Theme Issue Honoring Prof. Dr. Ludger Wessjohann's 60th Birthday: Natural Products in Modern Drug Discovery

Theme Issue Honoring Prof. Dr. Ludger Wessjohann's 60th Birthday: Natural Products in Modern Drug Discovery

Editor

Hidayat Hussain

MDPI • Basel • Beijing • Wuhan • Barcelona • Belgrade • Manchester • Tokyo • Cluj • Tianjin

Editor
Hidayat Hussain
Department of Bioorganic
Chemistry
Leibniz Institute of Plant
Biochemistry
Halle
Germany

Editorial Office
MDPI
St. Alban-Anlage 66
4052 Basel, Switzerland

This is a reprint of articles from the Special Issue published online in the open access journal *International Journal of Molecular Sciences* (ISSN 1422-0067) (available at: www.mdpi.com/journal/ijms/special_issues/Natural_Anticancer_Agents).

For citation purposes, cite each article independently as indicated on the article page online and as indicated below:

LastName, A.A.; LastName, B.B.; LastName, C.C. Article Title. *Journal Name* **Year**, *Volume Number*, Page Range.

ISBN 978-3-0365-4744-2 (Hbk)
ISBN 978-3-0365-4743-5 (PDF)

© 2022 by the authors. Articles in this book are Open Access and distributed under the Creative Commons Attribution (CC BY) license, which allows users to download, copy and build upon published articles, as long as the author and publisher are properly credited, which ensures maximum dissemination and a wider impact of our publications.

The book as a whole is distributed by MDPI under the terms and conditions of the Creative Commons license CC BY-NC-ND.

Contents

About the Editor vii

Preface to "Theme Issue Honoring Prof. Dr. Ludger Wessjohann's 60th Birthday: Natural Products in Modern Drug Discovery" ix

Hidayat Hussain
Prof. Ludger Wessjohann: A Lifelong Career Dedicated to a Remarkable Service in "Natural Products Sciences"
Reprinted from: *Int. J. Mol. Sci.* **2022**, 23, 5440, doi:10.3390/ijms23105440 1

Hidayat Hussain
Editorial to Special Issue "Theme Issue Honoring Prof. Dr. Ludger Wessjohann's 60th Birthday: Natural Products in Modern Drug Discovery"
Reprinted from: *Int. J. Mol. Sci.* **2022**, 23, 5835, doi:10.3390/ijms23105835 7

Johanna Voigt, Christoph Meyer and Frank Bordusa
Synthesis of Multiple Bispecific Antibody Formats with Only One Single Enzyme Based on Enhanced Trypsiligase [†]
Reprinted from: *Int. J. Mol. Sci.* **2022**, 23, 3144, doi:10.3390/ijms23063144 13

Ernest Oppong-Danquah, Martina Blümel, Silvia Scarpato, Alfonso Mangoni and Deniz Tasdemir
Induction of Isochromanones by Co-Cultivation of the Marine Fungus *Cosmospora* sp. and the Phytopathogen *Magnaporthe oryzae*
Reprinted from: *Int. J. Mol. Sci.* **2022**, 23, 782, doi:10.3390/ijms23020782 31

Shahrazad Sulaiman, Kholoud Arafat, Aya Mudhafar Al-Azawi, Noura Abdulraouf AlMarzooqi, Shamsa Nasser Ali Hussain Lootah and Samir Attoub
Butein and Frondoside-A Combination Exhibits Additive Anti-Cancer Effects on Tumor Cell Viability, Colony Growth, and Invasion and Synergism on Endothelial Cell Migration
Reprinted from: *Int. J. Mol. Sci.* **2021**, 23, 431, doi:10.3390/ijms23010431 51

Xunxun Wu, Xiaokun Li, Chunxue Yang and Yong Diao
Target Characterization of Kaempferol against Myocardial Infarction Using Novel In Silico Docking and DARTS Prediction Strategy
Reprinted from: *Int. J. Mol. Sci.* **2021**, 22, 12908, doi:10.3390/ijms222312908 69

Yen T. H. Lam, Manuel G. Ricardo, Robert Rennert, Andrej Frolov, Andrea Porzel and Wolfgang Brandt et al.
Rare Glutamic Acid Methyl Ester Peptaibols from *Sepedonium ampullosporum* Damon KSH 534 Exhibit Promising Antifungal and Anticancer Activity
Reprinted from: *Int. J. Mol. Sci.* **2021**, 22, 12718, doi:10.3390/ijms222312718 83

Katherine Yasmin M. Garcia, Mark Tristan J. Quimque, Gian Primahana, Andreas Ratzenböck, Mark Joseph B. Cano and Jeremiah Francis A. Llaguno et al.
COX Inhibitory and Cytotoxic Naphthoketal-Bearing Polyketides from *Sparticola junci*
Reprinted from: *Int. J. Mol. Sci.* **2021**, 22, 12379, doi:10.3390/ijms222212379 105

Fang-Fang You, Jing Zhang, Fan Cheng, Kun Zou, Xue-Qing Zhang and Jian-Feng Chen
ATG 4B Serves a Crucial Role in RCE-4-Induced Inhibition of the Bcl-2–Beclin 1 Complex in Cervical Cancer Ca Ski Cells
Reprinted from: *Int. J. Mol. Sci.* **2021**, 22, 12302, doi:10.3390/ijms222212302 119

Mahshid Deldar Abad Paskeh, Shafagh Asadi, Amirhossein Zabolian, Hossein Saleki, Mohammad Amin Khoshbakht and Sina Sabet et al.
Targeting Cancer Stem Cells by Dietary Agents: An Important Therapeutic Strategy against Human Malignancies
Reprinted from: *Int. J. Mol. Sci.* **2021**, 22, 11669, doi:10.3390/ijms222111669 135

Silke Schrom, Thomas Hebesberger, Stefanie Angela Wallner, Ines Anders, Erika Richtig and Waltraud Brandl et al.
MUG Mel3 Cell Lines Reflect Heterogeneity in Melanoma and Represent a Robust Model for Melanoma in Pregnancy
Reprinted from: *Int. J. Mol. Sci.* **2021**, 22, 11318, doi:10.3390/ijms222111318 185

Vladimir Maslivetc, Breana Laguera, Sunena Chandra, Ramesh Dasari, Wesley J. Olivier and Jason A. Smith et al.
Polygodial and Ophiobolin A Analogues for Covalent Crosslinking of Anticancer Targets
Reprinted from: *Int. J. Mol. Sci.* **2021**, 22, 11256, doi:10.3390/ijms222011256 205

Ridhima Kaul, Pradipta Paul, Sanjay Kumar, Dietrich Büsselberg, Vivek Dhar Dwivedi and Ali Chaari
Promising Antiviral Activities of Natural Flavonoids against SARS-CoV-2 Targets: Systematic Review
Reprinted from: *Int. J. Mol. Sci.* **2021**, 22, 11069, doi:10.3390/ijms222011069 217

Agnieszka Wojtkielewicz, Urszula Kiełczewska, Aneta Baj and Jacek W. Morzycki
Synthesis of Demissidine Analogues from Tigogenin via Imine Intermediates [†]
Reprinted from: *Int. J. Mol. Sci.* **2021**, 22, 10879, doi:10.3390/ijms221910879 267

Yue Yang, Ning Li, Tian-Ming Wang and Lei Di
Natural Products with Activity against Lung Cancer: A Review Focusing on the Tumor Microenvironment
Reprinted from: *Int. J. Mol. Sci.* **2021**, 22, 10827, doi:10.3390/ijms221910827 279

Yu Jin Kim, Nayeong Yuk, Hee Jeong Shin and Hye Jin Jung
The Natural Pigment Violacein Potentially Suppresses the Proliferation and Stemness of Hepatocellular Carcinoma Cells In Vitro
Reprinted from: *Int. J. Mol. Sci.* **2021**, 22, 10731, doi:10.3390/ijms221910731 303

Emmanuel Pina-Jiménez, Fernando Calzada, Elihú Bautista, Rosa María Ordoñez-Razo, Claudia Velázquez and Elizabeth Barbosa et al.
Incomptine A Induces Apoptosis, ROS Production and a Differential Protein Expression on Non-Hodgkin's
Lymphoma Cells [†]
Reprinted from: *Int. J. Mol. Sci.* **2021**, 22, 10516, doi:10.3390/ijms221910516 317

Haider N. Sultani, Ibrahim Morgan, Hidayat Hussain, Andreas H. Roos, Haleh H. Haeri and Goran N. Kaluerović et al.
Access to New Cytotoxic Triterpene and Steroidal Acid-TEMPO Conjugates by Ugi Multicomponent-Reactions [†]
Reprinted from: *Int. J. Mol. Sci.* **2021**, 22, 7125, doi:10.3390/ijms22137125 331

About the Editor

Hidayat Hussain

Hidayat Hussain obtained a Ph.D. degree in Medicinal Chemistry from University of Karachi Pakistan in 2004. He did post-doctorate (from 06-2004 to 09-2007) from University of Paderborn Germany and University of Maine France (from 10-2007 to 09-2009) along with senior scientist (from 01-2009 to 10-2010) at University of Paderborn Germany. Dr. Hidayat was also a Visiting Professor at Scripps Institution of Oceanography, San Diego USA from May 2017 to September 2017. He was also associate professor in University of Nizwa (Oman) from 2011 to 2017. Currently, Dr. Hidayat is associated with Leibniz Institute of Plant Biochemistry Halle, Germany. His research interests include the discovery of medicines from nature as synthesis of bioactive molecules for cancer, diabetes, malaria, and hepatitis, Alzheimer's disease. To date, he has authored and coauthored over 340 international publications with a cumulative impact factor of more than 1000 and given over 40 podium lectures at International Conferences. He published one book, over 10 book chapters and various review articles in International reputed Journals viz., Chemical Reviews and Natural Product Reports. He is a section Editor-in-Cheif of "Current Issues in Molecular Biology"Journal along with member of editorial board of several International Journals and also serving as a referee for over 100 international journals. Moreover he has over 7500 citations with h-index 40. According to the Stanford University database, his name is included in the top two percent of the most-cited scientists in various disciplines for 2020 and 2021.

Preface to "Theme Issue Honoring Prof. Dr. Ludger Wessjohann's 60th Birthday: Natural Products in Modern Drug Discovery"

This reprint is a specially designed reprint of the Special Issue in International Journal Molecular Sciences "Theme Issue Honoring Prof. Dr. Ludger Wessjohann's 60th Birthday: Natural Products in Modern Drug Discovery", which was dedicated to Prof. Dr. Ludger Wessjohann, on the occasion of his 60th Birthday.

Nature continuously produces biologically useful molecules and provides humankind with life-saving drugs or therapies. Natural products (NPs) offer a vast, unique and fascinating chemical diversity and these molecules have evolved for optimal interactions with biological macromolecules. Moreover, natural products feature pharmacologically active pharmacophores, which are pharmaceutically validated starting points for the development of new lead compounds. Over half of all approved (from 1981 to 2014) small-molecule drugs derived from NPs, including unaltered NPs, NPs synthetic derivatives and synthetic natural mimics, originated from a NPs pharmacophore or template. According to the FDA, NPs and their derivatives represent over one-third of all FDA-approved new drugs, in particular for anticancer/antibiotic lead compounds, which are remarkably enriched with NPs.

We wish to dedicate this Special Issue reprint version to Prof. Dr. Ludger Wessjohann, on the occasion of his 60th Birthday. He is a true research pioneer in the fields of Natural Product Chemistry and Medicinal Chemistry and made immense contribution to these fields.

Hidayat Hussain
Editor

Editorial

Prof. Ludger Wessjohann: A Lifelong Career Dedicated to a Remarkable Service in "Natural Products Sciences"

Hidayat Hussain

Department of Bioorganic Chemistry, Leibniz Institute of Plant Biochemistry, Weinberg 3, 06120 Halle, Germany; hidayat.hussain@ipb-halle.de

It is a great honor and a pleasure for me to serve as Guest Editor for this Issue of the *"International Journal of Molecular Sciences"*, dedicated to our mentor and colleague, Professor Dr. Ludger Wessjohann on the occasion of his 60th birthday. This Special Collection honoring Professor Wessjohann represents an excellent opportunity to celebrate not only a remarkable chemist, but also a great man.

Professor Wessjohann studied chemistry in Hamburg (Germany), Southampton (UK), and Oslo (Norway, Prof. Skattebøl). He earned his doctorate in 1990 with Prof. Armin de Meijere in Hamburg. After a short period as a lecturer in Brazil, he became a postdoctoral Feodor Lynen fellow of the Alexander von Humboldt Foundation with Prof. Paul Wender at Stanford University (USA), working on the total synthesis of Taxol®. After an assistant professorship in Munich (LMU, 1992–1998), he was appointed to the Chair of Bioorganic Chemistry at the Vrije Universiteit Amsterdam (NL), working on organometallic chemistry and biocatalysis. Since 2000, he has been the director of the Department of Bioorganic Chemistry at the Leibniz Institute of Plant Biochemistry (IPB) in Halle (Germany) and concurrently holds the chair of Natural Product Chemistry at the Martin Luther University of Halle-Wittenberg. From 2010–2017 he served as the Managing Director of the IPB (www.ipb-halle.de (accessed on 19 April 2022)).

Among others, he has been honored with the following awards: Leibniz Biotechnology Process of the Year (2019); Leibniz Bioactive Compound of the Year (2016 and 2018); Hugo Junkers Innovation Prize, State of Saxony-Anhalt, Germany (2014 and 2018); visiting Professor (invited) of King Saud University, Saudi Arabia; IQ Innovation Prize, City of Halle (2008); Microsoft IT-Founders Prize (Ontochem); winner of the business plan competition Sachsen-Anhalt (2006); and Feodor Lynen Fellowship (Alexander von Humboldt Foundation, 1990–1991). He is also a foreign member of the Brazilian Academy of Sciences, Brazil, and an honorary member of the Argentinean Society of Synthetic Organic Chemistry,

Argentina, and in 2019 was one of the thirteen foreign scientific advisors to the Colombian government's "Misíon des Sabios", with the task of drafting a future scientific agenda for the whole country.

He is a speaker at the Leibniz Science Campus Halle on "Plant-Based Bioeconomy" and the cofounder of the yearly "International Bioeconomy Conference" held at the National Academy of Sciences Leopoldina, in Halle. Additionally, he has been the organizer and co-organizer of several national and international conferences and conference series, e.g., the International Conference on the Chemistry of Selenium and Tellurium, the Multicomponent Reactions Conference, and the Brazilian Meeting of Organic Chemistry, to name but a few. Moreover, he is a member of steering committees and advisory boards of companies and organizations such as the national scientific advisory board of the Colombian government in the fields of bioeconomy, biotechnology, and environment; a board member of the European Federation of Biotechnology, Plant, Agriculture & Food Division; and the Dechema Fachgruppe Biotransformationen. Furthermore, Professor Wessjohann has served on numerous journal editorial boards, selection committees of foundations and science organizations (e.g., Alexander von Humboldt Foundation, DAAD, Finnish Academy of Sciences, Dutch Science Foundation).

He has authored or coauthored over 560 research articles (up to the end of 2021), including >20 reviews and >30 patent families issued or published. In addition, he has over 15000 citations, including a GS H-index of 63. Notably, according to the Stanford University database, his name is included in the top two percent of the most-cited scientists in various disciplines for 2020 and 2021. Moreover, he is the cofounder of six companies. He has been the major and/or thesis advisor for over 50 graduate students and has also directly supervised some 100 postdoctoral and visiting scholars. Furthermore, he has presented numerous research seminars at international and national scientific meetings in more than 30 countries. He has established very strong collaboration with scientists at several universities and academies in Cuba, Brazil, Colombia, Mexico, Vietnam, and some Arabic countries.

Professor Wessjohann's main research areas include molecular interactions in bioorganic and medicinal chemistry: (i) natural products—from isolation to total synthesis and bioactivity (incl. metabolomics, proteomics); (ii) biocatalysis and enzymatic reactions; (iii) new synthetic methods, combinatorial and medicinal chemistry (multicomponent reactions (MCR), pept(o)ides, macrocycles); and (iv) chemoinformatics. The main sources of his natural product discovery work are plants and higher fungi ("mushrooms"). His application covers the fields of anticancer agents, neuroactives, anti-infectives, and aroma compounds (flavor and fragrance), as well as plant protectants and phytoeffectors. He once said: "Coming originally from synthetic anticancer agents, natural product discovery quickly forced me to look more broadly. When you research organismic constituents, you do not know what they will be good for initially. Additionally, working in an institute with so many plant competent biologists, looking for compounds to increase plant productivity is a natural development—and it is a field underdeveloped in academic chemistry."

Professor Wessjohann's periods of service at IPB (since 2001) have been times of outstanding productivity and very dedicated service to the natural product research, academia, and health communities. Based on his remarkable research output, he is recognized as an internationally outstanding investigator and inspirational leader of collaborative research projects.

Prof. Dr. Wessjohann's research work, throughout his long career, has covered both plant-, fungi-, and coral-derived natural products developed toward drug candidates and agrochemical and food products, as can be seen from his extensive publication record. In addition, he has pioneered a number of synthetic methods for the synthesis of peptides, peptoids, and other natural products. However, beyond the analysis and discovery of phytochemicals, Ludger has pioneered many spectacular advances in the synthesis of fascinating natural products, intermediates, and derivatives. These scientific efforts led to the discovery of the second generation of tubulysins, the "tubugi" derivatives. Notably, tubugi-1 illustrated cytotoxic activity similar to that of tubulysin A and its activity was 30 fold higher than paclitaxel, with better stability and synthesizability than the parent compound [1]. Tubulysins are tetrapeptides, which constitute an intriguing natural product family with potent antimitotic properties [2]. Wessjohann's derivatives are especially suitable for targeting approaches as drug conjugates, e.g., with peptides and antibodies. Most excitingly, they enhance innate immunity against cancer and exhibit an atypical apoptosis mechanism. The synthetic tubulysin derivative, tubugi-1, improves the innate immune response by macrophage polarization, in addition to its direct cytotoxic effects in a murine melanoma model [3].

Total synthesis of natural products plays a remarkable role in achieving unambiguous structural confirmation, absolute configuration determination, and access to derivatives. Notably, his innovative syntheses enabled his teams to get sufficient natural product material for preliminary and detailed biological and pharmacological investigations, along with the synthesis of derivatives for (Q)SAR studies and improved application profiles. To pick out just a few, the Wessjohann group accomplished the total syntheses of hygrophorones, cordyheptapeptide A [4], or, in 2013, epothilone D [5]. Epothilones are anticancer macrolides, which possess various remarkable advantages in comparison to Taxol (paclitaxel) and Taxotere, which are among frontline anticancer agents. The Wessjohann group also reported the total first ever total synthesis of tubulysins (U and V) [6] and of tubulysin B [7]. Some other interesting molecules are (-)-julocrotine, viridic acid (tetrapeptide), selancins A and B (acylphloroglucinols), and empetrifranzinans A and C (acylphloroglucinols).

Notably, the Wessjohann lab is working on the applications of multicomponent reactions (MCRs) in the synthesis of natural and pseudo-natural products (in particular peptides and peptoids), which have been the subject of numerous publications throughout his career. It is noteworthy that the Wessjohann group is the world leader of synthesizing the peptoid backbones and macrocycles via Ugi and Passerini MCR reactions [8]. In addition, the group established the (cyclo-)-ligation and stapling of peptides with concomitant functionalization of the side chain [9] and backbone [10] via solution phase or on-resin MCR reactions. It is of note that the Wessjohann group established a MCR approach for the installation of turn-inducing moieties that facilitate the macrocyclization [11] along with ligation of peptides [12].

Prof. Wessjohann established an independent research program at IPB, involving the search for bioactive natural products, with well over a hundred new compounds described or patented, e.g., along with neuroscientists, he exploited the potential of *Rhodiola rosea* against Alzheimer disease and isolated ferulic acid eicosyl ester as a memory enhancer. Notably, this natural product used as nutraceutical (dried root material from *Rhodiola rosea*) possesses significant learning and memory enhancement properties in larval Drosophila and also partially compensates for age-related memory decline in adult flies [13], mice, and men. Another exemplary topic concerns natural sweetening and taste modifying agents from plants. His group isolated two sweet-tasting dammarane-type glycosides, balansins A and B, from *Mycetia balansae* Drake (Rubiaceae). Both balansins A and B demonstrated sweetening properties and their sweetness potencies at 0.1% and 0.2%, respectively, were equal to that of sucrose [14].

At the IPB, the Wessjohann group developed a process which they named "reverse metabolomics", which was employed to ultimately link many biological effects (e.g., taste sensory data) with metabolic profiles. The process is based on the correlation of chromatographic and spectroscopic profiles of partial or total extracts, metabolite profiles in particular, with biological activity profiles by chemo-informatic protocols, using activity correlation analysis (AcorA) [15]. In addition, this esteemed research group developed the first comparative metabolomics approach for the assessment of secondary metabolites of common medicinal plants across variety and species borders in the context of their genetic diversity, phylogeny and growth habitat, or processing, so as to set a framework for its authentication and quality control analysis, at a time when metabolomics is otherwise concentrated on single species only [16]. His work on the multiplex metabolic profiles of licorice and *Glycyrrhiza* species became one of the most cited papers in medicinal/taste plant metabolomics [17]. He also was the first to use 2D-NMR in metabolic profiling (of hops) [18].

Recently, he applied smart data analyses and machine-learning to bio- and chemoinformatics to analyze the complete published knowledge of the plant natural products chemistry of a whole country (Indonesia; specifically, Java), with thousands of plants, habitats, and applications, based on over 750,000 papers and database entries for the region, thereby correlating phylogeny with bioactivities, pinpointing "hot" and "cold" clades useful for a certain purpose, e.g., antibiotic potential, and the "white spots" of current research [19]. Wessjohann was a pioneer in bringing difficult enzymatic reactions into organic synthesis, e.g., regioselective aromatic prenylation [20] or recently hydroxylation [21]. He established various biocatalysis protocols, including the ligation of coenzyme A-conjugates with cinnamic acids, and this methodology has found an application in biocatalytic cascades, e.g., to vanillin and its analogs, which is the world's most-used flavoring agent [22].

Indeed, Professor Wessjohann has always been a highly inventive, creative, and stimulating mentor in encouraging his young students, along with collaborators, to expand their knowledge and vision in cross-border science fields via seminars, conferences, and critical scientific discussions in group seminars. Of note is his great intuition and ability to discuss everything about science with competence, creativity, and intelligence, which has undoubtedly allowed him to build and establish a remarkable work environment, in which a large and highly international and interdisciplinary group of coworkers has arisen, who, after leaving the group, continue to work in reputed institutions and universities (Figure 1). Of comparable importance is his establishment of a large school with many alumni, some of his Ph.D. students, posdoctoral fellows, and guest scientists who are shown in his academic family tree in Figure 1. Many of his former students are now professors around the world, with more than a dozen in Brazil alone. Indeed, his passion for honest science, work, and intriguing scientific ideas have inspired entire generations of scientists around the globe, as illustrated by the contributions to this outstanding Special Collection.

We are most grateful to Glinda He and the editorial team of the "IJMS" for their help in preparing this Special Collection and to Ines Stein, Prof. Wessjohann's assistant, for their valuable help. Notably, we thank all the chemists who contributed to the collection, demonstrating their esteem and affection for Professor Wessjohann. On this special occasion, I would like to join Ludger's colleagues, students, coworkers, collaborators, and friends in congratulating him on this 60th birthday. We all wish for him that he both continues and enjoys his scientific endeavors in good health for the coming years.

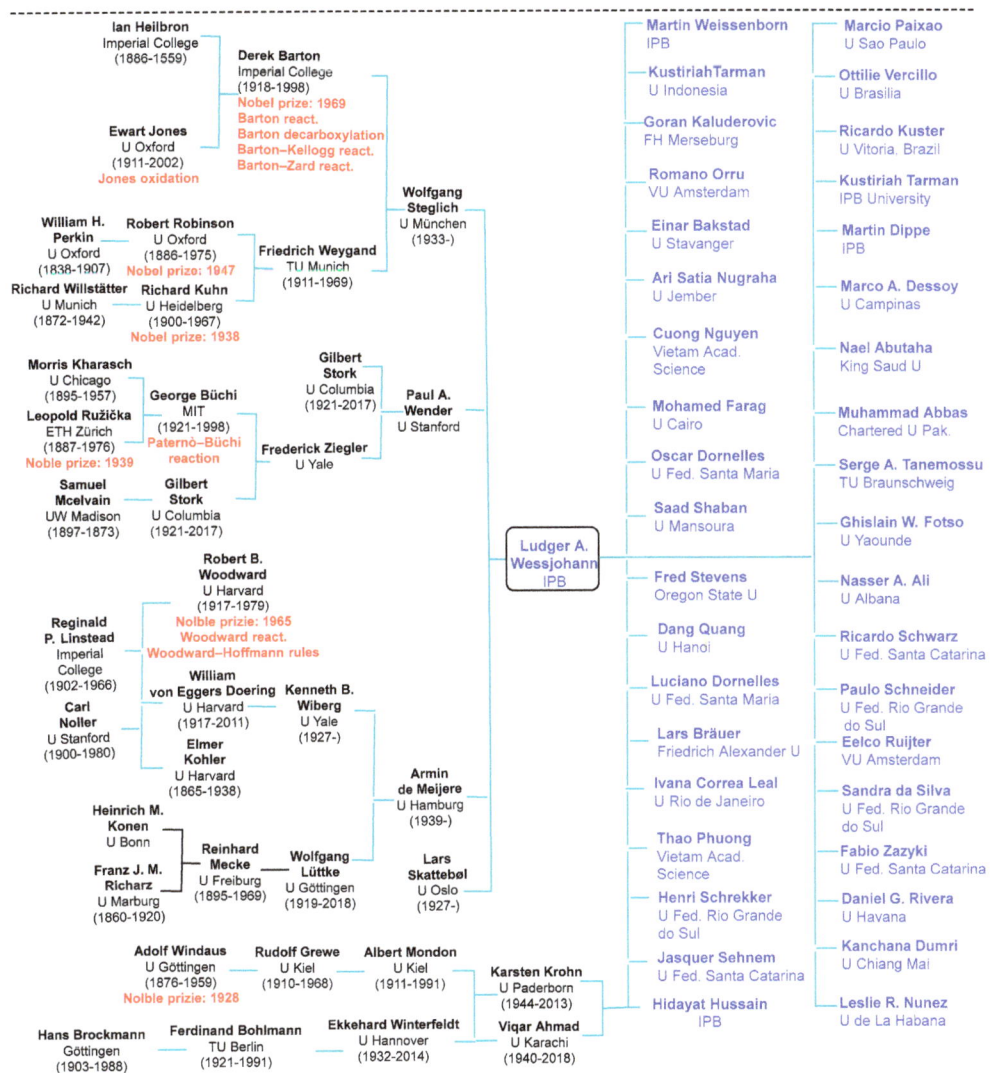

Figure 1. Ludger Wessjohann's academic family tree; only some of the Ph.D. students and postdoctoral fellows he has hosted and supervised and who eventually became university professors are considered; all researchers associated with him would exceed the limits of this compilation.

Funding: This research received no external funding.

Conflicts of Interest: The authors declare no conflict of interest.

References

1. Pando, O.; Stark, S.; Denkert, A.; Porzel, A.; Preusentanz, R.; Wessjohann, L.A. The Multiple Multicomponent Approach to Natural Product Mimics: Tubugis, N-Substituted Anticancer Peptides with Picomolar Activity. *J. Am. Chem. Soc.* **2011**, *133*, 7692–7695. [CrossRef] [PubMed]
2. Sasse, F.; Steinmetz, H.; Heil, J.; Höfle, G.; Reichenbach, H. Tubulysins, new cytostatic peptides from myxobacteria acting on microtubuli. Production, isolation, physico-chemical and biological properties. *J. Antibiot.* **2000**, *53*, 879–885. [CrossRef] [PubMed]

3. Drača, D.; Mijatović, S.; Krajnović, T.; Pristov, J.B.; Đukić, T.; Kaluđerović, G.N.; Wessjohann, L.A.; Maksimović-Ivanić, D. The synthetic tubulysin derivative, tubugi-1, improves the innate immune response by macrophage polarization in addition to its direct cytotoxic effects in a murine melanoma model. *Exp. Cell Res.* **2019**, *380*, 159–170. [CrossRef] [PubMed]
4. Puentes, A.R.; Neves Filho, R.A.W.; Rivera, D.G.; Wessjohann, L.A. Total synthesis of cordyheptapeptide A. *Synlett* **2017**, *28*, 1971–1974.
5. Wessjohann, L.A.; Scheid, G.O.; Eichelberger, U.; Umbreen, S. Total Synthesis of Epothilone D: The Nerol/Macroaldolization Approach. *J. Org. Chem.* **2013**, *78*, 10588–10595. [CrossRef]
6. Dömling, A.; Beck, B.; Eichelberger, U.; Sakamuri, S.; Menon, S.; Chen, Q.Z.; Lu, Y.; Wessjohann, L.A. Total Synthesis of Tubulysin U and V. *Angew. Chem. Int. Ed.* **2006**, *45*, 7235–7239, Erratum in *Angew. Chem. Int. Ed.* **2007**, *46*, 2347–2348. [CrossRef]
7. Pando, O.; Dorner, S.; Preusentanz, R.; Denkert, A.; Porzel, A.; Richter, W.; Wessjohann, L. First Total Synthesis of Tubulysin B. *Org. Lett.* **2009**, *11*, 5567–5569. [CrossRef]
8. Vercillo, O.E.; Andrade, C.K.Z.; Wessjohann, L.A. Design and Synthesis of Cyclic RGD Pentapeptoids by Consecutive Ugi Reactions. *Org. Lett.* **2008**, *10*, 205–208. [CrossRef]
9. Ricardo, M.G.; Llanes, D.; Wessjohann, L.A.; Rivera, D.G. Introducing the Petasis Reaction for Late-Stage Multicomponent Diversification, Labeling, and Stapling of Peptides. *Angew. Chem. Int. Ed.* **2019**, *58*, 2700–2704. [CrossRef]
10. Ricardo, M.G.; Marrrero, J.F.; Valdes, O.; Rivera, D.G.; Wessjohann, L.A. A peptide backbone stapling strategy enabled by the multicomponent incorporation of amide N-substituents. *Chem. A Eur. J.* **2019**, *25*, 769–774. [CrossRef] [PubMed]
11. Puentes, A.R.; Morejon, M.C.; Rivera, D.G.; Wessjohann, L.A. Peptide macrocyclization assisted by traceless turn inducers derived from Ugi peptide ligation with cleavable and resin-linked amines. *Org. Lett.* **2017**, *19*, 4022–4025. [CrossRef] [PubMed]
12. Vasco, A.V.; Ricardo, M.G.; Rivera, D.G.; Wessjohann, L.A. Ligation, Macrocyclization, and Simultaneous Functionalization of Peptides by Multicomponent Reactions (MCR). *Methods Mol. Biol.* **2022**, *2371*, 143–157.
13. Michels, B.; Zwaka, H.; Bartels, R.; Lushchak, O.; Franke, K.; Endres, T.; Fendt, M.; Song, I.; Bakr, M.; Budragchaa, T.; et al. Memory enhancement by ferulic acid ester across species. *Sci. Adv.* **2018**, *4*, eaat6994. [CrossRef]
14. Ley, J.; Reichelt, K.; Obst, K.; Wessjohann, L.; Wessjohann, S.; Van Sung, T.; Nguyen, T.A.; Van Trai, N. Orally consumed preparations comprising sweet-tasting triterpenes and triterpene glycosides. U.S. Patent 20140170083, 19 April 2014.
15. Degenhardt, A.; Wittlake, R.; Steilwind, S.; Liebig, M.; Runge, C.; Hilmer, J.M.; Krammer, G.; Gohr, A.; Wessjohann, L. Quantification of important flavour compounds in beef stocks and correlation to sensory results by "Reverse Metabolomics". In *Flavour Science*; Ferreira, V., Ed.; Elsevier: Amsterdam, The Netherlands, 2013; pp. 15–19.
16. Farag, M.A.; Porzel, A.; Al-Hammady, M.A.; Hegazy, M.E.; Meyer, A.; Mohamed, T.A.; Westphal, H.; Wessjohann, L.A. Soft Corals Biodiversity in the Egyptian Red Sea: A Comparative MS and NMR Metabolomics Approach of Wild and Aquarium Grown Species. *J. Proteome. Res.* **2016**, *15*, 1274–1287. [CrossRef] [PubMed]
17. Farag, M.A.; Porzel, A.; Wessjohann, L.A. Comparative metabolite profiling and fingerprinting of medicinal licorice roots using a multiplex approach of GC-MS, LC-MS and 1D-NMR techniques. *Phytochemistry* **2012**, *76*, 60–72. [CrossRef]
18. Farag, M.A.; Mahrous, E.A.; Lübken, T.; Porzel, A.; Wessjohann, L. Classification of commercial cultivars of *Humulus lupulus* L. (hop) by chemometric pixel analysis of two dimensional nuclear magnetic resonance spectra. *Metabolomics* **2014**, *10*, 21–32. [CrossRef]
19. Holzmeyer, L.; Hartig, A.K.; Franke, K.; Brandt, W.; Muellner-Riehl, A.N.; Wessjohann, L.A.; Schnitzler, J. Evaluation of plant sources for antiinfective lead compound discovery by correlating phylogenetic, spatial, and bioactivity data. *Proc. Natl. Acad. Sci. USA* **2020**, *117*, 12444–12451. [CrossRef]
20. Wessjohann, L.; Sontag, B. Prenylation of Benzoic Acid Derivatives Catalyzed by a Transferase from Escherichia coli Overproduction: Method Development and Substrate Specificity. *Angew. Chem. Int. Ed. Engl.* **1996**, *35*, 1697–1699. [CrossRef]
21. Dippe, M.; Herrmann, S.; Pecher, P.; Funke, E.; Pietzsch, M.; Wessjohann, L. Engineered bacterial flavin-dependent monooxygenases for the regiospecific hydroxylation of polycyclic phenols. *ChemBioChem* **2022**, *23*, e2021004.
22. Dippe, M.; Bauer, A.K.; Porzel, A.; Funke, E.; Müller, A.O.; Schmidt, J.; Beier, M.; Wessjohann, L.A. Coenzyme A-Conjugated Cinnamic Acids—Enzymatic Synthesis of a CoA-Ester Library and Application in Biocatalytic Cascades to Vanillin Derivatives. *Adv. Synth. Cat.* **2019**, *361*, 5346–5350. [CrossRef]

Editorial

Editorial to Special Issue "Theme Issue Honoring Prof. Dr. Ludger Wessjohann's 60th Birthday: Natural Products in Modern Drug Discovery"

Hidayat Hussain

Department of Bioorganic Chemistry, Leibniz Institute of Plant Biochemistry, Weinberg 3, 06120 Halle (Saale), Germany; hidayat.hussain@ipb-halle.de

Citation: Hussain, H. Editorial to Special Issue "Theme Issue Honoring Prof. Dr. Ludger Wessjohann's 60th Birthday: Natural Products in Modern Drug Discovery". *Int. J. Mol. Sci.* **2022**, 23, 5835. https://doi.org/10.3390/ijms23105835

Received: 5 May 2022
Accepted: 17 May 2022
Published: 23 May 2022

Publisher's Note: MDPI stays neutral with regard to jurisdictional claims in published maps and institutional affiliations.

Copyright: © 2022 by the author. Licensee MDPI, Basel, Switzerland. This article is an open access article distributed under the terms and conditions of the Creative Commons Attribution (CC BY) license (https://creativecommons.org/licenses/by/4.0/).

Nature continuously produces biologically useful molecules and provides mankind with life-saving drugs or therapies. Natural products (NPs) offer a vast, unique and fascinating chemical diversity, and these molecules have evolved for optimal interactions with biological macromolecules. Moreover, natural products feature pharmacologically active pharmacophores, which are pharmaceutically validated starting points for the development of new lead compounds. Over half of all approved (from 1981 to 2014) small-molecule drugs derived from NPs, including unaltered NPs, NPs synthetic derivatives and synthetic natural mimics, originated from an NP's pharmacophore or template. According to the FDA, NPs and their derivatives represent over one-third of all FDA-approved new drugs, in particular for anticancer/antibiotic lead compounds, which are remarkably enriched with NPs.

Many scientists have contributed to this Special Issue (SI), which includes 16 papers, original articles along with review articles that give the readers of the *International Journal of Molecular Sciences* an updated and new perspective about natural products in drug discovery.

Multicomponent reactions (MCRs) are reactions in which three or more reagents are allowed to undergo a reaction in a one-pot fashion in which almost all the atoms from the starting materials are incorporated into the final product. Furthermore, these MCRs feature various advantages over linear synthetic reactions, viz., fewer steps, no purification of reaction intermediates (fewer purification steps), ease of automation, and the creation of a bioactive compound library in a short time. Notably, these MCRs have great potential in Medicinal Chemistry in order to establish lead compounds in a short period of time as well as on an industrial scale. The Westermann group [1] has employed MCRs, in particular the Ugi-four component reaction (U-4CR), to prepare semi-synthetic analogs of triterpene acids (betulinic acid and fusidic acid), and steroids (cholic acid) conjugated with TEMPO (nitroxide). Notably, the nitroxide labelled betulinic and fusidic acid derivatives illustrated much better cytotoxic effects on prostate cancer (PC3) and colon cancer (HT29). The authors investigated the mechanism of the active molecules, which showed that these semi-synthetic compounds increased the level of caspase-3 significantly, which indicated the induction of apoptosis by activation of the caspase pathway.

Sesquiterpene lactones are a group of natural products reported from numerous plant species and are abundant in the Asteraceae family. These molecules possess secondary metabolites with pharmaceutical applications for cancer therapy. Sesquiterpene lactones illustrated numerous fascinating biological activities viz., antiamoebic, trypanocidal, anti-giardial, antibacterial, antidiabetic, antitumor, cytotoxic, and phytotoxic. The García-Hernández group [2] isolated the sesquiterpene lactone, incomptine A (IA) from *Decachaeta incompta*. Cytotoxic studies revealed that IA illustrates potent activity toward lymphoma cancer (U-937). The mechanistic investigation demonstrated that IA significantly enhances intracellular ROS levels along with the apoptotic activity. Moreover, in the proteomic investigation, 1548 proteins were differentially expressed. Of these 1548 proteins, 961 possessed a fold-change ≤ 0.67 and 587 possessed a fold-change ≥ 1.5. Notably, the majority of these proteins are involved in oxidative stress, apoptosis, and glycolytic metabolism.

Hepatocellular carcinoma (HCC) is a primary liver cancer that has high mortality and incidence worldwide. Furthermore, chemotherapeutic resistance is the major obstacle in HCC treatment. Violacein is an alkaloid featuring two indole units coupled via a pyrrolidinone ring. This alkaloid is produced by various bacterial strains, such as *Chromobacterium violaceum*, and demonstrates various biological effects, including anticancer effects. Kim et al. [3] revealed that violacein inhibits the stemness and proliferation of Hep3B and Huh7 HCC cells. Further study revealed that this molecule inhibited the proliferation of Hep3B and Huh7 HCC cells via inducing cell cycle arrest at the sub-G1 phase along with apoptotic cell death induction. Additionally, violacein induced MMP, enhanced ROS, upregulated p21 and p53, and activated the caspase (caspase-3, caspase-9) and PARP along with downregulation of ERK1/2 and AKT signalling. Violacein reduced the expression of CD133, Nanog, Oct4, and Sox2, via inhibiting STAT3/AKT/ERK pathways.

Solanidine and demissidine are leading compounds of the class "solanidane alkaloids" mainly found in the glycoside form in potato species, such as *Solanum tuberosum*, *S. acaule*, and *S. demissum*. Some members of the solanidane alkaloids illustrate potent antiproliferation properties and induce apoptosis in liver, cervical, stomach, and lymphoma cancer cells. Notably, the cytotoxic effects of α-chaconine toward HepG2 cells (hepatocellular cancer) are higher than the standards—camptothecin and doxorubicin. Wojtkielewicz et al. [4] established an approach for the synthesis of solanidanes from the spirostane sapogenin tigogenin. Moreover, the indolizidine core within the solanidane alkaloids was assembled in five steps starting from spirostane. In addition, numerous demissidine derivatives have been prepared from the alkaloid tigogenin through various imine intermediates.

Polygodial, a dialdehyde sesquiterpenoid, was reported from *Tasmannia lanceolata*. The literature revealed that ophiobolin A demonstrates potent effects toward apoptosis-resistant glioblastoma cells via the induction of a non-apoptotic cell death pathway. Maslivetc et al. [5] prepared dimers and trimers of polygodial and aphiobolin A in order to discover any metabolites that could crosslink biological primary amine-containing targets. The authors demonstrated that such molecules keep the pyrrolylation ability and illustrate increased single-digit micromolar potencies against apoptosis-resistant cancer cells.

The Macabeo research group [6] investigated the rice culture of *Sparticola junci* and produced the naphthoketal-bearing polyketides, sparticatechol A, sparticolin H, and sparticolin A. The absolute configurations of these natural products were established using ECD together with TDDFT. All three of these secondary metabolites proved to be highly active toward COX-2 and COX-1 enzymes, with sparticatechol A possessing the highest effects, which were higher than the standard Celecoxib. Furthermore, this molecule possesses preferential binding towards COX-2. Sparticolins H and A demonstrated moderate cytotoxic effects toward K-562 cells (myelogenous leukemia) and weak cytotoxicity toward mouse fibroblast cells (HeLa).

The genus *Sepedonium* is a tremendous source of natural products with intriguing chemical diversity. Nobert and his co-workers [7] investigated *Sepedonium* ampullosporum (strain KSH534) and isolated two new peptaibols, ampullosporin F and ampullosporin G, along with five known natural products. The structures of these molecules were established via extensive spectroscopic techniques. Moreover, the total synthesis of ampullosporins F and G was accomplished via a solid-phase strategy in order to establish the absolute configuration of all the chiral amino acids. Additionally, these two molecules possessed potent antifungal effects towards *Phytophthora infestans* and *Botrytis cinerea*. Furthermore, ampullosporins F and G demonstrated potent anticancer effects toward cell viability assays on PC-3 (prostate cancer) and HT-29 (colorectal cancer).

Microbial co-cultivation is a fascinating strategy employed in order to activate biosynthetic gene clusters. Furthermore, based on a comparative metabolomics approach and the anti-phytopathogenic effects of the co-cultures, Oppong-Danquah et al. [8] investigated *Magnaporthe oryzae* and the marine *Cosmospora* sp. Phytochemical investigation of *M. oryzae* and *Cosmospora* sp. produced five isochromanones, soudanones A, E, D, H, and I, along with the isochromans, pseudoanguillosporins A and B, naphtho-γ-pyrones, ustilaginoidin G, and cephalochromin. The basic structures of these compounds were established via NMR and absolute configuration through ECD together with Mosher's ester reaction. Among the tested compounds, only soudanones E and D illustrated antimicrobial effects towards *Phytophthora infestans* and *M. oryzae*, while pseudoanguillosporin A possessed potent anti-phytopathogenic effects towards *Xanthomonas campestris*, *Pseudomonas syringae*, *P. infestans*, and *M. oryzae*.

Despite the remarkable advances in immuno- and targeted therapies, breast and lung cancer are among the leading causes of cancer death. Combination therapy is an innovative strategy where a mixture of different drugs is used to treat diseases. Sulaiman et al. [9] explored such a combination therapy by using butein together with an anticancer flavonoid on lung (A549) and breast cancer (MDA-MB-231) cells together with another anticancer triterpene, frondoside-A. The authors demonstrated that butein was able to reduce the two cancer cell colony's growth and viability. In addition, this combination therapy decreases tumour growth on the chick embryo chorioallantoic membrane (CAM) in vivo. The authors further demonstrated that the anti-cancer effects of butein are due to significant inhibition of STAT3 phosphorylation, leading to PARP cleavage. This combination therapy was found to lead to synergistic effects on the inhibition of HUVEC migration.

Target identification is a challenging and important strategy for identifying drug lead development. In this regard, Src tyrosine kinase has been extensively developed as a factor in tumorigenesis via differentiation, regulating cell growth, survival, and adhesion. Wu et al. [10] developed a novel in silico docking strategy for the target identification of kaempferol. These results were further validated via TargetHunter and PharmMapper server protocols. From computational studies, it has been confirmed that Src is a validated target for kaempferol, and this validation was additionally verified by kaempferol cardioprotective potential in vitro and in vivo screening.

Cervical cancer is the fourth most common cancer in women and a leading cause of death for women in the US. The steroidal saponin, RCE-4 [(1β, 3β, 5β, 25S)-spirostan-1,3-diol-1-[α-L-rhamnopyranosyl-(1→2)-β-D-xylopyranoside] is produced by *Reineckia carnea*. The Chen research group [11] investigated the mode of action of RCE-4 towards cervical cancer and targeted the Bcl-2–Beclin 1 complex, an essential programmed cell death (PCD) regulator. Notably, the results of the Chen group illustrate that this molecule inhibited the formation of the Bcl-2–Beclin 1 complex and that the ATG 4B proteins served as a crucial co-factor. In addition, the sensitivity of RCE-4 to Ca Ski cells was increased remarkably by inhibiting the expression of the ATG 4B.

Melanoma is one of the most heterogeneous and aggressive cancers possessing a strong capability to evolve resistance towards different therapeutic strategies. Melanoma remains a most challenging cancer to treat because of foetal safety and balancing maternal needs. Schrom et al. [12] cultured four lymph node metastasis-derived pigmented and non-pigmented sections. Furthermore, the four cultures possessed different genotypic, phenotypic, and tumorigenic properties. For the treatment protocol, synthetic human lactoferricin-derived peptides were screened. Based on their scientific data, the authors claimed that these anti-tumour peptides could be employed as an optional alternative innovative therapeutic treatment during pregnancy.

Lung cancer also remains one of the leading cancer-related deaths that continues to challenge researchers involved in searching for lead compounds to treat this cancer. The tumour microenvironment (TME) is considered one of the leading signs of epithelial cancers, such as most lung cancers, and is associated with progression, tumorigenesis, metastasis, and invasion. Yang et al. [13] published a review that emphasised the important part of the TME and comprehensively described the antitumour properties and mode of action of numerous natural products that target the TME. Furthermore, the authors discussed combination therapy, i.e., the synergistic properties of various natural products employed in combination with other anticancer molecules. The authors also emphasised the use of nanotechnology previously used to increase the anticancer potentials of natural products.

Cancer stem cells (CSCs) are crucial factors for tumour stemness by increasing proliferation, colony formation along with metastasis. Furthermore, CSCs may also be employed as a potential therapy resistance protocol. The presence of CSCs can cause cancer recurrence, and their complete elimination can have significant therapeutic benefits. The review of Paskeh et al. [14] emphasises natural products-based targeting CSCs in cancer therapy. Numerous dietary aspects have been covered in this review on CSCs, including flavones, flavonols, chalcones, isoflavones, caffeic acid, cartenoids, and ginsenosides. Various molecular pathways such as the Sonic Hedgehog, Wnt/β-catenin, STAT3, NF-κB, and Gli1 that follow these molecules in suppressing CSC are also featured. The authors reported that upon exposure to these natural products, a potential decrease occurs in the CSC markers' levels, such as CD133, CD44, Oct4, and ALDH1, in order to impair cancer stemness.

The ongoing COVID-19 pandemic situation created by SARS-CoV-2 has become a leading health issue globally over the past two years. Kaul et al. [15] published a review that focussed on research data from in vitro screening of flavonoids on key SARS-CoV-2 targets. They analysed 27 research papers that included over 69 flavones for their anti-SARS-CoV-2 targets. The combination of flavonoids with other synthetic drugs demonstrated promising results. They further emphasised the importance of in silico studies of flavonoids towards SARS-CoV-2 and highlighted the clinical studies. The authors claimed that flavonols, myricetin, quercetin, baicalein, baicalin, and the flavan-3-ol EGCG, along with tannic acid, have great potential for in vivo evaluation and clinical studies.

Bispecific antibodies (bsAbs) were developed in the 1960s and are now considered to be a leading group of immunotherapies to treat cancer. Furthermore, numerous different bsAbs have been documented in the last decade, mainly generated genetically. Bordusa et al. [16] published a novel chemo-enzymatic protocol for generating bsAbs fragments through the covalent fusion of two functional antibody Fabs. They initially modified the single Fabs site through click anchors employing an enhanced Trypsiligase variant (eTl), and this was later followed by conversion into the heterodimers through click chemistry. In later stages, the authors employed the inverse electron-demand Diels–Alder reaction and strain-promoted alkyne-azide cycloaddition protocols, which are well-established strategies for their diminished side reactions and fast reaction kinetics. The authors also developed enzymatic C-C and C-N terminal coupling protocols of the two Fabs through peptide linkages. The resulting bsFabs illustrate cytotoxic effects on breast cancer cells.

Conflicts of Interest: The author declares no conflict of interest.

References

1. Sultani, H.N.; Morgan, I.; Hussain, H.; Roos, A.H.; Haeri, H.H.; Kaluđerović, G.N.; Hinderberger, D.; Westermann, B. Access to New Cytotoxic Triterpene and Steroidal Acid-TEMPO Conjugates by Ugi Multicomponent-Reactions. *Int. J. Mol. Sci.* **2021**, *22*, 7125. [CrossRef] [PubMed]
2. Pina-Jiménez, E.; Calzada, F.; Bautista, E.; Ordoñez-Razo, R.M.; Velázquez, C.; Barbosa, E.; García-Hernández, N. Incomptine A Induces Apoptosis, ROS Production and a Differential Protein Expression on Non-Hodgkin's Lymphoma Cells. *Int. J. Mol. Sci.* **2021**, *22*, 10516. [CrossRef] [PubMed]
3. Kim, Y.J.; Yuk, N.; Shin, H.J.; Jung, H.J. The Natural Pigment Violacein Potentially Suppresses the Proliferation and Stemness of Hepatocellular Carcinoma Cells in Vitro. *Int. J. Mol. Sci.* **2021**, *22*, 10731. [CrossRef] [PubMed]
4. Wojtkielewicz, A.; Kiełczewska, U.; Baj, A.; Morzycki, J.W. Synthesis of Demissidine Analogues from Tigogenin via Imine Intermediates. *Int. J. Mol. Sci.* **2021**, *22*, 10879. [CrossRef] [PubMed]
5. Maslivetc, V.; Laguera, B.; Chandra, S.; Dasari, R.; Olivier, W.J.; Smith, J.A.; Bissember, A.C.; Masi, M.; Evidente, A.; Mathieu, V.; et al. Polygodial and Ophiobolin A Analogues for Covalent Crosslinking of Anticancer Targets. *Int. J. Mol. Sci.* **2021**, *22*, 11256. [CrossRef] [PubMed]
6. Garcia, K.Y.M.; Quimque, M.T.J.; Primahana, G.; Ratzenböck, A.; Cano, M.J.B.; Llaguno, J.F.A.; Dahse, H.-M.; Phukhamsakda, C.; Surup, F.; Stadler, M.; et al. COX Inhibitory and Cytotoxic Naphthoketal-Bearing Polyketides from Sparticola junci. *Int. J. Mol. Sci.* **2021**, *22*, 12379. [CrossRef] [PubMed]

7. Lam, Y.T.H.; Ricardo, M.G.; Rennert, R.; Frolov, A.; Porzel, A.; Brandt, W.; Stark, P.; Westermann, B.; Arnold, N. Rare Glutamic Acid Methyl Ester Peptaibols from *Sepedonium ampullosporum* Damon KSH 534 Exhibit Promising Antifungal and Anticancer Activity. *Int. J. Mol. Sci.* **2021**, *22*, 12718. [CrossRef] [PubMed]
8. Oppong-Danquah, E.; Blümel, M.; Scarpato, S.; Mangoni, A.; Tasdemir, D. Induction of Isochromanones by Co-Cultivation of the Marine Fungus *Cosmospora* sp. and the Phytopathogen Magnaporthe oryzae. *Int. J. Mol. Sci.* **2022**, *23*, 782. [CrossRef] [PubMed]
9. Sulaiman, S.; Arafat, K.; Al-Azawi, A.M.; AlMarzooqi, N.A.; Lootah, S.N.A.H.; Attoub, S. Butein and Frondoside-A Combination Exhibits Additive Anti-Cancer Effects on Tumor Cell Viability, Colony Growth, and Invasion and Synergism on Endothelial Cell Migration. *Int. J. Mol. Sci.* **2022**, *23*, 431. [CrossRef] [PubMed]
10. Wu, X.; Li, X.; Yang, C.; Diao, Y. Target Characterization of Kaempferol against Myocardial Infarction Using Novel in Silico Docking and DARTS Prediction Strategy. *Int. J. Mol. Sci.* **2021**, *22*, 12908. [CrossRef] [PubMed]
11. You, F.-F.; Zhang, J.; Cheng, F.; Zou, K.; Zhang, X.-Q.; Chen, J.-F. ATG 4B Serves a Crucial Role in RCE-4-Induced Inhibition of the Bcl-2–Beclin 1 Complex in Cervical Cancer Ca Ski Cells. *Int. J. Mol. Sci.* **2021**, *22*, 12302. [CrossRef] [PubMed]
12. Schrom, S.; Hebesberger, T.; Wallner, S.A.; Anders, I.; Richtig, E.; Brandl, W.; Hirschmugl, B.; Garofalo, M.; Bernecker, C.; Schlenke, P.; et al. MUG Mel3 Cell Lines Reflect Heterogeneity in Melanoma and Represent a Robust Model for Melanoma in Pregnancy. *Int. J. Mol. Sci.* **2021**, *22*, 11318. [CrossRef] [PubMed]
13. Yang, Y.; Li, N.; Wang, T.-M.; Di, L. Natural Products with Activity against Lung Cancer: A Review Focusing on the Tumor Microenvironment. *Int. J. Mol. Sci.* **2021**, *22*, 10827. [CrossRef] [PubMed]
14. Deldar Abad Paskeh, M.; Asadi, S.; Zabolian, A.; Saleki, H.; Khoshbakht, M.A.; Sabet, S.; Naghdi, M.J.; Hashemi, M.; Hushmandi, K.; Ashrafizadeh, M.; et al. Targeting Cancer Stem Cells by Dietary Agents: An Important Therapeutic Strategy against Human Malignancies. *Int. J. Mol. Sci.* **2021**, *22*, 11669. [CrossRef] [PubMed]
15. Kaul, R.; Paul, P.; Kumar, S.; Büsselberg, D.; Dwivedi, V.D.; Chaari, A. Promising Antiviral Activities of Natural Flavonoids against SARS-CoV-2 Targets: Systematic Review. *Int. J. Mol. Sci.* **2021**, *22*, 11069. [CrossRef] [PubMed]
16. Voigt, J.; Meyer, C.; Bordusa, F. Synthesis of multiple bispecific antibody formats with only one single enzyme based on enhanced Trypsiligase. *Int. J. Mol. Sci.* **2022**, *23*, 3144. [CrossRef] [PubMed]

International Journal of *Molecular Sciences*

Article

Synthesis of Multiple Bispecific Antibody Formats with Only One Single Enzyme Based on Enhanced Trypsiligase [†]

Johanna Voigt, Christoph Meyer and Frank Bordusa *

Department of Naturstoffbiochemie, Institute for Biochemistry and Biotechnology, Martin-Luther-Universität Halle-Wittenberg, Kurt-Mothes-Straße 3a, 06120 Halle (Saale), Germany; johanna.hohgardt@biochemtech.uni-halle.de (J.V.); christoph.meyer@biochemtech.uni-halle.de (C.M.)
* Correspondence: frank.bordusa@biochemtech.uni-halle.de; Tel.: +49-345/5524801
† This paper is dedicated to L. A. Wessjohann on the occasion of his 60th birthday.

Abstract: Bispecific antibodies (bsAbs) were first developed in the 1960s and are now emerging as a leading class of immunotherapies for cancer treatment with the potential to further improve clinical efficacy and safety. Many different formats of bsAbs have been established in the last few years, mainly generated genetically. Here we report on a novel, flexible, and fast chemo–enzymatic, as well as purely enzymatic strategies, for generating bispecific antibody fragments by covalent fusion of two functional antibody Fab fragments (Fabs). For the chemo–enzymatic approach, we first modified the single Fabs site-specifically with click anchors using an enhanced Trypsiligase variant (eTl) and afterward converted the modified Fabs into the final heterodimers via click chemistry. Regarding the latter, we used the strain-promoted alkyne-azide cycloaddition (SPAAC) and inverse electron-demand Diels–Alder reaction (IEDDA) click approaches well known for their fast reaction kinetics and fewer side reactions. For applications where the non-natural linkages or hydrophobic click chemistry products might interfere, we developed two purely enzymatic alternatives enabling C- to C- and C- to N-terminal coupling of the two Fabs via a native peptide bond. This simple system could be expanded into a modular system, eliminating the need for extensive genetic engineering. The bispecific Fab fragments (bsFabs) produced here to bind the growth factors ErbB2 and ErbB3 with similar K_D values, such as the sole Fabs. Tested in breast cancer cell lines, we obtained biologically active bsFabs with improved properties compared to its single Fab counterparts.

Keywords: bispecific antibody; Trypsiligase; click chemistry; biorthogonal chemistry; antibody engineering

Citation: Voigt, J.; Meyer, C.; Bordusa, F. Synthesis of Multiple Bispecific Antibody Formats with Only One Single Enzyme Based on Enhanced Trypsiligase. *Int. J. Mol. Sci.* 2022, 23, 3144. https://doi.org/10.3390/ijms23063144

Academic Editors: Hidayat Hussain and Menotti Ruvo

Received: 1 December 2021
Accepted: 12 March 2022
Published: 15 March 2022

Publisher's Note: MDPI stays neutral with regard to jurisdictional claims in published maps and institutional affiliations.

Copyright: © 2022 by the authors. Licensee MDPI, Basel, Switzerland. This article is an open access article distributed under the terms and conditions of the Creative Commons Attribution (CC BY) license (https://creativecommons.org/licenses/by/4.0/).

1. Introduction

Bispecific antibodies (bsAbs) form a heterogeneous family of biological therapeutics. This fast-growing class of therapeutics holds, among others, promise mainly for the treatment of cancer and inflammatory diseases [1–3]. In 2020, a total of three bispecific antibodies were approved and more than 100 are being studied in clinical trials alone for the treatment of cancer [4].

In contrast to conventional monoclonal antibodies (mAbs), bsAbs simultaneously bind to two different types of antigens or, in their biparatopic form, two different epitopes on the same antigen, e. g., a cell target with a receptor on an immune cell or several targets on a cell surface to increase the cytotoxic potential and achieve improved efficacy with lower target expression [2]. In addition to full-length bsAbs, small bsAbs are of similar therapeutic interest due to the absence of the Fc-region. Better tissue penetration, the lack of Fc-mediated antibody effector functions, and easier production are only a few of the advantages of using antibody fragments [1,3].

The development of bsAbs accelerated in the 1980s with the progress of the synthesis techniques for their production. In particular, quadroma technology should be mentioned, which led to non-uniform bsAbs due to incorrect chain pairings, but nevertheless made numerous innovative bsAb formats possible [5–7]. One example is the small bsAb BiTE [8]. An important synthetic innovation to prevent mismatches of the heavy chains was the development of the knob-into-hole strategy [9]. At the same time, the first approaches were developed to suppress the mismatch of the light chains as well. In 2007, DVD (dual-variable-domain)-IgGs were developed that even had multiple fused antigen-binding domains [10]. Although relatively easy to generate by heterologous expression, their function can be influenced by the direct *N*- to *C*-terminal fusion of the domains. Antigen binding of the inner domain may be particularly affected. In addition, the increasing size of the bsAbs may also reduce the expression yield or homogeneity of the products [11–13]. Classical *C*- to *N*-terminal topology in bsAbs, therefore, requires considerable empirical control in the development process.

Posttranslational linkage of the single antigen-binding domains of mAbs or Fabs could address these limitations and simultaneously increase the flexibility regarding the individual structure of the final bsAbs/bsFabs. In addition to the classical *N*- to *C*-fusion of the antigen-binding domains, posttranslational domain coupling also allows more flexible orientations, e.g., *C*- to *C*-terminal domain linkage with free *N*-termini of the two binding domains, as is the case with native mAbs. Enzymatic coupling approaches seem to be particularly interesting for this purpose due to their inherent regio- and stereospecificity [14,15]. Nevertheless, only a few exceptions for purely enzymatic *C*- to *C*-terminal domain linkages are currently known [16,17]. In the most prominent case of the transpeptidase SortaseA (SrtA), a biorthogonal strategy was developed only with the help of a second enzyme, i.e., butelase1, using two different linker sequences, each specific for one enzyme [17]. A method that in principle only requires one enzyme was developed for transglutaminase [18]. The limitation of this approach, however, is that no human multichain Fabs or Abs were ligated, but instead camelidea single-chain domain antibodies were, which were also identical in their structure and thus, also specificity, and, therefore, led to the formation of monospecifics. Furthermore, in addition to the formation of the monospecific homodimers, these reactions also led to the simultaneous formation of inhomogeneous homomultimers with variable numbers of monomeric antibody domains. Chemo–enzymatic approaches have been developed as an alternative to installing a click anchor by enzyme catalysis to the proteins of interest first and subsequently coupling both moieties by click chemistry. Despite these promising strategies, there are, so far, only two enzymes with which the chemo–enzymatic generation of bsAbs (or analogs) has been initially demonstrated, i.e., SrtA and the formylglycine generating enzyme (FGE). In the case of SrtA, two full-length antibodies were chemo–enzymatically ligated, one with the strain-promoted alkyne-azide cycloaddition (SPAAC) *C*- to *C*-terminally to form a bsAb and, second, an antibody-(scFv)$_2$ conjugate using the inverse electron-demand Diels–Alder reaction (IEDDA) [19]. A general disadvantage regarding the combination with click chemistry may result from the catalytically active cysteine of SrtA which was shown to lower the yield in the modification step with SPAAC linkers. In the case of FGE, the cysteine to formylglycine conversion activity of the enzyme was used to form an aldehyde tag (CXPXR) in the proteins of interest [20,21]. The resulting formylglycines were subsequently reacted in a two-step process with a biofunctionalized linker by Pictet–Spengler ligation, resulting in the heterodimeric product. Unfortunately, the enzyme needs reducing reaction conditions to be catalytically active. This should be the reason why only single-chain nanobodies instead of disulfide-bridged antibodies have been used so far for bsAbs synthesis [16].

As an alternative to SrtA and FGE, we developed Trypsiligase (Tl), a trypsin-based biocatalyst by rational enzyme design [22]. Tl recognizes the short, hydrophilic amino acid tag YRH and catalyzes peptide bond formation via reverse proteolysis [22,23]. The synthesis reactions take place after the tyrosine in the YRH-motif. During synthesis, the enzyme allows the covalent coupling of functional groups attached to the enzyme-specific, RH-containing nucleophile, such as fluorophores, toxins, drugs, or click anchors [23,24]. Recently, we generated next-generation Trypsiligases via evolutionary enzyme engineering [25]. These variants are characterized by an increased synthesis potential compared to the original

Tl. Enhanced Trypiligase (eTl), representing the most efficient next-generation Trypsiligase so far, reaches product yields close to the thermodynamic maximum of the reaction type catalyzed.

While enzyme catalysis guarantees a regioselective reaction process, click chemistry opens the possibility of a flexible and modular synthesis principle including C- to N-, C- to C-terminal, and N- to N-terminal linkages of antigen-binding domains as well. Like enzyme synthesis, the click reactions are also biorthogonal and can be conducted in aqueous systems [26]. SPAAC and the IEDDA have been proven to be particularly suitable for biological applications. Both click reactions are fast, proceed entirely without or with only a few side reactions, and require only small substrate excesses [27,28]. In some cases, hydrophobic, sometimes bulky product structures may lead to undesirable problems with hydrophobicity, yield, or functionality.

In the present study, a synthesis approach for bsAbs is presented that only requires a single enzyme and can optionally be combined with click chemistry. It enables simple and rapid post-translational domain shuffling of antigen-binding domains to assess the optimal architecture of the bsAb. The function of the approach was evaluated using the example of the synthesis of bsFabs consisting of the antigen-binding domains of the growth factors ErbB2 and ErbB3. Both domains were covalently linked in a flexible orientation, purely enzyme-catalyzed or chemo–enzymatically with the enzymatic linkage of click anchors and downstream click chemistry. All bsFabs bind with similar K_D values as the individual Fabs. Tests in breast cancer cell lines demonstrated improved properties in the biological activity of the bsFabs compared to their single Fab counterparts.

2. Results and Discussion

The concept of posttranslational bsAbs assembling via eTl was investigated by fusing the antigen-binding moieties of two distinct Fabs, i.e., anti-ErbB2- and anti-ErbB3-Fab. Both Fabs target epitopes on growth factor receptors of the EGFR family. ErbB2 is abundant in many tissue tumors, such as breast cancer [29,30]. At the same time, ErbB3 is the preferred interaction partner of ErbB2. Once they interact, they form the most robust signal within this family of receptors [31,32]. The ErbB2-specific Fab used is derived from the well-known antibody Trastuzumab, whereas the anti-ErbB3 counterpart is derived from CDX3379, a therapeutic antibody from Celldex therapeutics [33,34].

Initially, the Fabs were equipped with a nucleotide sequence encoding either the C-terminal YRH- or N-terminal RH-motif on the genomic level. Both enzyme recognition sequences were inserted into the respective termini of the heavy chain of the two Fabs. Biosynthesis and purification of the constructs were performed according to established protocols [23,33]. Final assembly of bsFabs was evaluated either by direct coupling of a C-terminally tagged anti-ErbB3-Fab-YRH with the N-terminally modified anti-ErbB2-RH-Fab via eTl (Scheme 1A) or indirectly via eTl-coupling of click anchors to the anti-ErbB2- and anti-ErbB3-Fabs followed by click chemistry-mediated bsFab formation (Scheme 1B). Direct coupling via eTl according to Scheme 1A resulted in a C- to N- linkage of anti-ErbB3- and anti-ErbB2-Fab. On the contrary, C- to C-terminal linker structure was realized by click chemistry due to the eTl-coupling of the click anchors to the C-terminus of the two Fabs. The individual structure of the IEDDA and SPAAC based click linkers are shown in Figure 1. Previous studies showed that Tl also catalyzes the attachment of artificial functionalities to the N-terminus of proteins with high yields [24]. Finally, we tried to assemble a C- to C-linked anti-ErbB3-anti-ErbB2-bsFab by a purely enzymatic approach using the branched linker **5** (Figure 1) for Fab-coupling featuring two eTl-specific N-termini (Scheme 1C). C- to C-linkage was achieved by a two-step enzymatic reaction, ligating the first linker **5** to anti-ErbB2-Fab-YRH followed by the coupling of anti-ErbB3-Fab-YRH. The central lysine moiety in **5** allows the incorporation of a third functionality in addition to the two enzyme recognition sites. In our case, we used an azide to enable click chemistry with spectroscopic labels.

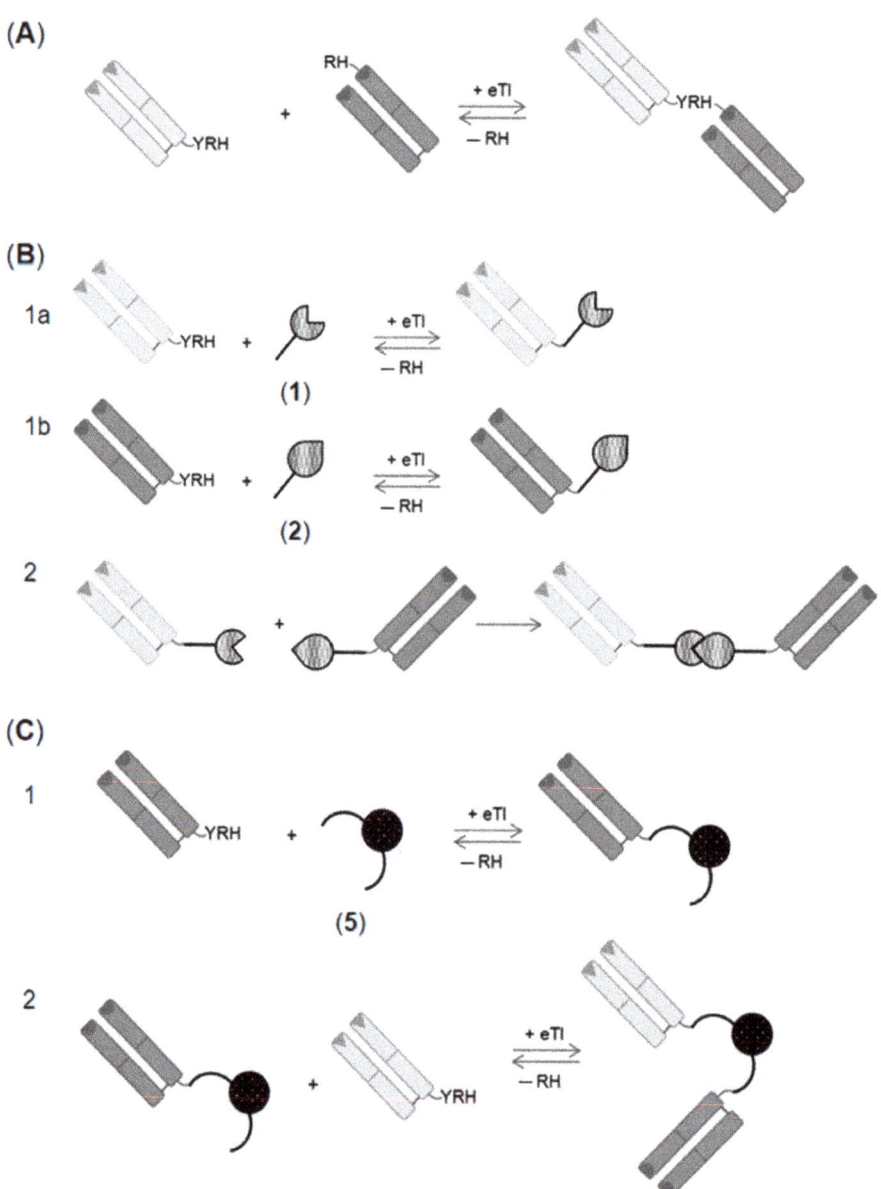

Scheme 1. Synthesis and structure of anti-ErbB2-anti-Erb3-bsFab formats. (**A**) Enzymatic synthesis of C- to N-linked anti-ErbB3-YRH-anti-ErbB2-bsFab via eTl-catalysis from anti-ErbB3-Fab-YRH and anti-ErbB2-RH-Fab; (**B**) Chemo–enzymatic synthesis of C- to C-linked anti-ErbB3-anti-ErbB2-bsFab via eTl-mediated coupling of click anchors to both Fabs (reaction 1a and 1b), followed by click coupling of the purified intermediate Fab products (reaction 2); (**C**) Enzymatic synthesis of C- to C-linked anti-ErbB3-anti-ErbB2-bsFab via a two-step eTl-catalysis initiated by the enzymatic coupling of anti-ErbB2-Fab-YRH to linker **5**, followed by a second enzymatic coupling of the purified intermediate to anti-ErbB3-Fab-YRH. Light grey Fab: anti-ErbB3-Fab-YRH, dark grey Fab: anti-ErbB2-Fab; eTl: enhanced Trypsiligase.

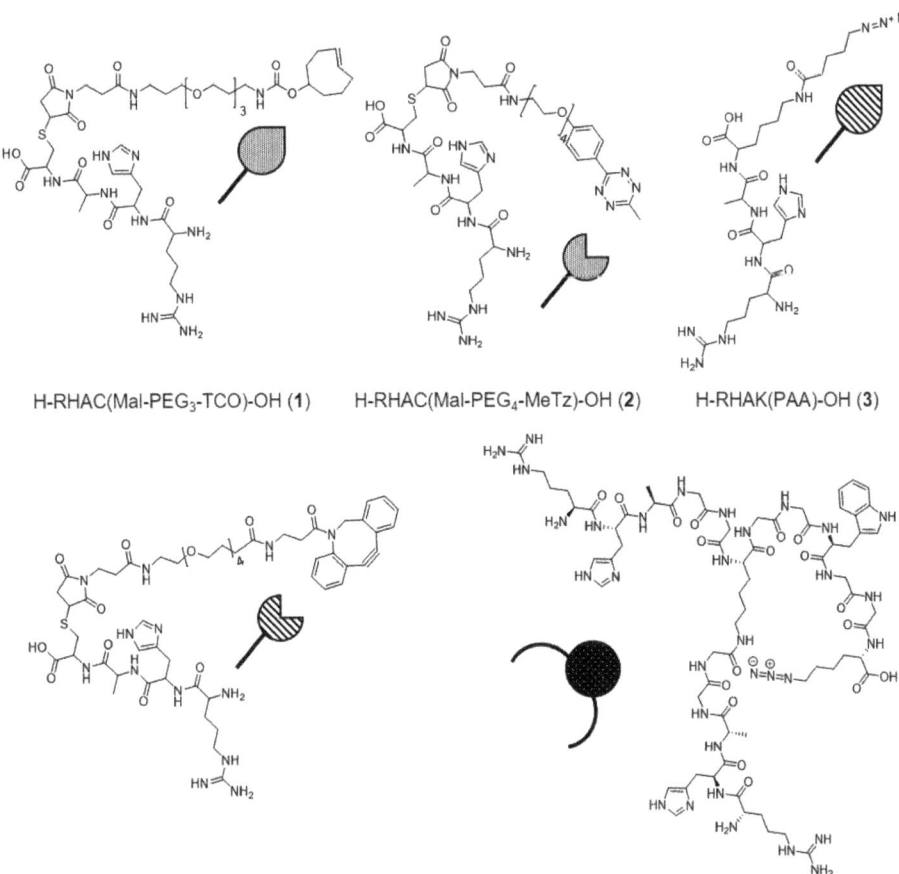

H-RHAC(Mal-PEG$_3$-TCO)-OH (1) H-RHAC(Mal-PEG$_4$-MeTz)-OH (2) H-RHAK(PAA)-OH (3)

H-RHAC(Mal-PEG$_4$-DBCO)-OH (4) H-RHAGGK(H-RHAGG)GGWGGK(N$_3$)-OH (5)

Figure 1. Structures of click linkers used. H-RHAC(Mal-PEG$_3$-TCO)-OH (1) and H-RHAC(Mal-PEG$_4$-MeTz)-OH (2) were used for IEDDA; H-RHAK(PSA)-OH (3) and H-RHAC(Mal-PEG$_4$-DBCO)-OH (4) were used for SPAAC; H-RHAGGK(H-RHAGG)GGWGGK(N$_3$)-OH (5) was used for enzymatic C- to C-terminal Fab coupling. Mal: maleiimide, TCO: *trans*-cyclooctene, MeTz: methyltetrazine, PAA: pentanoic acid azide, DBCO: dibenzocyclooctyne.

2.1. Enzymatic Synthesis of C- to N-Linked Anti-ErbB3-Anti-ErbB2-bsFab

The eTl-catalyzed direct coupling of anti-ErbB3-Fab-YRH and anti-ErbB2-RH-Fab, leading to the respective C- to N-linked bsFab conjugate, allows a simple one-step reaction regime, with a maximum yield of 60% of the bsFab product after reaction times of about 90 min at the reaction conditions used (Figure 2). After isolation by size exclusion chromatography (SEC), a single homogeneous bsFab conjugate was verified by LC-MS (Figure 2C), which was subsequently tested for biological functionality (Section 2.3). It should be mentioned that besides the main product and starting Fab substrates, only one further reaction product was found corresponding to the anti-ErbB3-Fab-Y-OH species in which the last two amino acid residues (RH) of the recognition sequence were missing. This indicated a certain enzymatic hydrolysis activity by eTl at the Tyr-Arg site, which is, however, usually negligible for this enzyme. The slightly reduced product yields indicated somewhat higher hydrolysis rates which may be due to limited accessibility of the enzyme recognition sequence at the anti-ErbB2-RH-Fab.

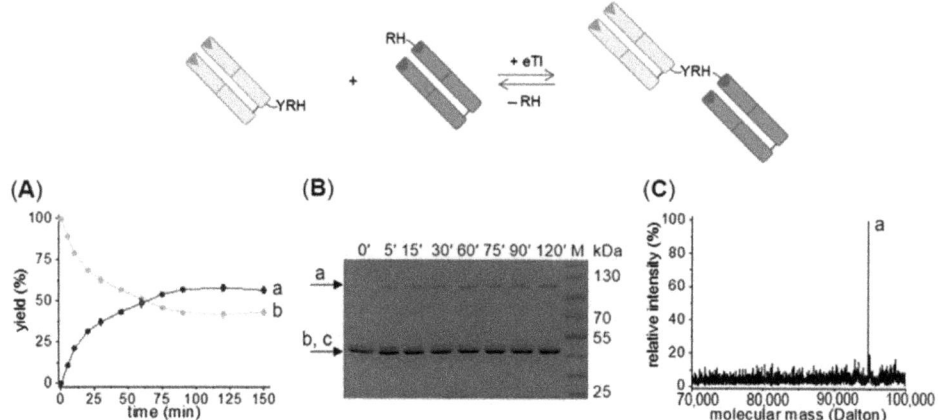

Figure 2. Synthesis of *C*- to *N*-terminal conjugated bsFab via direct eTl-coupling of anti-ErbB3-Fab-YRH and anti-ErbB2-RH-Fab. (**A**) Kinetics of product formation with reaction yields of about 60%, a: anti-ErbB3-YRH-anti-ErbB2-bsFab, b: anti-ErbB2-RH-Fab; (**B**) Time-resolved SDS-PAGE analysis of coupling reaction, c: anti-ErbB3-Fab-YRH, M: molecular marker; (**C**) MS analysis of the final bsFab product after purification by SEC, anti-ErbB3-YRH-anti-ErbB2-bsFab M_{calcd}: 94,804 Da, M_{found}: 94,805 Da. Light grey Fab: anti-ErbB3-Fab-YRH, dark grey Fab: anti-ErbB2-Fab-RH; eTl: enhanced Trypsiligase. Reaction conditions: 20 µM anti-ErbB2-RH-Fab, 100 µM anti-ErbB3-Fab-YRH, 2.5 µM eTl, 50 µM $ZnCl_2$, 100 mM HEPES/NaOH, pH 7.8, 100 mM NaCl, 10 mM $CaCl_2$.

2.2. Generation of C- to C-Linked Anti-ErbB2-Anti-ErbB3-bsFab

Posttranslational synthesis of *C*- to *C*-linked anti-ErbB2-anti-ErbB3-bsFab was evaluated by chemo–enzymatic and purely enzymatic approaches as well. Both are two-step reaction processes in which the Fab substrates were first site-specifically modified and then converted into the final bsFabs.

2.2.1. Chemo–Enzymatic Synthesis of C- to C-Linked Anti-ErbB2-Anti-ErbB3-bsFab

Chemo–enzymatic assembly of *C*- to *C*-linked anti-ErbB2-anti-ErbB3-bsFab begins with the site-specific coupling of click anchors to the starting Fabs by eTl-catalysis. For IEDDA click chemistry, the *trans*-cyclooctene (TCO) based click linker **1** was enzymatically coupled to anti-ErbB2-Fab-YRH, while the methyltetrazine (MeTz) anchor **2** was linked to anti-ErbB3-Fab-YRH. In the case of SPAAC chemistry (Figure S1), pentanoic acid azide (PAA) (compound **3**) instead of MeTz was used, and dibenzocyclooctyne (DBCO) (compound **4**) instead of TCO. Regardless of the nature of the click reagent, similar reaction conditions were used for all couplings.

The typical course of the eTl-mediated coupling reactions is shown in Figure 3 as an example of the synthesis of anti-ErbB2-Fab-TCO and anti-ErbB3-Fab-MeTz. According to Figure 3A, product yields higher than 72% were reached after approximately 75 min of reaction time. The reactions proceeded cleanly and resulted in an equilibrium between the reaction product and starting substrate at the end of the reaction (Figure 3B). Undesired by-products did not occur apart from traces of partially hydrolyzed anti-ErbB2-Y-OH (Figure 3C). With respect to former studies, such reaction processes are rather typical for eTl-catalyzed transamidation reactions [25]. Comparable yields for the SPAAC-based products, i.e., PAA and DBCO, were found in reactions with linkers **3** and **4** (Figure S1A). Following synthesis, the reaction products were isolated by affinity chromatography mainly to remove the excess click anchor substrates and eTl. The remaining quantities of Fab substrates, on the other hand, did not interfere with the further course of synthesis.

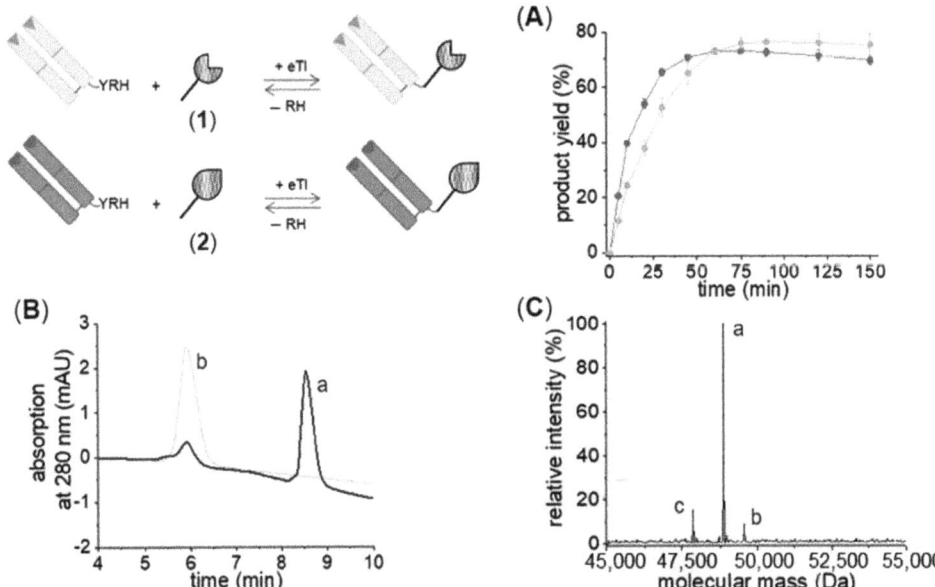

Figure 3. Coupling of anti-ErbB2-Fab-YRH and anti-ErbB3-Fab-YRH to click linkers **1** and **2**, respectively, via eTl-catalysis. (**A**) Reaction kinetics with maximum product yields higher than 72%, light grey line: anti-ErbB2-Fab-TCO, grey line: anti-ErbB3-Fab-MeTz; (**B**) HPLC analysis of the reaction mixture of anti-ErbB2-Fab-YRH with TCO click linker **1** at 0 min (light grey) and 75 min (dark grey) reaction times; (**C**) Mass spectrogram of the reaction mixture of anti-ErbB2-Fab-YRH with TCO click linker **1** after 75 min reaction time and after separation of excess click anchor via affinity chromatography (a: anti-ErbB2-Fab-TCO M_{calcd}: 48,860 Da, M_{found}: 48,861 Da, b: anti-ErbB2-Fab-YRH M_{calcd}: 49,548 Da, M_{found}: 49,549 Da, c: anti-ErbB2-Fab-Y-OH M_{calcd}: 47,870 Da, M_{found}: 47,871 Da). Light grey Fab: anti-ErbB3-Fab-YRH, dark grey Fab: anti-ErbB2-Fab-YRH; eTl: enhanced Trypsiligase. Reaction conditions: 100 µM anti-ErbB2-Fab-YRH and anti-ErbB3-Fab-YRH, 500 µM click linker **1** and **2**, 10 µM eTl, 100 µM $ZnCl_2$, 100 mM HEPES/NaOH, pH 7.8, 100 mM NaCl, 10 mM $CaCl_2$.

The subsequent SPAAC- and IEDDA-based click reactions were performed according to established protocols (Section 3.5). The course and analysis of the reactions are shown in Figure 4 using the IEDDA reaction as an example. As for the SPAAC reaction (Figure S1), a quantitative product yield could also be obtained for the IEDDA coupling at a stoichiometry of 1:2 of the starting substrates. The only differences were in the reaction times, which ranged from several minutes to a few hours. In fact, the IEDDA-based click reaction was completed within about 30 min (Figure 4A). The SPAAC reaction, on the other hand, took about 4 h to reach complete conversion (Figure S1B,C). Regardless of the individual reaction time, in all cases, the formation of the desired conjugates (Figures 4A,B and S1B,C) could be detected after only a few seconds (Figures 4C and S1D). The remaining bands after complete conversion at 40–55 kDa in the SDS-PAGE of Figure 4C (and Figure S1D, respectively) corresponded to unseparated, unmodified Fab species from the enzymatic reaction. These and the excess click component were finally separated by SEC in a one-step purification, yielding a final bispecific product of high purity and homogeneity (Figures 4D and S1E).

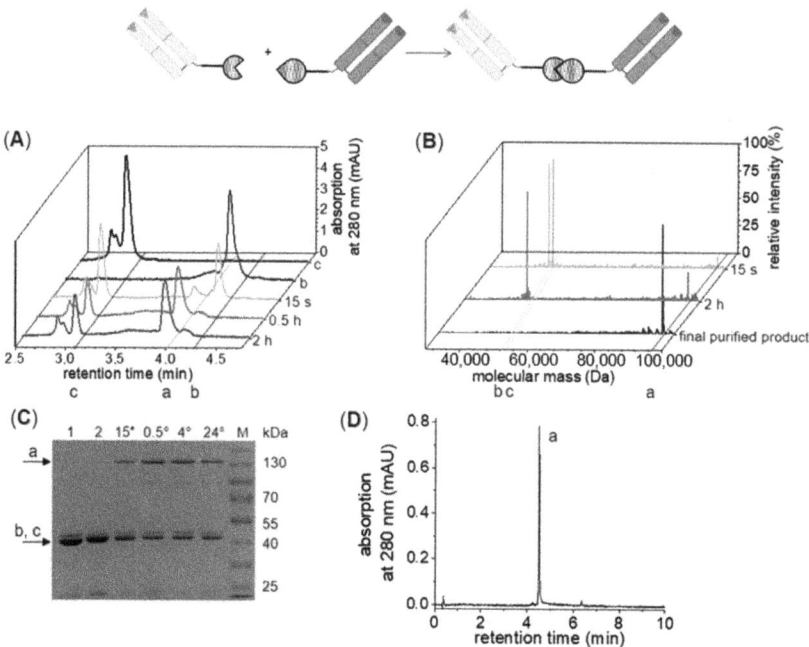

Figure 4. Course and analysis of the click reaction of eTl synthesized anti-ErbB3-Fab-MeTz with anti-ErbB2-Fab-TCO forming anti-ErbB3-Fab-IEDDA-anti-ErbB2-bsFab. (**A**) Time-resolved UPLC analysis of the click reaction showing a complete conversion within 0.5 h of the reaction time, a: anti-ErbB3-Fab-IEDDA-anti-ErbB2-bsFab, b: anti-ErbB3-Fab-MeTz, c: anti-ErbB2-Fab-TCO; (**B**) MS analysis of the click reaction after reaction start, 2 h reaction time and, of the final conjugation product (a: anti-ErbB3-Fab-IEDDA-anti-ErbB2-bsFab M_{calcd}: 96,536 Da, M_{found}: 96,537 Da, b: anti-ErbB3-Fab-MeTz M_{calcd}: 47,704 Da, M_{found}: 47,705 Da, c: anti-ErbB2-Fab-TCO M_{calcd}: 48,860 Da, M_{found}: 48,861 Da); (**C**) Time-resolved SDS-PAGE analysis of the click reaction (1: anti-ErbB3-Fab-MeTz, 2: anti-ErbB2-Fab-TCO, M: molecular marker); (**D**) UPLC profile of anti-ErbB3-Fab-IEDDA-anti-ErbB2-bsFab conjugate after purification via SEC. Light grey Fab: anti-ErbB3-Fab-MeTz, dark grey Fab: anti-ErbB2-Fab-TCO. Reaction conditions: 30 µM anti-ErbB3-Fab-MeTz, 60 µM anti-ErbB2-Fab-TCO in PBS, (**A**): 30–50% acetonitrile/ddH$_2$O in 10 min, (**D**): 10–80% acetonitrile/ddH$_2$O in 10 min.

2.2.2. Enzymatic Synthesis of C- to C-Linked Anti-ErbB2-Anti-ErbB3-bsFab

The eTl-catalyzed C- to C-linkage of two Fabs necessarily requires a special linker structure equipped with two nucleophilic recognition sequences for the biocatalyst (RH-motifs). Linker 5 obviously fulfilled this requirement. Furthermore, two Fab substrates were required, each carrying the recognition sequence YRH for eTl at the C-terminus. The reaction setting consisting of linker 5 and anti-ErbB2-Fab-YRH and anti-ErbB3-Fab-YRH fulfilled both requirements but bore the general risk that, in addition to the bispecific product, the respective homodimers are simultaneously formed from anti-ErbB2-Fab or anti-ErbB3-Fab. A sequential reaction mode, starting first with the coupling of anti-ErbB2-Fab-YRH with the linker and second, the coupling of anti-ErbB3-Fab-YRH to the resulting intermediate could minimize this risk, especially if the nucleophilic component (linker 5) was used in excess. Since the latter was the standard case in all previously performed eTl reactions, the reaction conditions were maintained. The results of both reactions are shown in Figure 5.

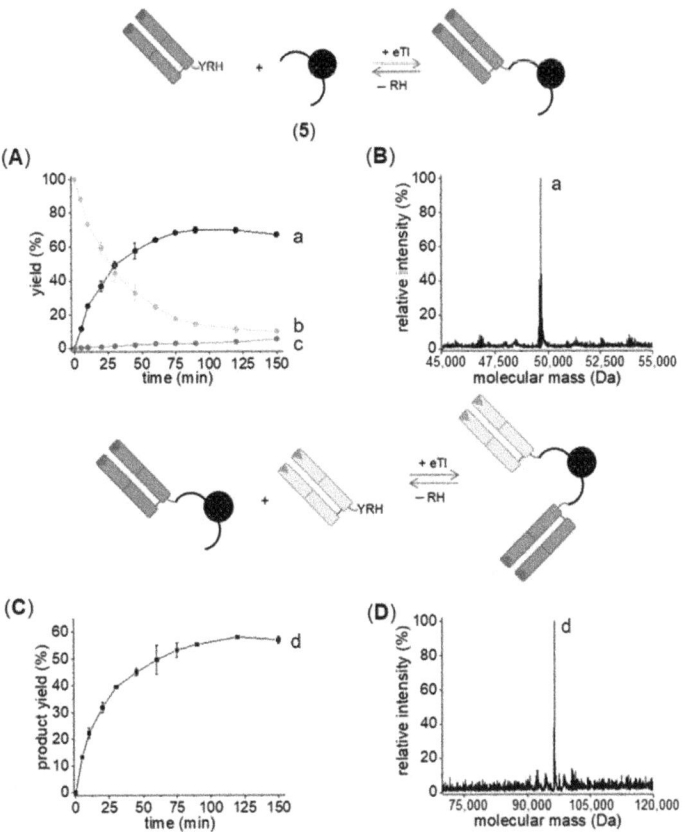

Figure 5. Results of the enzymatic C- to C-terminal coupling of anti-ErbB2-Fab-YRH and anti-ErbB3-Fab-YRH via linker **5**. (**A**) Reaction kinetics: the single modified product (black line) was mainly formed while only traces of the homodimeric product (dark grey line) were generated. a: anti-ErbB2-Fab-linker **5**, b: anti-ErbB2-Fab-YRH, c: homodimeric anti-ErbB2-Fab-linker **5**-anti-ErbB2-Fab; (**B**) MS analysis of the product purified by HIC, a: anti-ErbB2-Fab-linker **5** M_{calcd}: 49,538 Da, M_{found}: 49,539 Da; (**C**) Reaction course of anti-ErbB2-linker **5**-anti-ErbB3-bsFab synthesis with yields of about 60% analyzed by HIC, d: anti-ErbB2-linker **5**-anti-ErbB3-bsFab; (**D**) MS analysis of the purified bsFab, anti-ErbB2-linker **5**-anti-ErbB3-bsFab M_{calcd}: 96,244 Da, M_{found}: 96,245 Da. Reaction conditions: (**A**): 100 µM anti-ErbB2-Fab-YRH, 500 µM linker **5**, 10 µM eTI, 100 µM $ZnCl_2$; (**C**): 20 µM anti-ErbB2-Fab-linker **5**, 100 µM anti-ErbB3-Fab-YRH, 2.5 µM eTI, 50 µM $ZnCl_2$; (**A,C**): 100 mM HEPES/NaOH, pH 7.8, 100 mM NaCl, 10 mM $CaCl_2$.

As it can be seen in Figure 5A, for the first reaction of the two-step procedure, the coupling of anti-ErbB2-Fab-YRH with linker **5**, a product yield of higher 70% was obtained after approximately 75 min of reaction time. This corresponded to the results obtained with linkers **1** to **4**. Interestingly, only about 3% of the homodimeric anti-ErbB2-linker **5**-anti-ErbB2-Fab was formed (Figure 5A). Apparently, the 5-fold excess of the nucleophilic linker used was already sufficient to almost completely prevent unwanted homodimerization. After the monomeric anti-ErbB2-Fab-linker **5** product was purified by means of hydrophobic interaction chromatography (HIC) (Figure 5B), the second eTI-catalyzed coupling step was carried out by adding anti-ErbB3-Fab-YRH under again analogous reaction conditions. The result of this second enzymatic reaction is shown in Figure 5C,D. Corresponding to the course of the reaction shown in Figure 5C, a product yield of approx. 60% of the desired anti-ErbB2-Fab-linker **5**-anti-ErbB3-bsFab could be obtained in a reaction time of approx.

120 min. This yield corresponded to the C- to N-terminal coupling of the two Fabs by eTl (Section 2.1). Finally, the bispecific product was isolated and purified by SEC (Figure 5D) and subsequently used for functional studies (Section 2.3). Noticeably, with this method, it was rather impossible to control which Fab was attached to which site of the linker, which is, however, without relevance if the linker structure is symmetric. In cases where control is essential, two orthogonal Trypsiligase variants could be used in a one-step procedure to address this problem [25].

2.3. Analysis of In Vitro Functionality of the Generated bsFabs

2.3.1. Surface Plasmon Resonance (SPR)-Based Activity Assay

The in vitro function of all synthesized and purified anti-ErbB2-anti-ErbB3-bsFabs was first analyzed regarding their epitope-binding behavior. For this purpose, the dissociation constants (K_D) were investigated using an SPR-based activity assay compared to the single ErbB2- and ErbB3-Fabs. The K_D values obtained were determined sequentially by immobilizing the ectodomains of the receptors ErbB2 and ErbB3 on separate sensor chip surfaces and are summarized in Table 1. The complete SPR sensorgrams as well as the values for k_{on} and k_{off} are shown in Figure S6 and Figure S7. First, for the single anti-ErbB2- and anti-ErbB3-Fabs, it was noted that neither showed any cross-reactivity in their antigen-binding properties (Table 1). Second, it is clear from Table 1 that the additional N-terminal RH-motif in the anti-ErbB2-Fab did not affect the dissociation constant. As a consequence, the K_D values were in a narrow range and correspond to those described in the literature [23,33]. Third, the results further showed that all bsFabs synthesized retained their binding functionality. The determined dissociation constants were comparable to those of the individual single Fabs. Fourth, even the binding behavior of the inner domain of the C- to N-terminal linked bsFab format (anti-ErbB2-domain) did not appear to be affected by the outer anti-ErbB3-domain. Thus, it can be concluded that the type of linkage in which the two individual ErbB2 and ErbB3 Fabs are fused (at least as it is in the synthesized formats), has no significant influence on the binding properties to the antigens.

Table 1. Comparison of dissociation constants of single and bispecific ErbB2- and ErbB3-Fabs determined by SPR. Biotinylated ErbB2 or ErbB3 ectodomains were immobilized separately on streptavidin sensor chips; no binding was detected. SPAAC: strain-promoted alkyne-azide cycloaddition, IEDDA: inverse electron-demand Diels–Alder reaction.

	ErbB3-ECD K_D (pM)	ErbB2-ECD K_D (pM)
anti-ErbB3-Fab-YRH	99 ± 22	-
anti-ErbB2-Fab-YRH	-	139 ± 31
anti-ErbB2-RH-Fab	-	137 ± 43
anti-ErbB2-SPAAC-anti-ErbB3	73 ± 21	130 ± 28
anti-ErbB3-IEDDA-anti-ErbB2	63 ± 19	113 ± 20
anti-ErbB3-YRH-anti-ErbB2	95 ± 45	138 ± 46
anti-ErbB2-linker 5-anti-ErbB3	97 ± 33	108 ± 20

2.3.2. Receptor Internalization Assay

The internalization studies were conducted with three breast cancer cell lines. These cell lines differ in the number of ErbB2 and ErbB3 receptors expressed on their cell surfaces. SKBR-3 cells show high levels of both receptors, while HCC-1954 cells are known for high ErbB2 and low ErbB3 receptor expression. In contrast, the reverse holds true for MCF-7 cells. They are characterized by low ErbB2 and high ErbB3 receptor levels [35]. It was expected that the bsFab formats, in addition to binding the respective specific antigens, would also exhibit synergistic binding in the presence of both antigens on the cell surface.

First, all single and bispecific Fabs to be measured were modified non-specifically with an AlexaFluor568 NHS ester for the internalization studies. The resulting dye loading varied from 2 to 11 depending on the protein used. SDS-PAGE and UV/Vis absorption spectroscopic analyses showed no evidence of aggregate formation, especially at higher dye loadings (Figure S8). In addition, a lysosomal dye was used to follow the path of the proteins in the cells. The progression of internalization is shown for anti-ErbB3-IEDDA-anti-ErbB2-bsFab and SKBR-3 cells as an example of all bsFabs in Figure 6. In addition, the course of internalization of every single anti-ErbB2- and anti-ErbB3-Fab is shown. The complete data sets for all bsFabs and cell lines can be found in Figures S2–S5 in the Supplementary Materials. In general, little fluorescence intensities for all Fabs/bsFabs were observed for the MCF-7 cell line (Figure S3). In contrast, for SKBR-3 cells, significant fluorescence signals of all added proteins corresponding to that of the lysosomes were found, indicating internalization (Figure 6). Remarkably, the fluorescence signals of the bsFabs in the case of the SKBR-3 cells were much stronger than that of the single Fabs alone. The fluorescence intensity of the single anti-ErbB3-Fab was worse compared to that of the single anti-ErbB2 counterpart. A more in-depth quantification of these tendencies based on the performed fluorescence microscopic analyses was, however, not possible with certainty. Additional flow cytometric studies are recommended for this purpose, but these were beyond the focus of this work.

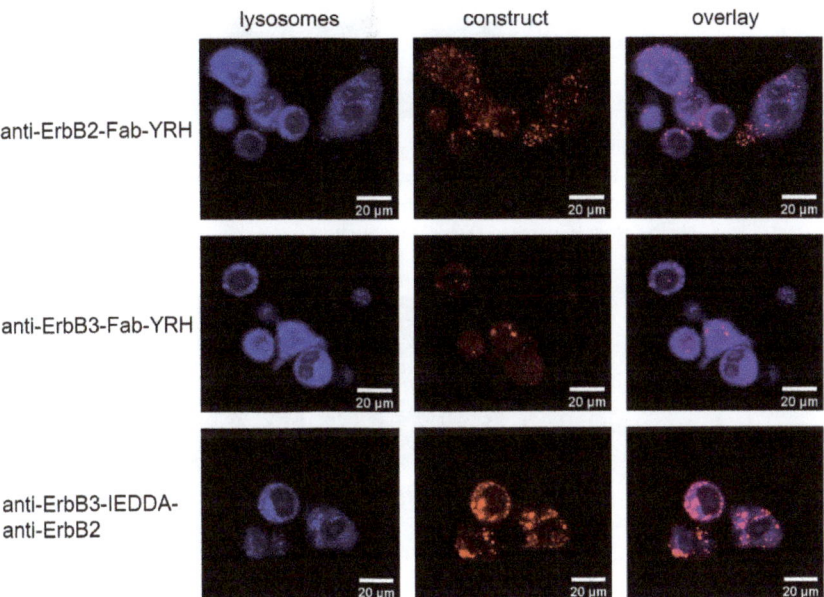

Figure 6. Internalization of anti-ErbB3-IEDDA-anti-ErbB2-bsFab compared to the single anti-ErbB2- and anti-ErbB3-Fab by lysosome-stained SKBR-3 cells after 24 h. Lysosomes were stained with LysoBriteBlue, single and bispecific Fabs were modified with AlexaFluor568 NHS ester. IEDDA: inverse electron-demand Diels–Alder reaction.

In the case of HCC-1954 cells, a similar trend to SKBR-3 cells was observed (Figure 7). While the fluorescence spots for the single Fabs were still congruent with the lysosomes, this was only partially the case for the bsFabs. A plausible explanation could be that the bsFab-receptor complexes are translocated into the nucleus via the non-canonical pathway [36–38]. To test this hypothesis initially, we performed the same experiment with nuclear instead of lysosome staining (Figure 8). While the fluorescence signal of the single anti-ErbB2-Fab-YRH was mainly found distant from that of the nucleus, the signals of all bsFabs could be found both distant and superimposed on the nucleus. This finding may indicate a different distribution of mono- and bispecifics in HCC-1954 cells. Whether the bsFabs can actually be found on or even in the cell nucleus cannot be concluded with certainty from this finding.

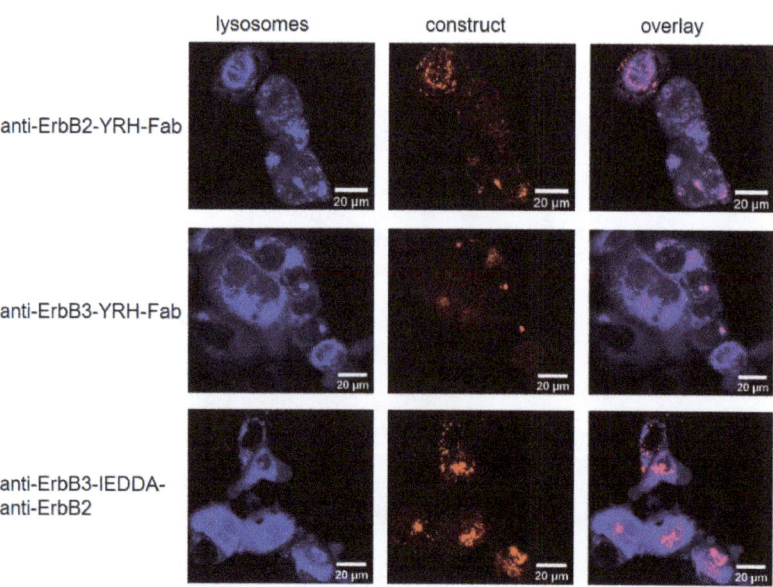

Figure 7. Internalization of anti-ErbB3-IEDDA-anti-ErbB2-bsFab compared to the single anti-ErbB2- and anti-ErbB3-Fab by lysosome-stained HCC-1954 cells after 24 h. Lysosomes were stained with LysoBriteBlue, single and bispecific Fabs were modified with AlexaFluor568 NHS ester. IEDDA: inverse electron-demand Diels–Alder reaction.

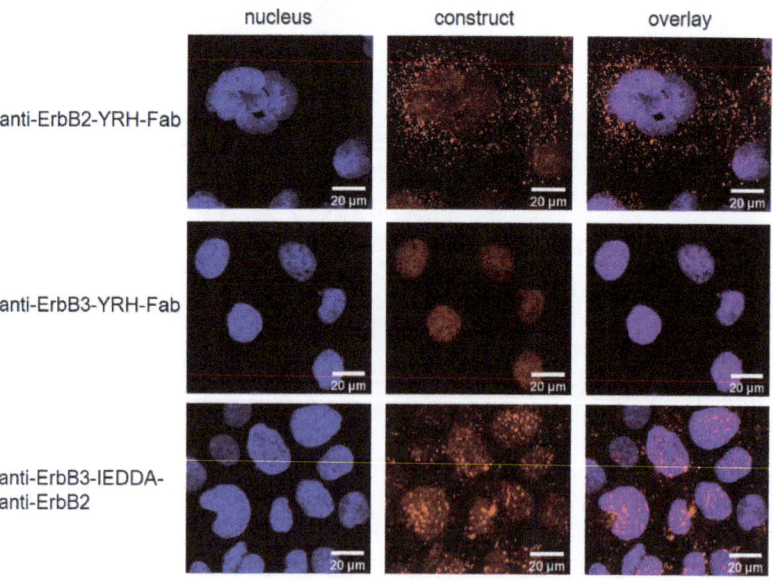

Figure 8. Internalization of anti-ErbB3-IEDDA-anti-ErbB2-bsFab compared to the single anti-ErbB2- and anti-ErbB3-Fab by cell nucleus-stained HCC-1954 cells after 24 h. Substantial portions of the bsFabs could be found in the cell nucleus. Cell nuclei were stained with HOECHST33342, single and bispecific Fabs were modified with AlexaFluor568 NHS ester. IEDDA: inverse electron-demand Diels–Alder reaction.

3. Materials and Methods

3.1. Chemicals and Peptide Synthesis

All reagents were purchased in the highest quality available at Sigma–Aldrich (St. Louis, MO, USA). Peptides were synthesized following standard procedures using Fmoc/tBu strategy [24,39,40]. Amino acids and reagents for peptide synthesis were purchased at IRIS Biotech (Marktredwitz, Germany). Click reagents were purchased at Jena Bioscience (Jena, Germany) and IRIS Biotech (Marktredwitz, Germany).

H-RHAC-OH peptide and maleimide click reagents were dissolved in a phosphate buffer (100 mM NaH$_2$PO$_4$/NaOH, 150 mM NaCl, pH 7.5) with a maximum of 10% (v/v) DMF. Reactants were mixed in a ratio of 1:1.1 (peptide to click component) and were incubated for 2 h at room temperature. Purification was done by preparative HPLC (waters system (Milford, MA, USA), XSelect® Peptide CSH™ C18) with a linear gradient of 5–95% acetonitrile/ddH$_2$O in 40 min. The identity of the synthesis products was verified by electrospray ionization mass spectrometry (waters Micromass® ZQ™ (Milford, MA, USA)) and the purity was determined by UPLC (5–95% acetonitrile/ddH$_2$O in 4 min at 220 nm; waters system (Milford, MA, USA), ACQUITY BEH 130, C18, 1.7 µm, 2.1 × 100 mm).

Analysis: H-RHAC(Mal-PEG$_3$-TCO)-OH: LC-MS (ESI$^+$) m/z: 1008.5 [M]$^+$, calcd. for [C$_{44}$H$_{72}$N$_{12}$O$_{13}$S]$^+$ m/z: 1009.3, purity: 94.1%; H-RHAC(Mal-PEG$_4$-MeTz)-OH: LC-MS (ESI$^+$) m/z: 999.8 [M]$^+$, calcd. for [C$_{42}$H$_{61}$N$_{15}$O$_{12}$S]$^+$ m/z: 999.4, purity: 97.7%; H-RHAC(Mal-PEG$_4$-DBCO)-OH: LC-MS (ESI$^+$) m/z: 1160.8 [M]$^+$, calcd. for [C$_{54}$H$_{73}$N$_{13}$O$_{14}$S]$^+$ m/z: 1159.5, purity: 97.1%; H-RHAK(PAA)-OH: LC-MS (ESI$^+$) m/z: 636.1 [M]$^+$, calcd. for [C$_{26}$H$_{45}$N$_{13}$O$_6$]$^+$ m/z: 635.4, purity: 91.5%; H-RHAGGK(H-RHAGG)GGWGGK(N$_3$)-OH: LC-MS (ESI$^+$) m/z: 1671.5 [M]$^+$, calcd. for [C$_{69}$H$_{106}$N$_{32}$O$_{18}$]$^+$ m/z: 1670.8, purity: 95.3%.

3.2. Production of eTl

The eTl encoding plasmid (pPICZaA) was linearized by *SacI*-HF (NEB (Ipswich, MA, USA) and purified by agarose gel electrophoresis (NucleoSpin Gel and PCR Clean-up-Kit (Macherey and Nagel (Düren, Germany)).

Pichia pastoris X-33 cells were electroporated according to the protocol described in [41]. The selection of positive clones was performed by Zeocin (200 µg/mL). Cells were cultivated for 24 h, 30 °C in YNB-medium (13.4 g/L YNB (w/o aa), 50 mM MES/NaOH; pH 6.0, 10 mM CaCl$_2$ with 20% (m/v) glucose and then pelleted and transferred in YNB-medium with 1% (v/v) methanol. After 3 days, the supernatant was harvested and diluted in IEX buffer (1:1) (20 mM Na-acetate/acetic acid; pH 4.0). After loading on a cation exchange column (HiPrep SP FF, Cytiva (Chalfont St Giles, UK)), the protein was eluted with 100 mM HEPES/NaOH; pH 7.8, 200 mM NaCl, 10 mM CaCl$_2$. Final purification was done by SEC (S75 pg, 16/600, Cytiva (Chalfont St Giles, UK)) in HEPES-buffer (100 mM HEPES/NaOH, pH 7.8, 100 mM NaCl, 10 mM CaCl$_2$).

Analysis: Calculated molecular mass: M$_{calcd}$: 23,758 Da, M$_{found}$: 23,759 Da.

3.3. Production of Fabs

The anti-ErbB3-Fab was designed according to the protocol described in [33]. Anti-ErbB3-Fab-YRH and anti-ErbB2-Fab-YRH have a prolonged C-terminal protein sequence at the heavy chain with the sequence: PGGYRHAAGEQKLISEEDL [25]. The N-terminally extended anti-ErbB2-RH-Fab was prolonged at the heavy chain with the amino acid sequence RHA, while the RH-motif represents the enzyme recognition tag, an A serves as an additional spacer amino acid. The pASK-anti-ErbB2-Fab plasmid and the pET-anti-ErbB3-Fab plasmid were transformed into *E. coli* BL21 (DE3) cells. Cultures were grown at 37 °C with either 100 µg/mL ampicillin or 50 µg/mL kanamycin until OD$_{600nm}$ of 0.8 was reached. Induction was done by adding 0.1 µg/mL anhydrotetracycline and 0.5 mM IPTG, respectively, for 4 h, 30 °C. Cell pellets were resuspended in phosphate buffer (50 mM NaH$_2$PO$_4$; pH 7.0) and lysed by sonification (amplitude 30%). The Fabs were purified via Protein G or Protein A affinity chromatography (HiTrap Protein G or A HP, Cytiva (Chalfont St Giles,

UK)) and SEC (S75 pg, 16/600) in HEPES-buffer (100 mM HEPES/NaOH, pH 7.8, 100 mM NaCl, 10 mM CaCl$_2$).

Analysis: Calculated molecular masses for Fabs: anti-ErbB3-Fab-YRH M$_{calcd}$: 48,401 Da, M$_{found}$: 48,402 Da; anti-ErbB2-Fab-YRH M$_{calcd}$: 49,548 Da, M$_{found}$: 49,549 Da; anti-ErbB2-RH-Fab M$_{calcd}$: 48,099 Da, M$_{found}$: 48,100 Da.

3.4. C-Terminal Modification of Fabs by eTl

The following standard reaction conditions were used for eTl catalyzed transamidation reactions: 100 µM Fab, 100 µM ZnCl$_2$ in HEPES buffer (100 mM HEPES/NaOH, pH 7.8, 100 mM NaCl, 10 mM CaCl$_2$) in a reaction volume of 100–500 µL. The nucleophile was freshly dissolved in buffer and added to Fab in 5-fold molar excess. The reaction was initiated by adding 10 µM of eTl. Incubation was performed at 30 °C, 550 rpm for 150 min. Reaction kinetics were examined by HIC according to [24] with a linear gradient from 95% equilibration buffer (1.5 M ammonium sulfate, 0.05 HEPES/NaOH; pH 7.5) to 0% in 15 min, SDS-PAGE [23,24,42] (molecular marker: PageRuler Plus pre-stained protein ladder (ThermoFisher Scientific (Waltham, MA, USA)) and LC-MS analysis [24].

3.5. Chemo–Enzymatic Production of bsFabs

A standard reaction mixture was prepared: 100 µM Fab, 500 µM linker **1–4**, 10 µM eTl, 100 µM ZnCl$_2$ in HEPES buffer (100 mM HEPES/NaOH, pH 7.8, 100 mM NaCl, 10 mM CaCl$_2$). When reaching the product maximum, the mixture was diluted in phosphate buffer (50 mM NaH$_2$PO$_4$; pH 7.0) (1:5). Anti-ErbB2- and anti-ErbB3-Fab mixtures were loaded onto a Protein G and A Spin Column, respectively (NAb Protein G or A Spin Columns, ThermoFisher Scientific (Waltham, MA, USA)). Purification was performed by 5 washing steps with phosphate buffer (50 mM NaH$_2$PO$_4$; pH 7.0) and 4 elution steps (100 mM glycine/HCl; pH 2.7). The elution fractions were pooled, and the buffer was exchanged by cross filtration to phosphate-buffered saline (PBS, 137 mM NaCl, 2.7 mM KCl, 10 mM Na$_2$HPO$_4$, 1.8 mM KH$_2$PO$_4$; pH 7.4). Click anchor modified Fabs were mixed together in a ratio of 1:2 according to the manufacturer's instructions (Jena Bioscience (Jena, Germany)) for the respective click reaction type and monitored by 10% SDS-PAGE, UPLC (waters system (Milford, MA, USA), Aeris^{TM} 3.6 µm WIDEPORE XB-C18 column, Phenomenex (Torrance, CA, USA)) and LC-MS. Purification of the bsFabs was done by SEC (S200 pg, 16/600, Cytiva (Chalfont St Giles, UK)) in PBS.

3.6. Enzymatic Production of C- to C- and C- to N-Terminal bsFabs

For the C- to C-terminal Fab couplings, a standard reaction mixture was prepared: 100 µM anti-ErbB2-Fab-YRH, 500 µM linker **5**, 10 µM eTl, 100 µM ZnCl$_2$ in HEPES buffer (100 mM HEPES/NaOH, pH 7.8, 100 mM NaCl, 10 mM CaCl$_2$). When reaching the maximum product yield, the mixture was purified by HIC. The single modified product was concentrated, buffer exchanged by cross filtration to HEPES buffer (100 mM HEPES/NaOH, pH 7.8, 100 mM NaCl, 10 mM CaCl$_2$) and used for the following modification step. A standard mixture was prepared to generate either the C- to C- or C- to N-terminal bsFab (100 µM anti-ErbB3-Fab-YRH, 20 µM modified product or anti-ErbB2-RH-Fab, 2.5 µM eTl, 50 µM ZnCl$_2$ in HEPES buffer (100 mM HEPES/NaOH, pH 7.8, 100 mM NaCl, 10 mM CaCl$_2$)). Upon reaching the maximum product yield, the reaction mixture was diluted in 100 mM glycine/HCl; pH 2.7 and incubated for 10 min. Purification was done by SEC (S200 pg, 16/600) in PBS.

3.7. SPR-Based Activity Assay

SPR spectroscopy was performed on a BIAcore X instrument (BIAcore (Uppsala, Sweden)). The biotinylated ErbB2 or ErbB3 ectodomains (BioCat, (Heidelberg, Germany)) were immobilized, each on a separate SAHC 200 M sensor chip (XanTec bioanalytics (Düsseldorf, Germany)), resulting in a surface density of approximately 1600 RU. The Fabs or bsFabs were applied in a dilution series using PBS as running buffer. Complex formation was observed at a continuous flow rate of 30 µL/min for 120 s. Kinetic parameters were determined by fitting the data to the 1:1 Langmuir binding model with the BIAevaluation software (BIAcore (Uppsala, Sweden)). After each injection, the surface was regenerated by injecting 10 µL 10 mM NaOH, 1 M NaCl.

3.8. Receptor Internalization Assay

Fluorescence labeling of the proteins was performed according to the AlexaFluor568 carboxylic acid NHS ester manufacturer's instructions (ThermoFisher Scientific (Waltham, MA, USA)). SKBR-3-Luc (JCRB1627.1), MCF-7-Luc (JCRB1372), and HCC-1954-Luc (JCRB1476) were purchased from JCRB (Tokyo, Japan). For internalization 5000 cells per well were seeded in a µ-slide 8-well chamber coverslip (ibidi (Gräfeling, Germany)) and incubated for 48 h at 37 °C, 5% CO_2 (RPMI1640, 10% (v/v) fetal bovine serum (FBS), ThermoFisher Scientific (Waltham, MA, USA)). The medium was replaced by FluoroBrite DMEM medium with 10% (v/v) FBS (ThermoFisher Scientific (Waltham, MA, USA)) and 50 nM labeled Fab or bsFab was added and incubated for 24 h. Before starting microscopy, cells were washed with FluoroBrite DMEM medium without FBS. Cells were mixed with equivalent amounts of FluoroBrite DMEM medium and life cell staining buffer (supplemented with either LysoBriteBlue (1:500), AATBioquest (Sunnyvale, CA, USA), or HOECHST33342 (1:2000), ThermoFisher Scientific (Waltham, MA, USA)) and incubated prior to microscopy for 0.5–2 h and 10 min, respectively (Eclipse TE2000-E, CFI Plan Fluor 40x oil lens, Nikon (Tokyo, Japan)). Lysosome and nucleus staining were excited at 405 nm (filter 450/35), fluorescent proteins at 568 nm (filter 650LP).

4. Conclusions

Currently, more than one hundred formats of bsAbs have been described in the scientific literature. Not infrequently, the format determines the therapeutic efficacy. In this study, we presented an approach that can generate numerous amounts of such formats with only a single enzyme. By combining this with click chemistry techniques, a modular synthesis kit was created that allowed rapid and flexible shuffling of the individual antigen-binding domains in different but well-defined arrangements. The formats generated in this way can then be used to screen potentially suitable candidates for the respective application. The heterologous expression of each individual format or the use of different coupling procedures with individual reaction conditions, including the need for two distinct enzymes, as is currently the case, is not necessary. We were able to show that both the purely enzymatic synthesis and the chemo–enzymatic reactions enable high product yields with only short reaction times. The products were homogeneous and showed a uniform architecture. The formation of multimeric synthesis products, as found with the use of transglutaminase [18], could not be observed with eTl. We were also able to initially demonstrate the biological function of all bispecific constructs. As expected, we could show that the synthesized and fluorescence-labeled anti-ErbB2-anti-ErbB3-bsFabs exhibited improved fluorescence intensities in mammalian breast cancer cell lines compared with the single Fabs alone. Our results suggest that the bispecific formats produced might follow a partially different endocytotic pathway after internalization in certain cell lines than the individual Fabs from which they were constructed. In particular, the findings with labeled bsFabs, which in the case of HCC-1954 cells lead to additional fluorescence signals overlapping with those of the nuclei, give rise to further studies. The background for this is the clinical finding that the diagnosis of EGF receptors localized in the cell nucleus is associated with poor patient prognosis in the case of severe cancer progression [43,44]. On the basis of these findings, it would be very promising to investigate whether the transport of DNA-damaging toxins in the direction of the cell nucleus mediated by the bsFabs may lead to additional effects on the cancer cell compared to the monospecific Fabs. With the trifunctional linker already used in this study, this would be possible without any problems from a synthetic point of view and with high flexibility in terms of the active substance. Studies in this direction are presently in progress.

Supplementary Materials: The following material is available online at https://www.mdpi.com/article/10.3390/ijms23063144/s1.

Author Contributions: J.V. designed and executed all experiments; C.M. did the majority of the chemical syntheses and helped with the experimental chemo–enzymatic design; F.B. initiated and designed the project; J.V. and F.B. analyzed and interpreted the data and drafted the paper. All authors have read and agreed to the published version of the manuscript.

Funding: This work was supported by DFG project "Chemoselective reactions for the synthesis and application of functional proteins" [SPP1623] and the Aninstitut für Technische Biochemie at the Martin-Luther-University Halle-Wittenberg.

Acknowledgments: We would like to thank DBC Andreas Simon for the support in cell culture and Noelle Schwerdtner for help in microscopy analysis. We are grateful to Cordelia Schiene-Fischer and Sandra Liebscher for their support in Biacore measurements. Furthermore, we thank Sandra Liebscher and DBC Andreas Simon for their helpful discussions. We thank BioPharma Translationsinstitut Dessau Forschungs GmbH for providing the plasmid of the eTI and Marcus Böhme for his support regarding Fab construction.

Conflicts of Interest: The authors declare no conflict of interest.

References

1. Labrijn, A.F.; Janmaat, M.L.; Reichert, J.M.; Parren, P.W.H.I. Bispecific antibodies: A mechanistic review of the pipeline. *Nat. Rev. Drug Discov.* **2019**, *18*, 585–608. [CrossRef] [PubMed]
2. Brinkmann, U.; Kontermann, R.E. The making of bispecific antibodies. *mAbs* **2017**, *9*, 182–212. [CrossRef]
3. Ma, J.; Mo, Y.; Tang, M.; Shen, J.; Qi, Y.; Zhao, W.; Huang, Y.; Xu, Y.; Qian, C. Bispecific Antibodies: From Research to Clinical Application. *Front. Immunol.* **2021**, *12*, 626616. [CrossRef] [PubMed]
4. Lim, S.M.; Pyo, K.H.; Soo, R.A.; Cho, B.C. The promise of bispecific antibodies: Clinical applications and challenges. *Cancer Treat. Rev.* **2021**, *99*, 102240. [CrossRef] [PubMed]
5. Milstein, C.; Cuello, A. Hybrid hybridomas and their use in immunohistochemistry. *Nature* **1983**, *305*, 537–540. [CrossRef] [PubMed]
6. Milstein, C.; Cuello, A. Hybrid hybridomas and the production of bi-specific monoclonal antibodies. *Immunol. Today* **1984**, *5*, 299–304. [CrossRef]
7. Suresh, M.R.; Cuello, A.C.; Milstein, C. Advantages of bispecific hybridomas in one-step immunocytochemistry and immunoassays. *Proc. Natl. Acad. Sci. USA* **1986**, *83*, 7989–7993. [CrossRef] [PubMed]
8. Mack, M.; Riethmüller, G.; Kufer, P. A small bispecific antibody construct expressed as a functional single-chain molecule with high tumor cell cytotoxicity. *Proc. Natl. Acad. Sci. USA* **1995**, *92*, 7021–7025. [CrossRef]
9. Ridgway, J.B.; Presta, L.G.; Carter, P. 'Knobs-into-holes' engineering of antibody CH3 domains for heavy chain heterodimerization. *Protein Eng. Des. Sel.* **1996**, *9*, 617–621. [CrossRef]
10. Wu, C.; Ying, H.; Grinnell, C.; Bryant, S.; Miller, R.; Clabbers, A.; Bose, S.; McCarthy, D.; Zhu, R.-R.; Santora, L.; et al. Simultaneous targeting of multiple disease mediators by a dual-variable-domain immunoglobulin. *Nat. Biotechnol.* **2007**, *25*, 1290–1297. [CrossRef]
11. Wu, P.; Shui, W.; Carlson, B.L.; Hu, N.; Rabuka, D.; Lee, J.; Bertozzi, C.R. Site-specific chemical modification of recombinant proteins produced in mammalian cells by using the genetically encoded aldehyde tag. *Proc. Natl. Acad. Sci. USA* **2009**, *106*, 3000–3005. [CrossRef] [PubMed]
12. DiGiammarino, E.L.; Harlan, J.E.; Walter, K.A.; Ladror, U.S.; Edalji, R.P.; Hutchins, C.W.; Lake, M.R.; Greischar, A.J.; Liu, J.; Ghayur, T.; et al. Ligand association rates to the inner-variable-domain of a dual-variable-domain immunoglobulin are significantly impacted by linker design. *mAbs* **2011**, *3*, 487–494. [CrossRef] [PubMed]
13. Steinmetz, A.; Vallée, F.O.; Beil, C.; Lange, C.; Baurin, N.; Beninga, J.; Capdevila, C.; Corvey, C.; Dupuy, A.; Ferrari, P.; et al. CODV-Ig, a universal bispecific tetravalent and multifunctional immunoglobulin format for medical applications. *mAbs* **2016**, *8*, 867–878. [CrossRef] [PubMed]
14. Falck, G.; Müller, K.M. Enzyme-Based Labeling Strategies for Antibody–Drug Conjugates and Antibody Mimetics. *Antibodies* **2018**, *7*, 4. [CrossRef] [PubMed]
15. Park, J.; Lee, S.; Kim, Y.; Yoo, T.H. Methods to generate site-specific conjugates of antibody and protein. *Bioorg. Med. Chem.* **2021**, *30*, 115946. [CrossRef] [PubMed]
16. Zang, B.; Ren, J.; Li, D.; Huang, C.; Ma, H.; Peng, Q.; Ji, F.; Han, L.; Jia, L. Freezing-assisted synthesis of covalent C-C linked bivalent and bispecific nanobodies. *Org. Biomol. Chem.* **2019**, *17*, 257–263. [CrossRef] [PubMed]
17. Harmand, T.J.; Bousbaine, D.; Chan, A.; Zhang, X.; Liu, D.R.; Tam, J.P.; Ploegh, H.L. One-Pot Dual Labeling of IgG 1 and Preparation of C-to-C Fusion Proteins Through a Combination of Sortase A and Butelase 1. *Bioconjug. Chem.* **2019**, *29*, 3245–3249. [CrossRef] [PubMed]

18. Plagmann, I.; Chalaris, A.; Kruglov, A.A.; Nedospasov, S.; Rosenstiel, P.; Rose-John, S.; Scheller, J. Transglutaminase-catalyzed covalent multimerization of camelidea anti-human TNF single domain antibodies improves neutralizing activity. *J. Biotechnol.* **2009**, *142*, 170–178. [CrossRef]
19. Bartels, L.; Ploegh, H.L.; Spits, H.; Wagner, K. Preparation of bispecific antibody-protein adducts by site-specific chemo-enzymatic conjugation. *Methods* **2019**, *154*, 93–101. [CrossRef] [PubMed]
20. Berteau, O.; Guillot, A.; Benjdia, A.; Rabot, S. A new type of bacterial sulfatase reveals a novel maturation pathway in prokaryotes. *J. Biol. Chem.* **2006**, *281*, 22464–22470. [CrossRef] [PubMed]
21. Holder, P.G.; Jones, L.C.; Drake, P.M.; Barfield, R.M.; Banas, S.; de Hart, G.W.; Baker, J.; Rabuka, D. Reconstitution of formylglycine-generating enzyme with copper (II) for aldehyde tag conversion. *J. Biol. Chem.* **2015**, *290*, 15730–15745. [CrossRef] [PubMed]
22. Liebscher, S.; Schöpfel, M.; Aumüller, T.; Sharkhuukhen, A.; Pech, A.; Höss, E.; Parthier, C.; Jahreis, G.; Stubbs, M.T.; Bordusa, F. N-terminal protein modification by substrate-activated reverse proteolysis. *Angew. Chem. Int. Ed. Engl.* **2014**, *53*, 3024–3028. [CrossRef] [PubMed]
23. Liebscher, S.; Kornberger, P.; Fink, G.; Trost-Gross, E.M.; Höss, E.; Skerra, A.; Bordusa, F. Derivatization of antibody Fab fragments: A designer enzyme for native protein modification. *Chembiochem* **2014**, *15*, 1096–1100. [CrossRef] [PubMed]
24. Meyer, C.; Liebscher, S.; Bordusa, F. Selective Coupling of Click Anchors to Proteins via Trypsiligase. *Bioconjug. Chem.* **2016**, *27*, 47–53. [CrossRef] [PubMed]
25. Wartner, R.; Böhme, M.; Bordusa, F.; Simon, A.H.; Richter, T. Trypsin Variants with Improved Enzymatic Properties. WO 2020/127808 A1, 25 June 2020.
26. Kolb, H.C.; Finn, M.G.; Sharpless, K.B. Click Chemistry: Diverse Chemical Function from a Few Good Reactions. *Angew. Chem. Int. Ed. Engl.* **2001**, *40*, 2004–2021. [CrossRef]
27. Oliveira, B.L.; Guo, Z.; Bernardes, G.J.L. Inverse electron demand Diels-Alder reactions in chemical biology. *Chem. Soc. Rev.* **2017**, *46*, 4895–4950. [CrossRef] [PubMed]
28. Dommerholt, J.; Rutjes, F.P.J.T.; van Delft, F.L. Strain-Promoted 1,3-Dipolar Cycloaddition of Cycloalkynes and Organic Azides. *Top. Curr. Chem.* **2016**, *374*, 16. [CrossRef] [PubMed]
29. Slamon, D.J.; Clark, G.M.; Wong, S.G.; Levin, W.J.; Ullrich, A.; McGuire, W.L. Human breast cancer: Correlation of relapse and survival with amplification of the HER-2/neu oncogene. *Science* **1987**, *235*, 177–182. [CrossRef] [PubMed]
30. Slamon, D.J.; Godolphin, W.; Jones, L.A.; Holt, J.A.; Wong, S.G.; Keith, D.E.; Levin, W.J.; Stuart, S.G.; Udove, J.; Ullrich, A. Studies of the HER-2/neu proto-oncogene in human breast and ovarian cancer. *Science* **1989**, *244*, 707–712. [CrossRef] [PubMed]
31. Wallasch, C.; Weiss, F.; Niederfellner, G.; Jallal, B.; Issing, W.; Ullrich, A. Heregulin-dependent regulation of HER2/neu oncogenic signaling by heterodimerization with HER3. *EMBO J.* **1995**, *14*, 4267–4275. [CrossRef] [PubMed]
32. Pinkas-Kramarski, R.; Soussan, L.; Waterman, H.; Levkowitz, G.; Alroy, I.; Klapper, L.; Lavi, S.; Seger, R.; Ratzkin, B.J.; Sela, M. Diversification of Neu differentiation factor and epidermal growth factor signaling by combinatorial receptor interactions. *EMBO J.* **1996**, *15*, 2452–2467. [CrossRef] [PubMed]
33. Lee, S.; Greenlee, E.B.; Amick, J.R.; Ligon, G.F.; Lillquist, J.S.; Natoli, E.J., Jr.; Hadari, Y.; Alvarado, D.; Schlessinger, J. Inhibition of ErbB3 by a monoclonal antibody that locks the extracellular domain in an inactive configuration. *Proc. Natl. Acad. Sci. USA* **2015**, *112*, 13225–13230. [CrossRef] [PubMed]
34. Xiao, Z.; Carrasco, R.A.; Schifferli, K.; Kinneer, K.; Tammali, R.; Chen, H.; Rothstein, R.; Wetzel, L.; Yang, C.; Chowdhury, P.; et al. A Potent HER3 Monoclonal Antibody That Blocks Both Ligand-Dependent and -Independent Activities: Differential Impacts of PTEN Status on Tumor Response. *Mol. Cancer Ther.* **2016**, *15*, 689–701. [CrossRef]
35. Nusinow, D.P.; Szpyt, J.; Ghandi, M.; Rose, C.M.; McDonald, E.R., 3rd; Kalocsay, M.; Jané-Valbuena, J.; Gelfand, E.; Schweppe, D.K.; Jedrychowski, M.; et al. Quantitative Proteomics of the Cancer Cell Line Encyclopedia. *Cell* **2020**, *180*, 387–402.e316. [CrossRef] [PubMed]
36. Bertelsen, V.; Stang, E. The Mysterious Ways of ErbB2/HER2 Trafficking. *Membranes* **2014**, *4*, 424–446. [CrossRef] [PubMed]
37. Giri, D.K.; Ali-Seyed, M.; Li, L.-Y.; Lee, D.-F.; Ling, P.; Bartholomeusz, G.; Wang, S.-C.; Hung, M.-C. Endosomal transport of ErbB-2: Mechanism for nuclear entry of the cell surface receptor. *Mol. Cell. Biol.* **2005**, *25*, 11005–11018. [CrossRef] [PubMed]
38. Cordo Russo, R.I.; Béguelin, W.; Díaz Flaqué, M.C.; Proietti, C.J.; Venturutti, L.; Galigniana, N.; Tkach, M.; Guzmán, P.; Roa, J.C.; O'Brien, N.A.; et al. Targeting ErbB-2 nuclear localization and function inhibits breast cancer growth and overcomes trastuzumab resistance. *Oncogene* **2015**, *34*, 3413–3428. [CrossRef]
39. Merrifield, R.B. Solid phase peptide synthesis. I. The synthesis of a tetrapeptide. *J. Am. Chem. Soc.* **1963**, *85*, 2149–2154. [CrossRef]
40. Bycroft, B.W.; Chan, W.C.; Chhabra, S.R.; Hone, N.D. A novel lysine-protecting procedure for continuous flow solid phase synthesis of branched peptides. *J. Chem. Soc.* **1993**, *9*, 778–779. [CrossRef]
41. Lin-Cereghino, J.; Wong, W.W.; Xiong, S.; Giang, W.; Luong, L.T.; Vu, J.; Johnson, S.D.; Lin-Cereghino, G.P. Condensed protocol for competent cell preparation and transformation of the methylotrophic yeast Pichia pastoris. *BioTechniques* **2005**, *38*, 44–48. [CrossRef] [PubMed]
42. Laemmli, U.K. Cleavage of structural proteins during the assembly of the head of bacteriophage T4. *Nature* **1970**, *227*, 680–685. [CrossRef] [PubMed]

43. Lo, H.-W.; Xia, W.; Wei, Y.; Ali-Seyed, M.; Huang, S.-F.; Hung, M.-C. Novel Prognostic Value of Nuclear Epidermal Growth Factor Receptor in Breast Cancer. *Cancer Res.* **2005**, *65*, 338. [PubMed]
44. Schillaci, R.; Guzmán, P.; Cayrol, F.; Beguelin, W.; Díaz Flaqué, M.C.; Proietti, C.J.; Pineda, V.; Palazzi, J.; Frahm, I.; Charreau, E.H.; et al. Clinical relevance of ErbB-2/HER2 nuclear expression in breast cancer. *BMC Cancer* **2012**, *12*, 74. [CrossRef] [PubMed]

Article

Induction of Isochromanones by Co-Cultivation of the Marine Fungus *Cosmospora* sp. and the Phytopathogen *Magnaporthe oryzae*

Ernest Oppong-Danquah [1], Martina Blümel [1], Silvia Scarpato [2], Alfonso Mangoni [2] and Deniz Tasdemir [1,3,*]

1. GEOMAR Centre for Marine Biotechnology (GEOMAR-Biotech), Research Unit Marine Natural Product Chemistry, GEOMAR Helmholtz Centre for Ocean Research Kiel, Am Kiel-Kanal 44, 24106 Kiel, Germany; eoppong-danquah@geomar.de (E.O.-D.); mbluemel@geomar.de (M.B.)
2. Dipartimento di Farmacia, Università degli Studi di Napoli Federico II, via Domenico Montesano 49, 80131 Napoli, Italy; silvia.scarpato@unina.it (S.S.); alfonso.mangoni@unina.it (A.M.)
3. Faculty of Mathematics and Natural Science, Kiel University, Christian-Albrechts-Platz 4, 24118 Kiel, Germany
* Correspondence: dtasdemir@geomar.de; Tel.: +49-431-6004430

Abstract: Microbial co-cultivation is a promising approach for the activation of biosynthetic gene clusters (BGCs) that remain transcriptionally silent under artificial culture conditions. As part of our project aiming at the discovery of marine-derived fungal agrochemicals, we previously used four phytopathogens as model competitors in the co-cultivation of 21 marine fungal strains. Based on comparative untargeted metabolomics analyses and anti-phytopathogenic activities of the co-cultures, we selected the co-culture of marine *Cosmospora* sp. with the phytopathogen *Magnaporthe oryzae* for in-depth chemical studies. UPLC-MS/MS-based molecular networking (MN) of the co-culture extract revealed an enhanced diversity of compounds in several molecular families, including isochromanones, specifically induced in the co-culture. Large scale co-cultivation of *Cosmospora* sp. and *M. oryzae* resulted in the isolation of five isochromanones from the whole co-culture extract, namely the known soudanones A, E, D (**1-3**) and their two new derivatives, soudanones H-I (**4-5**), the known isochromans, pseudoanguillosporins A and B (**6, 7**), naphtho-γ-pyrones, cephalochromin and ustilaginoidin G (**8, 9**), and ergosterol (**10**). Their structures were established by NMR, HR-ESIMS, FT-IR, electronic circular dichroism (ECD) spectroscopy, polarimetry ($[\alpha]_D$), and Mosher's ester reaction. Bioactivity assays revealed antimicrobial activity of compounds **2** and **3** against the phytopathogens *M. oryzae* and *Phytophthora infestans*, while pseudoanguillosporin A (**6**) showed the broadest and strongest anti-phytopathogenic activity against *Pseudomonas syringae*, *Xanthomonas campestris*, *M. oryzae* and *P. infestans*. This is the first study assessing the anti-phytopathogenic activities of soudanones.

Keywords: marine fungi; *Cosmospora* sp.; soudanone; *Magnaporthe oryzae*; co-culture; phytopathogen; molecular networking; metabolomics

1. Introduction

Fungi are prolific producers of bioactive natural products that have found valuable applications in the agrochemical industry as pesticides and biocontrol agents [1]. There is an urgent need for new, natural agrochemicals against crop diseases, but the rediscovery of the known compounds poses a major hurdle to fungal natural product biodiscovery endeavours. Although recent fungal genomics have indexed huge numbers of biosynthetic gene clusters (BGCs), the majority of BGCs remain transcriptionally silent under standard laboratory conditions [2]. The cryptic BGCs encode the enzymatic machinery for the synthesis of a plethora of yet untapped secondary metabolites (SMs). Therefore, the application of culture-based approaches that awaken the silent BGCs promises the discovery of bioactive and novel compounds [3]. One of the effective strategies employed to enhance the expression of silent BGCs in fungal cultures is co-cultivation, which is based on the premise that two or more microorganisms growing within a confined environment respond to environmental cues, which trigger the activation of BGCs to produce, often new, bioactive SMs. Co-cultivation has already proven successful to yield new bioactive compounds or enhance the production of previously identified compounds.

Different strategies exist for designing co-cultivation experiments, such as the co-cultivation of microbes from the same habitat, or the co-cultivation of microorganisms unlikely to co-occur in the same habitat [4,5]. Phytopathogens are considered effective competitors in co-cultivation experiments as, in natural settings, they must overcome plant innate immunity, including beneficial epiphytes and endophytes prior to their colonization [6]. Upon contact with a host plant, the pathogen has to successfully outcompete host symbionts, regarding space and nutrients, which is usually achieved by direct antagonism [7]. Consequently, phytopathogens lend themselves as useful models to investigate competitive microbial interactions.

In a previous study, we obtained 123 marine fungal isolates belonging to 30 genera from the Baltic Sea [4]. Based on phylogenetic and metabolic diversity, 21 fungal isolates were selected and co-cultured with two bacterial (*Pseudomonas syringae* and *Ralstonia solanacearum*) and two fungal (*Magnaporthe oryzae* and *Botrytis cinerea*) phytopathogens on two different solid media, Saboroud Agar (SA) and Potato Dextrose Agar (PDA). Comparative metabolomics of the crude organic extracts of co-cultures and the axenic cultures using HRMS/MS-based molecular networking (MN) and in vitro anti-phytopathogenic activity assays led to the prioritization of the co-culture of *Cosmospora* sp. and *M. oryzae* in the PDA medium. This co-culture was selected for in-depth chemical analysis because of (i) its unique chemical diversity, (ii) its clear induction of several compounds in the co-culture, and (iii) its antimicrobial activity against numerous phytopathogens [4].

Herein, we have investigated the chemical composition of the upscaled co-cultures of *Cosmospora* and *M. oryzae* and their axenic-cultures grown on a PDA medium by MN-based metabolomics to observe the specific induction of the isochromanone type SMs, which were absent in the mono-cultures. Chromatographic separation of the CH_2Cl_2-soluble portion of the co-cultures led to the isolation of two new isochromanones, soudanones H-I (**4-5**), and the known soudanones A, E, D (**1-3**), together with the known isochromans, pseudoanguillosporins A and B (**6, 7**), the known naphtho-γ-pyrones, cephalochromin and ustilaginoidin G (**8, 9**), and ergosterol (**10**). Herein, we report on the in-depth metabolomics analyses, followed by the isolation and structure elucidation of the induced isochromanones (**1-5**) and the known compounds (**6-10**), as well as their anti-phytopathogenic activities.

2. Results

2.1. Co-Culture and Description of MN and Annotations

A UPLC chromatogram of the crude EtOAc extract of a 21-day PDA whole co-culture of *Cosmospora* sp. and *M. oryzae* revealed the induction of the compounds **1-5** (Figure 1B), which were absent in the monocultures (Figure 1C,D). These metabolites were accumulated in the confrontation zone (Figure 1A), i.e., at the site of direct interaction between two fungi. We hypothesized that compounds **1-5** were produced for competitive advantage and so they were prioritized for isolation from the whole co-cultures.

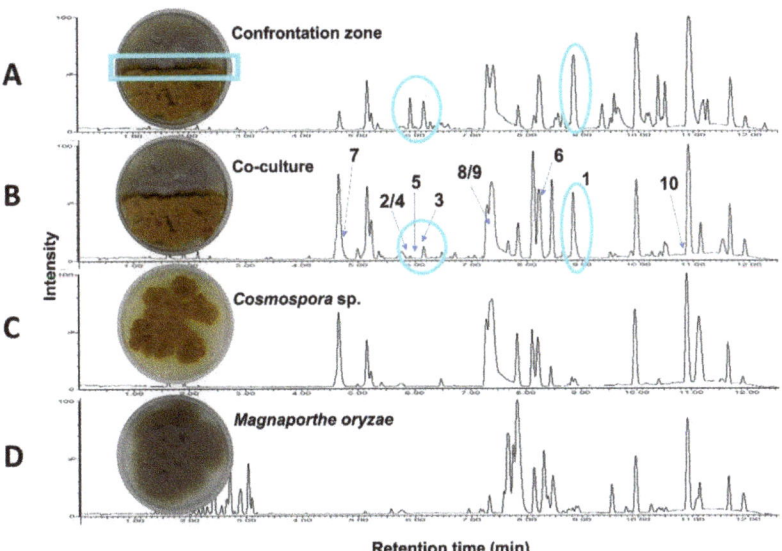

Figure 1. UPLC-MS chromatograms of the extracts obtained from (**A**) the confrontation zone (blue rectangle) excised from (**B**) the whole co-culture of *Cosmospora* sp. and *Magnaporthe oryzae*, and the monocultures of (**C**) *Cosmospora* sp., and (**D**) *M. oryzae*. Peak ions corresponding to compounds **1–5** (highlighted in blue) were induced in the whole co-culture (**B**) but of low intensities. They were however observed in higher intensities in the confrontation zone (**A**) (also highlighted in oval blue).

In order to investigate the metabolome and to obtain more information on the induced compounds (**1-5**), a UPLC-QToF-MS/MS based MN was generated with the crude extracts of the fungal co-culture and the corresponding axenic cultures (Figure 2). The MN analysis revealed that co-cultivation generally increased the chemical space of the fungi, indicated by a size increase of several clusters with co-culture induced compounds (blue only nodes), such as clusters C, D, E, I, K, L and M. Moreover, we observed five clusters (G, P, Q, R and T) that were exclusively induced in the co-cultures. The latter clusters remained unidentified as they showed no match to known compounds in several commercial and public databases. Unfortunately, no ion belonging to these clusters could be purified in sufficient quantity to allow chemical identification.

The biggest cluster, A, shared by the *Cosmospora* sp. and the co-culture extracts, was annotated to the naphtho-γ-pyrone class of compounds, which included ustilaginoidins A (*m/z* 515.1259 [M+H]$^+$), D (*m/z* 547.1599 [M+H]$^+$), E (*m/z* 533.1479 [M+H]$^+$) and V (*m/z* 535.1237 [M+H]$^+$) (Figure 2 and Supplementary Table S1) [8,9]. Two additional nodes (compounds), **8** (*m/z* 519.1284 [M+H]$^+$) and **9** (*m/z* 517.1150 [M+H]$^+$), belonging to this cluster were produced in significant amounts. They were purified and identified as the known naphtho-γ-pyrones, cephalochromin (**8**) and ustilaginoidin G (**9**), to confirm the cluster annotation.

Another small cluster annotation confirmed by purification was the isochroman class of compounds (cluster N). It contained two nodes originating from both *Cosmospora* sp. mono-culture and the co-culture extracts. One node, *m/z* 279.1939 [M+H]$^+$, was purified and identified as the known isochroman, pseudoanguillosporin A (**6**) [10]. The second node, *m/z* 279.0970 [M+H]$^+$, could not be assigned to any known compound in databases, so remains unidentified. Another node (singleton), *m/z* 295.1912 [M+H]$^+$, had a mass difference of 16 Da from **6**, being indicative of an additional oxygen atom. It was identified as the known isochroman, pseudoanguillosporin B (**7**), that bears a hydroxy substitution on the alkyl side chain [10]. Pseudoanguillosporin B did not cluster with the isochromans in cluster N because the product ions displayed a similarity score (cosine score) of <0.7, as estimated from the MN algorithm (refer to Section 4.4).

Cluster M was identified as a steroid family, based on the purification and identification of **10** as ergosterol *m/z* 419.3290 [M+Na]$^+$ [11,12]. Other putative annotations, based on manual dereplication, include the cyclodepsipeptides acuminatums B (*m/z* 888.5449 [M+H]$^+$) and C (*m/z* 874.5460 [M+H]$^+$) [13] in cluster J (nodes shared between *Cosmospora* sp. and co-culture), as well as a phospholipid (*m/z* 520.3740 [M+H]$^+$) [14] in cluster D, which was shared by *M. oryzae* and the co-culture. All annotations are listed in Table S1.

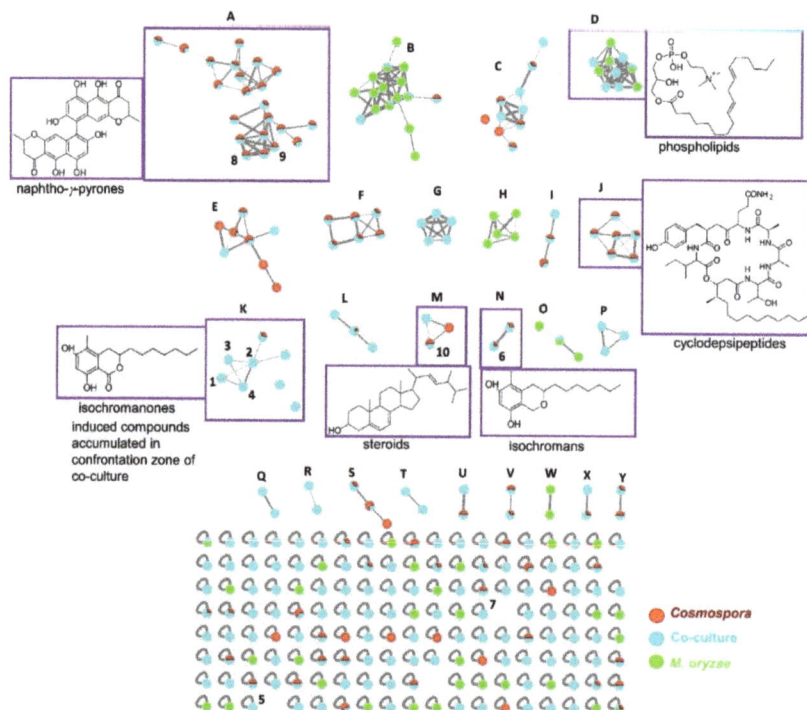

Figure 2. Molecular network (MN) of *Cosmospora* sp. (red), *M. oryzae* (green) and their co-culture (blue). Dereplicated clusters (A–Y) and representative structures are highlighted in purple next to each other. Compounds **1-4** are the induced isochromanones in the co-culture, while compounds **5** and **7** were displayed as singletons in the MN. Some nodes in clusters represent isotopic nodes with a mass difference of +1 Da from the neighbouring node.

MN identified the induced compounds **1-5** as belonging to a common molecular family, K (Figure 2), the isochromanones. The cluster K contained 8 nodes (Figure 3), of which 7 were exclusively observed in the co-culture. The first node, (**1**) *m/z* 293.1755 [M+H]$^+$, was isolated and identified as the known isochromanone, soudanone A. Another node, (**2**) *m/z* 309.1705 [M+H]$^+$, showed a mass difference of 16 Da from **1**, indicative of hydroxylation, and was identified as soudanone E. The sodiated adduct of **2** (*m/z* 331.1730 [M+Na]$^+$) was also observed. As indicated in the experimental section, the clusters were generated with a defined cosine score (similarity index) above 0.7 and the edges were modulated accordingly. The thick edge between nodes **2** and **4** indicates high structural similarity. As expected, these nodes shared highly similar MS/MS spectra, thus node **4** was predicted to be a positional isomer of **2**, the new soudanone H, visualized in the network as a dehydrated adduct (*m/z* 291.1600 [M-H$_2$O+H]$^+$. Node (**3**), *m/z* 307.1546 [M + H]$^+$, had a mass difference of 2 Da from nodes **2** and **4**, which suggested the oxidation of a hydroxyl group to a ketone.

Node 3 was identified as the known soudanone D by isolation. Other observed nodes in this cluster (K) were m/z 635.3860 [M+H]⁺, m/z 569.3860 [M+H]⁺, and m/z 553.2470 [M+H]⁺. Unfortunately, these nodes (compounds) representing potentially new isochromanones remain unidentified because they could not be purified in sufficient quantities. Notably, all nodes in this cluster originated from the co-culture (blue nodes) except m/z 553.2470 [M+H]⁺, which was also detected in the *Cosmospora* sp. monoculture (red node, Figure 3). In addition, another new isochromanone, (**5**) m/z 309.1341 [M + H]⁺, soudanone I, was identified as a single node in the MN (Figure 2). It did not cluster with the isochromanone molecular family, because it displayed slightly different product ions due to the shorter side chain and the terminal esterification. All identifications and annotations are listed in Table S1.

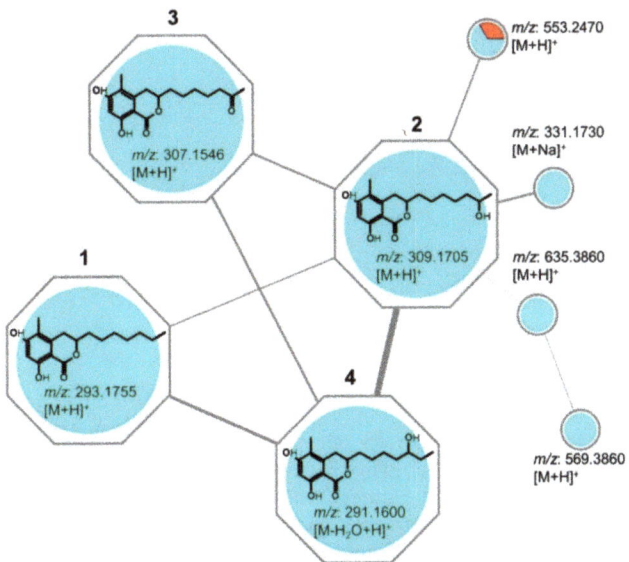

Figure 3. Close-up of molecular family cluster K containing the known compounds (**1, 2** and **3**) induced in the co-culture (and significantly observed in the confrontation zone). Likewise, the new compound **4** and the other nodes were induced in the co-culture, except m/z 553.2470 [M+H]⁺, which was also produced in the *Cosmospora* mono-culture. Nodes coloured blue originate from the co-culture, and red from *Cosmospora* monoculture. The compound **5** is observed as a singleton (Figure 2).

2.2. Isolation & Structure Elucidation

For isolation of the compounds, the crude EtOAc extract obtained from scaled-up solid co-cultures (100 PDA plates) of *Cosmospora* sp. and *M. oryzae* was subjected to a modified Kupchan solvent partitioning scheme [15] to yield *n*-hexane, CH₂Cl₂ and aqueous MeOH subextracts. Only the CH₂Cl₂ subextract exhibited activity against the phytopathogens *X. campestris*, *P. infestans* and *M. oryzae* (Table S2), and contained the prioritized isochromanones (**1-5**). The CH₂Cl₂ subextract was selected for further work-up and fractionated over a C18 SPE cartridge, followed by repeated RP-HPLC separations to afford the co-culture induced compounds **1-5**, as well as compounds **6-10** (Figure 4).

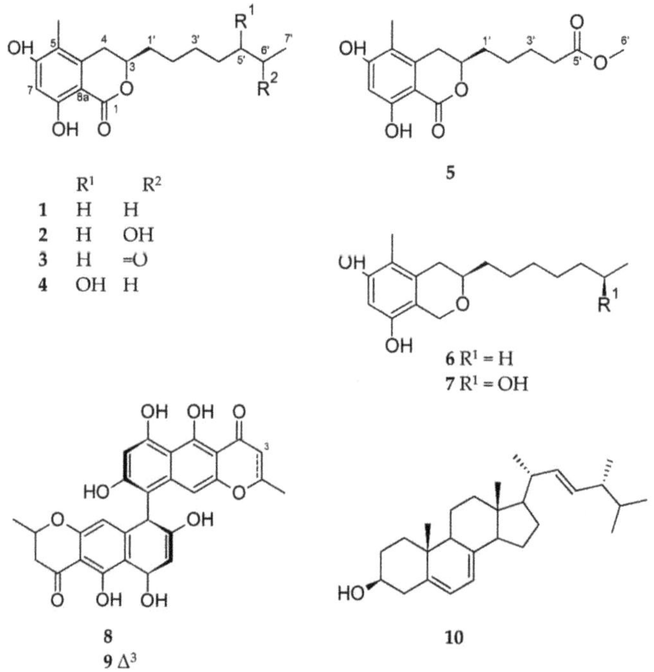

Figure 4. Chemical structures of compounds **1–10**.

Compound **1** was obtained as a yellowish amorphous film with the molecular formula $C_{17}H_{25}O_4$ deduced by HR-ESIMS (m/z 293.1755 ([M+H]$^+$)). 1D and 2D NMR data analyses (Tables 1 and 2, Figures S1–S10), as well as the FT-IR data (Figure S11), led to the identification of **1** as soudanone A, an isochromanone previously reported from a *Cadophora* sp. isolated from the Soudan Underground Mine [16]. Compound **1** exhibited the same sign and a similar magnitude of specific rotation value ($[\alpha]_D^{20}$-59, c 0.5, MeOH) as soudanone A ($[\alpha]_D^{23}$-43, c 0.065, MeOH) [16]. In addition, the experimental electronic circular dichroism (ECD) spectrum of **1** (Figure 5) was similar to that of soudanone A, in that it possessed an *R* configuration at C-3 [16]. Hence, compound **1** was unambiguously identified as soudanone A.

Compound **2** was assigned the molecular formula $C_{17}H_{25}O_5$, deduced by HR-ESIMS (m/z 309.1705 ([M+H]$^+$, Figure S18). With the knowledge of compound **2** and **1** clustering in the MS/MS molecular network, compound **2** was assigned as an oxygenated derivative of **1** with an increase of 16 mass units. The HRMS, FT-IR, ^{1}H and ^{13}C NMR, COSY, HMBC and NOESY spectra (Tables 1 and 2, Figures S12–S20) were consistent with the known soudanone E, that contains an OH group attached at a C-6' position on the alkyl chain [16]. The optical rotation of **2** ($[\alpha]_D^{20}$-50, c 0.05, MeOH) and soudanone E ($[\alpha]_D^{25}$-40, c 0.065, MeOH) were comparable in sign and magnitude. The measured ECD spectrum of **2** was similar to **1** (Figure 5) and not affected by the presence of a second chiral centre at C-6'. Hence, we assigned the absolute configuration of C-3 as *R*. Next, we attempted to determine the configuration at C-6', as yet uncharacterized in the literature [16], by comparing the NMR data of the side chain of **2** to those published for pseudoanguillosporin B (**7**) [10]. Pseudoanguillosporin B (**7**), a highly related isochroman type of molecule (also isolated in this study), contains a hydroxyl function at a C-6' position with *R* configuration, determined by Mosher's NMR method [10]. As shown in Table S3, compounds **2** and **7** showed almost identical ^{1}H/^{13}C NMR signals for positions in the side chain, particularly for H-6'/C-6' and H$_3$-7'/C-7', suggesting an *R* configuration for C-6'. To confirm this assignment, we performed Mosher's ester derivatization. Due to the very small amount of **2** available, we chose a nanomole-scale protocol using methoxyphenylacetic (MPA) acid as the chiral derivatizing agent [17]. The phenolic hydroxy groups were protected by reacting with (trimethylsilyl)diazomethane to give a dimethyl ether derivative, then the (*R*)-MPA and (*S*)-MPA esters were prepared (Figure S50). LC-MS analyses of the resulting reaction mixtures revealed compound **2** to be a mixture of two epimers at C-6′ in the 2:1 ratio (Figure S51; details can be found in the Supporting Information section). 1D-TOCSY experiments recorded on the crude reaction mixtures (Figure S52) were used to assign the ^{1}H NMR signals of the MPA esters. Due to the small amounts of esters obtained, only the chemical shift of the terminal methyl (H$_3$-7′) could be determined, but this was sufficient to assign configuration. For the most abundant epimer, H$_3$-7′ was deshielded in the (*R*)-MPA ester and shielded in the (*S*)-MPA ester. This indicated the 6′*R* configuration for the major epimer. Therefore, compound **2** was identified as an inseparable 2:1 mixture of (3*R*,6′*R*)-soudanone E and (3*R*,6′*S*)-soudanone E. It is worth noting that neither ^{1}H nor ^{13}C NMR provided any clue that compound **2** was in fact a mixture of diastereomers. This means that the spectra of the two diastereomers are identical, and indeed several examples have been reported where two diastereomers with chiral centers far from each other show indistinguishable NMR spectra [18].

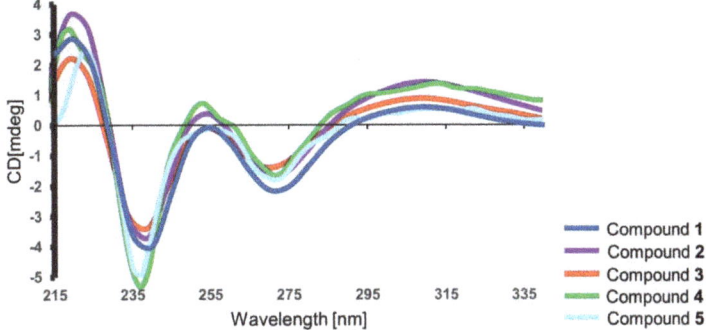

Figure 5. Experimental electronic circular dichroism (ECD) spectra of compounds **1** (dark blue), **2** (purple) and **3** (red), **4** (green) and **5** (light blue) in MeOH.

Table 1. ^{1}H NMR data of compounds **1–5** (^{a} acquired in CD$_3$OD at 500 MHz, ^{b} acquired in CDCl$_3$ at 600 MHz), δ in ppm, Mult (*J* in Hz).

Position	1 ^{a}	1 ^{b}	2 ^{b}	3 ^{b}	4 ^{b}	5 ^{b}
3	4.47 m	4.46 m	4.45 m	4.45 m	4.47 m	4.46 m
4	3.03 dd (16.7, 3.4) 2.69 dd (16.7, 11.5)	2.94 dd (16.5, 3.4) 2.70 dd (16.5, 11.6)	2.93 dd (16.5, 3.2) 2.69 dd (16.5, 11.8)	2.93 dd (16.5, 3.3) 2.69 dd (16.5, 11.7)	2.94 dd (16.5, 3.2) 2.71 dd (16.5, 11.8)	2.94 dd (16.5, 3.3) 2.70 dd (16.5, 11.7)
7	6.24 s	6.31 s	6.31 s	6.30 s	6.29 s	6.3 s
1′	1.83 m 1.76 m	1.88 m 1.73 m	1.87 m 1.73 m	1.86 m 1.76 m	1.90 m 1.75 m	1.90 m 1.76 m
2′	1.58 m 1.48 m	1.56 m 1.45 m	1.60 m 1.48 m	1.62 m 1.48 m	1.64 m 1.52 m	1.64 m 1.53 m
3′	1.35 m	1.34 m	1.37 m	1.36 m	1.52 m 1.42 m	1.70 m
4′	1.35 m	1.32 m	1.36 m 1.44 m	1.62 m	1.53 m 1.45 m	2.36 t (7.3)
5′	1.35 m	1.26 m	1.45 m	2.45 t (7.3)	3.54 m	
6′	1.35 m	1.29 m	3.81 m		1.52 m 1.44 m	3.68 s
7′	0.91 t (6.9)	0.88 t (6.9)	1.20 d (6.1)	2.15 s	0.95 t (7.5)	
5-Me	2.04 s	2.07 s	2.06 s	2.06 s	2.07 s	2.07 s
OH		11.26 s	11.26 s	11.18 s	11.26 s	11.26 s

Compound **3** was purified as a yellow amorphous solid with a molecular formula of $C_{17}H_{23}O_5$ on the basis of its HR-ESIMS data (m/z 307.1546 ([M+H]$^+$). Analysis of the 1D and 2D NMR data (Table 2, Figures S21–S29, including the FT-IR spectrum), as well as the comparison of its specific rotation value ($[\alpha]_D^{20}$-56, c 0.35, MeOH) with that of soudanone D ($[\alpha]_D^{23}$-40, c 0.05, MeOH), led to the identification of compound **3** as soudanone D [16]. Compound **3** also displayed a similar ECD spectrum (Figure 5) as **1**, hence was unambiguously identified as (3R)-soudanone D.

Table 2. ^{13}C NMR data of compounds **1-5** (150 MHz, CDCl$_3$).

Position	1	2	3	4	5
1	170.6, C	170.6, C	170.5, C	170.5, C	170.4, C
3	78.7, CH	78.6, CH	78.5, CH	78.5, CH	78.3, CH
4	30.6, CH$_2$	30.7, CH$_2$	30.7, CH$_2$	30.7, CH$_2$	30.6, CH$_2$
4a	139.5, C	139.4, C	139.4, C	139.5, C	139.4, C
5	113.1, C	113.2, C	113.0, C	112.9, C	113.0, C
6	160.9, C	161.0, C	160.7, C	160.5, C	160.6, C
7	101.6, CH	101.6, CH	101.6, CH	101.6, CH	101.7, CH
8	162.5, C	162.5, C	162.5, C	162.5, C	162.6, C
8a	102.1, C	102.0, C	102.1, C	102.2, C	102.2, C
1'	35.1, CH$_2$	35.0, CH$_2$	34.9, CH$_2$	35.1, CH$_2$	34.8, CH$_2$
2'	25.1, CH$_2$	25.0, CH$_2$	24.9, CH$_2$	25.6, CH$_2$	24.6, CH$_2$
3'	29.5, CH$_2$	29.4, CH$_2$	29.0, CH$_2$	25.2, CH$_2$	24.8, CH$_3$
4'	29.3, CH$_2$	25.7, CH$_2$	23.6, CH$_2$	36.8, CH$_2$	34.0, CH$_3$
5'	31.9, CH$_2$	39.2, CH$_2$	43.7, CH$_2$	73.3, CH	174.1, C
6'	22.8, CH$_2$	68.4, CH	209.3, C	30.4, CH$_2$	51.7, CH$_3$
7'	14.2, CH$_3$	23.7, CH$_3$	30.1, CH$_3$	10.0, CH$_3$	
5-Me	10.6, CH$_3$	10.6, CH$_3$	10.6, CH$_3$	10.6, CH$_3$	10.6, CH$_3$

Compound **4** was assigned the same molecular formula as compound **2**, $C_{17}H_{25}O_5$, deduced by HR-ESIMS (m/z 309.1703 ([M+H]$^+$, Figure S36). This indicated 6 degrees of unsaturation. The FT-IR absorption bands, observed at 3390 and 1646 cm^{-1}, indicated the presence of hydroxyl and carbonyl groups, respectively (Figure S38). The ^{13}C NMR spectrum contained 17 carbon resonances corroborating the predicted molecular formula from the HR-ESIMS. These included a carbonyl (δ_C 170.5), six olefinic carbons (δ_C 139.5, 112.9, 160.5, 101.6, 162.5 and 102.2), two oxymethine carbons (δ_C 78.5, 73.3) and two methyl groups (δ_C 10.0 and 10.6). ^{1}H NMR and HSQC spectra revealed an aromatic proton (δ_H 6.29 s, H-7), two oxymethine protons (δ_H 4.47 m, δ_H 3.54 m), an olefinic methyl (δ_H 2.07 s, H$_3$-4), a primary methyl group (δ_H 0.95 t, J = 7.5 Hz), and six aliphatic methylene protons between δ_H 1.26-2.94 (Tables 1 and 2, **Figures S30–S31**, Figure S33). This data was suggestive of the same isochromanone core structure found in compounds **1-3**, supported by whole set of 2D NMR experiments (Figures S32–S35). The COSY spectrum revealed only one large spin system (Figure 6 and Figure S32). It started from the diastereotopic methylene protons at δ_H 2.71 and 2.94 (H$_2$-4) that coupled with the oxymethine proton at δ_H 4.47 (m, H-3), which in turn coupled to a methylene proton (δ_H 1.75 and 1.90, H$_2$-1'). The spin network correlations further incorporated four methylenes (H$_2$-2'-H$_2$-6'), another oxymethine proton (δ_H 3.54 m, H-5') and a terminal methyl group (δ_H 0.95 t, J = 7.5 Hz, H$_3$-7'), resulting in an unbranched aliphatic chain similar to that observed in **2**. The major difference between **2** and **4** was represented by the attachment of the secondary OH function at C-5' in **4**, instead of C-6' in **2** (Figure 6). This assumption was verified on the basis of COSY correlations observed between H-5'/H-6' and H-5'/H-4', as well as by HMBC correlations between H-5'/C-7'and H-5'/C-3' (Figure 6). Thus, the side chain of **4** was confirmed as heptan-3-ol. The measured ECD spectrum of **4** was similar to that of **2** (Figure 5) and their optical rotation values were almost identical (**2** $[\alpha]_D^{20}$-50, c 0.05, MeOH; **4** $[\alpha]_D^{20}$-44, c 0.05, MeOH). The configuration of C-3 in **4** was thus deduced as (R). The absolute configuration of C-5' was investigated by Mosher's ester method mentioned above.

Due to the small amounts of sample, only the (R)-MPA ester was prepared (Figure S53), with the intention of using the variable-temperature NMR measurement to assign absolute configuration at C-5′ [17]. However, similar to compound 2, compound 4 was determined to be a mixture of two epimers at C-5′. Based on the observation of two peaks of nearly equal intensity in the LC-MS chromatogram of the (R)-MPA ester (Figure S54), in this case the epimers were determined to be in the 1:1 ratio. This made it unnecessary to use NMR to assign the configuration at C-5′. Therefore, compound 4 was identified as an inseparable 1:1 mixture of epimers, for which we propose the trivial names of (3R, 5′R)-soudanone H and (3R, 5′S)-soudanone H.

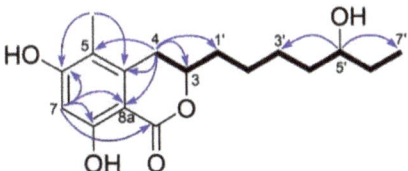

Figure 6. Key COSY (bold lines) and HMBC (blue arrows) correlations observed for compound 4.

Compound 5 was assigned the molecular formula $C_{16}H_{21}O_6$ based on the m/z 309.1341 ([M+H]$^+$) observed in the HR-ESIMS spectrum (Figure S45), indicating 7 degrees of unsaturation. In-depth analysis of the 1D and 2D NMR, and the FT-IR data of 5 (Figures S39–S47) provided evidence for the same isochromanone structure with a carbonyl function on the alkyl chain, as in 3. The main differences between compounds 5 and 3 rested in the absence of one methylene group (H_2-5′) in the proton spin system of the side chain of 5, and the replacement of the methyl ketone terminal group with a carboxymethyl ester attached at C-4′. Due to the ester formation, H_2-4′ was shifted to downfield and appeared as a triplet at δ_H 2.36 (J = 7.3 Hz). The presence of the carboxymethyl ester terminal in 5 was evident from the ^{1}NMR resonances of C-5′ (δ_C 174.1 s) and CH_3-6′ (δ_H 3.68 s; δ_C 51.7 q). COSY correlations on the side chain from H_2-1′ through H_2-4′ (Figure S41) and diagnostic HMBC correlations between H_3-6′/C-5′ and H_2-4′/C-5′ (Figure S43) confirmed that methyl pentanoate was the side chain of 5. The measured ECD spectrum (Figure 5) and the optical rotation value of 5 $[\alpha]_D^{20}$-20 (c 0.07, CH_3OH) were similar to compounds 1–4, hence the configuration at C-3 was also proposed as (R). We propose the trivial name (3R)-soudanone I for compound 5.

In addition to the five soudanones (1–5) described above, two known isochromans were isolated and identified as pseudoanguillosporins A (6) and B (7) by comparing their 1D NMR, HR-ESIMS and $[\alpha]_D$ data with those reported in the literature [10]. The other isolates 8-9 were identified as bis-naphtho-γ-pyrone type mycotoxins, cephalochromin (8) [10] and ustilaginoidin G (9) [8,19], while the compound 10 was characterized as ergosterol [11], based on their NMR, HR-ESIMS and $[\alpha]_D$ data.

2.3. Bioactivity of Pure Compounds

The inhibitory activities of the isolated compounds were tested against a panel of plant pathogens (Table 3). The new soudanones H and I (4 and 5) were not tested because they were obtained in very low amounts. Soudanone A (1) and ergosterol (10) were inactive against all phytopathogens, even at the highest test concentration (100 μg/mL), while soudanones E and D (2 and 3) exerted moderate activities against the oomycete *P. infestans*, the fungus *M. oryzae* and the bacterium *X. campestris* with IC$_{50}$ values between 12.8 and 71.5 μg/mL. Pseudoanguillosporin A (6) exhibited the strongest (IC$_{50}$ values 0.8–23.4 μg/mL) and broadest activity against all tested phytopathogens, except *E. amylovora*, which was not susceptible to any of the compounds. Cephalochromin (8) inhibited *P. infestans*, *R. solanacearum* and *X. campestris* (IC$_{50}$ values 2.3, 27.6 and 12.1 μg/mL, respectively). Ustilaginoidin G (9) exerted moderate inhibition against *P. infestans* (IC$_{50}$ value 7.2 μg/mL) and *X. campestris* (IC$_{50}$ value 21.7 μg/mL).

Table 3. Anti-phytopathogenic activity (IC_{50} values in µg/mL) of isolated compounds against phytopathogens. Compounds **4** and **5** were not tested due to their minute amounts. Positive controls for *Pseudomonas syringae* (Ps), *Xanthomonas campestris* (Xc) and *Erwinia amylovora* (Ea): chloramphenicol; *Ralstonia solanacearum* (Rs): tetracycline; *Phytophthora infestans* (Pi): cycloheximide, *Magnaporthe oryzae* (Mo): nystatin.

Compound	Bacteria				Oomycete	Fungus
	Ps	Xc	Ea	Rs	Pi	Mo
1	>100	>100	>100	>100	>100	>100
2	>100	71.5	>100	>100	27.6	12.8
3	>100	15.7	>100	>100	52.1	60.3
6	23.4	7.4	>100	>100	3.2	0.8
7	>100	67.1	>100	42.2	>100	>100
8	95.7	12.1	>100	27.6	2.3	>100
9	>100	21.7	>100	>100	7.2	>100
10	>100	>100	>100	>100	>100	>100
Control	0.4	0.5	0.2	1.0	0.6	0.4

2.4. Optimization Study to Enhance Production of the Soudanones

We intended to optimize the fermentation conditions to increase the production of the isochromanones to allow for further biological assays with pure compounds. Considering that the isochromanones were accumulated in the confrontation zone, i.e., the site of interaction between both co-cultivation partners, we tried co-cultivation by overlaying *Cosmospora* over pre-grown *M. oryzae* to examine if an enhanced contact area increases the production of the isochromanones (**1-5**). All five isochromanones were produced and observed in the resulting EtOAc crude extract (Figure S48). With the exception of **1**, whose production was evident from the UPLC-MS base peak chromatogram (Figure S48), soudanones **2-5** were still produced in small amounts (assessed by small peaks in UPLC-MS base peak chromatogram, Figure S48). In the next step, *Cosmospora* sp. and *M. oryzae* were co-cultured in a liquid culture regime using a potato dextrose broth (PDB) medium under aeration, which facilitates contact of competing strains and ensures constant oxygenation. However, co-cultivation in broth did not yield any of the isochromanones **1-5** (Figure S49). On the other hand, cephalochromin (**8**) and ustilaginoidin G (**9**) were already observed on day 3 of broth co-cultivation, peaking on day 6, with sustained levels of production as major peaks across all 21 days of cultivation (Figure S49). This result underscores the high complexity and unpredictability of fungal interactions. Although these approaches failed to increase further yields of **1-5**, other fermentation conditions, such as cultivation time, temperature, pH and media, may be further exploited to optimize the production of the isochromanones to enable further testing against a more diverse panel of pathogens.

3. Discussion

Enhancement of the chemical diversity of fungal metabolomes through co-cultivation is becoming a promising strategy for the discovery of new bioactive natural products. Recently, pathogens are emerging as effective model competitors to trigger or enhance the production of anti-pathogenic compounds (antibiotics) in microbial species which do not naturally co-exist with the pathogens. For example, Serrano et al. [5] showed the increased diversity of antifungal compounds following the interaction of the plant pathogen *B. cinerea* with a large panel of fungal strains isolated from different environmental niches, such as soils, leaf litters and rhizosphere. Several new molecules, identified by LC-MS, were induced in the co-cultures, supporting the hypothesis that co-cultivation with phytopathogens enhances chemical diversity [5]. Not only phytopathogens, but also human pathogens have been investigated as good competitors in co-cultivation experiments. The production of the antibiotics granaticin, granatomycin D and dihydrogranaticin B showed a multiple-fold increase when the tunicate derived actinomycete *Streptomyces* sp.

was co-cultured with human pathogens, such as *Bacillus subtilis* and methicillin-resistant *Staphylococcus aureus* (MRSA) [20].

In our previous study, we demonstrated that the use of economically-relevant phytopathogens as challengers in co-cultivation experiments can increase the chemical space of marine-derived fungi, since over 11% of all compounds detected exclusively originated from the co-cultures [4]. In addition, the MN-based metabolomics and anti-phytopathogenic bioactivity screening resulted in the prioritization of the co-culture of the marine fungus *Cosmospora* sp. and one of the most devastating rice pathogens *M. oryzae* in the PDA medium [4]. In the present study, large scale co-culture fermentation of *Cosmospora* and *M. oryzae* was conducted for natural product isolation. Five isochromanones (**1–5**) were biosynthesized only in the co-culture of *Cosmospora* sp. and *M. oryzae*, while the closely related isochromans, pseudoanguillosporins A (**6**) and B (**7**), and the naphtho-γ-pyrones, cephalochromin (**8**) and ustilaginoidin G (**9**), were detected also in the *Cosmospora* monoculture. Comparative metabolomics of the mono- and co-cultures by MN revealed the overall enhancement of chemical diversity in the co-culture. Five unannotated clusters (G, P, Q, R and T) were exclusively induced in the co-culture, whereas other clusters, such as the isochromanones (cluster K), increased in size by induction of new compounds.

Isochromanones are a family of benzopyranones widely distributed in nature. They represent an important family of secondary metabolites from plants and fungi [21,22] and display diverse biological activities, such as antibacterial, antiparasitic, anticancer, herbicidal and fungicidal [23,24]. Hence, it is reasonable to assume that the biosynthesis of isochromanones was induced in response to the fungal competition for the benefit of the producer fungus, as hypothesized [25]. To this end, the potential benefits of the induced compounds in the co-culture and constitutive compounds originating from the axenic cultures were assessed by their antimicrobial activities against a panel of phytopathogens. The isochromanones **2** and **3** showed varying degrees of inhibition against the test phytopathogens. Soudanone E (**2**), that contains an OH group at C-6' of the side chain, exhibited the highest activity against the oomycete *P. infestans*, the bacterium *X. campestris*, and the competing phytopathogen, *M. oryzae* (IC$_{50}$ values 27.6, 71.6 and 12.8 µg/mL, respectively). Soudanone D (**3**), bearing a keto function at the same position of the alkyl chain, showed only weak activity against *P. infestans* and *M. oryzae* (IC$_{50}$ values 52.1 and 60.3 µg/mL respectively), while soudanone A (**1**), with no oxygenation on the side chain, was devoid of any activity against the test pathogens. Thus, the hydroxylation on the aliphatic side chain, (at position C-6') seems to be a structural feature required for the anti-phytopathogenic activity of the isochromanones. Unfortunately, we were unable to assess the bioactivity of the new soudanones H-I (**4** and **5**), due to the minor amounts isolated, to provide additional insights into the structure-activity relationships of soudanones.

The highly related isochromans, pseudoanguillosporins A (**6**) and B (**7**), with the same type side chain displayed a more potent and broader range of activities against the phytopathogenic bacteria compared to the isochromanones. Pseudoanguillosporin A (**6**), that lacks oxygenation on the side chain, exhibited overall the most potent anti-phytopathogenic activity and its inhibitory potential against *M. oryzae* (IC$_{50}$ value 0.8 3.2 µg/mL) was comparable to that of the positive control nystatin (IC$_{50}$ value 0.4 µg/mL). Such biological activities were also reported by Kock et al. [10] where percentage-growth inhibition was evaluated. Interestingly, pseudoanguillosporin B (**7**), that bears an OH group at C-6', was inactive towards *M. oryzae* and *P. infestans* at 100 µg/mL (Table 3), representing an opposite trend for the anti-phytopathogenic activity of isochromans. Pseudoanguillosporin B (**7**) showed low activity against the phytopathogenic bacteria *X. campestris* and *R. solanacearum*. The naphtho-γ-pyrones, cephalochromin (**8**) and ustilaginoidin G (**9**), moderately inhibited the growth of *X. campestris*. This aligns with a previous study, in which cephalochromin (**8**) was shown selectively to inhibit the bacterial enoyl-acyl carrier protein, reductase FabI (a highly conserved enzyme from the type II fatty acid biosynthesis of many pathogens) in *S. aureus* and *E. coli* [26]. Anti-phytopathogenic activities were also reported for ustilaginoidin G (**9**) against *X. vesicatoria* (a bacterial spot of tomatoes) and *Agrobacterium tumefaciens* (a crown gall disease of several plants) [27]. Cephalochromin is biosynthesised by several fungal genera, such as *Cephalosporium*, *Pseudoanguillospora* and *Plenodomus* [28,29]. Interestingly, its production in *Plenodomus influorescens* (Dothideomycetes, Pleosporales) was notably enhanced upon interaction with a *Pyrenochaeta* sp. (Dothideomycetes, Pleosporales), both isolated from sediment [29]. Based on the bioactivities of the isolated functional metabolites, naphtho-γ-pyrones, isochromans and the induced isochromanones appear to inhibit the growth of several devastating crop pathogens.

As examined in our previous study, the co-cultures of *Cosmospora* and the other three plant pathogens (*B. cinerea*, *P. syringae* and *R. solanacearum*) did not induce the production of the isochromanones [4], except in their co-cultivation with *M. oryzae*. Therefore, the producer fungus could not be definitively identified. However, several lines of evidence support our hypothesis that the isochromanones **1–5** are produced by *Cosmospora* sp. Firstly, the isochromans, pseudoanguillosporins A (**6**) and B (**7**), produced in a *Cosmospora* monoculture, share many structural similarities with the isochromanones; both the heterocyclic ring and aliphatic side chains. Secondly, the molecular family cluster of the isochromanones (Figure 3) displayed one shared node between both the co-culture and the *Cosmospora* monoculture. Therefore, it is inferred that the isochromanones (soudanones) are likely to be expressed by *Cosmospora* sp. when triggered by *M. oryzae*. More so, it is reasonable to assume that *M. oryzae* would not produce antibiotics inhibiting its own growth in co-cultures without self-protection mechanisms such as detoxification [30]. It is, however, worth mentioning that *Pyricularia grisea* (in the same monophyletic clade as *M. oryzae* [31]) produced 6-hydroxymellein and 3-methoxy-6,8-dihydroxy-3-methyl-3,4-dihydroisocoumarin, which possess a similar isochromanone core structure to the soudanones [32]. Further work on genome mining would be necessary to confirm the real producer of the induced isochromanones.

Cosmospora is a teleomorph genus in the family Nectriaceae (Ascomycetes). Its asexual morphs are mostly affiliated to the genera *Acremonium* and *Fusarium*, which are well-studied for fungal secondary metabolites [33,34]. Despite their close relationship with well-studied genera, there is little insight into the biosynthetic machinery and secondary metabolite production in the genus *Cosmospora*, supporting the relevance of this study. Only five classes of compounds have been reported from *Cosmospora* spp.: sesquiterpene glycoside cosmosporasides with reported nitric oxide production-inhibitory effects in lipopolysaccharide-activated murine macrophage RAW264.7 cells [35]; isoxazolidinone, containing parnafungins with antifungal activity [36,37]; orcinol *p*-depside acquastatins with enzyme inhibitory activity against protein tyrosine phosphatase 1B [38]; naphtho-γ-pyrone cephalochromin with anticancer activity [28]; and the dichlororesorcinol cosmochlorins that increase osteoclast formation in RAW264.7 cells [39]. The current study is the first report on the production of the isochromans, pseudoanguillosporins, and soudanone type isochromanones from *Cosmospora* sp. Pseudoanguillosporins were previously isolated from *Pseudoanguillospora* sp and *Cadophora* sp. with anti-phytopathogenic and antifungal activities [10,16]. Currently, pseudoanguillosporin A is isolated from *Strobilurus* sp. commercially and marketed as an antifungal and a fungal mitochondrial inhibitor (https://adipogen.com, accessed on 01 December 2021). The soudanones were first characterized as antifungal agents produced in the axenic cultures of another Ascomycete *Cadophora* sp. [16]. Our study is the second report on the production of soudanones by fungi.

Although the optimization study was not entirely successful, the observation of the induced isochromanones in the overlaid co-cultivation on solid agar (Figure S48) and their accumulation in the confrontation zone (Figure 1), suggested that direct cell-to-cell contact between *Cosmospora* and *M. oryzae* on a PDA medium may be crucial for their production. Schroeckh et al. [40] demonstrated a similar phenomenon: only cell-to-cell contact between the fungus *Aspergillus nidulans* and the bacterium *Streptomyces rapamycinicus* resulted in a specific activation of a PKS gene to produce orsellinic acid and the depside lecanoric acid. Our data suggest a similar phenomenon and warrants further experiments to validate the hypothesis.

4. Materials and Methods

4.1. General Experimental Procedures

Optical rotations were measured with a monochromatic light source in MeOH at 20 °C on a Jasco P-2000 polarimeter (Jasco, Pfungstadt, Germany). FT-IR spectra were recorded on a PerkinElmer Spectrum Two FT-IR spectrometer (PerkinElmer, Boston, MA, USA). The 1D (^{1}H and ^{13}C) and 2D (COSY, HSQC, HMBC, NOESY and TOCSY) NMR spectra were obtained on a BRUKER AV 600 spectrometer (600 and 150 MHz for ^{1}H and ^{13}C NMR, respectively, Bruker®, Billerica, MA, USA) or a Bruker Avance III spectrometer (500 MHz for ^{1}H NMR, Bruker®, Billerica, MA, USA). All spectra were acquired in solvents as specified in the text with referencing to residual ^{1}H and ^{13}C signals in the deuterated solvent. HRMS/MS data were recorded on a Xevo G2-XS QToF Mass Spectrometer (Waters®, Milford, MA, USA) connected to an Acquity UPLC I-Class System (Waters®, Milford, MA, USA). Crude extracts were fractionated using Chromabond SPE C18 column cartridges (Macherey-Nagel, Düren, Germany). TLC analysis was done on silica gel 60 F254 plates (pre-coated aluminium sheets) cut into 10 × 10 cm (Macherey-Nagel, Düren, Germany). Semipreparative and Preparative HPLCs were performed using a VWR Hitachi Chromaster system (VWR International, Allison Park, PA, USA) consisting of a 5310 column oven, a 5260 autosampler, a 5110 pump and a 5430 diode array detector. Separation was achieved using an octadecyl silica gel semipreparative column (Onyx, 10 mm × 100 mm, Phenomenex, Torrance, CA, USA) equipped with a guard column. Further purification was achieved on a Synergi Polar-RP 80 Å column (250 × 4.6 mm, Phenomenex, Torrance, CA, USA). Circular dichroism spectroscopy was performed in MeOH on a J-810 CD spectrometer (Jasco, Pfungstadt, Germany) with a 0.5-mm cuvette path length. The averages of triplicate scans were acquired and the CD signal of the MeOH was subtracted subsequently.

4.2. Fungal Collection and Taxonomy

The fungus *Cosmospora* sp. (GenBank accession number MH79129) was previously isolated from the Baltic Sea environment [4]. The plant pathogen *M. oryzae* was obtained from Deutsche Sammlung für Mikroorganismen und Zellkulturen (DSMZ, Braunschweig, Germany). *Cosmospora* sp. And *M. oryzae* were individually maintained on PDA medium (potato infusion powder 4 g, glucose monohydrate 20 g, agar 15 g in 1 L) as pre-cultures for 14 days.

4.3. UPLC-QtoF-MS Analysis

Analysis of all crude extracts (0.1 mg/mL) and controls (media blanks and solvents) were performed on an UPLC I-Class system coupled to a Xevo G2-XS QToF Mass spectrometer (Waters®, Milford, MA, USA). Chromatography was achieved on an Acquity UPLC HSS T3 column (High Strength Silica C18, 1.8 µm, 100 × 2.1 mm, Waters®, Milford, MA, USA) maintained at 40 °C. A binary solvent system (A: 99.9% MilliQ®-water/0.1% formic acid ULC/MS grade and B: 99.9% acetonitrile ULC/MS grade/0.1% formic acid) was pumped at a flow rate of 0.4 mL/min with a gradient of 1% to 100% B in 11.5 min. The column was then washed for 3.5 min with 100% B, back to 1% B (initial condition) and reconditioned at 1% in 2.5 min. MS and MS/MS spectra were acquired in a data dependent analysis (DDA) mode with an electrospray ionization (ESI) source in the positive ionization mode. The *m/z* range was set to 50–1200 Da in the centroid data format. The QToF-MS source temperature was set to 150 °C, capillary voltage of 0.8 KV, cone and desolvation gas flow of 50 and 1200 L/h, respectively. Desolvation temperature was set to 550 °C with sampling cone and source offset at 40 and 80, respectively. Collision energy (CE) was ramped: low CE from 6-60 eV to high CE of 9-80 eV. As controls, solvent (MeOH) and non-inoculated medium were injected under the same conditions.

4.4. Molecular Networking and Annotations

A molecular family, as introduced by Nguyen et al. [41], describes structurally related compounds based on the similarity in mass spectral fragmentation patterns. A series of connected nodes in an MN typically represents a molecular family. UPLC-MS/MS RAW data files were converted to centroid .mzXML format using msconvert from the ProteoWizard suite version 3.0.10051 [42]. The converted .mzXML files were uploaded on Global Natural Products Social Molecular Networking (GNPS) platform with an open access File Transfer Protocol client WinSCP version 5.15.9. A molecular network was generated from the .mzXML files of the fungal cultures using the GNPS script suite [43]. The algorithm assumed precursor mass tolerance of 0.02 Da and a product ion tolerance of 0.02 Da to create consensus spectra. The network was generated using a minimum of 6 matched peaks and a defined cosine score above 0.7. The resulting MN was visualized using Cytoscape version 3.8.2 [44], and displayed with 'directed' style, with the edges modulated by cosine score. To simplify analysis of the network, only nodes that contained ions observed between retention times of 2-11.30 min were considered. Background nodes originating from the cultivation medium PDA and solvents (MeOH and EtoAc) were removed from the MN. Node colours were mapped based on the source of the spectra files, thus red for *Cosmospora* sp., green for *M. oryzae* and blue for the co-culture. The MN job on GNPS can be found at https://gnps.ucsd.edu/ProteoSAFe/status.jsp?task=2d3d3 01a8f3e43039432865c61f33028, accessed on 01 December 2021. Manual annotations were done using MassLynx® (version 4.2, Waters®, Milford, MA, USA). to predict the molecular formulae of the m/z [M+H]$^+$ or other adduct ions, e.g., m/z [M+Na]$^+$, and searching them against databases such us NPAtlas [45] and DNP [46]. Hits were considered based on retention times and fragmentation patterns of the ions and biological sources of the hits. Dereplication was also achieved by isolation and characterization by NMR and other data such as HR-ESIMS, IR and [α]$_D$. Identifications were also assigned confidence levels 1–4 as proposed by Sumner et al. [47].

4.5. Large Scale Co-Cultivation and Isolation of Compounds

About 1 cm plugs from each pre-culture, *Cosmospora* sp. and *M. oryzae*, were streaked and inoculated on opposite sides of the same agar plate. To obtain enough extract amount for compound isolation, 100 plates on PDA medium were inoculated. The co-culture plates were incubated at 22 °C in the dark for 21 days. Co-cultures were extracted twice with EtOAc (at a ratio of 4 plates: 200 mL of EtOAc) after homogenising with an Ultra-Turrax at 19,000 rpm. EtOAc extracts were washed with equal volumes of water (200 mL). The pooled organic phase was concentrated in vacuo to yield 1.01 g of dry organic extract, and then partitioned using a modified protocol originally designed by Kupchan and Tsou [15,29]. This resulted in three subextracts: *n*-hexane (KH, 314.1 mg), CH_2Cl_2 (KC, 646.7 mg), and aqueous MeOH (KM, 7.1 mg). The subextracts were analysed by UPLC-MS/MS, and ions of interest (compounds **1-5**) were annotated in KC. The KC extract was then fractionated over a Chromabond SPE C18 column cartridge (Macherey-Nagel, Düren, Germany) with a step gradient of 10% to 100% MeOH to afford 11 fractions (F1-F11). Each subfraction was checked for its composition by TLC using $CHCl_3$-CH_3OH (9.5:0.5 v/v) as mobile phase and vanillin sulfuric acid as visualization reagent. Fractions 7 and 8 (F7–8) were combined (250 mg) and purified by HPLC on a semi-preparative RP column (Phenomex, Onyx monolithic C18, 100 × 10 mm) with H_2O:MeCN mixture (70:30 to 0:100 in 17 min, flow rate 3.5 mL/min) as mobile phase to yield compounds **6** (4 mg, t_R 12.3 min) and **1** (2.8 mg, t_R 14.5 min). Combined fractions 5 and 6 (33 mg) were subjected to semi-preparative RP–HPLC, on the same column, eluting with H_2O:MeCN (64:36 to 0:100 in 11 min, flow rate 3.5 mL/min) to yield 6 subfractions (fraction F5–6.1 to 5–6.6). Subfraction 5–6.1 contained pure compound **7** (5 mg, t_R 3.6 min), while 5–6.3 contained pure compound **3** (0.6 mg, t_R 6.5 min). Subfraction 5–6.2 (2.4 mg, t_R 4–6 min) was re-chromatographed by RP-HPLC on a Xselect HSS T3 column (2.5 μm, 150 × 4.6 mm) using an isocratic mixture of H_2O:MeCN (50:50) and flow rate of 1 mL/min, to afford compounds **2** (1.5 mg, t_R 4.1 min), **4** (0.5 mg, t_R 5.2 min) and **5** (0.2 mg, t_R 5.9 min). Subfraction 5–6.5 was also purified by RP-HPLC on a Synergi 4u Polar-RP 80A (250 × 4.6 mm) using an isocratic mixture of H_2O:MeCN (35:65) and flow rate of 1 mL/min to yield **8** (2.2 mg, t_R 4.1 min) and **9** (1.2 mg, t_R 4.3 min).

In an effort to obtain sufficient amounts of **1**, KH was fractionated over modified silica SPE cartridge (CHROMOBOND®, Macherey-Nagel, Dueren, Germany) with a step gradient of EtOAc in *n*-hexane (0 to 100% EtoAc) to afford 11 fractions (H1-H11). Compound **10** (10 mg) was crystallised out of fraction H2 and purified by filtering.

Soudanone A (**1**): amorphous powder; $[\alpha]_D^{20}$-59 (*c* 0.5, MeOH); IR (film) ν_{max} 3251, 2928, 2856, 1645, 1614, 1463, 1386, 1246, 1142 cm^{-1}; ^{1}H (600 MHz) and ^{13}C (150 MHz) NMR data, see Tables 1 and 2; HR-ESIMS *m/z* 293.1755 [M+H]$^+$ (calcd for C$_{17}$H$_{25}$O$_4$, 293.1753).

Soudanone E (**2**): yellow amorphous powder; $[\alpha]_D^{20}$-50 (*c* 0.05, MeOH); IR (film) ν_{max} 3316, 2930, 2858, 1644, 1618, 1460, 1382, 1253, 1109 cm^{-1}; ^{1}H (600 MHz) and ^{13}C (150 MHz) NMR data, see Tables 1 and 2; HR-ESIMS *m/z* 309.1705 [M+H]$^+$ (calcd for C$_{17}$H$_{25}$O$_5$, 309.1702).

Soudanone D (**3**): yellow amorphous powder; $[\alpha]_D^{20}$-56 (*c* 0.35, MeOH); IR (film) ν_{max} 3252, 2930, 2861, 1705, 1655, 1618, 1465, 1379, 1254, 1174 cm^{-1}; ^{1}H (600 MHz) and ^{13}C (150 MHz) NMR data, see Tables 1 and 2; HR-ESIMS *m/z* 307.1546 [M+H]$^+$ (calcd for C$_{17}$H$_{23}$O$_5$, 307.1545).

Soudanone H (**4**): yellow amorphous powder; $[\alpha]_D^{20}$-44 (*c* 0.05, MeOH); IR (film) ν_{max} 3390, 2918, 2850, 1646, 1598, 1464, 1380, 1255, 1179 cm^{-1}; ^{1}H (600 MHz) and ^{13}C (150 MHz) NMR data, see Tables 1 and 2; HR-ESIMS *m/z* 309.1703 [M+H]$^+$ (calcd for C$_{17}$H$_{25}$O$_5$, 309.1702).

Soudanone I (**5**): yellow amorphous powder; $[\alpha]_D^{20}$-20 (*c* 0.07, MeOH); IR (film) ν_{max} 3311, 2948, 2864, 1729, 1648, 1616, 1382, 1253, 1174 cm^{-1}; ^{1}H (600 MHz) and ^{13}C (150 MHz) NMR data, see Tables 1 and 2; HR-ESIMS *m/z* 309.1341 [M+H]$^+$ (calcd for C$_{16}$H$_{21}$O$_6$, 309.1338).

4.6. Co-Cultivation by Overlaid Inoculation on PDA and Extraction

A 1 cm agar plug of *M. oryzae* pre-culture was streaked over the surface of an agar plate and incubated at 22 °C for 7 days until it formed a lawn on the agar surface. Agar plugs of *Cosmospora* sp. were then inoculated onto the *M. oryzae* lawn (overlaid inoculation) and incubated for 14 days, resulting in a total incubation period of 21 days. For extraction, the entire co-culture was chopped into pieces and extracted with EtOAc (2 × 20 mL) and extracted as previously described [48]. Briefly, the co-culture was homogenized by an Ultra-Turrax in 20 mL EtOAc, and the organic layer decanted into a separating funnel. The extraction with EtOAc was done twice to increase the efficiency. The pooled EtOAc layer was then washed twice with 20 mL of Milli-Q® water (Arium® Lab water systems, Sartorius). The EtOAc phase was collected and dried under vacuum, re-dissolved in methanol (3 mL of ULC/MS grade MeOH), filtered through a 0.2 µm PTFE membrane (VWR International, Darmstadt, Germany) into a pre-weighed vial, and dried under nitrogen. An aliquot (0.1 mg/mL in MeOH) of extract was prepared for analysis by UPLC-MS/MS, and remaining extract was stored at −20 °C.

4.7. Co-Cultivation in Liquid Broth (PDB)

About 3 cm agar plugs, each of *Cosmospora* sp. and *M. oryzae* precultures, were inoculated into a 2 L Erlenmeyer flask containing 500 mL of potato dextrose broth (PDB: potato infusion powder 4 g, glucose monohydrate 20 g in 1 L; pH 5.6) and incubated at 22 °C on an orbital shaker (VKS-75 control, Edmund Bühler, Hechingen, Germany) at 120 rpm. An aliquot of the broth (5 mL) was sampled each day with a sterile pipette for 21 days. The daily aliquots were extracted with EtOAc (10 mL) and washed twice with 10 mL of Milli-Q® water (Arium® Lab water systems, Sartorius). Aliquots (0.1 mg/mL in MeOH) of extracts were prepared for analysis by UPLC-MS/MS and remaining extracts were stored at −20 °C.

4.8. Biological Assays

Anti-phytopathogenic activity assays were performed using the broth dilution technique in 96-well microplates. Compounds **1-10** (except **4** and **5**) were tested for in vitro anti-phytopathogenic activity against six phytopathogens, i.e., four bacteria (*X. campestris, R. solanacearum, P. syringae,* and *E. amylovora*), a fungus (*M. oryzae*), and an oomycete (*P. infestans*). The selected plant pathogens are among the most widespread and devastating microbes and pose a direct impact on global food supply and forest products [49,50]. The protocol used was described in a previous publication [4]. Briefly, compounds were first dissolved in DMSO (effective conc. 0.5% (v/v)) and added to test organisms suspended in appropriate broth in 96-well microplates. Compounds were tested at varying concentrations (max test conc. 100 µg/mL) in triplicate. Chloramphenicol and tetracycline were used as positive controls in the antibacterial tests, nystatin in the antifungal test, and cycloheximide in the antioomycete test; growth media and 0.5% (v/v) DMSO were tested as negative controls. Optical density measurements were made at 600 nm using an Infinite M200 reader (TECAN Deutschland GmbH, Crailsheim, Germany) before and after incubation to obtain growth inhibition values. The IC_{50} values were estimated using the Microsoft Excel program.

4.9. Mosher's Esterification

Compounds **2** and **4** (approximately 0.2 mg and 0.1 mg, respectively) were treated with (trimethylsilyl)diazomethane to protect the phenolic OH groups prior to MPA derivatization. The reaction product was split into two aliquots, which were treated separately with a ten-fold excess of (*R*)- and (*S*)-MPA (methoxyphenylacetic acid), respectively, in the presence of EDC (1-ethyl-3-(3-dimethylaminopropyl)carbodiimide) and 4-DMAP (4-dimethylaminopyridine), to give the corresponding (*R*)- and (*S*)-MPA esters. The structure of the esterification products was confirmed by LC-MS, ^{1}H NMR and 1D-TOCSY experiments (detailed description in Supplementary Materials, Figures S50–S54).

5. Conclusions

Five isochromanones, including two new soudanones H-I (**4–5**), with varying anti-phytopathogenic activities were isolated and identified in the co-culture of a marine-derived fungus, *Cosmospora* sp., and the phytopathogen, *M. oryzae*, but were not produced in the axenic fungal cultures. This supports the hypothesis that using phytopathogens as model competitors in co-cultivation triggers the biosynthesis of a new bioactive and diverse chemistry in marine fungi. This is the first study to report on isochromans and isochromanones from *Cosmospora* and its co-culture, and the second on the chemical class of soudanones from fungi. MN was used as an essential metabolomics tool to unravel the putative producer of the induced compounds in the co-culture. It further facilitated the annotation and structural elucidation of the isolated compounds. Further investigation into the optimal fermentation conditions to enhance the production of the isochromanones (**1–5**), as well as to elucidate the BGC responsible for their biosynthesis will be the focus of our future work. This will expand our knowledge on the influence of the conditions of co-cultivation on the expression of BGCs and biosynthesis of metabolites in these two fungi.

Supplementary Materials: The following supporting information can be downloaded at: https://www.mdpi.com/article/10.3390/ijms23020782/s1.

Author Contributions: Conceptualization, D.T. and E.O.-D.; methodology and investigation, E.O.-D.; formal analysis, E.O.-D, D.T., S.S. and A.M.; writing—original draft preparation, E.O.-D, D.T. and M.B.; supervision, D.T. All authors have read and agreed to the published version of the manuscript.

Funding: This research received no external funding.

Institutional Review Board Statement: Not applicable.

Informed Consent Statement: Not applicable.

Data Availability Statement: Not applicable.

Acknowledgments: We thank Arlette Wenzel-Storjohann for her valuable assistance with bioassays, and Fengjie Li for his comments on structure elucidation. We are grateful to the Institute of Biochemistry, Kiel University, for providing the CD spectrometer for measurements. We acknowledge financial support by Land Schleswig-Holstein within the funding program, Open Access Publikationsfonds.

Conflicts of Interest: The authors declare no conflict of interest.

References

1. Nofiani, R.; de Mattos-Shipley, K.; Lebe, K.E.; Han, L.-C.; Iqbal, Z.; Bailey, A.M.; Willis, C.L.; Simpson, T.J.; Cox, R.J. Strobilurin biosynthesis in Basidiomycete fungi. *Nat. Commun.* **2018**, *9*, 1–11. [CrossRef] [PubMed]
2. Kumar, A.; Sørensen, J.L.; Hansen, F.T.; Arvas, M.; Syed, M.F.; Hassan, L.; Benz, J.P.; Record, E.; Henrissat, B.; Pöggeler, S.; et al. Genome sequencing and analyses of two marine fungi from the North Sea unraveled a plethora of novel biosynthetic gene clusters. *Sci. Rep.* **2018**, *8*, 1–16. [CrossRef] [PubMed]
3. Harwani, D.; Begani, J.; Lakhani, J. Co-cultivation strategies to induce de novo synthesis of novel chemical scaffolds from cryptic secondary metabolite gene clusters. In *Fungi and their Role in Sustainable Development: Current Perspectives*; Gehlot, P., Singh, J., Eds.; Springer: Singapore, 2018; pp. 617–631.
4. Oppong-Danquah, E.; Parrot, D.; Blümel, M.; Labes, A.; Tasdemir, D. Molecular networking-based metabolome and bioactivity analyses of marine-adapted fungi co-cultivated with phytopathogens. *Front. Microbiol.* **2018**, *9*, 1–20. [CrossRef] [PubMed]
5. Serrano, R.; González-Menéndez, V.; Rodríguez, L.; Martín, J.; Tormo, J.R.; Genilloud, O. Co-culturing of fungal strains against *Botrytis cinerea* as a model for the induction of chemical diversity and therapeutic agents. *Front. Microbiol.* **2017**, *8*, 1–15. [CrossRef]
6. Chaudhry, V.; Runge, P.; Sengupta, P.; Doehlemann, G.; Parker, J.E.; Kemen, E. Shaping the leaf microbiota: Plant–microbe–microbe interactions. *J. Exp. Bot.* **2020**, *72*, 36–56. [CrossRef]
7. De Souza, E.M.; Granada, C.E.; Sperotto, R.A. Plant pathogens affecting the establishment of plant-symbiont Interaction. *Front. Plant Sci.* **2016**, *7*, 1–5. [CrossRef]
8. Koyama, K.; Natori, S. Further characterization of seven bis (naphtho-γ-pyrone) congeners of ustilaginoidins, coloring matters of *Claviceps virens* (*Ustilaginoidea virens*). *Chem. Pharm. Bull.* **1988**, *36*, 146–152. [CrossRef]
9. Sun, W.; Wang, A.; Xu, D.; Wang, W.; Meng, J.; Dai, J.; Liu, Y.; Lai, D.; Zhou, L. New ustilaginoidins from rice false smut balls caused by *Villosiclava virens* and their phytotoxic and cytotoxic activities. *J. Agric. Food Chem.* **2017**, *65*, 5151–5160. [CrossRef]
10. Kock, I.; Draeger, S.; Schulz, B.; Elsässer, B.; Kurtán, T.; Kenéz, Á.; Rheinheimer, J. Pseudoanguillosporin A and B: Two new isochromans isolated from the endophytic fungus *Pseudoanguillospora* sp. *Eur. J. Org. Chem.* **2009**, *2009*, 1427–1434. [CrossRef]
11. Wang, Y.; Xu, L.; Ren, W.; Zhao, D.; Zhu, Y.; Wu, X. Bioactive metabolites from *Chaetomium globosum* L18, an endophytic fungus in the medicinal plant *Curcuma wenyujin*. *Phytomedicine* **2012**, *19*, 364–368. [CrossRef]
12. Heald, S.L.; Jeffs, P.W.; Wheat, R.W. The identification of ergosterol and Δ9(11)-dehydroergosterol from mycelia of *Coccidioides immitis* by reverse-phase high-performance liquid and gas chromatography and ultraviolet and mass spectrometry. *Exp. Mycol.* **1981**, *5*, 162–166. [CrossRef]
13. El-Elimat, T.; Figueroa, M.; Ehrmann, B.M.; Cech, N.B.; Pearce, C.J.; Oberlies, N.H. High-resolution MS, MS/MS, and UV database of fungal secondary metabolites as a dereplication protocol for bioactive natural products. *J. Nat. Prod.* **2013**, *76*, 1709–1716. [CrossRef]
14. Liu, H.; Zhao, X.; Guo, M.; Liu, H.; Zheng, Z. Growth and metabolism of *Beauveria bassiana* spores and mycelia. *BMC Microbiol.* **2015**, *15*, 1–12. [CrossRef]
15. Kupchan, S.M.; Tsou, G.; Sigel, C.W. Datiscacin, a novel cytotoxic cucurbitacin 20-acetate from *Datisca glomerata*. *J. Org. Chem.* **1973**, *38*, 1420–1421. [CrossRef]
16. Rusman, Y.; Held, B.W.; Blanchette, R.A.; Wittlin, S.; Salomon, C.E. Soudanones A-G: Antifungal isochromanones from the Ascomycetous fungus *Cadophora* sp. isolated from an iron mine. *J. Nat. Prod.* **2015**, *78*, 1456–1460. [CrossRef]
17. Seco, J.M.; Quiñoá, E.; Riguera, R. The assignment of absolute aonfiguration by NMR. *Chem. Rev.* **2004**, *104*, 17–117. [CrossRef]
18. Seki, M.; Mori, K. Synthesis of a prenylated and immunosuppressive marine galactosphingolipid with cyclopropane-containing alkyl chains: (2S,3R,11S,12R,2'''R,5'''Z,11'''S,12'''R)-plakoside A and its (2S,3R,11R,12S,2'''R,5'''Z,11'''R,12'''S) isomer. *Eur. J. Org. Chem.* **2001**, *2001*, 3797–3809. [CrossRef]
19. Matsumoto, M.; Minato, H.; Kondo, E.; Mitsugi, T.; Katagiri, K. Cephalochromin, dihydroisoustilaginoidin A, and iso-ustilaginoidin A from *Verticillium* sp. K-113. *J. Antibiot.* **1975**, *28*, 602–604. [CrossRef]
20. Sung, A.A.; Gromek, S.M.; Balunas, M.J. Upregulation and identification of antibiotic activity of a marine-derived *Streptomyces* sp. via co-cultures with human pathogens. *Mar. Drugs* **2017**, *15*, 250. [CrossRef]
21. Pal, S.; Chatare, V.; Pal, M. Isocoumarin and its derivatives: An overview on their synthesis and applications. *Curr. Org. Chem.* **2011**, *15*, 782–800. [CrossRef]
22. Noor, A.O.; Almasri, D.M.; Bagalagel, A.A.; Abdallah, H.M.; Mohamed, S.G.A.; Mohamed, G.A.; Ibrahim, S.R.M. Naturally occurring isocoumarins derivatives from endophytic fungi: Sources, isolation, structural characterization, biosynthesis, and biological activities. *Molecules* **2020**, *25*, 395. [CrossRef]

23. Qadeer, G.; Rama, N.; Fan, Z.-J.; Liu, B.; Liu, X.-F. Synthesis, herbicidal, fungicidal and insecticidal evaluation of 3-(dichlorophenyl)- isocoumarins and (±)-3-(dichlorophenyl)-3,4-dihydroisocoumarins. *J. Braz. Chem. Soc.* **2007**, *18*, 1176–1182. [CrossRef]
24. Shabir, G.; Saeed, A.; El-Seedi, H.R. Natural isocoumarins: Structural styles and biological activities, the revelations carry on. *Phytochemistry* **2021**, *181*, 1–23. [CrossRef]
25. Netzker, T.; Flak, M.; Krespach, M.K.C.; Stroe, M.C.; Weber, J.; Schroeckh, V.; Brakhage, A.A. Microbial interactions trigger the production of antibiotics. *Curr. Opin. Microbiol.* **2018**, *45*, 117–123. [CrossRef]
26. Zheng, C.J.; Sohn, M.J.; Lee, S.; Hong, Y.S.; Kwak, J.H.; Kim, W.G. Cephalochromin, a FabI-directed antibacterial of microbial origin. *Biochem. Biophys. Res. Commun.* **2007**, *362*, 1107–1112. [CrossRef]
27. Lu, S.; Sun, W.; Meng, J.; Wang, A.; Wang, X.; Tian, J.; Fu, X.; Dai, J.; Liu, Y.; Lai, D.; et al. Bioactive bis-naphtho-γ-pyrones from rice false smut pathogen *Ustilaginoidea virens*. *J. Agric. Food Chem.* **2015**, *63*, 3501–3508. [CrossRef]
28. Hsiao, C.-J.; Hsiao, G.; Chen, W.-L.; Wang, S.-W.; Chiang, C.-P.; Liu, L.-Y.; Guh, J.-H.; Lee, T.-H.; Chung, C.-L. Cephalochromin induces G0/G1 cell cycle arrest and apoptosis in A549 human non-small-cell lung cancer cells by inflicting mitochondrial disruption. *J. Nat. Prod.* **2014**, *77*, 758–765. [CrossRef]
29. Oppong-Danquah, E.; Budnicka, P.; Blümel, M.; Tasdemir, D. Design of fungal co-cultivation based on comparative metabolomics and bioactivity for discovery of marine fungal agrochemicals. *Mar. Drugs* **2020**, *18*, 73. [CrossRef]
30. Keller, N.P. Translating biosynthetic gene clusters into fungal armor and weaponry. *Nat. Chem. Biol.* **2015**, *11*, 671–677. [CrossRef]
31. Klaubauf, S.; Tharreau, D.; Fournier, E.; Groenewald, J.Z.; Crous, P.W.; de Vries, R.P.; Lebrun, M.H. Resolving the polyphyletic nature of *Pyricularia* (Pyriculariaceae). *Stud. Mycol.* **2014**, *79*, 85–120. [CrossRef]
32. Masi, M.; Santoro, E.; Clement, S.; Meyer, J.; Scafato, P.; Superchi, S.; Evidente, A. Further secondary metabolites produced by the fungus *Pyricularia grisea* isolated from buffelgrass (*Cenchrus ciliaris*). *Chirality* **2020**, *32*, 1234–1242. [CrossRef] [PubMed]
33. Herrera, C.S.; Rossman, A.Y.; Samuels, G.J.; Pereira, O.L.; Chaverri, P. Systematics of the *Cosmospora viliuscula* species complex. *Mycologia* **2015**, *107*, 532–557. [CrossRef] [PubMed]
34. Tian, J.; Lai, D.; Zhou, L. Secondary metabolites from *Acremonium* fungi: Diverse structures and bioactivities. *Mini-Rev. Med. Chem.* **2017**, *17*, 603–632. [CrossRef] [PubMed]
35. Lee, T.H.; Lu, C.K.; Wang, G.J.; Chang, Y.C.; Yang, W.B.; Ju, Y.M. Sesquiterpene glycosides from *Cosmospora joca*. *J. Nat. Prod.* **2011**, *74*, 1561–1567. [CrossRef] [PubMed]
36. Overy, D.; Calati, K.; Kahn, J.N.; Hsu, M.-J.; Martín, J.; Collado, J.; Roemer, T.; Harris, G.; Parish, C.A. Isolation and structure elucidation of parnafungins C and D, isoxazolidinone-containing antifungal natural products. *Bioorg. Med. Chem. Lett.* **2009**, *19*, 1224–1227. [CrossRef]
37. Parish, C.A.; Smith, S.K.; Calati, K.; Zink, D.; Wilson, K.; Roemer, T.; Jiang, B.; Xu, D.; Bills, G.; Platas, G.; et al. Isolation and structure elucidation of parnafungins, antifungal natural products that Inhibit mRNA polyadenylation. *J. Am. Chem. Soc.* **2008**, *130*, 7060–7066. [CrossRef]
38. Seo, C.; Sohn, J.H.; Oh, H.; Kim, B.Y.; Ahn, J.S. Isolation of the protein tyrosine phosphatase 1B inhibitory metabolite from the marine-derived fungus *Cosmospora* sp. SF-5060. *Bioorg. Med. Chem. Lett.* **2009**, *19*, 6095–6097. [CrossRef]
39. Shiono, Y.; Miyazaki, N.; Murayama, T.; Koseki, T.; Harizon; Katja, D.G.; Supratman, U.; Nakata, J.; Kakihara, Y.; Saeki, M.; et al. GSK-3β inhibitory activities of novel dichloresorcinol derivatives from *Cosmospora vilior* isolated from a mangrove plant. *Phytochem. Lett.* **2016**, *18*, 122–127. [CrossRef]
40. Schroeckh, V.; Scherlach, K.; Nutzmann, H.W.; Shelest, E.; Schmidt-Heck, W.; Schuemann, J.; Martin, K.; Hertweck, C.; Brakhage, A.A. Intimate bacterial-fungal interaction triggers biosynthesis of archetypal polyketides in *Aspergillus nidulans*. *Proc. Natl. Acad. Sci. USA* **2009**, *106*, 14558–14563. [CrossRef]
41. Nguyen, D.D.; Wu, C.-H.; Moree, W.J.; Lamsa, A.; Medema, M.H.; Zhao, X.; Gavilan, R.G.; Aparicio, M.; Atencio, L.; Jackson, C.; et al. MS/MS networking guided analysis of molecule and gene cluster families. *Proc. Natl. Acad. Sci. USA* **2013**, *110*, 2611–2620. [CrossRef]
42. Chambers, M.C.; Maclean, B.; Burke, R.; Amodei, D.; Ruderman, D.L.; Neumann, S.; Gatto, L.; Fischer, B.; Pratt, B.; Egertson, J. A cross-platform toolkit for mass spectrometry and proteomics. *Nat. Biotechnol.* **2012**, *30*, 918–920. [CrossRef]
43. Wang, M.; Carver, J.J.; Phelan, V.V.; Sanchez, L.M.; Garg, N.; Peng, Y.; Nguyen, D.D.; Watrous, J.; Kapono, C.A.; Luzzatto-Knaan, T.; et al. Sharing and community curation of mass spectrometry data with Global Natural Products Social Molecular Networking. *Nat. Biotechnol.* **2016**, *34*, 828–837. [CrossRef]
44. Shannon, P.; Markiel, A.; Ozier, O.; Baliga, N.S.; Wang, J.T.; Ramage, D.; Amin, N.; Schwikowski, B.; Ideker, T. Cytoscape: A software environment for integrated models of biomolecular interaction networks. *Genome Res.* **2003**, *13*, 2498–2504. [CrossRef]
45. Van Santen, J.A.; Jacob, G.; Singh, A.L.; Aniebok, V.; Balunas, M.J.; Bunsko, D.; Neto, F.C.; Castaño-Espriu, L.; Chang, C.; Clark, T.N.; et al. The Natural Products Atlas: An open access knowledge base for microbial natural products discovery. *ACS Cent. Sci.* **2019**, *5*, 1824–1833. [CrossRef]
46. Buckingham, J. Dictionary of Natural Products (On-Line Web Edition). Available online: http://www.chemnetbase.com (accessed on 1 December 2021).
47. Sumner, L.W.; Amberg, A.; Barrett, D.; Beale, M.H.; Beger, R.; Daykin, C.A.; Fan, T.W.-M.; Fiehn, O.; Goodacre, R.; Griffin, J.L.; et al. Proposed minimum reporting standards for chemical analysis. *Metabolomics* **2007**, *3*, 211–221. [CrossRef]

48. Oppong-Danquah, E.; Passaretti, C.; Chianese, O.; Blümel, M.; Tasdemir, D. Mining the metabolome and the agricultural and pharmaceutical potential of sea foam-derived fungi. *Mar. Drugs* **2020**, *18*, 128. [CrossRef]
49. Mansfield, J.; Genin, S.; Magori, S.; Citovsky, V.; Sriariyanum, M.; Ronald, P.; Toth, I.A.N. Top 10 plant pathogenic bacteria in molecular plant pathology. *Mol. Plant Pathol.* **2012**, *13*, 614–629. [CrossRef]
50. Dean, R.; Kan, J.A.V.; Pretorius, Z.A.; Hammond-Kosack, K.E.; Pietro, A.D.; Spanu, P.D.; Foster, G.D. The top 10 fungal pathogens in molecular plant pathology. *Mol. Plant Pathol.* **2012**, *13*, 414–430. [CrossRef]

Article

Butein and Frondoside-A Combination Exhibits Additive Anti-Cancer Effects on Tumor Cell Viability, Colony Growth, and Invasion and Synergism on Endothelial Cell Migration

Shahrazad Sulaiman [1], Kholoud Arafat [1], Aya Mudhafar Al-Azawi [1], Noura Abdulraouf AlMarzooqi [1], Shamsa Nasser Ali Hussain Lootah [1] and Samir Attoub [1,2,*]

[1] Department of Pharmacology & Therapeutics, College of Medicine & Health Sciences, United Arab Emirates University, Al Ain 17666, United Arab Emirates; sharazadjeffy@uaeu.ac.ae (S.S.); kholoud.arafat@uaeu.ac.ae (K.A.); 201870001@uaeu.ac.ae (A.M.A.-A.); 201311678@uaeu.ac.ae (N.A.A.); 201312535@uaeu.ac.ae (S.N.A.H.L.)
[2] Institut National de la Santé et de la Recherche Médicale (INSERM), 75013 Paris, France
* Correspondence: samir.attoub@uaeu.ac.ae

Citation: Sulaiman, S.; Arafat, K.; Al-Azawi, A.M.; AlMarzooqi, N.A.; Lootah, S.N.A.H.; Attoub, S. Butein and Frondoside-A Combination Exhibits Additive Anti-Cancer Effects on Tumor Cell Viability, Colony Growth, and Invasion and Synergism on Endothelial Cell Migration. *Int. J. Mol. Sci.* **2022**, *23*, 431. https://doi.org/10.3390/ijms23010431

Academic Editor: Hidayat Hussain

Received: 1 December 2021
Accepted: 29 December 2021
Published: 31 December 2021

Publisher's Note: MDPI stays neutral with regard to jurisdictional claims in published maps and institutional affiliations.

Copyright: © 2021 by the authors. Licensee MDPI, Basel, Switzerland. This article is an open access article distributed under the terms and conditions of the Creative Commons Attribution (CC BY) license (https://creativecommons.org/licenses/by/4.0/).

Abstract: Despite the significant advances in targeted- and immuno-therapies, lung and breast cancer are at the top list of cancer incidence and mortality worldwide as of 2020. Combination therapy consisting of a mixture of different drugs taken at once is currently the main approach in cancer management. Natural compounds are extensively investigated for their promising anti-cancer potential. This study explored the anti-cancer potential of butein, a biologically active flavonoid, on two major solid tumors, namely, A549 lung and MDA-MB-231 breast cancer cells alone and in combination with another natural anti-cancer compound, frondoside-A. We demonstrated that butein decreases A549 and MDA-MB-231 cancer cell viability and colony growth in vitro in addition to tumor growth on chick embryo chorioallantoic membrane (CAM) in vivo without inducing any noticeable toxicity. Additionally, non-toxic concentrations of butein significantly reduced the migration and invasion of both cell lines, suggesting its potential anti-metastatic effect. We showed that butein anti-cancer effects are due, at least in part, to a potent inhibition of STAT3 phosphorylation, leading to PARP cleavage and consequently cell death. Moreover, we demonstrated that combining butein with frondoside-A leads to additive effects on inhibiting A549 and MDA-MB-231 cellular viability, induction of caspase 3/7 activity, inhibition of colony growth, and inhibition of cellular migration and invasion. This combination reached a synergistic effect on the inhibition of HUVECs migration in vitro. Collectively, this study provides sufficient rationale to further carry out animal studies to confirm the relevance of these compounds' combination in cancer therapy.

Keywords: lung cancer; breast cancer; butein; frondoside-A; STAT3; angiogenesis; invasion; viability; tumor growth

1. Introduction

Lung and breast cancer are the most prevalent types of cancer worldwide [1], and this is despite the significant advances in cytotoxic-, targeted-, and immune-therapies. While these treatment approaches have led to an improvement in overall survival, they come with drawbacks limiting their success in providing a cure for patients. Challenges of co-lateral damage, acquired resistance, and limited efficacy to a small percentage of patients, in addition to high cost, have led to a renewed interest in natural compounds for cancer treatment [2–4]. These natural compounds may have fewer side effects and may be used as an adjuvant to reduce clinical cytotoxic drug dose or minimize their side effects [5]. More importantly, they may also lead to a better anti-cancer response when used in combination therapy.

Extensive pre-clinical research studies are investigating natural compounds for their potential anti-cancer effects. Butein (3, 4, 2', 4'-tetrahydroxychalcone) is one of these active ingredients, isolated from various plants including stem bark of cashews (*Semecarpus anacardium*), Rhus verniciflua Stokes, Caragana jubata, and the heartwood of Dalbergia odorifera [6]. The anti-cancer potential of butein has been demonstrated in leukemia and in various solid tumors investigated to date [6,7]. However, very few studies have investigated its anti-cancer effect in vivo or in combination with other natural agents such as frondoside-A, a triterpenoid glycoside isolated from the Atlantic cucumber, Cucumaria frondosa [8].

In this study, we determined the anti-cancer potential of butein on lung and breast cancer cell viability, colony growth, migration, and invasion in vitro and on tumor growth in vivo using the chick embryo CAM tumor xenograft model. The impact on angiogenesis was determined using HUVECs migration and tubules formation assays in vitro. In addition, the study investigated the anti-cancer impact of the combination of butein with the natural compound frondoside-A.

2. Results and Discussion

2.1. Butein Decreases Cellular Viability, Colony and Tumor Growth

The anti-cancer effect of butein was investigated on cell viability, colony and tumor growth in non-small cell lung cancer (NSCLC) cells A549 and triple-negative breast cancer (TNBC) cells MDA-MB-231. As shown in Figure 1, butein (1–100 µM) caused a concentration- and time-dependent decrease in the cellular viability of lung A549 and breast MDA-MB-231 cells (Figure 1A,B). The IC50 values at 72 h were 35.1 µM and 55.7 µM for A549 and MDA-MB-231 cells, respectively. Our results are in agreement with other studies showing that butein suppressed the survival of multiple myeloma cells U266 [9], prostate cancer cells "LNCaP, CWR22Rv1, and PC-3" [10], NSCLC cells A549 [11], cervical cells HeLa [12], C-33A and SiHa [13], breast cancer cells ER+ "MCF7, T47D, and ZR-75-1" and the TNBC cells "MDA-MB-231, Hs578T, BT-20, HCC-38, HCC-70 and MDA-MB-453" [14,15], hepatocarcinoma cells SK-HEP-1 [16], and acute lymphoblastic leukemia cells "RS4-11, CEM-C7, CEM-C1, and MOLT-4" [17].

Cleavage of PARP is widely accepted as a specific marker of apoptosis. In this study, we clearly demonstrated that butein (25 and 50 µM) induces a concentration- and time-dependent increase in the cleavage of the full-length PARP (116 kDa) to a large cleaved fragment (89 kDa) in both A549 and MDA-MB-231 cells (Figure 1C,D). These data are supported by previous reports demonstrating that 24 h treatment with butein induces PARP cleavage in HepG2 cells [18,19], LNCaP cells [10], and C-33A and SiHa [13].

To further confirm the anticancer potential of butein, we examined its ability to affect the growth capacity of pre-formed colonies using the colony growth assay. Toward this, lung (A549) and breast (MDA-MB-231) cells were grown for 14 days to form colonies and then treated with increasing concentration of butein for an additional seven days. Treatment with butein (10–100 µM) caused a significant concentration-dependent decrease in the number of colonies (Figure 2A–D). Similarly, it has been shown that butein decreases the colony formation of HeLa cells [12], and SMMC-7721 and HepG2 hepatocarcinoma cells [19].

To confirm the pharmacological relevance of our in vitro data, the anticancer activity of butein was investigated in vivo using the CAM tumor growth model. A549 and MDA-MB-231 cells grafted on the CAM formed tumors that were treated every 48 h with vehicle (DMSO) or butein (100 µM). At the end of the experiment (E17), tumors were recovered from the upper CAM and weighted. In line with our in vitro findings, we found that butein significantly inhibited tumor growth in vivo (Figure 3A–D). Butein showed no major toxic effect as there was no, or a small, difference in the number of surviving chick embryos between the control and the butein-treated eggs in A549 groups (85.7% vs. 86.7%) (Figure 3E) and MDA-MB-231 groups (78.6% vs. 93.3%) (Figure 3F). Similarly, it has been reported that butein decreases cervical HeLa [12], breast HER2+ BT-474 [15], and liver HepG2 [19,20] xenografts tumor growth in nude mice.

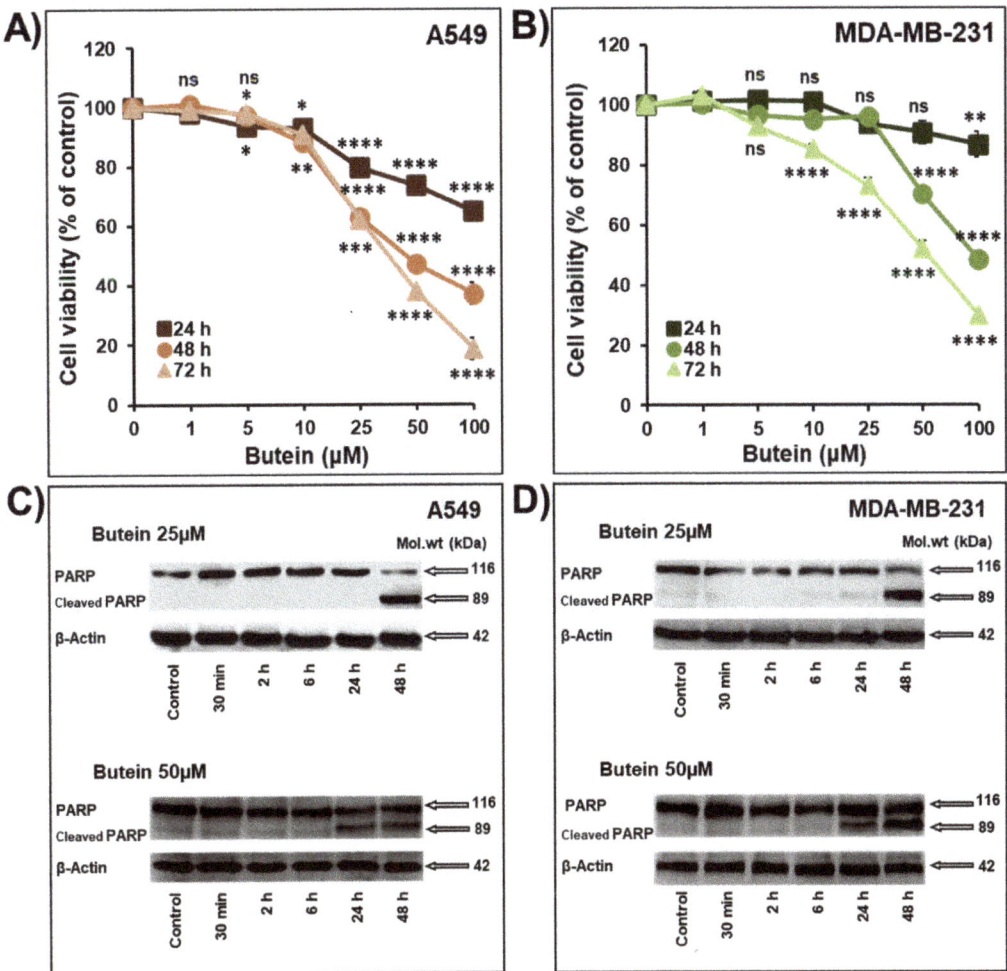

Figure 1. Inhibition of cellular viability associated PARP cleavage by butein. Exponentially growing A549 (**A**) and MDA-MB-231 (**B**) cells were treated with vehicle (0.1% DMSO) and the indicated concentrations of butein for 24, 48, and 72 h. Viable cells were determined using the CellTiter-Glo Luminescent Cell Viability Assay, based on ATP quantification, which indicates the presence of viable cells. Experiments were repeated at least three times. Western blot analysis shows PARP cleavage after butein (25 and 50 μM) treatment in A549 (**C**) and MDA-MB-231 (**D**) cancer cells. β-actin was used as a loading control. The data shown are representative of three in-dependent experiments. Shapes represent means; bars represent S.E.M. * Significantly different at $p < 0.05$, ** Significantly different at $p < 0.01$, *** Significantly different at $p < 0.001$, **** Significantly different at $p < 0.0001$. ns—non-significant.

2.2. Butein Decreases Lung and Breast Cancer Cell Migration and Invasion

Cancer cell migration and invasion are critical steps in the process of metastasis. To determine whether butein inhibited cell migration and invasion in vitro, lung A549 and breast MDA-MB-231 cancer cells were treated with low concentrations of butein (5 and 10 μM). As shown in Figure 4, butein induced a significant time- and concentration-dependent inhibition of A549 (Figure 4A) and MDA-MB-231 (Figure 4B) cancer cell migration. In addition, we demonstrated that butein was also able to decrease A549 and MDA-MB-231 cancer cell invasion in a concentration-dependent manner (Figure 4C,D).

A previous study also reported that butein at the concentration of 20 µM suppressed bladder cancer cells BLS-211 motility and invasion [20]. Similarly, butein at the high concentration of 50 µM suppressed CXCL12-induced cancer cell migration and invasion of breast cancer cells SKBr3 and the pancreatic cells AsPC-1 [21], respectively. Butein (15, 25, and 50 µM) also decreased HeLa cell migration and invasion [12], and at the concentrations of 50 and 75 µM inhibited the migration and invasion of SK-HEP-1 cells [16]. In comparison with our study, all previous studies used a much higher concentration of butein. In addition, it has been shown that butein reduces lung metastasis of mouse melanoma cells B16F10 [22] and the metastasis behavior of the hepatocellular carcinoma cells SK-Hep-1 [23].

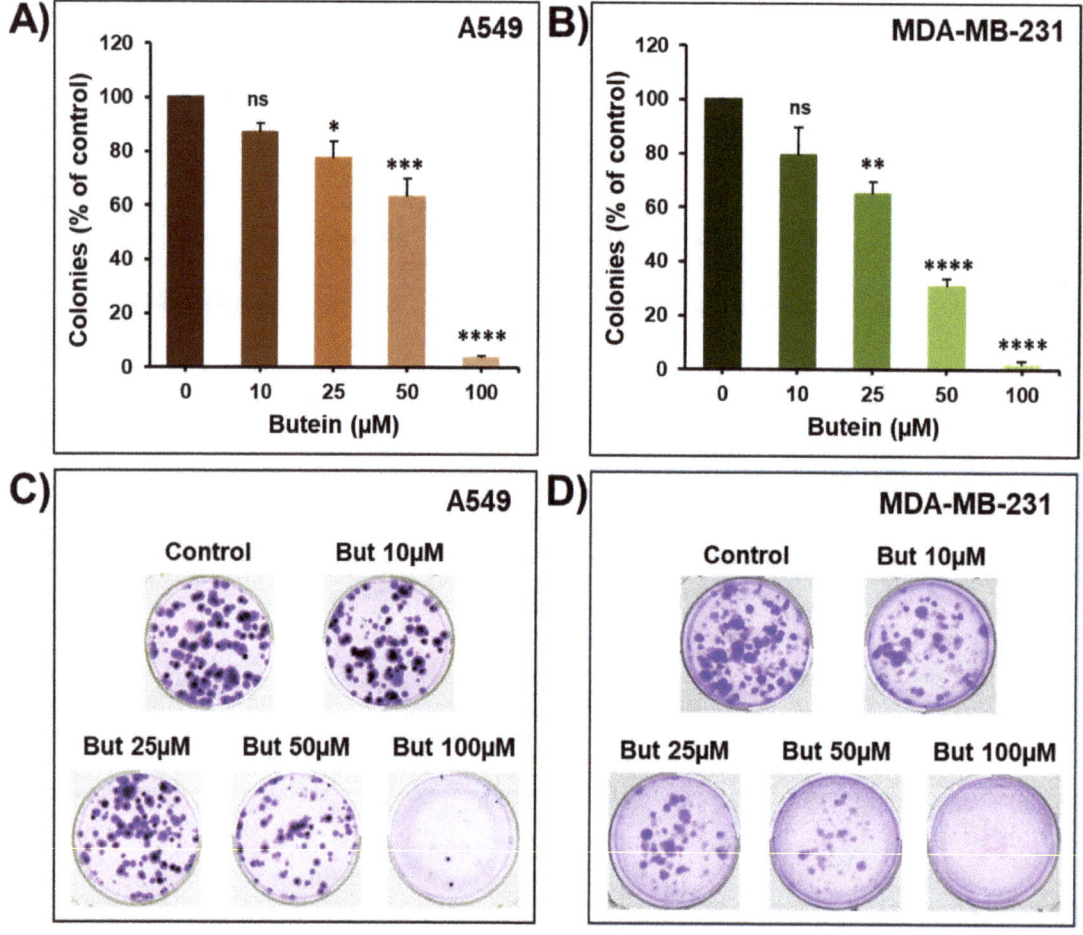

Figure 2. Effect of butein on colony growth. The growth of cancer cell-derived colonies from A549 (**A**) and MDA-MB-231 (**B**) cells was assessed by measuring the number of the colonies in control and butein-treated wells for seven days. (**C,D**) Representative pictures of the control and butein-treated colonies are shown for A549 and MDA-MB-231 cancer cells. Experiments were repeated at least three times. Columns represent means; bars represent S.E.M. * Significantly different at $p < 0.05$, ** Significantly different at $p < 0.01$, *** Significantly different at $p < 0.001$, **** Significantly different at $p < 0.0001$. ns—non-significant.

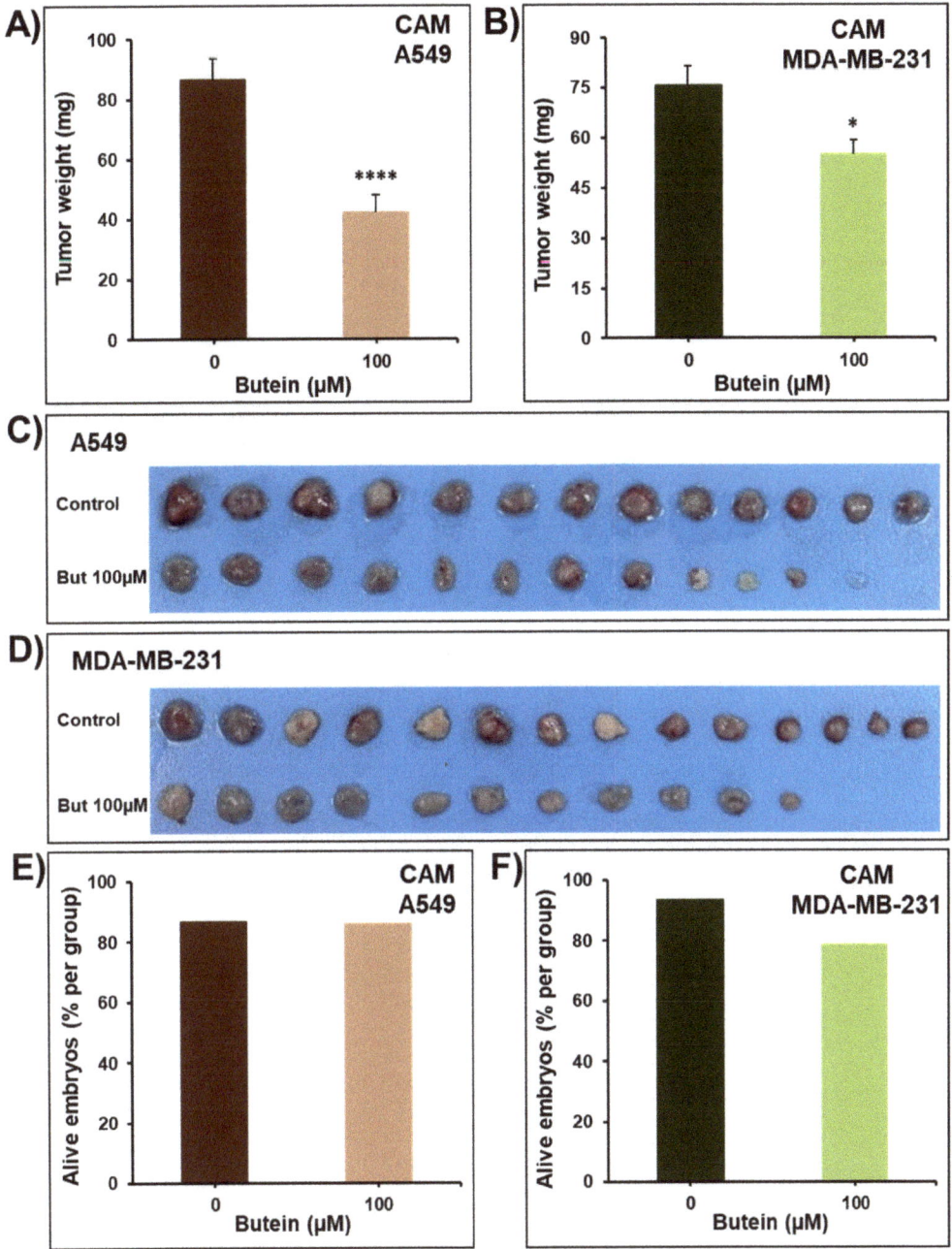

Figure 3. Impact of butein on tumor growth using the in vivo CAM tumor xenograft model. Volumes of 1×10^6 of A549 (**A,C**) and MDA-MB-231 (**B,D**) cells were grafted on the CAM of 9 days (E9) chick embryos. Tumors were treated with butein (100 µM) every 48 h for a total of 6 days. At E17, tumors were collected, weighed, and photographed (**C,D**). The viability of the chick embryos was assessed, and the percentage of alive embryos was determined (**E,F**). Columns are means; bars are S.E.M. * Significantly different at $p < 0.05$. **** Significantly different at $p < 0.0001$.

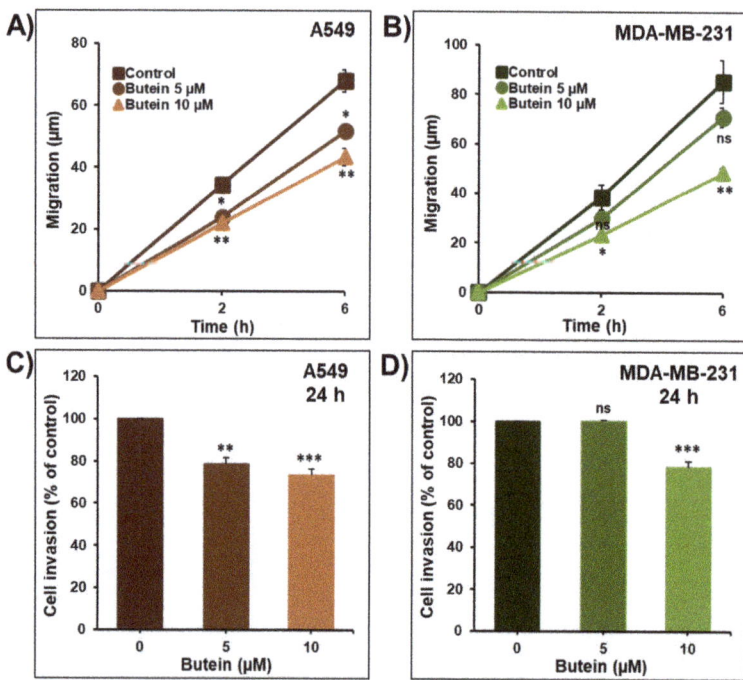

Figure 4. Butein impairs cancer cell migration and invasion. Wounds were introduced in A549 (**A**) and MDA-MB-231 (**B**) cells' confluent monolayers cultured in the presence or absence (control) of butein (5 and 10 µM). The mean distance that cells travelled from the edge of the scraped area after 2 and 6 h was measured using an inverted microscope. A549 (**C**) and MDA-MB-231 (**D**) cells were incubated for 24 h in the presence or absence of butein (5 and 10 µM). Cells that invaded the Matrigel and crossed the 8 µm pores insert were determined using the CellTiter-Glo Luminescent Cell Viability Assay. All experiments were repeated at least three times. Columns or shapes represent means; bars represent S.E.M. * Significantly different at $p < 0.05$, ** Significantly different at $p < 0.01$, *** Significantly different at $p < 0.001$. ns—non-significant.

2.3. Impact of Butein on STAT3 Phosphorylation

Signal Transducer and Activator of Transcription 3 (STAT3) is constitutively active in a wide variety of human cancers, including but not limited to breast, lung, and colorectal cancers [24,25]. Chronic STAT3 phosphorylation is associated with major cancer hallmarks, including survival, migration, invasion, and metastasis [26–29]. There is large community agreement that the activated STAT3 influences not only tumor growth but also the invasiveness of cancer cells [30]. However, STAT3 targeting for cancer therapy is still a major challenge [24,31–33]. In this context, we examined whether the observed anti-cancer effects of butein involve the STAT3 pathway. In this sense, the level of activated (phosphorylated) STAT3 was examined over time-period (0.5, 2, 6, 24, and 48 h) in the lung (A549) and breast (MDA-MB-231) cancer cells lines treated with 25 and 50 µM of butein. As shown in Figure 5, butein significantly decreased the level of phosphorylated STAT3 in A549 (Figure 5A,B) and MDA-MB-231 (Figure 5C,D). The inhibition of STAT3 phosphorylation was observed as early as 30 min post-treatment at both used concentrations and in both cell lines. This inhibition was maintained for almost 48 h. As expected, butein treatment had no effect on the level of total STAT3 (Figure 5E–H). Our results are in agreement with previous reports showing that butein inhibited the constitutive activation of STAT3 in multiple myeloma cells U266 [9] and in the hepatocellular carcinoma cell line HepG2 [18]. These two studies also demonstrate that this suppression of STAT3 phosphorylation was mediated through the inhibition of the upstream-activated c-src and JAK2 kinases. Butein also inhibited the constitutive active STAT3 in both head and neck squamous carcinoma SCC4 and in human prostate carcinoma DU145 cells [9]. Altogether, these data strongly suggest that butein mediates its anti-cancer effect, at least in part, through downregulation of the STAT3 signaling pathway.

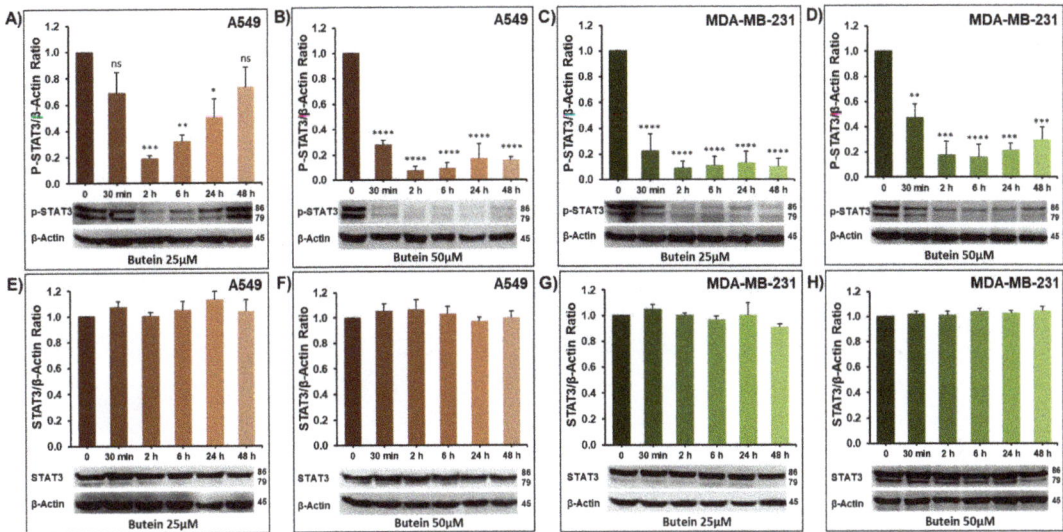

Figure 5. Western blot showing the inhibition of STAT3 phosphorylation by butein in A549 lung cancer cells (**A,B**), and MDA-MB-231 breast cancer cells (**C,D**). Effect of butein on total STAT3 (**E–H**). Each cell line was treated with 25 and 50 µM butein, and proteins were extracted at the indicated time-points (0.5, 2, 6, 24, and 48 h). β-actin was used as a loading control. The data shown are representative of three independent experiments. Columns represent means; bars represent S.E.M. * Significantly different at $p < 0.05$, ** Significantly different at $p < 0.01$, *** Significantly different at $p < 0.001$, **** Significantly different at $p < 0.0001$. ns—non-significant.

2.4. Butein in Combination Therapy with Frondoside-A Enhances Caspase 3/7 Inhibition of Cellular Viability

Frondoside-A has been previously reported to have strong anticancer activity against various types of cancer, including breast, lung, pancreas, prostate, and colon cancer [34–40]. The widespread effects of frondoside-A have been linked to various mechanisms, notably, inhibition of P21-activated kinase 1 (PAK1) [38,41]. Nguyen et al. (2017) reported that frondoside-A directly and specifically inhibits PAK1 activity in vitro with an IC50 equal to 1.2 µM [41]. PAK1 is overexpressed or overactivated in various cancer types, controlling cell growth, autophagy, angiogenesis, invasion, and metastasis [42]. Therefore, inhibiting such targets by frondoside-A could explain the wide range of activities frondoside-A exerts [38]. In current clinical oncology practices, combination therapy is the main approach in cancer management [43]. Therefore, we decided to explore the anticancer activity of butein combined with frondoside-A.

To investigate the therapeutic value of combining butein with frondoside-A, we used concentrations of butein (50 µM) and frondoside-A (2.5 µM) that induced a 50% decrease in cell viability of A549 cells at 48 h. However, for MDA-MB-231 cells, we combined concentrations of butein (50 µM) and frondoside-A (1 µM) that induced an almost 25% decrease in cell viability at 48 h.

We demonstrated that treatment of the A549 and MDA-MB-231 cells for 48 h with frondoside-A (1 and 2.5 µM) significantly enhance the inhibitory effects of butein 50 µM on cell viability (Figure 6A,B). This combination produced an inhibition of cell viability equal to the calculated additive effects of the drugs used alone (Figure 6C,D), demonstrating a clear additive effect between butein and frondoside-A in the inhibition of cellular viability.

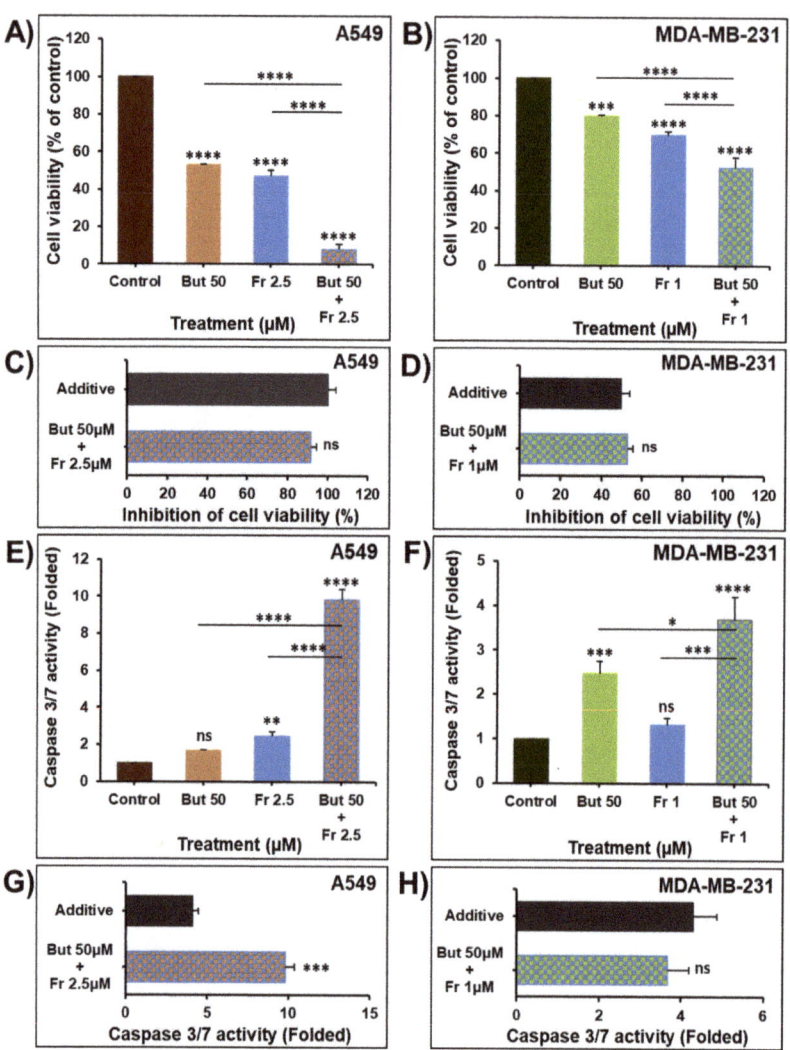

Figure 6. Effect of butein in combination with frondoside-A on the inhibition of cell viability of A549 (**A**) and MDA-MB-231 cells (**B**) after 48 h treatment. Effect of combinations of butein and frondoside-A on cell viability compared with the calculated additive effects of the two drugs alone (**C,D**). Induction of caspase 3/7 activity was also analyzed in A549 and MDA-MB-231 cells treated for 48 h with frondoside-A (2.5 and 1 µM, respectively), butein (50 µM) and their combination (**E,F**). Effect of combinations of butein and frondoside-A on caspase 3/7 activity compared with the calculated additive effects of the two drugs alone (**G,H**). Data were normalized to the number of viable cells per well and expressed as fold induction compared to the control group. All experiments were repeated at least three times. Columns are means; bars are S.E.M. The statistical significance is compared to the control except for the specified lines. * Significantly different at $p < 0.05$, ** Significantly different at $p < 0.01$, *** Significantly different at $p < 0.001$, **** Significantly different at $p < 0.0001$. ns—non-significant.

Caspase 3 activation induces the cleavage and consequently the inactivation of the downstream PARP events, leading to apoptosis [37]. The treatment of A549 and MDA-MB-231 cancer cells with butein (50 µM) for 48 h induced around two-fold increase in caspase 3/7 activity (Figure 6E,F). These data are in agreement with previous studies showing that treatment with butein increased caspase 3, 8, and 9 activities in HeLa cells [12], LNCaP cells [10], SKOV-3/PAX ovarian cancer cells [44], and cervical cancer cells C-33A and SiHa [13].

Despite the mild induction in caspase 3/7 activity when used alone, butein significantly synergizes with frondoside-A (2.5 µM) in the activation of caspase 3/7 in A549 cells (Figure 6E,G). In MDA-MB-231 cells, 48 h combination of butein (50 µM) and frondoside-A (1 µM) produced an increase in caspase 3/7 activity equal to the calculated additive effect of the drugs used alone (Figure 6F,H).

2.5. Impact of Butein in Combination with Frondoside-A on Colony Growth

To further assess the anti-cancer potential of combining butein with frondoside-A, the combined impact was investigated on the growth of pre-formed colonies of A549. Toward this, A549 cells were grown for ten days to form colonies that were treated with butein 50 µM, frondoside-A 1 µM, or a combination of both. As shown in Figure 7, combination treatment of the pre-formed colonies for two weeks significantly decreased the number of colonies compared to individual therapy. The combination caused additive effects by reducing the number of colonies by 98 ± 2%, similar to the calculated additive value of single treatments (104 ± 6%).

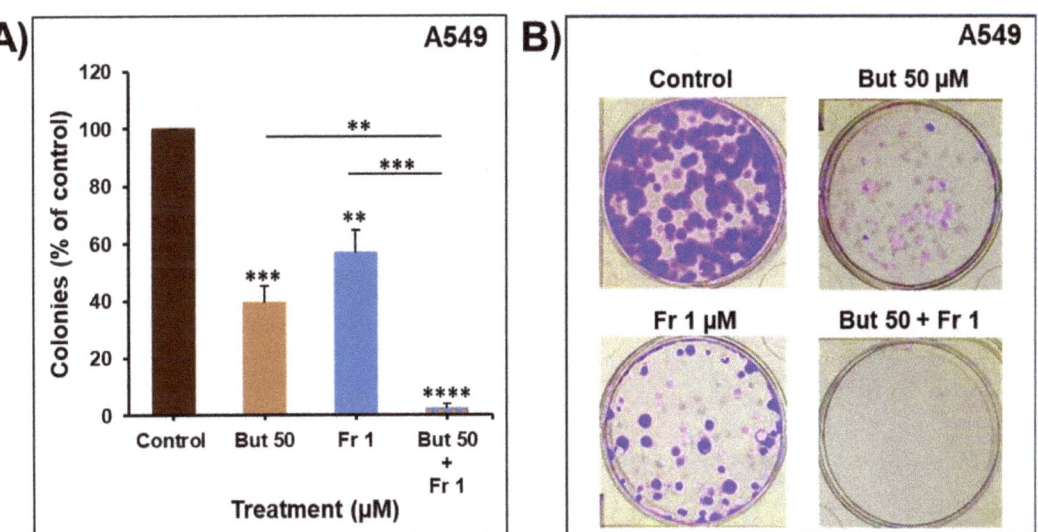

Figure 7. Impact of butein in combination with frondoside-A on A549 colony growth after 14 days of treatment (**A,B**). Experiments were repeated three times. Columns represent means; bars represent S.E.M. The statistical significance is compared to the control except for the specified lines. ** Significantly different at $p < 0.01$, *** Significantly different at $p < 0.001$, **** Significantly different at $p < 0.0001$.

2.6. Impact of Butein in Combination of Frondoside-A on Angiogenesis In Vitro

To further evaluate the therapeutic value of combining butein with frondoside-A, we investigated the impact of this combination on angiogenesis in vitro. First, the effect of the combination on HUVECs migration was determined for eight hours using transwell chambers. As shown in Figure 8A, 4% FBS increased HUVECs migration by eight-fold compared to 0% FBS. Treatment with butein (25 µM) or frondoside-A (0.5 µM) slightly decreased the HUVECs migration without reaching statistical significance. However, the combination of butein and frondoside-A led to a significant decrease in the FBS-induced HUVECs migration. This combination produced a more potent inhibition in the HUVECs migration than the calculated additive effects of the drugs used alone, demonstrating a clear synergism between butein and frondoside-A in the inhibition of HUVECs migration (Figure 8B).

The effect of this combination was next investigated on the ability of HUVECs to form capillary-like structures when seeded on Matrigel. As shown in Figure 8C–E, butein failed to inhibit the ability of HUVECs to form capillary-like structures, in contrast to frondoside-A, which significantly decreased the tube formation by approximately 40%. Combination of butein with frondoside-A slightly enhanced the frondoside-A inhibitory effect without reaching statistical significance. The impact of these treatments on HUVECs viability was determined at the end of the capillary-like-structure experiments. As you can see in Figure 8D, butein and frondoside-A did not affect HUVECs cell viability. However, their combination led to a slight, 7% decrease in cell viability (Figure 8D). Combining butein with frondoside-A could be a promising approach to target endothelial migration, an essential step in angiogenesis.

To the best of our knowledge, this is the first study demonstrating the synergistic impact of butein and frondoside-A on endothelial migration in vitro. The anti-angiogenic effect of butein in vitro was documented only once by Chung et al., (2013), who reported the ability of butein (1–20 µM) to inhibit the migration and tube formation of human endothelial progenitor cells in a concentration-dependent manner [45]. The differences in the endothelial cell type and experiment condition between the aforementioned report and this study might explain the variable results.

2.7. Additive Inhibition of Cellular Invasion by the Combination of Butein with Frondoside-A

To further evaluate the therapeutic value of combining butein with frondoside-A, we investigated whether the anti-invasive effect of butein could enhance the anti-invasive potential of frondoside-A. In this context, we have previously reported that frondoside-A possesses strong anti-invasiveness activity against breast [34] and lung [35] cancer cells. Treatment of A549 and MDA-MB-23 cells for 24 h with a low concentration of butein (5 and 10 µM, respectively) or frondoside-A (0.5 µM) (Figure 9A,B) led to a significant decrease in the invasiveness of the two cell lines using the Boyden chamber invasion assay. Similar to the cell viability data, the combination of butein and frondoside-A produced a decrease in cellular invasion of both cell lines equal to the calculated additive effects of the drugs used alone (Figure 9C,D). The anti-invasiveness additive effect of the combination of butein (10 µM) and frondoside-A (0.5 µM) for 24, 48, and 72 h was also confirmed in the ORIS Matrigel invasion assay using the MDA-MB-231 cells (Figure 9E).

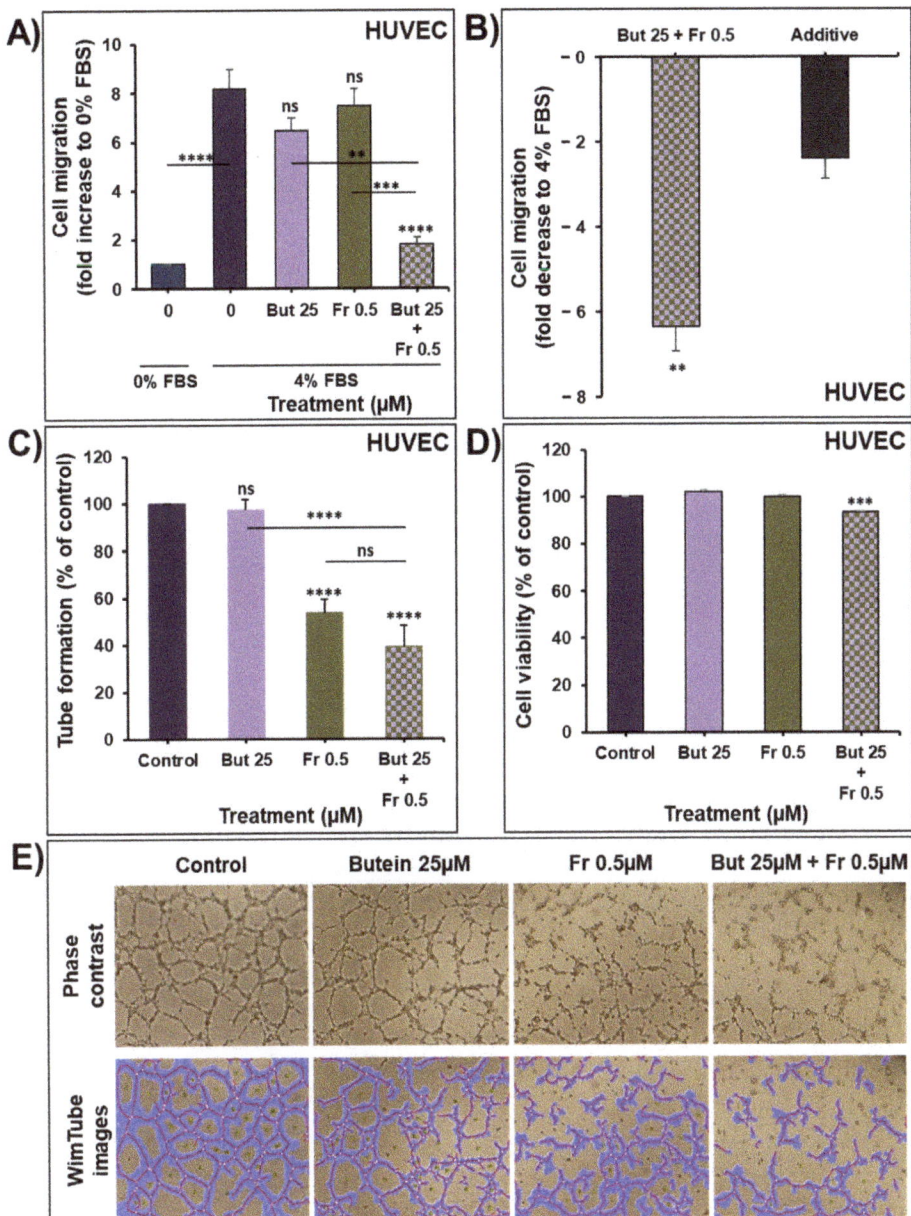

Figure 8. Effect of butein (25 µM) in combination with frondoside-A (0.5 µM) on HUVECs migration after 8 h of treatment (**A**). Effect of the combination on HUVECs migration compared with the calculated additive effects of the two drugs alone (**B**). Data were expressed as fold induction compared to the control group. (**C–E**) Impact of butein and frondoside-A alone and in combination on HUVECs capillary-like-structure formation and cell viability after 8 h of treatment. The statistical significance is compared to the control (4% FBS) except for the specified lines. All experiments were repeated at least three times. Columns are means; bars are S.E.M. ** Significantly different at $p < 0.01$, *** Significantly different at $p < 0.001$, **** Significantly different at $p < 0.0001$. ns—non-significant.

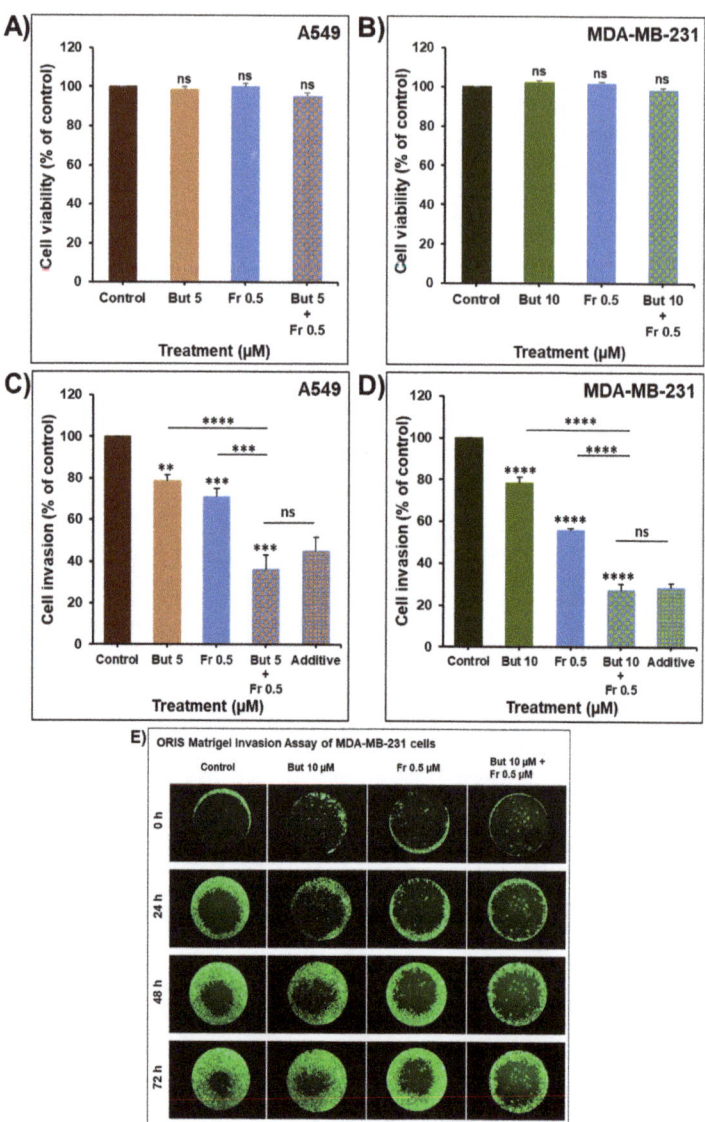

Figure 9. Impact of butein combination with frondoside-A on cellular invasion. A549 (**A**) and MDA-MB-231 (**B**) cells were treated for 24 h with butein (5 and 10 µM, respectively), frondoside-A (0.5 µM), and their combination. The effect on cell viability was determined as previously described. Using a Boyden chamber Matrigel invasion assay, A549 (**C**) and MDA-MB-231 (**D**) cells were incubated for 24 h with the non-toxic concentrations of butein and frondoside-A and their combination. Cells that invaded the Matrigel and crossed the 8 µm pores were determined using the CellTiter-Glo Luminescent Cell Viability Assay. (**E**) Oris Matrigel invasion assay showing an enhanced suppression of MDA-MB-231 cell invasion in the combination of butein with frondoside-A compared to drugs alone. All experiments were repeated at least three times. Columns are means; bars are S.E.M. The statistical significance is compared to the control except for the specified lines. ** Significantly different at $p < 0.01$, *** Significantly different at $p < 0.001$, **** Significantly different at $p < 0.0001$. ns–not significant.

3. Materials and Methods

3.1. Cell Culture and Reagents

Human NSCLC cells A549 were maintained in RPMI 1640 (Hyclone, Cramlington, UK), and human TNBC cells MDA-MB-231 were maintained in DMEM (Hyclone, Cramlington, UK). All media were supplemented with 1% of Penicillin-Streptomycin solution (Hyclone, Cramlington, UK) and with 10% fetal bovine serum (FBS; Hyclone, Cramlington, UK). Human Umbilical Vein Endothelial Cells (HUVECs) (Millipore, Temecula, CA, USA) were maintained in an EndoGRO™-VEGF complete media kit (Millipore, Temecula, CA, USA) in flasks coated with 0.2% Gelatin. The culture medium of all cells was changed every 3 days, and cells were passed once a week when the culture reached 95% confluency for cancer cells and 80% for HUVECs. In all experiments, cell viability was higher than 99% using trypan blue dye exclusion. Butein and frondoside-A were purchased from Sigma-Aldrich (Sigma-Aldrich, Saint Louis, MO, USA).

3.2. Cellular Viability

Cells were seeded at a density of 5000 cells/well into 96-well plates. After 24 h, cells were treated for another 24, 48, and 72 h with increasing concentrations of Butein (1–100 µM) in triplicate. Control cultures were treated with 0.1% DMSO (the drug vehicle). The effect of butein on cell viability was determined using the CellTiter-Glo Luminescent Cell Viability Assay (Promega Corporation, Madison; US), based on quantification of ATP, which indicates the presence of metabolically viable cells. The luminescent signal was measured using the GLOMAX Luminometer (Promega Corporation, Madison, WI, USA). Cellular viability was presented as a percentage (%) by comparing the butein-treated cells with the DMSO-treated cells, the viability of which is assumed to be 100%.

In the second set of experiments, cells were treated for 48 h with a combination of butein (50 µM) and frondoside-A (1 and 2.5 µM). The effects of these combinations on cell viability were presented as proportional cell viability (%) by comparing the drugs-treated cells with the DMSO-treated cells, the viability of which is assumed to be 100%.

3.3. Caspase 3/7 Activity

Cells were seeded at a density of 5000 cells/well into 96-well plate and treated with butein (50 µM) and frondoside-A (1 and 2.5 µM) for 48 h, in triplicate. Control cells were exposed to DMSO 0.1%. Caspase 3/7 activity was measured using a luminescent Caspase-Glo 3/7 assay kit, following the manufacturer's instructions (Promega Corporation, Madison, WI, USA). Caspase reagent was added, and the plate was mixed and incubated for 2.5 h at room temperature. Luminescence was measured using a GLOMAX Luminometer. Caspase 3/7 activity was normalized to the cellular viability and expressed as fold changes.

3.4. Clonogenic Assay

A549 and MDA-MB-231 cells were seeded into six-well plates at 100 cells/well. Cells were incubated for 14 days to form colonies and then treated every 3 days for another 7 days with increasing concentration of butein (10–100 µM). Colonies were then washed three times with PBS, fixed, and stained for 2 h with 0.5% crystal violet dissolved in (v/v) distilled water/methanol. Colonies were again washed three times with PBS, photographed, and counted. The percentages of colonies with more than 50 cells were determined and compared to the DMSO-treated colonies assumed to be 100%. The experiment was repeated three times. Data were presented as colonies percentage (%) by comparing the treated colonies with the control colonies. Colonies from representative experiments were photographed using an inverted phase-contrast microscope.

In the second set of experiments, A549 cells were kept for 7 days to form colonies and then were treated every 3 days for 14 days with a combination of butein (50 µM) and frondoside-A (1 µM). Data were presented as colonies percentage (%) by comparing the drug-treated colonies with the control colonies.

3.5. In Ovo Tumor Growth Assay

Fertilized White Leghorn eggs were incubated at 37.5 °C and 50% humidity. At the embryonic day 3 (E3), the CAM was dropped by drilling a small hole through the eggshell opposite to the round wide end followed by aspirating ~1.5–2 mL of albumin using a 5 mL syringe with an 18 G needle. Then, a small 1 cm^2 window was cut in the eggshell above the CAM using a delicate scissor and sealed with a semipermeable adhesive film (Suprasorb® F). At day 9 (E9), cancer cells were trypsinized, washed with complete medium, and suspended at a density of 1×10^6 cells/100 µL in 80% Matrigel Matrix (Corning, Bedford, UK). A 100 µL inoculum of cell suspension was added onto the CAM of each egg, for a total of 14–15 eggs per condition. Two days later, tumors were treated topically every second day at E11, E13, and E15, by dropping 100 µL of the vehicle (PBS with 0.1% of DMSO) or butein (100 µM). At the embryonic day 16 (E16), embryos were humanely euthanized by topical addition of 10–30 µL of Pentobarbitone Sodium (300 mg/mL, Jurox, Auckland, New Zealand). Tumors were carefully extracted from the upper CAM tissues, washed with PBS, and weighted to determine the effect of butein on tumor growth. Drug toxicity was assessed by comparing the percentage of alive embryos in the control and butein-treated groups at the end of the experiment. Alive embryos were determined by checking the voluntary movements of the embryos in addition to the integrity and pulsation of the blood vessels. The eggs were randomly assigned to the treatments, but the experimenter was not blinded to the identities of the groups. All data collected were used in statistical analysis. This assay was carried out according to the protocol approved by the animal ethics committee at the United Arab Emirates University. According to the European Directive 2010/63/EU on the protection of animals used for scientific purposes, experiments involving using chicken embryos on and before E18 do not require approval from the Institutional Animal Care and Use Committee (IACUC).

3.6. Scratch Wound Healing Migration Assay

A549 and MDA-MB-231 cells seeded at a density of 1.75×10^6 cells/well into a six-well plate reached confluence after 24 h. Then, a scrape was made through the confluent monolayer using a 200 µL tip. Afterwards, the dishes were washed twice and incubated at 37 °C in fresh medium containing 10% fetal bovine serum and two low concentrations of butein (5 and 10 µM). At the top side of each well, two random places were marked where the width of the wound was monitored using an inverted microscope at objective 4× (Olympus, Tokyo, Japan). Migration was expressed as the mean ± SEM of the wound difference between the measurements at time zero and the 2 and 6 h time-periods considered.

3.7. Boyden Chamber Matrigel Invasion Assay

The invasiveness of the lung cancer cells A549 and the breast cancer cells MDA-MB-231 was tested using a Corning BioCoat Matrigel Invasion Chamber (8 µm pore size) in a 24-well plate (Corning, Bedford, MA, USA), according to the manufacturer's protocol. Cells (1×10^5) in 0.5 mL of serum-free media were seeded into the upper chambers of the system with the indicated concentration of butein, frondoside-A, or butein in combination with frondoside-A. The bottom wells in the system were filled with the corresponding media supplemented with 10% fetal bovine serum as a chemoattractant and then incubated at 37 °C for 24 h. Non-invasive cells were removed from the upper surface of the filter by gently rubbing the area with a cotton swab. Cells that invaded the Matrigel and passed through the 8 µm pores of the insert were detected using CellTiter-Glo® Luminescent Cell Viability assay (Promega Corporation, Madison, WI, USA). This was done by incubating the inserts into wells having CellTiter-Glo® reagent mixed with medium (1:1) for 10 min, after which the luminescence signal was measured as described in the cellular viability section. The effects of the treatments on cellular invasion were presented as a percentage (%) by comparing the invading cells in the presence of the treatments with the control condition.

3.8. The Oris™ Matrigel Cell Invasion Assay

The impact of butein and frondoside-A, respectively, compared to the combination butein/frondoside-A on the invasiveness of MDA-MB-231-GFP cells was also investigated using a three-dimensional extracellular Matrigel matrix (Corning, Bedford, UK). Cells were seeded at 100,000 cells/well and allowed to attach overnight onto a 96-well plate coated with Matrigel. Once the cells formed a confluent monolayer, the silicone stoppers were removed. Wells were washed twice with PBS and then the cells were covered with 40 µL of Matrigel at the concentration of 6 mg/mL, incubated at 37 °C in the incubator for 45 min, and then incubated in complete media with the indicated treatments for 24, 48, and 72 h. The impact of the treatments of the invasiveness of the MDA-MB-231 GFP cells was assessed using an Olympus fluorescence microscope (Olympus, Tokyo, Japan). Representative figures were taken at 0, 24, 48, and 72 h.

3.9. Western Blotting Assay

A549 and MDA-MB-231 cells were seeded in 60 mm dishes at 750,000 cells/dish for 24 h and then treated with two concentrations of butein (25 and 50 µM) for another 0.5, 2, 6, 24, and 48 h. Control cultures were treated with 0.1% DMSO (the drug vehicle). Total cellular proteins were isolated using RIPA buffer (25 mM Tris.HCl, pH 7.6; 1% Nonidet P-40; 1% sodium deoxycholate; 0.1% SDS; 0.5% protease inhibitor cocktail; 1% PMSF; 1% phosphatase inhibitor cocktail) from the DMSO- and drug-treated cells. The whole-cell lysates were recovered by centrifugation at 14,000 rpm for 20 min at 4 °C to remove insoluble material, and protein concentrations of lysates were determined using a BCA protein assay kit (Thermo Fisher Scientific, Waltham, MA, USA). Proteins (30 µg) were separated by SDS-PAGE gel to determine the expression of STAT3, the level of p-STAT3, and PARP cleavage. After electrophoresis, the proteins were transferred onto a nitrocellulose membrane, blocked for 1 h at room temperature with 5% non-fat milk in TBST (TBS and 0.05% Tween 20), and then probed with specific primary antibodies and β-actin overnight at 4 °C. Antibodies to STAT3 (124H6) (1:1000), phospho-STAT3 (Tyr705) (D3A7) XP® (1:600), and cleaved poly (ADP-ribose) polymerase (PARP) (1:500) were obtained from Cell Signaling Technology (Cell Signaling, Beverly, MA, USA). The β-actin antibody (1:9000) was obtained from Santa Cruz Biotechnology, Inc (Santa Cruz, CA, USA). Blots were washed and exposed to secondary antibodies. Immunoreactive bands were detected using ECL substrate (Thermo Fisher Scientific, Waltham, MA, USA) and chemiluminescence was detected using the LI-COR C-DiGit blot scanner (LI-COR Biotechnology, Lincoln, NE, USA). Densitometry analysis was performed using an HP Deskjet F4180 Scanner with ImageJ software. The intensities of the bands were normalized to the intensities of the corresponding β-actin bands.

3.10. HUVECs Migration Assay

HUVECs migration assay was performed using Boyden chambers with inserts of 8 µm pores (Corning, Bedford, MA, USA). The bottom chambers were filled with 0.75 mL of EndoGRO™-Basal Medium supplemented with 4% FBS. Sub-confluent cells were trypsinized, collected, and resuspended with EndoGRO™-Basal Medium supplemented with 0.1% FBS. Typically, 50,000 cells/0.5 mL, in the presence and absence of test compounds, were added to the top of each migration chamber and cells were allowed to migrate to the underside of the chamber in a humidified incubator at 37 °C and 5% CO_2 for 8 h. After that, the upper chambers' non-migrating cells were removed by gently rubbing the area with a cotton swab. The migrating cells were determined using CellTiter-Glo® Luminescent Cell Viability assay (Promega Corporation, Madison, WI, USA) previously described in the cellular viability section.

3.11. Vascular Tube Formation Assay

Matrigel Matrix (Corning, Bedford, UK) was thawed, and 40–50 µL was added to the wells of a 96-well plate for coating. In order for the Matrigel to solidify, the plate was kept in a humidified incubator at 37 °C and 5% CO_2 for 1 h. HUVECs were trypsinized and seeded on the coated plate at a density of 2.5×10^4 cells/100 µL/well in the absence and presence of the indicated low concentrations of butein, frondoside-A, or butein in combination with frondoside-A. After 8 h of incubation, the tube networks at the different wells were photographed using an inverted phase-contrast microscope. The impact of the treatments on the ability of HUVECs to form capillary-like structures was assessed by measuring the total lengths of the formed tubes in the control and drugs-treated wells. Total tube lengths were measured using online image analysis software developed by Wimasis (https://www.wimasis.com/en/products/13/WimTube, accessed on 11 November 2021). The impact of the different treatments on the viability of HUVECs was determined using CellTiter-Glo® Luminescent Cell Viability assay (Promega Corporation, Madison, WI, USA) as previously described in the cellular viability section.

3.12. Statistical Analysis

Each experiment was repeated at least three times, and results are expressed as means ± SEM of the indicated data. Statistical analysis was performed with GraphPad Prism7 (La Jolla, CA, USA). The difference between experimental and control values was assessed by ANOVA followed by Dunnett's multiple comparisons test. For the combination experiments, data were assessed by ANOVA followed by Tukey's multiple comparisons test. The unpaired t-test was used to assess the difference between two groups. * $p < 0.05$, ** $p < 0.01$, *** $p < 0.001$, and **** $p < 0.0001$ indicate a significant difference.

4. Conclusions

In conclusion, we demonstrate that butein decreases lung and breast cancer cell viability and colony growth, leading to a significant decrease in tumor growth in vivo. Butein also decreases cancer cell migration and invasion, suggesting its potential anti-metastatic effect. STAT3 is an oncogene constitutively activated in both A549 and MDA-MB-231 cells used in this study and has been reported to be associated with cancer cell viability/proliferation, migration, and invasion. The reported anti-cancer effects of butein are due, at least in part, to the potent inhibition of STAT3 phosphorylation.

Very few studies have investigated butein anti-cancer effects in combination therapy. In this context, we demonstrate that butein combined with frondoside-A has an additive impact on lung and breast cancer cell viability, colony growth and invasion, and synergistically decreases endothelial cell migration.

This study provides sufficient rationale to carry out pre-clinical research further to confirm the therapeutic potential of this combination therapy using butein and frondoside-A on tumor growth and metastasis in vivo in chick embryo CAM and nude mice tumor xenograft models.

Author Contributions: Conceptualization, S.A.; methodology, S.S., K.A., A.M.A.-A., N.A.A., S.N.A.H.L., S.A.; validation, S.S., K.A., A.M.A.-A., N.A.A., S.N.A.H.L., S.A.; formal analysis, A.M.A.-A., S.A.; investigation, S.S., K.A., A.M.A.-A., N.A.A., S.N.A.H.L., S.A.; data curation, S.A.; writing—original draft preparation, A.M.A.-A., S.A.; writing—review and editing, S.S., K.A., A.M.A.-A., N.A.A., S.N.A.H.L., S.A.; visualization, A.M.A.-A., S.A.; supervision, S.A.; project administration, S.A.; funding acquisition, S.A. All authors have read and agreed to the published version of the manuscript.

Funding: This research was funded by a grant from the College of Medicine and Health Sciences, United Arab Emirates University, No. 31M473.

Institutional Review Board Statement: The in vivo study was conducted according to the guidelines of the Declaration of Helsinki and approved by the Institutional Animal Ethics Committee of the United Arab Emirates University (protocol code ERA_2019_5895).

Informed Consent Statement: Not applicable.

Data Availability Statement: Not applicable.

Conflicts of Interest: The authors declare no conflict of interest. The funders had no role in the design of the study; in the collection, analyses, or interpretation of data; in the writing of the manuscript, or in the decision to publish the results.

References

1. Sung, H.; Ferlay, J.; Siegel, R.L.; Laversanne, M.; Soerjomataram, I.; Jemal, A.; Bray, F. Global Cancer Statistics 2020: GLOBOCAN Estimates of Incidence and Mortality Worldwide for 36 Cancers in 185 Countries. *CA Cancer J. Clin.* **2021**, *71*, 209–249. [CrossRef] [PubMed]
2. Cragg, G.M.; Pezzuto, J.M. Natural Products as a Vital Source for the Discovery of Cancer Chemotherapeutic and Chemopreventive Agents. *Med. Princ. Pract.* **2016**, *25*, 41–59. [CrossRef] [PubMed]
3. Balik, K.; Modrakowska, P.; Maj, M.; Kaźmierski, Ł.; Bajek, A. Limitations of molecularly targeted therapy. *Med. Res. J.* **2019**, *4*, 99–105. [CrossRef]
4. Tan, S.; Li, D.; Zhu, X. Cancer immunotherapy: Pros, cons and beyond. *Biomed. Pharmacother.* **2020**, *124*, 109821. [CrossRef]
5. Lin, S.R.; Chang, C.H.; Hsu, C.F.; Tsai, M.J.; Cheng, H.; Leong, M.K.; Sung, P.-J.; Chen, J.-C.; Weng, C.-F. Natural compounds as potential adjuvants to cancer therapy: Preclinical evidence. *Br. J. Pharmacol.* **2020**, *177*, 1409–1423. [CrossRef]
6. Padmavathi, G.; Roy, N.K.; Bordoloi, D.; Arfuso, F.; Mishra, S.; Sethi, G.; Bishayee, A.; Kunnumakkara, A.B. Butein in health and disease: A comprehensive review. *Phytomedicine* **2017**, *25*, 118–127. [CrossRef]
7. Jayasooriya, R.G.P.T.; Molagoda, I.M.N.; Park, C.; Jeong, J.W.; Choi, Y.H.; Moon, D.-O.; Kim, M.-O.; Kim, G.-Y. Molecular chemotherapeutic potential of butein: A concise review. *Food Chem. Toxicol.* **2018**, *112*, 1–10. [CrossRef]
8. Girard, M.; Bélanger, J.; ApSimon, J.W.; Garneau, F.-X.; Harvey, C.; Brisson, J.-R. Frondoside A. A novel triterpene glycoside from the holothurian Cucumaria frondose. *Can. J. Chem.* **1990**, *68*, 11. [CrossRef]
9. Pandey, M.K.; Bokyung, S.; Kwang, S.A.; Aggarwal, B.B. Butein suppresses constitutive and inducible signal transducer and activator of transcription (stat) 3 activation and stat3-regulated gene products through the induction of a protein tyrosine phosphatase SHP-1. *Mol. Pharmacol.* **2009**, *75*, 525–533. [CrossRef] [PubMed]
10. Khan, N.; Adhami, V.M.; Afaq, F.; Mukhtar, H. Butein induces apoptosis and inhibits prostate tumor growth in Vitro and in Vivo. *Antioxid. Redox Signal* **2012**, *16*, 1195–1204. [CrossRef] [PubMed]
11. Li, Y.; Ma, C.; Qian, M.; Wen, Z.; Jing, H.; Qian, D. Butein induces cell apoptosis and inhibition of cyclooxygenase-2 expression in A549 lung cancer cells. *Mol. Med. Rep.* **2014**, *9*, 763–767. [CrossRef]
12. Bai, X.; Ma, Y.; Zhang, G. Butein suppresses cervical cancer growth through the PI3K/AKT/mTOR pathway. *Oncol. Rep.* **2015**, *33*, 3085–3092. [CrossRef]
13. Yang, P.-Y.; Hu, D.-N.; Kao, Y.-H.; Lin, I.-C.; Liu, F.-S. Butein induces apoptotic cell death of human cervical cancer cells. *Oncol. Lett.* **2018**, *16*, 6615–6623. [CrossRef]
14. Yang, L.H.; Ho, Y.J.; Lin, J.F.; Yeh, C.W.; Kao, S.H.; Hsu, L.S. Butein inhibits the proliferation of breast cancer cells through generation of reactive oxygen species and modulation of ERK and p38 activities. *Mol. Med. Rep.* **2012**, *6*, 1126–1132. [CrossRef]
15. Cho, S.G.; Woo, S.M.; Ko, S.G. Butein suppresses breast cancer growth by reducing a production of intracellular reactive oxygen species. *J. Exp. Clin. Cancer Res.* **2014**, *33*, 51. [CrossRef]
16. Ma, C.Y.; Ji, W.T.; Chueh, F.S.; Yang, J.S.; Chen, P.Y.; Yu, C.C.; Chung, J.G. Butein inhibits the migration and invasion of SK-HEP-1 human hepatocarcinoma cells through suppressing the ERK, JNK, p38, and uPA signaling multiple pathways. *J. Agric. Food Chem.* **2011**, *59*, 9032–9038. [CrossRef]
17. Tang, Y.L.; Huang, L.B.; Lin, W.H.; Wang, L.N.; Tian, Y.; Shi, D.; Wang, J.; Qin, G.; Li, A.; Liang, Y.N.; et al. Butein inhibits cell proliferation and induces cell cycle arrest in acute lymphoblastic leukemia via FOXO3a/p27kip1 pathway. *Oncotarget* **2016**, *7*, 18651–18664. [CrossRef]
18. Rajendran, P.; Ong, T.H.; Chen, L.; Li, F.; Shanmugam, M.K.; Vali, S.; Abbasi, T.; Kapoor, S.; Sharma, A.; Kumar, A.P.; et al. Suppression of signal transducer and activator of transcription 3 activation by butein inhibits growth of human hepatocellular carcinoma in vivo. *Clin. Cancer Res.* **2011**, *17*, 1428–1439. [CrossRef]
19. Zhou, Y.; Wang, K.; Zhou, N.; Huang, T.; Zhu, J.; Li, J. Butein activates p53 in hepatocellular cancer cells via blocking MDM2-mediated ubiquitination. *OncoTargets Therapy* **2018**, *11*, 2007–2015. [CrossRef]
20. Zhang, L.R.; Chen, W.; Li, X. A novel anticancer effect of butein: Inhibition of invasion through the ERK1/2 and NF-κB signaling pathways in bladder cancer cells. *FEBS Lett.* **2008**, *582*, 1821–1828. [CrossRef]
21. Chua, A.W.L.; Hay, H.S.; Rajendran, P.; Shanmugam, M.K.; Li, F.; Bist, P.; Koay, E.S.C.; Lim, L.H.K.; Kumar, A.P.; Sethi, G. Butein downregulates chemokine receptor CXCR4 expression and function through suppression of NF-κB activation in breast and pancreatic tumor cells. *Biochem. Pharmacol.* **2010**, *80*, 1553–1562. [CrossRef]
22. Lai, Y.-W.; Wang, S.-W.; Chang, C.-H.; Liu, S.-C.; Chen, Y.-J.; Chi, C.-W.; Chiu, L.-P.; Chen, S.-S.; Chiu, A.W.; Chung, C.-H. Butein inhibits metastatic behavior in mouse melanoma cells through VEGF expression and translation-dependent signaling pathway regulation. *BMC Complement. Altern. Med.* **2015**, *15*, 445. [CrossRef]
23. Liu, S.-C.; Chen, C.; Chung, C.-H.; Wang, P.-C.; Wu, N.-L.; Cheng, J.-K.; Lai, Y.-W.; Sun, H.-L.; Peng, C.-Y.; Tang, C.-H.; et al. Inhibitory effects of butein on cancer metastasis and bioenergetic modulation. *J. Agric. Food Chem.* **2014**, *62*, 9109–9117. [CrossRef]
24. Siveen, K.S.; Sikka, S.; Surana, R.; Dai, X.; Zhang, J.; Kumar, A.P.; Tan, B.K.H.; Sethi, G.; Bishayee, A. Targeting the STAT3 signaling pathway in cancer: Role of synthetic and natural inhibitors. *Biochim. Biophys. Acta* **2014**, *1845*, 136–154. [CrossRef] [PubMed]

25. Song, L.; Turkson, J.; Karras, J.G.; Jove, R.; Haura, E.B. Activation of Stat3 by receptor tyrosine kinases and cytokines regulates survival in human non-small cell carcinoma cells. *Oncogene* **2003**, *22*, 4150–4165. [CrossRef] [PubMed]
26. Christine, R.; Sylvie, R.; Erik, B.; Geneviève, P.; Amélie, R.; Gérard, R.; Marc, B.; Christian, G.; Attoub, S. Implication of STAT3 signaling in human colonic cancer cells during intestinal trefoil factor 3 (TFF3)—And vascular endothelial growth factor-mediated cellular invasion and tumor growth. *Cancer Res.* **2005**, *65*, 195–202.
27. Al Kubaisy, E.; Arafat, K.; De Wever, O.; Hassan, A.H.; Attoub, S. SMARCAD1 knockdown uncovers its role in breast cancer cell migration, invasion, and metastasis. *Expert Opin. Ther. Targets* **2016**, *20*, 1035–1043. [CrossRef]
28. Aryappalli, P.; Al-Qubaisi, S.S.; Attoub, S.; George, J.A.; Arafat, K.; Ramadi, K.B.; Mohamed, Y.A.; Al-Dhaheri, M.M.; Al-Sbiei, A.; Fernandez-Cabezudo, M.J.; et al. The IL-6/STAT3 signaling pathway is an early target of manuka honey-induced suppression of human breast cancer cells. *Front. Oncol.* **2017**, *7*, 167. [CrossRef] [PubMed]
29. El Hasasna, H.; Saleh, A.; Al Samri, H.; Athamneh, K.; Attoub, S.; Arafat, K.; Benhalilou, N.; Alyan, S.; Viallet, J.; Al Dhaheri, Y.; et al. Rhus coriaria suppresses angiogenesis, metastasis and tumor growth of breast cancer through inhibition of STAT3, NFkB and nitric oxide pathways. *Sci. Rep.* **2016**, *6*, 21144. [CrossRef]
30. Schütz, A.; Röser, K.; Klitzsch, J.; Lieder, F.; Aberger, F.; Gruber, W.; Mueller, K.M.; Pupyshev, A.; Moriggl, R.; Friedrich, K. Lung adenocarcinomas and lung cancer cell lines show association of MMP-1 expression with STAT3 activation. *Transl. Oncol.* **2015**, *8*, 97–105. [CrossRef]
31. Miklossy, G.; Hilliard, T.S.; Turkson, J. Therapeutic modulators of STAT signalling for human diseases. *Nat. Rev. Drug Discov.* **2013**, *12*, 611–629. [CrossRef]
32. Wong, A.L.; Soo, R.A.; Tan, D.S.; Lee, S.C.; Lim, J.S.; Marban, P.C.; Kong, L.R.; Lee, Y.J.; Wang, L.Z.; Thuya, W.L.; et al. Phase I and biomarker study of OPB-51602, a novel signal transducer and activator of transcription (STAT) 3 inhibitor, in patients with refractory solid malignancies. *Ann. Oncol.* **2015**, *26*, 998–1005. [CrossRef]
33. Munoz, J.; Dhillon, N.; Janku, F.; Watowich, S.S.; Hong, D.S. STAT3 Inhibitors: Finding a Home in Lymphoma and Leukemia. *Oncologist* **2014**, *19*, 536–544. [CrossRef]
34. Al Marzouqi, N.; Iratni, R.; Nemmar, A.; Arafat, K.; Al Sultan, M.A.; Yasin, J.; Collin, P.; Mester, J.; Adrian, T.E.; Attoub, S. Frondoside A inhibits human breast cancer cell survival, migration, invasion and the growth of breast tumor xenografts. *Eur. J. Pharmacol.* **2011**, *668*, 25–34. [CrossRef]
35. Attoub, S.; Arafat, K.; Gélaude, A.; Al Sultan, M.A.; Bracke, M.; Collin, P.; Takahashi, T.; Adrian, T.E.; De Wever, T. Frondoside A Suppressive Effects on Lung Cancer Survival, Tumor Growth, Angiogenesis, Invasion, and Metastasis. *PLoS ONE* **2013**, *8*, e53087. [CrossRef]
36. Al Shemaili, J.; Mensah-Brown, E.; Parekh, K.; Thomas, S.A.; Attoub, S.; Hellman, B.; Nyberg, F.; Adem, A.; Collin, P.; Adrian, T.E. Frondoside A enhances the antiproliferative effects of gemcitabine in pancreatic cancer. *Eur. J. Cancer* **2014**, *50*, 1391–1398. [CrossRef] [PubMed]
37. Attoub, S.; Arafat, K.; Khalaf, T.; Sulaiman, S.; Iratni, R. Frondoside a enhances the anti-cancer effects of oxaliplatin and 5-fluorouracil on colon cancer cells. *Nutrients* **2018**, *10*, 560. [CrossRef] [PubMed]
38. Adrian, T.E.; Collin, P. The anti-cancer effects of frondoside A. *Marine Drugs* **2018**, *16*, 64. [CrossRef]
39. Dyshlovoy, S.A.; Menchinskaya, E.S.; Venz, S.; Rast, S.; Amann, K.; Hauschild, J.; Otte, K.; Kalinin, V.I.; Silchenko, A.S.; Avilov, S.A.; et al. The marine triterpene glycoside frondoside A exhibits activity in vitro and in vivo in prostate cancer. *Int. J. Cancer* **2016**, *138*, 2450–2465. [CrossRef] [PubMed]
40. Sajwani, F.H. Frondoside A is a potential anticancer agent from sea cucumbers. *J. Cancer Res. Ther.* **2019**, *15*, 953–960. [CrossRef]
41. Nguyen, B.C.Q.; Yoshimura, K.; Kumazawa, S.; Tawata, S.; Maruta, H. Frondoside A from sea cucumber and nymphaeols from Okinawa propolis: Natural anti-cancer agents that selectively inhibit PAK1 in vitro. *Drug Discov. Ther.* **2017**, *11*, 110–114. [CrossRef] [PubMed]
42. Yao, D.; Li, C.; Rajoka, M.; He, Z.; Huang, J.; Wang, J.; Zhang, J. P21-Activated Kinase 1: Emerging biological functions and potential therapeutic targets in Cancer. *Theranostics* **2020**, *10*, 9741–9766. [CrossRef]
43. Palmer, A.C.; Sorger, P.K. Combination Cancer Therapy Can Confer Benefit via Patient-to-Patient Variability without Drug Additivity or Synergy. *Cell* **2017**, *171*, 1678–1691. [CrossRef]
44. Choi, H.S.; Kim, M.K.; Choi, Y.K.; Shin, Y.C.; Cho, S.-G.; Ko, S.-G. Rhus verniciflua Stokes (RVS) and butein induce apoptosis of paclitaxel-resistant SKOV-3/PAX ovarian cancer cells through inhibition of AKT phosphorylation. *BMC Complement. Altern. Med.* **2016**, *16*. [CrossRef] [PubMed]
45. Chung, C.H.; Chang, C.H.; Chen, S.S.; Wang, H.H.; Yen, J.Y.; Hsiao, C.J.; Wu, N.-L.; Chen, Y.-L.; Huang, T.-F.; Wang, P.-C.; et al. Butein inhibits angiogenesis of human endothelial progenitor cells via the translation dependent signaling pathway. *Evid. Based Complement. Altern. Med.* **2013**, *2013*, 943187. [CrossRef] [PubMed]

International Journal of *Molecular Sciences*

Article

Target Characterization of Kaempferol against Myocardial Infarction Using Novel In Silico Docking and DARTS Prediction Strategy

Xunxun Wu [1], Xiaokun Li [1], Chunxue Yang [2,*] and Yong Diao [1,*]

1. School of Biomedical Science, Huaqiao University, Quanzhou 362021, China; wuxunxun2015@163.com (X.W.); 18014071010@stu.hqu.edu.cn (X.L.)
2. Department of Pathology, The University of Hong Kong, Hong Kong 999077, China
* Correspondence: cxyang@hku.hk (C.Y.); diaoyong@hqu.edu.cn (Y.D.); Tel.: +86-595-22692516 (Y.D.)

Abstract: Target identification is a crucial process for advancing natural products and drug leads development, which is often the most challenging and time-consuming step. However, the putative biological targets of natural products obtained from traditional prediction studies are also informatively redundant. Thus, how to precisely identify the target of natural products is still one of the major challenges. Given the shortcomings of current target identification methodologies, herein, a novel in silico docking and DARTS prediction strategy was proposed. Concretely, the possible molecular weight was detected by DARTS method through examining the protected band in SDS-PAGE. Then, the potential targets were obtained from screening and identification through the PharmMapper Server and TargetHunter method. In addition, the candidate target Src was further validated by surface plasmon resonance assay, and the anti-apoptosis effects of kaempferol against myocardial infarction were further confirmed by in vitro and in vivo assays. Collectively, these results demonstrated that the integrated strategy could efficiently characterize the targets, which may shed a new light on target identification of natural products.

Keywords: target identification; kaempferol; docking; DARTS; Src

1. Introduction

Natural products have historically served as a prolific and unsurpassed source for novel candidates in the search for new drugs [1]. Target identification of known bioactive compounds and novel analogs is pivotal to understanding the therapeutic effects and underlying mechanisms of natural products [2]. How to identify the therapeutic target from the huge number of compounds from natural products is a challenging and costly task. Even though the identification and validation strategies have been improved, such as affinity chromatography of immobilized probe, label-free methods including drug affinity responsive target stability (DARTS), and virtual screening techniques. However, limitations still exist in the progress of single-method target identification, such as target selection and lower-abundance targets identification [3,4].

DARTS is a robust method for the determination of target proteins for natural products without chemical modification. The concept of DARTS is that ligand-bound proteins show altered stability compared to ligand-unbound proteins in the case of proteolysis. However, the DARTS method is not sensitive in identifying low-abundance proteins and validating of proteolysis of a cell lysate [2,3]. In silico target identification could be performed independent of target abundance and could be useful complement for traditional bench experiments [5]. Chemical similarity is a key criterion for in silico target identification. Structurally similar compounds have similar physicochemical properties and can show possibly similar biological effects [6–8]. Based on the similarity to a biologically active template, the structurally similar strategy offers an alternative avenue for the exploration of ligand–target interactions with a high chance and hit rate [9,10]. As the saying goes: 'The best way to discover a new drug is to start with an old one', which demonstrates that conventional drugs or targets may have new uses [11].

The Src tyrosine kinase is a non-receptor tyrosine kinase, and is widely discussed as a factor in tumorigenesis through regulating cell growth, differentiation, adhesion, and survival [12]. In addition, inhibition of Src ameliorates myocardial ischemia reperfusion injury and arrhythmia [13,14]. This evidence raises the possibility that the regulation of Src could relieve myocardial infarction. However, Src inhibitors, such as dasatinib, caused coronary artery disease in the clinic [15]. Therefore, it is important to discover the regulatory mechanism of Src and develop safe and effective inhibitors. Kaempferol (Kae) is one of the most commonly used flavonoids from natural products, such as Ginkgo biloba. Kae has a variety of therapeutic effects, for example, anti-inflammatory, anticancer, and antibacterial properties [16]. In addition, recent studies demonstrated that Kae could protect rat brain against I/R-induced damage [17,18]. However, the specific target and molecular mechanism remain to be identified.

In this study, we proposed a novel strategy that combined the in silico target identification and DARTS prediction method to identify the direct target of Kae. Then, compound-receptor interactions were further confirmed by molecular docking and surface plasmon resonance (SPR) analysis. Furthermore, the cardioprotective effects against myocardial infarction were analyzed (Figure 1). This initial phase of target identification can be done rapidly by integrated approaches to identify the optimal lead from different structural isomers. This novel methodology may help uncover new therapeutic targets and molecular pathways for myocardial infarction therapy. In addition, this knowledge could shed light on identifying new therapeutic targets and molecular pathways for currently untreatable diseases.

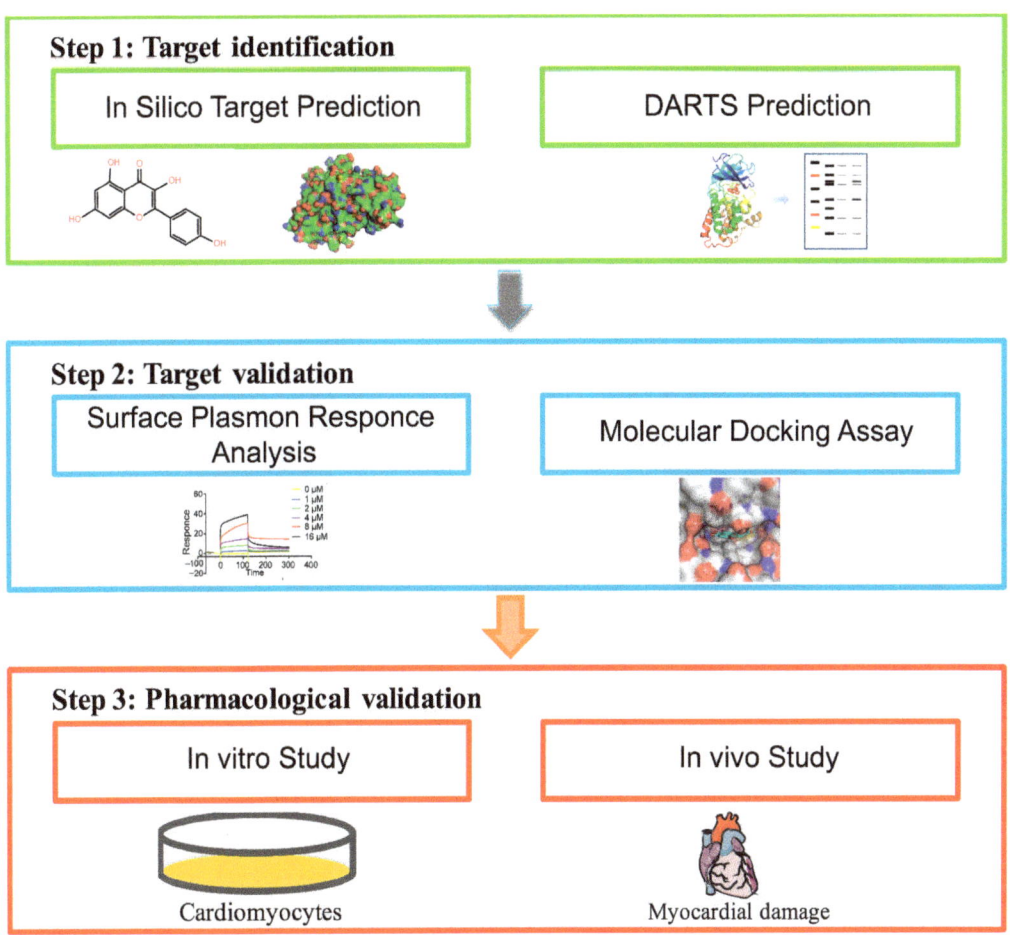

Figure 1. The flow diagram of target characterization of Kae against myocardial infarction using comprehensive in silico docking and DARTS prediction strategy.

2. Results

2.1. Target Identification of Kae by In Silico Docking and DARTS Prediction Strategy

The drug affinity responsive target stability (DARTS) approach is widely used for direct target protein identification. To reveal the direct target of Kae in cardioprotection, DARTS method was employed to find the potential candidates that could interact with Kae. Of particular note, to ensure the binding-promoted stability efficiency, the concentration of Kae used for DARTS assay was higher than that used in cell culture. Thus, after proteolytic digestion, high concentrations of Kae (100 and 200 µM) were added to H9C2 cells lysates and incubated for 1 h, respectively. Coomassie-blue-stained results showed that the abundance of the band at ~60 kDa was increased after the incubation with Kae, which may be caused by the resistance to pronase degradation (Figure 2C). PharmMapper is a web server for potential drug target identification with a comprehensive target pharmacophore database. On the other hand, TargetHunter is designed to search for target identification based on chemical similarity. In this study, PharmMapper Server and TargetHunter [19] tools were adopted for potential drug target identification (Figure 2A,B). The docking results obtained from PharmMapper were further ranked by z-score (Supplementary Table S1). From the top 10 target candidates of Kae identified by PharmMapper, tyrosine-protein kinase HCK and proto-oncogene tyrosine-protein kinase Src are ~60 kDa (Table 1). In addition, potential drug targets obtained from TargetHunter were ranked by the similarity to Kae (Supplementary Table S2). Tyrosine-protein kinase LCK, epidermal growth factor receptor erbB1, and tyrosine-protein kinase Src are published targets of compounds CHEMBL 115102 and CHEMBL116051, which are similar to Kae [20]. Among the three targets, tyrosine-protein kinase LCK and tyrosine-protein kinase Src are ~60 kDa (Table 2). Collectively, tyrosine-protein kinase Src has been identified as the best potential target of Kae through the integrated screening of DARTS and in silico target identification.

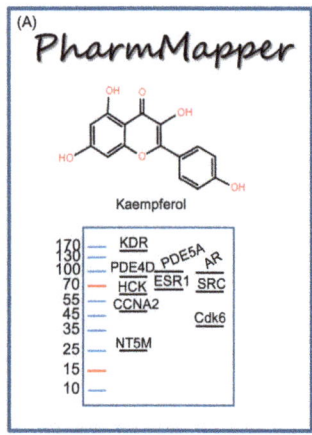 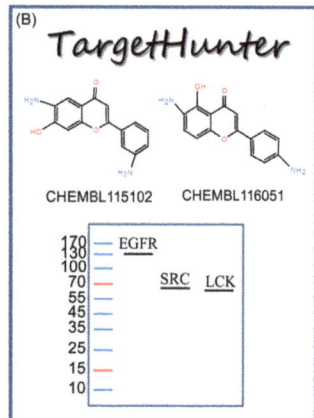 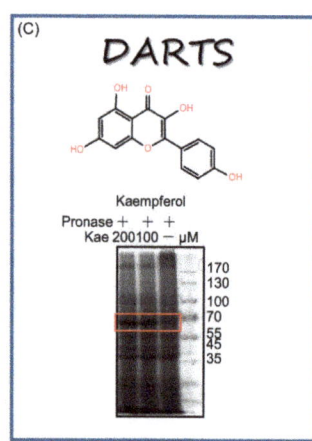

Figure 2. (**A**) The ranked list of hit targets of Kae from PharmMapper. (**B**) Target identification of Kae by TargetHunter using the 2D similarity compounds. (**C**) The DARTS assay was employed to detect the different bands, and a ~60 kD band increased upon Kae incubation. The red box indicated the detected bands at ~60 kD.

Table 1. Top 10 target candidates of Kae identified by PharmMapper.

Rank	Target	Z-Score	Name	Mass (Da)
1	PDE4D	4.59774	cAMP-specific 3,5-cyclic phosphodiesterase 4D	91,115
2	HCK	4.43052	Tyrosine-protein kinase HCK	59,600
3	Cdk6	4.36472	Cell division protein kinase 6	36,938
4	PDE5A	3.82585	cGMP-specific 3,5-cyclic phosphodiesterase	99,985
5	AR	3.29088	Androgen receptor	99,188
6	ESR1	3.27299	Estrogen receptor	66,216
7	SRC	3.26386	Proto-oncogene tyrosine-protein kinase Src	59,835
8	KDR	3.12717	VEGFR2 kinase	151,527
9	CCNA2	3.07088	Cyclin-A2	48,551
10	NT5M	2.97835	5(3)-deoxyribonucleotidase, mitochondrial	25,862

Table 2. Target candidates of Kae identified by TargetHunter.

Number.	Score	Name	Mass (Da)	Target
CHEMBL116051	0.74	Tyrosine-protein kinase LCK	58,001	LCK
		Epidermal growth factor receptor erbB1	134,277	EGFR
		Tyrosine-protein kinase Src	59,835	SRC
CHEMBL115102	0.71	Tyrosine-protein kinase LCK	58,001	LCK
		Epidermal growth factor receptor erbB1	134,277	EGFR
		Tyrosine-protein kinase Src	59,835	SRC

2.2. Src Is a Direct Target of Kae

To validate whether Kae was directly bound to Src, pronase was added to cell lysate and incubated with Kae. The results showed that Kae increased the resistance to pronase degradation and also promoted the stability of Src (Figure 3A). To further examine the interaction between Kae and Src, a CETSA assay was performed. As shown in Figure 3B, the addition of Kae to heat-denatured H9C2 cell lysates led to the stabilization of Src at different temperatures. Surface plasmon resonance (SPR), as a powerful technique, has been widely employed for the detection of protein–probe interactions [21]. SPR analysis revealed the potential interaction between Kae and Src (K_D = 8.666 µM) (Figure 3C). As predicted by molecular docking, there were several hydrogen bonds formed between Kae and the hinge region of Glu339 and Met341, as well as the gatekeeper residue of Thr338. H-bonding interactions are typical characteristics for the binding of ATP-competitive inhibitors with kinases [22]. The phenol moiety of Kae extends into the ATP back-site and makes one hydrogen bond with Glu310. In addition to polar interactions, the aromatic rings of Kae establish hydrophobic contact with lipophilic residues present in the ATP-site of the Src catalytic domain (Figure 3D). Collectively, these data suggest that Src is a direct target of Kae.

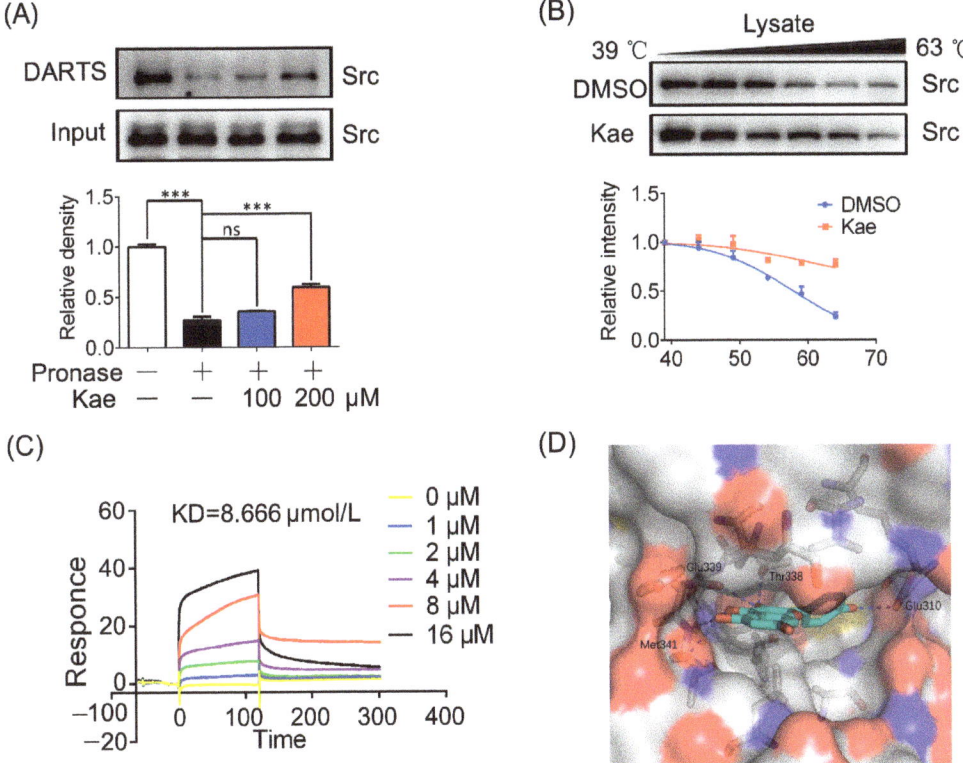

Figure 3. Identification and validation of Src as a direct target for Kae. (**A**) Immunoblot analysis of Src treated with DARTS assay (n = 3). (**B**) Immunoblot analysis of Src degradation insult by indicated temperature (n = 3). (**C**) Surface plasmon resonance (SPR) analysis of the interaction between Kae and Src: KD = 8.666 µmol/L (n = 3). (**D**) Predicted binding mode of Kae in the ATP-site of the Src catalytic domain. Protein is shown in surface, and key amino acids interacting with Kae are highlighted in stick representation. Kae is colored in cyan. Hydrogen bonds are depicted by blue dashed lines. *** $p < 0.001$ and ns, statistically not significant.

2.3. Kae Protects Cardiomyocytes against Oxidative Damage

To further verify the protective effects of Kae, we applied cardiac myoblast H9C2 cells for in vitro assays. H9C2 cells were maintained in medium supplied with different concentrations of Kae (10 and 20 μM). The results showed that the survival rate was increased with the addition of Kae, indicating that Kae protected cell survival against H_2O_2 insult (Figure 4A,B). Hoechst 33342 is an apoptotizing cell detection fluorescent probe that brightly stains chromatin in cell nuclei [23]. Apoptotizing cells exhibit apoptosis-related alterations in the chromatin state. In the H_2O_2 treated group, the nucleus showed extremely bright zones, which suggested abnormalities in the nucleus size or shape, and heterogeneous staining of chromatin. However, the cells treated with Kae were homogenously stained with Hoechst 33342, indicating that the addition of Kae protected the nucleus from oxidative damage (Figure 4C). Oxidative stress is a significant characteristic in myocardial infarction, and Kae treatment effectively suppressed H_2O_2-induced ROS generation (Figure 4D). In addition, the ROS level was detected, and the ROS level was reversed by the treatment of Kae, suggesting that Kae prevented mitochondrial fragmentation through the inhibition of oxidative stress (Figure 3E,F). In conclusion, these results indicate that Kae could protect against and prevent cell apoptosis via mitochondria protection.

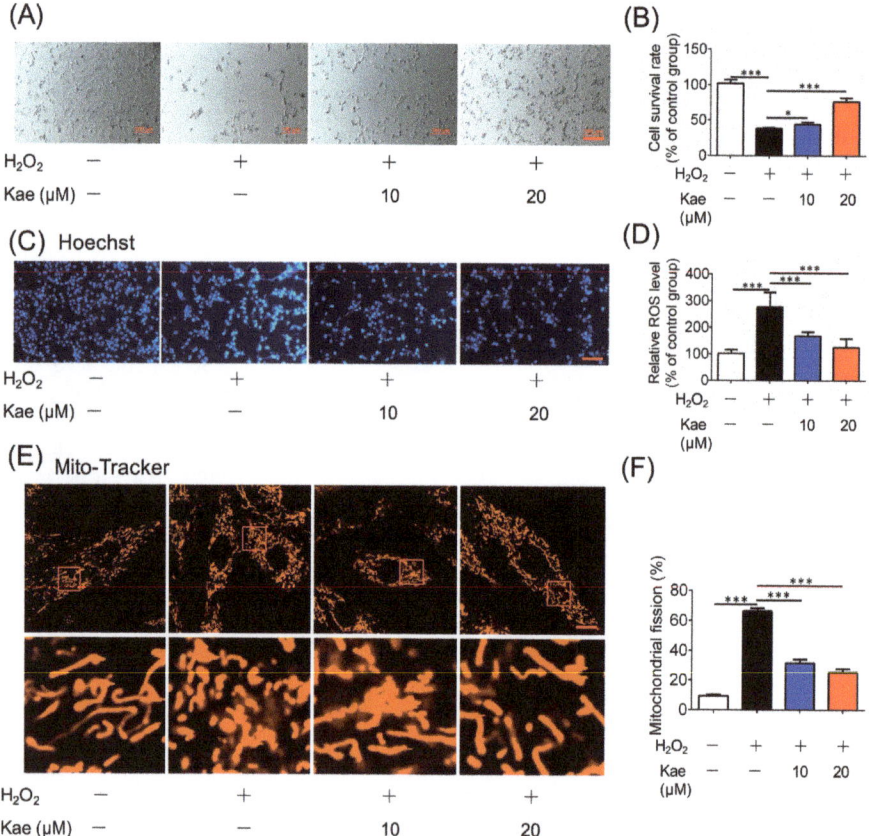

Figure 4. Kae protects cardiomyocytes. (**A**) Cell morphology analysis (scale bar, 100 μm). (**B**) Cell survival rate analysis (n = 6). (**C**) Kae-induced nuclear morphological changes were examined by using fluorescence (scale bar, 100 μm). (**D**) Intracellular ROS production (n = 6). (**E**) Mitochondrial fission imaging. (Scale bar, 5 μm). (**F**) Quantification of relative mitochondrial fission ratio (n = 6). * $p < 0.05$, *** $p < 0.001$.

2.4. Kae Protects the Heart against Ischemic Injury

To further examine the effects of Kae on cardioprotection, isoprenaline (ISO)-challenge-induced heart injury animal model was established [24] (Figure 5A). ISO treatment increase the leakage of creatine kinase (CK) and lactate dehydrogenase (LDH) in the rat blood, which was rescued by oral administration of Kae (30, 60 mg/kg) (Figure 5B,C). HE staining of heart tissue demonstrated that ISO induced cardiac injury with the color, texture, the presence of scar tissue and areas of softening or discoloration of septum, and broken fiber, but Kae administration reduced the cardiac infarct surface with structure normalization (Figure 5E,F). In addition, the ratio of heart weight and tibial length was increased by ISO insult, and oral administration of Kae at 60 mg/kg decreased the ratio (Figure 5D). Nuclear chromatin fragmentation is a hallmark apoptosis, which leads to an appearance of broken DNA strands [25]. We have applied terminal transferase dUTP nick-end labeling (TUNEL) assay to detect DNA degradation, which reveals a percentage of apoptosis cells. We found Kae treatment decreased the apoptosis cells to prevent cardiomyocyte apoptosis in mouse heart challenged with ISO (Figure 5G,H) compared with the control group. Therefore, Kae could alleviate heart injury induced by ISO via anti-apoptosis pathway.

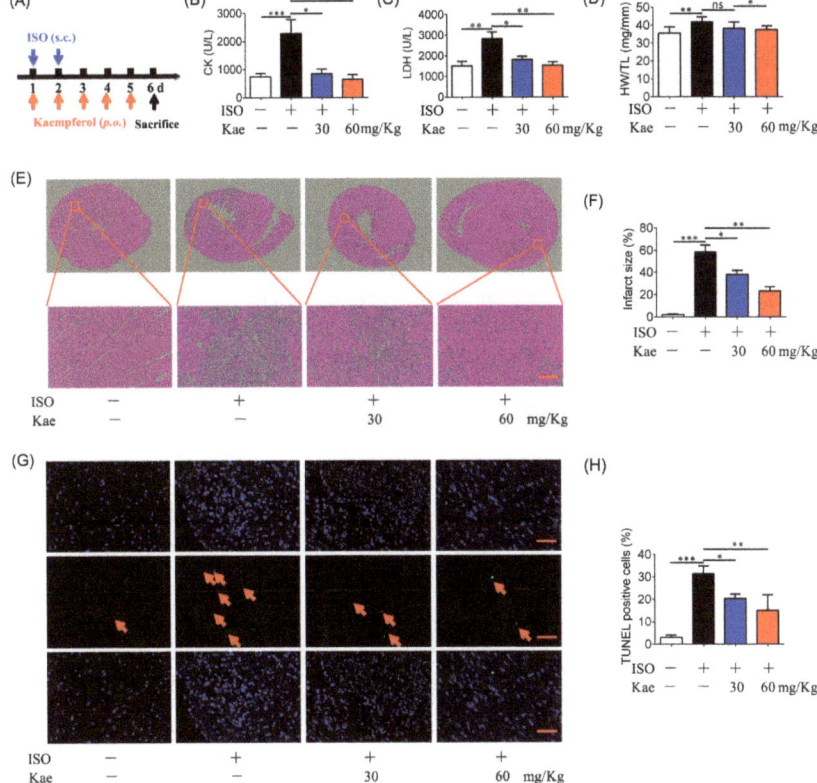

Figure 5. Kae protects the heart from ISO-induced myocardial damage in rats. (**A**) The myocardial damage in rats was established with isoprenaline (ISO) for 2 days, and Kae was orally administrated (30, 60 mg/kg, 5d). (**B**) The level of CK in the blood ($n = 6$). (**C**) The leakage of LDH in the blood ($n = 6$). (**D**) The ratio of heart weight to tibia length (HW/TL) ($n = 6$). (**E**) Representative HE-stained heart sections, the lower panel are enlarged from the red boxex in the upper panel ($n = 6$, scale bar, 50 μm). (**F**) Quantitative analysis of infarct size. (**G**) Apoptosis detection using TUNEL staining assay (scale bar, 50 μm). The red arrows indicate TUNEL positive cells. (**H**) Quantification of TUNEL positive cells ($n = 6$). * $p < 0.05$, ** $p < 0.01$, *** $p < 0.001$ and ns, statistically not significant.

3. Discussion

Target identification is used to study all binding targets that account for biological effects. Despite the technological advances in natural products development, the majority of their putative biological targets remain unknown [26,27]. In recent years, extensive strategies have been presented on target identification, such as affinity-based protein profiling, "label-free" methodology, and in silico docking [2,3,28]. However, these methods provide large numbers of candidate targets in a single experiment, which present us a new challenge for target selection. In the present study, a novel strategy that integrated of in silico docking and DARTS prediction was established, and Kae was selected as an example for investigation. With this strategy, the probable mass of the potential target can be identified by DARTS method, and in silico docking study can further help to predict the candidates. Therefore, using this strategy, Src protein was rapidly focused on as one of the candidate targets of Kae, in silico analogue, and pharmacophore docking. In addition, DARTS, CETSA, and SPR assays further identified that Kae was directly bound to Src. In vitro and in vivo study demonstrated that the cardioprotective effect of Kae was through anti-apoptosis pathway.

Traditionally, mass-spectrometry-based analysis would be employed followed by DARTS screen to further determine the candidate proteins of the specific gel band. However, the mass spectrometry data was informatively redundant for identifying the direct target of Kae. Recently, a large number of in silico target identification methods have been used and studied through open-source web servers, which are relatively fast and convenient, while it is still a challenge to reduce the number of false positives [29]. Therefore, the improvement of target identification strategies with high accuracy of target information is particularly important. Herein, a comprehensive in silico target identification and DARTS prediction strategy was established, which could quickly identify targets from a complex potential target pool.

The PharmMapper online tool is a convenient web server for potential drug target identification by reversed pharmacophore matching. Through target fishing by PharmMapper Server, Src was ranked the highest. TargetHunter is an in silico target identification tool for predicting targets based on chemical similarity searching, which is easy to operate and has high accuracy. When the structure of Kae was submitted to TargetHunter, CHEMBL115102 and CHEMBL116051 were identified as structurally similar compounds to Kae, and they share the target of Src. Nevertheless, the limits of both target identification methods were proven to be at least partially circumvented when the two different strategies were used in combination with each other. Indeed, several bands seemed to increase with Kae, such as the band located at ~60 kDa and 170 kDa. TargetHunter and PharmMapper methods are available online, and are free, fast, and convenient. However, compared to commercial software, the limited available database is the main obstacle ahead. In this study, TargetHunter and PharmMapper methods were performed, and a small number of targets with high mass are included in the database. Herein, the targets at 170 kDa were not top-ranked from the in silico study; thus, the band located at ~60 kDa was selected as an example to research. However, with the technological advances in machine learning and artificial intelligence methods, more comprehensive database and new methods for target identification are still on the way. Therefore, a constant improvement of target identification strategies is required for achieving more efficient and reliable targets.

Natural products have involved a large number of structural isomers and analogue compounds, with multiple structural isomers involved in different biological activities [30]. It was reported that the similar structures of compounds would tend to show similar biological characteristics and effects, and similar molecules are efficient for lead optimization [31]. With increasing knowledge of receptor–ligand interactions, the prediction method by similarity search is quite meaningful [31]. In recent years, the concept of molecular similarity has grown dramatically in the area of target identification. In addition, structure similarity search methods are especially applicable in natural products searches and useful in molecular mechanism elucidation [9]. Kae, a flavonoid, is structurally similar to other flavonoids. Thus, Kae was selected as a case study.

Myocardial infarction is a major cause of death in modern society, and Src family protein tyrosine kinase has been identified as a promising target for treating cardiovascular diseases, such as hypertension and ischemic heart disease [13]. NaKtide, a Na/K-ATPase-derived peptide Src inhibitor, ameliorates myocardial ischemia-reperfusion injury in vitro and in vivo [14]. In addition, Src inhibition improves arrhythmia through reducing the internalization and degradation of connexin 43 in the heart [32]. Interestingly, mitochondrial Src tyrosine kinase is inhibited by H/R from rat hearts. Inhibition of mitochondrial JNK/Sab/Src/ROS pathway could ameliorate H/R-associated oxidative stress [33]. However, dasatinib, a first-phase anti-acute myeloid neoplasms drug, leads a side effect of coronary artery disease [15]. Therefore, it is crucial to discover the endogenous cardiac arrhythmias regulatory mechanism of Src and develop safe and effective inhibitors. Kae, an ingredient isolated from natural products, has revealed a cardioprotective effect against myocardial infarction. Herein, using our novel strategy, the cardioprotective target of Kae was identified and verified.

However, several limitations still exist in this study. ISO-challenge-induced heart damage provides an easily operated model which produces myocardial damage similar to that seen in acute cardiac ischemia in humans [34]. To further evaluate the cardioprotective function of Kae on myocardial infarction, a coronary-artery-ligation-induced myocardial infarction model may be performed. In addition, creatine kinase (CK) and lactate dehydrogenase (LDH) were not specific to heart injury, and detection of CK-MB or troponin (T/I) in the blood might be a better choice. Echocardiography is the primary imaging modality for detecting cardiac functions. In short, further results about cardiac function require further study.

4. Materials and Methods

4.1. Reagents

Kaempferol (purity ≥98%) was obtained from Chengdu Biopurify Phytochemicals Ltd. (Chengdu, China). Mito-Tracker was obtained from Thermo Fisher Scientific (Xiamen, China). Isoprenaline hydrochloride (#I5627) was purchased from Sigma (St. Louis, MO, USA). Antibody against Src (#36D10) Rabbit mAb (#2109) was purchased from Cell Signaling Technology (Beverly, MA, USA). RIPA lysis buffer (#P0013D) and BCA assay (#P0010) were obtained from Beyotime (Suzhou, China).

4.2. Identification of Candidate Targets of Kae

The structure of Kae was drawn by ChemBio3D Ultra 14.0 software, and the mol2 format structure of Kae was uploaded. Then, results were presented after completing screening and scoring protocol for each target set by PharmMapper software [35] (http://www.lilab-ecust.cn/pharmmapper/, accessed date 2 October 2021). In addition, the Kae pharmacological targets were pooled with TargetHunter [19] (http://www.cbligand.org/TargetHunter, accessed date 2 October 2021).

4.3. Animals and Treatments

Sprague Dawley (SD) rats (Male, 200–220 g) were purchased from Wushi Animal Center (Fuzhou, China). All animal experiments were carried out in accordance with the National Institutes of Health guide for the care and use of laboratory animals, following protocols approved by ethics committee of Huaqiao University (no: A2020033). The mice were housed in cages with a constant temperature (20 ± 2 °C) and a 12-h light/dark cycle with free access to standard food and water.

The myocardial damage model on rats was established by ISO (65 mg/kg, *s.c.*, 2 days), as described previously [36], and Kae (30, 60 mg/kg, *p.o.*) was administrated for 5 consecutive days. Then, rats were euthanized, and the heart tissues were collected for further analysis.

4.4. Cell Culture

H9C2 cells were obtained from Cell Bank of Chinese Academy of Sciences and cultured in the medium of DMEM supplemented with 10% (v/v) FBS in an incubator with conditions of 37 °C with 5% CO_2 in air atmosphere.

For cell survival assay, H9C2 cells were seeded in a 96-well plate and incubated with Kae at given concentrations for 8 h with H_2O_2 (100 μM). The cells were observed by bright-field microscopy to determine the cytotoxicity. Cell survival was evaluated by Cell Counting Kit-8 (CCK8, APExBIO, Shanghai, China).

4.5. Immunoblotting Experiments

Cell lysates were harvested with RIPA lysis buffer, and the protein quantitation of all the samples was performed using BCA assay. Then, samples were separated by SDS-PAGE and transferred onto PVDF (0.45 μm) membrane. PVDF membranes were incubated with indicated primary antibody overnight at 4 °C followed by blocking for 1 h. The next day, membranes were washed and incubated with secondary antibodies for 2 h at room temperature. The protein bands were imaged using Tanon 500 system (Tanon, Shanghai, China).

4.6. Surface Plasmon Resonance (SPR) Analysis

Src protein was immobilized on a carboxymethylated 5 (CM5) sensor chip. Different concentrations of Src (1 to 16 μmol/L) were used for analysis using the Biacore T200 system (GE Healthcare Life Sciences, Uppsala, Sweden).

4.7. Molecular Docking

The 3D structure of the Src kinase catalytic domain in complex with the drug bosutinib was downloaded from the Protein Data Bank (PDB entry 4MXO). Upon removing bosutinib and water molecules, hydrogen atoms were added to the protein according to the protonation states of chemical groups at the physiological pH. The initial 3D conformer of Kae was generated using the ETKDG method implemented in RDKit (version 2020.09), and further minimized with MMFF94s force field. Kae was docked into the ATP site of the Src catalytic domain by the program LeDock [37].

4.8. The Assay of Mitochondrial Fission

After treatment, H9C2 cells were incubated with 50 nmol/L Mito Tracker for 30 min at 37 °C. Then, after washing with warm PBS three times, the mitochondrial fission was determined on confocal scanning microscopy (Zeiss, LSM 700).

For quantification of mitochondrial fission, the fluorescence images were performed using ImageJ software, as described previously [38]. Briefly, appropriate threshold for images were set up, and individual mitochondrion were analyzed for circularity ($4\pi \times area/perimeter^2$) and lengths of major and minor axes. The form factor (FF, the reciprocal of circularity value) and aspect ratio (AR, major axis/minor axis) were calculated. While the mitochondrion was a small perfect circle, parameters have a small value, and the values increase when it becomes elongated. The lower values of FF and AR indicate mitochondrial fission. In addition, fragmented or tubular mitochondria were counted by three experimenters.

4.9. Drug Affinity Responsive Target Stabilization Assay (DARTS)

The DARTS assay was conducted in H9C2 cells. In brief, approximately 1×10^7 cells were lysed on ice for 30 min. Indicated concentrations of Kae (diluted in $1 \times$ TNC buffer, 50 mmol/L Tris, 50 mmol/L NaCl, 10 mmol/L $CaCl_2$, pH = 7.4) were added into the aliquoted protein (5 mg/mL). Then, samples were gently mixed and incubated for 2 h at room temperature. Then, lysates were digested by pronase (1:400, w/w) for 30 min. Then, $1 \times$ loading buffer was added and boiled for 10 min. The samples were separated by sodium dodecyl sulfate polyacrylamide gel electrophoresis (SDS-PAGE) and stained with Coomassie blue.

4.10. Cellular Thermal Shift Assay (CETSA)

For cell lysate CETSA experiments, H9C2 cells were lysed with a freeze-thawed method using liquid nitrogen. Then, cell lysates were divided into two fractions, one incubated with the DMSO as the control group and the other incubated with Kae (200 µmol/L) for 30 min as the Kae-treated group (at room temperature). Then, the lysates from the two groups were aliquoted, respectively, followed by heating at sequentially increased temperature (39–63 °C with a 4 °C interval) for 5 min. After boiling for 10 min, immunoblotting assay was performed, and Src abundance level was analyzed.

4.11. Hoechst 33342 Staining

Cells were treated and harvested, followed by rapid staining, and fixed in 4% paraformaldehyde for 20 min at room temperature. Then, cells were washed with PBS for 5 min. After incubation with Hoechst 33342 (100 ng/mL) in the dark for 15 min, the cells were viewed using a fluorescence microscope.

4.12. Tissue Preparation, Hematoxylin Eosin and TUNEL Staining

First, tissue samples were fixated with 4% paraformaldehyde for 24 h, then also with cellular water. Graded alcohols (70%, 80%, 95%, and 100%) were used in dehydration for 1 h each. Next, 100% xylene was used for tissue clearing (1 h). After clearing, tissue sections were infiltrated with paraffin wax to support the tissue for thin sectioning. 5 µm thick heart cross-sections were cut, and the paraffin wax was removed, followed by tissue staining. Six serial cross-sections were collected and then placed in slide boxes and stored until use.

For H&E staining, the hematoxylin solution stains the nuclear chromatin and possibly other acidic cellular elements. Unbound hematoxylin is removed with water rinses, followed by an optional differentiation step using acid alcohol. Staining was observed using a light microscope.

TUNEL staining was performed using a One-Step TUNEL Apoptosis Assay Kit (Cat# C1086; Beyotime, Suzhou, China) according to the manufacturer's protocols with modifications. First, cross-sections were incubated with proteinase K (20 µg/mL) for 15 min. Following digestion, TUNEL reaction mix was added onto the slides and incubated at 37 °C for 1 h. After washing, staining was observed using a light microscope. All digital images were captured using a Nikon digital camera.

4.13. Determination of LDH and CK Leakage

Blood samples collected from mice were centrifuged at $3000 \times g$ for 30 min to obtain serum. LDH and CK leakage was determined by a colorimetric procedure with lactate dehydrogenase assay kit (A020-2-2) and creatine kinase assay kit (A032-1-1, Nanjing Jiancheng Bioengineering Institute, Nanjing, China) according to manufacturer's instructions, and the absorbance at 490 nm was measured on a microplate reader.

4.14. Statistical Analysis

All the data used were normally distributed. The difference between groups (>2 two groups) were performed using one-way ANOVA (Tukey's multiple comparisons test) with IBM SPSS Statistics 26 (IBM Corp., Armonk, NY, USA). All data are represented as mean $\pm$ SD, unless otherwise specified; * $p < 0.05$, ** $p < 0.01$, *** $p < 0.001$.

5. Conclusions

In conclusion, we have established a novel integrated strategy of in silico docking and DARTS prediction to efficiently identify the direct targets of Kae. Src was successfully identified and validated as a direct target of Kae, and we further verified its cardioprotective effect with in vitro and in vivo study. This study suggests that our strategy is convenient and efficient, and Kae might be a new potent cardioprotective drug candidate or a lead compound.

Supplementary Materials: The following are available online at https://www.mdpi.com/article/10.3390/ijms222312908/s1.

Author Contributions: Conceptualization, Y.D. and C.Y.; methodology, X.W. and X.L.; software, X.W.; validation, Y.D. and C.Y.; formal analysis, X.W.; investigation, X.W. and X.L.; resources, X.W.; data curation, X.W.; writing—original draft preparation, X.W. and C.Y.; writing—review and editing, Y.D.; visualization, X.W.; supervision, Y.D.; project administration, Y.D. and X.W.; funding acquisition, Y.D. and X.W. All authors have read and agreed to the published version of the manuscript.

Funding: This research was supported by the Scientific Research Funds of Huaqiao University, grant number 21BS126; The Science and the Technology Planning Projects of Quanzhou Municipal, grant number 2019N031; The Project of Science and Technology of Fujian Province of China, grant numbers 2019J05094, 2018J01127.

Institutional Review Board Statement: The study was conducted according to the guidelines of the Declaration of Helsinki, and approved by the Institutional Animal Ethics Committee of Huaqiao university via protocol no: A2020033.

Informed Consent Statement: Not applicable.

Data Availability Statement: The data generated during the current study are available with the corresponding author on reasonable request.

Conflicts of Interest: The authors declare no conflict of interest.

References

1. Zheng, X.; Ma, S.; Kang, A.; Wu, M.; Wang, L.; Wang, Q.; Wang, G.; Hao, H. Chemical dampening of Ly6C(hi) monocytes in the periphery produces anti-depressant effects in mice. *Sci. Rep.* **2016**, *6*, 19406. [CrossRef]
2. Pan, S.; Zhang, H.; Wang, C.; Yao, S.C.L.; Yao, S.Q. Target identification of natural products and bioactive compounds using affinity-based probes. *Nat. Prod. Rep.* **2016**, *33*, 612–620. [CrossRef]
3. Lomenick, B.; Hao, R.; Jonai, N.; Chin, R.M.; Aghajan, M.; Warburton, S.; Wang, J.; Wu, R.P.; Gomez, F.; Loo, J.A.; et al. Target identification using drug affinity responsive target stability (DARTS). *Proc. Natl. Acad. Sci. USA* **2009**, *106*, 21984–21989. [CrossRef]
4. Chen, X.; Wang, Y.; Ma, N.; Tian, J.; Shao, Y.; Zhu, B.; Wong, Y.; Liang, Z.; Zou, C.; Wang, J. Target identification of natural medicine with chemical proteomics approach: Probe synthesis, target fishing and protein identification. *Signal Transduct. Target. Ther.* **2020**, *5*, 72. [CrossRef]
5. Terstappen, G.; Reggiani, A. In silico research in drug discovery. *Trends Pharmacol. Sci.* **2001**, *22*, 23–26. [CrossRef]
6. Bender, A.; Young, D.W.; Jenkins, J.L.; Serrano, M.; Mikhailov, D.; Clemons, P.A.; Davies, J.W. Chemogenomic data analysis: Prediction of small-molecule targets and the advent of biological fingerprint. *Comb. Chem. High Throughput Screen.* **2007**, *10*, 719–731. [CrossRef] [PubMed]
7. Martin, Y.C.; Kofron, J.L.; Traphagen, L.M. Do structurally similar molecules have similar biological activity? *J. Med. Chem.* **2002**, *45*, 4350–4358. [CrossRef]
8. Schuffenhauer, A.; Floersheim, P.; Acklin, P.; Jacoby, E. Similarity metrics for ligands reflecting the similarity of the target proteins. *J. Chem. Inf. Comput. Sci.* **2003**, *43*, 391–405. [CrossRef] [PubMed]
9. Muegge, I.; Mukherjee, P. An overview of molecular fingerprint similarity search in virtual screening. *Expert Opin. Drug Discov.* **2016**, *11*, 137–148. [CrossRef]
10. Bender, A.; Glen, R.C. Molecular similarity: A key technique in molecular informatics. *Org. Biomol. Chem.* **2004**, *2*, 3204–3218. [CrossRef]
11. Raju, T. The Nobel chronicles—The first century. *Lancet* **2000**, *356*, 436. [CrossRef]
12. Cheng, M.; Huang, K.; Zhou, J.; Yan, D.; Tang, Y.-L.; Zhao, T.C.; Miller, R.J.; Kishore, R.; Losordo, D.W.; Qin, G. A critical role of Src family kinase in SDF-1/CXCR4-mediated bone-marrow progenitor cell recruitment to the ischemic heart. *J. Mol. Cell Cardiol.* **2015**, *81*, 49–53. [CrossRef] [PubMed]
13. Zhai, Y.; Yang, J.; Zhang, J.; Yang, J.; Li, Q.; Zheng, T. Src-family Protein Tyrosine Kinases: A promising target for treating Cardiovascular Diseases. *Int. J. Med. Sci.* **2021**, *18*, 1216–1224. [CrossRef]
14. Li, H.; Yin, A.; Cheng, Z.; Feng, M.; Zhang, H.; Xu, J.; Wang, F.; Qian, L. Attenuation of Na/K-ATPase/Src/ROS amplification signal pathway with pNaktide ameliorates myocardial ischemia-reperfusion injury. *Int. J. Biol. Macromol.* **2018**, *118*, 1142–1148. [CrossRef]
15. Redner, R.; Beumer, J.; Kropf, P.; Agha, M.; Boyiadzis, M.; Dorritie, K.; Farah, R.; Hou, J.; Im, A.; Lim, S.; et al. A phase-1 study of dasatinib plus all-trans retinoic acid in acute myeloid leukemia. *Leuk. Lymphoma* **2018**, *59*, 2595–2601. [CrossRef]
16. Calderón-Montaño, J.M.; Burgos-Morón, E.; Pérez-Guerrero, C.; López-Lázaro, M. A review on the dietary flavonoid kaempferol. *Mini Rev. Med. Chem.* **2011**, *11*, 298–344. [CrossRef] [PubMed]
17. Armstrong, S.C. Protein kinase activation and myocardial ischemia/reperfusion injury. *Cardiovasc. Res.* **2004**, *61*, 427–436. [CrossRef] [PubMed]

18. Zhou, M.; Ren, H.; Han, J.; Wang, W.; Zheng, Q.; Wang, D. Protective Effects of Kaempferol against Myocardial Ischemia/Reperfusion Injury in Isolated Rat Heart via Antioxidant Activity and Inhibition of Glycogen Synthase Kinase-3β. *Oxid. Med. Cell. Longev.* **2015**, *2015*, 481405. [CrossRef] [PubMed]
19. Wang, L.; Ma, C.; Wipf, P.; Liu, H.; Su, W.; Xie, X.Q. TargetHunter: An in silico target identification tool for predicting therapeutic potential of small organic molecules based on chemogenomic database. *AAPS J.* **2013**, *15*, 395–406. [CrossRef]
20. Cushman, M.; Zhu, H.; Geahlen, R.; Kraker, A. Synthesis and biochemical evaluation of a series of aminoflavones as potential inhibitors of protein-tyrosine kinases p56lck, EGFr, and p60v-src. *J. Med. Chem.* **1994**, *37*, 3353–3362. [CrossRef]
21. Luo, J.; Zhang, R.; Wang, X.; Hou, Z.; Guo, S.; Jiang, B. Binding properties of marine bromophenols with human protein tyrosine phosphatase 1B: Molecular docking, surface plasmon resonance and cellular insulin resistance study. *Int. J. Biol. Macromol.* **2020**, *163*, 200–208. [CrossRef]
22. Zhang, J.; Adrián, F.J.; Jahnke, W.; Cowan-Jacob, S.W.; Li, A.G.; Iacob, R.E.; Sim, T.; Powers, J.; Dierks, C.; Sun, F.; et al. Targeting Bcr-Abl by combining allosteric with ATP-binding-site inhibitors. *Nature* **2010**, *463*, 501–506. [CrossRef] [PubMed]
23. Crowley, L.C.; Marfell, B.J.; Waterhouse, N.J. Analyzing Cell Death by Nuclear Staining with Hoechst 33342. *Cold Spring Harb. Protoc.* **2016**, *2016*, 9. [CrossRef] [PubMed]
24. Garg, M.; Khanna, D. Exploration of pharmacological interventions to prevent isoproterenol-induced myocardial infarction in experimental models. *Ther. Adv. Cardiovasc. Dis.* **2014**, *8*, 155–169. [CrossRef] [PubMed]
25. Kyrylkova, K.; Kyryachenko, S.; Leid, M.; Kioussi, C. Detection of apoptosis by TUNEL assay. *Methods Mol. Biol.* **2012**, *887*, 41–47.
26. Chang, J.; Kim, Y.; Kwon, H.J. Advances in identification and validation of protein targets of natural products without chemical modification. *Nat. Prod. Rep.* **2016**, *33*, 719–730. [CrossRef]
27. Williams, D.E.; Andersen, R.J. Biologically active marine natural products and their molecular targets discovered using a chemical genetics approach. *Nat. Prod. Rep.* **2020**, *37*, 617–633. [CrossRef]
28. Dai, L.; Li, Z.; Chen, D.; Jia, L.; Guo, J.; Zhao, T.; Nordlund, P. Target identification and validation of natural products with label-free methodology: A critical review from 2005 to 2020. *Pharmacol. Ther.* **2020**, *216*, 107690. [CrossRef]
29. Galati, S.; Di Stefano, M.; Martinelli, E.; Poli, G.; Tuccinardi, T. Recent Advances in In Silico Target Fishing. *Molecules* **2021**, *26*, 5124. [CrossRef]
30. Lai, C.-J.-S.; Zha, L.; Liu, D.-H.; Kang, L.; Ma, X.; Zhan, Z.-L.; Nan, T.-G.; Yang, J.; Li, F.; Yuan, Y.; et al. Global profiling and rapid matching of natural products using diagnostic product ion network and in silico analogue database: Gastrodia elata as a case study. *J. Chromatogr. A* **2016**, *1456*, 187–195. [CrossRef]
31. Patterson, D.E.; Cramer, R.D.; Ferguson, A.M.; Clark, R.D.; Weinberger, L.E. Neighborhood behavior: A useful concept for validation of "molecular diversity" descriptors. *J. Med. Chem.* **1996**, *39*, 3049–3059. [CrossRef]
32. Rutledge, C.; Ng, F.; Sulkin, M.; Greener, I.; Sergeyenko, A.; Liu, H.; Gemel, J.; Beyer, E.; Sovari, A.; Efimov, I.; et al. c-Src kinase inhibition reduces arrhythmia inducibility and connexin43 dysregulation after myocardial infarction. *J. Am. Coll. Cardiol.* **2014**, *63*, 928–934. [CrossRef] [PubMed]
33. Chu, Q.; Zhang, Y.; Zhong, S.; Gao, F.; Chen, Y.; Wang, B.; Zhang, Z.; Cai, W.; Li, W.; Zheng, F.; et al. N-n-Butyl Haloperidol Iodide Ameliorates Oxidative Stress in Mitochondria Induced by Hypoxia/Reoxygenation through the Mitochondrial c-Jun N-Terminal Kinase/Sab/Src/Reactive Oxygen Species Pathway in H9c2 Cells. *Oxid Med. Cell. Longev.* **2019**, *2019*, 7417561. [CrossRef] [PubMed]
34. Wong, Z.W.; Thanikachalam, P.V.; Ramamurthy, S. Molecular understanding of the protective role of natural products on isoproterenol-induced myocardial infarction: A review. *Biomed. Pharmacother.* **2017**, *94*, 1145–1166. [CrossRef]
35. Wang, X.; Shen, Y.; Wang, S.; Li, S.; Zhang, W.; Liu, X.; Lai, L.; Pei, J.; Li, H. PharmMapper 2017 update: A web server for potential drug target identification with a comprehensive target pharmacophore database. *Nucleic Acids Res.* **2017**, *45*, W356–W360. [CrossRef]
36. Zhang, L.; Wei, T.-T.; Li, Y.; Li, J.; Fan, Y.; Huang, F.-Q.; Cai, Y.-Y.; Ma, G.; Liu, J.-F.; Chen, Q.-Q.; et al. Functional Metabolomics Characterizes a Key Role for -Acetylneuraminic Acid in Coronary Artery Diseases. *Circulation* **2018**, *137*, 1374–1390. [CrossRef] [PubMed]
37. Zhang, N.; Zhao, H. Enriching screening libraries with bioactive fragment space. *Bioorg. Med. Chem. Lett.* **2016**, *26*, 3594–3597. [CrossRef]
38. Chuang, J.-I.; Pan, I.L.; Hsieh, C.-Y.; Huang, C.-Y.; Chen, P.-C.; Shin, J.W. Melatonin prevents the dynamin-related protein 1-dependent mitochondrial fission and oxidative insult in the cortical neurons after 1-methyl-4-phenylpyridinium treatment. *J. Pineal. Res.* **2016**, *61*, 230–240. [CrossRef]

Article

Rare Glutamic Acid Methyl Ester Peptaibols from *Sepedonium ampullosporum* Damon KSH 534 Exhibit Promising Antifungal and Anticancer Activity

Yen T. H. Lam [1,2], Manuel G. Ricardo [1,3], Robert Rennert [1], Andrej Frolov [1,4], Andrea Porzel [1], Wolfgang Brandt [1], Pauline Stark [1], Bernhard Westermann [1] and Norbert Arnold [1,*]

1. Department of Bioorganic Chemistry, Leibniz Institute of Plant Biochemistry, D-06120 Halle (Saale), Germany; ThiHaiYen.Lam@ipb-halle.de (Y.T.H.L.); Manuel.GarciaRicardo@mpikg.mpg.de (M.G.R.); Robert.Rennert@ipb-halle.de (R.R.); Andrej.Frolov@ipb-halle.de (A.F.); Andrea.Porzel@ipb-halle.de (A.P.); Wolfgang.Brandt@ipb-halle.de (W.B.); Pauline.Stark@ipb-halle.de (P.S.); Bernhard.Westermann@ipb-halle.de (B.W.)
2. Department of Organic Chemistry, Faculty of Chemistry, Hanoi National University of Education, Hanoi 100000, Vietnam
3. Department of Biomolecular Systems, Max Planck Institute of Colloids and Interfaces, D-14476 Potsdam, Germany
4. Department of Biochemistry, Faculty of Biology, St. Petersburg State University, 199004 St. Petersburg, Russia
* Correspondence: Norbert.Arnold@ipb-halle.de; Tel.: +49-345-5582-1310

Abstract: Fungal species of genus *Sepedonium* are rich sources of diverse secondary metabolites (e.g., alkaloids, peptaibols), which exhibit variable biological activities. Herein, two new peptaibols, named ampullosporin F (**1**) and ampullosporin G (**2**), together with five known compounds, ampullosporin A (**3**), peptaibolin (**4**), chrysosporide (**5**), c(Trp-Ser) (**6**) and c(Trp-Ala) (**7**), have been isolated from the culture of *Sepedonium ampullosporum* Damon strain KSH534. The structures of **1** and **2** were elucidated based on ESI-HRMS^n experiments and intense 1D and 2D NMR analyses. The sequence of ampullosporin F (**1**) was determined to be Ac-Trp^1-Ala^2-Aib^3-Aib^4-Leu^5-Aib^6-Gln^7-Aib^8-Aib^9-Aib^{10}-GluOMe^{11}-Leu^{12}-Aib^{13}-Gln^{14}-Leuol^{15}, while ampullosporin G (**2**) differs from **1** by exchanging the position of Gln^7 with GluOMe^{11}. Furthermore, the total synthesis of **1** and **2** was carried out on solid-phase to confirm the absolute configuration of all chiral amino acids as L. In addition, ampullosporin F (**1**) and G (**2**) showed significant antifungal activity against *B. cinerea* and *P. infestans*, but were inactive against *S. tritici*. Cell viability assays using human prostate (PC-3) and colorectal (HT-29) cancer cells confirmed potent anticancer activities of **1** and **2**. Furthermore, a molecular docking study was performed in silico as an attempt to explain the structure-activity correlation of the characteristic ampullosporins (**1–3**).

Keywords: *Sepedonium ampullosporum*; peptaibols; ampullosporin; glutamic acid methyl ester; solid-phase peptide synthesis; antifungal; anticancer; molecular docking

1. Introduction

Filamentous fungi from saprophytic genera, e.g., *Acremonium*, *Trichoderma/Hypocrea* and *Sepedonium*, are known for a high abundance of nonribosomal-peptide-synthetase (NRPS)-derived metabolites, so-called pept"Aib"ols [1]. This class of compounds is defined as linear peptides with 5–20 amino acids (AA) residues including (*i*) a high proportion of the nonproteinogenic α,α-dialkylated amino acid α-aminoisobutyric acid (Aib); (*ii*) an N-acetyl terminus; (*iii*) and a C-terminal AA reduced into amino alcohol such as leucinol (Leuol) or phenylalaninol (Pheol) [1]. Peptaibols are intriguing not only because of the structural variability generated by varying amino acid building blocks, but also due to their broad range of bioactivities, i.e., cytotoxic [2–11], antibacterial [2,5,6,10–13], antiviral [14,15], antileishmanial [16], antifungal [5,6,17–19], plant root growth inhibiting [20], insecticidal [7], and anthelmintic activity [2]. Furthermore, ampullosporin A (3), a peptaibol isolated from *Sepedonium ampullosporum* Damon, has been shown to permit neuroleptic-like activity in mice [21,22], whereby it exhibited a more targeted interaction with the glutamatergic system, namely the *N*-methyl-D-aspartate (NMDA) receptor [23].

These biological activities appear to be related to the strong foldameric capacity of Aib so that peptaibols would adopt helical structures in artificial bilayers and natural membranes [24]. Consequently, the amphipathic voltage-dependent helices can act as ion channels in cell membranes, causing cytoplasmic leakage and cellular breakdown [24,25].

The genus *Sepedonium* (teleomorph *Hypocrea*, Ascomycota), which was established by H.F. Link and restricted to mold-like fungal parasites, is characterized by the occurrence of two synanamorphs, i.e., aleurioconidia (for persistence) and phialoconidia (for fast propagation). *Sepedonium* species were found settling on basidiocarps of Boletales s.l. [26]. Unlike the intensively studied genus *Trichoderma/Hypocrea*, there are only few reports on secondary metabolites isolated from *Sepedonium* spp. So far, compounds belonging to tropolones [27,28], anthraquinones (mono-, dimeric) [29], isoquinoline alkaloids [30], azaphilones [31], cyclopeptides [32,33], and peptaibols [15,17–19,21,22,34–37] are identified from species in this genus. Therefore, further studies on chemical components of *Sepedonium* spp. are promising for novel bioactive compounds, especially for new peptaibols.

During our ongoing work to study secondary metabolites from *Sepedonium* species, the culture broth and mycelial extract of *S. ampullosporum* Damon strain KSH 534 was investigated. So far, the chemical studies on *S. ampullosporum* strains have resulted in the isolation of characteristic 15-residue peptaibols, named ampullosprins A–D and E1–E4 [21,22], together with peptaibolin [35] and ampullosine [30].

The present paper describes the MS-guided isolation, structural elucidation, and total synthesis of two new linear 15-residue peptaibols, named ampullosporin F (1) and G (2), together with five known compounds including two linear peptaibols, ampullosporin A (3) and peptaibolin (4), as well as three cyclic peptides, chrysosporide (5), c(Trp-Ser) (6), and c(Trp-Ala) (7), from the semi-solid culture of *S. ampullosporum* Damon strain KSH 534. The absolute configuration of compounds 1 and 2 was determined by comparison to their synthetic counterparts. Moreover, the linear 15-residue peptaibols 1–3 were evaluated for their biological activity against the plant-pathogenic organisms *Botrytis cinerea* Pers., *Septoria tritici* Desm., and *Phytophthora infestans* (Mont.) de Bary, as well as their anticancer effects against human prostate (PC-3) and colorectal (HT-29) cancer cells. Additionally, the relationship between changes in amino acid sequence and activity of compounds 1–3 are discussed using a chemoinformatic molecular docking approach.

2. Results and Discussion

2.1. Isolation and Structural Elucidation of Compounds 1–7

The chromatographic separation of the culture broth and mycelial crude extract using Diaion HP 20 and Sephadex LH 20 in combination with (semi)preparative HPLC yielded seven compounds (1–7) (Figure 1).

Compound 1 was isolated as a white, amorphous solid. The amino acid sequence of 1 was determined based on positive and negative ion ESI-HRMS^n studies, which showed diagnostic fragments of the b and y series (Figures 2 and 3; Table 1; Table S1, Supplementary Materials).

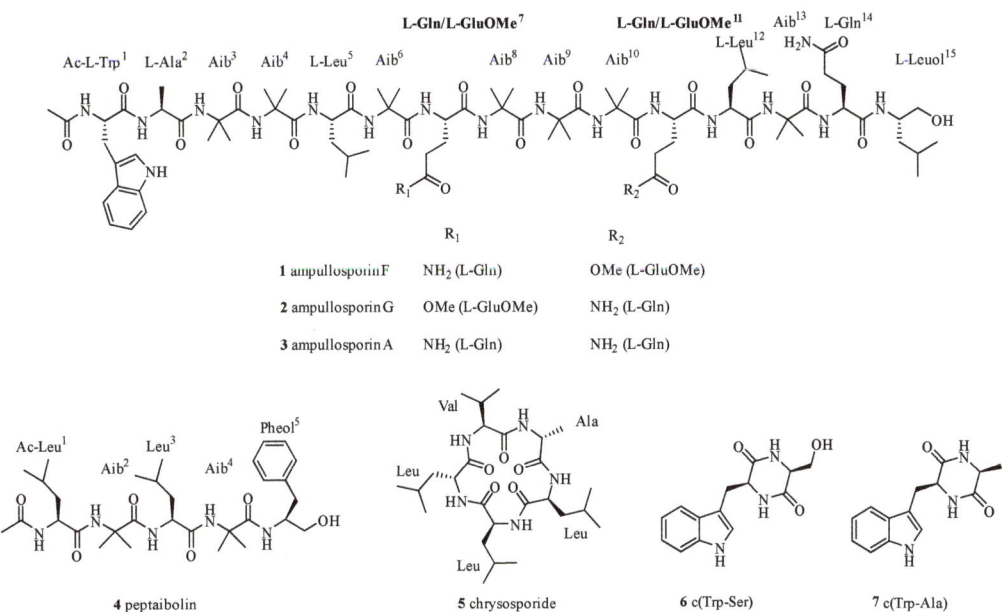

Figure 1. Structures of compounds 1–7.

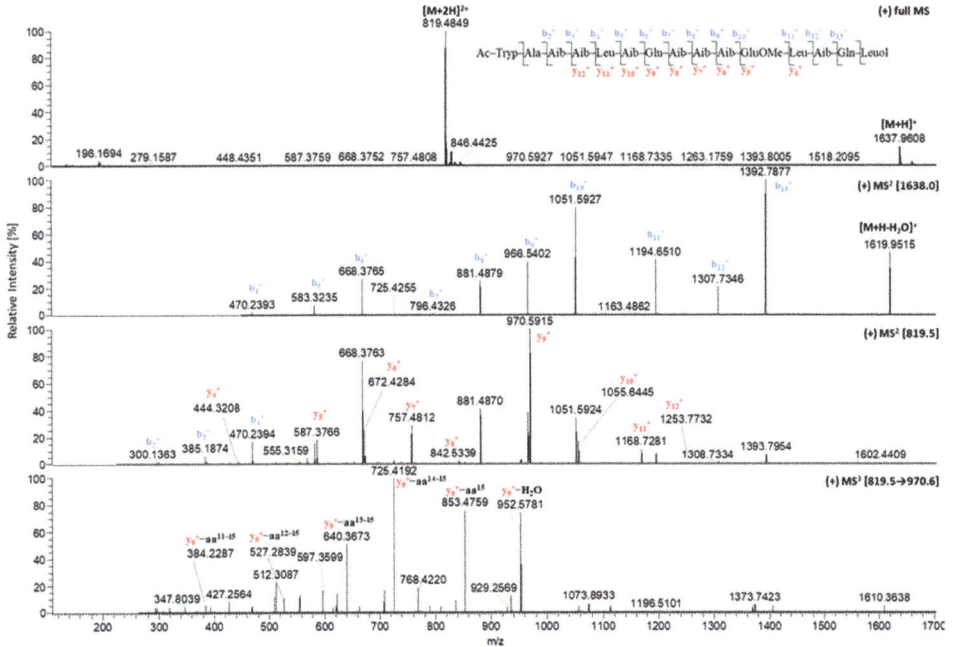

Figure 2. Positive ion ESI-HRMS^n spectra of ampullosporin F (**1**).

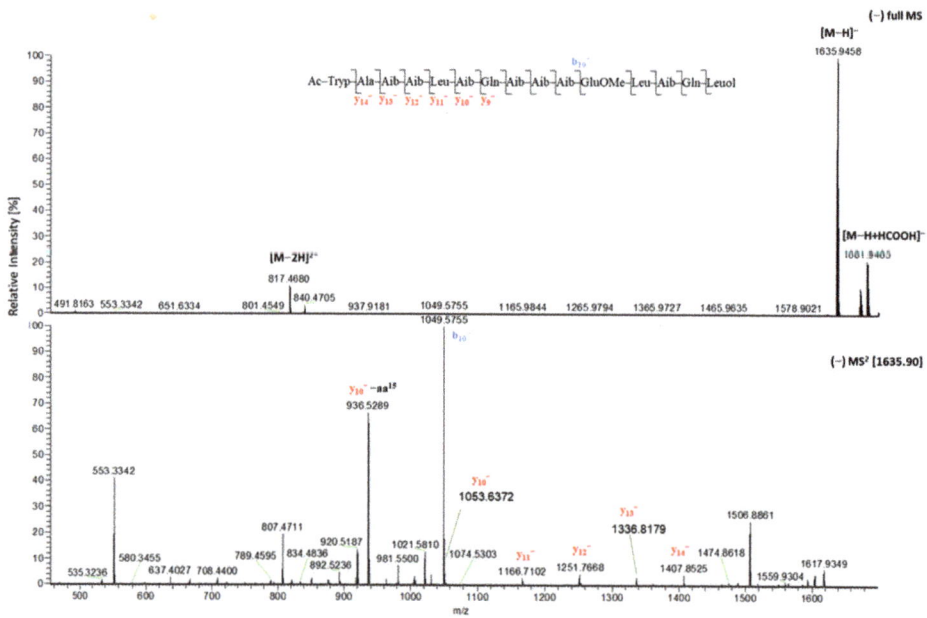

Figure 3. Negative ion ESI-HRMS^n spectra of ampullosporin F (1).

Table 1. Diagnostic fragment ions [m/z] of ampullosporin F (**1**) and ampullosporin G (**2**) from ESI-HRMS^n experiments in positive and negative ion modes.

	1	2		1	2
t_R (min)	11.62	11.54	y_{11}^+	1168.7280	1168.7294
[M + 2H]$^{2+}$	819.4849	819.4864	y_{12}^+	n.d.	n.d.
[M + H]$^+$	1637.9608	1637.9646	y_{13}^+	n.d.	n.d.
b_1^+	229.0977	229.0978	y_{14}^+	n.d.	n.d.
b_2^+	300.1361	300.1345	y_9^-	970.5923	970.5915
b_3^+	385.1870	385.1874	y_9^+-aa^{15}	853.4770	853.4781
b_4^+	470.2393	470.2400	y_9^+-aa$^{14-15}$	725.4192	725.4180
b_5^+	583.3235	583.3235	y_9^+-aa$^{13-15}$	640.3662	640.3654
b_6^+	668.3765	668.3763	y_9^+-aa$^{12-15}$	527.2811	527.2847
b_7^+	796.4326	811.4340	y_9^+-aa$^{11-15}$	n.d.	399.2281
b_8^+	881.4879	896.4871	y_9^+-aa$^{10-15}$	n.d.	n.d.
b_9^+	966.5402	981.5396	y_9^+-aa$^{9-15}$	n.d.	n.d.
b_{10}^+	1051.5927	1066.5923	y_9^+-aa$^{8-15}$	n.d.	n.d.
b_{11}^+	1194.6510	1194.6440	[M − H]$^-$	1635.9458	1635.9453
b_{12}^+	1307.7346	1307.7357	y_2^-	n.d.	n.d.
b_{13}^+	1392.7877	1392.7875	y_3^-	n.d.	n.d.
b_{14}^+	n.d.	n.d.	y_4^-	n.d.	n.d.
y_1^+	n.d.	n.d.	y_5^-	n.d.	570.3615
y_2^+	n.d.	246.1820	y_6^-	n.d.	655.4136
y_3^+	n.d.	n.d.	y_7^-	n.d.	740.4655
y_4^+	444.3214	444.3169	y_8^-	n.d.	825.5215
y_5^+	587.3766	572.3767	y_9^-	968.5731	968.5756
y_6^+	672.4281	657.4289	y_{10}^-	1053.6372	1053.6267
y_7^+	757.4816	742.4813	y_{11}^-	1166.7102	1166.7143
y_8^+	842.5358	827.5331	y_{12}^-	1251.7668	1251.7572
y_9^+	970.5923	970.5915	y_{13}^-	1336.8179	1336.8088
y_{10}^+	1055.6453	1055.6426	y_{14}^-	1407.8525	1407.8543

The signal of the [M + 2H]$^{2+}$ ion at m/z 819.4849 (calcd for $C_{78}H_{130}N_{18}O_{20}^{2+}$ 819.4849) was the most intense one in the positive mode full scan spectrum of 1, followed by the signal for the [M + H]$^{+}$ ion at m/z 1637.9608 (calcd for $C_{78}H_{129}N_{18}O_{20}^{+}$ 1637.9625), both corresponding to the molecular formula $C_{78}H_{128}N_{18}O_{20}$.

Fragmentation of the [M + H]$^{+}$ ion generated a series of product ions b_4^{+} to b_{13}^{+}, providing successive losses of Lxx^5, Aib^6, Gln^7, Aib^8, Aib^9, Aib^{10}, GluOMe^{11}, Lxx^{12} and Aib^{13}. The MS^2 spectrum of the [M + 2H]$^{2+}$ ion displayed further N-terminal b ions b_2^{+} and b_3^{+} corresponding to Aib^3 and Aib^4. The negative ion MS^2 spectrum of the [M − H]$^{-}$ yielded diagnostic fragment ions y_{13}^{-} and y_{14}^{-} representing Ala^2 and Ac-Trp^1 as the acetylated N-terminal amino acid. In addition, the observed series of C-terminal ions y_4^{+} to y_{12}^{+} and y_9^{-} to y_{12}^{-} fully supported the sequence deduced from b series. Thus, the N-terminal peptide part was shown to be Ac-Trp^1-Ala^2-Aib^3-Aib^4-Lxx^5-Aib^6-Gln^7-Aib^8-Aib^9-Aib^{10}-GluOMe^{11}-Lxx^{12}-Aib^{13}.

Furthermore, the MS^3 spectrum of the C-terminal ion y_9^{+} displayed the characteristic mass differences of 117 amu (y_9^{+}-aa^{15}, m/z 853.4770) and 245 amu (y_9^{+}-aa$^{14-15}$, m/z 725.4192) corresponding to Isoleucinol/Leucinol (Lxxol^{15}) as a C-terminal amino acid, and hence revealed the presence of Gln^{14}. Based on the mass spectrometric analyses, the tentative sequence of 1 was proposed as Ac-Trp^1-Ala^2-Aib^3-Aib^4-Lxx^5-Aib^6-Gln^7-Aib^8-Aib^9-Aib^{10}-GluOMe^{11}-Lxx^{12}-Aib^{13}-Gln^{14}-Lxxol^{15}.

The sequence of 1 resulting from mass spectrometry fragmentations was confirmed by 1D and 2D NMR data, which simultaneously specified the isomeric residues Leu and Ile.

The ^{1}H spectrum of 1 (Table 2) displayed resonances of twenty exchangeable amide protons in the range of 6.7–11 ppm, including one broad low-field singlet (δ_H 10.86, Trp^1), eight doublets, seven singlets representing seven Aib, as well as four broad singlets displaying two side-chain N-H$_2$ (δ_H 7.12 and 6.75, Gln^7; δ_H 7.13 and 6.73, Gln^{14}), which were assigned in combination with ^{1}H-^{15}N HSQC data. In addition, five aromatic protons (δ_H 7.22, 7.56, 7.33, 7.05, and 6.96, Trp^1) characterized the indole ring of tryptophan. In the high-field, multiple broad singlets between 1.30–1.50 ppm for Aib residues and a doublet for Ala^2 (δ_H 1.27, d, 7.4 Hz) were observed, while six characteristic doublets (δ_H 0.93, 0.86 (×2), 0.84, 0.83, and 0.81; each J = 6.6 Hz) were unambiguously assigned to CH$_3$ groups of Leu^5, Leu^{12} and Leuol^{15}.

As depicted in Figure 4, ^{1}H-^{1}H correlations from TOCSY and COSY spectra allowed defining eight N-H doublet peaks as part of eight spin systems, of which five correspond to one Ala, two Leu, and two Gln residues. Another leucine-related coupling system showing additional hydroxymethylene signals at δ_H 3.30/3.18, (δ_C 63.5) and a hydroxyl resonance at δ_H 4.50 bound to its methine at δ_H 3.78 (δ_C 48.2), demonstrated the presence of the reduced C-terminal Leuol residue. An additional moiety was pointed out by ^{1}H-^{1}H TOCSY correlations between the N-H amide group at δ_H 8.38, the methine at δ_H 4.40 (δ_C 54.4), and the methylene at δ_H 3.12/2.98, (δ_C 26.8). ^{1}H-^{13}C HMBC correlations demonstrated that this methylene carbon is linked to protons of the indole ring. Additionally, the ROESY correlation between the above amide at δ_H 8.38 and the acetyl CH$_3$ group at δ_H 1.86 (δ_C 22.2), which showed HMBC correlation to a carbonyl carbon at δ_C 170.2, clearly resulted in the characterization of an acetylated N-terminal Trp residue. Finally, the δ-methylene of a glutamine-related spin system exhibited HMBC correlation to a carbonyl carbon at δ_C 173.0, which itself showed a strong HMBC correlation with a methoxy group at δ_H 3.55 (δ_C 50.8), supporting the presence of a glutamic acid δ-methyl ester residue.

Table 2. NMR data of ampullosporin F (1) (600/150 MHz, DMSO-d6, δ in ppm).

Pos.	δ_H, Mult. J (Hz)	δ_C/δ_N	Pos.	δ_H, Mult. J (Hz)	δ_C/δ_N	Pos.	δ_H, Mult. J (Hz)	δ_C/δ_N
Ac			β	1.77 [a]; 1.58 [a]	38.4	C=O		173.0
CH₃	1.86 s	22.2	γ	1.72 [a]	23.8	α	3.95 m	54.4
C=O		170.2	δ₁	0.84 d 6.6	20.8	β	2.07 [a]; 2.00 [a]	25.3
Trp¹			δ₂	0.93 d 6.6	22.0	γ	2.60 m; 2.45 m	29.6
NH	8.38 d 6.5	125.3	Aib⁶			C=O		172.5
C=O		173.0	NH	7.87 s	127.9	O-CH₃	3.55 s	50.8
α	4.40	54.4	C=O		175.5	Leu¹²		
β	3.12 dd 14.6/3.6	26.8	α	1.48	55.4	NH	7.42 d 7.3	115.4
	2.98 dd 14.6/9.4		β	1.36 s	22.6	C=O		172.8
1-NH	10.86 br s	131.2	γ	1.48 s	26.1	α	4.08 m	52.5
2	7.22 d 2.4	123.4	Gln⁷			β	1.67 [a]; 1.59 [a]	38.4
3		109.8	NH	7.41 d 6.2	112.9	γ	1.73 [a]	23.7
3a		127.1	C=O		173.1	δ₁	0.86 d 6.6	22.2
4	7.56 d 8.1	117.9	α	3.82 m	55.7	δ₂	0.83 d 6.6	20.6
5	6.96 t 7.5	117.7	β	2.00 [a]; 1.91 [a]	25.8	Aib¹³		
6	7.05 [a]	120.5	γ	2.22 [a]; 2.07 [a]	31.0	NH	7.45 s	126.5
7	7.33 d 8.1	111.0	C=O		172.9	C=O		173.4
7a		136.0	N-H₂	7.12 br s	107.7	α		56.0
Ala²				6.75 br s		β	1.38 s	23.7
NH	8.52 [a]	121.9	Aib⁸			γ	1.41 s	25.4
C=O		174.0	NH	7.99 s	128.5	Gln¹⁴		
α	4.04 [a]	50.1	C=O		175.0	NH	7.19 d 7.7	109.2
β	1.27 d 7.4	15.5	α		55.1	C=O		173.1
Aib³		132.1	β	1.31 s	22.0	α	3.98 m	52.9
NH	8.46 s	174.6	γ	1.43 s	25.3	β	2.05 [a]; 1.81 [a]	26.7
C=O		55.4	Aib⁹			γ	2.18 [a]; 2.08 [a]	31.2
α		25.3	NH	7.96 s	125.2	C=O		173.5
β	1.38 s	22.6	C=O		175.0	N-H₂	7.13 br s	107.9
γ	1.35 s	22.6	α		55.3		6.73 br s	
Aib⁴			β	1.39 s	26.0	Leuol¹⁵		
NH	8.06 s	124.3	γ	1.32 s	22.0	NH	7.06 [a]	118.8
C=O		176.3	Aib¹⁰			α	3.78 m	48.2
α		55.3	NH	7.57 s	125.1	β	1.35 [a]	39.4
β	1.38 s	26.2	C=O		175.8	γ	1.66 [a]	23.5
γ	1.36 s	22.0	α		55.4	δ₁	0.86 d 6.6	23.2
Leu⁵			β	1.39 s	22.3	δ₂	0.81 d 6.6	21.2
NH	7.71 d 4.1	114.8	γ	1.49 s	26.1	β'	3.30 [a]; 3.18 m	63.5
C=O		173.7	GluOMe¹¹			O-H	4.50 t 6.0	
α	3.90 m	54.1	NH	7.76 d 6.6	111.2			

Chemical shifts of quaternary carbons were determined from ¹H, ¹³C HMBC correlation peaks; [a] overlapping signals, chemical shifts were determined from ¹H, ¹⁵N or ¹H, ¹³C HSQC correlation peaks.

Seven N-H singlets (δ_H 8.46, 8.06, 7.99, 7.96, 7.87, 7.57, and 7.45) were assigned for seven Aib residues due to their HMBC correlations to seven characteristic quaternary C-α carbons (δ_C 55.1–56.0) and seven carbonyl carbons (δ_C 173.4–176.3), as well as their ROESY interactions with methyl broad singlet peaks (δ_H 1.30–1.50/δ_C 22.0–26.2).

Detailed analysis of HMBC interactions of N-H resonances with carbonyl and C-α signals coupled with ROESY correlations between neighboring proton signals afforded the sequence establishment as in accordance with mass spectrometry fragmentations. Therefore, the structure of 1 was established as Ac-Trp¹-Ala²- Aib³-Aib⁴-Leu⁵-Aib⁶-Gln⁷-Aib⁸-Aib⁹-Aib¹⁰-GluOMe¹¹-Leu¹²-Aib¹³-Gln¹⁴-Leuol¹⁵, and named as ampullosporin F (1) consistent with the ampullosporin series reported by Ritzau et al. in 1997 [21] and Kronen et al. in 2001 [22].

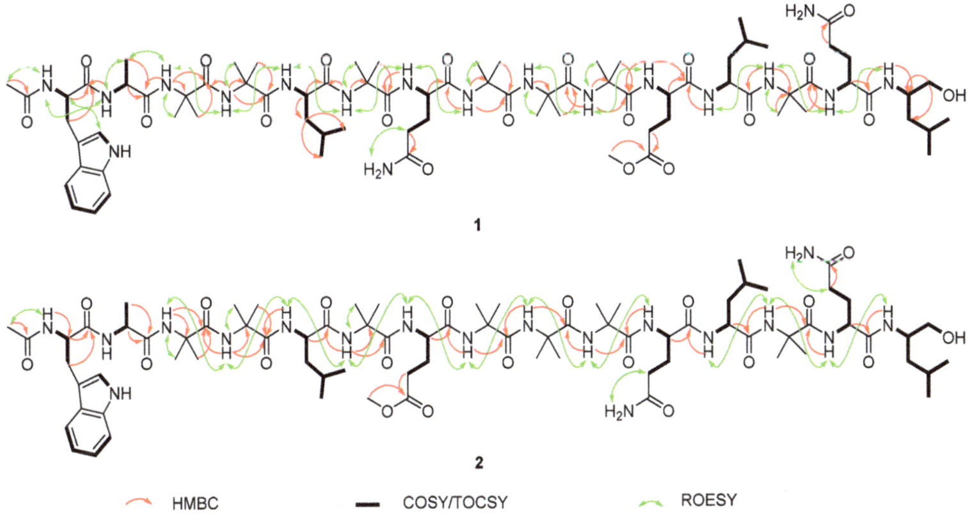

Figure 4. Key HMBC (H to C), COSY, and ROE correlations of ampullosporin F (**1**) and G (**2**).

Compound **2** was obtained as a white, amorphous solid. ESI-HRMS studies showed that **2** exhibited the same molecular formula ($C_{78}H_{128}N_{18}O_{20}$) as ampullosporin F (**1**). Positive and negative ion ESI-HRMS^n investigations (Table 1; Table S1, Supplementary Materials) revealed that **2** is an isomer of **1**. The only difference between **1** and **2** is the positional exchange of Gln^7/GluOMe^{11} in **1** by GluOMe^7/Gln^{11} in **2**. Likewise, NMR data of **2** (Table 3, Figure 4) are closely resemble that of **1**, including the strong correlation between the methoxy singlet at δ_H 3.54 (δ_C 51.12) and a carbonyl carbon at δ_C 172.20 confirming the glutamic acid methyl ester moiety. Thus, the structure of **2** was determined as Ac-Trp^1-Ala^2-Aib^3-Aib^4-Leu^5-Aib^6-GluOMe^7-Aib^8-Aib^9-Aib^{10}-Gln^{11}-Leu^{12}-Aib^{13}-Gln^{14}-Leuol^{15}, and trivially named ampullosporin G (**2**).

The δ-methyl ester of glutamic acid is rarely recognized in natural peptaibols, with only five examples out of the over 1450 peptaibiotics reported in the literature so far [5,8–11,14,38–50]. The first occurrence was reported for four peptaibols named Trichorzianines (TA) 1896, TA1924, TA1910, and TA1924a isolated from *Trichoderma atroviride* by Panizel et al. in 2013 [13]. The second example was described for Glu(OMe)^{18}-alamethicin F50 isolated from *Trichoderma arundinaceum* by Rivera-Chávez et al. in 2017 [8]. Consequently, the isolation and characterization of ampullosporin F (**1**) and G (**2**) represents the third report of glutamic acid methyl ester containing peptaibols isolated from natural sources, and the first occurrence in a *Sepedonium* species.

Compound **3**, isolated as a white, amorphous solid, possesses the molecular formula $C_{77}H_{127}N_{19}O_{19}$ as deduced from ESI-HRMS studies of the [M + H]$^+$ ion at m/z 1622.9628. The structure of **3** was identified as ampullosporin A (**3**, Ac-Trp^1-Ala^2-Aib^3-Aib^4-Leu^5-Aib^6-Gln^7-Aib^8-Aib^9-Aib^{10}-Gln^{11}-Leu^{12}-Aib^{13}-Gln^{14}-Leuol^{15}) based on MS studies in positive mode as well as 1D NMR experiments (Figures S29–S30, Table S2, Supplementary Materials).

Table 3. NMR data of ampullosporin G (**2**) (600/150 MHz, DMSO-d6, δ in ppm).

Pos.	δ_H, Mult. J (Hz)	δ_C/δ_N	Pos.	δ_H, Mult. J (Hz)	δ_C/δ_N	Pos.	δ_H, Mult. J (Hz)	δ_C/δ_N
Ac CH$_3$	1.85 br s	22.3	β	1.55 a; 1.76 a	38.7	α	3.93 m	55.0
C=O		170.4	γ	1.70 a	24.3	β	1.94 a; 2.00 a	26.3
Trp^1 NH	8.30 a	125.2	δ$_1$	0.833 d 6.6	21.1	γ	2.13 m; 2.34 m	31.5
C=O		173.0	δ$_2$	0.914 d 6.6	22.1	C=O		173.2
α	4.39 m	54.4	Aib^6			N-H$_2$	7.12 br s; 6.70 br s	107.6
β	3.11 dd 14.4/4.2; 2.98 dd 14.4/9.6	26.8	NH	7.85 s	127.8	Leu^{12}		
			C=O		176.0	NH	7.77 d 6.6	115.2
1-NH	10.84 br s	131.1	α		55.6	C=O		173.2
2	7.21 d 2.4	123.5	β	1.35 s	22.2	α	4.07 m	52.6
3		109.6	γ	1.45 s	26.2	β	1.58 a & 1.67 a	38.5
3a		126.9	GluOMe^7			γ	1.70 a	23.9
4	7.55 d 7.8	118.0	NH	7.47 d 6.0	111.8	δ$_1$	0.85 d 6.6	22.3
5	6.95 t-like 7.2	117.9	C=O		172.8	δ$_2$	0.82 d 6.6	20.7
6	7.06 t-like 7.8	120.7	α	3.87 m	55.2	Aib^{13}		
7	7.33 d 8.1	111.1	β	1.96 a; 2.03 a	25.4	NH	7.48 s	126.7
7a		135.9	γ	2.49 a; 2.39 m	29.7	C=O		173.7
Ala^2			C=O		172.2	α		56.0
NH	8.30 a	121.6	O-CH$_3$	3.54 s		β	1.37 s	23.8
C=O		174.0	Aib^8			γ	1.41 s	25.6
α	4.01 m	50.3	NH	7.97 s	128.7	Gln^{14}		
β	1.25 d 7.2	15.7	C=O		175.0	NH	7.22 d 8.4	109.3
Aib^3			α		55.4	C=O		170.6
NH	8.33 s	131.9	β	1.34 s	22.4	α	3.98 m	53.0
C=O		174.7	γ	1.42 s	25.4	β	1.80 a & 2.05 a	26.6
α		55.4	Aib^9			γ	2.09 a & 2.18 a	31.3
β	1.33 s	23.0	NH	7.87 s	125.0	C=O		173.8
γ	1.37 s	25.4	C=O		175.0	N-H$_2$	7.17 br s; 6.73 br s	108.2
Aib^4			α		55.4	Leuol^{15}		
NH	8.06 s	124.4	β	1.31 s	22.0	NH	7.07 d 8.4	119.1
C=O		176.4	γ	1.37 s	26.1	α	3.76 m	48.4
α		55.4	Aib^{10}			β	1.34 a	39.5
β	1.32 s	22.0	NH	7.56 s	125.1	γ	1.64 a	23.7
γ	1.37 s	26.5	C=O		175.9	δ$_1$	0.85 d 6.6	23.2
Leu^5			α		55.5	δ$_2$	0.81 d 6.6	21.3
NH	7.65 d 5.4	114.8	β	1.38 s	22.6	β'	3.18 m, 3.30 m	63.6
C=O		173.8	γ	1.48 s	26.2			
α	3.88 a	54.2	Gln^{11}			O-H	4.54 t-like	
			NH	7.74 d 5.4	111.9			
			C=O		173.5			

Chemical shifts of quaternary carbons were determined from ^{1}H, ^{13}C HMBC correlation peaks; [a] overlapping signals, chemical shifts were determined from ^{1}H, ^{15}N or ^{1}H, ^{13}C HSQC correlation peaks.

Compound **4** was obtained as a white, amorphous solid. On the basis of ESI-HRMS studies, the molecular formula $C_{31}H_{51}N_5O_6$ was determined from the [M + H]$^+$ ion's signal at m/z 590.3892. Mass fragmentation analyses coupled with 1D and 2D NMR investigations (Figures S31–S33, Table S2, Supplementary Materials) established **4** as peptaibolin (**4**, Ac-Leu^1-Aib^2-Leu^3-Aib^4-pheol^5), the shortest peptaibol with only five amino acid residues. Both compounds ampullosporin A (**3**) and peptaibolin (**4**) were previously isolated from cultures of *S. ampullosporum* HKI-0053 [21,35].

Compound **5** was obtained as a white, amorphous solid. ESI-HRMS analysis of the signal at m/z 510.3649 ([M + H]$^+$, calcd for $C_{26}H_{48}N_5O_5^+$ 510.3650) resulted in the molecular formula $C_{26}H_{47}N_5O_5$. Deduced from mass fragmentation analyses as well as 1D NMR investigations (Figure S34, Table S2, Supplementary Materials) compound **5** was identified as the cyclic pentapeptide chrysosporide (**5**).

Compound **6** was obtained as an amorphous solid and exhibited the molecular formula $C_{14}H_{15}N_3O_3$ based on ESI-HRMS analyses of the $[M - H]^-$ ion at m/z 272.1042 (Figure S35, Supplementary Materials). 1D NMR spectroscopic data (Figures S36–S37, Supplementary Materials) of **6** were in accordance with those of c(Trp-Ser) (**6**), which was first reported as a synthetic product in 2006 [51] and later isolated from several fungus like *Oidiodendron truncatum* GW3-13 [52], *Acrostalagmus luteoalbus* SCSIO F457 [53], and *Rheinheimera aquimaris* QSI02 [54].

Compound **7** was obtained as a white, amorphous solid. On the basis of ESI-HRMS studies, the molecular formula $C_{14}H_{15}N_3O_2$ was deduced from the $[M - H]^-$ ion at m/z 256.1089 (Figure S38, Supplementary Materials), indicating the absence of one oxygen atom in **7** compared to **6**. 1D NMR spectroscopic data (Figures S39 and S40, Supplementary Materials) of **7** were in agreement with those of c(Trp-Ala) (**7**), which was first synthesized in 1998 [55] and later recognized in different fungal sources such as *Eurotium* sp. [56] and *Eurotium chevalieri* MUT 2316 [57]. To the best of our knowledge, this is the first detection of c(Trp-Ser) (**6**) and c(Trp-Ala) (**7**) in a *Sepedonium* species.

2.2. In Situ Chemical Analysis

Because ampullosporin F (**1**) and G (**2**) were isolated as minor compounds with similar structures as ampullosporin A (**3**), the dominant component of *S. ampullosporum*, a question may arise with regard to the authenticity of these new compounds **1** and **2**. Are the isolated ampullosporins **1** and **2** biosynthesized by fungus itself or are they artefacts formed during extraction, fractionation and/or the purification processes using methanol?

In order to answer this question, a crude extract from cultivated *S. ampullosporum* was prepared using ethanol, instead of methanol, and investigated using LC-HRMS screening approach. From an enriched fraction of the ethanol crude extract of *S. ampullosporum*, the ampullosporin F (**1**) and G (**2**) as well as ampullosporin A (**3**) could be unambiguously detected by their characteristic mass (Figure S43, Supplementary Materials). Therefore, ampullosporin F (**1**) and G (**2**) are native constituents of *S. ampullosporum* Damon strain KSH 534.

2.3. Solid Phase Synthesis and Absolute Configuration of Ampullosporin F (**1**) and G (**2**)

The absolute configuration of **1** and **2** was established based on solid-phase peptide synthesis using Fmoc protected, L-configured amino acids (except the achiral Aib) and tetramethylfluoroform-amidinium hexafluorophosphate (TFFH) as a coupling reagent (Scheme 1).

The combined manual and automated synthesis was carried out on L-Leucinol 2-chlorotrityl polystyrene resin. The first four amino acids residues were incorporated by automated synthesis using the standard PyBOP (benzotriazol-1-yloxy-tripyrrolidinophosphonium hexafluorophosphate) strategy, while the rest of the synthesis was performed manually using TFFH or HATU (1-[bis(dimethylamino)methylene]-1H-1,2,3-triazolo[4,5-b]pyridinium 3-oxide hexafluorophosphate) activations. N-terminal acetylation was achieved after complete peptide assembly by treating the resin with Ac_2O, followed by solid-phase cleavage and global deprotection using TFA. The synthesized peptaibols **1** and **2** were purified by size exclusion column chromatography using Sephadex LH20 in combination with preparative HPLC. The ESI-HRMS^n, 1D NMR, and CD spectra of the synthetic peptaibols **1** and **2** were consistent with those of the natural ampullosporin F (**1**) and G (**2**) (Figures S17–S28, Supplementary Materials). Consequently, all chiral amino acids naturally present in **1** and **2** possess the L-configuration, which is typical for the particular class of ampullosporins [21,35].

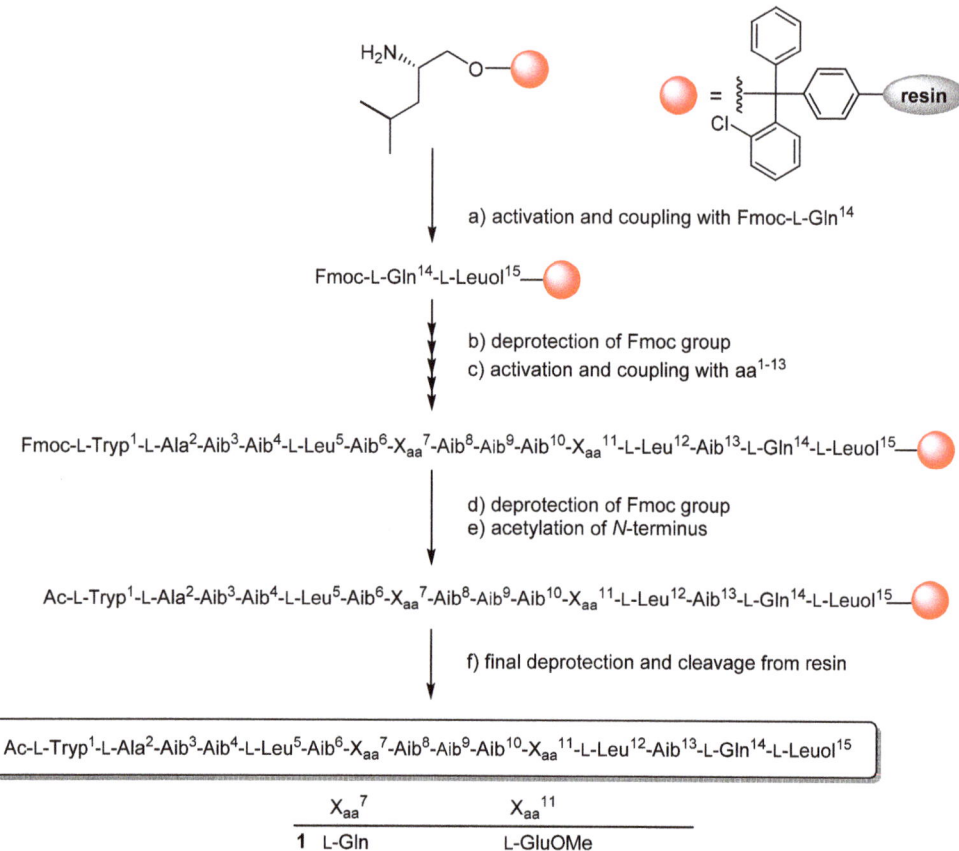

Scheme 1. Solid-phase peptide synthesis of ampullosporin F (**1**) and G (**2**).

2.4. Evaluation of Antifungal Activities

Isolated peptaibols **1–3** were examined for their antifungal activities against plant pathogenic ascomycetous fungi *Botrytis cinerea* (grey mold pathogen on many crops, e.g., strawberries and wine grapes) and *Septoria tritici* (causes septoria leaf blotch of wheat) as well as the oomycete *Phytophthora infestans* (causal agent of the late blight disease on potato and tomato) using a 96-well microtiter plate assay. The commercially available fungicides epoxiconazole and terbinafine were used as positive controls.

All three compounds **1–3** showed strong activity against *B. cinerea* and only slightly lower inhibitory effects against *P. infestans*, but were inactive against *S. tritici*. Interestingly, the growth-inhibitory effects against both *B. cinerea* and *P. infestans* increased with the presence of GluOMe in ampullosporin F (**1**) and G (**2**), compared to the ampullosporin A (**3**) (Table 4, Figure S41, Supplementary Materials).

Table 4. Biological activity of ampullosorin F (1), G (2) and A (3) (IC$_{50}$, µM).

Compound	Toxicity [a]		Antifungal Activity [b]		
	HT29	PC3	B. cinerea	S. tritici	P. infestans
1	5.90 ± 0.65	3.62 ± 0.49	7.27 ± 0.95	>125	14.75 ± 1.89
2	6.04 ± 0.64	3.35 ± 0.89	4.59 ± 0.34	>125	14.79 ± 2.84
3	10.45 ± 1.18	6.68 ± 1.67	11.41 ± 1.48	>125	19.49 ± 4.29
Epoxiconazole [c]			<1.5	<1.5	
Terbinafine [c]					48.30 ± 0.2

[a] Data represent biological quadruplicates (n = 4), each comprising technical triplicates. [b] The experiment was carried out in triplicates (n = 3), each comprising technical triplicates. [c] Used as positive control.

2.5. Evaluation of Anticancer Activities

The peptaibols 1–3 were tested for their effects on the viability of two different human cancer cell lines, namely prostate PC-3 adenocarcinoma cells and colorectal HT-29 adenocarcinoma cells. The cell viability and cytotoxicity assay was conducted by using resazurin and fluorometric read-out after 48 h cell treatment. The saponin digitonin (100 µM), a very potent permeabilizer of cell membranes, was used as positive control compromising the cells to yield 0% cell viability after 48 h. As negative control, representing 100% cell viability for data normalization, medium with 0.5% (v/v) DMSO supplementation (highest final DMSO concentration in test item samples) was measured. The peptaibols 1–3 were tested with concentrations in the range of 0.195–100 µM (factor 2 dilutions) in order to determine IC$_{50}$ values that have been calculated to be ~3–6 µM for both novel ampullosporins F (1) and G (2), with twofold higher activity in prostate PC-3 cancer cells compared to the colorectal cancer cells HT-29. Furthermore, in both cancer cell lines, 1 and 2 were found with twofold lower IC$_{50}$s compared to the already known ampullosporin A (3) (see Figure 5), indicating that the GluOMe modifications at the amino acid positions 7 and 11, respectively, enhance not only the antifungal but also the anticancer activity of the ampullosporins. Summarized results are shown in Table 4.

2.6. In Silico Molecular Docking

In 2009, Berek et al. investigated the neuroleptic-like activity of ampullosporin A (3) in mice. Thereby, they described a complete suppression of the effects of the N-methyl-D-aspartate (NMDA) receptor antagonist MK-801, and alteration of the activity of those glutamate receptors [23], indicating NMDA receptors as potential biological target molecules of ampullosporin A (3). Recently, Lu et al. (2017) investigated the interplay of MK-801 and NMDA receptor in more detail using cryo-EM structural analyses, explaining the allosteric antagonistic action of MK-801 (pdb-code 5UOW) [58]. Based on those results, we decided to proof the hypothesis of NMDA receptor binding of our ampullosporins (1–3) by using a chemoinformatic molecular docking approach based on two protein databank entries (5UOW and 6IRA). While 5UOW includes the MK-801 inhibitor but reflects the situation at nonhuman NMDA receptor proteins, 6IRA comprises the human GluN1/GluN2A ligand binding domain as the relevant part of the human NMDA receptor [59].

Based on the protein databank entry 5UOW, a putative binding site on the triheteromeric NMDA receptor GluN1/GluN2A/GluN2B was indicated by the cryo-EM-based localization of the inhibitor MK-801 in structure 5UOW (Figure 6). Moreover, our in silico docking approach based on 5UOW highlighted a very good docking of 3 in very close proximity to MK-801, as also shown in Figure 6. The indicated binding site is located in a hydrophobic cleft that is formed by several helices of both NMDA receptor subunits GluN1 and GluN2A. This theoretical co-localization of ampullosporin A (3) and the allosteric NMDA receptor antagonist MK-801 could explain the complete suppression of MK-801 effects caused by the peptaibol (3).

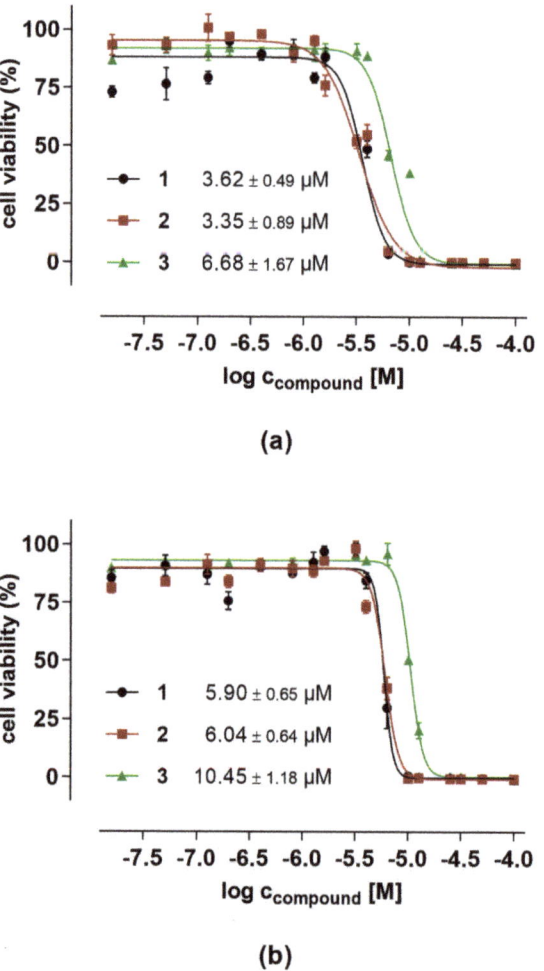

Figure 5. Cell viability of (**a**): prostate cancer PC-3 cells, and (**b**): colon cancer HT-29 cells treated for 48 h with the peptaibols **1** (●), **2** (■), and **3** (▲), respectively, as determined by resazurin-based fluorimetric assay. Data represent biological quadruplicates ($n = 4$), each comprising technical triplicates.

In addition, compounds **1–3** were tested for their antiproliferative effects in PC-3 prostate and HT-29 colorectal human cancer cells, which, to some extent, express the NMDA receptor proteins GluN1 and GluN2A (according to mRNA expression data; analyzed by using the Genevestigator software, data base HS_mRNASeq_HUMAN_GL-1; data not shown). Therefore, we also investigated the ampullosporins' binding in silico based on protein databank entry 6IRA representing the human GluN1/GluN2A NMDA receptor complex that, however, do not include MK-801. Indeed, also in our 6IRA-based in silico model the peptaibols **1–3** were docked with best scores at the same position in the hydrophobic cleft between GluN1 and GluN2A (Figure 7).

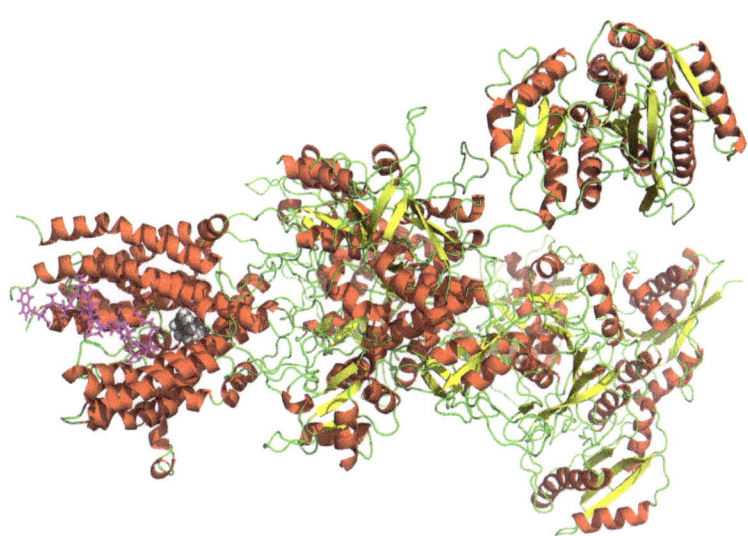

Figure 6. The complete structure of the nonhuman triheteromeric NMDA receptor GluN1/GluN2A/GluN2B (based on protein databank entry 5UOW) with docked ampullosporin A (3) (magenta atoms in pink, left site). The position of a proposed allosteric binding site is given by the cryo-EM structure-based position of the inhibitor MK-801 (drawn in grey) in structure of 5UOW.

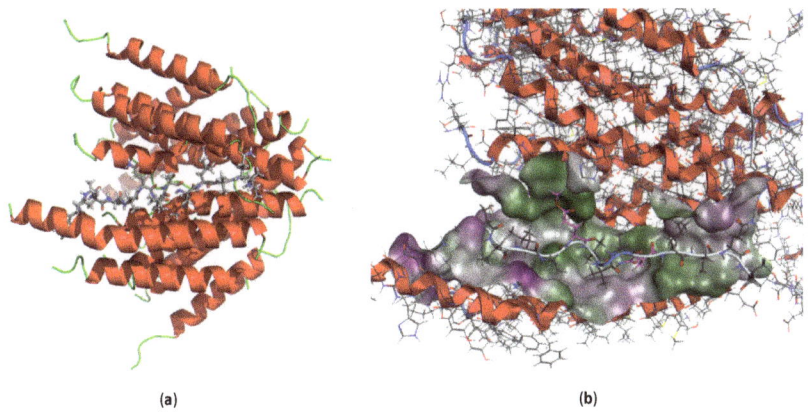

Figure 7. (a) Ampullosporins (1–3) dock with best docking scores into cleft formed by several helices of human NMDA receptor subunits GluN1 and GluN2A. (b) The hydrophobic potential blot—hydrophobic (in green)/hydrophilic (in lilac)—indicates the primarily hydrophobic character of the binding site.

Our in silico results, coupled with the observations by Berek et al. [23], led to the hypothesis that this hydrophobic cleft at the interphase between GluN1 and GluN2A could be the binding site for the very hydrophobic peptaibols **1–3**, causing their biological effects. To shed more light on the molecular ampullosporins binding mode and especially to investigate consequences of the sequence modifications in the novel ampullosporin F (**1**) and G (**2**), further docking studies were performed to validate this hydrophobic cleft as a potential binding site of compounds **1–3**.

As illustrated in Figure 8 and Figure S42 (Supplementary Materials), all compounds **1–3** fit nicely into the described hydrophobic cleft and seem to be stabilized by numerous hydrophobic interactions to hydrophobic receptor side chains. Accordingly, for ampullosporin A (**3**) an interaction energy of −92.0 kcal/mol was calculated. However, replacement of each of Gln residues (Gln[7] and Gln[11]) in ampullosporin A (**3**) by GluOMe resulted in additional hydrophobic interactions of ampullosporin F (**1**; GluOMe[11]) and ampullosporin G (**2**; GluOMe[7]) with hydrophobic receptor residues. In case of **1**, enhanced hydrophobic interactions to the receptor residues V820, I824 and F637 elevate the calculated interaction energy to be −94.5 kcal/mol. In case of **2**, the molecular docking indicates additional hydrophobic interactions of its GluOMe[7] modification with the hydrophobic receptor residues W608 and W611. The calculated interaction energy (−92.9 kcal/mol) for **2** is slightly enhanced compared to **3**. Based on these calculated interaction energies, the binding of both ampullosporin F (**1**) and G (**2**) should outperform the binding of ampullosporin A (**3**).

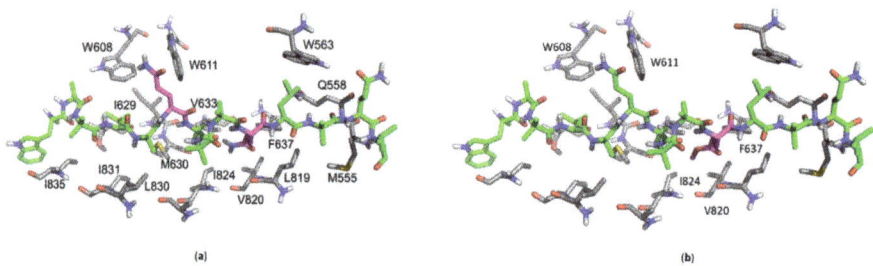

Figure 8. Docking arrangements of (**a**) ampullosporin A (**3**), and (**b**) ampullosporin F (**1**) in the proposed binding site in the hydrophobic cleft between NMDA receptor subunits GluN1 and GluN2A, as modeled based on protein databank entry 6IRA.

Interestingly, the determined antiproliferative activities of compounds **1**–**3** (Figure 5) correlate very well with the proposed enhanced binding of both novel ampullosporins (**1** and **2**). The antiproliferative IC$_{50}$ values in the anticancer assay were detected to be by twofold better for ampullosporin F (**1**) and G (**2**) than for ampullosporin A (**3**).

Whether or not the NMDA receptor is really the relevant molecular target explaining the observed antiproliferative effects of the ampullosporins on human cancer cells, further investigations are necessary. However, considering the well-published neuroleptic-like activities of ampullosporin A (**3**) [21,23,60] that can be easily attributed to NMDA receptor pathways, it would be of high interest to test our novel ampullosporins F (**1**) and G (**2**), with presumed improved NMDA receptor binding for their neuroleptic activity, e.g., in vivo in mice.

3. Materials and Methods

3.1. General Experimental Procedures

Column chromatography was carried out on Sephadex LH 20 (Fluka, Steinheim, Germany), while analytical TLC was performed on precoated silica gel F254 aluminum sheets (Merck, Darmstadt, Germany). Peptaibols were detected on TLC plates using ninhydrin reagent as described in the literature [17]. Diaion HP 20 was purchased from Supelco (Bellefonte, PA, USA). UV spectra were recorded on a Jasco V-770 UV-Vis/NIR spectrophotometer (Jasco, Pfungstadt, Germany), meanwhile CD spectra were obtained from a Jasco J-815 CD spectropolarimeter (Jasco, Pfungstadt, Germany). The specific rotation was measured with a Jasco P-2000 digital polarimeter (Jasco, Pfungstadt, Germany).

NMR spectra were obtained from an Agilent DD2-400 and an Agilent VNMRS 600 system (Varian, Palo Alto, CA, USA) using a 5-mm inverse detection cryoprobe. Compounds were dissolved in DMSO-d_6 (99.96% D for **1** and **2**, 99.80% D for **3**–**7**). The spectra were recorded at 399.82/599.83 MHz (^{1}H) and 100.54/150.84 (^{13}C), respectively. 2D NMR spectra were recorded using standard CHEMPACK 8.1 pulse sequences (s2pul, ^{1}H,^{1}H gDQCOSY,

^{1}H,^{1}H zTOCSY, ^{1}H,^{1}H ROESYAD, ^{1}H,^{13}C and ^{1}H,^{15}N gHSQCAD, ^{1}H,^{13}C gHMBCAD) implemented in Varian VNMRJ 4.2 spectrometer software (Varian, Palo Alto, CA, USA). The mixing time for the TOCSY experiments was set to 80 ms, for the ROESY experiments was set to 300 ms. The HMBC experiment was optimized for a long-range coupling of 8 Hz. ^{1}H and ^{13}C chemical shifts were referenced to internal DMSO-d_6 (δ_H 2.51 ppm and δ_C 39.5 ppm), whereas ^{15}N chemical shifts were given relative to liquid NH$_3$ (δ_N 0 ppm).

The solid phase peptide synthesis was partly carried out on a ResPep SL peptide synthesizer (Intavis Bioanalytical Instruments, Köln, Germany). The L-configured Fmoc-amino acids Fmoc-Trp(Boc)-OH, Fmoc-Ala-OH, Fmoc-Leu-OH, Fmoc-Gln(Trt)-OH and Fmoc-Glu(OMe)-OH were purchased from Carbolution Chemicals (Ingbert, Germany), while L-Leucinol 2-chlorotrityl polystyrene resin was obtained from Iris Biotech GmbH (Marktredwitz, Germany). TFFH was supplied Carbolution Chemicals (Ingbert, Germany). Piperidine, Ac$_2$O, DIPEA, and DMF were purchased from Sigma-Aldrich (Steinheim, Germany).

The high-resolution mass spectra in positive and negative modes were obtained from an Orbitrap Elite mass spectrometer (Thermofisher Scientific, Bremen, Germany) equipped with an ESI electrospray ion source (spray voltage 4.0 kV; capillary temperature 275 °C, source heater temperature 40 °C; FTMS resolution 60.000). Nitrogen was used as sheath gas. The sample solutions were introduced continuously via a 500 µL Hamilton syringe pump with a flow rate of 5 µL/min. The instrument was externally calibrated by the Pierce® LTQ Velos ESI positive ion calibration solution (product number 88323) and Pierce® ESI negative ion calibration solution (product number 88324) from Thermofisher Scientific (Rockford, IL, USA). The data were evaluated by the Xcalibur software 2.7 SP1 Thermofisher Scientific, Waltham, MA, USA). The collision induced dissociation (CID) MS^n measurements were performed using the relative collision energies given in Table S1 (Supplementary Materials).

The preparative HPLC was performed on a Shimadzu prominence system (Kyoto, Japan) which consists of a CBM-20A communications bus module, a SPD-M20A diode array detector, a FRC-10A fraction collector, a DGU-20A5R degassing unit, a LC-20AT liquid chromatograph, and a SIL-20A HT auto sampler, using either column 1 (ODS-A, 5 µm, 120 Å, 150 × 20 mm I.D; YMC, Devens, MA, USA) or column 2 (ODS-A, 5 µm, 120 Å, 150 × 10 mm I.D; YMC, Devens, MA, USA). The mobile phases were H$_2$O (A) and CH$_3$CN (B), with 0.1% formic acid contained in both solvents, using a gradient system.

3.2. Fungal Strain and Cultivation

The fungal strain *Sepedonium ampullosporum* Damon KSH 534 was isolated in August 1999, from *Boletus calopus* in Crista Acri near Cosenza, Italy (leg./det. C. Lavorato). A voucher specimen is deposited at the herbarium of the University Regensburg. The fungal culture of *S. ampullosporum* strain KSH 534 was stored on malt peptone agar (MPA) plates and transferred periodically. The upscaled semi-solid cultures, used for isolation, were grown in 31 Erlenmeyer flasks (size 1 L) each containing 1.5 g of cotton wool and 250 mL of malt peptone medium (2.5 g malt and 0.625 g peptone in 250 mL deionized water), resulting in a total volume of 7.75 L. Each culture flask was inoculated with a 10 × 10 mm agar plug of colonized fungus and incubated for 14 days at room temperature without agitation.

3.3. Extraction and Isolation

The mycelia were separated from the culture broth by vacuum filtration, frozen with liquid nitrogen and subsequently extracted with EtOAc (2 × 2 L) and MeOH (2 × 2 L) to yield two crude extracts (EtOAc, 2.20 g and MeOH, 8.98 g, respectively). Meanwhile, activated Diaion HP 20 (50 g) was added to the culture broth and agitated for 12 h at room temperature. Diaion HP 20 was then removed by vacuum filtration, washed with H$_2$O, eluted with MeOH to give a yellow solution, which was then evaporated *in vacuo* to dryness. This dryness (3.44 g) from culture broth was combined with the EtOAc extract from mycelia according to their LC-MS profiles. The resulting residue (5.64 g) was chromatographed on a Sephadex LH 20 column, using aq. MeOH 60% as eluent to afford 140 fractions

(8 mL each). Based on ESI-MS spectra, fractions 26–38 were combined (1.91 g), which were first separated by size exclusion column chromatography (using Sephadex LH 20, eluent: MeOH), then subjected to preparative HPLC using column 1 at a flow rate of 4.5 mL/min (0–30 min, 30–80% B; 31–50 min, 80–100% B) to afford 3 (ampullosporin A, t_R = 34.0 min, 100.9 mg), 1 (ampullosporin F, t_R = 37.5 min, 4.7 mg), and 2 (ampullosporin G, t_R = 40.5 min, 3.8 mg). In addition, the combination of fractions 39–48 (311.7 mg) was purified by semipreparative HPLC using column 2 at a flow rate of 2.2 mL/min (0–12 min, 30–100% B; 12–17 min, 100% B) to afford 4 (peptaibolin, t_R = 12.7 min, 1.4 mg) and 5 (chrysosporide, t_R = 14.5 min, 0.3 mg). Similarly, fractions 61–74 (85.4 mg) were combined and finally purified by semipreparative HPLC using column 2 at a flow rate of 2.2 mL/min (0–20 min, 10–100% B) to afford 6 (c(Trp-Ser), t_R = 10.6 min, 8.1 mg) and 7 (c(Trp-Ala), t_R = 11.2 min, 7.3 mg).

3.4. Sample Preparation for LC-MS Screening

For the preparation of the enriched fraction for the LC-HRMS screening, one stored deep frozen agar plate cultures of *S. ampullosporum* Damon strain KSH534 was crushed in small pieces and extracted with EtOH 96% (2 × 250 mL) in an ultrasonic bath at room temperature. The resulted yellow solution was evaporated *in vacuo* to dryness. The dried crude extract was redissolved in EtOH 96%/H$_2$O (1:2, v/v) to a concentration of 50 mg/mL. The resulting solution was separated on SPE cartridges Chromabond® C18 (loading 200 mg/3 mL, particle size 45 µm; Macherey-Nagel, Düren, Germany), targeted peptaibols 1–3 were eluted with EtOH 96%. After evaporation to dryness *in vacuo*, the enriched fraction was redissolved in CH$_3$CN and submitted to LC-HRMS.

3.5. Solid-Phase Peptide Synthesis

Compounds 1 and 2 were synthesized by combining manual and automated synthesis. The protocol was based on a Fmoc/t-butyl strategy in a 0.1 mmol scale starting from L-Leucinol 2-chlorotrityl resin (200–400 mesh, loading 0.67 mmol/g resin). The first four amino acid residues were incorporated by automated synthesis using the standard method based in PyBOP/NMM activation. The rest of the synthesis was performed manually using 4 equiv. of the amino acids and DMF as solvent. For all the Aib residues, the coupling cycle protocol was based on activation with TFFH (4 equiv.) and NMM (*N*-methylmorpholine, 4 equiv.) for 12 min and coupling time of 120 min. The rest of the amino acids were coupled using activation with HATU (4 equiv.) and NMM (4 equiv.) for 5 min and a coupling time of 120 min. Fmoc removals were carried out using a solution of 20% piperidine in DMF for two cycles of 10 min. *N*-terminal acetylation was performed after complete peptide assembly by treating the resin with Ac$_2$O (10 equiv) and DIPEA (10 equiv) in DMF for 30 min. Solid-phase cleavage and global deprotection was achieved by treating the resin with 5 mL of TFA/H$_2$O/TIPS (95:2.5:2.5, $v/v/v$) for 120 min. The cleavage mixture was concentrated under reduced pressure, suspended in a mixture of CH$_3$CN 50% in water and lyophilized.

For the peptide 1, the crude peptide mixture (183.8 mg) obtained after lyophilizing was redissolved in MeOH, and subjected to column chromatography (360 × 30 mm) using Sephadex LH20, eluting with MeOH to afford 30 fractions (8mL each). Fractions 6 and 7 were combined (78.6 mg) and purified by preparative HPLC using column 1 (0–2 min, 30–70% B; 3–15 min, 70–100% B; 10 mL/min) to give 1 in 7.0% total yield (t_R = 15.6 min, 11.5 mg).

For the peptide 2, the crude peptide mixture (201.7 mg) obtained after lyophilizing was redissolved in MeOH, and separated by column chromatography (360 × 30 mm) on Sephadex LH20, using MeOH as eluent to give 30 fractions (8mL each). Fractions 1–5 were combined (95.6 mg) and purified by preparative HPLC using column 1 (0–2 min, 30–80% B; 3–15 min, 80–100% B; 10 mL/min) to give 2 in 7.7% total yield (t_R = 16.2 min, 12.6 mg).

Natural ampullosporin F (1): white, amorphous solid; TLC R$_f$ 0.34 (*n*-BuOH/AcOH/H$_2$O 4:1:1); $[\alpha]_D^{23}$ -15.6 (*c* 0.100, MeOH); CD (MeOH) $[\theta]_{208}$-120119, $[\theta]_{224}$-103881 deg cm^2

× dmol^{-1}; UV (MeOH) λmax (log ε) 290 (23.56) nm; ^{1}H NMR and ^{13}C NMR see Table 2; ESI-HRMS m/z 819.4849 ([M + 2H]$^{2+}$, calcd for C$_{78}$H$_{130}$N$_{18}$O$_{20}$$^{2+}$ 819.4849); ESI-HRMS^n see Table 1, and Table S1 (Supplementary Materials).

Synthetic ampullosporin F (**1**): white, amorphous solid; CD (MeOH) [θ]$_{208}$-232819, [θ]$_{224}$-208346 deg cm^2 × dmol^{-1}; ^{1}H NMR and ^{13}C NMR in accordance with data of natural **1**; ESI-HRMS m/z 819.4852 ([M + 2H]$^{2+}$, calcd for C$_{78}$H$_{130}$N$_{18}$O$_{20}$$^{2+}$ 819.4849); ESI-HRMS^n in agreement with natural **1**.

Natural ampullosporin G (**2**): white, amorphous solid; TLC R$_f$ 0.44 (n-BuOH/AcOH/H$_2$O 4:1:1); [α]$_D^{23}$-17.1 (c 0.080, MeOH); CD (MeOH) [θ]$_{208}$-150953, [θ]$_{223}$-125977 deg cm^2 × dmol^{-1}; UV (MeOH) λmax (log ε) 281 (1.88), 290 (1.81) nm; ^{1}H NMR and ^{13}C NMR see Table 3; ESI-HRMS m/z 819.4860 ([M + 2H]$^{2+}$, calcd for C$_{78}$H$_{130}$N$_{18}$O$_{20}$$^{2+}$ 819.4849); ESI-HRMS^n see Table 1, and Table S1 (Supplementary Materials).

Synthetic ampullosporin G (**2**): white, amorphous solid; CD (MeOH) [θ]$_{207}$-246393, [θ]$_{222}$-205061 deg cm^2 × dmol^{-1}; ^{1}H NMR and ^{13}C NMR in accordance with data of natural **2**; ESI-HRMS m/z 819.4852 ([M + 2H]$^{2+}$, calcd for C$_{78}$H$_{130}$N$_{18}$O$_{20}$$^{2+}$ 819.4849); ESI-HRMS^n in agreement with natural **2**.

Natural ampullosporin A (**3**): white, amorphous solid; TLC R$_f$ 0.26 (n-BuOH/AcOH/H$_2$O 4:1:1); ^{1}H NMR and ^{13}C in agreement with data of Ritzau et al. (1997) [21]; ESI-HRMS m/z 811.9851 ([M + 2H]$^{2+}$, calcd for C$_{77}$H$_{129}$N$_{19}$O$_{19}$$^{2+}$ 811.9851); ESI-HRMS^n see Table S2, Supplementary Materials.

Natural peptaibolin (**4**): white, amorphous solid; TLC R$_f$ 0.88 (n-BuOH/AcOH/H$_2$O 4:1:1); ^{1}H NMR and ^{13}C in agreement with data of Hulsmann et al. (1998) [35]; ESI-HRMS m/z 590.3892 ([M + H]$^+$, calcd for C$_{31}$H$_{52}$N$_5$O$_6$$^+$ 590.3912); ESI-HRMS^n see Table S2, Supplementary Materials.

Natural chrysosporide (**5**): white, amorphous solid; TLC R$_f$ 0.68 (n-BuOH/AcOH/H$_2$O 4:1:1); ^{1}H NMR in agreement with data of Mitova et al. (2006) [32]; ESI-HRMS m/z 510.3649 ([M + H]$^+$, calcd for C$_{26}$H$_{48}$N$_5$O$_5$$^+$ 510.3650); ESI-HRMS^n see Table S2, Supplementary Materials.

Natural c(Trp-Ser) (**6**): white, amorphous solid; TLC R$_f$ 0.66 (n-BuOH/AcOH/H$_2$O 4:1:1); ^{1}H NMR and ^{13}C in agreement with data of Tullberg et al. (2006) [51]; ESI-HRMS m/z 272.1042 ([M − H]$^-$, calcd for C$_{14}$H$_{14}$N$_3$O$_3$$^-$ 272.1041).

Natural c(Trp-Ala) (**7**): white, amorphous solid; TLC R$_f$ 0.76 (n-BuOH/AcOH/H$_2$O 4:1:1); ^{1}H NMR and ^{13}C in agreement with data of Zhao et al. (2018) [56]; ESI-HRMS m/z 256.1089 ([M − H]$^-$, calcd for C$_{14}$H$_{14}$N$_3$O$_2$$^-$ 256.1092).

3.6. Antifungal Assay

Compounds **1–3** were tested in a 96-well microtiter plate assay against *Botrytis cinerea* Pers., *Septoria tritici* Desm., and *Phytophthora infestans* (Mont.) de Bary as described earlier [19]. The experiment was carried out in triplicates (n = 3), each comprising technical triplicates. For data analyses, GraphPad Prism version 8.0.2 (GraphPad Software, San Diego, CA, USA), SigmaPlot 14.0 (Systat Software, San Jose, CA, USA) and Microsoft Excel 2013 (Microsoft, Redmond, WA, USA) were used.

3.7. In Vitro Cell Proliferation Assay—Anticancer Activity

The prostate cancer cell line PC-3 (ATCC, Manassas, VA, USA) and the colon cancer cell line HT-29 (ATCC, Manassas, VA, USA) were cultured in RPMI 1640 medium supplemented with 2 mM L-glutamine and 10% heat-inactivated FCS. The cells were routinely grown in a humidified atmosphere with 5% CO$_2$ at 37 °C to reach subconfluency (~70–80%) prior to subsequent usage or subculturing. The adherent cells were rinsed with PBS and detached by using trypsin/EDTA (0.05% in PBS) prior to cell passaging and seeding. RPMI 1640 basal medium, FCS, L-glutamine, PBS and trypsin/EDTA for cell culturing were purchased from Capricorn Scientific GmbH (Ebsdorfergrund, Germany). The culture flasks, multi-well plates and further cell culture plastics were purchased from TPP (Trasadingen, Switzerland) and Greiner Bio-One GmbH (Frickenhausen, Germany), respectively.

Antiproliferative and cytotoxic effects, respectively, of the compounds **1**–**3** were investigated by performing a fluorimetric resazurin-based cell viability assay (Sigma-Aldrich, Taufkirchen, Germany). For that purpose, prostate PC-3 and colorectal HT-29 cancer cells were seeded in low densities into 96-well plates (3000–6000 cells per well; seeding confluency ~10%) and were allowed to adhere for 24 h. Subsequently, the cells were treated for 48 h with compound concentrations up to 100 µM. For control measurements, cells were treated in parallel with 0.5% DMSO (negative control, representing the final DMSO content of the highest concentrated test compound concentration) and 100 µM digitonin (positive control, for data normalization set to 0% cell viability), both in standard growth medium. As soon as the 48-h incubation was finished, the incubation medium was discarded, and cell were rinsed once with PBS. Resazurin solution in RPMI 1640 without phenol red and other supplements was prepared freshly prior to use, and added to the cells in a final resazurin concentration of 50 µM. Subsequently, the cells were incubated under standard growth conditions for further 2 h. Finally, the conversion of resazurin to resorufin by viable, metabolically active cells was measured with 540 nm excitation and 590 nm emission settings by using a SpectraMax M5 multiwell plate reader (Molecular Devices, San Jose, CA, USA). Data were determined in biological quatruplicates, each with technical triplicates. For data analyses GraphPad Prism version 8.0.2 (GraphPad Software, San Diego, CA, USA) and Microsoft Excel 2013 (Microsoft, Redmond, WA, USA) were used.

3.8. Computational Details

The X-ray structures of the human GluN1/GluN2A NMDA receptor in the glutamate/glycine-bound state at pH 7.8 (pdb-code 6IRA) [59] and nonhuman (frog and mouse) triheteromeric NMDA receptor GluN1/GluN2A/GluN2B in complex with glycine, glutamate, the uncompetitive NMDA receptor antagonist MK-801 and a GluN2B-specific Fab, at pH 6.5 (pdb-code 5UOW) [58] were downloaded from the protein databank [61] and used for in silico docking analyses of compounds **1**–**3**.

All theoretical investigations were performed using the molecular modeling software package MOE (molecular operating environment) [62]. The X-ray structure was prepared for docking studies by adding the missing hydrogen atoms using the 3D-protonate module implemented in MOE.

The putative active site of the enzyme is indicated by the co-crystallized inhibitor MK-801 (see Figure 6, left site). By closer inspection of this site, an almost complete hydrophobic cleft formed by several helices could be detected, which led to the hypothesis that this could be a perfect binding site for the very hydrophobic peptaibols due to multiple Aib amino acid containing residues. Therefore, docking studies were performed defining this as a binding site. For each of the three peptaibols, ampullosporin A (**3**), F (**1**), and G (**2**), 30 poses were generated using the triangle matcher for fast placement, the London dG as fitness function with subsequent induced fit relaxation of the binding site. The best scored docking poses are displayed and discussed.

4. Conclusions

In conclusion, the present study represents the chemical investigation of semi-solid culture of *Sepedonium ampullosporum* Damon strain KSH 534. There are seven constituents including two new 15-residue linear peptaibols, named ampullosporin F (**1**) and G (**2**), as well as two known linear peptaibols, ampullosporin A (**3**) and peptaibolin (**4**), together with three previously described cyclic peptides, chrysosporide (**5**), c(Trp-Ser) (**6**), and c(Trp-Ala) (**7**). The authenticity of peptaibols **1** and **2** was approved by using LC-HRMS approach. Additionally, the total synthesis of **1** and **2** was performed on solid-phase synthesis establishing the L-configuration of all chiral amino acids. Furthermore, peptaibols **1**–**3** showed significant anti-phytopathogenic activity against *B. cinerea* and *P. infestans*, but no activity against *S. tritici*. Moreover, compounds **1**–**3** exhibited strong anticancer activities against human prostate (PC-3) and colorectal (HT-29) cancer cells. Interestingly, for both antifungal and anticancer assays, the activities of ampullosporin F (**1**) and G

(2) were found to be twofold higher than the structurally similar ampullosporin A (3), demonstrating the effect of a GluOMe moiety on biological activity. Our molecular docking data on NMDA receptors suggests the better hydrophobic interaction of **1** and **2** than **3** with the hydrophobic cleft of the receptor, which might lead to the higher inhibitory effects of the new compounds **1** and **2** compared to **3**.

Supplementary Materials: Supplementary materials are available online at https://www.mdpi.com/article/10.3390/ijms222312718/s1.

Author Contributions: Conceptualization, N.A.; methodology, Y.T.H.L., M.G.R. and N.A.; software, R.R., A.F., A.P., P.S. and W.B.; formal analysis, Y.T.H.L., A.P. and R.R.; investigation, Y.T.H.L.; resources, N.A.; data curation, Y.T.H.L. and N.A.; writing—original draft preparation, Y.T.H.L. and R.R.; writing—review and editing, Y.T.H.L., R.R. and N.A.; supervision, B.W. and N.A.; project administration, N.A.; funding acquisition, B.W. and N.A. All authors have read and agreed to the published version of the manuscript.

Funding: The Ph.D. work of Yen T.H. Lam was funded by the Vietnamese Ministry of Education and Training (MOET), scholarship 911.

Institutional Review Board Statement: Not applicable.

Informed Consent Statement: Not applicable.

Data Availability Statement: The data presented in this study are available on request from the corresponding author.

Acknowledgments: The authors are indebted to G. Hahn for optical measurements, M. Brode for antifungal assays, M. Lerbs for a part of anticancer assays, and M. Saoud for detailed advice about data analysis using GraphPad.

Conflicts of Interest: The authors declare no conflict of interest.

References

1. Degenkolb, T.; Brückner, H. Peptaibiomics: Towards a Myriad of Bioactive Peptides Containing C^{α}-Dialkylamino Acids? *Chem. Biodivers.* **2008**, *5*, 1817–1843. [CrossRef] [PubMed]
2. Ayers, S.; Ehrmann, B.M.; Adcock, A.F.; Kroll, D.J.; Carcache de Blanco, E.J.; Shen, Q.; Swanson, S.M.; Falkinham, J.O., III; Wani, M.C.; Mitchell, S.M.; et al. Peptaibols from two unidentified fungi of the order Hypocreales with cytotoxic, antibiotic, and anthelmintic activities. *J. Pept. Sci.* **2012**, *18*, 500–510. [CrossRef]
3. Carroux, A.; Van Bohemen, A.I.; Roullier, C.; Robiou du Pont, T.; Vansteelandt, M.; Bondon, A.; Zalouk-Vergnoux, A.; Pouchus, Y.F.; Ruiz, N. Unprecedented 17-residue peptaibiotics produced by marine-derived *Trichoderma atroviride*. *Chem. Biodivers.* **2013**, *10*, 772–786. [CrossRef]
4. Liu, D.; Lin, H.; Proksch, P.; Tang, X.; Shao, Z.; Lin, W. Microbacterins A and B, new peptaibols from the deep sea actinomycete *Microbacterium sediminis* sp. nov. YLB-01(T). *Org. Lett.* **2015**, *17*, 1220–1223. [CrossRef] [PubMed]
5. Neumann, N.K.; Stoppacher, N.; Zeilinger, S.; Degenkolb, T.; Bruckner, H.; Schuhmacher, R. The peptaibiotics database—A comprehensive online resource. *Chem. Biodivers.* **2015**, *12*, 743–751. [CrossRef]
6. Mohamed-Benkada, M.; Francois Pouchus, Y.; Verite, P.; Pagniez, F.; Caroff, N.; Ruiz, N. Identification and biological activities of long-chain peptaibols produced by a marine-derived strain of *Trichoderma longibrachiatum*. *Chem. Biodivers.* **2016**, *13*, 521–530. [CrossRef] [PubMed]
7. Du, L.; Risinger, A.L.; Mitchell, C.A.; You, J.; Stamps, B.W.; Pan, N.; King, J.B.; Bopassa, J.C.; Judge, S.I.V.; Yang, Z.; et al. Unique amalgamation of primary and secondary structural elements transform peptaibols into potent bioactive cell-penetrating peptides. *Proc. Natl. Acad. Sci. USA* **2017**, *114*, E8957–E8966. [CrossRef] [PubMed]
8. Rivera-Chavez, J.; Raja, H.A.; Graf, T.N.; Gallagher, J.M.; Metri, P.; Xue, D.; Pearce, C.J.; Oberlies, N.H. Prealamethicin F50 and related peptaibols from *Trichoderma arundinaceum*: Validation of their authenticity via in situ chemical analysis. *RSC Adv.* **2017**, *7*, 45733–45751. [CrossRef] [PubMed]
9. Sica, V.P.; Rees, E.R.; Raja, H.A.; Rivera-Chavez, J.; Burdette, J.E.; Pearce, C.J.; Oberlies, N.H. In situ mass spectrometry monitoring of fungal cultures led to the identification of four peptaibols with a rare threonine residue. *Phytochemistry* **2017**, *143*, 45–53. [CrossRef] [PubMed]
10. van Bohemen, A.I.; Ruiz, N.; Zalouk-Vergnoux, A.; Michaud, A.; Robiou du Pont, T.; Druzhinina, I.; Atanasova, L.; Prado, S.; Bodo, B.; Meslet-Cladiere, L.; et al. Pentadecaibins I-V: 15-Residue Peptaibols Produced by a marine-derived *Trichoderma* sp. of the *Harzianum* clade. *J. Nat. Prod.* **2021**, *84*, 1271–1284. [CrossRef] [PubMed]
11. Wu, G.; Dentinger, B.T.M.; Nielson, J.R.; Peterson, R.T.; Winter, J.M. Emerimicins V-X, 15-Residue Peptaibols discovered from an *Acremonium* sp. through integrated genomic and chemical approaches. *J. Nat. Prod.* **2021**, *84*, 1113–1126. [CrossRef]

12. Kimonyo, A.; Bruckner, H. Sequences of metanicins, 20-residue peptaibols from the ascomycetous fungus CBS 597.80. *Chem. Biodivers.* **2013**, *10*, 813–826. [CrossRef] [PubMed]
13. Panizel, I.; Yarden, O.; Ilan, M.; Carmeli, S. Eight new peptaibols from sponge-associated *Trichoderma atroviride*. *Mar. Drugs* **2013**, *11*, 4937–4960. [CrossRef] [PubMed]
14. Kai, K.; Mine, K.; Akiyama, K.; Ohki, S.; Hayashi, H. Anti-plant viral activity of peptaibols, trichorzins HA II, HA V, and HA VI, isolated from *Trichoderma harzianum* HK-61. *J. Pestic. Sci.* **2018**, *43*, 283–286. [CrossRef] [PubMed]
15. Stadler, M.S.S.; Müller, H.; Henkel, T.; Lagojda, A.; Kleymann, G. New antiviral peptaibols from the mycoparasitic fungus *Sepedonium microspermum*. In *Book of Abstracts, 13*; Irseer Naturstofftage der DECHEMA: Irsee, Germany, 2001.
16. Fragiadaki, I.; Katogiritis, A.; Calogeropoulou, T.; Bruckner, H.; Scoulica, E. Synergistic combination of alkylphosphocholines with peptaibols in targeting *Leishmania infantum* in vitro. *Int. J. Parasitol. Drugs Drug Resist.* **2018**, *8*, 194–202. [CrossRef] [PubMed]
17. Otto, A.; Laub, A.; Porzel, A.; Schmidt, J.; Wessjohann, L.; Westermann, B.; Arnold, N. Isolation and total synthesis of Albupeptins A–D: 11-Residue peptaibols from the fungus *Gliocladium album*. *Eur. J. Org. Chem.* **2015**, *34*, 7449–7459. [CrossRef]
18. Otto, A.; Laub, A.; Haid, M.; Porzel, A.; Schmidt, J.; Wessjohann, L.; Arnold, N. Tulasporins A–D, 19-Residue peptaibols from the mycoparasitic fungus *Sepedonium tulasneanum*. *Nat. Prod. Commun.* **2016**, *11*, 1821–1824. [CrossRef]
19. Otto, A.; Laub, A.; Wendt, L.; Porzel, A.; Schmidt, J.; Palfner, G.; Becerra, J.; Kruger, D.; Stadler, M.; Wessjohann, L.; et al. Chilenopeptins A and B, Peptaibols from the Chilean *Sepedonium* aff. *chalcipori* KSH 883. *J. Nat. Prod.* **2016**, *79*, 929–938. [CrossRef]
20. Shi, W.L.; Chen, X.L.; Wang, L.X.; Gong, Z.T.; Li, S.; Li, C.L.; Xie, B.B.; Zhang, W.; Shi, M.; Li, C.; et al. Cellular and molecular insight into the inhibition of primary root growth of *Arabidopsis* induced by peptaibols, a class of linear peptide antibiotics mainly produced by *Trichoderma* spp. *J. Exp. Bot.* **2016**, *67*, 2191–2205. [CrossRef]
21. Ritzau, M.; Heinze, S.; Dornberger, K.; Berg, A.; Fleck, W.; Schlegel, B.; Hartl, A.; Grafe, U. Ampullosporin, a new peptaibol-type antibiotic from *Sepedonium ampullosporum* HKI-0053 with neuroleptic activity in mice. *J. Antibiot.* **1997**, *50*, 722–728. [CrossRef]
22. Kronen, M.; Kleinwachter, P.; Schlegel, B.; Hartl, A.; Grafe, U. Ampullosporines B,C,D,E1,E2,E3 and E4 from *Sepedonium ampullosporum* HKI-0053: Structures and biological activities. *J. Antibiot.* **2001**, *54*, 175–178. [CrossRef]
23. Berek, I.; Becker, A.; Schroder, H.; Hartl, A.; Hollt, V.; Grecksch, G. Ampullosporin A, a peptaibol from *Sepedonium ampullosporum* HKI-0053 with neuroleptic-like activity. *Behav. Brain Res.* **2009**, *203*, 232–239. [CrossRef] [PubMed]
24. Speckbacher, V.; Zeilinger, S. Secondary Metabolites of Mycoparasitic Fungi. In *Secondary Metabolites-Sources and Applications*, 1st ed.; Vijayakumar, R., Raja, S., Eds.; Intechopen: London, UK, 2018; pp. 37–55.
25. Milov, A.D.; Tsvetkov, Y.D.; Raap, J.; De Zotti, M.; Formaggio, F.; Toniolo, C. Conformation, self-aggregation, and membrane interaction of peptaibols as studied by pulsed electron double resonance spectroscopy. *Biopolymers* **2016**, *106*, 6–24. [CrossRef] [PubMed]
26. Sahr, T.A.H.; Besl, H.; Fischer, M. Infrageneric classification of the boleticolous genus *Sepedonium*: Species delimitation and phylogenetic relationships. *Mycologia* **1999**, *91*, 935–943. [CrossRef]
27. Divekar, P.V.; Vining, L.C. Reaction of anhydrosepedonin with alkali synthesis of a degradation product and some related dimethylhydroxybenzoic acids. *Can. J. Chem.* **1964**, *42*, 63–68. [CrossRef]
28. Divekar, P.V.; Raistrick, H.; Dobson, T.A.; Vining, L.C. Studies in the biochemistry of microorganisms part 117. Sepedonin, a tropolone metabolite of *Sepedonium chrysospermum* Fries. *Can. J. Chem.* **1965**, *43*, 1835–1848. [CrossRef]
29. Shibata, S.; Shoji, J.; Ohta, A.; Watanabe, M. Metabolic products of fungi. XI. Some observation on the occurrence of skyrin and rugulosin in mold metabolites, with a reference to structural relationship between penicilliopsin and skyrin. *Pharm. Bull* **1957**, *5*, 380–382. [CrossRef] [PubMed]
30. Quang, D.N.; Schmidt, J.; Porzel, A.; Wessjohann, L.; Haid, M.; Arnold, N. Ampullosine, a new isoquinoline alkaloid from *Sepedonium ampullosporum* (Ascomycetes). *Nat. Prod. Commun.* **2010**, *5*, 869–872. [CrossRef]
31. Closse, A.; Hauser, D. Isolierung und Konstitutionsermittlung von Chrysodin. *Helv. Chim. Acta.* **1973**, *56*, 2694–2698. [CrossRef]
32. Mitova, M.I.; Stuart, B.G.; Cao, G.H.; Blunt, J.W.; Cole, A.L.; Munro, M.H. Chrysosporide, a cyclic pentapeptide from a New Zealand sample of the fungus *Sepedonium chrysospermum*. *J. Nat. Prod.* **2006**, *69*, 1481–1484. [CrossRef]
33. Laub, A.; Lam, Y.T.H.; Mendez, Y.; Vidal, A.V.; Porzel, A.; Schmidt, J.; Wessjohann, L.A.; Westermann, B.; Arnold, N. Identification and total synthesis of two new cyclic pentapeptides from *Sepedonium microspermum* Besl. Manuscript in preparation.
34. Dornberger, K.; Ihn, W.; Ritzau, M.; Grafe, U.; Schlegel, B.; Fleck, W.F.; Metzger, J.W. Chrysospermins, new peptaibol antibiotics from *Apiocrea chrysosperma* Ap101. *J. Antibiot.* **1995**, *48*, 977–989. [CrossRef] [PubMed]
35. Hulsmann, H.; Heinze, S.; Ritzau, M.; Schlegel, B.; Grafe, U. Isolation and structure of peptaibolin, a new peptaibol from *Sepedonium* strains. *J. Antibiot.* **1998**, *51*, 1055–1058. [CrossRef]
36. Neuhof, T.; Berg, A.; Besl, H.; Schwecke, T.; Dieckmann, R.; von Dohren, H. Peptaibol production by *Sepedonium* strains parasitizing Boletales. *Chem. Biodivers.* **2007**, *4*, 1103–1115. [CrossRef] [PubMed]
37. Mitova, M.I.; Murphy, A.C.; Lang, G.; Blunt, J.W.; Cole, A.L.; Ellis, G.; Munro, M.H. Evolving trends in the dereplication of natural product extracts. 2. The isolation of chrysaibol, an antibiotic peptaibol from a New Zealand sample of the mycoparasitic fungus *Sepedonium chrysospermum*. *J. Nat. Prod.* **2008**, *71*, 1600–1603. [CrossRef] [PubMed]
38. Iijima, M.; Amemiya, M.; Sawa, R.; Kubota, Y.; Kunisada, T.; Momose, I.; Kawada, M.; Shibasaki, M. Acremopeptin, a new peptaibol from *Acremonium* sp. PF1450. *J. Antibiot.* **2017**, *70*, 791–794. [CrossRef] [PubMed]
39. Abdalla, M.A.; McGaw, L.J. Natural cyclic peptides as an attractive modality for therapeutics: A mini review. *Molecules* **2018**, *23*, 2080. [CrossRef] [PubMed] X

40. Jiao, W.H.; Khalil, Z.; Dewapriya, P.; Salim, A.A.; Lin, H.W.; Capon, R.J. Trichodermides A–E: New peptaibols isolated from the australian termite nest-derived fungus *Trichoderma virens* CMB-TN16. *J. Nat. Prod.* **2018**, *81*, 976–984. [CrossRef] [PubMed]
41. Marik, T.; Tyagi, C.; Racic, G.; Rakk, D.; Szekeres, A.; Vagvolgyi, C.; Kredics, L. New 19-residue peptaibols from *Trichoderma* Clade Viride. *Microorganisms* **2018**, *6*, 85. [CrossRef]
42. Ojo, O.S.; Nardone, B.; Musolino, S.F.; Neal, A.R.; Wilson, L.; Lebl, T.; Slawin, A.M.Z.; Cordes, D.B.; Taylor, J.E.; Naismith, J.H.; et al. Synthesis of the natural product descurainolide and cyclic peptides from lignin-derived aromatics. *Org. Biomol. Chem.* **2018**, *16*, 266–273. [CrossRef]
43. Singh, V.P.; Yedukondalu, N.; Sharma, V.; Kushwaha, M.; Sharma, R.; Chaubey, A.; Kumar, A.; Singh, D.; Vishwakarma, R.A. Lipovelutibols A–D: Cytotoxic lipopeptaibols from the himalayan cold habitat fungus *Trichoderma velutinum*. *J. Nat. Prod.* **2018**, *81*, 219–226. [CrossRef]
44. Touati, I.; Ruiz, N.; Thomas, O.; Druzhinina, I.S.; Atanasova, L.; Tabbene, O.; Elkahoui, S.; Benzekri, R.; Bouslama, L.; Pouchus, Y.F.; et al. Hyporientalin A, an anti-*Candida* peptaibol from a marine *Trichoderma orientale*. *World J. Microbiol. Biotechnol.* **2018**, *34*, 98. [CrossRef] [PubMed]
45. Katoch, M.; Singh, D.; Kapoor, K.K.; Vishwakarma, R.A. Trichoderma lixii (IIIM-B4), an endophyte of *Bacopa monnieri* L. producing peptaibols. *BMC Microbiol.* **2019**, *19*, 98. [CrossRef] [PubMed]
46. Momose, I.; Onodera, T.; Doi, H.; Adachi, H.; Iijima, M.; Yamazaki, Y.; Sawa, R.; Kubota, Y.; Igarashi, M.; Kawada, M. Leucinostatin Y: A Peptaibiotic produced by the entomoparasitic fungus *Purpureocillium lilacinum* 40-H-28. *J. Nat. Prod.* **2019**, *82*, 1120–1127. [CrossRef] [PubMed]
47. Kim, C.K.; Krumpe, L.R.H.; Smith, E.; Henrich, C.J.; Brownell, I.; Wendt, K.L.; Cichewicz, R.H.; O'Keefe, B.R.; Gustafson, K.R. Roseabol A, a new peptaibol from the fungus *Clonostachys rosea*. *Molecules* **2021**, *26*, 3594. [CrossRef]
48. Rawa, M.S.A.; Nogawa, T.; Okano, A.; Futamura, Y.; Nakamura, T.; Wahab, H.A.; Osada, H. A new peptaibol, RK-026A, from the soil fungus *Trichoderma* sp. RK10-F026 by culture condition-dependent screening. *Biosci. Biotechnol. Biochem.* **2021**, *85*, 69–76. [CrossRef] [PubMed]
49. Rawa, M.S.A.; Nogawa, T.; Okano, A.; Futamura, Y.; Wahab, H.A.; Osada, H. Zealpeptaibolin, an 11-mer cytotoxic peptaibol group with 3 Aib-Pro motifs isolated from *Trichoderma* sp. RK10-F026. *J. Antibiot.* **2021**, *74*, 485–495. [CrossRef]
50. Zhang, S.H.; Yang, J.; Ma, H.; Yang, Y.; Zhou, G.F.; Zhao, X.; Xu, R.; Nie, D.; Zhang, G.G.; Shan, J.J.; et al. Longibramides A–E, Peptaibols isolated from a mushroom derived fungus *Trichoderma longibrachiatum* Rifai DMG-3-1-1. *Chem. Biodivers.* **2021**, *18*, 2100128. [CrossRef]
51. Tullberg, M.; Grøtli, M.; Luthman, K. Efficient synthesis of 2,5-diketopiperazines using microwave assisted heating. *Tetrahedron* **2006**, *62*, 7484–7491. [CrossRef]
52. Li, L.; Li, D.; Luan, Y.; Gu, Q.; Zhu, T. Cytotoxic metabolites from the antarctic psychrophilic fungus *Oidiodendron truncatum*. *J. Nat. Prod.* **2012**, *75*, 920–927. [CrossRef]
53. Wang, F.Z.; Huang, Z.; Shi, X.F.; Chen, Y.C.; Zhang, W.M.; Tian, X.P.; Li, J.; Zhang, S. Cytotoxic indole diketopiperazines from the deep sea-derived fungus *Acrostalagmus luteoalbus* SCSIO F457. *Bioorg. Med. Chem. Lett.* **2012**, *22*, 7265–7267. [CrossRef]
54. Sun, S.; Dai, X.; Sun, J.; Bu, X.; Weng, C.; Li, H.; Zhu, H. A diketopiperazine factor from *Rheinheimera aquimaris* QSI02 exhibits anti-quorum sensing activity. *Sci. Rep.* **2016**, *6*, 39637. [CrossRef] [PubMed]
55. Caballero, E.; Avendaño, C.; Menéndez, J.C. Stereochemical issues related to the synthesis and reactivity of pyrazino[2′,1′-5,1]pyrrolo[2,3-b]indole-1,4-diones. *Tetrahedron Asymmetry* **1998**, *9*, 967–981. [CrossRef]
56. Zhao, D.; Cao, F.; Guo, X.-J.; Zhang, Y.-R.; Kang, Z.; Zhu, H.-J. Antibacterial Indole alkaloids and anthraquinones from a sewage-derived fungus *Eurotium* sp. *Chem. Nat. Compd.* **2018**, *54*, 399–401. [CrossRef]
57. Bovio, E.; Garzoli, L.; Poli, A.; Luganini, A.; Villa, P.; Musumeci, R.; McCormack, G.P.; Cocuzza, C.E.; Gribaudo, G.; Mehiri, M.; et al. Marine fungi from the sponge *Grantia compressa*: Biodiversity, chemodiversity, and biotechnological potential. *Mar. Drugs* **2019**, *17*, 220. [CrossRef] [PubMed]
58. Lu, W.; Du, J.; Goehring, A.; Gouaux, E. Cryo-EM structures of the triheteromeric NMDA receptor and its allosteric modulation. *Science* **2017**, *355*, 6331. [CrossRef] [PubMed]
59. Zhang, J.B.; Chang, S.; Xu, P.; Miao, M.; Wu, H.; Zhang, Y.; Zhang, T.; Wang, H.; Zhang, J.; Xie, C.; et al. Structural basis of the proton sensitivity of human GluN1-GluN2A NMDA receptors. *Cell Rep.* **2018**, *25*, 3582–3590. [CrossRef]
60. Nguyen, H.-H.; Imhof, D.; Kronen, M.; Schlegel, B.; Ha, A.; Gera, L.; Reissmann, S. Synthesis and biological evaluation of analogues of the peptaibol ampullosporin A. *J. Med. Chem.* **2002**, *45*, 2781–2787. [CrossRef]
61. Berman, H.M.; Westbrook, J.; Feng, Z.; Gilliland, G.; Bhat, T.N.; Weissig, H.; Shindyalov, I.N.; Bourne, P.E. The Protein Data Bank. *Nucleic Acids Res.* **2000**, *28*, 235–242. [CrossRef] [PubMed]
62. Chemical Computing Group Inc. *Molecular Operating Environment v2019.0101*; Chemical Computing Group Inc.: Montreal, QC, Canada, 2019.

International Journal of *Molecular Sciences*

Article

COX Inhibitory and Cytotoxic Naphthoketal-Bearing Polyketides from *Sparticola junci*

Katherine Yasmin M. Garcia [1,2], Mark Tristan J. Quimque [1,2,3], Gian Primahana [4,5,6], Andreas Ratzenböck [7], Mark Joseph B. Cano [1,2], Jeremiah Francis A. Llaguno [2], Hans-Martin Dahse [8], Chayanard Phukhamsakda [9,10], Frank Surup [4,5], Marc Stadler [4,5] and Allan Patrick G. Macabeo [2,*]

1. The Graduate School, University of Santo Tomas, España Blvd., Manila 1015, Philippines; katherineyasmin.garcia.gs@ust.edu.ph (K.Y.M.G.); marktristan.quimque@g.msuiit.edu.ph (M.T.J.Q.); markjoseph.cano.gs@ust.edu.ph (M.J.B.C.)
2. Laboratory for Organic Reactivity, Discovery and Synthesis (LORDS), Research Center for the Natural and Applied Sciences, University of Santo Tomas, España Blvd., Manila 1015, Philippines; jeremiahfrancis.llaguno.sci@ust.edu.ph
3. Chemistry Department, College of Science and Mathematics, Mindanao State University-Iligan Institute of Technology, Tibanga, Iligan City 9200, Philippines
4. Department of Microbial Drugs, Helmholtz Centre for Infection Research and German Centre for Infection Research (DZIF), Partner Site Hannover/Braunschweig, Inhoffenstrasse 7, 38124 Braunschweig, Germany; gian.primahana@helmholtz-hzi.de (G.P.); frank.surup@helmholtz-hzi.de (F.S.); marc.stadler@helmholtz-hzi.de (M.S.)
5. Institute of Microbiology, Technische Universität Braunschweig, Spielmannstraße 7, 38106 Braunschweig, Germany
6. Research Center for Chemistry, National Research and Innovation Agency (BRIN), Kawasan Puspitek, Serpong, Tangerang Selatan 15314, Indonesia
7. Institut für Organische Chemie, Universität Regensburg, Universitätstrasse 31, 93053 Regensburg, Germany; andreas.ratzenboeck@chemie.uni-regensburg.de
8. Leibniz-Institute for Natural Product Research and Infection Biology, Hans-Knöll-Institute (HKI), 07745 Jena, Germany; hans-martin.dahse@hki-jena.de
9. Center of Excellence in Fungal Research, Mae Fah Luang University, Chiang Rai 57100, Thailand; chayanard91@gmail.com
10. Institute of Plant Protection, College of Agriculture, Jilin Agricultural University, Changchun 130118, China
* Correspondence: agmacabeo@ust.edu.ph

Citation: Garcia, K.Y.M.; Quimque, M.T.J.; Primahana, G.; Ratzenböck, A.; Cano, M.J.B.; Llaguno, J.F.A.; Dahse, H.-M.; Phukhamsakda, C.; Surup, F.; Stadler, M.; et al. COX Inhibitory and Cytotoxic Naphthoketal-Bearing Polyketides from *Sparticola junci*. *Int. J. Mol. Sci.* **2021**, *22*, 12379. https://doi.org/10.3390/ijms222212379

Academic Editor: Silvie Rimpelová

Received: 16 October 2021
Accepted: 12 November 2021
Published: 17 November 2021

Publisher's Note: MDPI stays neutral with regard to jurisdictional claims in published maps and institutional affiliations.

Copyright: © 2021 by the authors. Licensee MDPI, Basel, Switzerland. This article is an open access article distributed under the terms and conditions of the Creative Commons Attribution (CC BY) license (https://creativecommons.org/licenses/by/4.0/).

Abstract: Axenic fermentation on solid rice of the saprobic fungus *Sparticola junci* afforded two new highly oxidized naphthalenoid polyketide derivatives, sparticatechol A (**1**) and sparticolin H (**2**) along with sparticolin A (**3**). The structures of **1** and **2** were elucidated on the basis of their NMR and HR-ESIMS spectroscopic data. Assignment of absolute configurations was performed using electronic circular dichroism (ECD) experiments and Time-Dependent Density Functional Theory (TDDFT) calculations. Compounds **1**–**3** were evaluated for COX inhibitory, antiproliferative, cytotoxic and antimicrobial activities. Compounds **1** and **2** exhibited strong inhibitory activities against COX-1 and COX-2. Molecular docking analysis of **1** conferred favorable binding against COX-2. Sparticolin H (**2**) and A (**3**) showed a moderate antiproliferative effect against myelogenous leukemia K-562 cells and weak cytotoxicity against HeLa and mouse fibroblast cells.

Keywords: *Sparticola junci*; structure elucidation; ECD-TDDFT; COX inhibitory; molecular docking; antiproliferative; cytotoxic

1. Introduction

Plant-associated fungi constitutes a myriad and relatively less explored repository of biologically active natural products that may serve as key starting points for pharmaceutical drug development, biotechnology and agrochemical applications [1–3]. The Dothideomycetes, which comprise the largest taxa in the Ascomycota, are known to produce fungal secondary metabolites, possessing distinct chemical structures associated with biological activities that have gained considerable interest. Several genera of Dothideomycetes elaborate oxidized bisnaphthalenoids that consist of 1,8-dihydroxynaphthalene-derived units and a decalin moiety bridged through spiroketal linkages [4]. Since the first discovery of the antibacterial diketo-bisnaphthalenoid natural product MK 3018 isolated from the fungal culture of *Tetraploa aristata* [5], a growing number of naphthospiroketal derivatives have been reported, originating from a diverse group of fungi and considered to be potential drug leads exhibiting a wide range of biological properties [6–11]. MK 3018 exhibited potent in vitro cytotoxicity against P388 murine leukemia cell line [6].

The 1,8-dihydroxynaphthalene polyketide *spiro*-mamakone is the first example of *nor*-spirodioxynaphthalene derivative featuring a *spiro*-nonaphthadiene skeleton. Its isolation spurred biosynthetic studies which have illustrated oxidative coupling and rearrangement of naphthalene subunits followed by decarboxylation and ring closure in the biogenetic pathways of *nor*- and *nor-seco* derivatives [12]. Our initial efforts in exploring the secondary metabolites of *Sparticola junci* (family Sporormiaceae), an ascomycetous saprophytic species previously isolated from the decomposing branches of the Spanish broom, *Spartium junceum* (family Fabaceae), has led to the isolation of additional antimicrobial and cytotoxic *nor-seco* congeners. These biologically active spirodioxynaphthalene derivatives showcase unprecedented structural frameworks that bear carboxyalkylidene–cyclopentanoid, carboxyl-functionalized oxabicyclo [3.3.0]octane, annelated 2-cyclopentenone/δ-lactone, and precursor catechol-bearing sub-structures [13,14].

In our continuing search of biologically active secondary metabolites from Dothideomycetes fungi [13–16], we herein report the isolation and structure identification of two new spirodioxynaphthalene derivatives, hitherto referred to as sparticatechol A (**1**) and sparticolin H (**2**) along with the known polyketide, sparticolin A (**3**) from the fermented solid rice culture and evaluation of their cyclooxygenase inhibitory, antiproliferative, cytotoxic and antimicrobial activities. To complement the observed biological studies, in silico molecular docking simulations on cyclooxygenases, and determination of drug-likeness, adsorption, distribution, metabolism, excretion, and toxicity (ADME-Tox) of the bioactive naphthoketal derivatives **1–3** were also carried out.

2. Results and Discussion

The ethyl acetate (EtOAc) extract of *Sparticola junci* obtained after fermentation on a solid rice medium was partitioned between *n*-heptane and 10% aqueous MeOH. The resulting aqueous methanolic crude extract was purified using gradient elution vacuum liquid chromatography followed by semi-preparative HPLC to afford metabolites **1–3** (Figure 1). This study highlights the structure elucidation and biological activity determination of new naphthoketals **1** and **2** based on NMR spectroscopic data, HR-ESIMS and ECD-TDDFT calculations. The known *nor-seco* spirodioxynapthalenoid, sparticolin A (**3**) was identified by comparing its physicochemical and NMR spectroscopic data with those reported in the literature. Compound **3** was previously obtained from the submerged fermentation culture of *S. junci* and exhibited weak antimicrobial and cytotoxic activities.

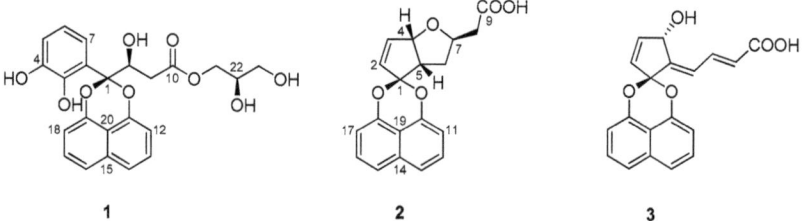

Figure 1. Dioxynaphthalenoids **1–3** from *Sparticola junci*.

Sparticatechol A (**1**) was obtained as an optically active reddish-brown syrup. The molecular formula $C_{23}H_{22}O_9$, indicating 13 degrees of unsaturation, was established based on the protonated molecular ion peak at *m/z* 443.1330 [M + H]$^+$ of its positive-ion HRESIMS. This is consistent with the number of proton and carbon peaks detected in its ^{1}H and ^{13}C NMR spectroscopic data (Table 1). Detailed examination of the ^{1}H, ^{13}C, and HSQC-DEPT NMR spectroscopic data revealed the presence of a carboxylic acid carbon, a ketal carbon, seven non-protonated aromatic carbons (four oxygenated and three non-oxygenated), nine aromatic methines, three sp^3 methylene and two oxygenated sp^3 methine carbon.

Table 1. NMR Spectroscopic Data of **1** and **2** in MeOH-d_4.

Position	1		2	
	δ_H ^{a} (Mult., *J* in Hz)	δ_C ^{a}, Type	δ_H ^{b} (Mult., *J* in Hz)	δ_C ^{c}, Type
1	-	105.1, C	-	111.0, C
2	-	123.5, C	5.88 (ddd, 5.7, 1.0, 0.4)	130.1, CH
3	-	145.1, C	6.25 (dd, 5.7, 1.7)	139.3, CH
4	-	147.3, C	5.07 (ddd, 7.0, 1.8, 1.0)	85.7, CH
5	6.59 (dd, 7.9, 1.5)	117.0, CH	3.14 (td, 8.5, 7.1)	51.1, CH
6	6.42 (dd, 7.9, 1.5)	119.8, CH	1.86–1.73 (m)	32.7, CH_2
7	6.80 (dd, 7.9, 1.5)	121.6, CH	4.24 (dtd, 12.1, 6.1, 5.6)	78.7, CH
8	4.81 (dd, 14.0, 3.1)	73.9, CH	2.52 (d, 6.3)	39.9, CH_2
9a	2.68 (dd, 15.7, 9.4)	38.0, CH_2	-	173.5, C
9b	2.97 (dd, 15.7, 3.2)			
10	-	173.1, C	-	148.2, C
11	-	148.7, C	6.87 (dd, 8.9, 7.4)	108.6, CH
12	7.02 (dd, 8.3, 1.3)	110.2, CH	7.40 (dd, 8.9, 7.4)	127.1, CH
13	7.37 (dd, 8.3)	121.6, CH	7.47 (dd, 8.9, 7.4)	120.1, CH
14	7.39 (dd, 8.3, 1.3)	128.2, CH	-	134.5, C
15	-	135.5, C	7.47 (dd, 8.9, 7.4)	120.1, CH
16	7.39 (dd, 8.3, 1.3)	128.2, CH	7.40 (dd, 8.9, 7.4)	127.1, CH
17	7.37 (dd, 8.3)	121.6, CH	6.87 (dd, 8.9, 7.4)	108.6, CH
18	7.02 (dd, 8.3, 1.3)	110.2, CH	-	148.4, C
19	-	148.9, C	-	114.1, C
20	-	115.5, C	-	-
21a	4.10 (td, 11.4, 6.3)	66.9, CH_2	-	-
21b	4.19 (td, 11.4, 4.3)		-	-
22	3.80 (dt, 15.2, 6.7)	71.1, CH	-	-
23	3.56 (ddd, 7.3, 5.7, 2.3)	64.0, CH_2	-	-

^{a} Recorded at 600 MHz, ^{b} Recorded at 500 MHz, ^{c} Recorded at 125 MHz; Carbon multiplicities were deduced from HSQC-DEPT-135 spectra. δ_H: proton chemical shift; δ_C: carbon chemical shift.

Analysis of the COSY spectrum identified two isolated three distinct spin systems of δ_H 7.37 (2H, dd, *J* = 8.3 Hz, H-13/H-17) and δ_H 7.02 (2H, dd, *J* = 8.3, 1.3 Hz, H-12/H-18), and of δ_H 7.39 (2H, dd, *J* = 8.3, 1.3 Hz, H-14/H-16) corresponding to C-12 to C-14 and to C-16 to C-18 naphthalene subunits of **1**. These aromatic protons are mutually *ortho*-coupled to each other (*J* = 8.3 Hz). HMBC correlations of H-14/H-16 with two non-protonated sp^2 carbons δ_C 135.5 (C-15) and δ_C 115.5 (C-20), along with long range correlations of H-12 and H-13 with δ_C 148.7 (C-11), and H-17 and H-18 with δ_C 148.9 (C-19), suggested that subunits C-12 to C-14 and C-16 to C-18 are linked to C-15 and C-20, leading to the construction of a naphthalene fragment. The midfield chemical shifts observed with C-11 and C-19 proposed 1,8-dioxygenation in the naphthalene unit. The remaining portion of **1** was constructed through analysis of COSY and HSQC-DEPT spectroscopic data, revealing aromatic signals representative of a 1,2,3-trisubstituted benzene ring at δ_H 6.59 (1H, dd, *J* = 7.9, 1.5 Hz, H-5), δ_H 6.42 (1H, t, *J* = 7.9 Hz, H-6), δ_H 6.80 (1H, dd, *J* = 7.9, 1.5 Hz, H-7), a two-proton spin system consisting of the oxygenated sp^3 methine δ_H 4.81 (1H, dd, *J* = 14.0, 3.1, H-8) in the β-carbon coupled to the diastereotopic methylene protons at δ_H 2.68 (1H, dd, *J* = 15.6, 9.4 Hz, H-9a), 2.97 (1H, dd, *J* = 15.7, 3.2 Hz, H-9b) connected to the β-carbon, and a three-proton spin system at δ_H 4.10 (1H, td, *J* = 11.4, 6.3 Hz, H-21a),

4.19 (1H, td, J = 11.0, 4.3 Hz, H-21b), δ_H 3.80 (2H, dd, J = 18.3, 10.4 Hz, H-22), δ_H 3.56 (2H, dt, J = 5.7, 4.8, 2.3 Hz, H-23) (Figure 2) arising from the glycerol subunit. HMBC correlations of δ_H 6.59 (H-5) with δ_C 145.1 (C-3) and δ_C 147.3 (C-4) and of δ_H 6.80 (H-7) with δ_C 105.1 (C-1) and C-3 allowed the identification of a substituted catechol fragment. In addition, the correlations of δ_H 4.10, 4.19 (H$_2$-21) with an oxygenated sp^3 methine δ_C 71.1 (C-22) and an oxygenated sp^3 methylene δ_C 64.0 (C-23) suggested the presence of dihydroxypropyl residue. Further HMBC correlations of δ_H 4.10, 4.19 (H$_2$-21), δ_H 4.81 (H-8) and δ_H 2.68, 2.97 (H$_2$-9) with the carboxylic carbon δ_C 173.1 (C-10) established the connectivities of the glycerol residue and the α-hydroxy carboxylic acid fragment, allowing for the construction of a glyceryl-β-hydroxybutanoate motif. Finally, HMBC correlations of H-7, H-8, and H$_2$-9 with the *spiro*-ketal carbon δ_C 105.1 (C-1) led to the attachment of the catechol moiety and the glyceryl-β-hydroxybutanoate motif to the 1,8-dioxygenated naphthalene subunit. On the basis of the data discussed above, the planar structure of **1** was identified as 2,3-dihydroxypropyl-3-(2-(2,3-dihydroxyphenyl)naphtho[1,8-de][1,3]dioxin-2-yl)-3-hydroxypropanoate.

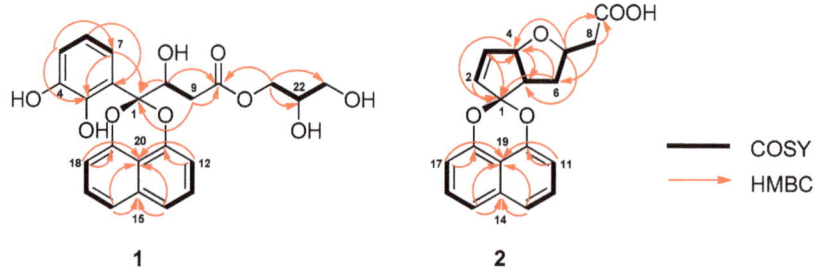

Figure 2. COSY and HMBC correlations in **1** and **2**.

TDDFT-ECD calculations were performed to determine the absolute configuration of sparticatechol A (**1**) unambiguously. Prior to ECD calculations, a conformational analysis was done using an MMFF94 force field to search for conformers with room-temperature equilibrium population > 1%. Geometry re-optimization via DFT calculations at a B3LYP/6-31G(d) basis set afforded two stable conformers with a configurational assignment arbitrarily chosen as (8S,22R) based on the previously determined configuration of the catechol-bearing naphthoketal [13] and the naturally occurring (2R)-glyceraldehyde-3-phosphate [17,18]. The two stable conformers differed in the orientation, mainly due to the flexibility of the ester chain, particularly along the C8-C9 and C22-C23 axes. Conformer 1A (95.03%) adapted almost an anti-staggered conformation with respect to the C7 *spiro*-ketal carbon and the C10 carbonyl carbon about the C8-C9 axis with a C7-C8-C9-C10 dihedral angle of 170.35°. Conformer 1B (4.97%), on the other hand, adapted a gauche configuration about the said axis; however, this was compensated by strong intramolecular hydrogen bonding between two hydroxy groups, C6(OH) and C22(OH). Moreover, along the C22-C23 axis, conformer 1B adapted an anti-staggered conformation. The 1A and 1B conformations were also supported by the ROESY correlations between H-8/H-7, H-8/H-9b, H-8/H-18, and H-6/H-22, respectively. Based on the harmonic vibrational frequency calculations, these conformers are confirmed stable. TDDFT calculations were carried out on both conformers at the following levels of theory/basis sets: B3LYP/6-31G(d), WB97XD/DGDZVP, and B97D/TZVP in the gas phase and polarizable continuum model (PCM) for acetonitrile. The experimental ECD spectrum of compound **1** showed two negative cotton effects (CE's)– one at 206 nm and a more prominent peak at 246 nm. In the latter CE peak, the calculated ECD spectrum at B3LYP/6-31G(d) in gas phase exhibited a blue shift of about 10 nm. The use of other levels of theory such as WB97XD and B97D only increased the said blue shift. Overall, the Boltzmann-averaged ECD spectrum of sparticatechol A (Figure 3) calculated at B3LYP/6-31G(d) provided good agreement with the experimental ECD spectrum. Thus, the absolute configuration of sparticatechol A (**1**) is (8*S*,22*R*).

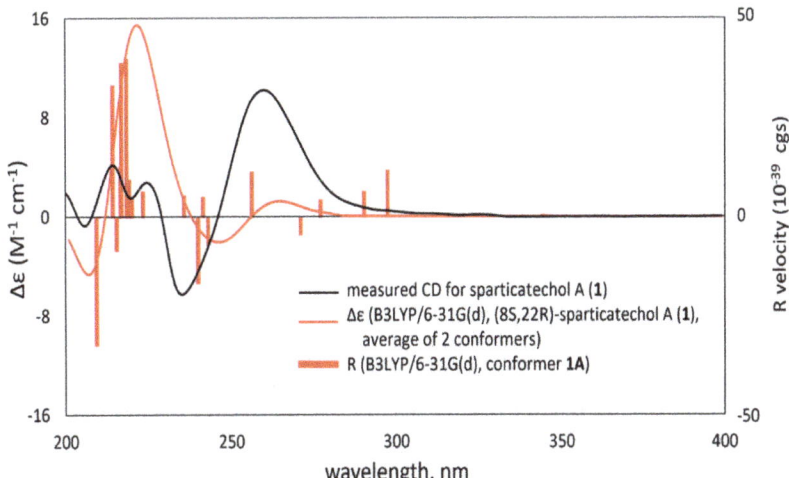

Figure 3. Experimental ECD spectrum of sparticatechol A (**1**, black solid curve) compared with B3LYP/6-31G(d)-calculated ECD spectra (red solid curve) for the B3LYP/6-31G(d)-optimized conformers of (8*S*,22*R*)-**1**.

Sparticolin H (**2**) was isolated as a dark brown syrup. The HRESIMS of **2** showed a sodiated molecular ion peak at *m/z* 347.0889, corresponding to the molecular formula $C_{19}H_{16}NaO_5$, indicating an index of hydrogen deficiency of 12. In addition to signals resonating for a 1,8-dioxynaphthalene subunit, detailed investigation of the ^1H, ^{13}C, (Table 1) and HSQC-DEPT NMR spectroscopic data revealed the presence of additional signals corresponding to a carboxylic acid, a ketal carbon, two sp^3 methylene carbons, two olefinic methines, and three sp^3 methines (two oxygenated and one non-oxygenated). The ^1H-^1H COSY NMR showed the presence of a seven-proton system corresponding to the C-2 to C-8 cyclopentanoid units in (Figure 2). Key HMBC correlations of δ_H 5.88 (H-2), δ_H 4.81 (H-3), δ_H 5.07 (H-4), δ_H 3.14 (H-5) with the *spiro*-ketal carbon δ_C 111.0 (C-1) allowed the identification of C-4/C-5 disubstituted 2-cyclopentene motif connected to the dioxygenated naphthalene fragment. Further HMBC correlations of H-4 with δ_C 78.7 (C-7) and H-7 with δ_C 51.1 (C-5) suggested the annelation of a tetrahydrofuran moiety in the cyclopentene ring, establishing a 1-oxabicyclo[3.3.0]octane subunit. Finally, the HMBC correlations of H-7 and H-8 to the carboxylic carbon δ_C 173.5 (C-9) allowed the determination of the gross structure of **2**. The vicinal coupling constant of 7 Hz ($^3J_{H-4,H-5}$) between H-4 and H-5 suggested a *cis* geometry. These NMR data correspond to a similar structural framework observed for sparticolin E, as described in our previous study [12]. In the NOESY spectrum, however, an NOE effect was only observed between H-4 and H-5. This analysis suggested that H-7 is spatially different from the two protons rendering an assignment of relative configurations of the chiral carbons as 4*S**, 5*R**, and 7*R**. Thus, the structure of compound **2** was deduced as the C-7 epimer of sparticolin E.

Sparticolin H (**2**) was also subjected to TDDFT-ECD calculations to confirm its absolute configuration. Due to its rigid structure, conformational analysis of **2** yielded only one prevalent conformer with an equilibrium population of 99.62% with a pre-assigned (4*S*,5*S*,7*R*) configuration with two other low-energy conformers. All conformers are confirmed stable per harmonic vibrational frequency calculations. The theoretically obtained Boltzmann-averaged ECD spectrum of **2** at WB97XD/DGDZVP (PCM/MeCN) showed a good correlation with the experimental data (Figure 4), which strongly suggests that the absolute configuration of sparticolin H is (4S,5S,7R).

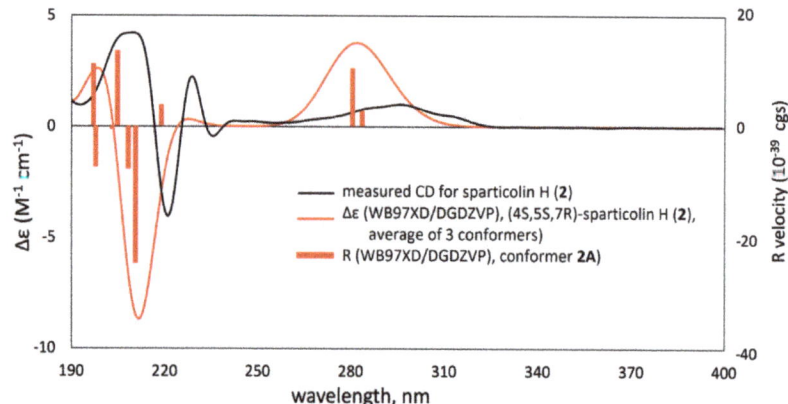

Figure 4. Experimental ECD spectrum of sparticolin H (**2**, black solid curve) compared with WB97XD/DGDZVP (PCM/MeCN)-calculated ECD spectra (red solid curve) for the WB97XD/DGDZVP-optimized conformers of (4S,5S,7R)-**2**.

Compounds 1–3 were also evaluated for their cytotoxic activity against mouse fibroblasts L-929 (ACC 2), human cell lines like HeLa cells (ACC 57), KB-3-1 (ACC 158), squamous carcinoma A-431 (ACC 91), lung carcinoma A-549 (ACC 107), ovarian carcinoma SK-OV-3 (HTB-77), prostate cancer PC-3 (ACC 465), and breast adenocarcinoma MCF-7 (ACC 115). Among the isolated compounds, sparticolins H (**2**) and A (**3**) exhibited a moderate antiproliferative effect against myelogenous leukemia (Table 2). In addition, compound **2** displayed moderate cytotoxic properties against HeLa cervical cancer cell line and mouse fibroblast cell line L-929. All compounds were weakly inhibitory against *Candida albicans* (DSM 1665), *Mucor hiemalis* (DSM 2656), *Pichia anomala* (DSM 6766), *Rhodoturula glutinis* (DSM 10134), *Schizosaccharomyces pombe* (DSM 70572), *Bacillus subtilis* (DSM 10), *Chromobacterium violaceum* (DSM 30191), *Escherichia coli* (DSM 1116), *Acinetobacter baumannii* (DSM 30008), *Mycobacterium smegmatis* (ATCC 700084), *Pseudomonas aeruginosa* (DSM PA14), *Staphylococcus aureus* (DSM 346) and *Mycobacterium tuberculosis* $H_{37}Rv$.

Table 2. Antiproliferative and cytotoxicity of compounds 1–3 against mammalian cancer cell lines.

Compound	Antiproliferative Effect GI_{50} (µM)		Cytotoxicity CC_{50} (µM)	Cytotoxicity IC_{50} (µM)	
	HUVEC	K-562	HeLa	L929	KB3.1
1	>50	97.5	85.3	–	–
2	>50	80.8	124.7	22.9	21.8
3	>50	91.3	119.8	–	–
Imatinib	18.5	0.17	65.8	N.D.	N.D.
Epothilone B	N.D.	N.D.	N.D.	1.57×10^{-3}	3.9×10^{-3}

N.D.: Not determined; Em dash (–): no observed cytotoxic activity.

The anti-inflammatory properties of **1–3** were assessed in vitro via their inhibitory activities against COX-1 and COX-2 (Table 3). All compounds were found to have high COX-2 and COX-1 inhibitory activities, with compound **1** exhibiting the highest activity. The interesting COX inhibitory activity of **1** prompted us to investigate further its binding behavior, particularly the protein-ligand interactions, against COX-2 and COX-1 via in silico molecular docking analysis (Figure 5). All calculations were done on AutoDock Vina with COX-2 (PDB ID: 3LN1) and COX-1 (PDB ID: 3KK6) as receptors. Validation of the docking protocol was done through redocking of the native ligand Celecoxib at the conserved active site for non-steroidal anti-inflammatory drugs (NSAID), which resulted in RMSD values of 0.448 Å for COX-2 and 0.668 Å for COX-1; both RMSDs are less than 2 Å, which indicates that the computed ligand–protein conformations are good [19–21]. Sparticatechol A (**1**) exhibited a high binding affinity of −9.2 kcal/mol against COX-2. Attempts to direct the docking of the compound to other plausible binding sites of COX-2, such as the hydrophilic side chain and the POX region, provided weaker binding, and thus reveals a preferential binding of the compound to the NSAID active site. At the active site, compound **1** is stabilized through various intermolecular forces at different regions of the molecule. The dihydroxyphenyl fragment exhibited two conventional H-bonding against Arg106 and Tyr341, while the aromatic ring interacted with the following residues: Ala 513 via pi-sigma and Leu517 via pi-alkyl. The naphthoketal moiety, on the other hand, is attached via pi-alkyl interplay with Val335, Leu338, and Val509. Four carbon–hydrogen bond interactions (against Gln178, Val335, Leu338, and Ser339) can also be observed for the β-hydroxyester chain. Meanwhile, sparticatechol A (**1**) is bound to the active site of COX-1 with a binding energy of -7.5 kcal/mol. The naphthalene fragment of **1** displayed two prominent pi interactions, particularly pi-donor hydrogen bonds with Ser530 and pi-alkyl attraction with Leu352 and Val 349. Several aromatic pi-alkyl interactions can also be observed between the dihydroxyphenyl fragment against Val349, Ala527, and Leu531. Moreover, three conventional hydrogen bonds helped strengthen **1**'s attachment to the enzyme, specifically between the following: C8(OH) and Val349, C22(OH) and Gln192, and C23(OH) and Ser353.

Table 3. In vitro Cyclooxygenase inhibitory activity of compounds **1–3**.

Compound	COX-1 IC_{50} (µM)	COX-2 IC_{50} (µM)
1	8.8×10^{-3}	0.3
2	1.8	1.5
3	1.4	2.3
Celecoxib	18.09	0.656

IC_{50} values represented as mean based on triplicate measurements.

Compounds **1–3** were assessed for their adsorption, metabolism, distribution, and excretion (ADME) properties in silico to provide a prediction of their potential pharmacokinetic behavior (Table S1). Lipinski's 'rule of five' (LRo5) was used to evaluate the pharmacokinetic profile of the compounds, which considers the following key physiochemical parameters: molecular weight (less than 500 Da), lipophilicity (MLogP less than 5), number of hydrogen bond donor (no more than 5), and number of hydrogen bond acceptors (no more than 10). As per LRo5, all three compounds exhibited good bioavailability and drug-likeness, with no violations against Lipinski parameters. Additionally, the compounds were subjected to a predictive brain or intestinal permeation modelling (BOILED-Egg) based on two functions, lipophilicity and total polar surface area [22]. From the BOILED-Egg plot (Figure 6), compounds **1** and **2**, located at the yellow region (yolk), are predicted to have a high propensity for blood–brain barrier (BBB) permeation. Compound **3**, on the hand, is predicted to be a non-BBB permeant, but is considered to have low gastrointestinal absorption, as it is slightly off the white region of the plot. Osiris Property Explorer was used to predict the toxicities of the compounds, specifically their potential mutagenicity, tumorigenicity, irritant effect, and reproductive toxicity (Table S2). All compounds were predicted to demonstrate mutagenic effects, which is due to the presence of a naphthalene fragment in their structures. However, structural manipulation, such as derivatization around the naphthalene core, may improve the compounds' toxicity profiles.

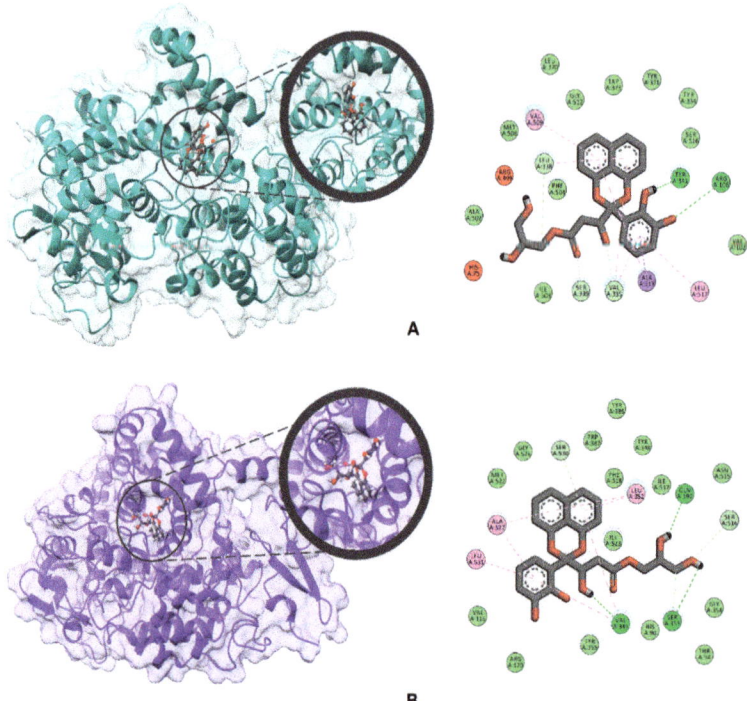

Figure 5. 3D and 2D docked poses of sparticatechol A (**1**) against (**A**) COX-2 (PDB ID: 3LN1) and (**B**) COX-1 (PDB ID: 3KK6).

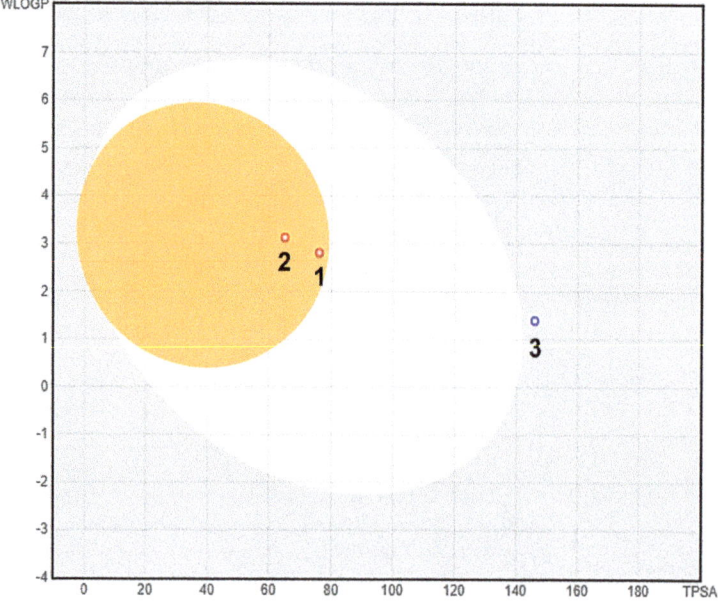

Figure 6. Prediction of GI tract and brain permeation of compounds **1–3** by brain or the intestinal estimated permeation predictive model (BOILED-Egg) method.

3. Materials and Methods

3.1. General Experimental Procedures

Specific optical rotations ($[\alpha]_D$) were measured on a PerkinElmer 241 polarimeter in a in 2.0 mm × 100 mm cell at 20 °C. UV-vis spectra were obtained on a Shimadzu UV-2450 spectrophotometer with a 1 cm quartz cell. IR spectra were measured either on a PerkinElmer Spectrum Two FT-IR spectrophotometer or a Shimadzu Prestige-21 spectrophotometer coupled with Diffuse Reflectance Spectroscopy in KBr. Nuclear magnetic resonance (NMR) spectra were obtained either on a Varian VNMRS-500 MHz or a Agilent DD2 MR Varian- 500 MHz (^{1}H 500 MHz, ^{13}C 125 MHz). Spectra were acquired at 25 °C in MeOH-d_4, with referencing to residual ^{1}H or ^{13}C signals in the deuterated solvent. HR-ESI mass spectra were measured on the use of the Agilent 6200 series TOF and 6500 series Q-TOF LC/MS system. The HPLC-DAD purification was carried out on a Shimadzu Prominence Liquid Chromatograph LC-20AT coupled with a SPD-M20A Photodiode Array Detector (Shimadzu Corp., Tokyo, Japan) and a semi-preparative reversed phase C_{18} column, Inertsil ODS-3 (10 mm I.D. × 250 mm, 5 μm, G.L. Sciences, Tokyo, Japan). The mobile phase was composed of ultrapure water (Milli-Q, Millipore, Schwalbach, Germany) as solvent A and acetonitrile (HPLC grade) as solvent B. Normal phase gravity column chromatography utilized silica gel 60 (Merck Art. 7734 and 9835). Aluminum TLC sheets pre-coated with silica gel 60 F254 (Merck Art. 1.05557) were used for routine analysis. TLC spots were visualized under UV light (254 nm and 365 nm) followed by spraying with a vanillin-sulfuric acid reagent.

3.2. Fungal Material and Fermentation

The voucher specimen is deposited at the Mae Fah Luang University Culture Collection, Chiang Rai, Thailand with designation numbers, MFLU 15-0030 and MFLUCC 15-1405 for ex-type culture and holotype specimen, respectively [13].

The fungus was cultivated on solid rice media (70 g brown rice, 0.3 g peptone, 0.1 g corn syrup, and 100 mL ultrapure water) in 15 × 1000 mL sterilized Fernbach culture flasks, followed by autoclaving (120 °C, 20 min). Five agar blocks of well-grown fungal culture in a malt extract agar plate were inoculated in the culture flasks and incubated under static conditions in a dark room at 25–30 °C, which lasted for 12 weeks until the fungal hyphae proliferated and the rice medium turned black in color.

3.3. Extraction and Isolation

The rice cultures were homogenized using a sterile metal spatula. Fermentation was terminated by the addition of EtOAc (3 × 300 mL) and the combined extracts were concentrated in a rotary evaporator to afford the crude extract (10.2 g). The crude EtOAc extract was reconstituted with 300 mL 10% aqueous MeOH and partitioned with n-heptane (3 × 100 mL). The combined organic layer was concentrated in vacuo to afford 2.73 g methanolic crude extract.

The resulting crude methanolic extract was fractionated, eluting initially with petroleum ether–EtOAc (1:1 and 2:3) followed by 20% increments of dichloromethane in methanol and finally methanol to afford 5 fractions. Fraction 1 (2.22 g) was chromatographed using petroleum ether–EtOAc (6:1, 3:1 and 1:1), EtOAc–MeOH (4:1 and 9:1) to yield 5 subfractions. Subfraction 1.5 was further purified twice using petroleum ether–EtOAc (1:12), EtOAc–MeOH (12:1, 7:3 and 1:1) followed by semi-preparative reversed phase HPLC by gradient elution using 40% solvent B for 5 min, 40%–100% solvent B for 20 min, 100% solvent B for 5 min and 100%–40% solvent B for 5 min to afford compound **1** (4.3 mg, t_R = 16.8 min). Compound **2** was obtained from the first subfraction of 1.5, which was purified by silica gel column chromatography using pure dichloromethane and dichloromethane–MeOH (13:1) followed by semi-preparative reversed phase HPLC using similar gradient conditions used for the purification of **1** to afford **2** (13.4 mg, t_R = 22.8 min). A subfraction 1.2 was column-chromatographed using a petroleum ether and a mixture of petroleum ether– CH_2Cl_2 (1:4,

1:8 and 1:9) and CH$_2$Cl$_2$–MeOH (9:1), followed by semi-preparative reversed phase HPLC employing similar gradient solvent system to afford compound 3 (3.5 mg, t_R = 23.1 min).

Sparticatechol A (**1**): brown syrup; $[\alpha]_D^{25}$ +120 (c 0.1, MeOH); TLC (CH$_2$Cl$_2$:MeOH, 9:1 v/v): R$_f$ = 0.17; UV (c 0.1, ACN) λ_{max} (log ε) 226 (5.0), 254 (4.7), 300 (4.5), 314 (4.4), 328 (4.4) nm; ^{1}H NMR (CD$_3$OD, 600 MHz), and ^{13}C NMR (CD$_3$OD, 600 MHz) see Table 1; IR (KBr): 3040, 1610, 1380, 1275, 1160, 1110, 1060, 955, 910, 870, 835, 755, 710 cm^{-1} HR-ESI-MS m/z: [M + H]$^+$ calcd for C$_{23}$H$_{23}$O$_9$, 443.1342; found, 443.1330; m/z [M + H – H$_2$O]$^+$ calcd for C$_{23}$H$_{21}$O$_8$, 425.1236; found, 425.1226; m/z [M + Na]$^+$ calcd for C$_{23}$H$_{22}$NaO$_9$, 465.1162; found, 465.1152. IUPAC nomenclature: (8S)-(22R)-2,3-dihydroxypropyl-3-(2-(2,3-dihydroxyphenyl)naphtho[1,8-de][1,3]dioxin-2-yl)-3-hydroxypropanoate.

Sparticolin H (**2**): dark brown syrup; $[\alpha]_D^{25}$ +104 (c 0.1, MeOH); TLC (CH$_2$Cl$_2$:MeOH, 9:1 v/v): R$_f$ = 0.43; UV (c 0.1, ACN) λ_{max} (log ε) 226 (4.9), 300 (4.3), 314 (4.3), 328 (4.2) nm; ^{1}H NMR (CD$_3$OD, 500 MHz), and ^{13}C NMR (CD$_3$OD, 125 MHz) see Table 1; IR (KBr): 1712, 1636, 1607, 1584, 1411, 1380, 1273, 1143, 1073, 1037, 957, 895, 820, 794, 756, 666 cm^{-1}; HR-ESI-MS m/z: [M + H]$^+$ calcd for C$_{19}$H$_{17}$O$_5$, 325.1071; found, 325.1066; m/z [M + Na]$^+$ calcd for C$_{19}$H$_{16}$NaO$_5$, 347.0890; found, 347.0889. IUPAC nomenclature: 2-((2R,3aR,6aS)-2,3,3a,6a-tetrahydrospiro[cyclopenta[b]furan-4,2'-naphtho[1,8-de][1,3]dioxin]-2-yl)acetic acid.

3.4. Biological Assays

Anti-cyclooxygenase (COX) Assay. The cyclooxygenase inhibitory potential of compounds **1–3** were assessed using the COX Activity Assay Kit (Catalog No. 760151, Cayman Chemical, Ann Arbor, MI, USA). The assay kit quantifies the peroxidase activity of cyclooxygenases by observing colorimetrically the oxidation of N,N,N',N'-tetramethyl-p-phenylenediamine (TMPD) at 590 nm. The samples were dissolved in DMSO with concentrations ranging from 0.1 to 10 µg/mL. Celecoxib was used as a reference compound. The assay was carried out according to the protocol provided by the manufacturer and the half-maximal inhibitory concentration of the compounds were determined using concentration-response curve in triplicate measurements.

Anti-proliferation and Cytotoxicity Assay. Compounds **1–3** were assayed against human umbilical vein endothelial cells (HUVEC, ATCC CRL-1730) and K562 human chronic myeloid leukemia cells (DSM ACC 10) for their antiproliferative effects (GI$_{50}$). Cytotoxicity properties were also assessed against several mammalian cancer cell lines including mouse fibroblast L929 (ACC 2), HeLa (KB3.1), human breast adenocarcinoma (MCF-7, ACC 115), adenocarcinomic human alveolar basal epithelial cells (A549, ACC 107), human prostate cancer cells (PC-3, ACC 465), ovarian carcinoma (SKOV-3, HTB-77) and squamous cell carcinoma (A431, ACC 91), which was expressed as IC$_{50}$ by MTT Assay [23] and against HeLa human cervix carcinoma cells (DSM ACC 57), expressed as CC$_{50}$ by Cell Titer Blue Assay [24]. Inhibitory concentrations are provided as 50% half-maximal inhibitory concentration (IC$_{50}$, concentration of the substance where a specific biological process is reduced by half), 50% inhibition of cell growth (GI$_{50}$, the concentration needed to reduce the growth of treated cells to half that of untreated cells) or 50% cytotoxic concentration (CC$_{50}$, the concentration that kills 50% of treated cells).

Antimicrobial Assay. The antimicrobial activities of **1–3** were evaluated against various fungal and bacterial strains. The minimum inhibitory concentration (MIC) of the tested compounds were determined in a sterile 96-well plates by the broth microdilution method according to our previously described procedures [25]. Gentamicin (MIC vs. *P. aeruginosa* = 0.21 µg/mL), kanamycin (MIC vs. *M. smegmatis* = 1.7 µg/mL), nystatin (MIC vs. *R. glutinis* = 2.1 µg/mL; MIC vs. *C. albicans*, *P. anomala*, *M. hiemalis*, *S. pombe* = 4.2 µg/mL), ciprobay (MIC vs. *A. baumannii* = 0.26 µg/mL) and oxytetracycline hydrochloride (MIC vs. *S. aureus* = 0.21 µg/mL; MIC vs. *C. violaceum* = 0.83 µg/mL; MIC vs. *E. coli* = 3.3 µg/mL; MIC vs. *B. subtilis* = 8.3 µg/mL). Nystatin was used as a positive control against fungi while ciprobay, gentamicin, kanamycin and oxytetracyclin were used as a positive control against Gram-positive and Gram-negative bacteria, respectively.

Antituberculosis Assay. Antituberculosis inhibitory activity of the compounds against replicating Mycobacterium tuberculosis $H_{37}Rv$ (American Type Culture Collection, Rockville, MD, USA) was determined using a fluorescence reading at 530 nm excitation wavelength and 590 nm emission wavelength in the Microplate Alamar Blue Assay (MABA) [26]. The MIC values was determined as the lowest concentration exhibiting a 90% fluorescence inhibition compared to the untreated bacterial control. Rifampin (MIC = 0.15 µg/mL), isoniazid (MIC = 0.63 µg/mL), and streptomycin (MIC = 0.83 µg/mL) were used as positive drug controls.

3.5. Computational Calculations

Theoretical ECD Calculations. A conformational analysis on both sparticatechol A (**1**) and sparticol H (**2**) was performed using the Avogadro (version 1.1.1) platform, which included a search for low-energy conformations using the MMFF94 molecular mechanics force field and conformer optimization following the steepest descent algorithm. The geometries of all stable conformers were re-optimized via density functional theory calculations at a B3LYP/6-31G(d) basis set using acetonitrile as a solvent model on a polarizable continuum model (PCM). Boltzmann population distribution was estimated for each conformer based on the calculated energies, taken as the sum of electronic and zero-point energies. The optimized geometries were subjected time-dependent DFT (TDDFT) using B3LYP/6-31G(d), WB97XD/DGDZVP, and B97D/TZVP basis sets at gas phase and acetonitrile PCM solvent model. A Gaussian distribution function was used to generate the ECD curve from the calculated rotatory strength values with 3000 cm^{-1} half-height width. All DFT calculations were carried out using Gaussian 16W while the visualization of results was done on GaussView 6.0 [27].

Molecular docking studies. Compound **1** was subjected to molecular docking analysis against COX-2 (PDB ID: 3LN1). All molecular docking experiments were performed on UCSF Chimera platform (version 1.14.1) (University of California-San Francisco, CA, United States) [28]. The three-dimensional structure of COX-2 was added to the docking platform as PDB format. The receptor was prepared by Each protein crystal structure removing all co-crystallized ligands and water molecules. Meanwhile, compounds **1–3** as ligands, were added to the docking platform as the SYBYL mol2 file, which were pre-optimized using the using the MMFF94 force field via the steepest descent algorithm using Avogadro software. Minimization and dock-prepping of ligand and protein structures were done using Antechamber [29], and molecular docking was performed using the BFGS algorithm of AutoDock Vina (version 1.1.2) [30]. The conformational protein–ligand structure was visualized and analyzed using Biovia Discovery Studios (version 4.1).

ADMET Profiling. In silico prediction of the adsorption, distribution, metabolism, and excretion (ADME) properties of compounds **1–3** was carried out using SwissADME software (Molecular Modeling Group, Swiss Institute of Bioinformatics, Lausanne, Switzerland, 2019) [22,31]. Pharmacokinetic profiles of the compounds were assessed based on the Lipinski's 'rule of five', which predicts a compound's oral druggability. Additionally, OSIRIS property explorer program (Thomas Sander, Idorsia Pharmaceuticals Ltd., Allschwil, Switzerland, 2017) was employed for the in silico prediction of toxicity prediction, particularly mutagenicity, tumorigenicity, irritant effects, and reproductive toxicity of the metabolites [19].

4. Conclusions

This study investigated the biologically active chemical constituents of the rice culture of *Sparticola junci*, where two highly oxygenated naphthoketal-bearing polyketides, sparticatechol A (**1**), sparticolin H (**2**), along with sparticolin A (**3**), were isolated and identified. Compounds **1** and **2** inhibited the two cyclooxygenase isozymes, COX-1 and COX-2 illustrating its anti-inflammatory potentials. In silico molecular docking analysis of **1** showed preferential binding in the NSAID active site. In addition, **1** and **2** also demonstrated weak antiproliferative effects against myeloid leukemia K-562 cell lines and cytotoxic activity against HeLa cervical carcinoma cell lines. Compounds **1–3** exhibited good bioavailability and drug-likeness, with no violations against Lipinski parameters. Our findings in general establish the potential of naphthoketal derivatives **1–3** as a possible drug inspiration for discovering new anti-inflammatory and cancer agents.

Supplementary Materials: The following are available online at https://www.mdpi.com/article/10.3390/ijms222212379/s1.

Author Contributions: Conceptualization, K.Y.M.G., M.T.J.Q., M.S., F.S. and A.P.G.M.; methodology, K.Y.M.G., M.T.J.Q., G.P., A.R., M.J.B.C., J.F.A.L., H.-M.D. and C.P.; software, M.T.J.Q.; validation, K.Y.M.G., M.T.J.Q., M.S., H.-M.D., F.S. and A.P.G.M.; formal analysis, K.Y.M.G., M.T.J.Q., G.P., A.R., M.J.B.C., J.F.A.L., H.-M.D., C.P., F.S. and A.P.G.M.; investigation, K.Y.M.G., M.T.J.Q., G.P., A.R., M.J.B.C., J.F.A.L., H.-M.D., C.P., F.S. and A.P.G.M.; resources, K.Y.M.G., M.T.J.Q., M.S., A.R., H.-M.D., F.S. and A.P.G.M.; data curation, K.Y.M.G., M.T.J.Q., M.J.B.C., J.F.A.L.; writing—original draft preparation, K.Y.M.G., M.T.J.Q., M.J.B.C., J.F.A.L., C.P., F.S. and A.P.G.M.; writing—review and editing, M.S., H.-M.D., F.S. and A.P.G.M.; visualization, K.Y.M.G. and M.T.J.Q.; supervision, M.S., F.S. and A.P.G.M.; project administration, A.P.G.M.; funding acquisition, K.Y.M.G. and A.P.G.M. All authors have read and agreed to the published version of the manuscript.

Funding: Department of Science and Technology (DOST)/Accelerated Science & Technology Human Resource Development Program (ASTHRDP).

Institutional Review Board Statement: Not applicable.

Informed Consent Statement: Not applicable.

Data Availability Statement: NMR and HRMS spectroscopic data are available at the Research Center for the Natural and Applied Sciences, University of Santo Tomas, Philippines and Institute of Organic Chemistry, University of Regensburg, Germany (AR), while the MIC vs pathogens and IC_{50} or GI_{50} or CC_{50} vs cancer cell-lines data are available at the Helmholtz Center for Infection Research, Germany and Leibniz Institute of Natural Products, Germany. IC_{50} and binding energy values vs COX enzymes are available at the Research Center for the Natural and Applied Sciences, University of Santo Tomas, Philippines.

Acknowledgments: KYMG acknowledges scholarship grant from the Department of Science and Technology (DOST) thru its Accelerated Science & Technology Human Resource Development Program (ASTHRDP). GP thanks the Deutscher Akademischer Austauschdienst (DAAD) for his graduate scholarship. We also thank Wera Collisi and Christel Kakoschke for expert technical assistance.

Conflicts of Interest: The authors declare no conflict of interest.

References

1. Kuephadungphan, W.; Macabeo, A.P.G.; Luangsa-ard, J.J.; Stadler, M. Discovery of novel biologically active secondary metabolites from Thai mycodiversity with anti-infective potential. *CRBIOT* **2004**, *3*, 160–172. [CrossRef]
2. Hyde, K.D.; Xu, J.; Rapior, S.; Jeewon, R.; Lumyong, S.; Niego, A.G.T.; Abeywickrama, P.D.; Aluthmuhandiram, J.V.S.; Brahamanage, R.S.; Brooks, S.; et al. The amazing potential of fungi: 50 ways we can exploit fungi industrially. *Fungal Divers.* **2019**, *97*, 1–136.
3. Bills, G.F.; Gloer, J.B. Biologically active secondary metabolites from the fungi. *Microbiol. Spectr.* **2016**, *4*, 1–32. [CrossRef] [PubMed]
4. Cai, Y.-S.; Guo, Y.-W.; Krohn, K. Structure, bioactivities, biosynthetic relationships and chemical synthesis of the spirodioxynaphthalenes. *Nat. Prod. Rep.* **2010**, *27*, 1840–1870. [CrossRef]
5. Ogishi, H.; Chiba, N.; Mikawa, T.; Sasaki, T.; Miyaji, S.; Sezaki, M. Mitsubishi Kasei Corp., JP 01294686. *Chem. Abstr.* **1990**, *113*, 38906q.
6. Petersen, F.; Moerker, T.; Vanzanella, F.; Peter, H.H. Production of cladospirone bisepoxide, a new fungal metabolite. *J. Antibiot.* **1994**, *47*, 1098–1103. [CrossRef]
7. Thiergardt, R.; Rihs, G.; Hug, P.; Peter, H.H. Cladospirone bisepoxide: Definite structure assignment including absolute configuration and selective chemical transformations. *Tetrahedron* **1995**, *51*, 733–742. [CrossRef]
8. Krohn, K.; Michel, A.; Flörke, U.; Aust, H.-J.; Draeger, S.; Schulz, B. Palmarumycins C_1–C_{16} from *Coniothyrium* sp.: Isolation, structure elucidation and biological activity. *Liebigs Ann. Chem.* **1994**, *1994*, 1099–1108. [CrossRef]
9. Jiao, P.; Swenson, D.C.; Gloer, J.B.; Campbell, J.; Shearer, C.A. Decaspirones A–E, Bioactive Spirodioxynaphthalenes from the freshwater aquatic fungus *Decaisnella thyridioides*. *J. Nat. Prod.* **2006**, *69*, 1667–1671. [CrossRef]

10. Dai, J.; Krohn, K.; Elsässer, B.; Flörke, U.; Draeger, S.; Schulz, B.; Pescitelli, G.; Salvadori, P.; Antus, S.; Kurtán, T. Metabolic products of the endophytic fungus *Microsphaeropsis* sp. from *Larix decidua*. *Eur. J. Org. Chem.* **2007**, *29*, 4845–4954. [CrossRef]
11. Chu, M.; Truumees, I.; Patel, M.G.; Gullo, V.P.; Pai, J.-K.; Das, P.R.; Puar, M.S. Two new phospholipase d inhibitors, Sch 49211 and Sch 49212, produced by the fungus *Nattrasia mangiferae*. *Bioorg. Med. Chem. Lett.* **1994**, *4*, 1539–1542. [CrossRef]
12. Van der Sar, S.A.; Lang, G.; Mitova, M.; Blunt, J.; Cole, A.L.J.; Cummings, N.; Ellis, G.; Munro, M.H.G. Biosynthesis of spiro-mamakone A, a structurally unprecedented fungal metabolite. *J. Org. Chem.* **2008**, *73*, 8635–8638. [CrossRef] [PubMed]
13. Phukhamsakda, C.; Macabeo, A.P.G.; Huch, V.; Cheng, T.; Hyde, K.D.; Stadler, M. Sparticolins A–G, Biologically active oxidized pirodioxynaphthalene derivatives from the ascomycete *Sparticola junci*. *J. Nat. Prod.* **2019**, *82*, 2878–2885. [CrossRef] [PubMed]
14. Garcia, K.Y.M.; Phukhamsakda, C.; Quimque, M.T.J.; Hyde, K.D.; Stadler, M.; Macabeo, A.P.G. Catechol-bearing polyketide derivatives from *Sparticola junci*. *J. Nat. Prod.* **2021**, *84*, 2053–2058. [CrossRef]
15. Macabeo, A.P.G.; Pilapil, L.A.E.; Garcia, K.Y.M.; Quimque, M.T.J.; Phukhamsakda, C.; Cruz, A.J.C.; Hyde, K.D.; Stadler, M. Alpha-Glucosidase- and Lipase-Inhibitory Phenalenones from a new species of *Pseudolophiostoma* originating from Thailand. *Molecules* **2020**, *25*, 965. [CrossRef]
16. Phukhamsakda, C.; Macabeo, A.P.G.; Yuyama, K.T.; Hyde, K.D.; Stadler, M. Biofilm inhibitory abscisic acid derivatives from the plant-associated Dothideomycete fungus, *Roussoella* sp. *Molecules* **2018**, *23*, 2190. [CrossRef] [PubMed]
17. Walton, L.J.; Corre, C.; Challis, G.L. Mechanisms for incorporation of glycerol-derived precursors into polyketide metabolites. *J. Ind. Microbiol.* **2005**, *33*, 105–120. [CrossRef]
18. Yan, Y.; Zang, X.; Jamieson, C.S.; Lin, H.-C.; Houk, K.N.; Zhou, J.; Tang, Y. Biosynthesis of the fungal glyceraldehyde-3-phosphate dehydrogenase inhibitor heptelidic acid and mechanism of self-resistance. *Chem. Sci.* **2020**, *11*, 9554–9562. [CrossRef]
19. Quimque, M.T.; Notarte, K.I.; Letada, A.; Fernandez, R.A.; Pilapil, D.Y.; Pueblos, K.R.; Agbay, J.C.; Dahse, H.-M.; Wenzel-Storjohann, A.; Tasdemir, D.; et al. Potential cancer- and Alzheimer's Disease-targeting phosphodiesterase inhibitors from *Uvaria alba*: Insights from *in vitro* and consensus virtual screening. *ACS Omega* **2021**, *6*, 8403–8417. [CrossRef] [PubMed]
20. Acharya, R.; Chacko, S.; Bose, P.; Lapenna, A.; Pattanayak, S.P. Structure based multitargeted molecular docking analysis of selected furanocoumarins against breast cancer. *Sci. Rep.* **2019**, *9*, 1–13. [CrossRef]
21. Quimque, M.T.J.; Liman, R.A.; Agbay, J.C.; Macabeo, A.P.G.; Corpuz, M.J.-A.; Wang, Y.-M.; Lu, T.-T.; Lin, C.H.; Villaflores, O.B. Computational and experimental assessments of magnolol as a neuroprotective agent and utilization of UiO-66(Zr) as Its drug delivery system. *ACS Omega* **2021**, *6*, 24382–24396.
22. Daina, A.; Michielin, O.; Zoete, V. SwissADME: A free web tool to evaluate pharmacokinetics, drug-likeness and medicinal chemistry friendliness of small molecules. *Sci. Rep.* **2017**, *7*, 1–13. [CrossRef] [PubMed]
23. Karuth, J.; Dahse, H.-M.; Rüttinger, H.-H.; Frohberg, P. Synthesis and characterization of novel 1,2,4-triazine derivatives with antiproliferative activity. *Bioorg. Med. Chem.* **2010**, *18*, 1816–1821. [CrossRef]
24. Becker, K.; Wessel, A.C.; Luangsa-ard, J.J.; Stadler, M. Viridistratins A–C, antimicrobial and cytotoxic benzo[j]fluoranthenes from stromata of *Annulohypoxylon viridistratum* (Hypoxylaceae, Ascomycota). *Biomolecules* **2020**, *10*, 805. [CrossRef] [PubMed]
25. Kuephadungphan, W.; Macabeo, A.P.G.; Luangsa-ard, J.J.; Tasanathai, K.; Thanakitpipattana, D.; Phongpaichit, S.; Yuyama, K.; Stadler, M. Studies on the biologically active secondary metabolites of the new spider parasitic fungus *Gibellula gamsii*. *Mycol. Prog.* **2019**, *18*, 135–146. [CrossRef]
26. Collins, L.; Franzblau, S.G. Microplate alamar blue assay versus BACTEC 460 system for high-throughput screening of compounds against *Mycobacterium tuberculosis* and *Mycobacterium avium*. *Antimicrob. Agents Chemother.* **1997**, *41*, 1004–1009. [CrossRef]
27. Frisch, M.J.; Trucks, G.W.; Schlegel, H.B.; Scuseria, G.E.; Robb, M.A.; Cheeseman, J.R.; Scalmani, G.; Barone, V.M.; Petersson, G.A.; Nakatsuji, H.; et al. *Gaussian 16, Revision A. 03*; Gaussian Inc.: Wallingford, CT, USA, 2016.
28. Pettersen, E.F.; Goddard, T.D.; Huang, C.C.; Couch, G.S.; Greenblatt, D.M.; Meng, E.C.; Ferrin, T.E. UCSF Chimera: A visualization system for exploratory research and analysis. *J. Comput. Chem.* **2004**, *25*, 1605–1612. [CrossRef]
29. Wang, J.; Wang, W.; Kollman, P.A.; Case, D.A. Automatic atom type and bond type perception in molecular mechanical calculations. *J. Mol. Graph. Modell.* **2006**, *25*, 247–260. [CrossRef]
30. Trott, O.; Olson, A.J. AutoDock Vina: Improving the speed and accuracy of docking with a new scoring function, efficient optimization, and multithreading. *J. Comput. Chem.* **2009**, *31*, 455–461. [CrossRef]
31. Daina, A.; Zoete, V. A BOILED-egg to predict gastrointestinal absorption and brain penetration of small molecules. *ChemMedChem* **2016**, *11*, 1117–1121. [CrossRef]

Article

ATG 4B Serves a Crucial Role in RCE-4-Induced Inhibition of the Bcl-2–Beclin 1 Complex in Cervical Cancer Ca Ski Cells

Fang-Fang You, Jing Zhang, Fan Cheng, Kun Zou, Xue-Qing Zhang and Jian-Feng Chen *

Hubei Key Laboratory of Natural Products Research and Development, College of Biological and Pharmaceutical Sciences, China Three Gorges University, Yichang 443002, China; fangfangyou2020@163.com (F.-F.Y.); zj031020@163.com (J.Z.); chengf@ctgu.edu.cn (F.C.); kzou@ctgu.edu.cn (K.Z.); happy.xueqing@163.com (X.-Q.Z.)
* Correspondence: chenjianfeng@ctgu.edu.cn; Tel.: +86-071-763-97478

Abstract: RCE-4, a steroidal saponin isolated from *Reineckia carnea*, has been studied previously and has exhibited promising anti-cervical cancer properties by inducing programmed cell death (PCD) of Ca Ski cells. Considering the cancer cells developed various pathways to evade chemotherapy-induced PCD, there is, therefore, an urgent need to further explore the potential mechanisms underlying its actions. The present study focused on targeting the Bcl-2–Beclin 1 complex, which is known as the key regulator of PCD, to deeply elucidate the molecular mechanism of RCE-4 against cervical cancer. The effects of RCE-4 on the Bcl-2–Beclin 1 complex were investigated by using the co-immunoprecipitation assay. In addition, autophagy-related genes (ATG) were also analyzed due to their special roles in PCD. The results demonstrated that RCE-4 inhibited the formation of the Bcl-2–Beclin 1 complex in Ca Ski cells via various pathways, and ATG 4B proteins involved in this process served as a key co-factor. Furthermore, based on the above, the sensitivity of RCE-4 to Ca Ski cells was significantly enhanced by inhibiting the expression of the ATG 4B by applying the ATG 4B siRNA plasmid.

Keywords: RCE-4; PCD; ATG 4B; the Bcl-2–Beclin 1 complex

1. Introduction

Cervical cancer is the third most common gynecological malignant tumor [1]. Clinically, chemotherapy remains the primary therapeutic regime, but the drug resistance and serious side effects often lead to poor treatment expectations. Natural medicines have attracted extensive attention due to their multiple targets and low toxicity. In our previous studies, RCE-4 (Figure 1), a natural candidate drug for cervical cancer isolated from *Reineckia carnea* [2–5], could induce PCD characterized by apoptosis and autophagy for cervical cancer Ca Ski cells selectively with an IC50 of 4.71 µmol/L. In addition, the tumor inhibition rate for a human cervical cancer xenograft in nude mice attained 69.1% with the extremely low toxicity to normal tissues [6]. These discoveries highlighted the tremendous value of RCE-4 for treating cervical cancer. Now, RCE-4 has been included as the typical ingredient of *Reineckia carnea* in the local Chinese medicinal material standards of Hubei Province [7]. However, some shortcomings of RCE-4 have also been exposed, such as the relatively high dosage and the molecular mechanism underlying its anti-cervical cancer actions not being fully clarified.

Most current anticancer chemotherapy drugs primarily act by activating programmed cell death (PCD) pathways including apoptosis and autophagy in cancer cells [8]. Targeting the process of PCD with different small-molecule compounds has become a promising therapeutic strategy over the last few decades [9,10]. However, tumor cells have developed novel mechanisms for evading chemotherapy-induced PCD, this could be associated with both the autophagy and inhibition of the more common apoptosis cell death pathways [11]. One of the hallmarks of human cancers is the intrinsic or acquired resistance to apoptosis. Evasion of apoptosis can be part of a cellular stress response to ensure the cell's survival upon exposure to stressful stimuli. Autophagy is also a protective mechanism of tumor cells that can promote the growth of established tumors. Autophagy-related stress tolerance can enable cell survival by maintaining energy production that can lead to tumor growth and therapeutic resistance. Hence, it was necessary to further explore PCD signaling pathways that are drug-induced, which are not only able to give us new insights into the pathogenesis of tumors but also to aid the development of new targeted therapeutic strategies.

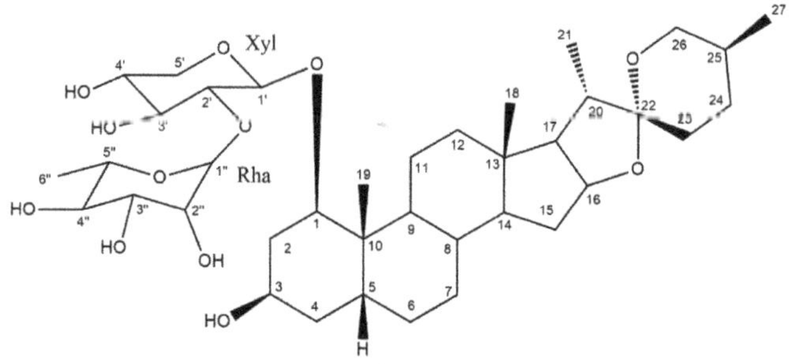

Figure 1. The structure of RCE-4 (1β, 3β, 5β, 25S)-spirostan-1,3-diol1-[α-L-rhamnopyranosyl-(1→2) -β-D-xylopyranoside].

With regards to apoptosis and autophagy as the main two forms of PCD, although there are obvious distinctions in intracellular processes, studies have proved the complex interplay between them [12–14]. Bcl-2–Beclin 1 complex, a macro-molecular protein consisting of Bcl-2 and Beclin 1 via the BH3 domain, was confirmed to be the core regulator of crosstalk between apoptosis and autophagy and the "master switch" of PCD [15,16]. The dynamic changes in intracellular Bcl-2–Beclin 1 complex molecules regulate the whole process of PCD, including the initiation, progress and termination [17,18]. The process of PCD can be changed when the binding of Bcl-2 and Beclin 1 is blocked or the dissociation of the complex is promoted, accompanied by the content of the intracellular Bcl-2–Beclin 1 complex decreasing. Numerous studies have shown that the Bcl-2–Beclin 1 complex is stable under nutrient-rich conditions, but when there is a lack of nutrients or other external existing stimuli, some signal molecules participate in the regulation of the complex [19,20]. For example, JNK1 could mediate the rapid phosphorylation of Bcl-2 [21,22], Mst1 could regulate the phosphorylation of Beclin 1 and Dapper 1 could promote the formation of the other competitive Beclin 1-related complexes [23,24], and these could lead to the inhibition of the formation of the Bcl-2–Beclin 1 complex. Ultimately, the apoptosis and autophagic flux processes would all be changed accordingly.

All in all, regulation of the Bcl-2–Beclin 1 complex is of great significance for drug development and the treatment of certain diseases, such as tumors, cerebral ischemia and neurodegenerative disease, etc. [25,26]. However, relevant reports focused on targeting the Bcl-2–Beclin 1 complex for developing anticancer drugs or studying the molecular mechanism of anticancer drugs have not been seen yet. Therefore, the present study focuses on the Bcl-2–Beclin 1 complex and takes RCE-4 as a model drug to deeply clarify the molecular mechanisms underlying its anti-cervical cancer effects.

In addition, it was expected to find effective measures to improve the sensitivity on the basis of understanding the molecular mechanisms, and this might provide theoretical support for the future clinical application of RCE-4.

2. Results

2.1. RCE-4 Induces Time- and Concentration-Dependent Apoptosis and Autophagy in Ca Ski Cells

RCE-4 was found to induce PCD of cervical cancer Ca Ski cells in our previous studies. In the present study, we further confirmed that RCE-4 could induce time- and concentration-dependent apoptosis and autophagy in Ca Ski cells. As shown in Figure 2A,B, Ca Ski cells were treated with RCE-4 for 6, 12 and 24 h. Compared with the control group, the expression levels of the apoptosis-related proteins cleaved caspase-3/-7/-9 and Bax were increased, caspase-3/-9 and Bcl-2 were reduced ($p < 0.05$ and $p < 0.01$). Additionally, the protein expression levels of the autophagy related proteins, LC3 II and Beclin1 were increased, and the expression of P62 was decreased. As shown in Figure 2C,E, cells treated with RCE-4 were stained with AO/EB, and an obviously nuclear contraction and rupture were observed in a trapezoidal pattern ($p < 0.01$). In addition, the results of flow cytometry also revealed that RCE-4 induced, significantly, apoptosis in Ca Ski cells (Figure 2D,F, $p < 0.01$).

2.2. RCE-4 Inhibits the Formation of Bcl-2–Beclin 1 Complex in Ca Ski Cells

Co-IP assay was used to analyze the relative content of the Bcl-2–Beclin 1 complex in Ca Ski cells. Bcl-2 was used as a bait and anti-Bcl-2 monoclonal antibody as IP, Beclin 1 was analyzed as the target protein to evaluate the relative content of the Bcl-2–Beclin 1 complex. In this experiment, Input was used as the positive control and homologous IgG was used as the negative control to exclude nonspecific binding. Thus, the content of Beclin 1 protein reflected the relative content of the Bcl-2–Beclin 1 complex compared with the blank group. When Beclin 1 was used as a bait, the reverse held true as well, the experimental design was similar. Ca Ski cells were treated with RCE-4 of 8 μmol/L for 6, 12 and 24 h, When using anti-Bcl-2 monoclonal antibody as IP, the expression level of Beclin 1 was significantly reduced, indicating that RCE-4 significantly inhibits the formation of the Bcl-2–Beclin 1 complex in Ca Ski cells. The experimental results when using anti-Beclin 1 monoclonal antibody as IP also confirmed the effect of RCE-4 in inhibiting the formation of the complex (Figure 3).

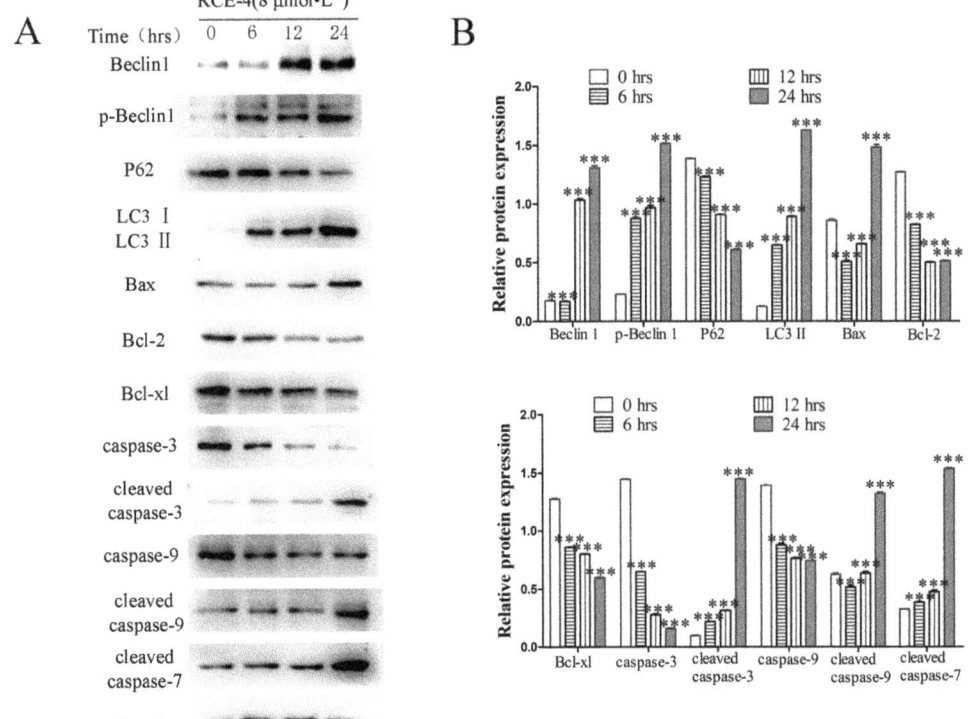

Figure 2. *Cont.*

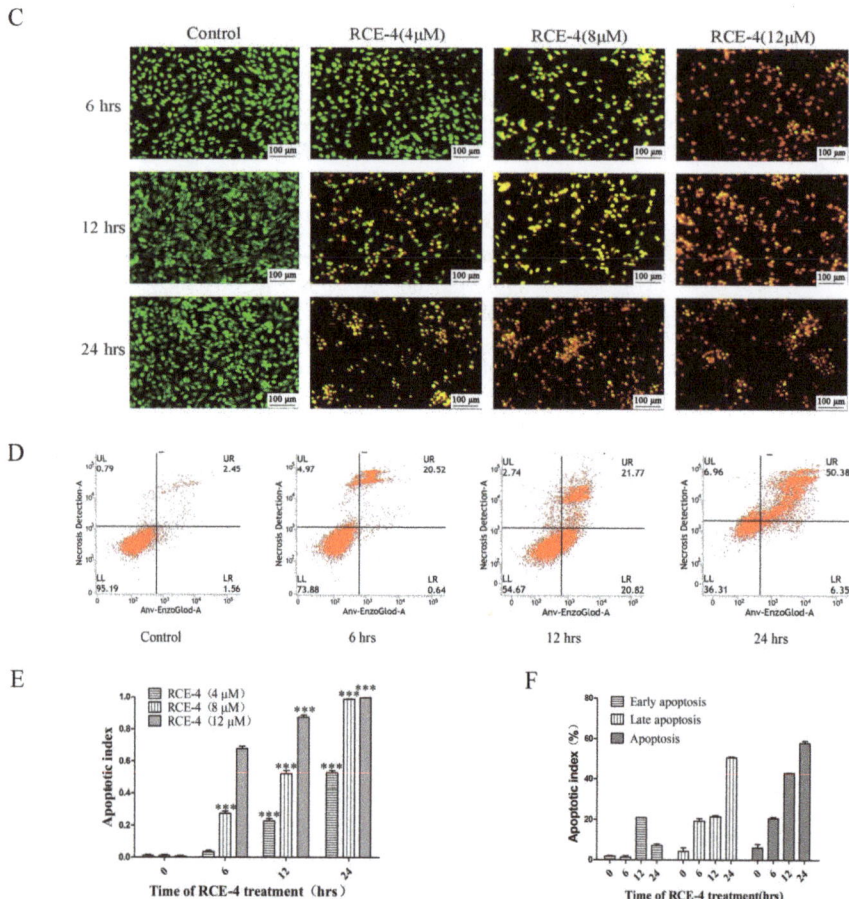

Figure 2. RCE-4 induces time- and concentration-dependent apoptosis and autophagy in Ca Ski cells. (**A**) The protein expression levels of apoptosis associated proteins in Ca Ski cells treated with RCE-4 at different times was analyzed by western blotting. β-actin was used as the loading control. (**B**) Densitometry analysis of the expression of proteins. (**C**) and (**E**) AO/EB Staining was used to detect apoptosis in Ca Ski cells. Scale bar: 100 μM. Viable cells, bright green chromatin with organized structure; viable apoptotic cells, bright green chromatin, which is highly condensed or fragmented; non-apoptotic nonviable cells, bright orange chromatin with organized structure; nonviable apoptotic cells, bright orange chromatin, which is highly condensed or fragmented. Data are presented as the mean ± standard deviation of three independent experiments. (**D,F**) Ca Ski cells were treated with RCE-4 for 24 h, and apoptosis was detected by flow cytometry. Data are presented as the mean ± standard deviation of three independent repeats, *** $p < 0.001$ vs. Control. AO, acridine orange.

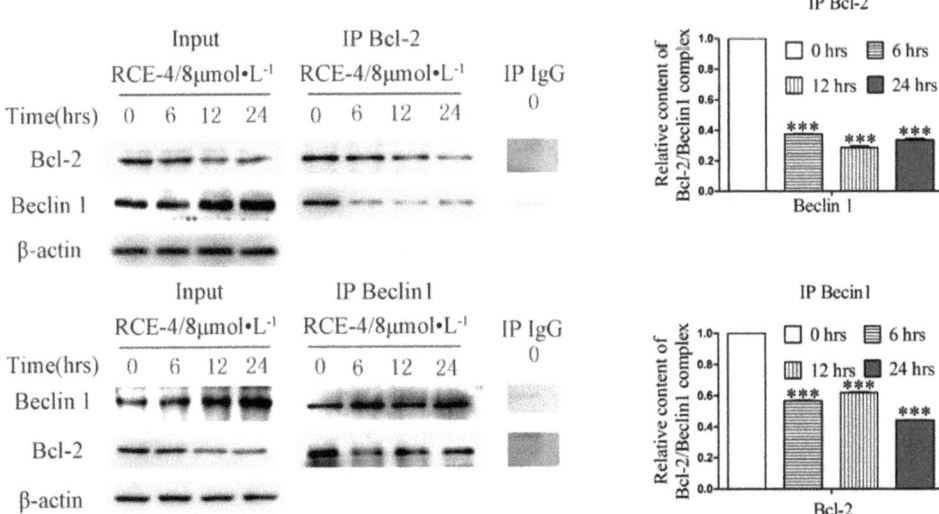

Figure 3. Effect of RCE-4 on the Bcl-2–Beclin 1 complex in Ca Ski cells. Co-IP was used to confirm the effects of RCE-4 on the Bcl-2–Beclin 1 complex in Ca Ski cells. The Ca Ski cells were treated with RCE-4 of 8 μmol/L for 6, 12 and 24 h. Whole cell lysates were purified using Protein A/G PLUS-Sepharose, and then mixed with Bcl-2, Beclin 1 or control IgG antibodies. The immunoprecipitate was captured on Protein A/G PLUS-Agarose and analyzed using western blotting with antibodies against Bcl-2 and Beclin 1. Relative expression of Beclin 1 and Bcl-2 expressed as a percentage of the untreated group, and relative content of the Bcl-2–Beclin 1 complex in Ca Ski cells. Data are presented as the mean ± standard deviation of three independent experiments. *** $p < 0.001$ vs. Control.

2.3. RCE-4 Inhibits the Formation of Bcl-2–Beclin 1 Complex via Various Pathways

Supportive evidence has gradually revealed that various signal pathways co-regulate the formation of the intracellular Bcl-2–Beclin 1 complex. As shown in Figure 4, under external stimulation, intracellular signal molecule JNK1 was activated and phosphorylated, leading to the phosphorylation of Bcl-2 between the BH4 and BH3 domains on multiple residues, including Thr69, Ser70 and Ser87, which resulted in the dissociation of the Bcl-2–Beclin 1 complex [21,22,27]. In addition, Mst1, a pro-apoptotic protein kinase, could promote the formation of the Bcl-2–Beclin 1 complex by phosphorylating Beclin 1 [23]. On the other side, some core complexes involved in the autophagy regulatory network, such as Beclin 1–HMGB-1 complex [28–30], Beclin 1– ATG14–Vps34–Vps15 complex [24,31], Beclin 1–Vps34–Vps15–UVRAG complex and Beclin 1–Vps34–Vps15–UVRAG–Rubicon complex [32,33], would competitively bind to Beclin 1, thus resulting in the reduction of free Beclin 1 molecules, thereby preventing the formation of the Bcl-2–Beclin 1 complex.

The results of western blot (Figure 5A,B) showed that the expression of Dapper1, p-Beclin 1, p-Bcl-2 and p-JNK1 significantly increased and Mst 1 obviously decreased after Ca Ski cells treated with RCE-4 of 8 μmol/L for 6, 12 and 24 h, this inhibited the formation of the Bcl-2–Beclin 1 complex ($p < 0.001$). Additionally, the relative contents of Beclin 1–HMGB-1 complex, Beclin 1–ATG 14–Vps34–Vps15 complex, Beclin 1–Vps34–Vps15–UVRAG complex and Beclin 1–Vps34–Vps15–UVRAG–Rubicon complex were all observed to be increased by the Co-IP assay, thus resulting in the inhibition of the combination of Bcl-2 and Beclin 1 molecules (Figure 5C–E; $p < 0.001$).

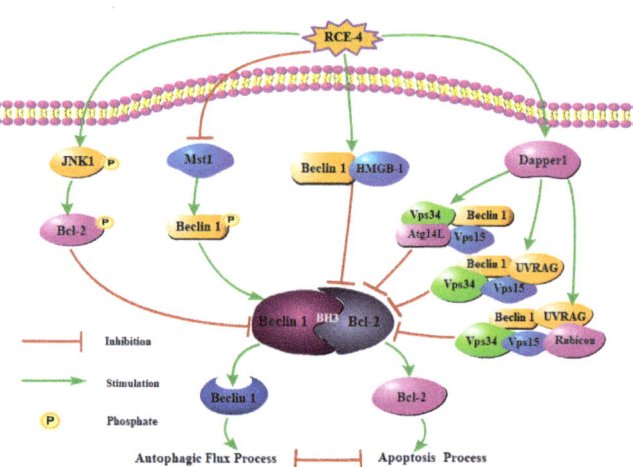

Figure 4. Various signal pathways co-regulated the formation of intracellular Bcl-2–Beclin 1 complex.

2.4. ATG 4B Plays a Critical Role in the Inhibition of Bcl-2–Beclin 1 Complex Induced by RCE-4

The Co-IP assay was used to analyze the effect of ATG family proteins on the Bcl-2–Beclin 1 complex in Ca Ski cells treated with RCE-4. Input reflected the regulation of RCE-4 on ATG family proteins. IP Bcl-2 reflected that ATG family proteins bound to the Bcl-2–Beclin 1 complex owing to no reports being seen that ATG molecules could bind to Bcl-2 directly. However, IP Beclin 1 was uncertain because some special ATG proteins can bind to Beclin 1 and form other complexes. For example, ATG14 molecules could bind to Beclin 1 and form the Beclin 1–ATG14–Vps34–Vps15 complex. Thus, for IP Beclin 1, it should be analyzed based on the actual conditions.

As shown in Figure 6A–D, (1) for untreated Ca Ski cells (RCE-4 of 0 µmol/L), when both IP Bcl-2 and IP Beclin 1 could capture the specific protein, this indicated that this protein was involved in the formation of the Bcl-2–Beclin 1 complex. Our results showed that the formation of Bcl-2–Beclin 1 complex might require ATG 3/4 B/5/12/14/16 L1 as co-factors, while ATG 7/13 were irrelevant to the formation of the Bcl-2–Beclin 1 complex ($p < 0.001$). (2) For ATG 3/5/12/16L1, the trend of the Input results was that IP Bcl-2 and IP Beclin 1 were all reduced, which indicated that although ATG 3/5/12/16L1 participated in the formation of the Bcl-2–Beclin 1 complex, it had less effect on the formation of the complex in Ca Ski cells treated with RCE-4 ($p < 0.001$). (3) For ATG14, the trend of the Input results was increased, IP Bcl-2 was reduced but IP Beclin 1 was increased. These results indicated that RCE-4 promoted the expression of ATG 14; however, the increased ATG 14 molecules did not participate in the formation of the Bcl-2–Beclin 1 complex, whereas they were competitively bound to Beclin 1 and formed the Beclin 1–ATG 14–Vps34–Vps15 complex ($p < 0.001$). (4) For ATG 4B, the trend of the Input results was increased, contrary to ATG 14, it was interesting to see that IP Bcl-2 was increased but IP Beclin 1 was reduced. First of all, when Ca Ski cells were stimulated by RCE-4, the expression of ATG 4B significantly increased. Secondly, because no reports had been seen that ATG 4B molecules could bind to Bcl-2 or Beclin 1 directly, so the increased ATG 4B molecules were most likely to bind with the Bcl-2–Beclin 1 complex. In addition, considering that the formation of the Bcl-2–Beclin 1 complex was inhibited in Ca Ski cells treated with RCE-4, if ATG 4B molecules could bind to the Bcl-2–Beclin 1 complex directly, we could conclude that the single Bcl-2–Beclin 1 complex bound more ATG 4B molecules compared with the untreated cells, and its purpose might be to maintain the stability of the complex. In short, these results indicated that, whatever the possibility, ATG 4B not only participated in the formation of the Bcl-2–Beclin 1 complex but also played a key role in protecting the complex from dissociation.

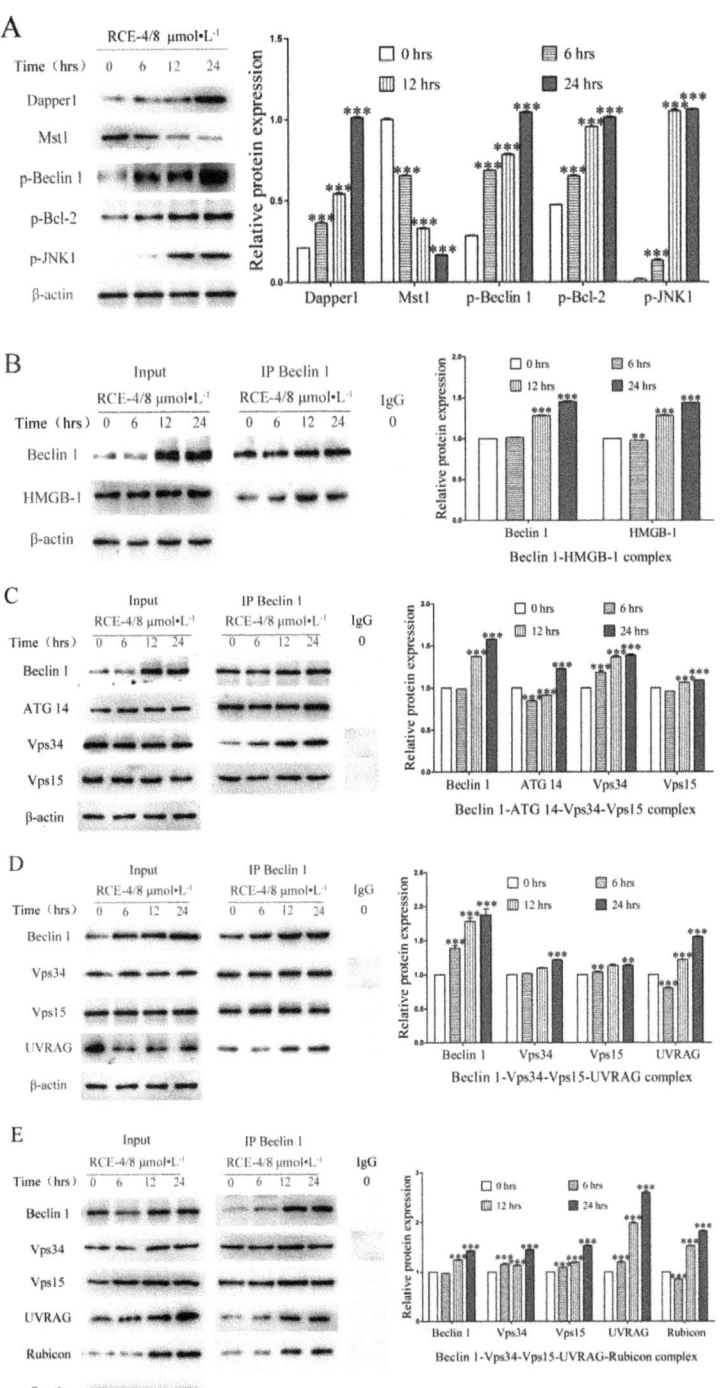

Figure 5. Pathway by which RCE-4 inhibits the formation of the Bcl-2–Beclin 1 complex. (**A**) Protein expression levels in Ca

Ski cells treated with RCE-4 at different time points were analyzed by western blotting. β-actin was used as the loading control. (**B**) The content of Beclin 1–HMGB-1 complex was detected by Co-IP assay. Data are presented as the mean ± standard deviation of three independent experiments. *** $p < 0.001$ vs. Control. (**C**) The content of Beclin 1–ATG 14–Vps34–Vps15 complex was detected by Co-IP assay. Data are presented as the mean ± standard deviation of three independent experiments. *** $p < 0.001$ vs. Control. (**D**) The content of Beclin 1–Vps34–Vps15–UVRAG complex was detected by Co-IP assay. Data are presented as the mean ± standard deviation of three independent experiments. ** $p < 0.01$, *** $p < 0.001$ vs. Control. (**E**) The content of Beclin 1–Vps34–Vps15–UVRAG–Rubicon complex was detected by Co-IP assay. Data are presented as the mean ± standard deviation of three independent experiments. *** $p < 0.001$ vs. Control.

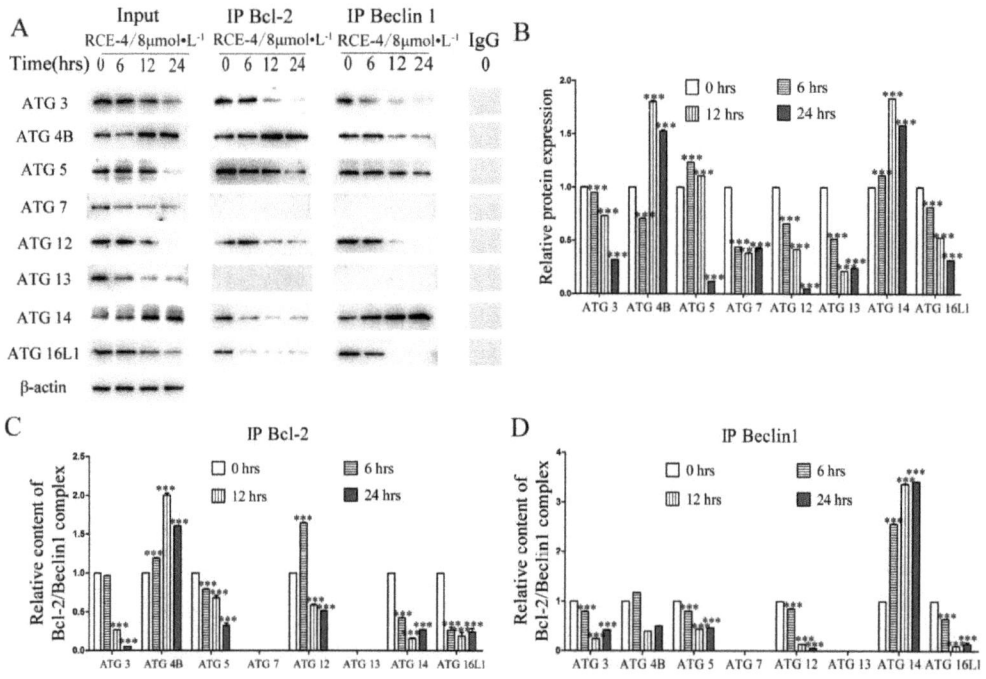

Figure 6. Analysis of the roles of ATG family proteins in the inhibition of Bcl-2–Beclin 1 complex induced by RCE-4. (**A**) The Ca Ski cells were treated with RCE-4 of 8 μmol/L for 6, 12, and 24 h. Following, the Co-IP experiments were performed. (**B–D**) Densitometry analysis of the expression of proteins. Data are presented as the mean ± standard deviation of three independent experiments. *** $p < 0.001$ vs. Control.

2.5. RCE-4 Combined with ATG siRNA Enhances the Sensitivity to Ca Ski Cells

Based on the discovery of the important role of ATG 4B for the Bcl-2–Beclin 1 complex, we tried to improve the sensitivity of RCE-4 to Ca Ski cells by regulating the expression of ATG 4B using ATG 4B siRNA. As shown in Figure 7A, RCE-4 combined with ATG 4B siRNA significantly enhanced the proliferation inhibition and greatly improved the sensitivity of RCE-4 to Ca Ski cells, with IC50 from 4.67 μmol/L reduced to 1.37 μmol/L ($p < 0.001$). In addition, in line with our expectations, the results of Co-IP (Figure 7B,C) declared that the knock out of ATG 4B enhanced the inhibition of RCE-4 on the formation of the Bcl-2–Beclin 1 complex ($p < 0.001$). Furthermore, apoptosis and the depolarization of MMP in Ca Ski cells were detected. As illustrated in Figure 7D–G, the knock out of ATG 4B significantly enhanced the RCE-4-induced apoptosis and depolarization of MMP, compared with only RCE-4-treated cells ($p < 0.001$).

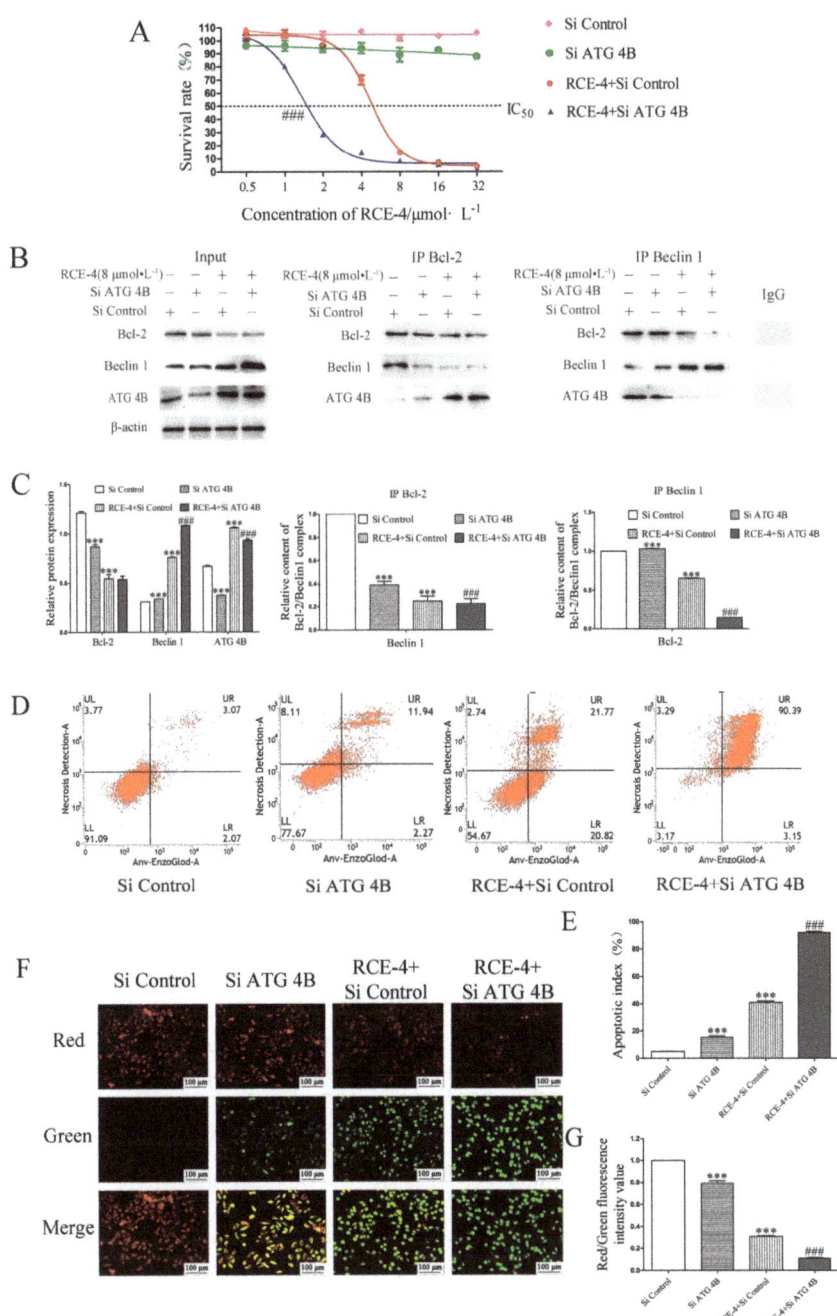

Figure 7. RCE-4 combined with ATG 4B siRNA enhanced the sensitivity of Ca Ski cells. (**A**) RCE-4 combined with or without ATG 4B siRNA inhibited the proliferation of Ca Ski cells. Data are presented as the mean ± standard deviation of

three independent experiments. ### $p < 0.001$ vs. RCE-4. (**B**) Co-IP was used to confirm the effects of RCE-4 combined with ATG 4B siRNA on the Bcl-2–Beclin 1 complex in Ca Ski cells. (**C**) Relative expression of Beclin 1 and Bcl-2 expressed as a percentage of the untreated group, and relative content of the Bcl-2–Beclin 1 complex in Ca Ski cells. Data are presented as the mean ± standard deviation of three independent experiments. *** $p < 0.001$ vs. Control. ### $p < 0.001$ vs. RCE-4. (**D**) and (**E**) The effect of RCE-4 combined with ATG 4B siRNA on the apoptosis of Ca Ski cells. Data are presented as the mean ± standard deviation of three independent experiments. *** $p < 0.001$ vs. Control. ### $p < 0.001$ vs. RCE-4. (**F,G**) The effect of RCE-4 combined with ATG 4B siRNA on the mitochondrial transmembrane potential of Ca Ski cells. In normal mitochondria, JC-1 aggregates in the mitochondrial matrix to form polymers, which emit a strong red fluorescence. In damaged mitochondria, JC-1 only exists as a monomer in the cytoplasm and exhibits green fluorescence due to the decrease in or loss of membrane potential. Data are presented as the mean ± standard deviation of three independent experiments. *** $p < 0.001$ vs. Control. ### $p < 0.001$ vs. RCE-4.

3. Discussion

Autophagy and apoptosis play significant physiological roles in cellular survival, stress adaptation and the development of tumors [10,25]. They can be regulated by a variety of regulatory elements (such as sphingolipids, MAPk, etc.) and signaling pathways (such as PI3K/Akt), while the Bcl-2–Beclin 1 complex plays a "toggle switch" role in the occurrence and development of apoptosis and autophagy, which determines whether the cell enters the apoptosis or initiates the autophagy process [15]. When the formation of the Bcl-2–Beclin 1 complex is inhibited or disrupted, Bcl-2 and Beclin 1 molecules are released and enter mitochondria and endoplasmic reticulum, respectively, and regulate the process of apoptosis and autophagy. In recent years, more and more studies have focused on this complex. Álvaro F. Fernández et al. found that disruption of the Bcl-2–Beclin 1 complex is an effective mechanism to increase autophagy, prevent premature ageing, improve health span and promote longevity in mammals [34]. Some studies have also found that drug-induced tumor cell death is related to the Bcl-2–Beclin 1 complex pathway [18,26,35,36].

In the present study, we demonstrated the effect of RCE-4 on the Bcl-2–Beclin 1 complex using Co-IP assay. The results showed that the relative content of the Bcl-2–Beclin1 complex in Ca Ski cells treated with RCE-4 was significantly reduced, compared with the control group, which indicated that RCE-4 could inhibit the formation of the Bcl-2–Beclin1 complex. The downside was that whether RCE-4 could disrupt the Bcl-2–Beclin 1 complex directly had not been identified due to the limitation of the test conditions.

Furthermore, we explored the mechanism by which RCE-4 inhibited the formation of the Bcl-2–Beclin 1 complex. Multiple pathways were involved in the regulation, including: (1) RCE-4 induced phosphorylation of JNK1, which in turn led to phosphorylation of multiple residues of Bcl-2 located between the BH4 and BH3 domains, thereby blocking the binding to Beclin 1 via the BH3 domain. (2) RCE-4 inhibited the expression of pro-apoptotic protein kinase Mst1, thus the phosphorylation of Beclin 1 was promoted, in the same way, the binding of the two was blocked. (3) Some important molecules (such as HMGB-1, UVRAG, ATG14, Vps15/34, etc.) involved in the autophagy process competitively bound to Beclin 1 to form other complexes, such as Beclin 1–HMGB–1 complex, Beclin 1–ATG 14–Vps34–Vps15 complex, Beclin 1–Vps34–Vps15–UVRAG complex and Beclin 1–Vps34–Vps15–UVRAG–Rubicon complex, which also affected the formation of the Bcl-2–Beclin 1complex. These findings indicated the complexity of the anti-cervical cancer molecular mechanism of RCE-4.

Multiple protein molecules are involved in the crosstalk between apoptosis and autophagy as co-regulatory factors, especially ATG family members [37–39] such as ATG 4, which plays a central role in the LC3 lipid conjugation system, essential for the late step of autophagosome formation [40–42]. ATG 5 participates in the external activation of apoptosis and can cause cell death with the participation of FADD-mediated caspase enzymes [43–45]. ATG 3/7 can be cleaved by the caspase enzyme and lose its ability to induce autophagy, but it can also affect apoptosis by translocating to mitochondria, making cells more sensitive to apoptosis [46,47]. This progress suggests that the ATG family members play important roles in the process of PCD characterized by apoptosis and autophagy, but no previous studies have found whether ATG family proteins are involved in the formation of the Bcl-2–Beclin 1 complex. In this regard, we tried to make a preliminary exploration of this. Our results indicated that ATG family proteins, such as ATG 3/4B/5/12/14/16L1, participated in the formation of the Bcl-2–Beclin 1 complex as co-factors, while ATG 7/13 were not involved; among them ATG 4B played a key role in maintaining the stability of the complex. A more interesting finding was that when Ca Ski cells were stimulated by RCE-4, the expression of ATG 4B significantly increased, which led to more ATG 4B molecules binding to the Bcl-2–Beclin 1 complex to maintain the stability of the complex, this might be a self-protection mechanism of cells under stress conditions. In previous studies, scholars have found that ATG 4B was complicated related to tumor progression. For example, the expression of ATG 4B is significantly increased in colorectal cancer cells, suggesting that ATG 4B may be important for cancer biology [48]. Debra Akin et al. found that ATG 4B had a positive impact on the tumor growth of Saos-2 cells, and osteosarcoma Saos-2 cells lacking ATG 4B failed to survive under amino acid starvation conditions and had attenuated tumor growth in mice [49]. These findings show the possibility of ATB 4B as a novel target for cancer therapy, but the research into the deep mechanism underlying its actions had not been clarified, our findings could provide a new explanation for this.

Based on the above research, we tried to enhance the sensitivity of RCE-4 to Ca Ski cells via knocking out ATG 4B by siRNA. The results demonstrated that RCE-4 combined with ATG 4B siRNA significantly enhanced the proliferation inhibition of Ca Ski cells by inhibiting RCE-4-induced autophagy, enhancing RCE-4-induced apoptosis and strengthening the inhibition of RCE-4 on the formation of the Bcl-2–Beclin 1 complex in Ca Ski cells. In our previous studies, we found that RCE-4-induced autophagy was protective for Ca Ski cells, which could protect Ca Ski cells from apoptosis, and inhibiting autophagy could enhance RCE-4-induced apoptosis and the sensitivity of RCE-4 to Ca Ski cells; this was consistent with the results of this experiment. Debra Akin et al. also showed similar results, the ATG 4B antagonist named NSC185058 effectively inhibited ATG 4B activity in vitro and in cells, inhibited autophagy and had a negative impact on the growth of osteosarcoma tumors [48]. In another study, the inhibition of ATG 4B by siRNA enhanced lupulone derivatives-induced apoptosis in prostate cancer cells [40]. Moreover, blocking autophagy by inhibiting ATG 4B sensitized several types of resistant carcinoma cells, including MDA-MB-231 and A549 cell lines, to radiation therapy [50].

4. Materials and Methods

4.1. Cell Culture

Human cervical cancer Ca Ski cells were obtained from The Cell Bank of Type Culture Collection of The Chinese Academy of Sciences and maintained in RPMI-1640 culture medium (Gibco/Thermo Fisher Scientific, Waltham, MA, USA) supplemented with 10% fetal bovine serum (Zhejiang Tianhang Biological Technology Co., Ltd, Hangzhou, Zhejiang, China), 0.2% HEPES and 2% double antibody at 37 °C with 5% CO_2 in a humidified incubator.

4.2. Reagent and Antibodies

The RCE-4 used in the present study was a spiral steranol saponin extracted and purified from *Reineckia carnea* [51]. The RCE-4 stock solution of 50 mmol·L^{-1} was prepared using DMSO and diluted to the desired concentrations in RPMI-1640 medium. Rabbit monoclonal antibodies against Beclin 1 (#3495S), Bcl-2 (#4223S), p-Bcl-2 (#2875T; #2827T), Mst 1 (#14946S), ATG 3 (#3415S), ATG 4B (#13507S), ATG 5 (#12994T), ATG 7 (#8558T), ATG 12 (#4180S), ATG 13 (#13468S), ATG 14 (#96752S), ATG 16L1 (#8089S), HMGB-1 (#MAB1690-SP), UVRAG (#NBP2-24482SS), Vps34 (#3358T), Rubicon (#8465S) and β-actin (#8457S) (Cell Signaling Technology, Boston, USA; R&D Systems, Minneapolis, MN, USA) were used at a dilution of 1:1000. HRP-conjugated secondary antibody (1:5000) was purchased fromSanta Cruz Biotechnology, Inc., Dallas, TX, USA.

4.3. Cytotoxicity Assay

An MTT assay was used to evaluate the effects of RCE-4 extract on cell growth, as previously described. Ca Ski cells (1×10^5/mL) were seeded into 96-well plates, incubated at 37 °C for 12 h. The Ca Ski cells were pretreated with ATG 4B siRNA for 6 h, and then treated with RCE-4 (0, 0.5, 1, 2, 4, 8, 16 and 32 μmol/L) for an additional 48 h. Subsequently, 20 μL 5 mg/mL MTT (BS0328; Amersco, Spokane, WA, USA) reagent was added per well. After 4 h, the media was gently removed and 150 μL DMSO (Sigma-Aldrich, St. Louis, MO, USA) was added. The absorbance was measured using a microplate reader at 490 nm, subtracting the baseline reading.

4.4. Flow Cytometry Assay

Flow cytometry was used to detect the effects of RCE-4 on apoptosis of Ca Ski cells. The GFP-CERTIED® Apoptosis/Necrosis Detection kit (cat. no. ENZ-51002-100; Enzo Life Sciences, New York, NY, USA) allows for easy differentiation of early apoptosis from late apoptosis or necrosis; thus allowing for analysis of the separate death pathways in detail. A blank control was set; Ca Ski cells were pretreated with or without ATG 4B siRNA and then treated with RCE-4 for 24 h. Cells were collected, centrifuged, separated from the supernatant, washed twice with PBS, then sat to dry. The cell suspension was prepared with a buffer/water ratio of 1:9. The appropriate amounts of apoptotic and necrotic staining reagents were added according to the ratio of suspension–Staining solution = 100:1, then mixed gently. The mixture was stained for 15 min at room temperature in the dark and flow cytometry was used to detect staining.

4.5. Acridine Orange/Ethidium Bromide (AO/EB) Double Fluorescent Staining

AO/EB double fluorescence staining was used to detect the effects of RCE-4 on the apoptosis of Ca Ski cells. AO can penetrate through the cell membrane and bind to DNA, fluorescing green upon doing so. EB can only bind with DNA if the cell membrane is damaged, giving an orange-red appearance, and the orange-red brightness is higher than the green color of AO. Apoptotic cells are unevenly stained due to the high concentration of chromatin, whereas the non-apoptotic cells have a normal structure and uniform coloring. Observed under a fluorescent microscope, four cell states can be seen: viable cells, bright green chromatin with organized structure; viable apoptotic cells, bright green chromatin, which is highly condensed or fragmented; non-apoptotic nonviable cells, bright orange chromatin with organized structure; nonviable apoptotic cells, bright orange chromatin, which is highly condensed or fragmented. After treatment, the Ca Ski cells were collected and washed with PBS and AO Stain Buffer (1x) once. An appropriate quantity of AO Stain Buffer (1x) was used to resuspend the cells, mixed according to a cell suspension ratio of AO Stain–cell suspension of 19:1, and then an appropriate quantity of AO/EB stain was added and indicated in the dark for 20 min. A fluorescence microscope was used to observe staining and data was analyzed using Image-Pro-Plus version 7.0.

4.6. MMP Assay

The effect of RCE-4 on MMP in Ca Ski cells was assessed using JC-1 staining (cat. no. C2006; Beyotime Institute of Biotechnology, Songjiang, Shanghai, China). JC-1 is an ideal and widely used fluorescent probe for the detection of MMP. In normal mitochondria, JC-1 aggregates in the mitochondrial matrix to form polymers, which emits a strong red fluorescence. In damaged mitochondria, JC-1 only exists as a monomer in the cytoplasm and exhibits green fluorescence due to the decrease in or loss of membrane potential. JC-1 can be used not only for qualitative detection but also for quantitative detection as the change in its color can directly reflect the change in MMP since the degree of mitochondrial depolarization can be measured by the ratio of red to green fluorescence intensity. After treatment, Ca Ski cells were incubated with JC-1 dye for 20 min at room temperature, and then washed twice with dyeing buffer. Cells were treated with carbonyl cyanide m-chlorophenylhydrazone CCCP (0.1 μmol/L), and those treated with mitochondrial inhibitors were used as the positive controls. A fluorescence microscope was used to observe the staining. Data were analyzed using Image-Pro-Plus.

4.7. Western Blotting

Ca Ski cells were treated with RCE-4 of 8 µmol/L for 6, 12, and 24 h. Total protein was extracted from the cells using a protein extraction kit. Following protein quantification, protein lysates were added to the sample buffer and boiled for 10 min. Subsequently, samples were loaded on a 10% SDS-gel and resolved using SDS-PAGE (electrophoresis at 80 V and 30 mA for 2.5 h). Subsequently, proteins were transferred at 200 mA to PVDF membranes. After blocking at room temperature for 2 h, membranes were incubated with the primary antibodies at 4 °C overnight in 5% skimmed milk. Membranes were washed five times using tributyltin compound plus 0.05% Tween-20, and incubated with the secondary antibody for 1.5 h at 37 °C. Signals were visualized using ECL (Beyotime Institute of Biotechnology, Songjiang, Shanghai, China), and developed using Kodak film and produced using a Tanon 5200 luminescence imaging system (Tanon Technology, Shanghai, China). Densitometry analysis was performed using ImageJ version 2.1 (National Institutes of Health, Bethesda, MD, USA) using β-actin as the control load.

4.8. Plasmid Small Interfering (si)RNA Transfection

The plasmids ATG 4B siRNA (h) (cat. no. sc-72584; Santa Cruz Biotechnology, Inc., Dallas, TX, USA), ATG 4B (h)-PR (sc-72584-PR; Santa Cruz Biotechnology, Inc.), and Control siRNAs including Control (cat. no. sc-37007; Santa Cruz Biotechnology, Inc.) and Control siRNA (FITC Conjugate)-A (cat. no. sc-36869; Santa Cruz Biotechnology, Inc.) were obtained from Youning Life Support Technology. The plasmids were transfected into Ca Ski cells according to specifications.

4.9. Co-Immunoprecipitation (Co-IP)

After Ca Ski cells were treated with RCE-4 of 8 µmol/L, lysis buffer (cat. no. abs9116-100 ML; Absin, Shanghai, China) was used to lyse the Ca Ski cells. Protein A/G PLUS-Agarose (cat. no. sc-2003; Santa Cruz Biotechnology, Inc.) was used to purify the lysates, which were subsequently mixed with Bcl-2 (cat. no. 4223S; Cell Signaling Technology, Boston, MA, USA) or Normal Rabbit IgG (cat. no. 2729S; Cell Signaling Technology, Inc.) and then shaken for 2 h at room temperature. The immunoprecipitants were captured using protein A/G agarose beads and analyzed by western blotting.

4.10. Statistical Analysis

Data are presented as the mean ± standard deviation. Data were analyzed using GraphPad Prism version 5.0 (GraphPad Software Inc, San Diego, CA, USA). Comparisons between groups were assessed using a Student's t-test or a one-way ANOVA. $p < 0.05$ was considered to indicate a statistically significant difference.

5. Conclusions

This study demonstrated that inhibition of ATG 4B significantly reduced RCE-4-induced autophagy, enhanced RCE-4-induced apoptosis and strengthened the inhibition of RCE-4 on the formation of the Bcl-2–Beclin 1 complex in Ca Ski cells, thereby enhancing the sensibility of RCE-4 to Ca Ski cells.

Author Contributions: J.-F.C. conceived and designed the research, supplied reagents, materials, analysis tools and designed certain experimental methods. F.-F.Y. performed the experiments, analyzed the data and wrote the manuscript. F.C., K.Z., J.Z. and X.-Q.Z. assisted in the analysis of data and provided key commentary. All authors have read and agreed to the published version of the manuscript.

Funding: This research was funded by National Natural Science Foundation of China, grant number 81773952.

Institutional Review Board Statement: Not applicable.

Informed Consent Statement: Not applicable.

Data Availability Statement: The original data of the graphics in the paper can be obtained by contacting the main author.

Acknowledgments: We thank the Medical College of China Three Gorges University for providing the required instruments for the experiments.

Conflicts of Interest: The authors declare no competing interest.

References

1. Hu, Z.; Ma, D. The precision prevention and therapy of HPV-related cervical cancer: New concepts and clinical implications. *Cancer Med.* **2018**, *7*, 5217–5236. [CrossRef]
2. Xiang, W.; Zhang, R.; Jin, G.; Tian, L.; Cheng, F.; Wang, J.; Xing, X.; Xi, W.; Tang, S.; Chen, J. RCE-4, a potential anti-cervical cancer drug isolated from *Reineckia carnea*, induces autophagy via the dual blockade of PI3K and ERK pathways in cervical cancer CaSki cells. *Int. J. Mol. Med.* **2019**, *45*, 245–254. [CrossRef]
3. Bai, C.; Yang, X.; Zou, K.; He, H.; Wang, J.; Qin, H.; Yu, X.; Liu, C.; Zheng, J.; Cheng, F.; et al. Anti-proliferative effect of RCE-4 from *Reineckia carnea* on human cervical cancer HeLa cells by inhibiting the PI3K/Akt/mTOR signaling pathway and NF-κB activation. *Naunyn-Schmiedeberg's Arch. Pharmacol.* **2016**, *389*, 573–584. [CrossRef]
4. Wang, G.; Huang, W.; He, H.; Fu, X.; Wang, J.; Zou, K.; Chen, J. Growth inhibition and apoptosis-inducing effect on human cancer cells by RCE-4, a spirostanol saponin derivative from natural medicines. *Int. J. Mol. Med.* **2012**, *31*, 219–224. [CrossRef]
5. Yan, W.; Zou, K.; He, H.; Zhang, Y.; Li, X.; LI, X.; Yang, X.; Wang, J.; Deng, Z. Effects of steroidal saponin RCE-4 from *Reineckia carnea* (Andr.) Kunth on Ras/Erk and p16/cyclin D1/CDK4 signaling pathways in the human cervix cancer Ca Ski cells. *Chin. J. Clin. Pharmacol. Ther.* **2018**, *23*, 247–254.
6. Yang, X.J.; Bai, C.H.; Zou, K.; He, H.B.; Yu, X.Q. Steroidal saponin RCE-4 from *Reineckia carnea* (Andr.) Kunth inhibits growth of human cervical cancer xenograft in nude mice. *J. Third Mil. Med. Univ.* **2016**, *5*, 476–482. (In Chinese)
7. Hubei Institute for Drug Control. *Quality Standards for Traditional Chinese Medicines in Hubei Province*; Hubei Science and Technology Press: Wuhan, China, 2009. (In Chinese)
8. Ke, B.; Tian, M.; Li, J.; Liu, B.; He, G. Targeting programmed cell death using small-molecule compounds to improve potential cancer therapy. *Med. Res. Rev.* **2016**, *36*, 983–1035. [CrossRef] [PubMed]
9. Choi, A.M.; Ryter, S.W.; Levine, B. Autophagy in Human Health and Disease. *N. Engl. J. Med.* **2013**, *368*, 651–662. [CrossRef] [PubMed]
10. Friedlander, R.M. Apoptosis and Caspases in Neurodegenerative Diseases. *N. Engl. J. Med.* **2003**, *348*, 1365–1375. [CrossRef]
11. Mishra, A.P.; Salehi, B.; Sharifi-Rad, M.; Pezzani, R.; Kobarfard, F.; Sharifi-Rad, J.; Nigam, M. Programmed Cell Death, from a Cancer Perspective: An Overview. *Mol. Diagn. Ther.* **2018**, *22*, 281–295. [CrossRef]
12. Gordy, C.; He, Y. The crosstalk between autophagy and apoptosis: Where does this lead? *Protein Cell* **2012**, *3*, 17–27. [CrossRef]
13. Lin, Y.; Jiang, M.; Chen, W.; Zhao, T.; Wei, Y. Faculty Opinions recommendation of Cancer and ER stress: Mutual crosstalk between autophagy, oxidative stress and inflammatory response. *Biomed. Pharmacother.* **2020**, *118*, 109249. [CrossRef]
14. Kapuy, O.; Vinod, P.K.; Mandl, J.; Bánhegyi, G. A cellular stress-directed bistable switch controls the crosstalk between autophagy and apoptosis. *Mol. BioSyst.* **2013**, *9*, 296–306. [CrossRef]
15. Decuypere, J.-P.; Parys, J.B.; Bultynck, G. Regulation of the Autophagic Bcl-2/Beclin 1 Interaction. *Cells* **2012**, *1*, 284–312. [CrossRef]
16. Marquez, R.T.; Xu, L. Bcl-2-Beclin 1 complex: Multiple, mechanisms regulating autophagy/apoptosis toggle switch. *Am. J. Cancer Res.* **2012**, *2*, 214–221. [PubMed]
17. Nopparat, C.; Porter, J.E.; Ebadi, M.; Govitrapong, P. 1-Methyl-4-phenylpyridinium-induced cell death via autophagy through a Bcl-2/Beclin 1 complex-dependent pathway. *Neurochem. Res.* **2014**, *39*, 225–232. [CrossRef] [PubMed]
18. Rahman, A.; Bishayee, K.; Habib, K.; Sadra, A.; Huh, S.-O. 18α-Glycyrrhetinic acid lethality for neuroblastoma cells via de-regulating the Beclin-1/Bcl-2 complex and inducing apoptosis. *Biochem. Pharmacol.* **2016**, *117*, 97–112. [CrossRef] [PubMed]
19. He, C.; Zhu, H.; Li, H.; Zou, M.-H.; Xie, Z. Dissociation of Bcl-2-Beclin1 complex by activated AMPK enhances cardiac autophagy and protects against cardiomyocyte apoptosis in diabetes. *Diabetes* **2012**, *62*, 1270–1281. [CrossRef]
20. Li, D.; Wang, J.; Hou, J.; Fu, J.; Chang, D.; Bensoussan, A.; Liu, J. Ginsenoside Rg1 protects starving H9c2 cells by dissociation of Bcl-2-Beclin1 complex. *BMC Complement. Altern. Med.* **2016**, *16*, 1–12. [CrossRef]
21. Ke, D.; Ji, L.; Wang, Y.; Fu, X.; Chen, J.; Wang, F.; Zhao, D.; Xue, Y.; Lan, X.; Hou, J. JNK1 regulates RANKL-induced osteoclastogenesis via activation of a novel Bcl-2-Beclin1-autophagy pathway. *FASEB J.* **2019**, *33*, 11082–11095. [CrossRef]
22. Wei, Y.; Pattingre, S.; Sinha, S.; Bassik, M.; Levine, B. JNK1-Mediated Phosphorylation of Bcl-2 regulates starvation-induced autophagy. *Mol. Cell* **2008**, *30*, 678–688. [CrossRef] [PubMed]
23. Maejima, Y.; Kyoi, S.; Zhai, P.; Liu, T.; Li, H.; Ivessa, A.; Sciarretta, S.; Del Re, D.P.; Zablocki, D.K.; Hsu, C.-P.; et al. Mst1 inhibits autophagy by promoting the interaction between Beclin1 and Bcl-2. *Nat. Med.* **2013**, *19*, 1478–1488. [CrossRef]

24. Ma, B.; Cao, W.; Li, W.; Gao, C.; Qi, Z.; Zhao, Y.; Du, J.; Xue, H.; Peng, J.; Wen, J.; et al. Dapper1 promotes autophagy by enhancing the Beclin1-Vps34-Atg14L complex formation. *Cell Res.* **2014**, *24*, 912–924. [CrossRef] [PubMed]
25. Menzies, F.M.; Fleming, A.; Rubinsztein, D.C. Compromised autophagy and neurodegenerative diseases. *Nat. Rev. Neurosci.* **2015**, *16*, 345–357. [CrossRef]
26. Li, Z.; Li, Q.; Lv, W.; Jiang, L.; Geng, C.; Yao, X.; Shi, X.; Liu, Y.; Cao, J. The interaction of Atg4B and Bcl-2 plays an important role in Cd-induced crosstalk between apoptosis and autophagy through disassociation of Bcl-2-Beclin1 in A549 cells. *Free Radic. Biol. Med.* **2019**, *130*, 576–591. [CrossRef]
27. Wei, Y.; Sinha, S.C.; Levine, B. Dual Role of JNK1-mediated phosphorylation of Bcl-2 in autophagy and apoptosis regulation. *Autophagy* **2008**, *4*, 949–951. [CrossRef] [PubMed]
28. Tang, D.; Kang, R.; Livesey, K.M.; Cheh, C.-W.; Farkas, A.M.; Loughran, P.; Hoppe, G.; Bianchi, M.E.; Tracey, K.J.; Zeh, H.J.; et al. Endogenous HMGB1 regulates autophagy. *J. Cell Biol.* **2010**, *190*, 881–892. [CrossRef] [PubMed]
29. Tang, D.; Kang, R.; Cheh, C.-W.; Livesey, K.M.; Liang, X.; Schapiro, N.E.; Benschop, R.; Sparvero, L.J.; Amoscato, A.; Tracey, K.J.; et al. HMGB1 release and redox regulates autophagy and apoptosis in cancer cells. *Oncogene* **2010**, *29*, 5299–5310. [CrossRef]
30. Kong, Q.; Xu, L.-H.; Xu, W.; Fang, J.-P.; Xu, H.-G. HMGB1 translocation is involved in the transformation of autophagy complexes and promotes chemoresistance in leukaemia. *Int. J. Oncol.* **2015**, *47*, 161–170. [CrossRef]
31. Obara, K.; Ohsumi, Y. Atg14: A Key Player in Orchestrating Autophagy. *Int. J. Cell Biol.* **2011**, *2011*, 1–7. [CrossRef]
32. Matsunaga, K.; Saitoh, T.; Tabata, K.; Omori, H.; Satoh, T.; Kurotori, N.; Maejima, I.; Shirahama-Noda, K.; Ichimura, T.; Isobe, T.; et al. Two Beclin 1-binding proteins, Atg14L and Rubicon, reciprocally regulate autophagy at different stages. *Nat. Cell Biol.* **2009**, *11*, 385–396. [CrossRef]
33. Zhong, Y.; Wang, Q.; Li, X.; Yan, Y.; Backer, J.M.; Chait, B.T.; Heintz, N.; Yue, Z. Distinct regulation of autophagic activity by Atg14L and Rubicon associated with Beclin 1-phosphatidylinositol-3-kinase complex. *Nat. Cell Biol.* **2009**, *11*, 468–476. [CrossRef]
34. Fernández, F.; Sebti, S.; Wei, Y.; Zou, Z.; Shi, M.; McMillan, K.L.; He, C.; Ting, T.; Liu, Y.; Chiang, W.-C.; et al. Disruption of the beclin 1–BCL2 autophagy regulatory complex promotes longevity in mice. *Nature* **2018**, *558*, 136–140. [CrossRef] [PubMed]
35. Chen, Y.; Zhang, W.; Guo, X.; Ren, J.; Gao, A. The crosstalk between autophagy and apoptosis was mediated by phosphorylation of Bcl-2 and beclin1 in benzene-induced hematotoxicity. *Cell Death Dis.* **2019**, *10*, 1–15. [CrossRef]
36. Lian, J.; Karnak, D.; Xu, L. The Bcl-2-Beclin 1 interaction in (−)-gossypol-induced autophagy versus apoptosis in prostate cancer cells. *Autophagy* **2010**, *6*, 1201–1203. [CrossRef]
37. Mizushima, N.; Yoshimori, T.; Ohsumi, Y. The role of ATG proteins in autophagosome formation. *Annu. Rev. Cell Dev. Biol.* **2011**, *27*, 107–132. [CrossRef] [PubMed]
38. Wu, S.; Su, J.; Qian, H.; Guo, T. SLC27A4 regulate ATG4B activity and control reactions to chemotherapeutics-induced autophagy in human lung cancer cells. *Tumor Biol.* **2015**, *37*, 6943–6952. [CrossRef]
39. Vezenkov, L.; Honson, N.S.; Kumar, N.S.; Bosc, D.; Kovacic, S.; Nguyen, T.G.; Pfeifer, T.A.; Young, R.N. Development of fluorescent peptide substrates and assays for the key autophagy-initiating cysteine protease enzyme, ATG4B. *Bioorg. Med. Chem.* **2015**, *23*, 3237–3247. [CrossRef]
40. Zhang, L.; Li, J.; Ouyang, L.; Liu, B.; Cheng, Y. Unraveling the roles of Atg4 proteases from autophagy modulation to targeted cancer therapy. *Cancer Lett.* **2016**, *373*, 19–26. [CrossRef] [PubMed]
41. Betin, V.M.; MacVicar, T.D.; Parsons, S.F.; Anstee, D.J.; Lane, J.D. A cryptic mitochondrial targeting motif in Atg4D links caspase cleavage with mitochondrial import and oxidative stress. *Autophagy* **2012**, *8*, 664–676. [CrossRef]
42. Betin, V.M.; Lane, J.D. Atg4D at the interface between autophagy and apoptosis. *Autophagy* **2009**, *5*, 1057–1059. [CrossRef] [PubMed]
43. Pyo, J.O.; Jang, M.H.; Kwon, Y.K.; Lee, H.J.; Jun, J.I.; Woo, H.N.; Cho, D.H.; Choi, B.; Lee, H.; Kim, J.H.; et al. Essential roles of Atg5 and FADD in autophagic cell death: Dissection of autophagic cell death into vacuole formation and cell death. *J. Biol. Chem.* **2005**, *280*, 20722–20729. [CrossRef] [PubMed]
44. Codogno, P.; Meijer, A.J. Atg5: More than an autophagy factor. *Nat. Cell Biol.* **2006**, *8*, 1045–1047. [CrossRef] [PubMed]
45. Miller, B.; Zhao, Z.; Stephenson, L.M.; Cadwell, K.; Pua, H.H.; Lee, H.K.; Mizushima, N.; Iwasaki, A.; He, Y.-W.; Swat, W.; et al. The autophagy geneATG5plays an essential role in B lymphocyte development. *Autophagy* **2008**, *4*, 309–314. [CrossRef]
46. Cao, Q.-H.; Liu, F.; Yang, Z.-L.; Fu, X.-H.; Yang, Z.-H.; Liu, Q.; Wang, L.; Wan, X.-B.; Fan, X.-J. Prognostic value of autophagy related proteins ULK1, Beclin 1, ATG3, ATG5, ATG7, ATG9, ATG10, ATG12, LC3B and p62/SQSTM1 in gastric cancer. *Am. J. Transl. Res.* **2016**, *8*, 3831–3847.
47. Oral, O.; Oz-Arslan, D.; Itah, Z.; Naghavi, A.; Deveci, R.; Karacali, S.; Gozuacik, D. Cleavage of Atg3 protein by caspase-8 regulates autophagy during receptor-activated cell death. *Apoptosis* **2012**, *17*, 810–820. [CrossRef]
48. Fu, Y.; Hong, L.; Xu, J.; Zhong, G.; Gu, Q.; Gu, Q.; Guan, Y.; Zheng, X.; Dai, Q.; Luo, X.; et al. Discovery of a small molecule targeting autophagy via ATG4B inhibition and cell death of colorectal cancer cells in vitro and in vivo. *Autophagy* **2019**, *15*, 295–311. [CrossRef]
49. Akin, D.; Wang, S.K.; Habibzadegah-Tari, P.; Law, B.; Ostrov, D.; Li, M.; Yin, X.-M.; Kim, J.-S.; Horenstein, N.; Dunn, W.A., Jr. A novel ATG4B antagonist inhibits autophagy and has a negative impact on osteosarcoma tumors. *Autophagy* **2014**, *10*, 2021–2035. [CrossRef] [PubMed]

50. Apel, A.; Herr, I.; Schwarz, H.; Rodemann, H.P.; Mayer, A. Blocked autophagy sensitizes resistant carcinoma cells to radiation therapy. *Cancer Res.* **2008**, *68*, 1485–1494. [CrossRef]
51. Wang, Q.; Cheng, F.; Liu, C.X.; Zou, K. Content determination of a steroid saponin from rhizome of *Reineckia carnea*. *J. Huazhong Norm. Univ.* **2013**, *47*, 60–62, 77. (In Chinese)

Review

Targeting Cancer Stem Cells by Dietary Agents: An Important Therapeutic Strategy against Human Malignancies

Mahshid Deldar Abad Paskeh [1,2,†], Shafagh Asadi [3,†], Amirhossein Zabolian [4], Hossein Saleki [5], Mohammad Amin Khoshbakht [5], Sina Sabet [5], Mohamad Javad Naghdi [5], Mehrdad Hashemi [1,2], Kiavash Hushmandi [6], Milad Ashrafizadeh [7,8], Sepideh Mirzaei [9], Ali Zarrabi [8,10] and Gautam Sethi [11,12,*]

[1] Department of Genetics, Faculty of Advanced Science and Technology, Tehran Medical Sciences, Islamic Azad University, Tehran 1916893813, Iran; deldar_mahshid@yahoo.com (M.D.A.P.); mhashemi@iautmu.ac.ir (M.H.)
[2] Farhikhtegan Medical Convergence Sciences Research Center, Farhikhtegan Hospital Tehran Medical Sciences, Islamic Azad University, Tehran 1916893813, Iran
[3] Asu Vanda Gene Industrial Research Company, Tehran 1533666398, Iran; Shaphagh@icloud.com
[4] Resident of Orthopedics, Department of Orthopedics, School of Medicine, 5th Azar Hospital, Golestan University of Medical Sciences, Gorgan 4934174515, Iran; ah_zabolian@student.iautmu.ac.ir
[5] Young Researchers and Elite Club, Tehran Medical Sciences, Islamic Azad University, Tehran 1916893813, Iran; h.saleki@student.iautmu.ac.ir (H.S.); Khoshbakht.ma1378@gmail.com (M.A.K.); Sinasabet771124@gmail.com (S.S.); M.j.naghdi1378@gmail.com (M.J.N.)
[6] Department of Food Hygiene and Quality Control, Division of Epidemiology, Faculty of Veterinary Medicine, University of Tehran, Tehran 1419963114, Iran; Kiavash.hushmandi@gmail.com
[7] Faculty of Engineering and Natural Sciences, Sabanci University, Orta Mahalle, Üniversite Caddesi No. 27, Orhanlı, Istanbul 34956, Turkey; milad.ashrafizadeh@sabanciuniv.edu
[8] Sabanci University Nanotechnology Research and Application Center (SUNUM), Istanbul 34956, Turkey; alizarrabi@sabanciuniv.edu
[9] Department of Biology, Faculty of Science, Islamic Azad University, Science and Research Branch, Tehran 1477893855, Iran; Sepideh.mirzaei@srbiau.ac.ir
[10] Department of Biomedical Engineering, Faculty of Engineering and Natural Sciences, Istinye University, Sariyer, Istanbul 34396, Turkey
[11] Department of Pharmacology, Yong Loo Lin School of Medicine, National University of Singapore, Singapore 117600, Singapore
[12] Cancer Translational Research Programme, Yong Loo Lin School of Medicine, National University of Singapore, Singapore 117600, Singapore
* Correspondence: phcgs@nus.edu.sg
† These authors have participated equally to manuscript preparation.

Abstract: As a multifactorial disease, treatment of cancer depends on understanding unique mechanisms involved in its progression. The cancer stem cells (CSCs) are responsible for tumor stemness and by enhancing colony formation, proliferation as well as metastasis, and these cells can also mediate resistance to therapy. Furthermore, the presence of CSCs leads to cancer recurrence and therefore their complete eradication can have immense therapeutic benefits. The present review focuses on targeting CSCs by natural products in cancer therapy. The growth and colony formation capacities of CSCs have been reported can be attenuated by the dietary agents. These compounds can induce apoptosis in CSCs and reduce tumor migration and invasion via EMT inhibition. A variety of molecular pathways including STAT3, Wnt/β-catenin, Sonic Hedgehog, Gli1 and NF-κB undergo down-regulation by dietary agents in suppressing CSC features. Upon exposure to natural agents, a significant decrease occurs in levels of CSC markers including CD44, CD133, ALDH1, Oct4 and Nanog to impair cancer stemness. Furthermore, CSC suppression by dietary agents can enhance sensitivity of tumors to chemotherapy and radiotherapy. In addition to in vitro studies, as well as experiments on the different preclinical models have shown capacity of natural products in suppressing cancer stemness. Furthermore, use of nanostructures for improving therapeutic impact of dietary agents is recommended to rapidly translate preclinical findings for clinical use.

Keywords: medicinal herbs; cancer treatment; cancer stem cells; drug resistance; metastasis; proliferation

1. Introduction

The cancer is the second leading cause of death worldwide after cardiovascular diseases [1]. Based on the new estimates published by Siegel and co-authors, prostate cancer is the most common cancer in males, while breast cancer is the most common cancer in females. Noteworthy, lung cancer is the most aggressive cancer in both sexes and causes highest death among other tumors [2]. Regardless of cancer incidence rate and cell deaths, there have been efforts in developing novel therapeutics for tumor treatment [3–6]. Surgery or tumor resection is beneficial in early stages of cancer and when tumor cells diffuse into various tissues in body, it is impossible to eliminate cancer by using surgery [7,8]. Radiotherapy is less invasive compared to surgery, but it has its own problems including side effects and risk of resistance [9]. Immunotherapy is a new emerging therapeutic modality in cancer and uses checkpoint inhibitors in impairing cancer progression [10–12]. Another strategy is tumor treatment is chemotherapy that is most common compared to other modalities, but resistance and adverse impacts reduce its potential [13,14]. In order to prevent therapy resistance, combination cancer therapy has been utilized. In this strategy, a combination of chemotherapy and radiotherapy, or chemotherapy and immunotherapy are applied to suppress cancer progression and inhibit resistance [15,16]. Other kinds of treatments including photothermal therapy induced by nanoparticles in ablating tumor progression, inducing DNA damage and preventing cell cycle progression [17–19]. Besides, gene therapy using small interfering RNA (siRNA), short hairpin RNA (shRNA) and CRISPR/Cas9 system can be applied in cancer suppression [14,20–22]. These strategies have been partially advantageous in improving overall survival and prognosis of cancer patients. However, cancer is still a challenge for healthcare providers and new attempts should be made in this case [23–30].

The plant derived-natural products have been under attention in recent years in field of cancer therapy [31–33]. These agents have great therapeutic activities that anti-tumor activity is among them [34,35]. Due to the potential of phytochemicals in apoptosis induction, cell cycle arrest, metastasis inhibition and multitargeting capacity, they are able to inhibit cancer progression [20,36–42]. Clinical trials have shown that plant derived-natural products are generally well-tolerated in cancer patients [43,44]. Therefore, they can be extracted for developing commercialized drugs in cancer therapy [45]. Experiments have shown that natural occurring compounds are beneficial in sensitizing tumor cells to therapeutic modalities [46,47]. For instance, curcumin can increase sensitivity of cancer cells to cisplatin, docetaxel, paclitaxel and doxorubicin [31,48,49]. A combination of resveratrol and radiotherapy is advantageous in suppressing cancer progression [50]. Hence, application of natural products alone or in combination with other therapeutic strategies can pave the way towards effective tumor treatment [51–53]. The present review article focuses on using dietary agents in cancer therapy via targeting cancer stem cells (CSCs). For this purpose, we first provide a summary of CSCs, their analysis, markers and metabolism. Then, we show how CSCs can enhance stemness and progression of tumor cells. Finally, we mechanistically discuss how each phytochemical can be beneficial in cancer suppression via targeting CSCs.

2. Cancer Stem Cells

The stem cells have self-renewal capacity and can develop colonies [54,55]. The stem cells exist in various phases of life from embryonic phase to adulthood and are able to differentiation in forming various organs and tissues of body [56,57]. A kind of cells with characteristics similar to stem cells was found to be involved in carcinogenesis and called CSCs [58–60]. The CSCs are abundantly found in the tumor microenvironment and due to their self-renewal capacity, they can preserve population of tumor cells and mediating tumorigenesis. Besides, CSCs can differentiate into different cell kinds, enhancing cancer progression [61,62]. Overall, there are two concepts for carcinogenesis. At the first model, known as stochastic model, tumor cells are similar and have the same potential in tumorigenesis. Based on this model, the accumulation of mutations has resulted in carcinogenesis [61,63]. However, upon discovery of CSCs, a new concept of tumorigenesis was introduced, known as hierarchical model that CSCs are responsible for cancer development, maintaining and tumor seeding [64].

The identification of CSCs occurred in nineteenth century, when it was found that there are dormant cells in adult tissues that can be activated by stimuli and have capacity of proliferation and generating large masses of cells [65]. Although this was a great idea showing a special function for stem cells in cancer progression, it was ignored until in 1994 that Lapidot and colleagues isolated CSCs from leukemia cells and confirmed their presence [58]. The isolated CSCs were injected in mice and they showed potential in tumor initiation and development. After the isolation of CSCs from breast cancer in 2003 [60], more investigation was performed to isolated CSCs from other kinds of tumors including brain tumors, colorectal cancer and liver cancer [66–68].

Overall, CSCs have three distinct features from normal cells including differentiation, self-renewal capacity and homeostasis control [69,70]. Regardless of tumor stage, CSCs can be abundantly found in TME and a variety of techniques for isolation and enrichment of CSCs are utilized such as side population detection of cells with ability for Hoechst 33,342 exclusion, sphere forming capacity and aldehyde dehydrogenase (ALDH) measurement [71]. The Oct4, SOX2, Nanog, c-Myc and KLF4 are able to regulate CSC features and related signaling networks are Wnt, Notch, Hedgehog and PI3K/Akt, among others. Furthermore, complicated conditions in TME such as hypoxia, stromal cells, growth factors and extracellular matrix are able to control CSC characteristics in tumor cells [72–77].

One of the important aspects of CSCs is their metabolism. Based on experiments, CSCs rely on glycolysis, mitochondrial oxidative phosphorylation and other metabolic pathways that can be targeted therapeutically for suppressing CSCs [78–80]. It has been reported that CSCs are able to induce glycolysis via upregulating glucose transporters (GLUTs), hexokinases (HKs), monocarboxylate transporters and pyruvate dehydrogenase kinase 1 [81–85]. The CD133+ cells that exist in pancreatic cancer and glioma, have the ability of oxidative phosphorylation and enhancing expression level of genes involved in tricarboxylic acid (TCA) cycle [86,87]. Noteworthy, the levels of reactive oxygen species (ROS) can also affect CSC metabolism. The CSCs that are in dormant conditions can preserve low levels of ROS via glycolysis or stimulating antioxidant defense system. However, CSCs want to proliferate and differentiate, they induce ROS overgeneration via oxidative phosphorylation [88]. Therefore, metabolism, growth and differentiation of CSCs have a close association that should be considered.

3. Cancer Stem Cells in Oncology

The presence of CSCs is in favor of tumor progression. Thanks to experiments performed recently to shed some light on the role of CSCs. It seems that presence of CSCs in TME results in immune evasion and immunosuppression [89]. The CSCs are able to induce drug resistance feature of tumor cells. It has been reported that STAT3 signaling enhances stemness and CSC features in ovarian tumor cells to mediate their resistance to cisplatin and paclitaxel chemotherapy [90]. An interesting study has provided new insight about process of CSC generation in breast cancer and enhancing carcinogenesis. In this case, adipose-derived stem cells and breast cancer cells (MDA-MB-231 cells) fuse to produce CSCs. This process is mediated by CD44 [91]. As it was mentioned, CSCs involve in triggering drug resistance in tumor cells. Noteworthy, CSCs can also mediate radioresistance features. An experiment on nasopharyngeal cancer demonstrated that hTERT promotes CSCs features in nasopharyngeal cancer and mediates radioresistance. Knockdown of hTERT is correlated with reduced CSC characteristics and enhanced sensitivity to radiotherapy [92].

A variety of molecular pathways can modulate CSC features in tumors. For instance, microRNA (miRNA)-326 is suggested to be a tumor-suppressor in cervical cancer. The miRNA-326 binds to 3'-UTR of transcription factor 4 (TCF4) to diminish its expression. Upon TCF4 down-regulation, a significant decrease occurs in CSC features via down-regulating CD44 and SOX4 expression levels [93]. On the other hand, tumor-promoting factors pave the way for increasing CSC features in tumors. For instance, DUSP9 undergoes overexpression in triple-negative breast cancer and down-regulates ERK1/2 expression to enhance levels of CSC markers including SOX2, Oct4 and ALDH1 [94]. Therefore, recapitulation of CSC niche can promote tumor progression and mediate drug resistance feature [95]. Identification of such factors and reducing their expression can pave the way to drug sensitivity. For instance, silencing RAD51AP1 is associated with impairment in self-renewal capacity of CSCs and inducing drug sensitivity in colorectal cancer [96]. Another example is musashi-1 that enhances glioblastoma progression. Silencing musashi-1 diminishes CSC features in glioblastoma [97].

Each experiment has focused on a certain molecular pathway that leads to cancer progression via enhancing CSC features. The lung cancer stemness and CSC features can be mediated via JAK2/STAT3 axis. As upstream mediator, aryl hydrocarbon receptor induces JAK2/STAT3 axis to promote lung cancer stemness [98]. Another experiment reveals that long non-coding RNA (lncRNA)-WDFY3-AS2 reduces miRNA-139-5p expression to promote SCD4 expression. Then, ovarian cancer stemness enhances via promoting CSC features and resistance to cisplatin is mediated [99]. The NEDD4 expression undergoes upregulation in breast cancer and preserves stemness via promoting CSC features [100].

One of the important aspects of cancer progression is the role of extracellular vesicles (EVs), especially exosomes. Briefly, exosomes can provide cell–cell communication via transferring proteins, lipids and nucleic acids [101–103]. A recent study has shown that EVs can stimulate generation of CSCs from stem or progenitor cells [104]. The exosomes containing lncRNA UCA1 enhances SOX2 expression via miRNA-122-5p down-regulation to enhance differentiation and self-renewal capacity of CSCs [105]. It is worth mentioning that CSCs can also secrete exosomes in cancer progression. For instance, exosomes derived from CSCs contain lncRNA DOCK9-AS2 that can enhance growth, metastasis and stemness of thyroid tumor via inducing Wnt/β-catenin axis [106]. Overall, experiments highlight the fact that CSCs play a significant role in progression of tumors and their targeting is of importance in cancer therapy. Furthermore, a variety of molecular pathways including PCGF1, circFAM73A, CXCL1 and NUMB are able to affect CSC features. The growth, metastasis and therapy response are mainly regulated by CSC characteristics [107–115].

The interesting point is the role of CSC markers as prognostic factors in tumor [116]. The clinical studies have confirmed this statement. The overexpression of BMI-1 and CD44 as CSC markers occurs in head and neck squamous carcinoma to promote cancer progression and mediate undesirable prognosis [117]. The CD133 and CXCR4 as other CSC markers also demonstrate alterations in osteosarcoma patients. The CD133 upregulation occurs in 26% of patients, while CXCR4 demonstrates overexpression in 36% of cases. The overexpression of aforementioned CSC markers provides undesirable prognosis and survival of osteosarcoma patients [118].

4. Search Strategy

Various databases including Google scholar, Web of Science and Pubmed were used to search and collect articles. The names of dietary agents discussed in this article and other words including "cancer" and "cancer stem cell" were searched to find the relevant articles. Furthermore, there were many phytochemicals found whose impact on CSCs have not been evaluated yet but can be considered in future studies.

5. Dietary Agents and Cancer Stem Cells

5.1. Flavonoids

5.1.1. Flavones

Nobiletin

The nobiletin is a potent anti-tumor agent capable of suppressing tumor migration via EMT inhibition [119]. Nobiletin stimulates apoptosis and DNA damage to impair progression of oral cancer cells [120]. Nobiletin suppresses breast cancer progression in a dose-dependent manner. Nobiletin enhances miRNA-200b expression to elevate apoptosis and pyroptosis in breast cancer cells [121]. Three experiments have shown role of nobiletin in affecting CSCs in tumor therapy. The Wnt/β-catenin signaling is a possible target of nobiletin in impairing CSC characteristics [122]. The invasion and angiogenesis are suppressed by nobiletin via targeting CSCs. Nobiletin (100 and 200 µM) reduces STAT3 expression via binding to CD36 to inhibit NF-κB signaling, leading to a significant decrease in migration and metastasis of CSCs [123]. In order to potentiate efficacy of nobiletin in CSC suppression, its co-administration with xanthohumol is recommended. This combination suppresses migration of CSCs and decreases CD44v6 expression. Furthermore, they induce apoptosis and cycle arrest at G2/M phase. This combination impairs progression of colorectal CSCs and enhances their sensitivity to oxaliplatin and 5-flouroruacil chemotherapy [124].

Chrysin

The chrysin is a new emerging anti-tumor agent capable of suppressing growth and invasion of tumor cells, and promoting their sensitivity to chemotherapy [125]. A combination of chrysin and daidzein inhibits colorectal cancer progression via suppressing ERK and Akt molecular pathways [126]. Chrysin-loaded nanostructures are capable of enhancing apoptosis via triggering p53 expression. Furthermore, chrysin-loaded nanoparticles suppress PI3K/JNK axis and inhibit tumor growth in vivo [127]. Chrysin enhances miRNA-let-7a expression, while it reduces H19 and COPB2 expression levels to impair progression of gastric cancer cells [128].

An experiment has applied micellar nanoparticles for co-delivery of chrysin and docetaxel in cancer therapy. The application of micelles promotes therapeutic effect of both docetaxel and chrysin. The micelles were biodegradable and capable of docetaxel and chrysin co-delivery in synergistic cancer chemotherapy. This combination enhanced ROS levels to induce apoptosis in colon CSCs. Furthermore, docetaxel- and chrysin-loaded micelles inhibit migration and invasion of CSCs, impairing colon cancer metastasis [129]. Another study has focused on a derivative of chrysin, known as CHM-04 that has 3.2-fold higher anti-tumor activity compared to chrysin. The CHM-04 suppresses colony formation capacity and invasion of breast CSCs and induces apoptotic cell death [130].

Apigenin

Similar to chrysin, apigenin is a potent anti-tumor agent against various cancers including breast cancer, lung cancer and gastric cancer [7]. Apigenin impairs progression of multiple myeloma cells in a dose-dependent manner. Apigenin is able to down-regulate STAT1 expression in suppressing COX-2/iNOS axis [131]. A combination of apigenin and hesperidin prevent DNA repair in breast tumor to potentiate DOX activity in cancer suppression [132]. Furthermore, apigenin reduces activity of ABCG2 and ABCC4 as drug efflux transporters to enhance internalization of doxorubicin in breast cancer and mediate apoptosis [133].

The activation of YAP/TAZ axis is responsible for CSC features in triple negative breast cancer. The apigenin administration (0–64 µM) decreases colony formation and self-renewal capacity of CSCs. Furthermore, apigenin reduces number of CD44+ cells in breast cancer. These anti-tumor activities were mediated via suppressing YAP/TAZ axis [134]. The presence of CD133 cells decreases potential of cisplatin in lung cancer suppression. The apigenin administration (10–30 µM) induces apoptosis in CSCs via p53 upregulation and enhances cisplatin cytotoxicity against lung tumors [135]. The stimulation of tumor-promoting factors enhances CSC features in tumors. For instance, activation of PI3K/Akt signaling leads to CSC features in prostate cancer via inducing NF-κB signaling. The apigenin administration (0–100 µM) suppresses PI3K/Akt/NF-κB axis to reduce Oct3/4 levels, as CSC markers in prostate cancer. Apigenin reduces survival of CSCs and enhances p21 and p27 upregulation. For decreasing viability of CSCs in prostate cancer therapy, apigenin induces both intrinsic and extrinsic apoptosis. Furthermore, apigenin disrupts invasion and metastasis of CSCs [136]. When prostate cancer cells obtain stemness and CSC features, they can easily achieve resistance to cisplatin chemotherapy. Similar to previous experiment, apigenin (15 µM) suppressed phosphorylation of PI3K, Akt and NF-κB in impairing CSC features. Furthermore, apigenin inhibited cell cycle progression of CSCs in prostate cancer via upregulating p21, CDK2, CDK4 and CDK6. By reducing Snail expression, apigenin impaired progression and invasion of CSCs. Apigenin enhanced capase-8, Apaf-1 and p53 levels, while it decreased Bcl-2, sharpin and survivin levels in triggering apoptosis in CSCs in prostate cancer. These impacts of apigenin promote sensitivity of prostate cancer cells to cisplatin chemotherapy [137]. For disrupting cancer stemness, apigenin (40 µM) decreases expression levels of CD44, CD105, Nanog, Oct4, VEGF and REX-1 [138]. These studies demonstrate that apigenin is a potent inhibitor of CSCs in tumor treatment that further experiments can focus on revealing more signaling pathways affected by apigenin [139].

Baicalein

The baicalein is another anti-tumor agent that can induce apoptosis in tumors via upregulating caspase-3, -8 and -9 levels. Baicalein reduces MMP-2 and MMP-9 expression levels to impair metastasis of cancer cells [140]. The baicalein prevents SNO-induced ezrin tension and reduces iNOS levels to impair progression of lung tumors [141]. By suppressing Akt and Nrf2 molecular pathways, baicalein induce both apoptosis and autophagy in gastric tumor cells [142]. Furthermore, baicalein mediates proteasomal degradation of MAP4K3 to impair progression of lung tumor cells [143].

The members of Sonic Hedgehog signaling including SHH, SMO and Gli2 undergo upregulation to mediate CSC features in pancreatic cancer. The overexpression of Sonic leads to upregulation of SOX2 and Oct4 as CSC markers in pancreatic cancer. The baicalein administration (0–300 µM) suppresses Sonic signaling to reduce SOX2 expression and impair CSC features in pancreatic cancer [144].

Wogonin

The wogonin has demonstrated great therapeutic impacts in pre-clinical experiments [145]. The wogonin is able to induce senescent in breast tumors via down-regulating TXNRD2 expression [146]. Wogonin diminishes expression levels of Notch1 at mRNA and protein levels to suppress growth and metastasis of skin cancer cells [147]. This section focuses on wogonin impact on CSCs.

As it was mentioned, natural products can enhance ROS levels to induce apoptosis in CSCs [148]. A same strategy is followed by wogonin in osteosarcoma therapy. For this purpose, wogonin (0–80 µM) enhances ROS levels to reduce expression of factors responsible for CSC features such as STAT3, Akt and Notch1 [149]. In addition to triggering apoptosis and reducing survival of CSCs in osteosarcoma, wogonin is able to affect invasion of CSCs. In this way, wogonin administration (0–80 µM) decreases MMP-9 expression to impair migration and invasion of CSCs in osteosarcoma (Figure 1) [150]. Interestingly, studies have only focused on osteosarcoma and more experiments on other tumor models should be performed to shed more light on anti-tumor activities of wogonin via targeting CSCs. Overall, flavones are potential agents in suppressing stemness and CSC features in tumors that have been summarized in Table 1.

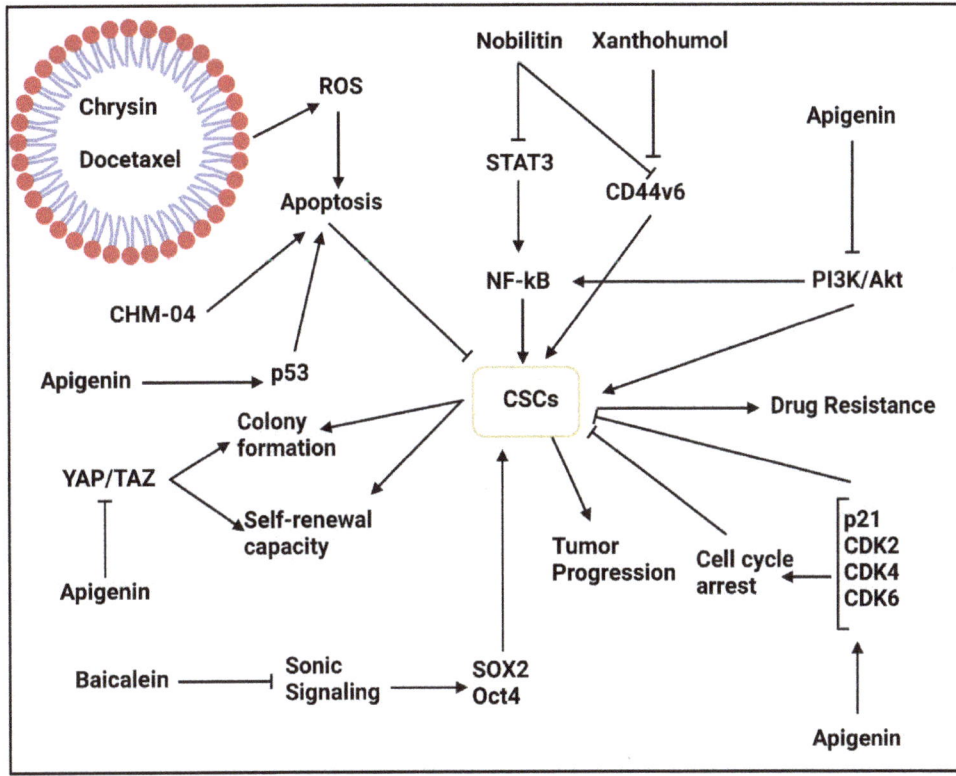

Figure 1. The potential of flavones in suppressing CSCs. Abbreviations: CSCs, cancer stem cells; NF-κB, nuclear factor-kappaB; CDK, cyclin-dependent kinase; Oct4, octamer-4; YAP, Yes-associated protein, SOX2, sex determining region Y-box 2; TAZ, Transcriptional coactivator with PDZ-binding motif; CHM-04, a chrysin derivative; STAT3, signal transducer and activator of transcription 3; Akt, protein kinase-B; PI3K, phosphoinositide 3-kinase; ROS, reactive oxygen species.

Table 1. Flavones in targeting CSCs for tumor suppression.

Anti-Tumor Agent	Cancer Type	In Vitro/ In Vivo	Cell Line/Animal Model	Study Design	Remarks	Refs
Chrysin	Colorectal cancer	In vitro	HT-29 cells	-	The chrysin- and docetaxel-loaded micelles exert synergistic therapy in suppressing growth and invasion of CSCs Enhancing ROS production to mediate cell death	[129]
CHM-04 (chrysin derivative)	Breast cancer	In vitro	MCF-7 and MDA-MB-231 cells	10 µM	Higher anti-tumor potential (3.2-fold increase) compared to chrysin Inducing apoptosis and reducing migration Suppressing colony formation of CSCs	[130]
Apigenin	Breast cancer	In vitro In vivo	MDA-MB-231 cells Nude mice	0–64 µM	Suppressing CSC features in breast cancer via inhibiting YAP/TAZ axis	[134]

Table 1. Cont.

Anti-Tumor Agent	Cancer Type	In Vitro/In Vivo	Cell Line/Animal Model	Study Design	Remarks	Refs
Apigenin	Lung cancer	In vitro	A549 and H1299 cells	10, 20 and 30 µM	Enhancing p53 expression and impairing CSC features Promoting sensitivity of cancer cells to cisplatin chemotherapy	[151]
Apigenin	Prostate cancer	In vitro	PC3 cells	25 µM	Reducing viability of CSCs and triggering apoptosis via p21 and p27 upregulation Triggering extrinsic pathway of apoptosis Suppressing PI3K/Akt/NF-κB axis	[136]
Apigenin	Prostate cancer	In vitro	PC3 cells	15 µM	Inducing apoptosis and cell cycle arrest in CSCs Suppressing migration and invasion of CSCs Potentiating anti-tumor activity of cisplatin Inhibiting PI3K/Akt and NF-κB pathways	[137]
Apigenin	Head and neck cancer	In vitro	HN-8, HN-30, and HSC-3 cells	0–100 µM	Decreasing viability of cancer cells in a dose-dependent manner Suppressing CSC features via reducing expression levels of CD44, Nanog and CD105	[138]
Apigenin	Prostate cancer	In vitro	PC3, LNCaP, or isolated CD44+ CD133+ and CD44+ stem cells	0–2 µM	Inducing apoptosis in CSCs Suppressing proliferation and viability of CSCs Upregulating expression levels of p21, p27, Bax, Bid, caspase-3 and caspase-8 Inhibiting ERK, PARP and NF-κB expression levels	[139]
Baicalein	Pancreatic cancer	In vitro	PANC-1, BxPC-3, and SW1990 cells	0–300 µM	Reducing SOX-2 expression and impairing CSC features via down-regulating Sonic expression	[144]
Wogonin	Osteosarcoma	In vitro	CD133+ Cal72 cells	0–80 µM	Triggering cell death in cancer cells via enhancing ROS levels	[149]
Wogonin	Osteosarcoma	In vitro	CD133+ CAL72 cells	0–80 µM	Triggering apoptosis in CSCs Inhibiting self-renewal capacity Reducing migration and invasion of CSCs via reducing MMP-9 expression	[150]

5.1.2. Flavanones

Naringenin administration (100 µM) prevents colony formation, metastasis and EMT in breast tumor, while it induces apoptosis via upregulating p53 and ERα at mRNA level, as tumor-suppressor factors [152].

Naringin administration (300 µM) impairs CSC features in esophageal cancer and inhibit viability of CSCs [153].

Hesperetin (50–200 µM) also demonstrated potential in triggering apoptosis in breast CSCs. Hesperetin impairs invasion of breast CSCs and stimulates cell cycle arrest in breast CSCs. The hesperetin enhances p53 expression, while it down-regulates Notch1 expression in impairing stemness in breast CSCs [154].

The experiments evaluating role of flavanones in targeting CSCs in tumor suppression are limited and more studies are required to reveal true potential of these natural products in cancer therapy.

5.1.3. Flavonols

Fisetin

The fisetin is a natural flavonol that has demonstrated high anti-tumor activity and capacity in chemoprevention [155,156]. Fisetin mediates histone demethylation to induce DNA damage and impair progression of pancreatic tumor cells [157]. To date, just one experiment has evaluated role of fisetin in targeting CSCs that is included here. The proliferation, metastasis, angiogenesis and carcinogenesis of renal CSCs undergo inhibition by fisetin. The in vitro and in vivo experiments have shown role of fisetin in suppressing renal CSCs. The fisetin is able to decrease expression levels of cyclin Y and CDK16 via inhibiting 5hmC modification in CpG islands. Furthermore, fisetin can reduce TET1 levels in renal CSCs. Therefore, a significant decrease occurs in growth of renal CSCs and their angiogenesis and migration are suppressed [158].

Epigallocatechin 3-Gallate

The epigallocatechin 3-gallate (EGCG) is another naturally occurring compound that has demonstrated high potential in cancer therapy. The EGCG enhances Beclin and LC3 levels to induce autophagy in bladder cancer cells. Furthermore, EGCG induces apoptosis in bladder cancer via upregulating Bax, caspase-3 and caspase-9 levels [159]. The EGCG reduces Sonic Hedgehog expression to suppress PI3K/Akt axis, leading to apoptosis in colon cancer cells [160]. The EGCG impairs migration and metastasis of cervical tumor cells via down-regulating TGF-β and subsequent inhibition of EMT mechanism [161]. Therefore, EGCG can be considered as a promising agent in cancer therapy [162]. Noteworthy, EGCG targets CSCs in affecting cancer progression. The lung tumor cells demonstrate high expression level of CLOCK to improve their CSC features. Noteworthy, EGCG (0–40 µM) reduces mRNA and protein levels of CLOCK to impair CSCs features and suppress self-renewal capacity of CSCs in lung tumor therapy [163]. The miRNAs are considered as important modulators of CSCs in cancer [164–167]. The expression level of hsa-miRNA-485-5p undergoes down-regulation in serum of lung tumor patients, showing tumor-suppressor role of this miRNA. Restoring miRNA-485-5p expression impairs lung tumor proliferation and stimulates apoptosis via down-regulating RXRα expression. The EGCG administration (0–40 µM) enhances miRNA-485-5p expression to reduce RXRα expression, leading to a decrease in expression levels of CD133 and CD44 as CSC markers [168]. The CSC features in lung cancer mainly depend on upregulation of β-catenin and its inhibition suppresses lung tumor progression. An experiment has shown that EGCG (0–100 µM) inhibits Wnt/β-catenin axis to impair CSC features in lung cancer [169].

Similar to lung cancer, a number of experiments have focused on anti-tumor activity of EGCG in colorectal cancer via targeting CSCs. In this way, EGCG (0–40 µM) suppresses Wnt/β-catenin axis to reduce expression levels of CD133, CD44, ALDHA1, Nanog and Oct4 in impairing CSC features in colorectal tumor [170]. The colorectal tumor cells demonstrate resistance to various chemotherapeutic agents including oxaliplatin and

5-flourouracil [171,172]. The EGCG (0–400 μM) decreases expression levels of Notch1, Bmi1, Suz12 and EZH2, while it enhances expression levels of miRNA-34a, miRNA-145 and miRNA-200c in impairing CSC features in colorectal cancer and enhancing 5-flourouracil sensitivity [173]. The bladder cancer cells rely on Sonic Hedgehog signaling to enhance their CSC features and mediate their progression. Noteworthy, EGCG suppresses Sonic signaling to decrease CD44, CD133, Nanog, Oct4 and ALDH1 in bladder cancer therapy [174]. Therefore, EGCG is a potent inhibitor of CSCs in various cancers and for this purpose, it targets various molecular pathways including Notch and NF-κB signaling pathways (Table 2) [175–177].

Table 2. Gallate in suppressing CSCs.

Cancer Type	In Vitro/In Vivo	Cell Line/Animal Model	Study Design	Remarks	Refs
Lung cancer	In vitro In vivo	A549 and H1299 cell lines Xenograft model	0–40 μM 20 mg/kg	Suppressing self-renewal capacity of tumor cells CLOCK down-regulation Suppressing tumor growth in vivo	[163]
Lung cancer	In vitro	A549, H460, H1299, and HEK-293T cells	0–40 μM	Suppressing CSC features in lung cancer Promoting miRNA-485-5p expression Down-regulating PXRα expression	[168]
Lung cancer	In vitro	A549 and H1299 cells	0–100 μM	The enrichment of CSC features in lung cancer Wnt/β-catenin inhibition prevents CSC characteristics	[169]
Colorectal cancer	In vitro	DLD-1 and SW480 cells	0–60 μM	Proliferation inhibition Apoptosis induction Impairing CSC features via inhibiting Wnt/β-catenin axis	[170]
Colorectal cancer	In vitro	HCT116 cells	0–200 μM	Inhibiting self-renewal capacity of CSCs Reducing expression levels of Notch1, Suz12 and EZH2 Increasing sensitivity of tumor cells to 5-flourouracil chemotherapy	[173]
Bladder cancer	In vitro	EJ and UM-UC-3 cells	0–90 μM	Inhibiting Sonic Hedgehog signaling Reducing CSC features Apoptosis induction in CSCs Impairing proliferation of CSCs	[174]
Head and neck cancer	In vitro In vivo	K3, K4 and K5 cells Xenograft model	0–10 μM	Impairing CSC features via suppressing Notch signaling	[176]
Nasopharyngeal cancer	In vitro In vivo	CNE2 and C666-1 cells Xenograft model	0–50 μM	Inhibiting self-renewal capacity of CSCs Suppressing migration via EMT inhibition Down-regulating NF-κB expression	[177]

5.1.4. Chalcones
Isoliquiritigenin

The isoliquiritigenin (ISL) is derived from licorice root and it is a modulator of molecular pathways in cancer suppression [178]. The ISL suppresses TGF-β signaling to reduce Smad3 expression, resulting in EMT inhibition and decreased endometrial cancer migration [179]. Furthermore, ISL enhances miRNA-200c expression to inhibit β-catenin signaling, leading to EMT suppression in triple-negative breast cancer [180]. A combination of ISL and docosahexaenoic acid induces apoptosis in colorectal cancer via enhancing ROS levels and mediating phosphorylation of JNK and ERK [181]. The ISL-loaded liposomes suppress colorectal tumor progression via inhibiting AMPK-mediated glycolysis [182]. This section focuses on anti-tumor activity of ISL via targeting CSCs. The ISL administration (25 μM) suppresses self-renewal and multidifferential capacities of CSCs in breast cancer. Furthermore, ISL inhibit growth and colony formation of breast CSCs. The ISL suppresses β-catenin signaling and decreases ABCG2 expression to enhance drug sensitivity of breast cancer cells and suppress CSC features [183]. The expression levels of CD44 and ALDH1 undergo down-regulation upon ISL administration (0–50 μM) in oral cancer. The ISL suppresses colony formation and invasion of CSCs in oral tumor. In suppressing CSCs, ISL also decreases expression level of GRP78 at mRNA and protein levels. Furthermore, by reducing ABCG2 expression, ISL enhances sensitivity of oral CSCs to cisplatin chemotherapy [184]. Another experiment reveals that ISL administration (25 and 50 μM) enhances SIF1 expression via demethylation to suppress DNMT1 methyltransferase. Furthermore, ISL simultaneously suppresses β-catenin signaling to eradicate CSCs in breast cancer therapy [185].

5.1.5. Isoflavonoids
Daidzein

An experiment has focused on a derivative of daidzein, known as N-t-boc-Daidzein. This derivative is able to suppress CSC features in ovarian cancer in concentration (0–10 μM)- and time (0–70 h)-dependent manner. This derivative induces apoptosis in CSCs via upregulating caspase-3, -8 and -9 levels. Suppressing growth and survival of CSCs is mediated by N-t-boc-Daidzein. In addition to triggering caspase cascade, N-t-boc-Daidzein mediates mitochondrial depolarization and decreases Akt expression via degradation to impair CSC features in ovarian tumor [186].

Genistein

The genistein is an isoflavonoid compound that has potent anti-tumor activity. The genistein administration decreases chance of mammary cancer recurrence [187]. The glioblastoma cells exposed to radiation, demonstrate enhanced metastasis. The genistein is able to suppress Akt2 pathway in reducing invasion and migration of glioblastoma cells [188]. The genistein administration reduces hsa-circ-0031250 expression to promote miRNA-873-5p expression, leading to lung cancer suppression and decreased proliferation rate [189]. Both inflammation and GSK-3 are suppressed by genistein to inhibit ovarian tumor progression in vivo [190]. In order to improve genistein capacity in cancer suppression, nanoparticles have been developed for its targeted delivery [191,192].

The Sonic Hedgehog signaling enhances CSC features in nasopharyngeal cancer. The administration of genistein (0–100 μM) inhibits sonic signaling and reduces expression level of CSC markers including CD44, ALDH1, Oct4 and Nanog to impair nasopharyngeal progression [193]. A same strategy occurs in renal cancer, so that genistein administration (0–90 μM) suppresses Sonic Hedgehog signaling to impair growth and colony formation of renal CSCs and stimulate apoptotic cell death [194]. The genistein is able to mediate differentiation of CSCs in a paracrine mechanism to suppress cancer progression. An experiment on breast tumor has shown genistein administration (2 μM or 40 nM) stimulates PI3K/Akt and MEK/ERK molecular pathways to induce differentiation of CSCs in breast cancer [195]. Another experiment also reveals that low doses of genistein (15 μM) inhibits self-renewal capacity of gastric CSCs in vitro and in vivo [196]. The capacity of genistein in suppressing CSC features in gastric cancer is attributed to down-regulating FOXM1 expression that subsequently, reduces expression levels of CD133, CD44 and Nanog. Furthermore, by inhibiting FOXM1 expression, genistein reduces Twist1 expression that is in favor of inhibiting EMT in CSCs via reducing N-cadherin levels and enhancing E-cadherin levels [197]. In addition to aforementioned molecular pathways, genistein suppresses Gli1 expression to impair CSC features in gastric cancer [198]. Some of the molecular pathways are similar among various cancers. For instance, previous research demonstrated that genistein suppresses Gli1 pathway in reversing CSC features in gastric cancer. Another experiment in prostate cancer reveals that genistein is able to suppress Gli1 pathway in inhibiting CSC characteristics [199]. Furthermore, it was mentioned that genistein suppresses nasopharyngeal CSCs via inhibiting Sonic signaling. A similar study reveals Hedgehog signaling inhibition and subsequent Gli1 down-regulation by genistein in prostate cancer stemness suppression [199]. Genistein also suppresses Hedgehog signaling in impairing CSC features in breast cancer [200]. Overall, studies highlight the fact that genistein is a potent inhibitor of cancer progression via targeting CSCs (Figure 2, Table 3) [200–204].

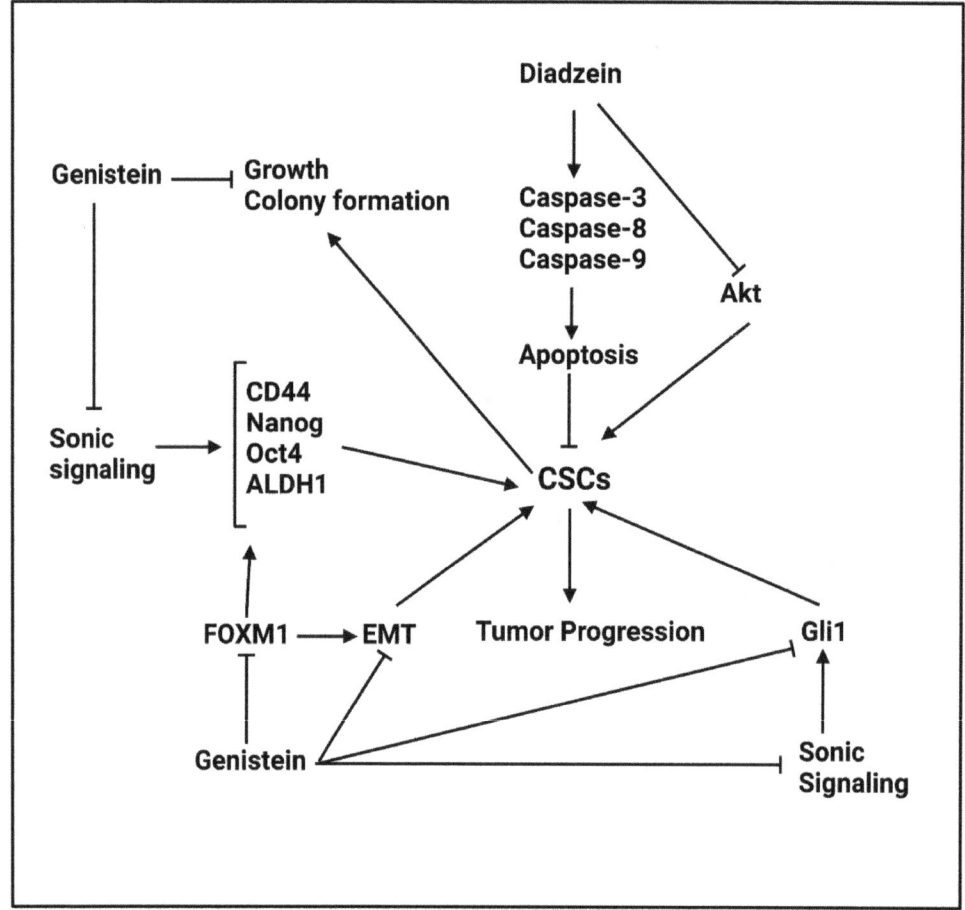

Figure 2. The potential of isoflavonoids in suppressing CSCs.

Table 3. The potential of genistein in cancer suppression via targeting CSCs.

Cancer Type	In Vitro/In Vivo	Cell Line/Animal Model	Study Design	Remarks	Refs
Nasopharyngeal cancer	In vitro	CNE2 and HONE1 cells	0–100 µM	Reducing expression level of CSC markers including CD44 and CD133 via suppressing Sonic signaling	[193]
Renal cancer	In vitro	786-O and ACHN cell lines	0–90 µM	Suppressing growth and colony formation capacities of CSCs Triggering apoptosis and reducing CSC marker expression Inhibiting Sonic Hedgehog signaling	[205]
Breast cancer	In vitro	MCF-7 and MDA-MB-231 cells	2 µM or 40 nM	Decreasing number of CSCs Inducing PI3K/Akt and MEK/ERK pathways in a paracrine manner and mediating differentiation of CSCs	[195]
Gastric cancer	In vitro	MGC-803 and SGC-7901 cells	15 µM	Suppressing colony formation and self-renewal capacities Inhibiting chemoresistance via down-regulating ERK1/2 and ABCG2	[196]
Gastric cancer	In vitro	SGC-7901 cells	0–10 µmol/L	Decreasing FoxM1 expression to prevent self-renewal capacity and migration of CSCs Decreasing expression levels of CSC markers including CD44, CD133 and ALDH1	[197]
Gastric cancer	In vitro	AGS and MKN45 cells	10 µg/ml	Decreasing expression level of CD44 via suppressing Sonic Hedgehog signaling	[198]
Prostate cancer	In vitro In vivo	22RV1, DU145 cells Xenograft animal model	15 and 30 µM 10 mg/kg	Suppressing Hedgehog/Gli1 axis to impair stemness and CSC features in prostate cancer	[199]

5.2. Pomegranate and Its Bioactive Compounds

5.2.1. Ellagic Acid

Ellagic acid is a polyphenolic compound that can suppress cancer progression. The ellagic acid increases sensitivity of bladder cancer cells via inhibiting drug transporters and EMT mechanism [206]. The p21 and p53 levels undergo upregulation by ellagic acid to induce apoptosis in prostate cancer cells [207]. The ellagic acid stimulates autophagy via Akt down-regulation and AMPK upregulation to reduce viability and survival of ovarian cancer cells [208]. The ellagic acid is a potent agent in suppressing drug resistance via down-regulating P-glycoprotein (P-gp_levels [209]. This section focuses on anti-tumor activity of ellagic acid based on targeting CSCs.

Ellagic acid is able to target CSCs in tumor eradication. An experiment has shown role of ellagic acid in breast cancer therapy via targeting CSCs. The β-catenin stabilization is responsible for CSC features in breast cancer and ACTN4 functions as upstream mediator. Silencing ACTN4 impairs growth, migration and colony formation capacities in breast cancer. The ellagic acid administration (0–50 µM) reduces ACTN4 expression to inhibit β-catenin signaling in CSCs, impairing breast cancer progression [210]. The ALDH level as an inducer of drug resistance and CSC marker undergoes down-regulation by a combination of ellagic acid and urolithins in colon CSCs [211]. Although just two experiments have evaluated role of ellagic acid in targeting CSCs, these obviously demonstrate role of ellagic acid in tumor treatment and more studies are required in this case.

5.2.2. Caffeic Acid

Caffeic acid (CA) is a phenolic acid compound that is isolated from natural sources such as tea and has various therapeutic activities. The anti-tumor activity of CA and its derivatives have been under attention, summarized in this review [212]. The CA can mediate ROS overgeneration to induce cell death in cervical cancer cells [213]. Furthermore, CA suppresses P-gp function to mediate chemosensitivity of tumors [214]. Noteworthy, CA derivatives are also able to induce apoptosis via survivin down-regulation and caspase-3 and -9 upregulation [215,216].

The colorectal CSCs that have high ability in mediating radioresistance feature. The CD44+ and CD133+ CSCs in colorectal cancer have self-renewal capacity and carcinogenesis impact. The CA administration inhibits PI3K/Akt axis that is responsible for CSCs features in colorectal cancer, leading to enhanced radiosensitivity and reduced carcinogenesis impact [217]. The keratinocytes demonstrate high expression level of NF-κB that is beneficial for upregulating Snail expression and mediating migration and CSCs features. The CA administration (0–100 µM) enhances p38 expression to reduce potential of NF-κB in binding to Snail. Therefore, invasion and CSC features undergo inhibition [218]. The CA has a naturally occurring derivative, known as caffeic acid phenethyl ester (CAPE) that can suppress self-renewal capacity and progenitor formation in breast cancer in a dose-dependent manner [219]. The expression of miRNAs is mainly affected by CA in cancer therapy. The TGF-β/Smad2 axis mediate CSC features in tumors. The CA administration (20 µM) increases miRNA-184a expression via DNA methylation, as a tumor-suppressor factor. Then, overexpressed miRNA-184a inhibits Smad2 expression by binding to its 3′-UTR, impairing CSC features [220]. Hence, CA and its derivative CAPE are potential anti-tumor agents in cancer treatment via targeting CSCs.

5.2.3. Luteolin

The luteolin is a natural flavonoid that can suppress cancer progression via regulating autophagy [221]. The luteolin suppresses Akt/mTOR axis to reduce MMP-9 expression, impairing metastasis and invasion of breast cancer cells [222]. The luteolin suppresses EMT mechanism via down-regulating YAP/TAZ, resulting in a significant decrease in breast cancer invasion [223]. Furthermore, lutelon enhances death receptor 5 (DR5) expression and mediates mitochondrial fission to promote TRAIL sensitivity of lung tumors [224]. To date, just one experiment has evaluated role of luteolin in targeting CSCs. Mechanistically, IL-6 stimulates STAT3 signaling to promote CSC features in oral cancer and mediate their resistance to radiotherapy. The luteolin administration (0–40 µM) suppresses IL-6/STAT3 axis to impair self-renewal capacity of CSCs and reduce expression level of CSC markers including ALDH1 and CD44 [225].

5.2.4. Quercetin

Quercetin is a flavonol present in fruits and vegetables such as grape, anion, apple, berries and broccoli [226]. The diet intake of quercetin is estimated to be higher than 70% of all flavonol intake [227]. Quercetin has different pharmacological activities including immunomodulatory, hepatoprotective, neuroprotective and nephroprotective that can be mediated via regulating autophagy [228]. Furthermore, quercetin can be considered as a protective agent against ischemic/reperfusion injury [229]. Regardless of its protective impacts, quercetin is suggested to display significant anti-tumor activity. Quercetin reduces CDK6 expression to impair progression of breast and lung cancer cells [230]. Quercetin enhances reactive oxygen species (ROS) levels to induce ferroptosis. Furthermore, quercetin promotes cell death via lysosome activation [231]. Angiogenesis and metastasis of esophageal cancer cells was suppressed by quercetin via reducing expression levels of VEGF-A, MMP-2 and MMP-9 [159]. Quercetin reverses multidrug resistance and using nanoparticles for its delivery enhances its potential in cancer suppression [232]. This section focuses on CSC targeting by quercetin in cancer therapy.

The PI3K/Akt/mTOR axis is responsible for growth and progression of CSCs in breast cancer. The quercetin administration (0–200 μM) suppresses PI3K/Akt/mTOR axis in induce cell cycle arrest in G1 phase in breast cancer cells. The quercetin impairs viability, colony formation and mammosphere formation in CD44+ stem cells in breast cancer and triggers apoptosis [233]. Another experiment also reveals role of quercetin in eradicating CSCs in breast cancer. The quercetin administration (0–200 μM) suppresses growth, metastasis and self-renewal capacity of CSCs in breast cancer. For this purpose, quercetin reduces expression levels of ALDH1, CXCR4, mucin 1 and epithelial cell adhesion molecule (EpCAM) [234]. The nuclear translocation of Y-box binding protein 1 (YB-1) is responsible for CSCs features in breast cancer. For reversing multidrug resistance in breast cancer, quercetin decreases expression level of P-gp as an efflux transporter. More importantly, quercetin inhibits nuclear translocation of YB-1 to suppress CSCs features in breast cancer cells and enhance their sensitivity to doxorubicin, paclitaxel and vincristine [235]. It is worth mentioning that by down-regulating expression levels of P-gp, BCRP and MRP1, quercetin (0–2 μM) increases internalization of doxorubicin in breast cancer cells, resulting in eradication of CSCs [236].

Another cancer that can be affected by quercetin is pancreatic cancer. For enhancing potential of quercetin in suppressing CSCs, its combination therapy has been used. For this purpose, a combination of sulforaphane (SFN) (0–10 μM) and quercetin (20 μM) effectively inhibits self-renewal capacity of CSCs [237]. In pancreatic cancer, CSC divisions tend to be symmetric. The low expression of miRNA-200b-3p induces Notch signaling to make daughter cells be symmetric. However, quercetin administration (50 μM) enhances expression level of miRNA-200b-3p as tumor-suppressor to reduce Notch expression, resulting in asymmetric divisions in CSCs [238]. A combination of SFN, quercetin and catechins reduces ALDH1 expression in pancreatic cancer, as CSC marker. For this purpose, SFN, quercetin and catechins enhances miRNA-let-7 expression to down-regulate K-ras expression [239]. Hence, quercetin and its combination with anti-tumor agents synergistically suppress CSC features in pancreatic cancer [240].

The induction of Notch1 signaling and enhanced CSC features lead to radioresistance in colon cancer. The quercetin administration (20 μM) suppresses Notch1 signaling and reduces expression levels of CSC markers including SOX9, CD133 and CD44 [241]. The CD133+ colorectal cancer cells seem to be responsible for triggering drug resistance in colorectal cancer. The quercetin administration (10–100 μM) induces apoptosis and cycle arrest (G2/M phase) in CD133+ cells to enhance sensitivity of colorectal tumor cells to doxorubicin chemotherapy [242]. Therefore, quercetin targets CSCs to increase therapy response of colorectal cancer cells.

The prostate cancer (PCa) is the most common cancer and second malignant tumor in men [2,243]. The various molecular pathways are responsible for PCa progression and various anti-tumor compounds and genetic tools have been utilized for its treatment [20,244–247]. The quercetin has shown capacity in suppressing growth and invasion of PCa stem cells (CD44+/CD133+ cells). The midkin (MK) pathway is responsible for CSC features in PCa. Co-application of quercetin and siRNA-MK enhances potential in suppressing CSCs. Furthermore, this combination induces apoptosis and G1 arrest in CSCs. The molecular pathways including PI3K/Akt, ERK1/2, p38, ABCG2 and NF-κB undergo down-regulation by quercetin and siRNA-MK in inhibiting CSCs in PCa [248]. Furthermore, a combination of quercetin and EGCG induces apoptosis (via caspase-3/7 upregulation and down-regulating Bcl-2, survivin and XIAP) in CSCs and prevents their growth and invasion. The reduced metastasis of CSCs by quercetin and EGCG is attributed to suppressing epithelial-to-mesenchymal transition (EMT) via down-regulating Snail, vimentin, Slug and β-catenin [249]. Therefore, quercetin is a well-known compound in eradicating CSCs in preventing cancer progression (Figure 3, Table 4) [250,251].

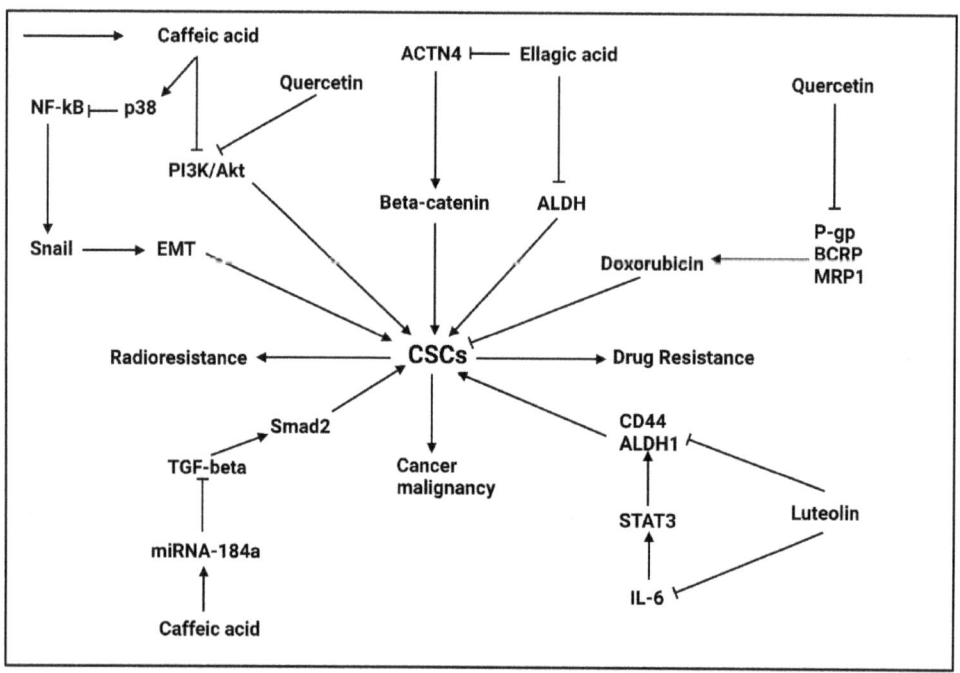

Figure 3. The potential of bioactive constituents of pomegranate in suppressing CSCs.

Table 4. The CSCs as promising targets of quercetin in tumor therapy.

Cancer Type	In Vitro/ In Vivo	Cell Line/Animal Model	Study Design	Remarks	Refs
Breast cancer	In vitro In vivo	MCF-7 cells Nude mice	0–200 µM	Reducing number of CSCs Impairing CSC features Suppressing PI3K/Akt/mTOR axis	[233]
Breast cancer	In vitro	MDA-MB-231 cells	0–200 µM	Disrupting tumor progression and CSC features via down-regulating ALDH1A1, CXCR4, MUC1 and EpCAM	[234]
Breast cancer	In vitro	MCF-7 cells	0–70 µM	Suppressing CSCs in breast cancer via preventing nuclear translocation of YB-1	[235]
Breast cancer	In vitro	MCF-10A, MCF-7, MDA-MB-231 and AC16 cells	0–2 µM	Enhancing intracellular accumulation of doxorubicin via down-regulating P-gp, MRP1 and BCRP Eliminating CSCs and potentiating doxorubicin's anti-tumor activity	[236]
Pancreatic cancer	In vitro	AsPC1 and PANC1 cells	50 µM	Suppressing self-renewal capacity of tumor cells via enhancing miRNA-200b expression	[238]

Table 4. Cont.

Cancer Type	In Vitro/In Vivo	Cell Line/Animal Model	Study Design	Remarks	Refs
Pancreatic cancer	In vitro In vivo	BxPc-3 and MIA-PaCa2 cells Nude mice and xenografts	100, 200 and 400 μM 50 mg/kg	Synergistic impact between quercetin and sulforaphane Down-regulating NF-κB signaling Impairing tumor growth in vitro and in vivo	[240]
Colon cancer	In vitro	DLD-1 and HT-29 cells	0–50 μM	Suppressing Notch signaling, impairing CSCs features and increasing sensitivity of tumor cells to radiotherapy	[241]
Colorectal cancer	In vitro	HT29 cells	0–100 μM	Triggering apoptosis and cell cycle arrest in CSCs Promoting sensitivity of tumor cells to doxorubicin chemotherapy	[242]
Prostate cancer	In vitro	PC3, LNCaP and ARPE-19 cells	40 μM	Queretin and midkine-siRNA co-application suppresses CSC features via dual inhibition of PI3K/Akt and MAPK/ERK molecular pathways	[248]

5.3. Carotenoids

5.3.1. Astaxanthin

The astaxanthin is derived from algae and has anti-tumor activity. Astaxanthin The astaxanthin impairs gastric cancer development and progression via suppressing inflammatory storm [252]. Astaxanthin induces apoptosis and promotes PARPγ expression to reduce esophageal cancer progression [253]. The pontin overexpression is responsible for CSC features in breast cancer. Suppressing pontin expression impairs colony formation and CSC characteristics in breast cancer. The astaxanthin administration (80 and 100 μM) reduces pontin expression and impairs CSC features via down-regulating Oct4, Nanog and mutp53 levels [254].

5.3.2. β-Carotene

The β-carotene is another member of carotenoids that is able to effectively suppress cancer progression. The β-carotene suppresses IL-6/STAT3 axis and inhibits M2 polarization of macrophages to impair colon tumor progression [255]. β-carotene induces apoptosis and reduces antioxidant markers to impair breast tumor progression [256]. A combination of β-carotene and lycopene suppresses esophageal cancer progression via down-regulating COX-2 and cyclin D1 expression levels [257]. For promoting potential of β-carotene in breast cancer suppression, lipid–polymer nanoparticles have been developed for its targeted delivery [258]. Compared to astaxanthin, more experiments have focused on role of β-carotene in targeting CSCs. The β-carotene administration (20 and 40 μM) reduces growth and colony formation capacities of colon CSCs. The β-carotene reduces DNMT3A expression at mRNA levels and prevents DNA methylation. Furthermore, β-carotene promotes histone H3 and H4 acetylation levels in impairing CSC progression [259]. The self-renewal capacity of CSCs undergoes a decrease by β-carotene in neuroblastoma cells and this leads to enhanced sensitivity of tumor cells to cisplatin chemotherapy [260]. Furthermore, in vivo experiment on xenograft model has shown role of β-carotene in suppressing CSC features in neuroblastoma [261]. The Oct3/4 and DLK1 undergo down-regulation by β-carotene in decreasing CSC features and stemness of neuroblastoma cells [262]. The potential of β-carotene in reducing DLK1 expression and impairing CSC features is mediated via retinoic acid receptor β [263].

5.4. Sulforaphane

The sulforaphane (SFN) is an isothiocyanate and is present in various vegetables including cabbage broccoli, cauliflower and Brussels. The SFN is generated upon glucoraphanin hydrolysis. The SNF application in cancer therapy has witnessed a growing increase [264]. A recent experiment demonstrates that SFN reduces expression level of H19, as a lncRNA to suppress growth and invasion of pancreatic cancer cells [265]. The SFN is able to suppress tubulin polymerization in triggering apoptosis and cell cycle arrest in glioblastoma cells [266]. The expression level of FAT-1 undergoes down-regulation by SFN to impair proliferation and invasion of bladder cancer cells [267]. The SFN administration is of interest in increasing sensitivity of breast cancer cells to doxorubicin chemotherapy via inhibiting myeloid-derived suppressor cells [268]. Therefore, SFN is a potent anti-tumor agent [269] and this section focuses on CSC targeting by SFN in cancer treatment.

The activation of Sonic Hedgehog signaling promotes stemness of gastric cancer cells. The SFN administration (0–10 μM) stimulates apoptosis in CSCs and inhibits their growth to suppress gastric cancer proliferation. For this purpose, SFN inhibits Sonic signaling [270]. The upregulation of TAp63α occurs in colorectal tumor spheres with CSC features. Noteworthy, TAp63α enhances self-renewal capacity and CSC markers in colorectal tumor. Investigation of molecular pathways reveals that SFN increases Lgr5 expression by binding to its promoter, leading to β-catenin signaling activation. The SFN administration (0–10 μM) reduces TAp63α expression to inhibit expression of CSC markers such as CD133, CD44, Nanog and Oct4 [224].

Human tumor necrosis factor (TNF)-related apoptosis ligand (TRAIL) is suggested to be involved in triggering apoptosis in tumor cells [271]. However, experiments have shown capacity of tumor cells in obtaining resistance to TRAIL-mediated apoptosis [272–275]. Furthermore, TRAIL can surprisingly enhance progression of cancer cells via inducing NF-κB signaling [276,277]. An experiment has shown synergistic impact between TRAIL and SFN in PCa therapy. Both TRAIL and SFN are able to suppress CSCs in PCa, and SFN has higher ability compared to TRAIL. However, TRAIL induces NF-κB signaling and promotes PCa progression. The SFN administration (10 μM) decreases expression level of NF-kB, CXCR4, Jagged1, Notch-1, SOX2, Nanog and ALDH1 to impairs CSC features and stemness in PCa [278]. It is worth mentioning that down-regulation of ALDH1, c-Rel and Nothc-1 as CSC markers by SFN, significantly elevates sensitivity of prostate and pancreas cancer cells to cisplatin, gemcitabine and 5-flourouracil chemotherapy [279]. The stimulation of NF-κB signaling is responsible for ALDH1 upregulation and enhanced CSC features in pancreatic cancer. Therefore, NF-κB inhibition by SFN (20 μmol/L) impairs CSC features in pancreatic cancer cells and enhances their sensitivity to sorafenib chemotherapy [280]. Overall, the experiments advocate the fact that SFN is a potent agent in targeting CSC and suppressing stemness of cancers (Figure 4, Table 5) [281–285].

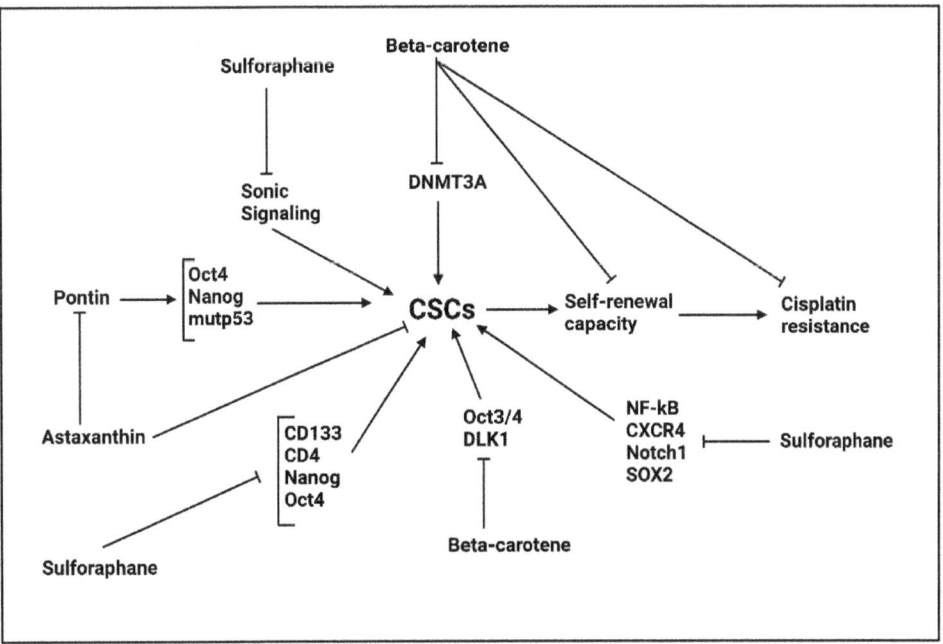

Figure 4. The potential of carotenoids in suppressing CSCs.

Table 5. The sulforaphane as a potential anti-tumor agent in cancer therapy.

Cancer Type	In vitro/In Vivo	Cell Line/Animal Model	Study Design	Remarks	Refs
Triple-negative breast cancer	In vitro	SUM149 and SUM159 cells	2.5 and 5 µM	Inducing apoptosis in CSCs and reducing ALDH expression as CSC marker	[286]
Melanoma	In vitro In vivo	A375 cells Xenograft	20 µM 10 µM/kg	The EZH2 promotes CSC features in melanoma and is suppressed by sulforaphane	[287]
Pancreatic cancer	In vivo	Mice	20 mg/kg	Suppressing Sonic/Gli1 axis to impair CSC features and self-renewal capacity in pancreatic cancer	[288]
Glioblastoma	In vitro In vivo	U87, U373, U118, and SF767 cells Mice	0–50 µM 100 mg/kg	Suppressing stem cell-like spheroids Inducing apoptosis and eliminating CSCs Promoting drug sensitivity	[289]
Leukemia	In vitro	KU812 cells	0–30 µM	Enhancing ROS levels Suppressing β-catenin signaling to reduce GSH levels Promoting potential of imatinib in tumor suppression	[290]
Lung cancer	In vitro	NSCLC PC9 cells	0–12 µM	Suppressing Sonic Hedgehog signaling Decreasing levels of CD133 and CD44 Inhibiting gefitinib resistance	[291]

Table 5. Cont.

Cancer Type	In vitro/In Vivo	Cell Line/Animal Model	Study Design	Remarks	Refs
Lung cancer	In vitro In vivo	A549 cells Nude mice	0–40 µM 25 and 50 mg/kg	Preventing tobacco-mediated CSC feature acquisition via suppressing IL-6/∆Np63α/Notch axis	[194]
Oral cancer	In vitro	SAS or GNM cells	0–50 µM	Upregulating miRNA-200c expression to impair CSC features and stemness in oral cancer	[292]
Epidermal squamous cell carcinoma	In vitro	SCC-13 cells	0–20 µM	The SCC-3 cells derived from CSCs have high sensitivity to a combination of sulforaphane and cisplatin This combination reduces viability of tumor cells and their capacity in colony formation	[293]

5.5. Curcumin

The *curcuma longa* is a medicinal plant belonging to family of Zingiberaceae family that has a well-known bioactive compound, curcumin that can be isolated from its rhizome [294,295]. The curcumin and other curcuminoids lead to yellow color of rhizome. Although content of curcuminoids in rhizome of *curcuma longa* is various, it seems that curcumin is the main component and comprises 77% [296]. The curcumin has demonstrated various therapeutic activities that among them, anti-cancer potential is of importance [297]. The curcumin administration is beneficial in suppressing proliferation and metastasis of cancer cells. A recent experiment has shown role of curcumin in down-regulating NF-κB, ERK, MMP-2 and MMP-9, while it can enhance p38 and JNK levels to induce cell death and prevent invasion of leukemia cells [298]. Recently, attention has been directed towards chemopreventive role of curcumin in cancer therapy [299]. The curcumin- and docetaxel-loaded micellar nanoparticles effectively suppress progression of ovarian cancer cells via suppressing angiogenesis and inducing apoptosis. The curcumin is able to sensitize ovarian cancer cells to docetaxel chemotherapy [300]. Due to poor bioavailability of curcumin, nanostructures are mainly applied for curcumin delivery alone or its combination with other anti-tumor agents to exert synergistic cancer therapy [31,301]. A fibrin matrix has been developed for prolonged release of curcumin and suppressing cancer metastasis. As studies have shown role of the nanoplatforms for promoting therapeutic activity of curcumin, future clinical trials can be performed [302]. The current section focuses on anti-tumor activity of curcumin based on targeting CSCs.

The capacity of curcumin in cancer suppression via targeting CSCs is attributed to affecting various molecular pathways. The activation of Wnt/β-catenin and Sonic Hedgehog molecular pathways results in CSC features in lung cancer and enhancing tumor progression. The curcumin administration (0–40 µM) induces apoptosis in CSCs and diminishes their proliferation. Furthermore, curcumin reduces CSC hallmarks such as Nanog, Oct4, CD133, CD44 and ALDH1A1. Mechanistically, these anti-tumor activities of curcumin are mediated via inhibiting Wnt and Hedgehog signaling pathways [303]. In addition, nuclear factor-kappaB (NF-κB) pathway also participates in enhancing survival of CSCs [304]. The down-regulation of NF-κB signaling impairs CSC features in bladder cancer [305]. In liver cancer cells, curcumin reduces their proliferation and colony formation capacities as well as inhibiting CSCs features. The curcumin capacity in suppressing CSCs features in liver cancer is pertained to inhibiting NF-κB signaling [306]. Hence, identification of molecular pathways affected by curcumin in suppressing CSC features can broaden our understanding towards underlying mechanisms of its anti-tumor activity.

The signal transducer and activator of transcription 3 (STAT3) is another tumor-promoting factor related to cancer progression [307–309]. The upregulation of STAT3 enhances growth and invasion of tumor cells and mediates their therapy resistance [310–316]. The activation of STAT3 signaling promotes CSC features in tumors and its inhibition, for instance by acetaminophen, reduces CSC markers and colony formation capacity [317,318]. The curcumin administration (0–40 µM) decreases levels of CD44, ALDH, SOX2, Nanog and c-Myc as CSC markers to impairs thyroid cancer progression and induce apoptosis. Furthermore, by reducing CSC features, curcumin enhances potential of cisplatin in thyroid cancer suppression [319]. Similar phenomenon occurs in bladder cancer, so that curcumin administration (0–50 µM) decreases CD44, CD133, ALDH1, Nanog and Oct4 to impair CSC characteristics in bladder cancer and induce apoptosis and proliferation inhibition. The investigation of molecular pathways reveals that these anti-tumor activities are mediated via sonic signaling inhibition [320]. The interesting point is the curcumin and CD44 coupling in cancer therapy. An experiment demonstrates that colon cancer cells overexpressing CD44 as CSC marker are more sensitive to curcumin compared to CD44- cells. The curcumin-CD44 coupling stimulates apoptosis in colorectal stem cells via preventing influx of glutamine into cancer cells and reducing its intracellular accumulation [321].

In order to potentiate anti-tumor activity of curcumin, its combination with other tumor-suppressor agents is performed. A combination of curcumin and quinacrine can be beneficial in impairing breast cancer progression via targeting CSCs. The quinacrine is able to induce DNA damage in cancer cells, but its intracellular accumulation should be improved. In this case, curcumin is helpful and by suppressing ABCG2 activity via occupying its ligand-binding site, enhances internalization of quinacrine, resulting in cell death and DNA damage in breast CSCs [322]. Furthermore, curcumin also has capacity of inducing DNA damage in CSCs [323]. The curcumin can negatively affect colony formation capacity of lung cancer cells. It is suggested that curcumin prevents self-renewal capacity of CSCs to suppress colony formation of lung cancer cells [323]. Therefore, curcumin is a potent suppressor of CSCs in lung cancer and for this purpose, it can inhibit JAK2/STAT3 axis to impair CSC features [324]. Based on previously discussed molecular pathways, both STAT3 and NF-κB pathways are suppressed by curcumin in targeting CSCs. Noteworthy, STAT3 can function as upstream mediator of NF-κB and mediates its activation via enhancing IKKα stability [325]. A combination of curcumin (10 µM) and epigallocatechin gallate (10 µM) suppresses STAT3/NF-κB axis to impair CSC features and reduce number of CD44+ cells [326]. In addition to curcumin, its derivatives have been also capable of targeting CSCs. The colon cancer cells positive for ALDH and CD133 demonstrate activation of STAT3 signaling and its phosphorylation. An analogue of curcumin, known as GO-Y030 can suppress STAT3 phosphorylation to impair CSC features in colon cancer and prevent its progression [327]. Therefore, curcumin and its derivatives suppress proliferation and invasion of tumors via targeting CSCs. Furthermore, they can enhance sensitivity of cancer cells to therapies by targeting CSCs [328]. Interestingly, curcumin can be advantageous in enhancing sensitivity of cancer cells to radiotherapy [329]. The breast cancer stem-like cells have the ability of obtaining radioresistance. A combination of curcumin and glucose gold nanoparticles induces apoptosis in CSCs and reduces expression levels of hypoxia inducible factor-1α (HIF-1α) and HSP90 in sensitizing breast cancer cells to radiotherapy [330]. Therefore, curcumin is a versatile agent in suppressing CSC features in cancer and various molecular pathways such as Wnt, Sonic, STAT3 and NF-κB, among others, are affected to reduce CSC markers such as CD44, ALDH and CD133 (Figure 5, Table 6) [331–338].

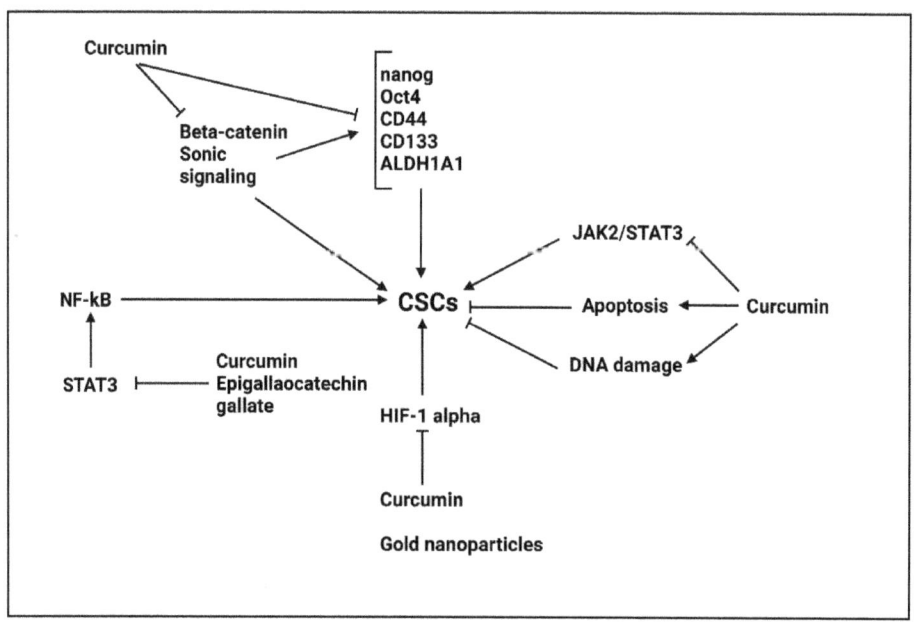

Figure 5. Curcumin in inhibiting CSCs.

Table 6. Curcumin potential of suppressing CSCs in tumor treatment.

Cancer Type	In Vitro/In Vivo	Cell Line/Animal Model	Study Design	Remarks	Refs
Lung cancer	In vitro	A549 and H1299 cells	0–40 µM	Inducing apoptosis Reducing levels of CSC markers Inhibiting Wnt/β-catenin and Sonic signaling pathways	[303]
Bladder cancer	In vitro	UM-UC-3 and EJ cells	0–50 µM	Apoptosis induction Decreasing expression levels of CD44, CD133, Oct4, Nanog and ALDH1 Inhibiting Sonic Hedgehog signaling	[320]
Breast cancer	In vitro	MCF-10A and MCF-7 cells	8 µM	A combination of curcumin and quinacrine induces DNA damage and reduces ABCG2 expression to impair cancer progression	[322]
Breast cancer	In vitro	SUM159 and MCF7 cell lines	0–40 µM	Reducing colony formation capacity of CSCs Down-regulating expression levels of CD44, ALDH1, Nanog and Oct4 Inhibiting Wnt/β-catenin axis	[333]
Breast cancer	In vitro	MDA-MB-231 and MCF-7 cells	5 µmol/L	Inducing apoptosis via Bcl-2 down-regulation Enhancing sensitivity of breast CSCs to mitomycin C	[328]

Table 6. Cont.

Cancer Type	In Vitro/In Vivo	Cell Line/Animal Model	Study Design	Remarks	Refs
Brain cancer	In vitro	U87MG cells	-	The curcumin-loaded nanoparticles efficiently penetrate into BBB to induce apoptosis in CSCs and reduce tumor progression	[332]
Pancreatic cancer	In vitro In vivo	BxPC3, MiaPaCa2 and Panc1 PDAC cells Nude mice	0–20 µM 100 mg/kg	Suppressing self-renewal capacity of tumor cells Retarding tumor growth in vivo Enhancing sensitivity to gemcitabine chemotherapy	[339]
Prostate cancer	In vitro	DU145 cells	-	Inhibiting growth and invasion of prostate CSCs Overexpression of miRNA-770-5p and miRNA-1247	[337]
Colorectal cancer	In vitro	HCT116 and DLD1 cells	0–20 µM	Reducing expression level of CSC markers including CD44, Oct4 and ALDH1 Inhibiting STAT3 signaling	[338]

5.6. Resveratrol

The resveratrol (Res) is a well-known compound in treatment of different ailments including diabetes, cancer and neurological disorders [340,341]. The recent years have witnessed an increase in attention towards Res application in cancer therapy. Res induces apoptosis via impairing mitochondrial function in cancer cells. Furthermore, tumor-promoting factors such as NF-κB, COX-2 and PI3K undergo down-regulation by Res in tumor suppression [342]. A combination of Res and radiation stimulates apoptosis in breast cancer via Bcl-2 down-regulation and Bax upregulation [50]. Res suppresses phosphorylation of STAT3 at tyrosine 705 to impair metastasis of cervical cancer cells [343]. Down-regulation of HPV E6 and E7 by Res leads to apoptosis and cell cycle arrest in cervical cancer [344]. Furthermore, Res inhibits progression and growth of PCa cells via suppressing PI3K/Akt signaling [345]. The c-Myc as downstream target of PI3K/Akt axis is suppressed by Res in lung cancer therapy [346]. Loading Res on nanostructures significantly promotes its potential in thyroid cancer suppression. Furthermore, Res-loaded nanoparticles decrease tumor volume up to 55% in thyroid cancer in vivo [347].

The tumor sphere formation capacity in renal cancer is mediated by CSCs and Sonic signaling plays a significant role in this case. The Res administration (0–30 µM) down-regulates Sonic signaling to inhibit CSCs features in renal cancer, leading to apoptosis and proliferation inhibition [348]. As it was mentioned earlier, Res is loaded on nanoparticles to promote its anti-tumor activity [349–351]. Another promising strategy for enhancing bioavailability of Res in synthesizing analogues with superior activities. The pterostilbene is a derivative of Res that has higher bioavailability compared to Res. The pterostilbene is able to reduce expression levels of CD133, Oct4, SOX2, Nanog and STAT3 in impairing CSC features and retarding progression of cervical cancer [352]. Furthermore, Res is able to suppress Wnt/β-catenin in reducing CSC features in breast cancer. It is worth mentioning that Res enhances LC-3II, Beclin-1 and ATG7 expression levels in triggering autophagy in CSCs and reducing breast cancer progression [353]. It is worth mentioning that autophagy has both tumor-promoting and tumor-suppressor roles in cancer [221,354,355] and its induction by Res requires more clarification.

The STAT3 signaling is considered as a tumor-promoting factor in osteosarcoma. The upregulation of STAT3 by VEGFR2 results in metastasis of osteosarcoma cells [356]. Furthermore, induction of JAK2/STAT3 axis by exosomal LCP1 leads to carcinogenesis and migration of osteosarcoma cells [357]. The Res administration (0–3 µM) decreases cytokine generation and suppresses JAK2/STAT3 axis to impair CSCs features in osteosarcoma via CD133 down-regulation [358]. Another molecular pathway that is responsible for CSCs features is nutrient-deprivation autophagy factor-1 (NAF-1). The Res inhibits NAF-1 signaling to reduce CSC features including SOX2, Oct4 and Nanog in pancreatic cancer treatment [359]. The interesting point is the interaction between tumor microenvironment (TME) components and molecular pathways in providing CSCs features. The T lymphocytes and fibroblasts cells present in TME can effectively induce NF-κB signaling and upon nuclear translocation, NF-κB significantly enhances growth, metastasis and survival of CSCs in colorectal cancer. The TNF-β and TGF-β3 are secreted by T lymphocytes and fibroblasts in inducing NF-κB signaling. Noteworthy, Res (0–10 µM) prevents secretion of TNF-β and TGF-β3 by lymphocytes and fibroblasts in reducing CSC features in colorectal cancer [360]. Another experiment reveals that cancer-associated fibroblasts (CAFs) existing in TME are able to reduce SOX2 expression and suppress stemness in breast cancer via reducing CSC features [361].

Another strategy that can be followed by Res, is the induction of endothelial differentiation. For this purpose, Res and sulindac stimulate vascular endothelial cadherin (VE-cadherin) and von Willebrand factor (vWF) in triggering trans-differentiation in CSCs and mediating their transformation into endothelial lineage [362]. Recently, attention has been directed towards using siRNA and natural compounds in synergistic cancer therapy [312]. An experiment has shown that RAD51 has oncogenic role and enhances CSC features in cervical cancer. A combination of Res and RAD51-siRNA induces apoptosis in HeLa cells and prevents cervical cancer progression [363]. Overall, experiments reveal role of Res in suppressing CSCs in cancer progression (Figure 6, Table 7) [364–367].

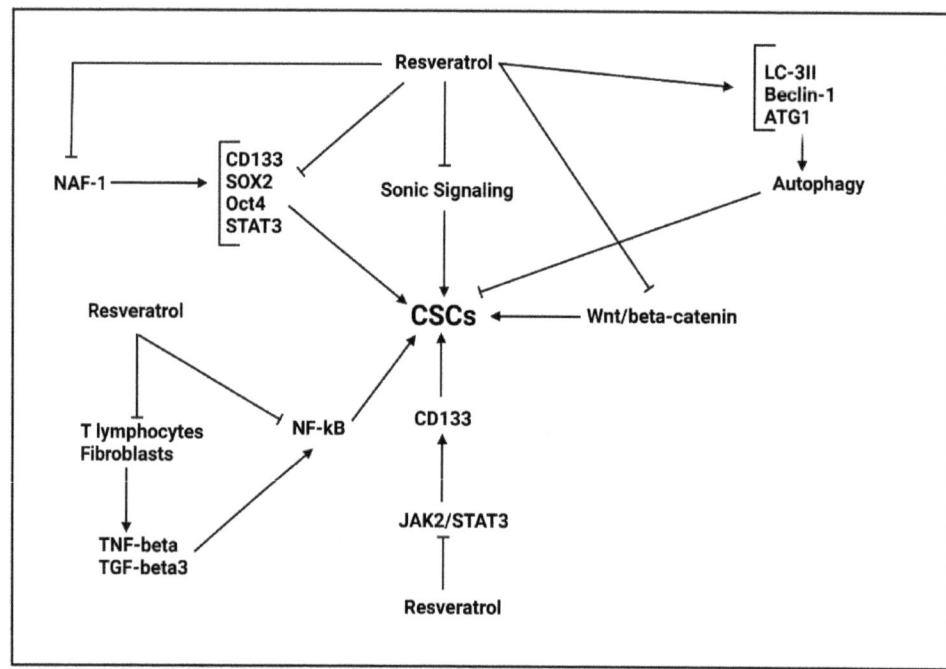

Figure 6. Res in suppressing CSCs in tumor therapy.

Table 7. Determining resveratrol potential in targeting CSCs and suppressing tumor progression.

Cancer Type	In Vitro/In Vivo	Cell Line/Animal Model	Study Design	Remarks	Refs
Pancreatic cancer	In vitro	MiaPaCa-2 and Panc-1 cells	50 μmol/L	Promoting gemcitabine sensitivity Suppressing lipid metabolism Down-regulation of SREBP1 Inhibiting stemness Suppressing colony formation and CSC features	[368]
Pancreatic cancer	In vitro In vivo	CSCs and mice	0–30 μM	Inhibiting tumor growth and development in vivo Inducing apoptosis in CSCs via upregulating caspase-3 and -7, and down-regulating Bcl-2 and XIAP Decreasing expression levels of Nanog, SOX2, c-Myc and Oct4 as CSC markers Reducing ABCG2 expression as an inducer of drug resistance in CSCs	[369]
Medulloblastoma	In vitro	CSCs derived from medulloblastoma	150 μM	Suppressing proliferation of CSCs and increasing sensitivity to radiotherapy	[370]
Ovarian cancer	In vitro	A2780 cells	0–50 μM	Suppressing self-renewal capacity of CSCs Triggering cell death in CSCs via mediating ROS overgeneration	[371]
Breast cancer	In vitro	MCF-7 cells	0–500 μM	Preventing proliferation of CSCs Triggering apoptosis Mediating oxidative damage Caspase cascade activation and inducing PARP cleavage Reducing SOD, MnSOD and catalase levels	[372]
Nasopharyngeal carcinoma	In vitro	TW01, TW06, and HONE-1 cells	0–100 μM	Suppressing self-renewal capacity and migration of CSCs P53 overexpression EMT inhibition	[373]
Colorectal cancer	In vitro	HCT116 cells	5 μM	Upregulation of CD133, CD44 and ALDH1 as CSC markers by TGF-β Suppressing CSC features by resveratrol and triggering apoptosis in CSCs by caspase-3 upregulation Promoting drug sensitivity	[374]

5.7. Berberine

The berberine (BBR) is an isoquinoline alkaloid mainly derived from *Coptis chinensis* with great therapeutic impacts [375]. In addition, BBR can be isolated from other *Berberis* plants such as Berberis julianae and Scutellaria baicalensis [376]. Although BBR has pharmacological activities including antioxidant, anti-inflammatory, anti-diabetes, hepatoprotective and renoproective, among others, much attention has been directed towards its anti-tumor activities [377–380]. The BBR induces apoptosis, autophagy and cycle arrest (G1 phase) in colon cancer cells. Mechanistically, BBR enhances PTEN expression, while it reduces expression of PI3K, Akt and mTOR [151]. Furthermore, BBR suppresses fatty acid metabolism and diminishes extracellular vesicle production in favor of cancer growth suppression [381]. The BBR is able to decrease expression level of IGF2BP3 in triggering cycle arrest (G0/G1 phase) in colorectal cancer cells [382]. The current section emphasizes on BBR function in affecting CSCs.

The pancreatic cancer treatment is an increasing challenge for physicians, as the tumor cells have high growth and migration capabilities [383,384]. Furthermore, they can develop drug resistance and risk of recurrence is present in pancreatic cancer patients. The CSCs play a significant role in aforementioned processes [385–388]. An experiment has shown that BBR (15 µM) reduces expression levels of SOX2, Nanog and POU5F1 as CSC markers to impair resistance of pancreatic cancer cells to gemcitabine and suppress their progression [389]. One of the targets of BBR in cancer therapy is miRNAs. Overall, the expression level of tumor-promoting miRNAs undergoes down-regulation by BBR, while an increase occurs in expression profile of tumor-suppressor miRNAs [390–392]. The oral CSCs have self-renewal capacity and demonstrate high colony formation and migration abilities that are in favor of triggering drug resistance. The miRNA-21 overexpression plays a tumor-promoting role in oral CSCs. The BBR (10 µM) reduces expression level of miRNA-21 to impair self-renewal capacity of CSCs and decrease ALDH1 expression. Then, a significant increase occurs in sensitivity of oral CSCs to cisplatin and 5-flourouracil chemotherapy [393].

The glioma-associated oncogene-1 (Gli1) is a tumor-promoting factor that its upregulation enhances growth and invasion of colorectal cancer cells [394]. Noteworthy, Gli1 participates in enhancing CSCs features and mediating tumor progression [395]. Therefore, its inhibition by BBR can exert anti-tumor activities. A recent experiment has shown that chemotherapy enhances ovarian cancer progression via enhancing CSCs features. The Gli1 and its downstream target BMI1 are involved in promoting CSCs features in ovarian cancer. The BBR administration down-regulates Gli1 expression to inhibit BMI1, leading to a decrease in CSC features [396]. Overall, CD133, β-catenin, n-Myc, nestin, SOX2 and Notch2 undergo down-regulation by BBR in suppressing CSC features [397]. Furthermore, BBR co-administration with Dodecyl-TPP (d-TPP) can synergistically suppress CSC features in breast cancer [205]. Moreover, to selectively target CSCs, nanostructures can be utilized [398,399]. In an effort, BBR was loaded on liposomal nanocarriers and results demonstrated potential of these nanocarriers in crossing over CSC membrane. Then, they can reduce expression level of ABCC1, ABCC2, ABCC3 and ABCG2. Furthermore, BBR-loaded liposomes selectively internalize in mitochondria and induce apoptosis via triggering mitochondrial membrane potential loss, enhancing Bax expression and reducing Bcl-2 level. Then, cytochrome C (cyt C) release occurs and promotes caspase-3 and -9 expressions to mediate apoptosis in CSCs and suppress breast cancer progression. Furthermore, in vivo experiment on xenografts in nude mice also has shown potential of BBR-loaded liposomes in suppressing tumor growth (Table 8) [400].

Table 8. The BBR in targeting CSCs for tumor suppression.

Cancer Type	In Vitro/In Vivo	Cell Line/Animal Model	Study Design	Remarks	Refs
Pancreatic cancer	In vitro	PANC-1 and MIA PaCa-2 cells	15 µM	Reducing population of CSCs Decreasing expression level of CSC markers including SOX2, ALDH1, Nanog and POU5F1 Promoting sensitivity of tumor cells to gemcitabine chemotherapy	[389]

Table 8. Cont.

Cancer Type	In Vitro/ In Vivo	Cell Line/Animal Model	Study Design	Remarks	Refs
Oral cancer	In vitro	SAS and OECM-1 cells	0–40 µM	Suppressing tumor progression in a dose-dependent manner Reducing colony formation, migration and self-renewal capacity of tumor cells Reducing miRNA-21 expression to impair CSC features	[393]
Ovarian cancer	In vitro	SKOV3 and A2780 cells	5 µM	Chemotherapy promotes Gli1 expression and facilitates CSC features Suppressing Gli1 pathway by berberine Inhibiting EMT-mediated metastasis and reducing CSC features	[396]
Neuroblastoma	In vitro	N2a cells	10 and 20 µg/mL	Inducing apoptosis and cell cycle arrest in tumor cells Suppressing metastasis via EMT inhibition Inhibiting CSCs features	[397]
Breast cancer	In vitro In vivo	MCF-7 cells Xenografts	40 µM of liposomal berberine	The berberine-loaded liposomes selectively target CSCs and induce apoptosis Enhancing berberine accumulation in tumor cells Retarding tumor growth in vivo	[400]

5.8. Ginseng and Its Derivatives

The ginsenosides are derivatives from ginseng that have therapeutics impacts including anti-diabetes, anti-inflammatory and neuroprotective, and among them, anti-tumor activity of ginsenosides is of importance [401–403]. The ginsenoside Rg3 is able to suppress progression of colorectal tumor cells and enhance their sensitivity to oxaliplatin and 5-flouroruacil chemotherapy. Mechanistically, ginsenoside Rg3 (200 µM) reduces levels of CD24, CD44 and EpCAM to inhibit stemness and population of CSCs [404]. The ginsenoside Rh2 also suppress skin cancer progression in a concentration-dependent manner (0–1 mg/mL) and induces autophagy to suppress β-catenin signaling and impair CSC features in skin cancer. Autophagy inhibition abrogates capacity of ginsenoside Rh2 in suppressing β-catenin signaling and CSC features in skin cancer. Therefore, autophagy induction by ginsenoside Rh2 is of importance for inhibiting β-catenin-mediated CSC features in skin cancer [405]. However, a special attention should be directed towards autophagy, as it has both tumor-promoting and tumor-suppressor roles in cancer [355]. An experiment demonstrates that ginsenoside F2 (0–120 µM) stimulates apoptosis and autophagy in breast CSCs. Noteworthy, autophagy plays as a pro-survival mechanism and autophagy inhibition increases ginsenoside F2 capacity in mediating apoptosis in CSCs [406]. The ginsenoside Rb1 and compound K are able to suppress self-renewal capacity of CSCs in ovarian cancer. Furthermore, Rb1 and compound K enhance sensitivity of ovarian cancer cells to cisplatin and paclitaxel chemotherapy via suppressing CSCs. The drug sensitivity activity of Rb1 and compound K are mediated via suppressing β-catenin signaling and subsequent down-regulation of ABCG2 and P-gp. Furthermore, Rb1 and compound K suppress CSC migration and invasion via EMT inhibition [407]. For suppressing self-renewal capacity of CSCs, ginsenoside Rg3 (0–100 µM) suppresses Akt signaling. Furthermore, ginsenoside Rg3 reduces HIF-1α expression to down-regulate

SOX-2 and Bmi-1 expression levels, leading to decreased stemness and CSC features in breast cancer [408]. Overall, ginsenosides similar to genistein, are able to suppress cancer progression via affecting CSCs [409–411]. Besides, ginsenosides are derived from ginseng extract and it has been reported that ginseng extract is also capable of suppressing CSC features in cancer (Figure 7, Table 9) [412,413]. Figures 8 and 9 show the chemical structures of selected phytochemicals discussed in the current review.

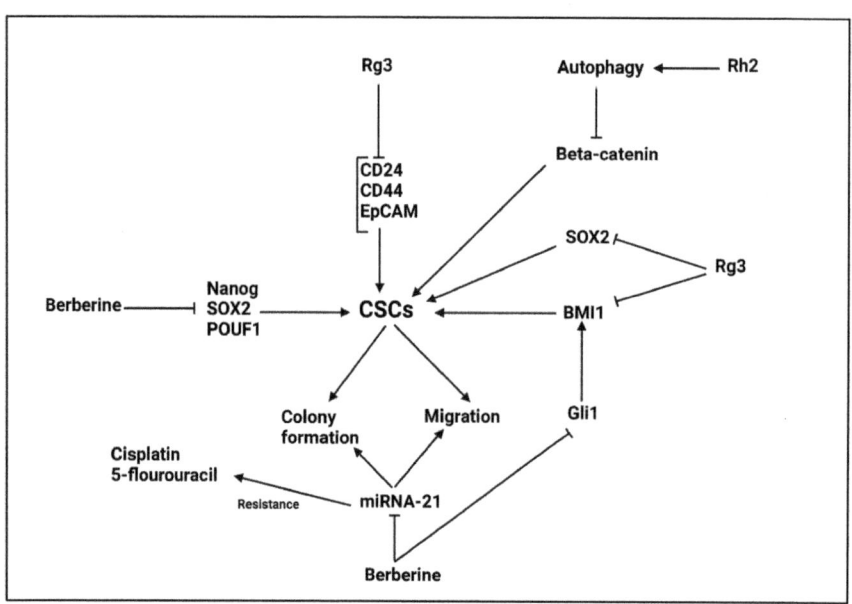

Figure 7. The berberine and ginsenosides as modulators of CSCs in tumor eradication.

Table 9. Ginsenosides as potential inhibitors of CSCs in different tumors.

Anti-Tumor Agent	Cancer Type	In Vitro/ In Vivo	Cell Line/Animal Model	Study Design	Remarks	Refs
Ginsenoside Rg3	Colorectal cancer	In vitro In vivo	LoVo, SW620 and HCT116 cells Nude mice	-	Promoting anti-tumor potential of oxaliplatin and 5-flourouracil Suppressing stemness and CSC features	[404]
Ginsenoside Rh2	Skin cancer	In vitro	A431 cells	0–1 mg/mL	Inhibiting viability of tumor cells Autophagy induction and β-catenin signaling inhibition	[405]
Ginsenoside Rb1	Ovarian cancer	In vitro In vivo	SKOV-3 and HEYA8 cells Mice	0–500 nM 50 mg/kg	Suppressing Wnt signaling and reversing EMT to impair cancer stemness and inhibit drug resistance	[407]
Gisnenoside Rg3	Breast cancer	In vitro	MCF-7 and MDA-MB-231 cells	0–100 µM	Inhibiting Akt/HIF-1α axis Reducing expression levels of SOX2 and Bmi-1 Inhibiting CSC features and self-renewal capacities	[408]

Table 9. Cont.

Anti-Tumor Agent	Cancer Type	In Vitro/ In Vivo	Cell Line/Animal Model	Study Design	Remarks	Refs
Ginsenoside F2	Breast cancer	In vitro	MCF-7 cells	0–140 µM	Apoptosis induction in CSCs Inhibiting autophagy promotes potential of ginsenoside F2 in tumor suppression	[406]
Ginsenoside Rh2	Hepatocellular carcinoma	In vitro In vivo	HepG2 and Huh7 cells Mice	0–1 mg/mL 1 mg/kg	Inhibiting CSC features in a dose-dependent manner Autophagy induction and suppressing β-catenin signaling	[409]

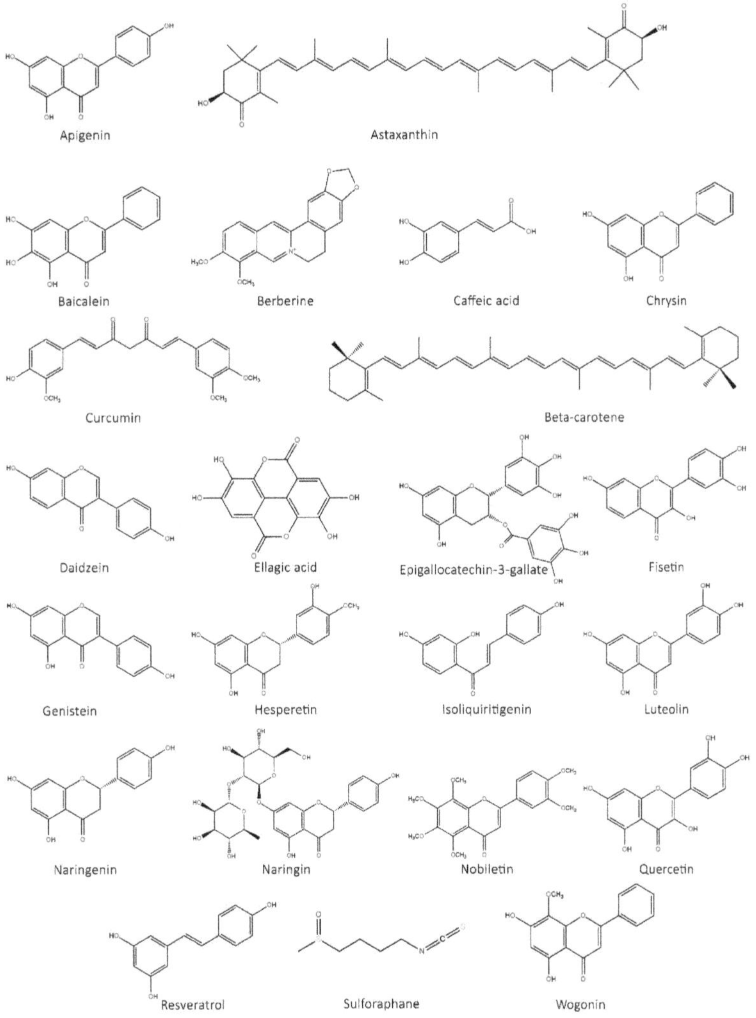

Figure 8. The chemical structures of selected phytochemicals discussed in the review.

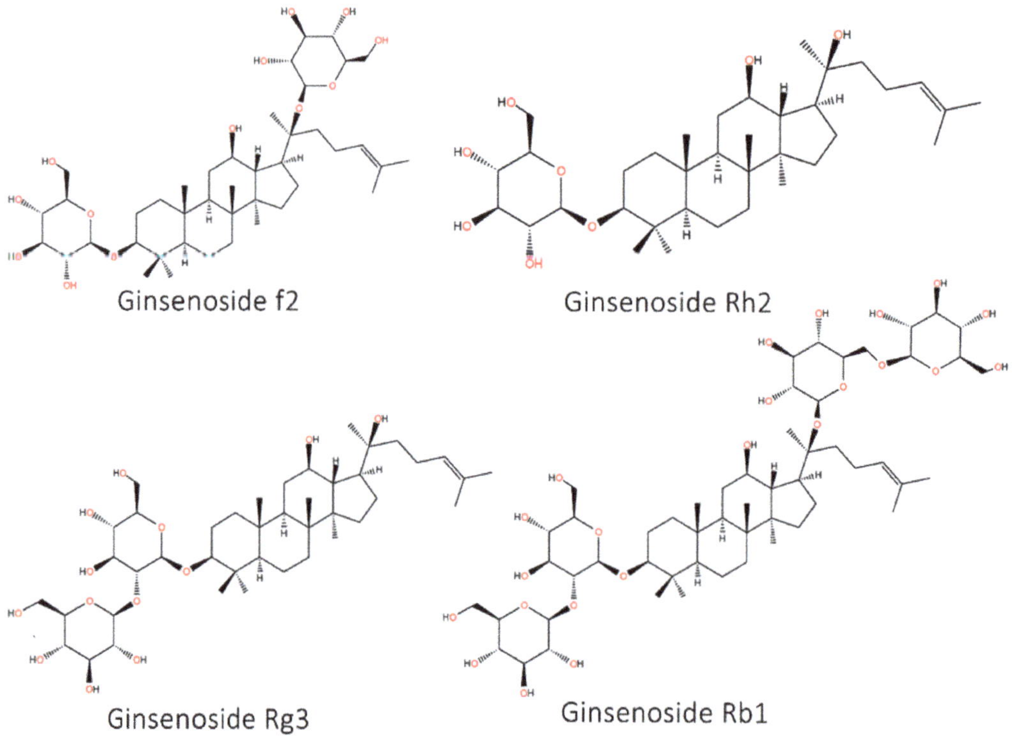

Figure 9. Chemical structures of ginsenosides.

6. Conclusions and Remarks

A variety of factors have been found to be responsible for tumor progression. The CSCs role in tumor progression has been confirmed in different experiments. In addition, molecular pathways participating in CSC feature suppression/induction have been shown. The CSCs enhance stemness of tumor cells and can mediate their growth and migration. Furthermore, CSCs are able to reduce sensitivity of tumor cells to chemotherapy and radiotherapy. The CD44, CD33, ALDH1, Nanog, SOX2 and Oct2 are the most well-known CSC markers. The overexpression of these factors significantly enhances stemness of tumors. Various molecular pathways are able to enhance cancer progression and stemness that include STAT3, NF-κB, Wnt/β-catenin signaling and Sonic Hedgehog, among others. Such molecular pathways were discussed in main text and it was shown that their inhibition can reduce expression of CSC markers and mediate tumor suppression.

Different strategies can be employed for suppressing CSCs. One of the potential methods is application of gene therapy. As signaling networks involved in regulating CSC features have been recognized, gene therapy can inactivate/activate molecular pathways and modulate CSC markers. However, there are some drawbacks with gene therapy, such as high cost. Therefore, plant derived-natural products are of importance due to their multitargeting capacity. The present review attempted to reveal the possible role of selected dietary agents in suppressing CSCs. As it was mentioned in main text and Tables, various doses of these anti-cancer compounds have been utilized and most of them are less than 100 µM. However, each natural product has its own IC_{50} and based on various experiments performed in this case, it is possible to determine optimal dose for using in cancer therapy. One of the limitations of experiments is lack of using nanostructures for delivery of dietary agents and enhancing their capacity in CSC suppression. Hence, future experiments can focus on this aspect and different nanoparticles such as lipid-based nanoparticles, carbon-based nanoparticles and polymeric nanoparticles, among others, can be utilized for targeted delivery. Furthermore, nanostructures overcome poor bioavailability of dietary agents in CSC suppression.

Noteworthy, the molecular pathways inducing CSC marker overexpression, can be targeted by natural products. The PI3K/Akt, Wnt, STAT3, NF-κB and Sonic signaling pathways undergo down-regulation by natural products to reduce expression level of CSC markers including CD44, CD133, Notch, Oct4 and Nanog, impairing cancer stemness and progression. The dietary agents induce apoptosis via both extrinsic and intrinsic pathways to reduce viability of CSCs. These compounds decrease expression level of MMPs and suppress EMT to reduce migration and invasion of CSCs. All these impacts are vital for preventing cancer progression. By suppressing CSC features, natural products enhance sensitivity of tumors to radiotherapy and chemotherapy. However, there are limited experiments in this case. Furthermore, there is lack of study evaluating potential of natural products in targeting CSCs and their interaction with immune system. One of the limitations related that was noted in these studies is that most of them have been performed under in vitro settings and have evaluated the potential impact of dietary agents in targeting CSCs in different cellular models. Hence, additional in vivo studies should be performed to analyze the exact potential of the natural products in targeting CSCs, suppressing tumor progression and investigating possible limitations associated with their use. There are no clinical trials performed so far to evaluate the role of natural products in targeting CSCs in cancer patients. However, there are clinical studies on using dietary agents in treatment of cancer patients (NCT03769766, NCT03980509, NCT01042938; clinicaltrials.gov, accessed on 22 October 2021). Therefore, still enormous work remains to be done to analyze the role of natural products in specifically targeting CSCs and providing effective treatment for cancer patients.

Author Contributions: M.D.A.P., S.A. and A.Z. (Amirhossein Zabolian), H.S., M.A.K., S.S., M.J.N., M.H., K.H. and M.A. prepared first draft and participated in manuscript writing. S.M., A.Z. (Ali Zarrabi) and G.S. collected articles, supervised and edited English language. All authors have read and agreed to the published version of the manuscript.

Funding: This work was supported by Singapore Ministry of Education Tier 1 Grant [R-184-000-301-114] to GS.

Institutional Review Board Statement: Not applicable.

Informed Consent Statement: Not applicable.

Data Availability Statement: Not applicable.

Conflicts of Interest: The authors declare no conflict of interest.

Abbreviations

siRNA: Small interfering RNA; shRNA, short hairpin RNA; CSCs, cancer stem cells; ALDH, aldehyde dehydrogenase; GLUTs, glucose transporters; HKs, hexokinases; ROS, reactive oxygen species; TCA, tricarboxylic acid; miRNA, microRNA; TCF4, transcription factor 4; EVs, extracellular vesicles; EGCG, epigallocatechin 3-gallate; ISL, isoliquiritigenin; P-gp, P-glycoprotein; CA, caffeic acid; CAPE, caffeic acid phenyl ester; DR5, death receptor 5; EpCAM, epithelial cell adhesion molecule; YB-1, Y-box binding protein 1; COX-2, cyclooxygenase-2; STAT3, signal transducer and activator of transcription 3; EMT, epithelial-to-mesenchymal transition; NF-κB, nuclear factor-kappaB; SFN, sulforaphane; TGF-β, transforming growth factor-β; TNF, tumor necrosis factor; TRAIL, TNF-related apoptosis ligand; HIF-α, hypoxia inducible factor-α; PCa, prostate cancer; MK, midkin; Res, resveratrol; NAF-1, nutrient-deprivation autophagy factor-1; TME, tumor microenvironment; VE, vascular endothelial; vWF, von Willebrand factor; BBR, berberine; Gli1, glioma-associated oncogene-1; d-TPP, Dodecyl-TPP; cyt C, cytochrome C.

References

1. Abadi, A.J.; Mirzaei, S.; Mahabady, M.K.; Hashemi, F.; Zabolian, A.; Hashemi, F.; Raee, P.; Aghamiri, S.; Ashrafizadeh, M.; Aref, A.R.; et al. Curcumin and its derivatives in cancer therapy: Potentiating antitumor activity of cisplatin and reducing side effects. *Phytother Res.* **2021**. [CrossRef] [PubMed]
2. Siegel, R.L.; Miller, K.D.; Fuchs, H.E.; Jemal, A. Cancer Statistics, 2021. *CA Cancer J. Clin.* **2021**, *71*, 7–33. [CrossRef]
3. Ashrafizade, M.; Delfi, M.; Hashemi, F.; Zabolian, A.; Saleki, H.; Bagherian, M.; Azami, N.; Farahani, M.V.; Omid Sharifzadeh, S.; Hamzehlou, S. Biomedical application of chitosan-based nanoscale delivery systems: Potential usefulness in siRNA delivery for cancer therapy. *Carbohydr. Polym.* **2021**, *260*, 117809. [CrossRef] [PubMed]
4. Kirtonia, A.; Gala, K.; Fernandes, S.G.; Pandya, G.; Pandey, A.K.; Sethi, G.; Khattar, E.; Garg, M. Repurposing of drugs: An attractive pharmacological strategy for cancer therapeutics. *Semin. Cancer Biol.* **2021**, *68*, 258–278. [CrossRef]
5. Merarchi, M.; Sethi, G.; Shanmugam, M.K.; Fan, L.; Arfuso, F.; Ahn, K.S. Role of Natural Products in Modulating Histone Deacetylases in Cancer. *Molecules* **2019**, *24*, 1047. [CrossRef] [PubMed]
6. Chopra, P.; Sethi, G.; Dastidar, S.G.; Ray, A. Polo-like kinase inhibitors: An emerging opportunity for cancer therapeutics. *Expert Opin. Investig. Drugs* **2010**, *19*, 27–43. [CrossRef] [PubMed]
7. Ashrafizadeh, M.; Bakhoda, M.R.; Bahmanpour, Z.; Ilkhani, K.; Zarrabi, A.; Makvandi, P.; Khan, H.; Mazaheri, S.; Darvish, M.; Mirzaei, H. Apigenin as tumor suppressor in cancers: Biotherapeutic activity, nanodelivery, and mechanisms with emphasis on pancreatic cancer. *Front. Chem.* **2020**, *8*, 829. [CrossRef] [PubMed]
8. Antognelli, C.; Mandarano, M.; Prosperi, E.; Sidoni, A.; Talesa, V.N. Glyoxalase-1-Dependent Methylglyoxal Depletion Sustains PD-L1 Expression in Metastatic Prostate Cancer Cells: A Novel Mechanism in Cancer Immunosurveillance Escape and a Potential Novel Target to Overcome PD-L1 Blockade Resistance. *Cancers* **2021**, *13*, 2965. [CrossRef]
9. Ashrafizadeh, M.; Farhood, B.; Eleojo Musa, A.; Taeb, S.; Najafi, M. The interactions and communications in tumor resistance to radiotherapy: Therapy perspectives. *Int. Immunopharmacol.* **2020**, *87*, 106807. [CrossRef] [PubMed]
10. Cao, J.; Sun, X.; Zhang, X.; Chen, D. 6PGD Upregulation is Associated with Chemo- and Immuno-Resistance of Renal Cell Carcinoma via AMPK Signaling-Dependent NADPH-Mediated Metabolic Reprograming. *Am. J. Med. Sci.* **2020**, *360*, 279–286. [CrossRef]
11. Gupta, B.; Sadaria, D.; Warrier, V.U.; Kirtonia, A.; Kant, R.; Awasthi, A.; Baligar, P.; Pal, J.K.; Yuba, E.; Sethi, G.; et al. Plant lectins and their usage in preparing targeted nanovaccines for cancer immunotherapy. *Semin. Cancer Biol.* **2020**. [CrossRef] [PubMed]
12. Warrier, V.U.; Makandar, A.I.; Garg, M.; Sethi, G.; Kant, R.; Pal, J.K.; Yuba, E.; Gupta, R.K. Engineering anti-cancer nanovaccine based on antigen cross-presentation. *Biosci. Rep.* **2019**, *39*, BSR20193220. [CrossRef] [PubMed]
13. Mirzaei, S.; Zarrabi, A.; Asnaf, S.E.; Hashemi, F.; Zabolian, A.; Hushmandi, K.; Raei, M.; Goharrizi, M.A.S.B.; Makvandi, P.; Samarghandian, S.; et al. The role of microRNA-338-3p in cancer: Growth, invasion, chemoresistance, and mediators. *Life Sci.* **2021**, *268*, 119005. [CrossRef]
14. Dehshahri, A.; Ashrafizadeh, M.; Ghasemipour Afshar, E.; Pardakhty, A.; Mandegary, A.; Mohammadinejad, R.; Sethi, G. Topoisomerase inhibitors: Pharmacology and emerging nanoscale delivery systems. *Pharmacol. Res.* **2020**, *151*, 104551. [CrossRef] [PubMed]
15. Wang, C.; Tang, H.; Geng, A.; Dai, B.; Zhang, H.; Sun, X.; Chen, Y.; Qiao, Z.; Zhu, H.; Yang, J.; et al. Rational combination therapy for hepatocellular carcinoma with PARP1 and DNA-PK inhibitors. *Proc. Natl. Acad. Sci. USA* **2020**, *117*, 26356–26365. [CrossRef] [PubMed]
16. Galván Morales, M.A.; Barrera Rodríguez, R.; Santiago Cruz, J.R.; Teran, L.M. Overview of New Treatments with Immunotherapy for Breast Cancer and a Proposal of a Combination Therapy. *Molecules* **2020**, *25*, 5685. [CrossRef] [PubMed]
17. Kong, Q.; Wei, D.; Xie, P.; Wang, B.; Yu, K.; Kang, X.; Wang, Y. Photothermal Therapy via NIR II Light Irradiation Enhances DNA Damage and Endoplasmic Reticulum Stress for Efficient Chemotherapy. *Front. Pharmacol.* **2021**, *12*, 670207. [CrossRef] [PubMed]
18. Zhou, H.; Zeng, X.; Li, A.; Zhou, W.; Tang, L.; Hu, W.; Fan, Q.; Meng, X.; Deng, H.; Duan, L.; et al. Upconversion NIR-II fluorophores for mitochondria-targeted cancer imaging and photothermal therapy. *Nat. Commun.* **2020**, *11*, 6183. [CrossRef] [PubMed]
19. Jo, G.; Lee, B.Y.; Kim, E.J.; Park, M.H.; Hyun, H. Indocyanine Green and Methyl-β-Cyclodextrin Complex for Enhanced Photothermal Cancer Therapy. *Biomedicines* **2020**, *8*, 476. [CrossRef] [PubMed]
20. Ashrafizadeh, M.; Hushmandi, K.; Rahmani Moghadam, E.; Zarrin, V.; Hosseinzadeh Kashani, S.; Bokaie, S.; Najafi, M.; Tavakol, S.; Mohammadinejad, R.; Nabavi, N.; et al. Progress in delivery of siRNA-based therapeutics employing nano-vehicles for treatment of prostate cancer. *Bioengineering* **2020**, *7*, 91. [CrossRef]
21. Mohammadinejad, R.; Dehshahri, A.; Sassan, H.; Behnam, B.; Ashrafizadeh, M.; Samareh Gholami, A.; Pardakhty, A.; Mandegary, A. Preparation of carbon dot as a potential CRISPR/Cas9 plasmid delivery system for lung cancer cells. *Minerva Biotecnol.* **2020**, *32*, 106–113. [CrossRef]
22. Mohammadinejad, R.; Sassan, H.; Pardakhty, A.; Hashemabadi, M.; Ashrafizadeh, M.; Dehshahri, A.; Mandegary, A. ZEB1 and ZEB2 gene editing mediated by CRISPR/Cas9 in A549 cell line. *Bratisl. Lek. Listy* **2020**, *121*, 31–36. [CrossRef] [PubMed]
23. Antognelli, C.; Frosini, R.; Santolla, M.F.; Peirce, M.J.; Talesa, V.N. Oleuropein-Induced Apoptosis Is Mediated by Mitochondrial Glyoxalase 2 in NSCLC A549 Cells: A Mechanistic Inside and a Possible Novel Nonenzymatic Role for an Ancient Enzyme. *Oxidative Med. Cell. Longev.* **2019**, *2019*, 8576961. [CrossRef] [PubMed]

24. Mirzaei, S.; Gholami, M.H.; Hashemi, F.; Zabolian, A.; Farahani, M.V.; Hushmandi, K.; Zarrabi, A.; Goldman, A.; Ashrafizadeh, M.; Orive, G. Advances in understanding the role of P-gp in doxorubicin resistance: Molecular pathways, therapeutic strategies, and prospects. *Drug Discov. Today* **2021**. [CrossRef] [PubMed]
25. Kirtonia, A.; Ashrafizadeh, M.; Zarrabi, A.; Hushmandi, K.; Zabolian, A.; Bejandi, A.K.; Rani, R.; Pandey, A.K.; Baligar, P.; Kumar, V.; et al. Long noncoding RNAs: A novel insight in the leukemogenesis and drug resistance in acute myeloid leukemia. *J. Cell. Physiol.* **2021**. [CrossRef]
26. Ashrafizadeh, M.; Zarrabi, A.; Mirzaei, S.; Hashemi, F.; Samarghandian, S.; Zabolian, A.; Hushmandi, K.; Ang, H.L.; Sethi, G.; Kumar, A.P.; et al. Gallic acid for cancer therapy: Molecular mechanisms and boosting efficacy by nanoscopical delivery. *Food Chem. Toxicol.* **2021**, *157*, 112576. [CrossRef]
27. Ashrafizadeh, M.; Mirzaei, S.; Gholami, M.H.; Hashemi, F.; Zabolian, A.; Raei, M.; Hushmandi, K.; Zarrabi, A.; Voelcker, N.H.; Aref, A.R.; et al. Hyaluronic acid-based nanoplatforms for Doxorubicin: A review of stimuli-responsive carriers, co-delivery and resistance suppression. *Carbohydr. Polym.* **2021**, *272*, 118491. [CrossRef]
28. Paskeh, M.D.A.; Mirzaei, S.; Orouei, S.; Zabolian, A.; Saleki, H.; Azami, N.; Hushmandi, K.; Baradaran, B.; Hashmi, M.; Aref, A.R.; et al. The role of miRNA-489 as a new onco-suppressor factor in different cancers based on pre-clinical and clinical evidence. *Int. J. Biol. Macromol.* **2021**, *191*, 727–732. [CrossRef]
29. Mirzaei, S.; Gholami, M.H.; Hashemi, F.; Zabolian, A.; Hushmandi, K.; Rahmanian, V.; Entezari, M.; Girish, Y.R.; Kumar, K.S.S.; Aref, A.; et al. Employing siRNA tool and its delivery platforms in suppressing cisplatin resistance: Approaching to a new era of cancer chemotherapy. *Life Sci.* **2021**, *277*, 119430. [CrossRef]
30. Mirzaei, S.; Mahabady, M.K.; Zabolian, A.; Abbaspour, A.; Fallahzadeh, P.; Noori, M.; Hashemi, F.; Hushmandi, K.; Daneshi, S.; Kumar, A.P.; et al. Small interfering RNA (siRNA) to target genes and molecular pathways in glioblastoma therapy: Current status with an emphasis on delivery systems. *Life Sci.* **2021**, *275*, 119368. [CrossRef]
31. Ashrafizadeh, M.; Najafi, M.; Makvandi, P.; Zarrabi, A.; Farkhondeh, T.; Samarghandian, S. Versatile role of curcumin and its derivatives in lung cancer therapy. *J. Cell. Physiol.* **2020**, *235*, 9241–9268. [CrossRef] [PubMed]
32. Sethi, G.; Shanmugam, M.K.; Warrier, S.; Merarchi, M.; Arfuso, F.; Kumar, A.P.; Bishayee, A. Pro-Apoptotic and Anti-Cancer Properties of Diosgenin: A Comprehensive and Critical Review. *Nutrients* **2018**, *10*, 645. [CrossRef] [PubMed]
33. Kashyap, D.; Tuli, H.S.; Yerer, M.B.; Sharma, A.; Sak, K.; Srivastava, S.; Pandey, A.; Garg, V.K.; Sethi, G.; Bishayee, A. Natural product-based nanoformulations for cancer therapy: Opportunities and challenges. *Semin. Cancer Biol.* **2021**, *69*, 5–23. [CrossRef]
34. Patel, S.M.; Nagulapalli Venkata, K.C.; Bhattacharyya, P.; Sethi, G.; Bishayee, A. Potential of neem (Azadirachta indica L.) for prevention and treatment of oncologic diseases. *Semin. Cancer Biol.* **2016**, *40–41*, 100–115. [CrossRef]
35. Liu, C.; Ho, P.C.; Wong, F.C.; Sethi, G.; Wang, L.Z.; Goh, B.C. Garcinol: Current status of its anti-oxidative, anti-inflammatory and anti-cancer effects. *Cancer Lett.* **2015**, *362*, 8–14. [CrossRef]
36. Samec, M.; Liskova, A.; Koklesova, L.; Mersakova, S.; Strnadel, J.; Kajo, K.; Pec, M.; Zhai, K.; Smejkal, K.; Mirzaei, S. Flavonoids targeting HIF-1: Implications on cancer metabolism. *Cancers* **2021**, *13*, 130. [CrossRef]
37. Koklesova, L.; Liskova, A.; Samec, M.; Zhai, K.; Abotaleb, M.; Ashrafizadeh, M.; Brockmueller, A.; Shakibaei, M.; Biringer, K.; Bugos, O.; et al. Carotenoids in Cancer Metastasis—Status Quo and Outlook. *Biomolecules* **2020**, *10*, 1653. [CrossRef] [PubMed]
38. Koklesova, L.; Liskova, A.; Samec, M.; Buhrmann, C.; Samuel, S.M.; Varghese, E.; Ashrafizadeh, M.; Najafi, M.; Shakibaei, M.; Büsselberg, D.; et al. Carotenoids in cancer apoptosis—the road from bench to bedside and back. *Cancers* **2020**, *12*, 2425. [CrossRef]
39. Mishra, S.; Verma, S.S.; Rai, V.; Awasthee, N.; Chava, S.; Hui, K.M.; Kumar, A.P.; Challagundla, K.B.; Sethi, G.; Gupta, S.C. Long non-coding RNAs are emerging targets of phytochemicals for cancer and other chronic diseases. *Cell. Mol. Life Sci.* **2019**, *76*, 1947–1966. [CrossRef]
40. Tewari, D.; Nabavi, S.F.; Nabavi, S.M.; Sureda, A.; Farooqi, A.A.; Atanasov, A.G.; Vacca, R.A.; Sethi, G.; Bishayee, A. Targeting activator protein 1 signaling pathway by bioactive natural agents: Possible therapeutic strategy for cancer prevention and intervention. *Pharmacol. Res.* **2018**, *128*, 366–375. [CrossRef]
41. Shanmugam, M.K.; Ong, T.H.; Kumar, A.P.; Lun, C.K.; Ho, P.C.; Wong, P.T.; Hui, K.M.; Sethi, G. Ursolic acid inhibits the initiation, progression of prostate cancer and prolongs the survival of TRAMP mice by modulating pro-inflammatory pathways. *PLoS ONE* **2012**, *7*, e32476. [CrossRef]
42. Li, F.; Shanmugam, M.K.; Chen, L.; Chatterjee, S.; Basha, J.; Kumar, A.P.; Kundu, T.K.; Sethi, G. Garcinol, a polyisoprenylated benzophenone modulates multiple proinflammatory signaling cascades leading to the suppression of growth and survival of head and neck carcinoma. *Cancer Prev. Res.* **2013**, *6*, 843–854. [CrossRef]
43. Choi, Y.H.; Han, D.H.; Kim, S.W.; Kim, M.J.; Sung, H.H.; Jeon, H.G.; Jeong, B.C.; Seo, S.I.; Jeon, S.S.; Lee, H.M.; et al. A randomized, double-blind, placebo-controlled trial to evaluate the role of curcumin in prostate cancer patients with intermittent androgen deprivation. *Prostate* **2019**, *79*, 614–621. [CrossRef]
44. Howells, L.M.; Iwuji, C.O.O.; Irving, G.R.B.; Barber, S.; Walter, H.; Sidat, Z.; Griffin-Teall, N.; Singh, R.; Foreman, N.; Patel, S.R.; et al. Curcumin Combined with FOLFOX Chemotherapy Is Safe and Tolerable in Patients with Metastatic Colorectal Cancer in a Randomized Phase IIa Trial. *J. Nutr.* **2019**, *149*, 1133–1139. [CrossRef]
45. Hsieh, Y.S.; Yang, S.F.; Sethi, G.; Hu, D.N. Natural bioactives in cancer treatment and prevention. *BioMed Res. Int.* **2015**, *2015*, 182835. [CrossRef]

46. Manu, K.A.; Shanmugam, M.K.; Ramachandran, L.; Li, F.; Siveen, K.S.; Chinnathambi, A.; Zayed, M.E.; Alharbi, S.A.; Arfuso, F.; Kumar, A.P.; et al. Isorhamnetin augments the anti-tumor effect of capecitabine through the negative regulation of NF-κB signaling cascade in gastric cancer. *Cancer Lett.* **2015**, *363*, 28–36. [CrossRef]
47. Manu, K.A.; Shanmugam, M.K.; Li, F.; Chen, L.; Siveen, K.S.; Ahn, K.S.; Kumar, A.P.; Sethi, G. Simvastatin sensitizes human gastric cancer xenograft in nude mice to capecitabine by suppressing nuclear factor-kappa B-regulated gene products. *J. Mol. Med.* **2014**, *92*, 267–276. [CrossRef] [PubMed]
48. Hussain, Y.; Islam, L.; Khan, H.; Filosa, R.; Aschner, M.; Javed, S. Curcumin-cisplatin chemotherapy: A novel strategy in promoting chemotherapy efficacy and reducing side effects. *Phytother. Res.* **2021**. [CrossRef] [PubMed]
49. Ashrafizadeh, M.; Zarrabi, A.; Hashemi, F.; Moghadam, E.R.; Hashemi, F.; Entezari, M.; Hushmandi, K.; Mohammadinejad, R.; Najafi, M. Curcumin in cancer therapy: A novel adjunct for combination chemotherapy with paclitaxel and alleviation of its adverse effects. *Life Sci.* **2020**, *256*, 117984. [CrossRef]
50. Amini, P.; Nodooshan, S.J.; Ashrafizadeh, M.; Eftekhari, S.-M.; Aryafar, T.; Khalafi, L.; Musa, A.E.; Mahdavi, S.R.; Najafi, M.; Farhood, B. Resveratrol induces apoptosis and attenuates proliferation of MCF-7 cells in combination with radiation and hyperthermia. *Curr. Mol. Med.* **2021**, *21*, 142–150. [CrossRef] [PubMed]
51. Moballegh Nasery, M.; Abadi, B.; Poormoghadam, D.; Zarrabi, A.; Keyhanvar, P.; Khanbabaei, H.; Ashrafizadeh, M.; Mohammadinejad, R.; Tavakol, S.; Sethi, G. Curcumin Delivery Mediated by Bio-Based Nanoparticles: A Review. *Molecules* **2020**, *25*, 689. [CrossRef] [PubMed]
52. Dai, X.; Ahn, K.S.; Wang, L.Z.; Kim, C.; Deivasigamni, A.; Arfuso, F.; Um, J.Y.; Kumar, A.P.; Chang, Y.C.; Kumar, D.; et al. Ascochlorin Enhances the Sensitivity of Doxorubicin Leading to the Reversal of Epithelial-to-Mesenchymal Transition in Hepatocellular Carcinoma. *Mol. Cancer Ther.* **2016**, *15*, 2966–2976. [CrossRef] [PubMed]
53. Ong, S.K.L.; Shanmugam, M.K.; Fan, L.; Fraser, S.E.; Arfuso, F.; Ahn, K.S.; Sethi, G.; Bishayee, A. Focus on Formononetin: Anticancer Potential and Molecular Targets. *Cancers* **2019**, *11*, 611. [CrossRef]
54. Bhuvanalakshmi, G.; Gamit, N.; Patil, M.; Arfuso, F.; Sethi, G.; Dharmarajan, A.; Kumar, A.P.; Warrier, S. Stemness, Pluripotentiality, and Wnt Antagonism: sFRP4, a Wnt antagonist Mediates Pluripotency and Stemness in Glioblastoma. *Cancers* **2018**, *11*, 25. [CrossRef]
55. Mandal, S.; Arfuso, F.; Sethi, G.; Dharmarajan, A.; Warrier, S. Encapsulated human mesenchymal stem cells (eMSCs) as a novel anti-cancer agent targeting breast cancer stem cells: Development of 3D primed therapeutic MSCs. *Int. J. Biochem. Cell Biol.* **2019**, *110*, 59–69. [CrossRef] [PubMed]
56. Carvalho, L.S.; Gonçalves, N.; Fonseca, N.A.; Moreira, J.N. Cancer Stem Cells and Nucleolin as Drivers of Carcinogenesis. *Pharmaceuticals* **2021**, *14*, 60. [CrossRef]
57. Ma, Z.; Wang, Y.Y.; Xin, H.W.; Wang, L.; Arfuso, F.; Dharmarajan, A.; Kumar, A.P.; Wang, H.; Tang, F.R.; Warrier, S.; et al. The expanding roles of long non-coding RNAs in the regulation of cancer stem cells. *Int. J. Biochem. Cell Biol.* **2019**, *108*, 17–20. [CrossRef] [PubMed]
58. Lapidot, T.; Sirard, C.; Vormoor, J.; Murdoch, B.; Hoang, T.; Caceres-Cortes, J.; Minden, M.; Paterson, B.; Caligiuri, M.A.; Dick, J.E. A cell initiating human acute myeloid leukaemia after transplantation into SCID mice. *Nature* **1994**, *367*, 645–648. [CrossRef] [PubMed]
59. Bonnet, D.; Dick, J.E. Human acute myeloid leukemia is organized as a hierarchy that originates from a primitive hematopoietic cell. *Nat. Med.* **1997**, *3*, 730–737. [CrossRef] [PubMed]
60. Al-Hajj, M.; Wicha, M.S.; Benito-Hernandez, A.; Morrison, S.J.; Clarke, M.F. Prospective identification of tumorigenic breast cancer cells. *Proc. Natl. Acad. Sci. USA* **2003**, *100*, 3983–3988. [CrossRef] [PubMed]
61. Koren, E.; Fuchs, Y. The bad seed: Cancer stem cells in tumor development and resistance. *Drug Resist. Updates* **2016**, *28*, 1–12. [CrossRef] [PubMed]
62. Warrier, S.; Patil, M.; Bhansali, S.; Varier, L.; Sethi, G. Designing precision medicine panels for drug refractory cancers targeting cancer stemness traits. *Biochim. Biophys. Acta. Rev. Cancer* **2021**, *1875*, 188475. [CrossRef] [PubMed]
63. Plaks, V.; Kong, N.; Werb, Z. The cancer stem cell niche: How essential is the niche in regulating stemness of tumor cells? *Cell Stem Cell* **2015**, *16*, 225–238. [CrossRef] [PubMed]
64. Cao, J.; Bhatnagar, S.; Wang, J.; Qi, X.; Prabha, S.; Panyam, J. Cancer stem cells and strategies for targeted drug delivery. *Drug Deliv. Transl. Res.* **2020**, *11*, 1779–1805. [CrossRef] [PubMed]
65. Capp, J.-P. Cancer stem cells: From historical roots to a new perspective. *J. Oncol.* **2019**, *2019*, 5189232. [CrossRef] [PubMed]
66. Singh, S.K.; Hawkins, C.; Clarke, I.D.; Squire, J.A.; Bayani, J.; Hide, T.; Henkelman, R.M.; Cusimano, M.D.; Dirks, P.B. Identification of human brain tumour initiating cells. *Nature* **2004**, *432*, 396–401. [CrossRef] [PubMed]
67. Ricci-Vitiani, L.; Lombardi, D.G.; Pilozzi, E.; Biffoni, M.; Todaro, M.; Peschle, C.; De Maria, R. Identification and expansion of human colon-cancer-initiating cells. *Nature* **2007**, *445*, 111–115. [CrossRef] [PubMed]
68. Ma, S.; Chan, K.W.; Hu, L.; Lee, T.K.W.; Wo, J.Y.H.; Ng, I.O.L.; Zheng, B.J.; Guan, X.Y. Identification and characterization of tumorigenic liver cancer stem/progenitor cells. *Gastroenterology* **2007**, *132*, 2542–2556. [CrossRef] [PubMed]
69. Elgendy, S.M.; Alyammahi, S.K.; Alhamad, D.W.; Abdin, S.M.; Omar, H.A. Ferroptosis: An emerging approach for targeting cancer stem cells and drug resistance. *Crit. Rev. Oncol./Hematol.* **2020**, *155*, 103095. [CrossRef] [PubMed]
70. Hiremath, I.S.; Goel, A.; Warrier, S.; Kumar, A.P.; Sethi, G.; Garg, M. The multidimensional role of the Wnt/β-catenin signaling pathway in human malignancies. *J. Cell. Physiol.* **2021**. [CrossRef]

71. Duan, J.-J.; Qiu, W.; Xu, S.-L.; Wang, B.; Ye, X.-Z.; Ping, Y.-F.; Zhang, X.; Bian, X.-W.; Yu, S.-C. development. Strategies for isolating and enriching cancer stem cells: Well begun is half done. *Stem Cells Dev.* **2013**, *22*, 2221–2239. [CrossRef] [PubMed]
72. Batlle, E.; Clevers, H. Cancer stem cells revisited. *Nat. Med.* **2017**, *23*, 1124–1134. [CrossRef] [PubMed]
73. Prager, B.C.; Xie, Q.; Bao, S.; Rich, J.N. Cancer Stem Cells: The Architects of the Tumor Ecosystem. *Cell Stem Cell* **2019**, *24*, 41–53. [CrossRef] [PubMed]
74. Yang, L.; Shi, P.; Zhao, G.; Xu, J.; Peng, W.; Zhang, J.; Zhang, G.; Wang, X.; Dong, Z.; Chen, F.; et al. Targeting cancer stem cell pathways for cancer therapy. *Signal Transduct. Target. Ther.* **2020**, *5*, 8. [CrossRef] [PubMed]
75. Nallanthighal, S.; Heiserman, J.P.; Cheon, D.J. The Role of the Extracellular Matrix in Cancer Stemness. *Front. Cell Dev. Biol.* **2019**, *7*, 86. [CrossRef]
76. Ma, Z.; Wang, L.Z.; Cheng, J.T.; Lam, W.S.T.; Ma, X.; Xiang, X.; Wong, A.L.; Goh, B.C.; Gong, Q.; Sethi, G.; et al. Targeting Hypoxia-Inducible Factor-1-Mediated Metastasis for Cancer Therapy. *Antioxid. Redox Signal.* **2021**, *34*, 1484–1497. [CrossRef] [PubMed]
77. Ong, P.S.; Wang, L.Z.; Dai, X.; Tseng, S.H.; Loo, S.J.; Sethi, G. Judicious Toggling of mTOR Activity to Combat Insulin Resistance and Cancer: Current Evidence and Perspectives. *Front. Pharmacol.* **2016**, *7*, 395. [CrossRef] [PubMed]
78. De Francesco, E.M.; Sotgia, F.; Lisanti, M.P. Cancer stem cells (CSCs): Metabolic strategies for their identification and eradication. *Biochem. J.* **2018**, *475*, 1611–1634. [CrossRef] [PubMed]
79. Jagust, P.; de Luxán-Delgado, B.; Parejo-Alonso, B.; Sancho, P. Metabolism-Based Therapeutic Strategies Targeting Cancer Stem Cells. *Front. Pharmacol.* **2019**, *10*, 203. [CrossRef] [PubMed]
80. Snyder, V.; Reed-Newman, T.C.; Arnold, L.; Thomas, S.M.; Anant, S. Cancer Stem Cell Metabolism and Potential Therapeutic Targets. *Front. Oncol.* **2018**, *8*, 203. [CrossRef] [PubMed]
81. Peng, F.; Wang, J.H.; Fan, W.J.; Meng, Y.T.; Li, M.M.; Li, T.T.; Cui, B.; Wang, H.F.; Zhao, Y.; An, F.; et al. Glycolysis gatekeeper PDK1 reprograms breast cancer stem cells under hypoxia. *Oncogene* **2018**, *37*, 1062–1074. [CrossRef]
82. Ciavardelli, D.; Rossi, C.; Barcaroli, D.; Volpe, S.; Consalvo, A.; Zucchelli, M.; De Cola, A.; Scavo, E.; Carollo, R.; D'Agostino, D.; et al. Breast cancer stem cells rely on fermentative glycolysis and are sensitive to 2-deoxyglucose treatment. *Cell Death Dis.* **2014**, *5*, e1336. [CrossRef] [PubMed]
83. Liu, P.P.; Liao, J.; Tang, Z.J.; Wu, W.J.; Yang, J.; Zeng, Z.L.; Hu, Y.; Wang, P.; Ju, H.Q.; Xu, R.H.; et al. Metabolic regulation of cancer cell side population by glucose through activation of the Akt pathway. *Cell Death Differ.* **2014**, *21*, 124–135. [CrossRef] [PubMed]
84. Hur, W.; Ryu, J.Y.; Kim, H.U.; Hong, S.W.; Lee, E.B.; Lee, S.Y.; Yoon, S.K. Systems approach to characterize the metabolism of liver cancer stem cells expressing CD133. *Sci. Rep.* **2017**, *7*, 45557. [CrossRef]
85. Shen, Y.A.; Wang, C.Y.; Hsieh, Y.T.; Chen, Y.J.; Wei, Y.H. Metabolic reprogramming orchestrates cancer stem cell properties in nasopharyngeal carcinoma. *Cell Cycle* **2015**, *14*, 86–98. [CrossRef]
86. Janiszewska, M.; Suvà, M.L.; Riggi, N.; Houtkooper, R.H.; Auwerx, J.; Clément-Schatlo, V.; Radovanovic, I.; Rheinbay, E.; Provero, P.; Stamenkovic, I. Imp2 controls oxidative phosphorylation and is crucial for preserving glioblastoma cancer stem cells. *Genes Dev.* **2012**, *26*, 1926–1944. [CrossRef] [PubMed]
87. Sancho, P.; Burgos-Ramos, E.; Tavera, A.; Bou Kheir, T.; Jagust, P.; Schoenhals, M.; Barneda, D.; Sellers, K.; Campos-Olivas, R.; Graña, O.; et al. MYC/PGC-1α Balance Determines the Metabolic Phenotype and Plasticity of Pancreatic Cancer Stem Cells. *Cell Metab.* **2015**, *22*, 590–605. [CrossRef]
88. Tanabe, A.; Sahara, H. The metabolic heterogeneity and flexibility of cancer stem cells. *Cancers* **2020**, *12*, 2780. [CrossRef]
89. Jain, S.; Annett, S.L.; Morgan, M.P.; Robson, T. The Cancer Stem Cell Niche in Ovarian Cancer and Its Impact on Immune Surveillance. *Int. J. Mol. Sci.* **2021**, *22*, 4091. [CrossRef]
90. Escalona, R.M.; Bilandzic, M.; Western, P.; Kadife, E.; Kannourakis, G.; Findlay, J.K.; Ahmed, N. TIMP-2 regulates proliferation, invasion and STAT3-mediated cancer stem cell-dependent chemoresistance in ovarian cancer cells. *BMC Cancer* **2020**, *20*, 960. [CrossRef]
91. Chan, Y.W.; So, C.; Yau, K.L.; Chiu, K.C.; Wang, X.; Chan, F.L.; Tsang, S.Y. Adipose-derived stem cells and cancer cells fuse to generate cancer stem cell-like cells with increased tumorigenicity. *J. Cell. Physiol.* **2020**, *235*, 6794–6807. [CrossRef]
92. Chen, K.; Li, L.; Qu, S.; Pan, X.; Sun, Y.; Wan, F.; Yu, B.; Zhou, L.; Zhu, X. Silencing hTERT attenuates cancer stem cell-like characteristics and radioresistance in the radioresistant nasopharyngeal carcinoma cell line CNE-2R. *Aging* **2020**, *12*, 25599–25613. [CrossRef] [PubMed]
93. Zhang, J.; He, H.; Wang, K.; Xie, Y.; Yang, Z.; Qie, M.; Liao, Z.; Zheng, Z. miR-326 inhibits the cell proliferation and cancer stem cell-like property of cervical cancer in vitro and oncogenesis in vivo via targeting TCF4. *Ann. Transl. Med.* **2020**, *8*, 1638. [CrossRef]
94. Jimenez, T.; Barrios, A.; Tucker, A.; Collazo, J.; Arias, N.; Fazel, S.; Baker, M.; Halim, M.; Huynh, T.; Singh, R.; et al. DUSP9-mediated reduction of pERK1/2 supports cancer stem cell-like traits and promotes triple negative breast cancer. *Am. J. Cancer Res.* **2020**, *10*, 3487–3506. [PubMed]
95. Moon, S.; Ok, Y.; Hwang, S.; Lim, Y.S.; Kim, H.Y.; Na, Y.J.; Yoon, S. A Marine Collagen-Based Biomimetic Hydrogel Recapitulates Cancer Stem Cell Niche and Enhances Progression and Chemoresistance in Human Ovarian Cancer. *Mar. Drugs* **2020**, *18*, 498. [CrossRef] [PubMed]

96. Bridges, A.E.; Ramachandran, S.; Tamizhmani, K.; Parwal, U.; Lester, A.; Rajpurohit, P.; Morera, D.S.; Hasanali, S.L.; Arjunan, P.; Jedeja, R.N.; et al. RAD51AP1 Loss Attenuates Colorectal Cancer Stem Cell Renewal and Sensitizes to Chemotherapy. *Mol. Cancer Res.* **2021**, *19*. [CrossRef]
97. Yarmishyn, A.A.; Yang, Y.P.; Lu, K.H.; Chen, Y.C.; Chien, Y.; Chou, S.J.; Tsai, P.H.; Ma, H.I.; Chien, C.S.; Chen, M.T.; et al. Musashi-1 promotes cancer stem cell properties of glioblastoma cells via upregulation of YTHDF1. *Cancer Cell Int.* **2020**, *20*, 597. [CrossRef]
98. Xiong, J.; Zhang, X.; Zhang, Y.; Wu, B.; Fang, L.; Wang, N.; Yi, H.; Chang, N.; Chen, L.; Zhang, J. Aryl hydrocarbon receptor mediates Jak2/STAT3 signaling for non-small cell lung cancer stem cell maintenance. *Exp. Cell Res.* **2020**, *396*, 112288. [CrossRef]
99. Wu, Y.; Wang, T.; Xia, L.; Zhang, M. LncRNA WDFY3-AS2 promotes cisplatin resistance and the cancer stem cell in ovarian cancer by regulating hsa-miR-139-5p/SDC4 axis. *Cancer Cell Int.* **2021**, *21*, 284. [CrossRef]
100. Jeon, S.A.; Kim, D.W.; Lee, D.B.; Cho, J.Y. NEDD4 Plays Roles in the Maintenance of Breast Cancer Stem Cell Characteristics. *Front. Oncol.* **2020**, *10*, 1680. [CrossRef]
101. Ashrafizaveh, S.; Ashrafizadeh, M.; Zarrabi, A.; Husmandi, K.; Zabolian, A.; Shahinozzaman, M.; Aref, A.R.; Hamblin, M.R.; Nabavi, N.; Crea, F. Long non-coding RNA in the doxorubicin resistance of cancer cells. *Cancer Lett.* **2021**, *508*, 104–114. [CrossRef] [PubMed]
102. Wee, I.; Syn, N.; Sethi, G.; Goh, B.C.; Wang, L. Role of tumor-derived exosomes in cancer metastasis. *Biochim. Biophys. Acta Rev. Cancer* **2019**, *1871*, 12–19. [CrossRef] [PubMed]
103. Weng, J.; Xiang, X.; Ding, L.; Wong, A.L.; Zeng, Q.; Sethi, G.; Wang, L.; Lee, S.C.; Goh, B.C. Extracellular vesicles, the cornerstone of next-generation cancer diagnosis? *Semin. Cancer Biol.* **2021**, *74*, 105–120. [CrossRef] [PubMed]
104. Afify, S.M.; Hassan, G.; Yan, T.; Seno, A.; Seno, M. Cancer Stem Cell Initiation by Tumor-Derived Extracellular Vesicles. In *Methods in Molecular Biology*; Springer: New York, NY, USA, 2021. [CrossRef]
105. Gao, Z.; Wang, Q.; Ji, M.; Guo, X.; Li, L.; Su, X. Exosomal lncRNA UCA1 modulates cervical cancer stem cell self-renewal and differentiation through microRNA-122-5p/SOX2 axis. *J. Transl. Med.* **2021**, *19*, 229. [CrossRef] [PubMed]
106. Dai, W.; Jin, X.; Han, L.; Huang, H.; Ji, Z.; Xu, X.; Tang, M.; Jiang, B.; Chen, W. Exosomal lncRNA DOCK9-AS2 derived from cancer stem cell-like cells activated Wnt/β-catenin pathway to aggravate stemness, proliferation, migration, and invasion in papillary thyroid carcinoma. *Cell Death Dis.* **2020**, *11*, 743. [CrossRef] [PubMed]
107. Sun, L.; Huang, C.; Zhu, M.; Guo, S.; Gao, Q.; Wang, Q.; Chen, B.; Li, R.; Zhao, Y.; Wang, M.; et al. Gastric cancer mesenchymal stem cells regulate PD-L1-CTCF enhancing cancer stem cell-like properties and tumorigenesis. *Theranostics* **2020**, *10*, 11950–11962. [CrossRef] [PubMed]
108. Ji, G.; Zhou, W.; Du, J.; Zhou, J.; Wu, D.; Zhao, M.; Yang, L.; Hao, A. PCGF1 promotes epigenetic activation of stemness markers and colorectal cancer stem cell enrichment. *Cell Death Dis.* **2021**, *12*, 633. [CrossRef]
109. Xia, Y.; Lv, J.; Jiang, T.; Li, B.; Li, Y.; He, Z.; Xuan, Z.; Sun, G.; Wang, S.; Li, Z.; et al. CircFAM73A promotes the cancer stem cell-like properties of gastric cancer through the miR-490-3p/HMGA2 positive feedback loop and HNRNPK-mediated β-catenin stabilization. *J. Exp. Clin. Cancer Res.* **2021**, *40*, 103. [CrossRef]
110. Ciummo, S.L.; D'Antonio, L.; Sorrentino, C.; Fieni, C.; Lanuti, P.; Stassi, G.; Todaro, M.; Di Carlo, E. The C-X-C Motif Chemokine Ligand 1 Sustains Breast Cancer Stem Cell Self-Renewal and Promotes Tumor Progression and Immune Escape Programs. *Front. Cell Dev. Biol.* **2021**, *9*, 689286. [CrossRef]
111. Wang, X.; Cai, J.; Zhao, L.; Zhang, D.; Xu, G.; Hu, J.; Zhang, T.; Jin, M. NUMB suppression by miR-9-5P enhances CD44(+) prostate cancer stem cell growth and metastasis. *Sci. Rep.* **2021**, *11*, 11210. [CrossRef]
112. Pan, X.W.; Zhang, H.; Xu, D.; Chen, J.X.; Chen, W.J.; Gan, S.S.; Qu, F.J.; Chu, C.M.; Cao, J.W.; Fan, Y.H.; et al. Identification of a novel cancer stem cell subpopulation that promotes progression of human fatal renal cell carcinoma by single-cell RNA-seq analysis. *Int. J. Biol. Sci.* **2020**, *16*, 3149–3162. [CrossRef] [PubMed]
113. Lin, X.; Chen, W.; Wei, F.; Xie, X. TV-circRGPD6 Nanoparticle Suppresses Breast Cancer Stem Cell-Mediated Metastasis via the miR-26b/YAF2 Axis. *Mol. Ther. J. Am. Soc. Gene Ther.* **2021**, *29*, 244–262. [CrossRef] [PubMed]
114. Iwamoto, K.; Takahashi, H.; Okuzaki, D.; Osawa, H.; Ogino, T.; Miyoshi, N.; Uemura, M.; Matsuda, C.; Yamamoto, H.; Mizushima, T.; et al. Syntenin-1 promotes colorectal cancer stem cell expansion and chemoresistance by regulating prostaglandin E2 receptor. *Br. J. Cancer* **2020**, *123*, 955–964. [CrossRef] [PubMed]
115. Li, M.; Pan, M.; Wang, J.; You, C.; Zhao, F.; Zheng, D.; Guo, M.; Xu, H.; Wu, D.; Wang, L.; et al. miR-7 Reduces Breast Cancer Stem Cell Metastasis via Inhibiting RELA to Decrease ESAM Expression. *Mol. Ther. Oncolytics* **2020**, *18*, 70–82. [CrossRef]
116. Ko, C.C.H.; Chia, W.K.; Selvarajah, G.T.; Cheah, Y.K.; Wong, Y.P.; Tan, G.C. The Role of Breast Cancer Stem Cell-Related Biomarkers as Prognostic Factors. *Diagnostics* **2020**, *10*, 721. [CrossRef] [PubMed]
117. Jakob, M.; Sharaf, K.; Schirmer, M.; Leu, M.; Küffer, S.; Bertlich, M.; Ihler, F.; Haubner, F.; Canis, M.; Kitz, J. Role of cancer stem cell markers ALDH1, BCL11B, BMI-1, and CD44 in the prognosis of advanced HNSCC. *Strahlenther. Onkol.* **2021**, *197*, 231–245. [CrossRef] [PubMed]
118. Mardani, A.; Gheytanchi, E.; Mousavie, S.H.; Madjd Jabari, Z.; Shooshtarizadeh, T. Clinical Significance of Cancer Stem Cell Markers CD133 and CXCR4 in Osteosarcomas. *Asian Pac. J. Cancer Prev.* **2020**, *21*, 67–73. [CrossRef]
119. Ashrafizadeh, M.; Zarrabi, A.; Saberifar, S.; Hashemi, F.; Hushmandi, K.; Hashemi, F.; Moghadam, E.R.; Mohammadinejad, R.; Najafi, M.; Garg, M. Nobiletin in cancer therapy: How this plant derived-natural compound targets various oncogene and onco-suppressor pathways. *Biomedicines* **2020**, *8*, 110. [CrossRef]

120. Yang, J.; Yang, Y.; Wang, L.; Jin, Q.; Pan, M. Nobiletin selectively inhibits oral cancer cell growth by promoting apoptosis and DNA damage in vitro. *Oral Surg. Oral Med. Oral Pathol. Oral Radiol.* **2020**, *130*, 419–427. [CrossRef] [PubMed]
121. Wang, J.G.; Jian, W.J.; Li, Y.; Zhang, J. Nobiletin promotes the pyroptosis of breast cancer via regulation of miR-200b/JAZF1 axis. *Kaohsiung J. Med. Sci.* **2021**, *37*, 572–582. [CrossRef] [PubMed]
122. Hermawan, A.; Putri, H. Bioinformatics Studies Provide Insight into Possible Target and Mechanisms of Action of Nobiletin against Cancer Stem Cells. *Asian Pac. J. Cancer Prev.* **2020**, *21*, 611–620. [CrossRef]
123. Sp, N.; Kang, D.Y.; Kim, D.H.; Park, J.H.; Lee, H.G.; Kim, H.J.; Darvin, P.; Park, Y.M.; Yang, Y.M. Nobiletin Inhibits CD36-Dependent Tumor Angiogenesis, Migration, Invasion, and Sphere Formation Through the Cd36/Stat3/Nf-Kb Signaling Axis. *Nutrients* **2018**, *10*, 772. [CrossRef] [PubMed]
124. Turdo, A.; Glaviano, A.; Pepe, G.; Calapà, F.; Raimondo, S.; Fiori, M.E.; Carbone, D.; Basilicata, M.G.; Di Sarno, V.; Ostacolo, C.; et al. Nobiletin and Xanthohumol Sensitize Colorectal Cancer Stem Cells to Standard Chemotherapy. *Cancers* **2021**, *13*, 3927. [CrossRef]
125. Moghadam, E.R.; Ang, H.L.; Asnaf, S.E.; Zabolian, A.; Saleki, H.; Yavari, M.; Esmaeili, H.; Zarrabi, A.; Ashrafizadeh, M.; Kumar, A.P. Broad-Spectrum Preclinical Antitumor Activity of Chrysin: Current Trends and Future Perspectives. *Biomolecules* **2020**, *10*, 1374. [CrossRef] [PubMed]
126. Salama, A.A.A.; Allam, R.M. Promising targets of chrysin and daidzein in colorectal cancer: Amphiregulin, CXCL1, and MMP-9. *Eur. J. Pharmacol.* **2021**, *892*, 173763. [CrossRef]
127. Kim, K.M.; Jung, J. Upregulation of G Protein-Coupled Estrogen Receptor by Chrysin-Nanoparticles Inhibits Tumor Proliferation and Metastasis in Triple Negative Breast Cancer Xenograft Model. *Front. Endocrinol.* **2020**, *11*, 560605. [CrossRef]
128. Chen, L.; Li, Q.; Jiang, Z.; Li, C.; Hu, H.; Wang, T.; Gao, D.; Wang, D. Chrysin Induced Cell Apoptosis Through H19/let-7a/COPB2 Axis in Gastric Cancer Cells and Inhibited Tumor Growth. *Front. Oncol.* **2021**, *11*, 651644. [CrossRef]
129. Ghamkhari, A.; Pouyafar, A.; Salehi, R.; Rahbarghazi, R. Chrysin and Docetaxel Loaded Biodegradable Micelle for Combination Chemotherapy of Cancer Stem Cell. *Pharm. Res.* **2019**, *36*, 165. [CrossRef] [PubMed]
130. Debnath, S.; Kanakaraju, M.; Islam, M.; Yeeravalli, R.; Sen, D.; Das, A. In silico design, synthesis and activity of potential drug-like chrysin scaffold-derived selective EGFR inhibitors as anticancer agents. *Comput. Biol. Chem.* **2019**, *83*, 107156. [CrossRef]
131. Adham, A.N.; Abdelfatah, S.; Naqishbandi, A.M.; Mahmoud, N.; Efferth, T. Cytotoxicity of apigenin toward multiple myeloma cell lines and suppression of iNOS and COX-2 expression in STAT1-transfected HEK293 cells. *Phytomed. Int. J. Phytother. Phytopharm.* **2021**, *80*, 153371. [CrossRef]
132. Korga-Plewko, A.; Michalczyk, M.; Adamczuk, G.; Humeniuk, E.; Ostrowska-Lesko, M.; Jozefczyk, A.; Iwan, M.; Wojcik, M.; Dudka, J. Apigenin and Hesperidin Downregulate DNA Repair Genes in MCF-7 Breast Cancer Cells and Augment Doxorubicin Toxicity. *Molecules* **2020**, *25*, 4421. [CrossRef] [PubMed]
133. Sudhakaran, M.; Parra, M.R.; Stoub, H.; Gallo, K.A.; Doseff, A.I. Apigenin by targeting hnRNPA2 sensitizes triple-negative breast cancer spheroids to doxorubicin-induced apoptosis and regulates expression of ABCC4 and ABCG2 drug efflux transporters. *Biochem. Pharmacol.* **2020**, *182*, 114259. [CrossRef]
134. Li, Y.W.; Xu, J.; Zhu, G.Y.; Huang, Z.J.; Lu, Y.; Li, X.Q.; Wang, N.; Zhang, F.X. Apigenin suppresses the stem cell-like properties of triple-negative breast cancer cells by inhibiting YAP/TAZ activity. *Cell Death Discov.* **2018**, *4*, 105. [CrossRef] [PubMed]
135. Li, Y.; Chen, X.; He, W.; Xia, S.; Jiang, X.; Li, X.; Bai, J.; Li, N.; Chen, L.; Yang, B. Apigenin Enhanced Antitumor Effect of Cisplatin in Lung Cancer via Inhibition of Cancer Stem Cells. *Nutr. Cancer* **2021**, *73*, 1489–1497. [CrossRef] [PubMed]
136. Erdogan, S.; Doganlar, O.; Doganlar, Z.B.; Serttas, R.; Turkekul, K.; Dibirdik, I.; Bilir, A. The flavonoid apigenin reduces prostate cancer CD44(+) stem cell survival and migration through PI3K/Akt/NF-κB signaling. *Life Sci.* **2016**, *162*, 77–86. [CrossRef] [PubMed]
137. Erdogan, S.; Turkekul, K.; Serttas, R.; Erdogan, Z. The natural flavonoid apigenin sensitizes human CD44(+) prostate cancer stem cells to cisplatin therapy. *Biomed. Pharmacother.* **2017**, *88*, 210–217. [CrossRef]
138. Ketkaew, Y.; Osathanon, T.; Pavasant, P.; Sooampon, S. Apigenin inhibited hypoxia induced stem cell marker expression in a head and neck squamous cell carcinoma cell line. *Arch. Oral Biol.* **2017**, *74*, 69–74. [CrossRef] [PubMed]
139. Erdogan, S.; Turkekul, K.; Dibirdik, I.; Doganlar, Z.B.; Doganlar, O.; Bilir, A. Midkine silencing enhances the anti-prostate cancer stem cell activity of the flavone apigenin: Cooperation on signaling pathways regulated by ERK, p38, PTEN, PARP, and NF-κB. *Investig. New Drugs* **2020**, *38*, 246–263. [CrossRef] [PubMed]
140. Tuli, H.S.; Aggarwal, V.; Kaur, J.; Aggarwal, D.; Parashar, G.; Parashar, N.C.; Tuorkey, M.; Kaur, G.; Savla, R.; Sak, K.; et al. Baicalein: A metabolite with promising antineoplastic activity. *Life Sci.* **2020**, *259*, 118183. [CrossRef]
141. Zhang, X.; Ruan, Q.; Zhai, Y.; Lu, D.; Li, C.; Fu, Y.; Zheng, Z.; Song, Y.; Guo, J. Baicalein inhibits non-small-cell lung cancer invasion and metastasis by reducing ezrin tension in inflammation microenvironment. *Cancer Sci.* **2020**, *111*, 3802–3812. [CrossRef]
142. Li, P.; Hu, J.; Shi, B.; Tie, J. Baicalein enhanced cisplatin sensitivity of gastric cancer cells by inducing cell apoptosis and autophagy via Akt/mTOR and Nrf2/Keap 1 pathway. *Biochem. Biophys. Res. Commun.* **2020**, *531*, 320–327. [CrossRef] [PubMed]
143. Li, J.; Yan, L.; Luo, J.; Tong, L.; Gao, Y.; Feng, W.; Wang, F.; Cui, W.; Li, S.; Sun, Z. Baicalein suppresses growth of non-small cell lung carcinoma by targeting MAP4K3. *Biomed. Pharmacother.* **2021**, *133*, 110965. [CrossRef]
144. Song, L.; Chen, X.; Wang, P.; Gao, S.; Qu, C.; Liu, L. Effects of baicalein on pancreatic cancer stem cells via modulation of sonic Hedgehog pathway. *Acta Biochim. Biophys. Sin.* **2018**, *50*, 586–596. [CrossRef] [PubMed]

145. Sharifi-Rad, J.; Herrera-Bravo, J.; Salazar, L.A.; Shaheen, S.; Abdulmajid Ayatollahi, S.; Kobarfard, F.; Imran, M.; Imran, A.; Custódio, L.; Dolores López, M.; et al. The Therapeutic Potential of Wogonin Observed in Preclinical Studies. *Evid.-Based Complement. Altern. Med.* **2021**, *2021*, 9935451. [CrossRef]
146. Yang, D.; Guo, Q.; Liang, Y.; Zhao, Y.; Tian, X.; Ye, Y.; Tian, J.; Wu, T.; Lu, N. Wogonin induces cellular senescence in breast cancer via suppressing TXNRD2 expression. *Arch. Toxicol.* **2020**, *94*, 3433–3447. [CrossRef] [PubMed]
147. Xin, N.J.; Han, M.; Gao, C.; Fan, T.T.; Shi, W. Wogonin suppresses proliferation and invasion of skin epithelioid carcinoma cells through Notch1. *Cell. Mol. Biol.* **2020**, *66*, 29–35. [CrossRef] [PubMed]
148. Kirtonia, A.; Sethi, G.; Garg, M. The multifaceted role of reactive oxygen species in tumorigenesis. *Cell. Mol. Life Sci.* **2020**, *77*, 4459–4483. [CrossRef]
149. Koh, H.; Sun, H.N.; Xing, Z.; Liu, R.; Chandimali, N.; Kwon, T.; Lee, D.S. Wogonin Influences Osteosarcoma Stem Cell Stemness Through ROS-dependent Signaling. *In Vivo* **2020**, *34*, 1077–1084. [CrossRef]
150. Huynh, D.L.; Kwon, T.; Zhang, J.J.; Sharma, N.; Gera, M.; Ghosh, M.; Kim, N.; Kim Cho, S.; Lee, D.S.; Park, Y.H.; et al. Wogonin suppresses stem cell-like traits of CD133 positive osteosarcoma cell via inhibiting matrix metallopeptidase-9 expression. *BMC Complement. Altern. Med.* **2017**, *17*, 304. [CrossRef]
151. Li, G.; Zhang, C.; Liang, W.; Zhang, Y.; Shen, Y.; Tian, X. Berberine regulates the Notch1/PTEN/PI3K/AKT/mTOR pathway and acts synergistically with 17-AAG and SAHA in SW480 colon cancer cells. *Pharm. Biol.* **2021**, *59*, 21–30. [CrossRef]
152. Hermawan, A.; Ikawati, M.; Jenie, R.I.; Khumaira, A.; Putri, H.; Nurhayati, I.P.; Angraini, S.M.; Muflikhasari, H.A. Identification of potential therapeutic target of naringenin in breast cancer stem cells inhibition by bioinformatics and in vitro studies. *Saudi Pharm. J. Off. Publ. Saudi Pharm. Soc.* **2021**, *29*, 12–26. [CrossRef]
153. Tajaldini, M.; Samadi, F.; Khosravi, A.; Ghasemnejad, A.; Asadi, J. Protective and anticancer effects of orange peel extract and naringin in doxorubicin treated esophageal cancer stem cell xenograft tumor mouse model. *Biomed. Pharmacother.* **2020**, *121*, 109594. [CrossRef] [PubMed]
154. Hermawan, A.; Ikawati, M.; Khumaira, A.; Putri, H.; Jenie, R.I.; Angraini, S.M.; Muflikhasari, H.A. Bioinformatics and In Vitro Studies Reveal the Importance of p53, PPARG and Notch Signaling Pathway in Inhibition of Breast Cancer Stem Cells by Hesperetin. *Adv. Pharm. Bull.* **2021**, *11*, 351–360. [CrossRef]
155. Farooqi, A.A.; Naureen, H.; Zahid, R.; Youssef, L.; Attar, R.; Xu, B. Cancer chemopreventive role of fisetin: Regulation of cell signaling pathways in different cancers. *Pharmacol. Res.* **2021**, *172*, 105784. [CrossRef]
156. Imran, M.; Saeed, F.; Gilani, S.A.; Shariati, M.A.; Imran, A.; Afzaal, M.; Atif, M.; Tufail, T.; Anjum, F.M. Fisetin: An anticancer perspective. *Food Sci. Nutr.* **2021**, *9*, 3–16. [CrossRef]
157. Ding, G.; Xu, X.; Li, D.; Chen, Y.; Wang, W.; Ping, D.; Jia, S.; Cao, L. Fisetin inhibits proliferation of pancreatic adenocarcinoma by inducing DNA damage via RFXAP/KDM4A-dependent histone H3K36 demethylation. *Cell Death Dis.* **2020**, *11*, 893. [CrossRef] [PubMed]
158. Si, Y.; Liu, J.; Shen, H.; Zhang, C.; Wu, Y.; Huang, Y.; Gong, Z.; Xue, J.; Liu, T. Fisetin decreases TET1 activity and CCNY/CDK16 promoter 5hmC levels to inhibit the proliferation and invasion of renal cancer stem cell. *J. Cell. Mol. Med.* **2019**, *23*, 1095–1105. [CrossRef]
159. Yin, Z.; Li, J.; Kang, L.; Liu, X.; Luo, J.; Zhang, L.; Li, Y.; Cai, J. Epigallocatechin-3-gallate induces autophagy-related apoptosis associated with LC3B II and Beclin expression of bladder cancer cells. *J. Food Biochem.* **2021**, *45*, e13758. [CrossRef] [PubMed]
160. Ding, F.; Yang, S. Epigallocatechin-3-gallate inhibits proliferation and triggers apoptosis in colon cancer via the hedgehog/phosphoinositide 3-kinase pathways. *Can. J. Physiol. Pharmacol.* **2021**, *99*. [CrossRef] [PubMed]
161. Panji, M.; Behmard, V.; Zare, Z.; Malekpour, M.; Nejadbiglari, H.; Yavari, S.; Nayerpour Dizaj, T.; Safaeian, A.; Maleki, N.; Abbasi, M.; et al. Suppressing effects of green tea extract and Epigallocatechin-3-gallate (EGCG) on TGF-β- induced Epithelial-to-mesenchymal transition via ROS/Smad signaling in human cervical cancer cells. *Gene* **2021**, *794*, 145774. [CrossRef]
162. Mokhtari, H.; Yaghmaei, B.; Sirati-Sabet, M.; Jafari, N.; Mardomi, A.; Abediankenari, S.; Mahrooz, A. Epigallocatechin-3-gallate Enhances the Efficacy of MicroRNA-34a Mimic and MicroRNA-93 Inhibitor Co-transfection in Prostate Cancer Cell Line. *Iran. J. Allergy Asthma Immunol.* **2020**, *19*, 612–623. [CrossRef] [PubMed]
163. Jiang, P.; Xu, C.; Zhang, P.; Ren, J.; Mageed, F.; Wu, X.; Chen, L.; Zeb, F.; Feng, Q.; Li, S. Epigallocatechin-3-gallate inhibits self-renewal ability of lung cancer stem-like cells through inhibition of CLOCK. *Int. J. Mol. Med.* **2020**, *46*, 2216–2224. [CrossRef]
164. Xia, L.; Li, F.; Qiu, J.; Feng, Z.; Xu, Z.; Chen, Z.; Sun, J. Oncogenic miR-20b-5p contributes to malignant behaviors of breast cancer stem cells by bidirectionally regulating CCND1 and E2F1. *BMC Cancer* **2020**, *20*, 949. [CrossRef] [PubMed]
165. Wei, Y.; Li, H.; Qu, Q. miR-484 suppresses endocrine therapy-resistant cells by inhibiting KLF4-induced cancer stem cells in estrogen receptor-positive cancers. *Breast Cancer* **2021**, *28*, 175–186. [CrossRef] [PubMed]
166. Zhou, Q.; Cui, F.; Lei, C.; Ma, S.; Huang, J.; Wang, X.; Qian, H.; Zhang, D.; Yang, Y. ATG7-mediated autophagy involves in miR-138-5p regulated self-renewal and invasion of lung cancer stem-like cells derived from A549 cells. *Anti-Cancer Drugs* **2021**, *32*, 376–385. [CrossRef]
167. Razi, S.; Sadeghi, A.; Asadi-Lari, Z.; Tam, K.J.; Kalantari, E.; Madjd, Z. DCLK1, a promising colorectal cancer stem cell marker, regulates tumor progression and invasion through miR-137 and miR-15a dependent manner. *Clin. Exp. Med.* **2021**, *21*, 139–147. [CrossRef]

168. Jiang, P.; Xu, C.; Chen, L.; Chen, A.; Wu, X.; Zhou, M.; Haq, I.U.; Mariyam, Z.; Feng, Q. Epigallocatechin-3-gallate inhibited cancer stem cell-like properties by targeting hsa-mir-485-5p/RXRα in lung cancer. *J. Cell. Biochem.* **2018**, *119*, 8623–8635. [CrossRef] [PubMed]
169. Zhu, J.; Jiang, Y.; Yang, X.; Wang, S.; Xie, C.; Li, X.; Li, Y.; Chen, Y.; Wang, X.; Meng, Y.; et al. Wnt/β-catenin pathway mediates (-)-Epigallocatechin-3-gallate (EGCG) inhibition of lung cancer stem cells. *Biochem. Biophys. Res. Commun.* **2017**, *482*, 15–21. [CrossRef] [PubMed]
170. Chen, Y.; Wang, X.Q.; Zhang, Q.; Zhu, J.Y.; Li, Y.; Xie, C.F.; Li, X.T.; Wu, J.S.; Geng, S.S.; Zhong, C.Y.; et al. (-)-Epigallocatechin-3-Gallate Inhibits Colorectal Cancer Stem Cells by Suppressing Wnt/β-Catenin Pathway. *Nutrients* **2017**, *9*, 572. [CrossRef]
171. Park, M.; Sundaramoorthy, P.; Sim, J.J.; Jeong, K.Y.; Kim, H.M. Synergistically Anti-metastatic Effect of 5-Flourouracil on Colorectal Cancer Cells via Calcium-mediated Focal Adhesion Kinase Proteolysis. *Anticancer Res.* **2017**, *37*, 103–114. [CrossRef] [PubMed]
172. Ashrafizadeh, M.; Zarrabi, A.; Hushmandi, K.; Hashemi, F.; Hashemi, F.; Samarghandian, S.; Najafi, M. MicroRNAs in cancer therapy: Their involvement in oxaliplatin sensitivity/resistance of cancer cells with a focus on colorectal cancer. *Life Sci.* **2020**, *256*, 117973. [CrossRef] [PubMed]
173. Toden, S.; Tran, H.M.; Tovar-Camargo, O.A.; Okugawa, Y.; Goel, A. Epigallocatechin-3-gallate targets cancer stem-like cells and enhances 5-fluorouracil chemosensitivity in colorectal cancer. *Oncotarget* **2016**, *7*, 16158–16171. [CrossRef] [PubMed]
174. Sun, X.; Song, J.; Li, E.; Geng, H.; Li, Y.; Yu, D.; Zhong, C. (-)-Epigallocatechin-3-gallate inhibits bladder cancer stem cells via suppression of sonic hedgehog pathway. *Oncol. Rep.* **2019**, *42*, 425–435. [CrossRef] [PubMed]
175. Abd El-Rahman, S.S.; Shehab, G.; Nashaat, H. Epigallocatechin-3-Gallate: The Prospective Targeting of Cancer Stem Cells and Preventing Metastasis of Chemically-Induced Mammary Cancer in Rats. *Am. J. Med. Sci.* **2017**, *354*, 54–63. [CrossRef]
176. Lee, S.H.; Nam, H.J.; Kang, H.J.; Kwon, H.W.; Lim, Y.C. Epigallocatechin-3-gallate attenuates head and neck cancer stem cell traits through suppression of Notch pathway. *Eur. J. Cancer* **2013**, *49*, 3210–3218. [CrossRef] [PubMed]
177. Li, Y.J.; Wu, S.L.; Lu, S.M.; Chen, F.; Guo, Y.; Gan, S.M.; Shi, Y.L.; Liu, S.; Li, S.L. (-)-Epigallocatechin-3-gallate inhibits nasopharyngeal cancer stem cell self-renewal and migration and reverses the epithelial-mesenchymal transition via NF-κB p65 inactivation. *Tumour Biol. Int. Soc. Oncodev. Biol. Med.* **2015**, *36*, 2747–2761. [CrossRef]
178. Wang, K.L.; Yu, Y.C.; Hsia, S.M. Perspectives on the Role of Isoliquiritigenin in Cancer. *Cancers* **2021**, *13*, 115. [CrossRef]
179. Chen, H.Y.; Chiang, Y.F.; Huang, J.S.; Huang, T.C.; Shih, Y.H.; Wang, K.L.; Ali, M.; Hong, Y.H.; Shieh, T.M.; Hsia, S.M. Isoliquiritigenin Reverses Epithelial-Mesenchymal Transition Through Modulation of the TGF-β/Smad Signaling Pathway in Endometrial Cancer. *Cancers* **2021**, *13*, 1236. [CrossRef]
180. Peng, F.; Tang, H.; Du, J.; Chen, J.; Peng, C. Isoliquiritigenin Suppresses EMT-Induced Metastasis in Triple-Negative Breast Cancer through miR-200c/C-JUN/[Formula: See text]-Catenin. *Am. J. Chin. Med.* **2021**, *49*, 505–523. [CrossRef]
181. Jin, H.; Kim, H.S.; Yu, S.T.; Shin, S.R.; Lee, S.H.; Seo, G.S. Synergistic anticancer effect of docosahexaenoic acid and isoliquiritigenin on human colorectal cancer cells through ROS-mediated regulation of the JNK and cytochrome c release. *Mol. Biol. Rep.* **2021**, *48*, 1171–1180. [CrossRef]
182. Wang, G.; Yu, Y.; Wang, Y.Z.; Yin, P.H.; Xu, K.; Zhang, H. The effects and mechanisms of isoliquiritigenin loaded nanoliposomes regulated AMPK/mTOR mediated glycolysis in colorectal cancer. *Artif. Cells Nanomed. Biotechnol.* **2020**, *48*, 1231–1249. [CrossRef]
183. Wang, N.; Wang, Z.; Peng, C.; You, J.; Shen, J.; Han, S.; Chen, J. Dietary compound isoliquiritigenin targets GRP78 to chemosensitize breast cancer stem cells via β-catenin/ABCG2 signaling. *Carcinogenesis* **2014**, *35*, 2544–2554. [CrossRef] [PubMed]
184. Hu, F.W.; Yu, C.C.; Hsieh, P.L.; Liao, Y.W.; Lu, M.Y.; Chu, P.M. Targeting oral cancer stemness and chemoresistance by isoliquiritigenin-mediated GRP78 regulation. *Oncotarget* **2017**, *8*, 93912–93923. [CrossRef] [PubMed]
185. Wang, N.; Wang, Z.; Wang, Y.; Xie, X.; Shen, J.; Peng, C.; You, J.; Peng, F.; Tang, H.; Guan, X.; et al. Dietary compound isoliquiritigenin prevents mammary carcinogenesis by inhibiting breast cancer stem cells through WIF1 demethylation. *Oncotarget* **2015**, *6*, 9854–9876. [CrossRef]
186. Green, J.M.; Alvero, A.B.; Kohen, F.; Mor, G. 7-(O)-Carboxymethyl daidzein conjugated to N-t-Boc-hexylenediamine: A novel compound capable of inducing cell death in epithelial ovarian cancer stem cells. *Cancer Biol. Ther.* **2009**, *8*, 1747–1753. [CrossRef] [PubMed]
187. Andrade, F.O.; Liu, F.; Zhang, X.; Rosim, M.P.; Dani, C.; Cruz, I.; Wang, T.T.Y.; Helferich, W.; Li, R.W.; Hilakivi-Clarke, L. Genistein Reduces the Risk of Local Mammary Cancer Recurrence and Ameliorates Alterations in the Gut Microbiota in the Offspring of Obese Dams. *Nutrients* **2021**, *13*, 201. [CrossRef] [PubMed]
188. Liu, X.; Wang, Q.; Liu, B.; Zheng, X.; Li, P.; Zhao, T.; Jin, X.; Ye, F.; Zhang, P.; Chen, W.; et al. Genistein inhibits radiation-induced invasion and migration of glioblastoma cells by blocking the DNA-PKcs/Akt2/Rac1 signaling pathway. *Radiother. Oncol. J. Eur. Soc. Ther. Radiol. Oncol.* **2021**, *155*, 93–104. [CrossRef]
189. Yu, Y.; Xing, Y.; Zhang, Q.; Zhang, Q.; Huang, S.; Li, X.; Gao, C. Soy isoflavone genistein inhibits hsa_circ_0031250/miR-873-5p/FOXM1 axis to suppress non-small-cell lung cancer progression. *IUBMB Life* **2021**, *73*, 92–107. [CrossRef] [PubMed]
190. Erten, F.; Yenice, E.; Orhan, C.; Er, B.; Demirel Öner, P.; Defo Deeh, P.B.; Şahin, K. Genistein suppresses the inflammation and GSK-3 pathway in an animal model of spontaneous ovarian cancer. *Turk. J. Med. Sci.* **2021**, *51*, 1465–1471. [CrossRef] [PubMed]
191. Dev, A.; Sardoiwala, M.N.; Kushwaha, A.C.; Karmakar, S.; Choudhury, S.R. Genistein nanoformulation promotes selective apoptosis in oral squamous cell carcinoma through repression of 3PK-EZH2 signalling pathway. *Phytomed. Int. J. Phytother. Phytopharm.* **2021**, *80*, 153386. [CrossRef]

192. Vodnik, V.V.; Mojić, M.; Stamenović, U.; Otoničar, M.; Ajdžanović, V.; Maksimović-Ivanić, D.; Mijatović, S.; Marković, M.M.; Barudžija, T.; Filipović, B.; et al. Development of genistein-loaded gold nanoparticles and their antitumor potential against prostate cancer cell lines. *Mater. Sci. Eng. C Mater. Biol. Appl.* **2021**, *124*, 112078. [CrossRef] [PubMed]
193. Zhang, Q.; Cao, W.S.; Wang, X.Q.; Zhang, M.; Lu, X.M.; Chen, J.Q.; Chen, Y.; Ge, M.M.; Zhong, C.Y.; Han, H.Y. Genistein inhibits nasopharyngeal cancer stem cells through sonic hedgehog signaling. *Phytother. Res.* **2019**, *33*, 2783–2791. [CrossRef]
194. Xie, C.; Zhu, J.; Jiang, Y.; Chen, J.; Wang, X.; Geng, S.; Wu, J.; Zhong, C.; Li, X.; Meng, Z. Sulforaphane Inhibits the Acquisition of Tobacco Smoke-Induced Lung Cancer Stem Cell-Like Properties via the IL-6/ΔNp63α/Notch Axis. *Theranostics* **2019**, *9*, 4827–4840. [CrossRef] [PubMed]
195. Liu, Y.; Zou, T.; Wang, S.; Chen, H.; Su, D.; Fu, X.; Zhang, Q.; Kang, X. Genistein-induced differentiation of breast cancer stem/progenitor cells through a paracrine mechanism. *Int. J. Oncol.* **2016**, *48*, 1063–1072. [CrossRef]
196. Huang, W.; Wan, C.; Luo, Q.; Huang, Z.; Luo, Q. Genistein-inhibited cancer stem cell-like properties and reduced chemoresistance of gastric cancer. *Int. J. Mol. Sci.* **2014**, *15*, 3432–3443. [CrossRef] [PubMed]
197. Cao, X.; Ren, K.; Song, Z.; Li, D.; Quan, M.; Zheng, Y.; Cao, J.; Zeng, W.; Zou, H. 7-Difluoromethoxyl-5,4′-di-n-octyl genistein inhibits the stem-like characteristics of gastric cancer stem-like cells and reverses the phenotype of epithelial-mesenchymal transition in gastric cancer cells. *Oncol. Rep.* **2016**, *36*, 1157–1165. [CrossRef] [PubMed]
198. Yu, D.; Shin, H.S.; Lee, Y.S.; Lee, D.; Kim, S.; Lee, Y.C. Genistein attenuates cancer stem cell characteristics in gastric cancer through the downregulation of Gli1. *Oncol. Rep.* **2014**, *31*, 673–678. [CrossRef]
199. Zhang, L.; Li, L.; Jiao, M.; Wu, D.; Wu, K.; Li, X.; Zhu, G.; Yang, L.; Wang, X.; Hsieh, J.T.; et al. Genistein inhibits the stemness properties of prostate cancer cells through targeting Hedgehog-Gli1 pathway. *Cancer Lett.* **2012**, *323*, 48–57. [CrossRef] [PubMed]
200. Fan, P.; Fan, S.; Wang, H.; Mao, J.; Shi, Y.; Ibrahim, M.M.; Ma, W.; Yu, X.; Hou, Z.; Wang, B.; et al. Genistein decreases the breast cancer stem-like cell population through Hedgehog pathway. *Stem Cell Res. Ther.* **2013**, *4*, 146. [CrossRef]
201. Ning, Y.X.; Li, Q.X.; Ren, K.Q.; Quan, M.F.; Cao, J.G. 7-difluoromethoxyl-5,4′-di-n-octyl genistein inhibits ovarian cancer stem cell characteristics through the downregulation of FOXM1. *Oncol. Lett.* **2014**, *8*, 295–300. [CrossRef] [PubMed]
202. Sekar, V.; Anandasadagopan, S.K.; Ganapasam, S. Genistein regulates tumor microenvironment and exhibits anticancer effect in dimethyl hydrazine-induced experimental colon carcinogenesis. *BioFactors* **2016**, *42*, 623–637. [CrossRef] [PubMed]
203. Ning, Y.; Xu, M.; Cao, X.; Chen, X.; Luo, X. Inactivation of AKT, ERK and NF-κB by genistein derivative, 7-difluoromethoxyl-5,4′-di-n-octylygenistein, reduces ovarian carcinoma oncogenicity. *Oncol. Rep.* **2017**, *38*, 949–958. [CrossRef] [PubMed]
204. Ning, Y.; Luo, C.; Ren, K.; Quan, M.; Cao, J. FOXO3a-mediated suppression of the self-renewal capacity of sphere-forming cells derived from the ovarian cancer SKOV3 cell line by 7-difluoromethoxyl-5,4′-di-n-octyl genistein. *Mol. Med. Rep.* **2014**, *9*, 1982–1988. [CrossRef] [PubMed]
205. De Francesco, E.M.; Ózsvári, B.; Sotgia, F.; Lisanti, M.P. Dodecyl-TPP Targets Mitochondria and Potently Eradicates Cancer Stem Cells (CSCs): Synergy With FDA-Approved Drugs and Natural Compounds (Vitamin C and Berberine). *Front. Oncol.* **2019**, *9*, 615. [CrossRef]
206. Wu, Y.S.; Ho, J.Y.; Yu, C.P.; Cho, C.J.; Wu, C.L.; Huang, C.S.; Gao, H.W.; Yu, D.S. Ellagic Acid Resensitizes Gemcitabine-Resistant Bladder Cancer Cells by Inhibiting Epithelial-Mesenchymal Transition and Gemcitabine Transporters. *Cancers* **2021**, *13*, 2032. [CrossRef]
207. Mohammed Saleem, Y.I.; Selim, M.I. MDM2 as a target for ellagic acid-mediated suppression of prostate cancer cells in vitro. *Oncol. Rep.* **2020**, *44*, 1255–1265. [CrossRef]
208. Elsaid, F.G.; Alshehri, M.A.; Shati, A.A.; Al-Kahtani, M.A.; Alsheri, A.S.; Massoud, E.E.; El-Kott, A.F.; El-Mekkawy, H.I.; Al-Ramlawy, A.M.; Abdraboh, M.E. The anti-tumourigenic effect of ellagic acid in SKOV-3 ovarian cancer cells entails activation of autophagy mediated by inhibiting Akt and activating AMPK. *Clin. Exp. Pharmacol. Physiol.* **2020**, *47*, 1611–1621. [CrossRef]
209. Yoganathan, S.; Alagaratnam, A.; Acharekar, N.; Kong, J. Ellagic Acid and Schisandrins: Natural Biaryl Polyphenols with Therapeutic Potential to Overcome Multidrug Resistance in Cancer. *Cells* **2021**, *10*, 458. [CrossRef]
210. Wang, N.; Wang, Q.; Tang, H.; Zhang, F.; Zheng, Y.; Wang, S.; Zhang, J.; Wang, Z.; Xie, X. Direct inhibition of ACTN4 by ellagic acid limits breast cancer metastasis via regulation of β-catenin stabilization in cancer stem cells. *J. Exp. Clin. Cancer Res.* **2017**, *36*, 172. [CrossRef]
211. Núñez-Sánchez, M.; Karmokar, A.; González-Sarrías, A.; García-Villalba, R.; Tomás-Barberán, F.A.; García-Conesa, M.T.; Brown, K.; Espín, J.C. In vivo relevant mixed urolithins and ellagic acid inhibit phenotypic and molecular colon cancer stem cell features: A new potentiality for ellagitannin metabolites against cancer. *Food Chem. Toxicol.* **2016**, *92*, 8–16. [CrossRef]
212. Mirzaei, S.; Gholami, M.H.; Zabolian, A.; Saleki, H.; Farahani, M.V.; Hamzehlou, S.; Far, F.B.; Sharifzadeh, S.O.; Samarghandian, S.; Khan, H.; et al. Caffeic acid and its derivatives as potential modulators of oncogenic molecular pathways: New hope in the fight against cancer. *Pharmacol. Res.* **2021**, *171*, 105759. [CrossRef] [PubMed]
213. Tyszka-Czochara, M.; Bukowska-Strakova, K.; Kocemba-Pilarczyk, K.A.; Majka, M. Caffeic Acid Targets AMPK Signaling and Regulates Tricarboxylic Acid Cycle Anaplerosis while Metformin Downregulates HIF-1α-Induced Glycolytic Enzymes in Human Cervical Squamous Cell Carcinoma Lines. *Nutrients* **2018**, *10*, 841. [CrossRef] [PubMed]
214. Teng, Y.N.; Wang, C.C.N.; Liao, W.C.; Lan, Y.H.; Hung, C.C. Caffeic Acid Attenuates Multi-Drug Resistance in Cancer Cells by Inhibiting Efflux Function of Human P-glycoprotein. *Molecules* **2020**, *25*, 247. [CrossRef] [PubMed]
215. Colpan, R.D.; Erdemir, A. Co-delivery of quercetin and caffeic-acid phenethyl ester by polymeric nanoparticles for improved antitumor efficacy in colon cancer cells. *J. Microencapsul.* **2021**, *38*, 381–393. [CrossRef] [PubMed]

216. Sari, C.; Sümer, C.; Celep Eyüpoğlu, F. Caffeic acid phenethyl ester induces apoptosis in colorectal cancer cells via inhibition of survivin. *Turk. J. Biol.* **2020**, *44*, 264–274. [CrossRef]
217. Park, S.R.; Kim, S.R.; Hong, I.S.; Lee, H.Y. A Novel Therapeutic Approach for Colorectal Cancer Stem Cells: Blocking the PI3K/Akt Signaling Axis with Caffeic Acid. *Front. Cell Dev. Biol.* **2020**, *8*, 585987. [CrossRef]
218. Yang, Y.; Li, Y.; Wang, K.; Wang, Y.; Yin, W.; Li, L. P38/NF-κB/snail pathway is involved in caffeic acid-induced inhibition of cancer stem cells-like properties and migratory capacity in malignant human keratinocyte. *PLoS ONE* **2013**, *8*, e58915. [CrossRef]
219. Omene, C.O.; Wu, J.; Frenkel, K. Caffeic Acid Phenethyl Ester (CAPE) derived from propolis, a honeybee product, inhibits growth of breast cancer stem cells. *Investig. New Drugs* **2012**, *30*, 1279–1288. [CrossRef]
220. Li, Y.; Jiang, F.; Chen, L.; Yang, Y.; Cao, S.; Ye, Y.; Wang, X.; Mu, J.; Li, Z.; Li, L. Blockage of TGFβ-SMAD2 by demethylation-activated miR-148a is involved in caffeic acid-induced inhibition of cancer stem cell-like properties in vitro and in vivo. *FEBS Open Bio* **2015**, *5*, 466–475. [CrossRef]
221. Ashrafizadeh, M.; Ahmadi, Z.; Farkhondeh, T.; Samarghandian, S. Autophagy regulation using luteolin: New insight into its anti-tumor activity. *Cancer Cell Int.* **2020**, *20*, 537. [CrossRef] [PubMed]
222. Wu, H.T.; Lin, J.; Liu, Y.E.; Chen, H.F.; Hsu, K.W.; Lin, S.H.; Peng, K.Y.; Lin, K.J.; Hsieh, C.C.; Chen, D.R. Luteolin suppresses androgen receptor-positive triple-negative breast cancer cell proliferation and metastasis by epigenetic regulation of MMP9 expression via the AKT/mTOR signaling pathway. *Phytomed. Int. J. Phytother. Phytopharm.* **2021**, *81*, 153437. [CrossRef]
223. Cao, D.; Zhu, G.Y.; Lu, Y.; Yang, A.; Chen, D.; Huang, H.J.; Peng, S.X.; Chen, L.W.; Li, Y.W. Luteolin suppresses epithelial-mesenchymal transition and migration of triple-negative breast cancer cells by inhibiting YAP/TAZ activity. *Biomed. Pharmacother.* **2020**, *129*, 110462. [CrossRef] [PubMed]
224. Chen, Y.; Wang, M.H.; Zhu, J.Y.; Xie, C.F.; Li, X.T.; Wu, J.S.; Geng, S.S.; Han, H.Y.; Zhong, C.Y. TAp63α targeting of Lgr5 mediates colorectal cancer stem cell properties and sulforaphane inhibition. *Oncogenesis* **2020**, *9*, 89. [CrossRef] [PubMed]
225. Tu, D.G.; Lin, W.T.; Yu, C.C.; Lee, S.S.; Peng, C.Y.; Lin, T.; Yu, C.H. Chemotherapeutic effects of luteolin on radio-sensitivity enhancement and interleukin-6/signal transducer and activator of transcription 3 signaling repression of oral cancer stem cells. *J. Formos. Med. Assoc.* **2016**, *115*, 1032–1038. [CrossRef] [PubMed]
226. Almeida, A.F.; Borge, G.I.A.; Piskula, M.; Tudose, A.; Tudoreanu, L.; Valentová, K.; Williamson, G.; Santos, C.N. Bioavailability of quercetin in humans with a focus on interindividual variation. *Compr. Rev. Food Sci. Food Saf.* **2018**, *17*, 714–731. [CrossRef]
227. Xiao, L.; Luo, G.; Tang, Y.; Yao, P. Quercetin and iron metabolism: What we know and what we need to know. *Food Chem. Toxicol.* **2018**, *114*, 190–203. [CrossRef] [PubMed]
228. Ashrafizadeh, M.; Ahmadi, Z.; Farkhondeh, T.; Samarghandian, S. Autophagy as a molecular target of quercetin underlying its protective effects in human diseases. *Arch. Physiol. Biochem.* **2019**. [CrossRef] [PubMed]
229. Ashrafizadeh, M.; Samarghandian, S.; Hushmandi, K.; Zabolian, A.; Shahinozzaman, M.; Saleki, H.; Esmaeili, H.; Raei, M.; Entezari, M.; Zarrabi, A.; et al. Quercetin in attenuation of ischemic/reperfusion injury: A review. *Curr. Mol. Pharmacol.* **2020**. [CrossRef]
230. Yousuf, M.; Khan, P.; Shamsi, A.; Shahbaaz, M.; Hasan, G.M.; Haque, Q.M.R.; Christoffels, A.; Islam, A.; Hassan, M.I. Inhibiting CDK6 Activity by Quercetin Is an Attractive Strategy for Cancer Therapy. *ACS Omega* **2020**, *5*, 27480–27491. [CrossRef] [PubMed]
231. Wang, Z.X.; Ma, J.; Li, X.Y.; Wu, Y.; Shi, H.; Chen, Y.; Lu, G.; Shen, H.M.; Lu, G.D.; Zhou, J. Quercetin induces p53-independent cancer cell death through lysosome activation by the transcription factor EB and Reactive Oxygen Species-dependent ferroptosis. *Br. J. Pharmacol.* **2021**, *178*, 1133–1148. [CrossRef]
232. Liu, M.; Fu, M.; Yang, X.; Jia, G.; Shi, X.; Ji, J.; Liu, X.; Zhai, G. Paclitaxel and quercetin co-loaded functional mesoporous silica nanoparticles overcoming multidrug resistance in breast cancer. *Colloids Surf. B Biointerfaces* **2020**, *196*, 111284. [CrossRef]
233. Li, X.; Zhou, N.; Wang, J.; Liu, Z.; Wang, X.; Zhang, Q.; Liu, Q.; Gao, L.; Wang, R. Quercetin suppresses breast cancer stem cells (CD44(+)/CD24(-)) by inhibiting the PI3K/Akt/mTOR-signaling pathway. *Life Sci.* **2018**, *196*, 56–62. [CrossRef]
234. Wang, R.; Yang, L.; Li, S.; Ye, D.; Yang, L.; Liu, Q.; Zhao, Z.; Cai, Q.; Tan, J.; Li, X. Quercetin Inhibits Breast Cancer Stem Cells via Downregulation of Aldehyde Dehydrogenase 1A1 (ALDH1A1), Chemokine Receptor Type 4 (CXCR4), Mucin 1 (MUC1), and Epithelial Cell Adhesion Molecule (EpCAM). *Med. Sci. Monit. Int. Med. J. Exp. Clin. Res.* **2018**, *24*, 412–420. [CrossRef]
235. Li, S.; Zhao, Q.; Wang, B.; Yuan, S.; Wang, X.; Li, K. Quercetin reversed MDR in breast cancer cells through down-regulating P-gp expression and eliminating cancer stem cells mediated by YB-1 nuclear translocation. *Phytother. Res.* **2018**, *32*, 1530–1536. [CrossRef] [PubMed]
236. Li, S.; Yuan, S.; Zhao, Q.; Wang, B.; Wang, X.; Li, K. Quercetin enhances chemotherapeutic effect of doxorubicin against human breast cancer cells while reducing toxic side effects of it. *Biomed. Pharmacother.* **2018**, *100*, 441–447. [CrossRef]
237. Srivastava, R.K.; Tang, S.N.; Zhu, W.; Meeker, D.; Shankar, S. Sulforaphane synergizes with quercetin to inhibit self-renewal capacity of pancreatic cancer stem cells. *Front. Biosci. (Elite Ed.)* **2011**, *3*, 515–528. [CrossRef]
238. Nwaeburu, C.C.; Abukiwan, A.; Zhao, Z.; Herr, I. Quercetin-induced miR-200b-3p regulates the mode of self-renewing divisions in pancreatic cancer. *Mol. Cancer* **2017**, *16*, 23. [CrossRef]
239. Appari, M.; Babu, K.R.; Kaczorowski, A.; Gross, W.; Herr, I. Sulforaphane, quercetin and catechins complement each other in elimination of advanced pancreatic cancer by miR-let-7 induction and K-ras inhibition. *Int. J. Oncol.* **2014**, *45*, 1391–1400. [CrossRef]
240. Zhou, W.; Kallifatidis, G.; Baumann, B.; Rausch, V.; Mattern, J.; Gladkich, J.; Giese, N.; Moldenhauer, G.; Wirth, T.; Büchler, M.W.; et al. Dietary polyphenol quercetin targets pancreatic cancer stem cells. *Int. J. Oncol.* **2010**, *37*, 551–561. [CrossRef] [PubMed]

241. Li, Y.; Wang, Z.; Jin, J.; Zhu, S.X.; He, G.Q.; Li, S.H.; Wang, J.; Cai, Y. Quercetin pretreatment enhances the radiosensitivity of colon cancer cells by targeting Notch-1 pathway. *Biochem. Biophys. Res. Commun.* **2020**, *523*, 947–953. [CrossRef] [PubMed]
242. Atashpour, S.; Fouladdel, S.; Movahhed, T.K.; Barzegar, E.; Ghahremani, M.H.; Ostad, S.N.; Azizi, E. Quercetin induces cell cycle arrest and apoptosis in CD133(+) cancer stem cells of human colorectal HT29 cancer cell line and enhances anticancer effects of doxorubicin. *Iran. J. Basic Med. Sci.* **2015**, *18*, 635–643.
243. Sikka, S.; Chen, L.; Sethi, G.; Kumar, A.P. Targeting PPARγ Signaling Cascade for the Prevention and Treatment of Prostate Cancer. *PPAR Res.* **2012**, *2012*, 968040. [CrossRef]
244. Hussain, Y.; Mirzaei, S.; Ashrafizadeh, M.; Zarrabi, A.; Hushmandi, K.; Khan, H.; Daglia, M. Quercetin and Its Nano-Scale Delivery Systems in Prostate Cancer Therapy: Paving the Way for Cancer Elimination and Reversing Chemoresistance. *Cancers* **2021**, *13*, 1602. [CrossRef] [PubMed]
245. Soleymani, L.; Zarrabi, A.; Hashemi, F.; Zabolian, A.; Banihashemi, S.; Moghadam, S.; Hushmandi, K.; Samarghandian, S.; Ashrafizadeh, M.; Khan, H. Role of ZEB family members in proliferation, metastasis and chemoresistance of prostate cancer cells: Revealing signaling networks. *Curr. Cancer Drug Targets* **2021**. [CrossRef]
246. Zhang, J.; Ahn, K.S.; Kim, C.; Shanmugam, M.K.; Siveen, K.S.; Arfuso, F.; Samym, R.P.; Deivasigamanim, A.; Lim, L.H.; Wang, L.; et al. Nimbolide-Induced Oxidative Stress Abrogates STAT3 Signaling Cascade and Inhibits Tumor Growth in Transgenic Adenocarcinoma of Mouse Prostate Model. *Antioxid. Redox Signal.* **2016**, *24*, 575–589. [CrossRef] [PubMed]
247. Lee, J.H.; Kim, C.; Baek, S.H.; Ko, J.H.; Lee, S.G.; Yang, W.M.; Um, J.Y.; Sethi, G.; Ahn, K.S. Capsazepine inhibits JAK/STAT3 signaling, tumor growth, and cell survival in prostate cancer. *Oncotarget* **2017**, *8*, 17700–17711. [CrossRef]
248. Erdogan, S.; Turkekul, K.; Dibirdik, I.; Doganlar, O.; Doganlar, Z.B.; Bilir, A.; Oktem, G. Midkine downregulation increases the efficacy of quercetin on prostate cancer stem cell survival and migration through PI3K/AKT and MAPK/ERK pathway. *Biomed. Pharmacother.* **2018**, *107*, 793–805. [CrossRef]
249. Tang, S.N.; Singh, C.; Nall, D.; Meeker, D.; Shankar, S.; Srivastava, R.K. The dietary bioflavonoid quercetin synergizes with epigallocatechin gallate (EGCG) to inhibit prostate cancer stem cell characteristics, invasion, migration and epithelial-mesenchymal transition. *J. Mol. Signal.* **2010**, *5*, 14. [CrossRef] [PubMed]
250. Chen, S.F.; Nieh, S.; Jao, S.W.; Liu, C.L.; Wu, C.H.; Chang, Y.C.; Yang, C.Y.; Lin, Y.S. Quercetin suppresses drug-resistant spheres via the p38 MAPK-Hsp27 apoptotic pathway in oral cancer cells. *PLoS ONE* **2012**, *7*, e49275. [CrossRef]
251. Shen, X.; Si, Y.; Wang, Z.; Wang, J.; Guo, Y.; Zhang, X. Quercetin inhibits the growth of human gastric cancer stem cells by inducing mitochondrial-dependent apoptosis through the inhibition of PI3K/Akt signaling. *Int. J. Mol. Med.* **2016**, *38*, 619–626. [CrossRef] [PubMed]
252. Han, H.; Lim, J.W.; Kim, H. Astaxanthin Inhibits Helicobacter pylori-induced Inflammatory and Oncogenic Responses in Gastric Mucosal Tissues of Mice. *J. Cancer Prev.* **2020**, *25*, 244–251. [CrossRef] [PubMed]
253. Cui, L.; Li, Z.; Xu, F.; Tian, Y.; Chen, T.; Li, J.; Guo, Y.; Lyu, Q. Antitumor Effects of Astaxanthin on Esophageal Squamous Cell Carcinoma by up-Regulation of PPARγ. *Nutr. Cancer* **2021**. [CrossRef]
254. Ahn, Y.T.; Kim, M.S.; Kim, Y.S.; An, W.G. Astaxanthin Reduces Stemness Markers in BT20 and T47D Breast Cancer Stem Cells by Inhibiting Expression of Pontin and Mutant p53. *Mar. Drugs* **2020**, *18*, 577. [CrossRef]
255. Lee, N.Y.; Kim, Y.; Kim, Y.S.; Shin, J.H.; Rubin, L.P.; Kim, Y. β-Carotene exerts anti-colon cancer effects by regulating M2 macrophages and activated fibroblasts. *J. Nutr. Biochem.* **2020**, *82*, 108402. [CrossRef] [PubMed]
256. Sowmya Shree, G.; Yogendra Prasad, K.; Arpitha, H.S.; Deepika, U.R.; Nawneet Kumar, K.; Mondal, P.; Ganesan, P. β-carotene at physiologically attainable concentration induces apoptosis and down-regulates cell survival and antioxidant markers in human breast cancer (MCF-7) cells. *Mol. Cell. Biochem.* **2017**, *436*, 1–12. [CrossRef]
257. Ngoc, N.B.; Lv, P.; Zhao, W.E. Suppressive effects of lycopene and β-carotene on the viability of the human esophageal squamous carcinoma cell line EC109. *Oncol. Lett.* **2018**, *15*, 6727–6732. [CrossRef] [PubMed]
258. Jain, A.; Sharma, G.; Kushwah, V.; Garg, N.K.; Kesharwani, P.; Ghoshal, G.; Singh, B.; Shivhare, U.S.; Jain, S.; Katare, O.P. Methotrexate and beta-carotene loaded-lipid polymer hybrid nanoparticles: A preclinical study for breast cancer. *Nanomedicine* **2017**, *12*, 1851–1872. [CrossRef] [PubMed]
259. Kim, D.; Kim, Y.; Kim, Y. Effects of β-carotene on Expression of Selected MicroRNAs, Histone Acetylation, and DNA Methylation in Colon Cancer Stem Cells. *J. Cancer Prev.* **2019**, *24*, 224–232. [CrossRef]
260. Lee, H.A.; Park, S.; Kim, Y. Effect of β-carotene on cancer cell stemness and differentiation in SK-N-BE(2)C neuroblastoma cells. *Oncol. Rep.* **2013**, *30*, 1869–1877. [CrossRef] [PubMed]
261. Kim, Y.S.; Gong, X.; Rubin, L.P.; Choi, S.W.; Kim, Y. β-Carotene 15,15′-oxygenase inhibits cancer cell stemness and metastasis by regulating differentiation-related miRNAs in human neuroblastoma. *J. Nutr. Biochem.* **2019**, *69*, 31–43. [CrossRef]
262. Lim, J.Y.; Kim, Y.S.; Kim, K.M.; Min, S.J.; Kim, Y. B-carotene inhibits neuroblastoma tumorigenesis by regulating cell differentiation and cancer cell stemness. *Biochem. Biophys. Res. Commun.* **2014**, *450*, 1475–1480. [CrossRef]
263. Kim, Y.S.; Kim, E.; Park, Y.J.; Kim, Y. Retinoic acid receptor β enhanced the anti-cancer stem cells effect of β-carotene by down-regulating expression of delta-like 1 homologue in human neuroblastoma cells. *Biochem. Biophys. Res. Commun.* **2016**, *480*, 254–260. [CrossRef]
264. Khan, S.; Awan, K.A.; Iqbal, M.J. Sulforaphane as a potential remedy against cancer: Comprehensive mechanistic review. *J. Food Biochem.* **2021**, e13886. [CrossRef]

265. Luo, Y.; Yan, B.; Liu, L.; Yin, L.; Ji, H.; An, X.; Gladkich, J.; Qi, Z.; De La Torre, C.; Herr, I. Sulforaphane Inhibits the Expression of Long Noncoding RNA H19 and Its Target APOBEC3G and Thereby Pancreatic Cancer Progression. *Cancers* **2021**, *13*, 827. [CrossRef]
266. Li, J.; Zhou, Y.; Yan, Y.; Zheng, Z.; Hu, Y.; Wu, W. Sulforaphane-cysteine downregulates CDK4/CDK6 and inhibits tubulin polymerization contributing to cell cycle arrest and apoptosis in human glioblastoma cells. *Aging* **2020**, *12*, 16837–16851. [CrossRef] [PubMed]
267. Wang, F.; Liu, P.; An, H.; Zhang, Y. Sulforaphane suppresses the viability and metastasis, and promotes the apoptosis of bladder cancer cells by inhibiting the expression of FAT-1. *Int. J. Mol. Med.* **2020**, *46*, 1085–1095. [CrossRef] [PubMed]
268. Rong, Y.; Huang, L.; Yi, K.; Chen, H.; Liu, S.; Zhang, W.; Yuan, C.; Song, X.; Wang, F. Co-administration of sulforaphane and doxorubicin attenuates breast cancer growth by preventing the accumulation of myeloid-derived suppressor cells. *Cancer Lett.* **2020**, *493*, 189–196. [CrossRef] [PubMed]
269. Ezeka, G.; Adhikary, G.; Kandasamy, S.; Friedberg, J.S.; Eckert, R.L. Sulforaphane inhibits PRMT5 and MEP50 function to suppress the mesothelioma cancer cell phenotype. *Mol. Carcinog.* **2021**, *60*, 429–439. [CrossRef]
270. Ge, M.; Zhang, L.; Cao, L.; Xie, C.; Li, X.; Li, Y.; Meng, Y.; Chen, Y.; Wang, X.; Chen, J.; et al. Sulforaphane inhibits gastric cancer stem cells via suppressing sonic hedgehog pathway. *Int. J. Food Sci. Nutr.* **2019**, *70*, 570–578. [CrossRef] [PubMed]
271. Dai, X.; Zhang, J.; Arfuso, F.; Chinnathambi, A.; Zayed, M.E.; Alharbi, S.A.; Kumar, A.P.; Ahn, K.S.; Sethi, G. Targeting TNF-related apoptosis-inducing ligand (TRAIL) receptor by natural products as a potential therapeutic approach for cancer therapy. *Exp. Biol. Med.* **2015**, *240*, 760–773. [CrossRef]
272. Stöhr, D.; Schmid, J.O.; Beigl, T.B.; Mack, A.; Maichl, D.S.; Cao, K.; Budai, B.; Fullstone, G.; Kontermann, R.E.; Mürdter, T.E.; et al. Stress-induced TRAILR2 expression overcomes TRAIL resistance in cancer cell spheroids. *Cell Death Differ.* **2020**, *27*, 3037–3052. [CrossRef]
273. Bauer, J.A.; Lupica, J.A.; Didonato, J.A.; Lindner, D.J. Nitric Oxide Inhibits NF-κB-mediated Survival Signaling: Possible Role in Overcoming TRAIL Resistance. *Anticancer Res.* **2020**, *40*, 6751–6763. [CrossRef]
274. Watanabe, A.; Miyake, K.; Akahane, K.; Goi, K.; Kagami, K.; Yagita, H.; Inukai, T. Epigenetic Modification of Death Receptor Genes for TRAIL and TRAIL Resistance in Childhood B-Cell Precursor Acute Lymphoblastic Leukemia. *Genes* **2021**, *12*, 864. [CrossRef] [PubMed]
275. She, T.; Shi, Q.; Li, Z.; Feng, Y.; Yang, H.; Tao, Z.; Li, H.; Chen, J.; Wang, S.; Liang, Y.; et al. Combination of long-acting TRAIL and tumor cell-targeted photodynamic therapy as a novel strategy to overcome chemotherapeutic multidrug resistance and TRAIL resistance of colorectal cancer. *Theranostics* **2021**, *11*, 4281–4297. [CrossRef]
276. Stuckey, D.W.; Shah, K. TRAIL on trial: Preclinical advances in cancer therapy. *Trends Mol. Med.* **2013**, *19*, 685–694. [CrossRef]
277. Puar, Y.R.; Shanmugam, M.K.; Fan, L.; Arfuso, F.; Sethi, G.; Tergaonkar, V. Evidence for the Involvement of the Master Transcription Factor NF-κB in Cancer Initiation and Progression. *Biomedicines* **2018**, *6*, 82. [CrossRef] [PubMed]
278. Labsch, S.; Liu, L.; Bauer, N.; Zhang, Y.; Aleksandrowicz, E.; Gladkich, J.; Schönsiegel, F.; Herr, I. Sulforaphane and TRAIL induce a synergistic elimination of advanced prostate cancer stem-like cells. *Int. J. Oncol.* **2014**, *44*, 1470–1480. [CrossRef] [PubMed]
279. Kallifatidis, G.; Labsch, S.; Rausch, V.; Mattern, J.; Gladkich, J.; Moldenhauer, G.; Büchler, M.W.; Salnikov, A.V.; Herr, I. Sulforaphane increases drug-mediated cytotoxicity toward cancer stem-like cells of pancreas and prostate. *Mol. Ther. J. Am. Soc. Gene Ther.* **2011**, *19*, 188–195. [CrossRef] [PubMed]
280. Rausch, V.; Liu, L.; Kallifatidis, G.; Baumann, B.; Mattern, J.; Gladkich, J.; Wirth, T.; Schemmer, P.; Büchler, M.W.; Zöller, M.; et al. Synergistic activity of sorafenib and sulforaphane abolishes pancreatic cancer stem cell characteristics. *Cancer Res.* **2010**, *70*, 5004–5013. [CrossRef]
281. Huang, J.; Tao, C.; Yu, Y.; Yu, F.; Zhang, H.; Gao, J.; Wang, D.; Chen, Y.; Gao, J.; Zhang, G.; et al. Simultaneous Targeting of Differentiated Breast Cancer Cells and Breast Cancer Stem Cells by Combination of Docetaxel- and Sulforaphane-Loaded Self-Assembled Poly(D, L-lactide-co-glycolide)/Hyaluronic Acid Block Copolymer-Based Nanoparticles. *J. Biomed. Nanotechnol.* **2016**, *12*, 1463–1477. [CrossRef]
282. Zhu, J.; Wang, S.; Chen, Y.; Li, X.; Jiang, Y.; Yang, X.; Li, Y.; Wang, X.; Meng, Y.; Zhu, M.; et al. miR-19 targeting of GSK3β mediates sulforaphane suppression of lung cancer stem cells. *J. Nutr. Biochem.* **2017**, *44*, 80–91. [CrossRef] [PubMed]
283. Li, Y.; Zhang, T.; Korkaya, H.; Liu, S.; Lee, H.F.; Newman, B.; Yu, Y.; Clouthier, S.G.; Schwartz, S.J.; Wicha, M.S.; et al. Sulforaphane, a dietary component of broccoli/broccoli sprouts, inhibits breast cancer stem cells. *Clin. Cancer Res. Off. J. Am. Assoc. Cancer Res.* **2010**, *16*, 2580–2590. [CrossRef] [PubMed]
284. Gu, H.F.; Ren, F.; Mao, X.Y.; Du, M. Mineralized and GSH-responsive hyaluronic acid based nano-carriers for potentiating repressive effects of sulforaphane on breast cancer stem cells-like properties. *Carbohydr. Polym.* **2021**, *269*, 118294. [CrossRef]
285. Rodova, M.; Fu, J.; Watkins, D.N.; Srivastava, R.K.; Shankar, S. Sonic hedgehog signaling inhibition provides opportunities for targeted therapy by sulforaphane in regulating pancreatic cancer stem cell self-renewal. *PLoS ONE* **2012**, *7*, e46083. [CrossRef]
286. Burnett, J.P.; Lim, G.; Li, Y.; Shah, R.B.; Lim, R.; Paholak, H.J.; McDermott, S.P.; Sun, L.; Tsume, Y.; Bai, S.; et al. Sulforaphane enhances the anticancer activity of taxanes against triple negative breast cancer by killing cancer stem cells. *Cancer Lett.* **2017**, *394*, 52–64. [CrossRef]
287. Fisher, M.L.; Adhikary, G.; Grun, D.; Kaetzel, D.M.; Eckert, R.L. The Ezh2 polycomb group protein drives an aggressive phenotype in melanoma cancer stem cells and is a target of diet derived sulforaphane. *Mol. Carcinog.* **2016**, *55*, 2024–2036. [CrossRef] [PubMed]

288. Li, S.H.; Fu, J.; Watkins, D.N.; Srivastava, R.K.; Shankar, S. Sulforaphane regulates self-renewal of pancreatic cancer stem cells through the modulation of Sonic hedgehog-GLI pathway. *Mol. Cell. Biochem.* **2013**, *373*, 217–227. [CrossRef]
289. Bijangi-Vishehsaraei, K.; Reza Saadatzadeh, M.; Wang, H.; Nguyen, A.; Kamocka, M.M.; Cai, W.; Cohen-Gadol, A.A.; Halum, S.L.; Sarkaria, J.N.; Pollok, K.E.; et al. Sulforaphane suppresses the growth of glioblastoma cells, glioblastoma stem cell-like spheroids, and tumor xenografts through multiple cell signaling pathways. *J. Neurosurg.* **2017**, *127*, 1219–1230. [CrossRef]
290. Lin, L.C.; Yeh, C.T.; Kuo, C.C.; Lee, C.M.; Yen, G.C.; Wang, L.S.; Wu, C.H.; Yang, W.C.; Wu, A.T. Sulforaphane potentiates the efficacy of imatinib against chronic leukemia cancer stem cells through enhanced abrogation of Wnt/β-catenin function. *J. Agric. Food Chem.* **2012**, *60*, 7031–7039. [CrossRef] [PubMed]
291. Wang, F.; Wang, W.; Li, J.; Zhang, J.; Wang, X.; Wang, M. Sulforaphane reverses gefitinib tolerance in human lung cancer cells via modulation of sonic hedgehog signaling. *Oncol. Lett.* **2018**, *15*, 109–114. [CrossRef]
292. Liu, C.M.; Peng, C.Y.; Liao, Y.W.; Lu, M.Y.; Tsai, M.L.; Yeh, J.C.; Yu, C.H.; Yu, C.C. Sulforaphane targets cancer stemness and tumor initiating properties in oral squamous cell carcinomas via miR-200c induction. *J. Formos. Med. Assoc.* **2017**, *116*, 41–48. [CrossRef]
293. Kerr, C.; Adhikary, G.; Grun, D.; George, N.; Eckert, R.L. Combination cisplatin and sulforaphane treatment reduces proliferation, invasion, and tumor formation in epidermal squamous cell carcinoma. *Mol. Carcinog.* **2018**, *57*, 3–11. [CrossRef]
294. Trošelj, K.G.; Samaržija, I.; Tomljanović, M.; Kujundžić, R.N.; Đaković, N.; Mojzeš, A. Implementing Curcumin in Translational Oncology Research. *Molecules* **2020**, *25*, 5240. [CrossRef]
295. Deng, S.; Shanmugam, M.K.; Kumar, A.P.; Yap, C.T.; Sethi, G.; Bishayee, A. Targeting autophagy using natural compounds for cancer prevention and therapy. *Cancer* **2019**, *125*, 1228–1246. [CrossRef]
296. Goel, A.; Kunnumakkara, A.B.; Aggarwal, B.B. Curcumin as "Curecumin": From kitchen to clinic. *Biochem. Pharmacol.* **2008**, *75*, 787–809. [CrossRef]
297. Termini, D.; Den Hartogh, D.J.; Jaglanian, A.; Tsiani, E. Curcumin against Prostate Cancer: Current Evidence. *Biomolecules* **2020**, *10*, 1536. [CrossRef]
298. Zhu, G.; Shen, Q.; Jiang, H.; Ji, O.; Zhu, L.; Zhang, L. Curcumin inhibited the growth and invasion of human monocytic leukaemia SHI-1 cells in vivo by altering MAPK and MMP signalling. *Pharm. Biol.* **2020**, *58*, 25–34. [CrossRef]
299. Patra, S.; Pradhan, B.; Nayak, R.; Behera, C.; Rout, L.; Jena, M.; Efferth, T.; Bhutia, S.K. Chemotherapeutic efficacy of curcumin and resveratrol against cancer: Chemoprevention, chemoprotection, drug synergism and clinical pharmacokinetics. *Semin. Cancer Biol.* **2021**, *73*, 310–320. [CrossRef]
300. Hu, Y.; Ran, M.; Wang, B.; Lin, Y.; Cheng, Y.; Zheng, S. Co-Delivery of Docetaxel and Curcumin via Nanomicelles for Enhancing Anti-Ovarian Cancer Treatment. *Int. J. Nanomed.* **2020**, *15*, 9703–9715. [CrossRef]
301. Ashrafizadeh, M.; Zarrabi, A.; Hashemi, F.; Zabolian, A.; Saleki, H.; Bagherian, M.; Azami, N.; Bejandi, A.K.; Hushmandi, K.; Ang, H.L. Polychemotherapy with curcumin and doxorubicin via biological nanoplatforms: Enhancing antitumor activity. *Pharmaceutics* **2020**, *12*, 1084. [CrossRef]
302. Aravind, S.R.; Lakshmi, S.; Ranjith, S.; Krishnan, L.K. Sustained release of curcumin from fibrin matrix induces cancer cell death and immunomodulation. *Biomed. Pharmacother.* **2021**, *133*, 110967. [CrossRef] [PubMed]
303. Zhu, J.Y.; Yang, X.; Chen, Y.; Jiang, Y.; Wang, S.J.; Li, Y.; Wang, X.Q.; Meng, Y.; Zhu, M.M.; Ma, X.; et al. Curcumin Suppresses Lung Cancer Stem Cells via Inhibiting Wnt/β-catenin and Sonic Hedgehog Pathways. *Phytother. Res.* **2017**, *31*, 680–688. [CrossRef] [PubMed]
304. Windmöller, B.A.; Beshay, M.; Helweg, L.P.; Flottmann, C.; Beermann, M.; Förster, C.; Wilkens, L.; Greiner, J.F.W.; Kaltschmidt, C.; Kaltschmidt, B. Novel Primary Human Cancer Stem-Like Cell Populations from Non-Small Cell Lung Cancer: Inhibition of Cell Survival by Targeting NF-κB and MYC Signaling. *Cells* **2021**, *10*, 1024. [CrossRef] [PubMed]
305. Geng, H.; Guo, W.; Feng, L.; Xie, D.; Bi, L.; Wang, Y.; Zhang, T.; Liang, Z.; Yu, D. Diallyl trisulfide inhibited tobacco smoke-mediated bladder EMT and cancer stem cell marker expression via the NF-κB pathway in vivo. *J. Int. Med. Res.* **2021**, *49*, 0300060521992900. [CrossRef] [PubMed]
306. Marquardt, J.U.; Gomez-Quiroz, L.; Arreguin Camacho, L.O.; Pinna, F.; Lee, Y.H.; Kitade, M.; Domínguez, M.P.; Castven, D.; Breuhahn, K.; Conner, E.A.; et al. Curcumin effectively inhibits oncogenic NF-κB signaling and restrains stemness features in liver cancer. *J. Hepatol.* **2015**, *63*, 661–669. [CrossRef] [PubMed]
307. Garg, M.; Shanmugam, M.K.; Bhardwaj, V.; Goel, A.; Gupta, R.; Sharma, A.; Baligar, P.; Kumar, A.P.; Goh, B.C.; Wang, L.; et al. The pleiotropic role of transcription factor STAT3 in oncogenesis and its targeting through natural products for cancer prevention and therapy. *Med. Res. Rev.* **2020**, *41*, 1291–1336. [CrossRef]
308. Lee, J.H.; Chiang, S.Y.; Nam, D.; Chung, W.S.; Lee, J.; Na, Y.S.; Sethi, G.; Ahn, K.S. Capillarisin inhibits constitutive and inducible STAT3 activation through induction of SHP-1 and SHP-2 tyrosine phosphatases. *Cancer Lett.* **2014**, *345*, 140–148. [CrossRef]
309. Mohan, C.D.; Bharathkumar, H.; Bulusu, K.C.; Pandey, V.; Rangappa, S.; Fuchs, J.E.; Shanmugam, M.K.; Dai, X.; Li, F.; Deivasigamani, A.; et al. Development of a novel azaspirane that targets the Janus kinase-signal transducer and activator of transcription (STAT) pathway in hepatocellular carcinoma in vitro and in vivo. *J. Biol. Chem.* **2014**, *289*, 34296–34307. [CrossRef]
310. Ashrafizadeh, M.; Gholami, M.H.; Mirzaei, S.; Zabolian, A.; Haddadi, A.; Farahani, M.V.; Kashani, S.H.; Hushmandi, K.; Najafi, M.; Zarrabi, A.; et al. Dual relationship between long non-coding RNAs and STAT3 signaling in different cancers: New insight to proliferation and metastasis. *Life Sci.* **2021**, *270*, 119006. [CrossRef]

311. Mirzaei, S.; Gholami, M.H.; Mahabady, M.K.; Nabavi, N.; Zabolian, A.; Banihashemi, S.M.; Haddadi, A.; Entezari, M.; Hushmandi, K.; Makvandi, P.; et al. Pre-clinical investigation of STAT3 pathway in bladder cancer: Paving the way for clinical translation. *Biomed. Pharmacother.* **2021**, *133*, 111077. [CrossRef]
312. Ashrafizadeh, M.; Zarrabi, A.; Orouei, S.; Zarrin, V.; Rahmani Moghadam, E.; Zabolian, A.; Mohammadi, S.; Hushmandi, K.; Gharehaghajlou, Y.; Makvandi, P.; et al. STAT3 pathway in gastric cancer: Signaling, therapeutic targeting and future prospects. *Biology* **2020**, *9*, 126. [CrossRef]
313. Arora, L.; Kumar, A.P.; Arfuso, F.; Chng, W.J.; Sethi, G. The Role of Signal Transducer and Activator of Transcription 3 (STAT3) and Its Targeted Inhibition in Hematological Malignancies. *Cancers* **2018**, *10*, 327. [CrossRef]
314. Kim, S.M.; Lee, J.H.; Sethi, G.; Kim, C.; Baek, S.H.; Nam, D.; Chung, W.S.; Kim, S.H.; Shim, B.S.; Ahn, K.S. Bergamottin, a natural furanocoumarin obtained from grapefruit juice induces chemosensitization and apoptosis through the inhibition of STAT3 signaling pathway in tumor cells. *Cancer Lett.* **2014**, *354*, 153–163. [CrossRef]
315. Kim, C.; Lee, S.G.; Yang, W.M.; Arfuso, F.; Um, J.Y.; Kumar, A.P.; Bian, J.; Sethi, G.; Ahn, K.S. Formononetin-induced oxidative stress abrogates the activation of STAT3/5 signaling axis and suppresses the tumor growth in multiple myeloma preclinical model. *Cancer Lett.* **2018**, *431*, 123–141. [CrossRef]
316. Lee, J.H.; Kim, C.; Kim, S.H.; Sethi, G.; Ahn, K.S. Farnesol inhibits tumor growth and enhances the anticancer effects of bortezomib in multiple myeloma xenograft mouse model through the modulation of STAT3 signaling pathway. *Cancer Lett.* **2015**, *360*, 280–293. [CrossRef] [PubMed]
317. Chen, Y.; Shao, Z.; Jiang, E.; Zhou, X.; Wang, L.; Wang, H.; Luo, X.; Chen, Q.; Liu, K.; Shang, Z. CCL21/CCR7 interaction promotes EMT and enhances the stemness of OSCC via a JAK2/STAT3 signaling pathway. *J. Cell. Physiol.* **2020**, *235*, 5995–6009. [CrossRef] [PubMed]
318. Pingali, P.; Wu, Y.J.; Boothello, R.; Sharon, C.; Li, H.; Sistla, S.; Sankaranarayanan, N.V.; Desai, U.R.; Le, A.T.; Doebele, R.C.; et al. High dose acetaminophen inhibits STAT3 and has free radical independent anti-cancer stem cell activity. *Neoplasia* **2021**, *23*, 348–359. [CrossRef] [PubMed]
319. Khan, A.Q.; Ahmed, E.I.; Elareer, N.; Fathima, H.; Prabhu, K.S.; Siveen, K.S.; Kulinski, M.; Azizi, F.; Dermime, S.; Ahmad, A.; et al. Curcumin-Mediated Apoptotic Cell Death in Papillary Thyroid Cancer and Cancer Stem-Like Cells through Targeting of the JAK/STAT3 Signaling Pathway. *Int. J. Mol. Sci.* **2020**, *21*, 438. [CrossRef] [PubMed]
320. Wang, D.; Kong, X.; Li, Y.; Qian, W.; Ma, J.; Wang, D.; Yu, D.; Zhong, C. Curcumin inhibits bladder cancer stem cells by suppressing Sonic Hedgehog pathway. *Biochem. Biophys. Res. Commun.* **2017**, *493*, 521–527. [CrossRef] [PubMed]
321. Huang, Y.T.; Lin, Y.W.; Chiu, H.M.; Chiang, B.H. Curcumin Induces Apoptosis of Colorectal Cancer Stem Cells by Coupling with CD44 Marker. *J. Agric. Food Chem.* **2016**, *64*, 2247–2253. [CrossRef] [PubMed]
322. Nayak, D.; Tripathi, N.; Kathuria, D.; Siddharth, S.; Nayak, A.; Bharatam, P.V.; Kundu, C. Quinacrine and curcumin synergistically increased the breast cancer stem cells death by inhibiting ABCG2 and modulating DNA damage repair pathway. *Int. J. Biochem. Cell Biol.* **2020**, *119*, 105682. [CrossRef]
323. Mirza, S.; Vasaiya, A.; Vora, H.; Jain, N.; Rawal, R. Curcumin Targets Circulating Cancer Stem Cells by Inhibiting Self-Renewal Efficacy in Non-Small Cell Lung Carcinoma. *Anti-Cancer Agents Med. Chem.* **2017**, *17*, 859–864. [CrossRef] [PubMed]
324. Wu, L.; Guo, L.; Liang, Y.; Liu, X.; Jiang, L.; Wang, L. Curcumin suppresses stem-like traits of lung cancer cells via inhibiting the JAK2/STAT3 signaling pathway. *Oncol. Rep.* **2015**, *34*, 3311–3317. [CrossRef] [PubMed]
325. Hahn, Y.I.; Saeidi, S.; Kim, S.J.; Park, S.Y.; Song, N.Y.; Zheng, J.; Kim, D.H.; Lee, H.B.; Han, W.; Noh, D.Y.; et al. STAT3 Stabilizes IKKα Protein through Direct Interaction in Transformed and Cancerous Human Breast Epithelial Cells. *Cancers* **2020**, *13*, 82. [CrossRef] [PubMed]
326. Chung, S.S.; Vadgama, J.V. Curcumin and epigallocatechin gallate inhibit the cancer stem cell phenotype via down-regulation of STAT3-NFκB signaling. *Anticancer Res.* **2015**, *35*, 39–46.
327. Lin, L.; Liu, Y.; Li, H.; Li, P.K.; Fuchs, J.; Shibata, H.; Iwabuchi, Y.; Lin, J. Targeting colon cancer stem cells using a new curcumin analogue, GO-Y030. *Br. J. Cancer* **2011**, *105*, 212–220. [CrossRef]
328. Zhou, Q.M.; Sun, Y.; Lu, Y.Y.; Zhang, H.; Chen, Q.L.; Su, S.B. Curcumin reduces mitomycin C resistance in breast cancer stem cells by regulating Bcl-2 family-mediated apoptosis. *Cancer Cell Int.* **2017**, *17*, 84. [CrossRef] [PubMed]
329. Mansouri, K.; Rasoulpoor, S.; Daneshkhah, A.; Abolfathi, S.; Salari, N.; Mohammadi, M.; Rasoulpoor, S.; Shabani, S. Clinical effects of curcumin in enhancing cancer therapy: A systematic review. *BMC Cancer* **2020**, *20*, 791. [CrossRef] [PubMed]
330. Yang, K.; Liao, Z.; Wu, Y.; Li, M.; Guo, T.; Lin, J.; Li, Y.; Hu, C. Curcumin and Glu-GNPs Induce Radiosensitivity against Breast Cancer Stem-Like Cells. *BioMed Res. Int.* **2020**, *2020*, 3189217. [CrossRef] [PubMed]
331. Bano, N.; Yadav, M.; Das, B.C. Differential Inhibitory Effects of Curcumin Between HPV+ve and HPV-ve Oral Cancer Stem Cells. *Front. Oncol.* **2018**, *8*, 412. [CrossRef] [PubMed]
332. Kuo, Y.C.; Wang, L.J.; Rajesh, R. Targeting human brain cancer stem cells by curcumin-loaded nanoparticles grafted with anti-aldehyde dehydrogenase and sialic acid: Colocalization of ALDH and CD44. *Mater. Sci. Eng. C Mater. Biol. Appl.* **2019**, *102*, 362–372. [CrossRef]
333. Li, X.; Wang, X.; Xie, C.; Zhu, J.; Meng, Y.; Chen, Y.; Li, Y.; Jiang, Y.; Yang, X.; Wang, S.; et al. Sonic hedgehog and Wnt/β-catenin pathways mediate curcumin inhibition of breast cancer stem cells. *Anti-Cancer Drugs* **2018**, *29*, 208–215. [CrossRef] [PubMed]

334. Buhrmann, C.; Kraehe, P.; Lueders, C.; Shayan, P.; Goel, A.; Shakibaei, M. Curcumin suppresses crosstalk between colon cancer stem cells and stromal fibroblasts in the tumor microenvironment: Potential role of EMT. *PLoS ONE* **2014**, *9*, e107514. [CrossRef] [PubMed]
335. Siddappa, G.; Kulsum, S.; Ravindra, D.R.; Kumar, V.V.; Raju, N.; Raghavan, N.; Sudheendra, H.V.; Sharma, A.; Sunny, S.P.; Jacob, T.; et al. Curcumin and metformin-mediated chemoprevention of oral cancer is associated with inhibition of cancer stem cells. *Mol. Carcinog.* **2017**, *56*, 2446–2460. [CrossRef] [PubMed]
336. Yu, Y.; Kanwar, S.S.; Patel, B.B.; Nautiyal, J.; Sarkar, F.H.; Majumdar, A.P. Elimination of Colon Cancer Stem-Like Cells by the Combination of Curcumin and FOLFOX. *Transl. Oncol.* **2009**, *2*, 321–328. [CrossRef] [PubMed]
337. Zhang, H.; Zheng, J.; Shen, H.; Huang, Y.; Liu, T.; Xi, H.; Chen, C. Curcumin Suppresses In Vitro Proliferation and Invasion of Human Prostate Cancer Stem Cells by Modulating DLK1-DIO3 Imprinted Gene Cluster MicroRNAs. *Genet. Test. Mol. Biomark.* **2018**, *22*, 43–50. [CrossRef] [PubMed]
338. Chung, S.S.; Dutta, P.; Chard, N.; Wu, Y.; Chen, Q.H.; Chen, G.; Vadgama, J. A novel curcumin analog inhibits canonical and non-canonical functions of telomerase through STAT3 and NF-κB inactivation in colorectal cancer cells. *Oncotarget* **2019**, *10*, 4516–4531. [CrossRef] [PubMed]
339. Yoshida, K.; Toden, S.; Ravindranathan, P.; Han, H.; Goel, A. Curcumin sensitizes pancreatic cancer cells to gemcitabine by attenuating PRC2 subunit EZH2, and the lncRNA PVT1 expression. *Carcinogenesis* **2017**, *38*, 1036–1046. [CrossRef] [PubMed]
340. Ashrafizadeh, M.; Najafi, M.; Orouei, S.; Zabolian, A.; Saleki, H.; Azami, N.; Sharifi, N.; Hushmandi, K.; Zarrabi, A.; Ahn, K.S. Resveratrol modulates transforming growth factor-beta (tgf-β) signaling pathway for disease therapy: A new insight into its pharmacological activities. *Biomedicines* **2020**, *8*, 261. [CrossRef] [PubMed]
341. Ashrafizadeh, M.; Zarrabi, A.; Najafi, M.; Samarghandian, S.; Mohammadinejad, R.; Ahn, K.S. Resveratrol targeting tau proteins, amyloid-beta aggregations, and their adverse effects: An updated review. *Phytother. Res.* **2020**, *34*, 2867–2888. [CrossRef] [PubMed]
342. Ashrafizadeh, M.; Taeb, S.; Haghi-Aminjan, H.; Afrashi, S.; Moloudi, K.; Musa, A.E.; Najafi, M.; Farhood, B. Resveratrol as an Enhancer of Apoptosis in Cancer: A Mechanistic Review. *Anti-Cancer Agents Med. Chem.* **2020**. [CrossRef]
343. Sun, X.; Xu, Q.; Zeng, L.; Xie, L.; Zhao, Q.; Xu, H.; Wang, X.; Jiang, N.; Fu, P.; Sang, M. Resveratrol suppresses the growth and metastatic potential of cervical cancer by inhibiting STAT3(Tyr705) phosphorylation. *Cancer Med.* **2020**, *9*, 8685–8700. [CrossRef] [PubMed]
344. Sun, X.; Fu, P.; Xie, L.; Chai, S.; Xu, Q.; Zeng, L.; Wang, X.; Jiang, N.; Sang, M. Resveratrol inhibits the progression of cervical cancer by suppressing the transcription and expression of HPV E6 and E7 genes. *Int. J. Mol. Med.* **2021**, *47*, 335–345. [CrossRef] [PubMed]
345. Ye, M.; Tian, H.; Lin, S.; Mo, J.; Li, Z.; Chen, X.; Liu, J. Resveratrol inhibits proliferation and promotes apoptosis via the androgen receptor splicing variant 7 and PI3K/AKT signaling pathway in LNCaP prostate cancer cells. *Oncol. Lett.* **2020**, *20*, 169. [CrossRef]
346. Li, W.; Li, C.; Ma, L.; Jin, F. Resveratrol inhibits viability and induces apoptosis in the small-cell lung cancer H446 cell line via the PI3K/Akt/c-Myc pathway. *Oncol. Rep.* **2020**, *44*, 1821–1830. [CrossRef] [PubMed]
347. Xiong, L.; Lin, X.M.; Nie, J.H.; Ye, H.S.; Liu, J. Resveratrol and its Nanoparticle suppress Doxorubicin/Docetaxel-resistant anaplastic Thyroid Cancer Cells in vitro and in vivo. *Nanotheranostics* **2021**, *5*, 143–154. [CrossRef] [PubMed]
348. Sun, H.; Zhang, T.; Liu, R.; Cao, W.; Zhang, Z.; Liu, Z.; Qian, W.; Wang, D.; Yu, D.; Zhong, C. Resveratrol Inhibition of Renal Cancer Stem Cell Characteristics and Modulation of the Sonic Hedgehog Pathway. *Nutr. Cancer* **2021**, *73*, 1157–1167. [CrossRef] [PubMed]
349. Ahmadi, Z.; Mohammadinejad, R.; Ashrafizadeh, M. Drug delivery systems for resveratrol, a non-flavonoid polyphenol: Emerging evidence in last decades. *J. Drug Deliv. Sci. Technol.* **2019**, *51*, 591–604. [CrossRef]
350. Ashrafizadeh, M.; Javanmardi, S.; Moradi-Ozarlou, M.; Mohammadinejad, R.; Farkhondeh, T.; Samarghandian, S.; Garg, M. Natural products and phytochemical nanoformulations targeting mitochondria in oncotherapy: An updated review on resveratrol. *Biosci. Rep.* **2020**, *40*, BSR20200257. [CrossRef] [PubMed]
351. Ashrafizadeh, M.; Rafiei, H.; Mohammadinejad, R.; Farkhondeh, T.; Samarghandian, S. Anti-tumor activity of resveratrol against gastric cancer: A review of recent advances with an emphasis on molecular pathways. *Cancer Cell Int.* **2021**, *21*, 66. [CrossRef] [PubMed]
352. Shin, H.J.; Han, J.M.; Choi, Y.S.; Jung, H.J. Pterostilbene Suppresses both Cancer Cells and Cancer Stem-Like Cells in Cervical Cancer with Superior Bioavailability to Resveratrol. *Molecules* **2020**, *25*, 228. [CrossRef] [PubMed]
353. Fu, Y.; Chang, H.; Peng, X.; Bai, Q.; Yi, L.; Zhou, Y.; Zhu, J.; Mi, M. Resveratrol inhibits breast cancer stem-like cells and induces autophagy via suppressing Wnt/β-catenin signaling pathway. *PLoS ONE* **2014**, *9*, e102535. [CrossRef] [PubMed]
354. Ashrafizadeh, M.; Ahmadi, Z.; Farkhondeh, T.; Samarghandian, S. Modulatory effects of statins on the autophagy: A therapeutic perspective. *J. Cell. Physiol.* **2020**, *235*, 3157–3168. [CrossRef] [PubMed]
355. Ashrafizadeh, M.; Zarrabi, A.; Orouei, S.; Hushmandi, K.; Hakimi, A.; Zabolian, A.; Daneshi, S.; Samarghandian, S.; Baradaran, B.; Najafi, M. MicroRNA-mediated autophagy regulation in cancer therapy: The role in chemoresistance/chemosensitivity. *Eur. J. Pharmacol.* **2020**, *892*, 173660. [CrossRef] [PubMed]
356. Zheng, B.; Zhou, C.; Qu, G.; Ren, C.; Yan, P.; Guo, W.; Yue, B. VEGFR2 Promotes Metastasis and PD-L2 Expression of Human Osteosarcoma Cells by Activating the STAT3 and RhoA-ROCK-LIMK2 Pathways. *Front. Oncol.* **2020**, *10*, 543562. [CrossRef] [PubMed]

357. Ge, X.; Liu, W.; Zhao, W.; Feng, S.; Duan, A.; Ji, C.; Shen, K.; Liu, W.; Zhou, J.; Jiang, D.; et al. Exosomal Transfer of LCP1 Promotes Osteosarcoma Cell Tumorigenesis and Metastasis by Activating the JAK2/STAT3 Signaling Pathway. *Mol. Ther. Nucleic Acids* **2020**, *21*, 900–915. [CrossRef] [PubMed]
358. Peng, L.; Jiang, D. Resveratrol eliminates cancer stem cells of osteosarcoma by STAT3 pathway inhibition. *PLoS ONE* **2018**, *13*, e0205918. [CrossRef] [PubMed]
359. Qin, T.; Cheng, L.; Xiao, Y.; Qian, W.; Li, J.; Wu, Z.; Wang, Z.; Xu, Q.; Duan, W.; Wong, L.; et al. NAF-1 Inhibition by Resveratrol Suppresses Cancer Stem Cell-Like Properties and the Invasion of Pancreatic Cancer. *Front. Oncol.* **2020**, *10*, 1038. [CrossRef]
360. Buhrmann, C.; Shayan, P.; Brockmueller, A.; Shakibaei, M. Resveratrol Suppresses Cross-Talk between Colorectal Cancer Cells and Stromal Cells in Multicellular Tumor Microenvironment: A Bridge between In Vitro and In Vivo Tumor Microenvironment Study. *Molecules* **2020**, *25*, 4292. [CrossRef]
361. Suh, J.; Kim, D.H.; Surh, Y.J. Resveratrol suppresses migration, invasion and stemness of human breast cancer cells by interfering with tumor-stromal cross-talk. *Arch. Biochem. Biophys.* **2018**, *643*, 62–71. [CrossRef]
362. Pouyafar, A.; Rezabakhsh, A.; Rahbarghazi, R.; Heydarabad, M.Z.; Shokrollahi, E.; Sokullu, E.; Khaksar, M.; Nourazarian, A.; Avci, Ç.B. Treatment of cancer stem cells from human colon adenocarcinoma cell line HT-29 with resveratrol and sulindac induced mesenchymal-endothelial transition rate. *Cell Tissue Res.* **2019**, *376*, 377–388. [CrossRef]
363. Ruíz, G.; Valencia-González, H.A.; León-Galicia, I.; García-Villa, E.; García-Carrancá, A.; Gariglio, P. Inhibition of RAD51 by siRNA and Resveratrol Sensitizes Cancer Stem Cells Derived from HeLa Cell Cultures to Apoptosis. *Stem Cells Int.* **2018**, *2018*, 2493869. [CrossRef] [PubMed]
364. Pradhan, R.; Chatterjee, S.; Hembram, K.C.; Sethy, C.; Mandal, M.; Kundu, C.N. Nano formulated Resveratrol inhibits metastasis and angiogenesis by reducing inflammatory cytokines in oral cancer cells by targeting tumor associated macrophages. *J. Nutr. Biochem.* **2021**, *92*, 108624. [CrossRef] [PubMed]
365. Jhaveri, A.; Luther, E.; Torchilin, V. The effect of transferrin-targeted, resveratrol-loaded liposomes on neurosphere cultures of glioblastoma: Implications for targeting tumour-initiating cells. *J. Drug Target.* **2019**, *27*, 601–613. [CrossRef] [PubMed]
366. Buhrmann, C.; Yazdi, M.; Popper, B.; Kunnumakkara, A.B.; Aggarwal, B.B.; Shakibaei, M. Induction of the Epithelial-to-Mesenchymal Transition of Human Colorectal Cancer by Human TNF-β (Lymphotoxin) and its Reversal by Resveratrol. *Nutrients* **2019**, *11*, 704. [CrossRef] [PubMed]
367. Su, Y.C.; Li, S.C.; Wu, Y.C.; Wang, L.M.; Chao, K.S.; Liao, H.F. Resveratrol downregulates interleukin-6-stimulated sonic hedgehog signaling in human acute myeloid leukemia. *Evid.-Based Complement. Altern. Med.* **2013**, *2013*, 547430. [CrossRef]
368. Zhou, C.; Qian, W.; Ma, J.; Cheng, L.; Jiang, Z.; Yan, B.; Li, J.; Duan, W.; Sun, L.; Cao, J.; et al. Resveratrol enhances the chemotherapeutic response and reverses the stemness induced by gemcitabine in pancreatic cancer cells via targeting SREBP1. *Cell Prolif.* **2019**, *52*, e12514. [CrossRef]
369. Shankar, S.; Nall, D.; Tang, S.N.; Meeker, D.; Passarini, J.; Sharma, J.; Srivastava, R.K. Resveratrol inhibits pancreatic cancer stem cell characteristics in human and KrasG12D transgenic mice by inhibiting pluripotency maintaining factors and epithelial-mesenchymal transition. *PLoS ONE* **2011**, *6*, e16530. [CrossRef]
370. Lu, K.H.; Chen, Y.W.; Tsai, P.H.; Tsai, M.L.; Lee, Y.Y.; Chiang, C.Y.; Kao, C.L.; Chiou, S.H.; Ku, H.H.; Lin, C.H.; et al. Evaluation of radiotherapy effect in resveratrol-treated medulloblastoma cancer stem-like cells. *Childs Nerv. Syst.* **2009**, *25*, 543–550. [CrossRef]
371. Seino, M.; Okada, M.; Shibuya, K.; Seino, S.; Suzuki, S.; Takeda, H.; Ohta, T.; Kurachi, H.; Kitanaka, C. Differential contribution of ROS to resveratrol-induced cell death and loss of self-renewal capacity of ovarian cancer stem cells. *Anticancer Res.* **2015**, *35*, 85–96.
372. Dewangan, J.; Tandon, D.; Srivastava, S.; Verma, A.K.; Yapuri, A.; Rath, S.K. Novel combination of salinomycin and resveratrol synergistically enhances the anti-proliferative and pro-apoptotic effects on human breast cancer cells. *Apoptosis Int. J. Program. Cell Death* **2017**, *22*, 1246–1259. [CrossRef]
373. Shen, Y.A.; Lin, C.H.; Chi, W.H.; Wang, C.Y.; Hsieh, Y.T.; Wei, Y.H.; Chen, Y.J. Resveratrol Impedes the Stemness, Epithelial-Mesenchymal Transition, and Metabolic Reprogramming of Cancer Stem Cells in Nasopharyngeal Carcinoma through p53 Activation. *Evid.-Based Complement. Altern. Med.* **2013**, *2013*, 590393. [CrossRef]
374. Buhrmann, C.; Yazdi, M.; Popper, B.; Shayan, P.; Goel, A.; Aggarwal, B.B.; Shakibaei, M. Resveratrol Chemosensitizes TNF-β-Induced Survival of 5-FU-Treated Colorectal Cancer Cells. *Nutrients* **2018**, *10*, 888. [CrossRef]
375. Song, D.; Hao, J.; Fan, D. Biological properties and clinical applications of berberine. *Front. Med.* **2020**, *14*, 564–582. [CrossRef]
376. Zhou, M.; Deng, Y.; Liu, M.; Liao, L.; Dai, X.; Guo, C.; Zhao, X.; He, L.; Peng, C.; Li, Y. The pharmacological activity of berberine, a review for liver protection. *Eur. J. Pharmacol.* **2020**, *890*, 173655. [CrossRef]
377. Chuang, T.C.; Wu, K.; Lin, Y.Y.; Kuo, H.P.; Kao, M.C.; Wang, V.; Hsu, S.C.; Lee, S.L. Dual down-regulation of EGFR and ErbB2 by berberine contributes to suppression of migration and invasion of human ovarian cancer cells. *Environ. Toxicol.* **2021**, *36*, 737–747. [CrossRef] [PubMed]
378. Liu, J.; Luo, X.; Guo, R.; Jing, W.; Lu, H. Cell Metabolomics Reveals Berberine-Inhibited Pancreatic Cancer Cell Viability and Metastasis by Regulating Citrate Metabolism. *J. Proteome Res.* **2020**, *19*, 3825–3836. [CrossRef] [PubMed]
379. Mohammadinejad, R.; Ahmadi, Z.; Tavakol, S.; Ashrafizadeh, M. Berberine as a potential autophagy modulator. *J. Cell. Physiol.* **2019**, *234*, 14914–14926. [CrossRef] [PubMed]
380. Ashrafizadeh, M.; Najafi, M.; Mohammadinejad, R.; Farkhondeh, T.; Samarghandian, S. Berberine Administration in Treatment of Colitis: A Review. *Curr. Drug Targets* **2020**, *21*, 1385–1393. [CrossRef] [PubMed]

381. Gu, S.; Song, X.; Xie, R.; Ouyang, C.; Xie, L.; Li, Q.; Su, T.; Xu, M.; Xu, T.; Huang, D.; et al. Berberine inhibits cancer cells growth by suppressing fatty acid synthesis and biogenesis of extracellular vesicles. *Life Sci.* **2020**, *257*, 118122. [CrossRef]
382. Zhang, Y.; Liu, X.; Yu, M.; Xu, M.; Xiao, Y.; Ma, W.; Huang, L.; Li, X.; Ye, X. Berberine inhibits proliferation and induces G0/G1 phase arrest in colorectal cancer cells by downregulating IGF2BP3. *Life Sci.* **2020**, *260*, 118413. [CrossRef] [PubMed]
383. Jung, Y.Y.; Ko, J.H.; Um, J.Y.; Chinnathambi, A.; Alharbi, S.A.; Sethi, G.; Ahn, K.S. LDL cholesterol promotes the proliferation of prostate and pancreatic cancer cells by activating the STAT3 pathway. *J. Cell. Physiol.* **2021**, *236*, 5253–5264. [CrossRef] [PubMed]
384. Pandya, G.; Kirtonia, A.; Sethi, G.; Pandey, A.K.; Garg, M. The implication of long non-coding RNAs in the diagnosis, pathogenesis and drug resistance of pancreatic ductal adenocarcinoma and their possible therapeutic potential. *Biochim. Biophys. Acta Rev. Cancer* **2020**, *1874*, 188423. [CrossRef]
385. Suzuki, S.; Okada, M.; Sanomachi, T.; Togashi, K.; Seino, S.; Sato, A.; Yamamoto, M.; Kitanaka, C. Therapeutic targeting of pancreatic cancer stem cells by dexamethasone modulation of the MKP-1-JNK axis. *J. Biol. Chem.* **2020**, *295*, 18328–18342. [CrossRef] [PubMed]
386. Jagust, P.; Alcalá, S.; Sainz, B., Jr.; Heeschen, C.; Sancho, P. Glutathione metabolism is essential for self-renewal and chemoresistance of pancreatic cancer stem cells. *World J. Stem Cells* **2020**, *12*, 1410–1428. [CrossRef] [PubMed]
387. Valle, S.; Alcalá, S.; Martin-Hijano, L.; Cabezas-Sáinz, P.; Navarro, D.; Muñoz, E.R.; Yuste, L.; Tiwary, K.; Walter, K.; Ruiz-Cañas, L.; et al. Exploiting oxidative phosphorylation to promote the stem and immunoevasive properties of pancreatic cancer stem cells. *Nat. Commun.* **2020**, *11*, 5265. [CrossRef] [PubMed]
388. Alcalá, S.; Mayoral-Varo, V.; Ruiz-Cañas, L.; López-Gil, J.C.; Heeschen, C.; Martín-Pérez, J.; Sainz, B., Jr. Targeting SRC Kinase Signaling in Pancreatic Cancer Stem Cells. *Int. J. Mol. Sci.* **2020**, *21*, 7437. [CrossRef] [PubMed]
389. Park, S.H.; Sung, J.H.; Chung, N. Berberine diminishes side population and down-regulates stem cell-associated genes in the pancreatic cancer cell lines PANC-1 and MIA PaCa-2. *Mol. Cell. Biochem.* **2014**, *394*, 209–215. [CrossRef]
390. Zhu, C.; Li, J.; Hua, Y.; Wang, J.; Wang, K.; Sun, J. Berberine Inhibits the Expression of SCT through miR-214-3p Stimulation in Breast Cancer Cells. *Evid.-Based Complement. Altern. Med.* **2020**, *2020*, 2817147. [CrossRef] [PubMed]
391. Li, J.; Zhang, S.; Wu, L.; Pei, M.; Jiang, Y. Berberine inhibited metastasis through miR-145/MMP16 axis in vitro. *J. Ovarian Res.* **2021**, *14*, 4. [CrossRef] [PubMed]
392. Zhan, Y.; Han, J.; Xia, J.; Wang, X. Berberine Suppresses Mice Depression Behaviors and Promotes Hippocampal Neurons Growth Through Regulating the miR-34b-5p/miR-470-5p/BDNF Axis. *Neuropsychiatr. Dis. Treat.* **2021**, *17*, 613–626. [CrossRef] [PubMed]
393. Lin, T.H.; Hsieh, P.L.; Liao, Y.W.; Peng, C.Y.; Lu, M.Y.; Yang, C.H.; Yu, C.C.; Liu, C.M. Berberine-targeted miR-21 chemosensitizes oral carcinomas stem cells. *Oncotarget* **2017**, *8*, 80900–80908. [CrossRef] [PubMed]
394. Xu, M.; Wang, J.; Li, H.; Zhang, Z.; Cheng, Z. AIM2 inhibits colorectal cancer cell proliferation and migration through suppression of Gli1. *Aging* **2020**, *13*, 1017–1031. [CrossRef] [PubMed]
395. Doheny, D.; Sirkisoon, S.; Carpenter, R.L.; Aguayo, N.R.; Regua, A.T.; Anguelov, M.; Manore, S.G.; Arrigo, A.; Jalboush, S.A.; Wong, G.L.; et al. Combined inhibition of JAK2-STAT3 and SMO-GLI1/tGLI1 pathways suppresses breast cancer stem cells, tumor growth, and metastasis. *Oncogene* **2020**, *39*, 6589–6605. [CrossRef] [PubMed]
396. Zhao, Y.; Yang, X.; Zhao, J.; Gao, M.; Zhang, M.; Shi, T.; Zhang, F.; Zheng, X.; Pan, Y.; Shao, D.; et al. Berberine inhibits chemotherapy-exacerbated ovarian cancer stem cell-like characteristics and metastasis through GLI1. *Eur. J. Pharmacol.* **2021**, *895*, 173887. [CrossRef] [PubMed]
397. Naveen, C.R.; Gaikwad, S.; Agrawal-Rajput, R. Berberine induces neuronal differentiation through inhibition of cancer stemness and epithelial-mesenchymal transition in neuroblastoma cells. *Phytomed. Int. J. Phytother. Phytopharm.* **2016**, *23*, 736–744. [CrossRef]
398. Liufu, C.; Li, Y.; Lin, Y.; Yu, J.; Du, M.; Chen, Y.; Yang, Y.; Gong, X.; Chen, Z. Synergistic ultrasonic biophysical effect-responsive nanoparticles for enhanced gene delivery to ovarian cancer stem cells. *Drug Deliv.* **2020**, *27*, 1018–1033. [CrossRef]
399. Ertas, Y.N.; Abedi Dorcheh, K.; Akbari, A.; Jabbari, E. Nanoparticles for Targeted Drug Delivery to Cancer Stem Cells: A Review of Recent Advances. *Nanomaterials* **2021**, *11*, 1755. [CrossRef] [PubMed]
400. Ma, X.; Zhou, J.; Zhang, C.X.; Li, X.Y.; Li, N.; Ju, R.J.; Shi, J.F.; Sun, M.G.; Zhao, W.Y.; Mu, L.M.; et al. Modulation of drug-resistant membrane and apoptosis proteins of breast cancer stem cells by targeting berberine liposomes. *Biomaterials* **2013**, *34*, 4452–4465. [CrossRef] [PubMed]
401. Ashrafizadeh, M.; Ahmadi, Z.; Mohammadinejad, R.; Farkhondeh, T.; Samarghandian, S. MicroRNAs mediate the anti-tumor and protective effects of ginsenosides. *Nutr. Cancer* **2020**, *72*, 1264–1275. [CrossRef]
402. Ashrafizadeh, M.; Tavakol, S.; Mohammadinejad, R.; Ahmadi, Z.; Yaribeygi, H.; Jamialahmadi, T.; Johnston, T.P.; Sahebkar, A. Paving the Road toward Exploiting the Therapeutic Effects of Ginsenosides: An Emphasis on Autophagy and Endoplasmic Reticulum Stress. *Adv. Exp. Med. Biol.* **2021**, *1308*, 137–160. [PubMed]
403. Hashemi, F.; Zarrabi, A.; Zabolian, A.; Saleki, H.; Farahani, M.V.; Sharifzadeh, S.O.; Ghahremaniyeh, Z.; Bejandi, A.K.; Hushmandi, K.; Ashrafizadeh, M. Novel strategy in breast cancer therapy: Revealing the bright side of ginsenosides. *Curr. Mol. Pharmacol.* **2021**. [CrossRef] [PubMed]
404. Tang, Y.C.; Zhang, Y.; Zhou, J.; Zhi, Q.; Wu, M.Y.; Gong, F.R.; Shen, M.; Liu, L.; Tao, M.; Shen, B.; et al. Ginsenoside Rg3 targets cancer stem cells and tumor angiogenesis to inhibit colorectal cancer progression in vivo. *Int. J. Oncol.* **2018**, *52*, 127–138. [CrossRef] [PubMed]

405. Liu, S.; Chen, M.; Li, P.; Wu, Y.; Chang, C.; Qiu, Y.; Cao, L.; Liu, Z.; Jia, C. Ginsenoside rh2 inhibits cancer stem-like cells in skin squamous cell carcinoma. *Cell. Physiol. Biochem. Int. J. Exp. Cell. Physiol. Biochem. Pharmacol.* **2015**, *36*, 499–508. [CrossRef] [PubMed]
406. Mai, T.T.; Moon, J.; Song, Y.; Viet, P.Q.; Phuc, P.V.; Lee, J.M.; Yi, T.H.; Cho, M.; Cho, S.K. Ginsenoside F2 induces apoptosis accompanied by protective autophagy in breast cancer stem cells. *Cancer Lett.* **2012**, *321*, 144–153. [CrossRef]
407. Deng, S.; Wong, C.K.C.; Lai, H.C.; Wong, A.S.T. Ginsenoside-Rb1 targets chemotherapy-resistant ovarian cancer stem cells via simultaneous inhibition of Wnt/β-catenin signaling and epithelial-to-mesenchymal transition. *Oncotarget* **2017**, *8*, 25897–25914. [CrossRef] [PubMed]
408. Oh, J.; Yoon, H.J.; Jang, J.H.; Kim, D.H.; Surh, Y.J. The standardized Korean Red Ginseng extract and its ingredient ginsenoside Rg3 inhibit manifestation of breast cancer stem cell-like properties through modulation of self-renewal signaling. *J. Ginseng Res.* **2019**, *43*, 421–430. [CrossRef]
409. Yang, Z.; Zhao, T.; Liu, H.; Zhang, L. Ginsenoside Rh2 inhibits hepatocellular carcinoma through β-catenin and autophagy. *Sci. Rep.* **2016**, *6*, 19383. [CrossRef] [PubMed]
410. Lu, S.L.; Wang, Y.H.; Liu, G.F.; Wang, L.; Li, Y.; Guo, Z.Y.; Cheng, C. Graphene Oxide Nanoparticle-Loaded Ginsenoside Rg3 Improves Photodynamic Therapy in Inhibiting Malignant Progression and Stemness of Osteosarcoma. *Front. Mol. Biosci.* **2021**, *8*, 663089. [CrossRef]
411. Phi, L.T.H.; Wijaya, Y.T.; Sari, I.N.; Kim, K.S.; Yang, Y.G.; Lee, M.W.; Kwon, H.Y. 20(R)-Ginsenoside Rg3 Influences Cancer Stem Cell Properties and the Epithelial-Mesenchymal Transition in Colorectal Cancer via the SNAIL Signaling Axis. *OncoTargets Ther.* **2019**, *12*, 10885–10895. [CrossRef] [PubMed]
412. Park, J.W.; Park, J.H.; Han, J.W. Fermented Ginseng Extract, BST204, Suppresses Tumorigenesis and Migration of Embryonic Carcinoma through Inhibition of Cancer Stem Cell Properties. *Molecules* **2020**, *25*, 3128. [CrossRef] [PubMed]
413. Oh, J.; Jeon, S.B.; Lee, Y.; Lee, H.; Kim, J.; Kwon, B.R.; Yu, K.Y.; Cha, J.D.; Hwang, S.M.; Choi, K.M.; et al. Fermented red ginseng extract inhibits cancer cell proliferation and viability. *J. Med. Food* **2015**, *18*, 421–428. [CrossRef] [PubMed]

Article

MUG Mel3 Cell Lines Reflect Heterogeneity in Melanoma and Represent a Robust Model for Melanoma in Pregnancy

Silke Schrom [1], Thomas Hebesberger [1], Stefanie Angela Wallner [1], Ines Anders [1], Erika Richtig [2], Waltraud Brandl [3], Birgit Hirschmugl [3,4], Mariangela Garofalo [5], Claudia Bernecker [6], Peter Schlenke [6], Karl Kashofer [7], Christian Wadsack [3,4], Ariane Aigelsreiter [7], Ellen Heitzer [8], Sabrina Riedl [4,9,10], Dagmar Zweytick [4,9,10], Nadine Kretschmer [11], Georg Richtig [12] and Beate Rinner [1,4,*]

1. Division of Biomedical Research, Medical University of Graz, 8036 Graz, Austria; silke.schrom@medunigraz.at (S.S.); thomas.hebesberger@medunigraz.at (T.H.); stefaniewallner@a1.net (S.A.W.); ines.anders@medunigraz.at (I.A.)
2. Department of Dermatology, Medical University of Graz, 8036 Graz, Austria; erika.richtig@medunigraz.at
3. Department of Obstetrics and Gynecology, Medical University of Graz, 8036 Graz, Austria; waltraud.brandl@medunigraz.at (W.B.); birgit.hirschmugl@medunigraz.at (B.H.); christian.wadsack@medunigraz.at (C.W.)
4. BioTechMed-Graz, 8010 Graz, Austria; sabrina.riedl@uni-graz.at (S.R.); dagmar.zweytick@uni-graz.at (D.Z.)
5. Department of Pharmaceutical and Pharmacological Sciences, University of Padova, 35122 Padova, Italy; mariangela.garofalo@unipd.it
6. Department of Blood Group Serology and Transfusion Medicine, Medical University of Graz, 8036 Graz, Austria; c.bernecker@medunigraz.at (C.B.); peter.schlenke@medunigraz.at (P.S.)
7. Diagnostic and Research Institute of Pathology, Medical University of Graz, 8036 Graz, Austria; karl.kashofer@medunigraz.at (K.K.); ariane.aigelsreiter@medunigraz.at (A.A.)
8. Institute of Human Genetics, Diagnostic and Research Center for Molecular BioMedicine, Medical University of Graz, 8036 Graz, Austria; ellen.heitzer@medunigraz.at
9. Institute of Molecular Biosciences, Biophysics Division, University of Graz, 8010 Graz, Austria
10. BioHealth, 8010 Graz, Austria
11. Institute of Pharmaceutical Sciences, Department of Pharmacognosy, University of Graz, 8010 Graz, Austria; nadine.kretschmer@uni-graz.at
12. Division of Oncology, Medical University of Graz, 8036 Graz, Austria; georg.richtig@medunigraz.at
* Correspondence: beate.rinner@medunigraz.at; Tel.: +43-316-3857-3524

Abstract: Melanomas are aggressive tumors with a high metastatic potential and an increasing incidence rate. They are known for their heterogeneity and propensity to easily develop therapy-resistance. Nowadays they are one of the most common cancers diagnosed during pregnancy. Due to the difficulty in balancing maternal needs and foetal safety, melanoma is challenging to treat. The aim of this study was to provide a potential model system for the study of melanoma in pregnancy and to illustrate melanoma heterogeneity. For this purpose, a pigmented and a non-pigmented section of a lymph node metastasis from a pregnant patient were cultured under different conditions and characterized in detail. All four culture conditions exhibited different phenotypic, genotypic as well as tumorigenic properties, and resulted in four newly established melanoma cell lines. To address treatment issues, especially in pregnant patients, the effect of synthetic human lactoferricin-derived peptides was tested successfully. These new *BRAF*-mutated MUG Mel3 cell lines represent a valuable model in melanoma heterogeneity and melanoma pregnancy research. Furthermore, treatment with anti-tumor peptides offers an alternative to conventionally used therapeutic options—especially during pregnancy.

Keywords: melanoma; tumor heterogeneity; pregnancy; anti-tumor peptides; in vitro model

1. Introduction

Melanoma is one of the most aggressive and heterogeneous types of cancers, with a strong tendency to develop resistance against different therapy approaches [1,2]. Among cutaneous melanomas, approximately 35% are diagnosed during gestation or the postpartum period, forming a specific subset of patients with pregnancy-associated melanoma (PAM) [3]. This can be due to changing life situations; more women are having children at an older age and the incidence of pregnancy-related cancers increases generally with age [4].

Another challenge in the cure of melanoma is that melanoma presents with one of the highest mutation rates of all cancers [5]. In 80% of all cases, mutations occur within the RAS-RAF-MEK-ERK-MAP kinase pathway [6]. The *BRAF* gene is mutated with a hotspot mutation at codon 600 (mostly V600E) in 40% to 50% of all melanoma cases, followed by NRAS mutations (G12, G13, Q61) in 12% to 20% [7]. However, to date, FDA-approved targeted therapies in advanced cutaneous melanoma have been available only for *BRAF* mutations [8]. Several years ago, *BRAF* inhibitors such as vemurafenib, dabrafenib, and encorafenib revolutionized the treatment of melanoma, nowadays given in combination with MEK inhibitors like cobimetinib, trametinib and binimetinib. In addition, immunotherapeutic agents like the anti-CTLA-4 antibody ipilimumab and the anti-PD-1 antibodies pembrolizumab and nivolumab have improved disease outcomes in melanoma patients [9,10]. Melanoma responds well to targeted therapies and immune therapies; however, resistance occurs in a significant percentage of patients [1,11].

Consequent to tumor heterogeneity in advanced melanoma, resistances against therapies for *BRAF* and MEK inhibitors develop. Intratumoral heterogeneity therefore has important implications for treatment selection [12]. The relationship between genomic and immunological heterogeneity, as well as the link to tumor growth and therapy response, has not been extensively studied [13].

Furthermore, due to pigmentation, melanomas exhibit a diverse morphology. To date, the role of melanin in melanoma, particularly its effects on the ability to metastasize, is almost unknown. Sarna et al. showed in mouse experiments that increased melanin expression inhibits the invasive capabilities of melanoma cells in vitro based on cell elasticity and summarized that cells with melanin expression were able to spread less than non-pigmented cells [14].

To better understand the biology of melanoma and thus develop new therapies specifically targeted for pregnant patients, it is necessary to establish patient-derived cell line models that reflect the heterogeneity of melanoma. Most existing melanoma cell lines are cultured in fetal bovine serum (FBS) and result in one cell line. This only reflects a small clonal proportion of the heterogenic tumor cell pool. Human platelet lysate (hPL), an alternative to FBS, is rich in several nutrients that promote cell growth and proliferation. However, it is currently not widely used in cell culture [15,16].

In particular, the treatment of cancers during and shortly after pregnancy is a challenge. Two peptides: R-DIM-P-LF11-322 and R-DIM-P-LF11-334—derived from the human host defense peptide lactoferricin (hLFcin)—were used for innovative melanoma treatment. Both peptides were previously shown to exert antitumor activity on different melanoma cells in vitro and in vivo [17–21]. Riedl et al. were able to prove that the negatively charged phospholipid phosphatidylserine (PS) is exposed on the outer leaflet of the plasma membrane of melanoma cells [22–24]. This general marker of cancer cells can serve as a target for host defense peptides like R-DIM-P-LF11-322 and R-DIM-P-LF11-334, which have emerged as a potential new and alternative anti-cancer therapeutic [18,21].

In this study, we present a patient with pregnancy-associated lymph node metastasis (PALNM) from a *BRAF*-mutated melanoma. Thereby, the aim was to establish multiple well-characterized cell lines from one patient, preserving intratumoral heterogeneity. A pigmented and non-pigmented tumor fraction were considered, and different cultivation conditions were used. Lastly, the effects of innovative peptides for melanoma treatment, such as R-DIM-P-LF11-322 and R-DIM-P-LF11-334—which might be of particular interest for pregnant patients—were assessed.

2. Results

2.1. Cell Growth and Classical Melanoma Markers Confirm the Heterogenic Character of MUG Mel3 Cell Lines

The patient received no therapy nearly one year before re-developing a new lymph node metastasis during gestation. Cells from this lymph node metastasis were used to establish MUG Mel3 cell lines. After mechanical tumor dissociation, the cells were cultured under two different growth conditions: hPL and FBS, illustrated in Figure 1.

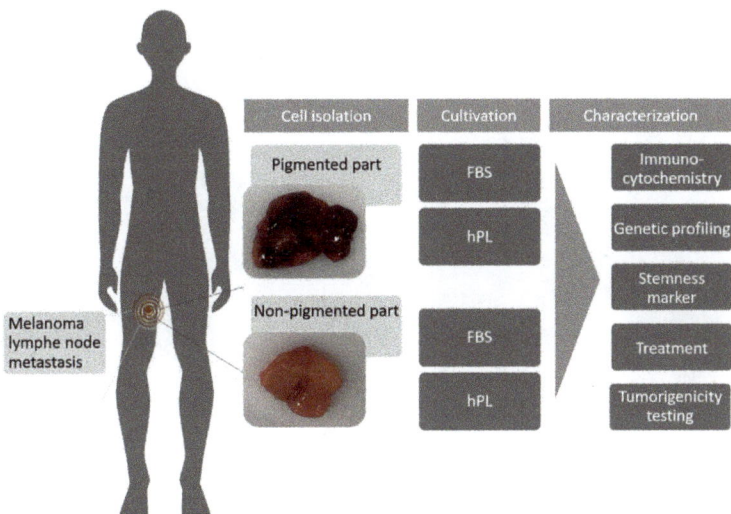

Figure 1. Overview of the experimental procedures. The location of the lymph node metastasis is illustrated, as well as the morphology of both areas (pigmented and non-pigmented). Furthermore, information about the cultivation conditions and individual steps in characterization are provided.

Cells with hPL started adhering to the cell culture flask's surface on the first day of cultivation, FBS cultures on the second day. Morphologically, cultures with hPL supplementation presented a semi-adherent growth behaviour at the start of primary cultures, marked by a more round-shaped phenotype and a high proportion of cells in suspension compared to well-adherent spindle-shaped cells in FBS cultures. No proliferation assays and statistical evaluation could be performed at the beginning of cultures due to insufficient amounts of cells, which is common in primary cell culture. Cells isolated from the pigmented part displayed a brownish staining immediately after the start of culture—especially in the hPL culture. However, this staining vanished after more than three passages. During the first two days, the focus of cultivation was to remove dead cells by centrifugation and media changes. On the third day, the first cell growth could be observed, and on day six and eight, representative pictures of the different conditions of primary cultures were taken (Figure 2).

Proliferation was recorded by passaging, starting at day 10 for MUG Mel3 PF, on day 13 for MUG Mel3 Ph, day 23 for MUG Mel3 NPF, and day 38 for MUG Mel3 NPh. Nomenclature of the pigmented part resulted only from first impressions of the tissue sample, since no pigmentation was visible after passaging the cells. Optimal hPL concentration was determined after several months in culture, as soon as enough cells were available for analysis by MTS cell proliferation assay. The highest OD values, which indicated the highest number of viable cells, were achieved between concentrations of 1.25% and 5% hPL (Supplementary Figure S1).

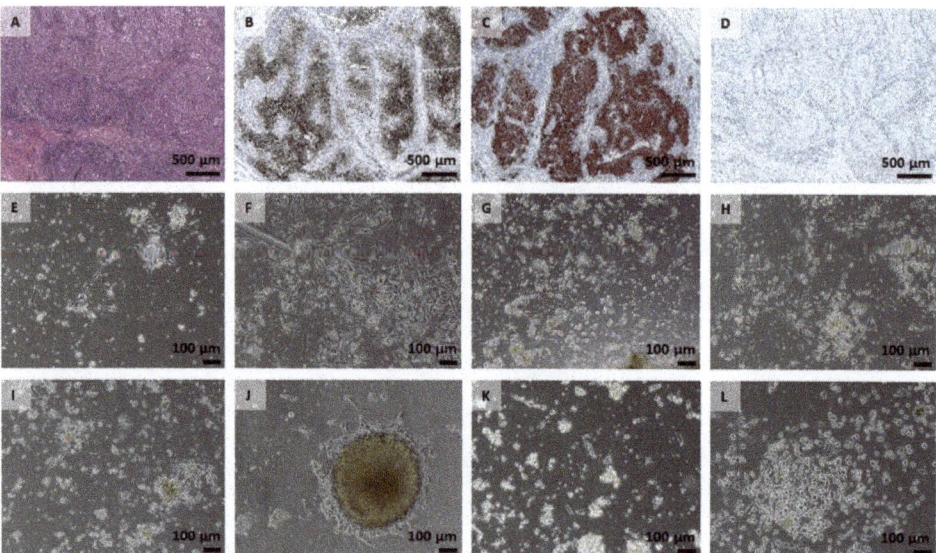

Figure 2. IHC of primary tumor tissue and morphological overview of cell growth. (**A–D**) Formalin fixed, paraffin embedded (FFPE) and stained tissue sections of the primary tumor tissue. (**A**) H&E, (**B**) HMB45; (**C**) Melan-A; (**D**) Tyrosinase. Stained tissue sections were scanned with a PANNORAMIC® 1000 (3DHISTECH, Budapest, Hungary) and pictures were analysed using CaseViewer 2.4 software (3DHISTECH, Budapest, Hungary). (**E–L**) Morphological presentation of the individual cultures observed over time; representative areas are illustrated. (**E**) MUG Mel3 PF (day six); (**F**) MUG Mel3 PF (day eight); (**G**) MUG Mel3 Ph (day six); (**H**) MUG Mel3 Ph (day eight); (**I**) MUG Mel3 NPF (day six); (**J**) MUG Mel3 NPF (day eight); (**K**) MUG Mel3 NPh (day six); (**L**) MUG Mel3 NPh (day eight).

MUG Mel3 Ph and MUG Mel3 NPh were then cultured in a 2.5% hPL medium. As soon as continuous growth was observed, ICC staining was performed. The melanoma marker MCSP was highly expressed in all four cell lines and did not markedly differ between the four treatment conditions.

However, differences could be observed for melanoma markers between the pigmented and the unpigmented sections. MUG Mel3 PF stained weak for HMB45, Melan-A, and tyrosinase. MUG Mel3 Ph stained weakly for tyrosinase as well, but contrary to MUG Mel3 PF, strongly for HMB45 and medium for Melan-A. Both MUG Mel3 NPF and MUG Mel3 NPh highly expressed each of the stained melanoma markers HMB45, Melan-A, Tyrosinase, and MCSP (Figure 3, Table 1).

Melan-A and SOX10, a marker used to identify metastatic melanoma [25], were assessed by qPCR, demonstrating significantly higher values for hPL cultured cells compared to FBS supplemented cells. Normalized to MUG Mel3 NPF (1 ± 0.11-fold expression), SOX10 mRNA was 0.93 ± 0.02-fold expressed in MUG Mel3 PF, 1.58 ± 0.76 in MUG Mel3 NPh, and 1.5 ± 0.24 in MUG Mel 3 Ph. Comparing cells of the same origin based on pigmentation but different cultivation methods, the analysis revealed that cells cultured from the pigmented part had a 1.6-fold higher SOX10 mRNA expression when established with hPL supplementation ($p = 0.015$) and a 1.58-fold higher expression for the non-pigmented part ($p = 0.258$). Melan-A mRNA expression was even more varied, with a 0.24 ± 0.3-fold expression of MUG Mel3 PF, 2.64 ± 0.3 of MUG Mel3 NPh, and 1.71 ± 0.26 of MUG Mel3 Ph normalized to MUG Mel3 NPF with 1 ± 0.34.

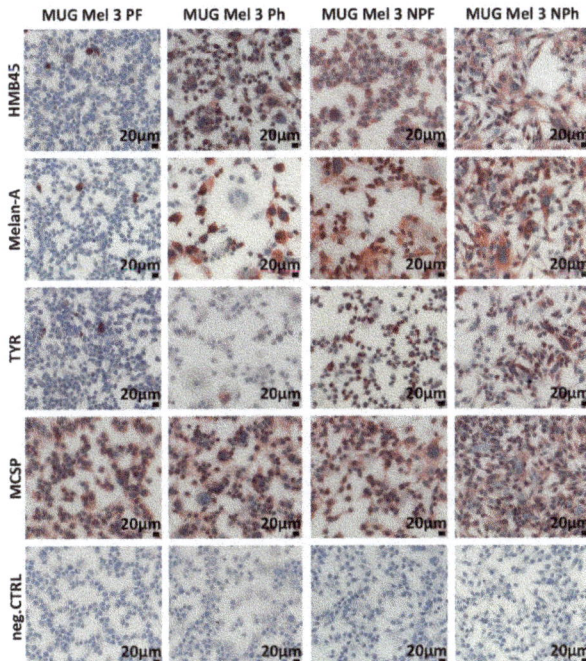

Figure 3. ICC for melanoma cell markers of cultivated MUG Mel3 cell lines. HMB45 staining: MUG Mel3 PF weak staining, MUG Mel3 Ph, MUG Mel3 NPF, and MUG Mel3 NPh strong staining. Melan A: MUG Mel3 PF weak staining, MUG Mel3 Ph, MUG Mel3 NPF, and MUG Mel3 NPh strong staining. Tyrosinase: MUG Mel3 PF and MUG Mel3 Ph weak staining, MUG Mel3 NPF and MUG Mel3 NPh strong staining. MCSP: highly expressed in all four cell lines. For each tested antigen, the corresponding IgG isotype control (negative control) was applied.

Table 1. Overview of melanoma marker. ICC markers HMB45, Melan-A, tyrosinase, and MCSP, and mRNA levels for SOX10, Melan-A, and % CD271 are summarized for all four cell lines. ICC data are depicted in Figure 3, qPCR data in Figure 4. For normalization of qPCR data, GAPDH, and β-Actin were used as housekeeping genes. qPCR and CD271 levels were obtained from three independent experiments.

	Pigmented		Non-Pigmented	
	FBS	hPL	FBS	hPL
Growth Behavior	Fast	Slow	Fast	Slow
Adhesion	++	Semi-adherent	++	Semi-adherent
ICC-HMB45	−	+	++	++
ICC-Melan-A	−	+	++	++
ICC-tyrosinase	−	−	+	++
ICC-MCSP	++	++	++	++
qPCR-Sox10	+	++	+	++
qPCR-Melan-A	−	++	+	++
CD271	13.94% ± 4.22	20.09% ± 6.71	3.66% ± 1.67	2.91% ± 1.40

The mRNA expression was greater for the pigmented and non-pigmented parts in hPL cultures, with a 7.26-fold higher expression for the pigmented part ($p = 0.003$) and a 2.64-fold higher expression for the non-pigmented part ($p = 0.002$). Moreover, we found significant differences in Melan-A mRNA expression levels when comparing cells derived from the pigmented and the non-pigmented part of the metastasis. The non-pigmented part had higher Melan-A expression levels independently of hPL or FBS supplementation, and was 4.25-fold higher in FBS cultures ($p = 0.04$) and 1.55-fold higher in hPL cultures ($p = 0.01$). However, SOX10 mRNA expression did not differ (for FBS cultivation $p = 0.364$ and for hPL cultured cells $p = 0.859$; Figure 4A,B, Table 1). Furthermore, CD271—a stemness marker associated with cell migration properties in melanoma—was measured in biological triplicates by flow cytometry. Different cell culture supplementation with either hPL or FBS did not affect CD271 expression. A higher percentage of CD271+ cells could be observed for cells derived from the pigmented part (Figure 4C).

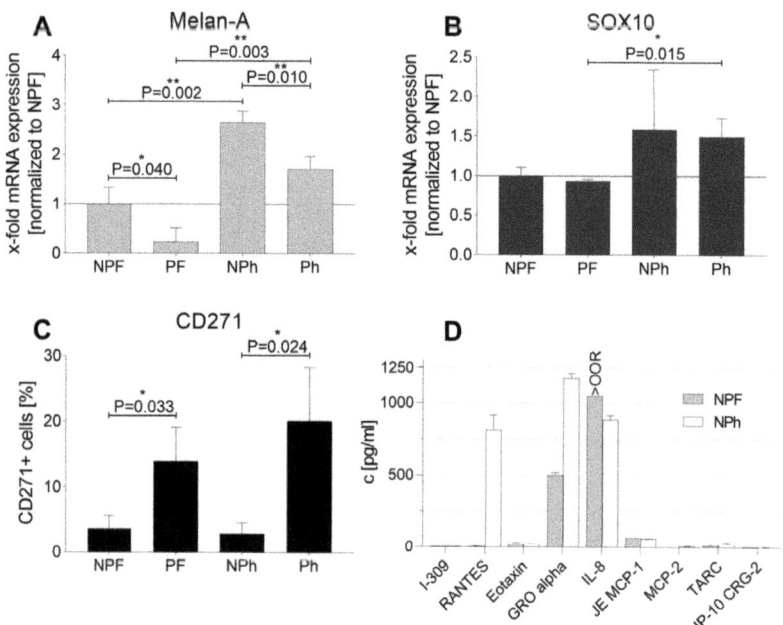

Figure 4. Melan-A and SOX10 mRNA, and percentage of CD271 expression levels as well as chemokine secretion in MUG Mel3 cultures. (**A**) Melan-A mRNA expression in all used cell lines, comparison normalized to MUG Mel3 NPF culture ($n = 3$); (**B**) SOX10 mRNA expression in all four cell lines, comparison normalized to MUG Mel3 NPF ($n = 3$) (**C**) CD271 surface marker expression determined by FACS analyses ($n = 3$); (**D**) secretion of chemokines in different culture conditions using FBS and hPL. The statistical analyses were conducted in GraphPad Prism 9.2.0 using a two-tailed student's t-tests to compare culture methods (FBS and hPL) within cell line origins (pigmented or non-pigmented) or to compare the origins of cells within the same cultivation method. A p-value ≤ 0.05 was considered statistically significant ($p \leq 0.05 = *, p \leq 0.01 = **$).

STR authentication was carried out for all four cell lines and confirmed origins of the pigmented and the non-pigmented part, with differences for both parts at locus D16S539. Losses could be observed for higher passages at locus D16S539 for MUG Mel3 NPh and at locus D7S820 for MUG Mel3 Ph (Supplementary Table S1).

2.2. Measurement of Growth Factors and Chemokines with Luminex Technology

Growth factors were measured at the very beginning of primary cell culture, before testing optimal amounts of hPL on cell proliferation. Therefore, supernatant from cells cultivated in RPMI supplemented with either 10% FBS or 10% hPL were compared for human chemokine expression (Supplementary Figure S2). Melanoma cells of the non-pigmented part, cultured with FBS or hPL, both secreted growth-regulated oncogene (GRO) alpha (FBS: 501.32 ± 15.66 pg/mL and hPL: 1172.46 ± 37.56 pg/mL) and IL-8

(FBS: higher than standard 1 at 1050 pg/mL and hPL: 885.83 ± 27.73 pg/mL) into the supernatant. Whereas RANTES was only excreted by MUG Mel3 supplemented with hPL (815.51 ± 100.45 pg/mL; Figure 4D).

2.3. Copy Number Profiling

To compare the copy number status of the cell lines and the primary tissues, a sWGS was performed. To this end, genome-wide copy number profiles were established from the pigmented and non-pigmented tissues as well as for two different passages of the FBS and hPL cultured cell lines (Figure 5).

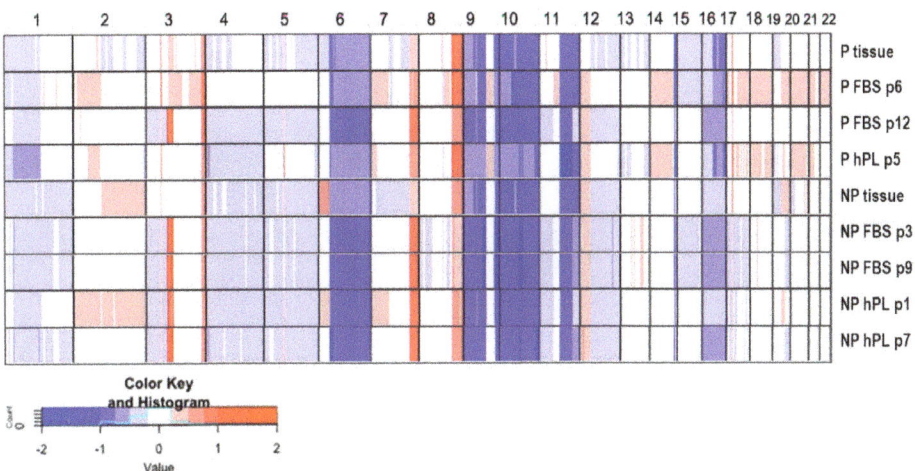

Figure 5. Genome-wide copy number profile of lymph node tissues and corresponding cell lines. Heat map depicting segmented log2-ratios pigmented (P) and non-pigmented (NP) tissues and various passages of their corresponding cell lines cultivated with FBS or hPL. Blue indicates loss of chromosomal material; red indicates gain of chromosomal material.

While some CNAs, such as losses at chr6q, chr9p, chr9q, chr10, chr11p, or gains at chr3q, chr7q, and chr8q were shared by both tumor tissue and cell lines—indicating a common origin of all samples—other CNAs (gains on chr2, 3,17,18) were observed only in a subset of samples. Interestingly, some changes (in particular on chr3) seemed to occur only after culturing and were not detected in either of the tissue samples. Cell lines derived from the non-pigmented tissue shared more CNAs with each other as compared to the pigmented tissue, with only a few differences between the FBS and hPL cultured cells. Moreover, CNAs were more stable in the non-pigmented tissue over passages regardless of supplementation with either FBS or hPL (Figure 5).

2.4. BRAF Mutation Analysis

A next-generation sequencing (NGS) analysis of pathologic lymph nodes in December 2018 revealed a *BRAF* V600E mutation (*BRAF*:NM_004333:exon15:c.T1799A:p.V600E) with variant allele frequencies (VAF) of 63.66 and 60.77% for the two replicates. No other mutation was identified in the analysed regions. *BRAF* V600E was assessed in pigmented and non-pigmented parts of the tissue as well as in the derived cultures using ddPCR. VAFs did not significantly differ between any of the samples with VAFs of 77.01% for the pigmented tissue, 73.98% for the non-pigmented tissue, 79.90% for MUG Mel3 PF, 85.24% for MUG Mel3 NPF, 79.85% for MUG Mel3 Ph, and 85.97% MUG Mel3 NPh.

2.5. Assessment of Tumorigenicity

To directly compare the tumorigenic potential of all four cell lines, we subcutaneously injected each of MUG Mel3 NPF, MUG Mel3 NPh, MUG Mel3 PF, and MUG Mel3 Ph into mice of two different strains. Thereby, we used CR ATH HO lacking T cells and NGX mice deficient in T cells, B cells, and natural killer cells. Twenty-five days post injection, all MUG Mel3 cell lines formed tumors, whereas cells derived from the non-pigmented part presented a higher tumor growth rate compared to cells derived from the pigmented part (Table 2 and Figure 6).

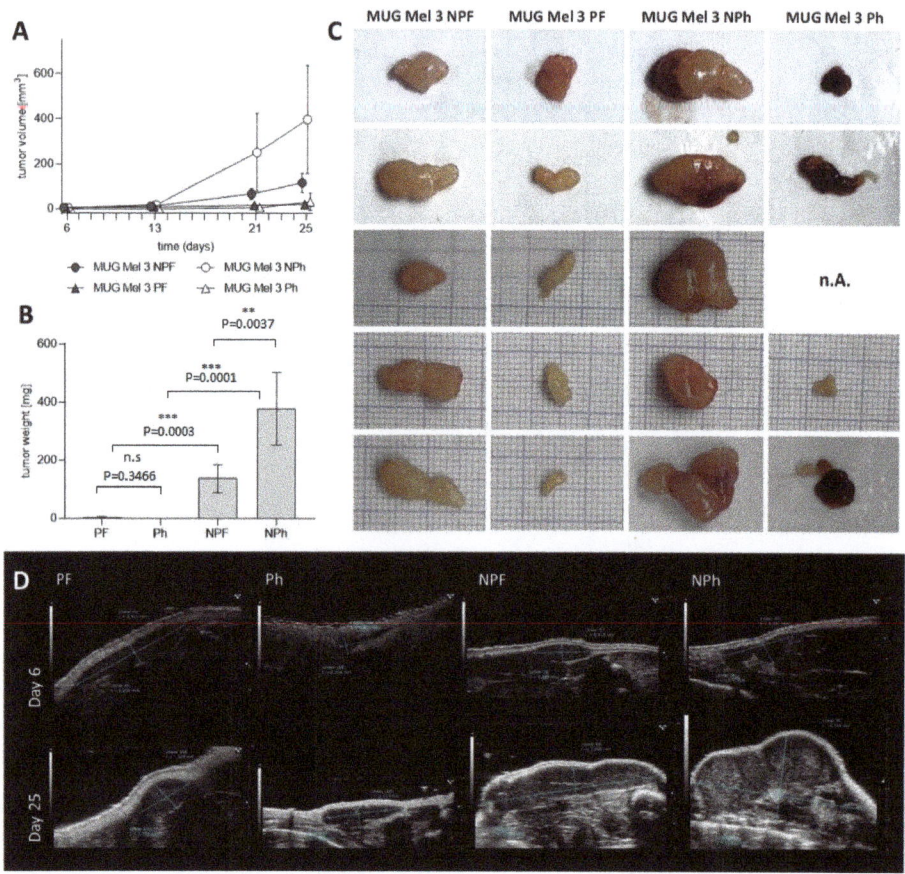

Figure 6. Tumorigenic profile of all four MUG Mel3 cell lines in CR ATH HO nude mice. Tumor volume [mm^3] over time—significance was tested by comparing culture conditions (FBS or hPL) within cell lines of origin (pigmented or non-pigmented) and the origins of cells within the same cultivation method on day of sacrifice: MUG Mel3 NPh/MUG Mel3 NPF p-value = 0.0319; MUG Mel3 NPh/ MUG Mel3 Ph p-value = 0.0056; MUG Mel3 PF/Ph p-value = 0.5024 and MUG Mel3 NPF/ MUG Mel3 PF p value = 0.0011—(**A**) and tumor weight [mg] on day of sacrifice (**B**) are illustrated ($p \leq 0.01 = $ **, $p \leq 0.001 = $ ***). Excised tumors of all mice are shown in (**C**) ($n = 5$). High frequency ultrasound (HF-US) images of one representative mouse are presented in (**D**). The images on day six were recorded with a 50 MHz (MX700) transducer; the images on day 25 with a 40 MHz (MX550D) transducer.

MUG Mel3 Ph recovered pigmentation in vivo (Figure 6C and Supplementary Figure S3C). IHC staining for the melanoma markers HMB45, Melan-A and tyrosinase were conducted and are depicted in Table 2. HMB45 and Melan-A were expressed in all xenograft tumors to a different extent. Tyrosinase expression could only be observed in hPl cultivated cells (Table 2, Supplementary Figure S4).

Table 2. Overview of melanoma markers expressed in xenografts.

CR ATH HO	MUG Mel3 PF	MUG Mel3 Ph	MUG Mel3 NPF	MUG Mel3 NPh
HMB45	++	++	++	++
Melan-A	+	++	++	++
Tyrosinase	−	+	−	+
NXG	**MUG Mel3 PF**	**MUG Mel3 Ph**	**MUG Mel3 NPF**	**MUG Mel3 NPh**
HMB45	+	++	++	++
Melan-A	+	++	++	++
Tyrosinase	−	+	−	−

2.6. PS Exposure and Peptide Treatment

Since new innovative therapies with low side effects are sought, and as the anti-tumor peptides used in this study have recently been tested promisingly on melanoma cell lines in vitro, they might represent a suitable approach for the treatment of pregnant women. PS exposure, serving as a target for anti-cancer peptides like R-DIM-P-LF11-322 and R-DIM-P-LF11-334, was quantified for MUG Mel3 cells using the Annexin V (AV) apoptosis kit. Cells cultivated with FBS presented with significantly (MUG Mel3 PF/MUG Mel3 NPF: p-value = 0.0250; MUG Mel3 PF/MUG Mel3 Ph p-value = 0.0002; MUG Mel3 NPh/ MUG Mel3 Ph: p-value = 0.0887 and finally MUG Mel3 NPF/MUG Mel3 NPh p-value = 0.0013) more PS on the outside of cells in both cultured parts, pigmented and non-pigmented (Supplementary Figure S5). The toxicity of anti-cancer peptides was measured by PI-uptake. Both tested peptides, R-DIM-P-LF11-322 and R-DIM-P-LF11-334, were also highly active against MUG Mel3 cells (Table 3).

Table 3. IC_{50} values [µM] of peptides R-DIM-P-LF11-322 and R-DIM-P-LF11-334 on all four MUG Mel3 cell lines. $IC_{50} \pm SD$ [µM] is summarized for all MUG Mel3 cell lines cultivated in 2.5% hPL or 10% FBS. p-values were calculated between the effects of R-DIM-P-LF11-322 and R-DIM-P-LF11-334.

	IC_{50} (PI) [µM]		
	R-DIM-P-LF11-322	**R-DIM-P-LF11-334**	**p-Value**
MUG Mel 3 PF (10% FBS)	14.4 ± 0.4	21.9 ± 1.7	0.0017
MUG Mel 3 Ph (2.5% hPL)	4.3 ± 0.3	6.6 ± 0.7	0.0064
MUG Mel 3 NPF (10% FBS)	22.1 ± 1.2	28.1 ± 1.3	0.0042
MUG Mel 3 NPh (2.5% hPL)	16.0 ± 0.9	26.9 ± 0.7	0.0001

Comparing both peptides, R-DIM-P-LF11-322 elicited cytotoxic effects at lower concentrations than R-DIM-P-LF11-334, ranging from 4.3 ± 0.3 µM for MUG Mel3 Ph up to 22.1 ± 1.2 µM for MUG Mel3 NPF. When focusing on the origin of cells, IC_{50}-values for R-DIM-P-LF11-322 and R-DIM-P-LF11-334 on cells cultured from the non-pigmented part of the lymph node metastasis were significantly higher compared to cells from the pigmented part, as analysed by the student's t test ($p < 0.001$, except the effect from R-DIM-P-LF1134 in MUG Mel3 PF/ MUG Mel3 NPF ($p < 0.01$)). In detail, the effect of R-DIM-P-LF11-322 compared for all cell lines: MUG Mel3 PF/ MUG Mel3 Ph p-value = 0.0001, MUG Mel3 PF/ MUG Mel3 NPF p-value = 0.0005; MUG Mel3 NPF/MUG Mel3 NPh p-value = 0.0021 and MUG Mel3 NPh/MUG Mel3 Ph p-value = 0.0001 and, in detail, the effect of R-DIM-P-LF11-334 compared for all cell lines: MUG Mel3 PF/ MUG Mel3 Ph p-value= 0.0001; MUG Mel3 PF/ MUG Mel3 NPF p-value = 0.0074, MUG Mel3 NPF/ MUG Mel3 NPh p-value = 0.232, MUG Mel3 NPH/MUG Mel3 Ph p-value = 0.0001 (Figure 7A–D).

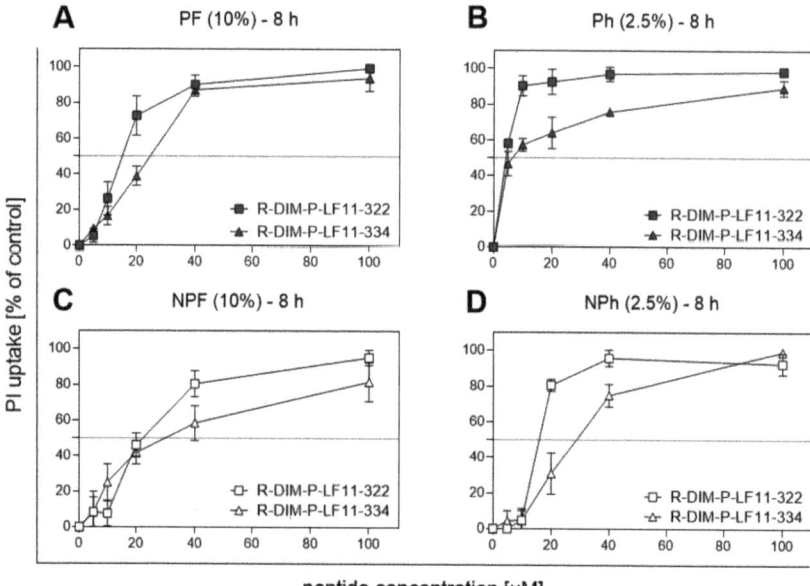

Figure 7. Cytotoxic capacity of R-DIM-P-LF11-322 and R-DIM-P-LF11-334 peptides on all four MUG Mel3 cell lines. Cytotoxicity was measured by PI-uptake in the presence of 10% FBS or 2.5% hPL after 8 h of incubation. (**A**) MUG Mel3 PF, (**B**) MUG Mel3 Ph, (**C**) MUG Mel3 NPF, and (**D**) MUG Mel3 NPh. Data for PI measurements from at least three independent experiments are presented as mean ± SD.

3. Discussion

Malignant melanoma is still one of the most aggressive and therapy-resistant tumors, caused by its highly intra- and intertumoral heterogenic character [1,26]. Furthermore, melanoma is one of the most prevalently diagnosed neoplasms in gestating women and is a major contributor to maternal morbidity and mortality. PAM is associated with 17% higher mortality compared with melanoma in non-pregnant female patients [27]. The incidence of melanoma development during pregnancy is increasing, largely related to an advanced maternal age [28,29]. Hyperpigmentation and increased melanocytic activity associated with melasma, as well as linea nigra, genital and areolar darkening, has been observed in pregnant women [30]. These conditions appear to correlate with elevated levels of oestrogen, progesterone, beta-endorphin, and beta- and alpha-melanocyte stimulating hormone [31]. For pregnancy-associated melanoma research, a melanoma cell line isolated directly from a pregnant patient, if possible without prior therapy, would be valuable. To our knowledge, no such cell lines are available for research currently. Research on biological mechanisms that foster PALNM and PAM could shed light on melanoma pathogenesis, leading to new therapeutic opportunities that may not only be related to pregnant patients. Particular emphasis was also placed on maintaining heterogeneity in our established cell lines, a major contributor associated with therapy resistances.

Previous studies have identified melanoma as the most heterogeneous type of tumor, consisting of multiple distinct subclonal populations of tumor cells [2,32,33]. Our data reflect melanoma tumor heterogeneity by expression of different genotypic and phenotypic characteristics for all four established MUG Mel3 cell lines, confirmed by qPCR data and chemokine expression. From the starting point of primary cultures, a distinct growth behaviour could be observed. Cell lines cultivated under hPL conditions presented a slower proliferation rate with a semi-adherent character. Staining for Melan-A, HMB45 and tyrosinase illustrated a diverse expression pattern in cell cultures and xenografts of all four MUG Mel3 cell lines. The only ICC marker uniformly expressed in culture was MCSP, a marker abundantly expressed in most human melanoma lesions, which has been associated with melanoma cell invasion [34,35].

Furthermore, mRNA levels for SOX10, a marker for melanoma sentinel lymph node metastasis [25], displayed higher levels in hPL supplemented culture. The same trend could be observed for mRNA levels of Melan-A, which were significantly higher expressed in hPL cultured cells as well. Furthermore, GRO alpha, initially identified by its growth stimulatory activity on melanoma cells as well as RANTES, which was described as being released in high amounts by melanoma cells in the presence of FBS, were secreted in larger quantities in hPL cultivated cells [36]. In murine xenograft models Payne and Cornelius were able to demonstrate that expression of RANTES in melanoma cells formed concentration-dependent aggressive tumors in nude mice and promoted tumor progression [37]. These findings correlate with our data, as MUG Mel3 NPh, the cell line excreting the highest RANTES levels, also illustrated the highest tumorigenic potential in both tested mouse models. Except for MUG Mel3 Ph, which caused engraftment in only 80% of mice, all other three cell lines induced 100% tumor engraftment. MUG Mel3 Ph was the only cell line that also caused re-pigmentation in the majority of formed tumors.

Contradictorily, CD271, a marker to detect stem cell-like melanoma progenitor cells, was significantly more highly expressed in cell lines derived from the pigmented part. CD271+ melanoma cells has been described as contributing to proliferation, tumorigenicity and plasticity [38–40], which does not coincide with our observations. This discrepancy could be explained by the previous work of Quintana et al., focusing on the usability of human melanoma cell lines. They compared genes expressed in commonly used cell lines to tumor tissues of origin and stated that the tumorigenic capacity in melanoma is not restricted to a small population. More likely, tumorigenic competence might reflect a reversible state in melanoma. They also concluded that so far, no markers are available that robustly distinguish tumorigenic from non-tumorigenic melanoma cells [41].

Regarding heterogeneity in melanoma, therapy-refractory subpopulations can cause tumor recurrence even in responding patients [42,43]. Even though highly specific drugs targeting BRAF mutations have improved overall response rate (ORR) and overall survival (OS) in the last decade, resistances still arise [8]. Generally, BRAF mutations are considered early oncogenic events and are therefore present in most tumor cells, resulting in a low intratumoral heterogeneity [1,44,45]. However, Chiappetta et al. used laser capture microdissection and showed that BRAF mutations cannot be detected in all areas of melanoma, indicating that therapy might not always be equally effective in patients [46]. All four established MUG Mel3 melanoma cell lines, as well as the tumor tissues, exhibited high VAFs for the BRAF V600E allele. Pregnant patients, harbouring a BRAF V600E mutation—the most abundant mutation in melanoma—were described as developing extensive metastatic melanoma during pregnancy or in the first year after birth, with poor survival regardless of therapeutic interventions [47].

Even though new interventions like targeted- and immunotherapies improved OS and progression-free survival (PFS), these therapies involve the risk of severe adverse events and are therefore often contraindicated during pregnancy [42,48].

New treatment strategies with minimal side effects are sought. We tested anti-tumor peptides derived from the human lactoferricin, a host defence peptide targeting negatively charged PS. PS, normally present on the inner leaflet of non-neoplastic cells, was proven to be transferred to the outer leaflet of MUG Mel3 cells, making them suitable for anti-tumor peptide treatment. In previous studies, both tested peptides R-DIM-P-LF11-322 and R-DIM-P-LF334 were shown to exert anti-tumoral activity against melanoma cells [17,18,20,21]. Both tested peptides were active against all four MUG Mel3 cell lines. R-DIM-P-LF11-322, previously described as being more active than R-DIM-P-LF11-334, exhibited lower IC50 values. Regarding serum supplementation, cells grown in the presence of FBS illustrated higher IC50 values than hPL cultured cells, indicating higher amounts of bivalent ions like Ca^{2+} present in FBS media. Bivalent ions were described as competing for negative charges present on the cell surface, lowering the cytotoxic capacity of peptides and thereby delaying the induction of cell death. Furthermore, R-DIM-P-LF11-322 was shown to be less affected by degradation excited by components present in serum [18], and exhibited lower IC50 values. Such new treatment approaches, targeting a common shared feature like PS exposure on the outside of cancerous cells, could help improve PFS by targeting subpopulations responsible for melanoma recurrence after conventional therapies.

All these findings illustrate the importance of maintaining melanoma heterogeneity to study new treatment options or combination therapies, and to further investigate resistance formation and therapy responses ex vivo.

4. Material and Methods

4.1. Patient History

A 26-year-old woman suffered from an ulcerated melanoma (tumor thickness 3.9 mm AJCC 2017) in January 2017 on her right thigh. Re-excision had been conducted in February 2017 without sentinel lymph node biopsy in an external hospital. The patient was referred to our university hospital initially in June 2017. The tumor board decided to perform adjuvant treatment with interferon-alpha, and the drug was given from July 2017 till December 2017. Then, the patient stopped due to side effects. In May 2018, she presented with a pathologic lymph node in the right groin and a solitary metastasis was excised in June 2018 (pN1b). At this point, adjuvant treatment was proposed to the patient, but was refused. In November 2018 she presented with pathologic lymph nodes distal to the right groin, while pregnant in the 13th week, which turned out to be PALNM (Pregnancy-associated lymph node metastasis). According to the tumor board, surgery was recommended and scheduled. Melanoma metastases were excised, and the patient was upgraded to pN3b, taking the former metastasis into account. The herein presented study was approved by the local ethics committee board of the Medical University of Graz (vote #31-457ex18/19; valid until 31.07.2020) in accordance with the Helsinki Declaration. The patient gave written informed consent for the study specific procedure. We confirm that all experiments were performed in accordance with relevant guidelines and regulations.

4.2. Pooled Human Platelet Lysate (hPL)

Informed consent was obtained from the donors and their blood was routinely tested according to Austrian and European Guidelines. hPL was produced as described [49]. In brief, four buffy coat units of blood group 0 were pooled with one plasma unit of blood group AB (male donor) to prevent interactions of cultured cells with AB0 antigens and isoagglutinins. The pools were centrifuged (461 g, 10 min, 22 °C) to get one unit of platelet-rich plasma (PRP). After leukodepletion by inline filtration, the PRP was frozen for at least 24 h (h) at −30 °C. A total of 10–13 units each were thawed at 37 °C and further pooled to one hPL batch. This summing-up of 50–65 donations achieves a minimization of individual donor variations. The pooled hPL was again frozen and stored at −30 °C. Before final proportioning, pooled hPL was centrifuged at 4000× g for 15 min at low brake speed to deplete any immunogenic platelet membrane fragments, and the supernatant was aliquoted and frozen at −30 °C until use. During the manufacturing process, two tests for microbiologic sterility (BacT/ALERT, bioMérieux, France) and mycoplasma contamination were performed.

4.3. Cell Culture

A part of the excised lymph nodes was available for cell culture. The material was morphologically divided into a pigmented (P) and non-pigmented (NP) part. Both parts were incubated for 10 min in a 10× concentrated penicillin (100 U/mL)/streptomycin (100 U/mL; pen-strep; Gibco, Life Technologies, Darmstadt, Germany) antibiotic bath in PBS to avoid possible contamination. Afterwards, the pieces were mechanically crushed and divided into different wells and cultivated with FBS or hPL. Four approaches with the following nomenclature were propagated as separate cell lines: MUG Mel3 PF (pigmented lymph node-part cultured in FBS), MUG Mel3 Ph (pigmented lymph node-part cultured in hPL), MUG Mel3 NPF (non-pigmented lymph node-part cultured in FBS), and MUG Mel3 NPh (non-pigmented lymph node-part cultured in hPL). Basic medium was RPMI (Gibco, Life Technologies, Darmstadt, Germany) supplemented with 10% FBS (M&B Stricker, Bernried, Germany) or 10% hPL (obtained from the Department of Blood Group Serology and Transfusion Medicine, Medical University of Graz), 2 mM L-glutamine (Gibco, Life Technologies, Darmstadt, Germany) and for the first passages, additional 1× pen-strep. Complete 10% hPL medium was prepared by addition of 2mM L-glutamine and 10% hPL to RPMI and was left at room temperature (RT) for 4 h. The formed clot was removed, and medium was incubated overnight (O/N) at 4° C. The next day, the medium was aliquoted into 50 mL tubes and warmed up to 37 °C in a water bath for 1 h, then centrifuged for 10 min, 3000× g at RT. Medium was filtered through a 0.22 µm filter unit (TPP, Biomedica, Trasadingen, Switzerland). After growth of the cells, an attempt was made to reduce hPL concentrations from 10% to 2.5% (0%; 1.25%; 2.5%; 5%; 7.5%; 10%). hPL cultivated cell lines were frozen in animal-origin free Synth-a-Freeze™ Cryopreservation Medium (Gibco, Life Technologies, Darmstadt, Germany). Cell lines between passages 0–25 were used for experiments.

The *BRAF* V600E mutated melanoma cell line A375 was used as a control and was cultured in DMEM supplemented with 10% FBS, 2 mM L-glutamine, and 1× pen-strep.

All cell lines were periodically checked for mycoplasma contamination (Minerva Biolabs, Berlin, Germany). Cell line authentication was conducted by short tandem repeat (STR) profiling (Promega, Madison, WI, USA; Table S1). Cell culture microscopic pictures were taken at RT on an Eclipse Ti2 inverted microscope (Nikon, Tokyo, Japan), 10× magnification, numerical aperture 0.30 with a DS-Fi2 camera (Nikon, Tokyo, Japan). Pictures were analysed with the NIS-Elements BR 5.02.00 software (Nikon, Tokyo, Japan).

4.4. Cell Line Authentication

DNA from tumor tissue and cells was prepared using the QIAamp DNA Mini kit (Qiagen, Hilden, Germany) in accordance with the manufacturer's protocol. After normalizing the DNA, 1 µL of each sample was amplified using the Power Plex® 16 HS System (Promega). One µL of the product was mixed with Hi-Di formamide (Applied Biosystems Inc., Foster City, CA, USA) and Internal Lane Standard (ILS600), denatured and fractionated on an ABI 3730 Genetic Analyzer (Applied Biosystems Inc.). The resulting data were processed and evaluated using ABI Genemapper 3.7.

4.5. CellTiter 96® AQueous Non-Radioactive Cell Proliferation Assay (MTS)

To determine optimal hPL concentration in growth medium, MUG Mel3 Ph and MUG Mel3 NPh cells were each seeded at 10^4 cells/100 µL medium in a clear 96-multiwell plate and incubated for 72 h at 5% CO_2 and 37 °C. Twenty µL of PMS/MTS solution (Promega) was added to each well. After 2 h of incubation at 5% CO_2 and 37 °C, OD values were measured at 490 nm and 720 nm for correction wavelength on a CLARIOstar photometer (BMG Labtech, Ortenberg, Germany). This experiment was performed only once in six replicates and was conducted to define optimal hPL concentration for culturing MUG Mel3 cells.

4.6. Immunocytochemistry (ICC)

Specific melanoma cell markers were selected for the characterization of isolated primary melanoma cells. Cells were seeded in glass chamber slides (7×10^4 per chamber) and grown in complete medium (10% FBS) to a confluency of 80%. Before fixation, cells were washed with 1× Hanks Balanced Salt Solution (HBSS, Thermo Fisher Scientific, Waltham, MA, USA) and dried for 1 h at RT. Cells were fixed by consecutive wash steps with 4% formaldehyde (Donauchem, Vienna, Austria), 1× PBS (Medicargo, Uppsala, Sweden), ice cold methanol (Merck, Darmstadt, Germany), and ice-cold acetone (Merck). All following procedure steps were performed at RT. All slides were rehydrated in 1× TBE pH 7.5 with

0.1% Tween (Sigma, St. Lois, MO, USA) buffer for 3 min, which was also used as washing buffer in-between steps. Endogenous peroxide activity was quenched by hydrogen peroxide block (Thermo Scientific, Rockford, IL, USA) for 10 min. Ultra V Block from Thermo Scientific was applied to reduce nonspecific background staining, by 5 min incubation. Selected monoclonal primary antibodies were diluted as shown in Table S2. Negative controls of the same isotype (Dako, Glostrup, Denmark) were applied and incubated for 1 h. After washing, primary enhancer was added for 30 min, followed by HRP Polymer (Thermo Scientific) and AEC chromogen (Thermo Scientific) incubation for 15 min in the dark, and for 5 min, respectively. Finally, cells were counterstained with hematoxylin and blued with running hot tap water each for 5 min. Slides were mounted with aquatex (Merck) and analysed by an Olympus BX53 reflected-light microscope (Hamburg, Germany), 10× magnification, numerical aperture: 0.30 with an Olympus U-TV1X-2 T7 camera (Tokyo, Japan) at RT. Pictures taken were evaluated by the CellSens Standard 2 software (Olympus, Hamburg, Germany).

4.7. qPCR

Melanoma cells (2.2×10^5 cells) were seeded onto six-well plates and incubated for 24 h. Subsequently, RNA isolation was performed with the Monarch® total RNA Miniprep Kit (New England Biolabs, Ipswich., MA, USA) according to the product manual. RNA quality and quantity were measured with a NanoDrop 2000 (Thermo Fisher Scientific). To remove melanin, RNA samples were further filtered using the OneStep PCR Inhibitor Removal Kit (Zymo Research, Irvine, CA, USA). cDNA was obtained from mRNA using the High-Capacity cDNA Reverse Transcription Kit (Thermo Fisher Scientific, Waltham, MA, USA) according to the manufacturer's manual. qPCR runs were performed on the CFX384 (Bio-Rad Laboratories Inc, Hercules, CA, USA) using primers against Melan-A and Sry-related HMG-Box gene 10 (SOX10; Table S3), and the 5X HOT FIREPol EvaGreen qPCR Supermix (Solis BioDyne, Tartu, Estonia). Results were analysed using $2^{-\Delta\Delta Ct}$ values. Means of GAPDH and β-Actin (ACTB) as reference genes were used for calculation of ΔCt values. Three independent biological experiments were conducted each in triplicates.

4.8. FACS Analysis

1×10^6 cells each were harvested from 80–90% confluent cultures and stained in 100 µL FACS buffer with 2.5 µL V450 mouse anti-Human CD271 antibody Clone C40-1457 (BD Horizon, Heidelberg, Germany) for 30 min at 4 °C. Cells were measured on a CytoFLEX S Flow Cytometer (Beckman Coulter, IN, USA) and further analysed using the Software CytExpert 2.3 (Beckman Coulter). The percentage of CD271 positive cells was calculated based on unstained controls and FSC/SSC gating was used to exclude debris, dead cells and doublets. The A375 cell line served as a positive control and Hela cells as negative controls (data not shown). Experiments were run in three independent experiments.

4.9. Growth Factors xMAP® Technology

Cytokine concentrations were determined from the non-pigmented lymph node part, using analyte-specific capture beads coated with target-specific capture antibodies according to the manufacturer's specifications. The analytes were detected by biotinylated analyte-specific antibodies. Following binding of the fluorescent detection label, the reporter fluorescent signal was measured with the Bio-Plex 200 multiplex suspension array system (Bio-Rad, Hercules, CA, USA) and detected with Bio-Plex 5.0 Software (Bio-Rad, Hercules, CA, USA). The sensitivity for the respective cytokines was as followed: CCL1/I-309 1.73–1260 pg/mL; CCL5/RANTES 8.27–6030 pg/mL; CCL11/Eotaxin 20.23–14,750 pg/mL; CXCL1/GRO alpha 15.30–11,160 pg/mL; IL-8/CXCL8 1.44–1050 pg/mL; CCL2/JE/MCP-1 11.87–8650 pg/mL; CCL8/MCP-2 4.66–3400 pg/mL; CCL17/TARC 30.62–22,320 pg/mL; CXCL10/IP 0.43–310 pg/mL. Analyte levels of complete growth medium mixed with the additives FBS or hPL served as background controls (Supplementary Figure S2).

4.10. DNA Extraction

Genomic DNA of the primary tumor and all four MUG Mel3 cell lines was isolated on a Maxwell MDxResearch System (Promega).

4.11. Copy Number Profiling

Genome-wide copy number alterations (CNA) were established using shallow whole-genome sequencing (sWGS). Shotgun libraries were prepared using the TruSeq DNA LT Sample preparation Kit (Illumina, San Diego, CA, USA). Briefly, 380 ng, 144 ng and 360 ng input DNA from all four MUG Mel3 cell lines and both parts of the lymph node were fragmented in 130 µL using the Covaris System (Covaris, Woburn, MA, USA). After concentrating the volume to 50 µL, end repair, A-tailing and adapter ligation were performed following the manufacturer's instructions. For selective amplification of the library fragments that have adapter molecules on both ends, 15 PCR cycles were used for higher concentration samples. Libraries were quality checked on an Agilent Bioanalyzer using a DNA 7500 Chip (Agilent Technologies, Santa Clara, CA, USA) and quantified using qPCR with a commercially available PhiX library (Illumina, San Diego, CA, USA) as a standard. Libraries were pooled equimolarily and sequenced on an Illumina MiSeq in a 150 bp single read run. On completion of the run data were base-call demultiplexed on the instrument (provided as Illumina). Genome-wide copy number calling was performed as previously described [50].

4.12. Mutation Analysis Using AmpliSeq (BRAF)

Highly multiplexed PCR was used to generate amplicon libraries to amplify 207 amplicons, covering approximately 2800 COSMIC mutations from 50 oncogenes and tumor suppressor genes. (Cancer Hotspot Panel v2, Cat. Nr. 4475346; Thermo Fisher Scientific, Waltham, MA, USA). All analyses were performed in duplicate. Libraries were prepared using the Ion AmpliSeq Library Kit 2.0 and sequencing was performed on an Ion Proton Sequencer (Thermo Fisher Scientific, Waltham, MA, USA). Emulsion PCR and sequencing runs were performed with the appropriate kits (Ion One Touch Template Kit version 2 and Ion Proton 200 Sequencing Kit; Thermo Fisher Scientific) using Ion PI chips. Sequencing length was set to 520 flows and yielded reads ranging from 70 to 150 bp, consistent with the expected amplicon size range. Initial data analysis was performed using the Ion Torrent Suite Software version 4.1 Plug-ins (Thermo Fisher Scientific).

4.13. Tumorigenicity Study

CR ATH HO mice (Crl:NU(NCr)-Foxn1nu, Charles River Laboratories, Kent, UK) and NXG mice (NOD-Prkdcscid-IL2rgTm1/Rj, Janvier Labs, Saint Berthevin Cedex, France) were maintained in-house (4–5 weeks of age, weight between 15 and 20 g). In accordance with a protocol approved by the committee for institutional animal care and use at the Austrian Federal Ministry of Science and Research (BMWFW); vote 66.010/0046-WF/V/3b/2016) animal work was carefully carried out. All mice were maintained under specific pathogen free (SPF) conditions in individually ventilated cages with ad libitum access to food and water. For cell injection and ultrasound imaging, mice were anaesthetized with constant administration of 2% isoflurane in a constant airflow of 2.5 L per minute. Animals were sacrificed 25 days post-injection.

Mice were divided into two groups ($n = 5$) according to strain. Animals were injected into each flank subcutaneously with 2.5×10^6 cells in 100 µL PBS cell suspension using a 27G needle. Injection was performed as described in the following: front right flank received MUG Mel3 NPh, back right flank received MUG Mel3 NPF, front left flank received MUG Mel3 Ph, back left flank received MUG Mel3 PF. Inoculation of the tumor was monitored daily and ultrasound imaging started on day six post-injection once a week. After sacrifice, histopathological examination was performed. The mice were dissected, and tumors extracted. All other organs were checked visually for structural changes. The tissue was fixed in 4% paraformaldehyde solution for 24 h, and were further embedded in paraffin.

4.14. Ultrasound Imaging

High-frequency ultrasound (HF-US) was performed using a Vevo3100 HF-US system with a 50 MHz (MX700) or a 40 MHz (MX550D) transducer (Fujifilm VisualSonics, Inc., Toronto, ON, Canada) reaching a spatial resolution of about 30–40 µm. Images in sagittal and transverse planes were obtained of the region of interest. Calculation of tumor volume was performed using VevoLab Software (Version 5.5.1, Fujifilm VisualSonics, Inc.) using the formula displayed in Equation (1):

$$V\left[\text{mm}^3\right] = \frac{3.141 \times \text{length [mm]} \times \text{width [mm]} \times \text{depth [mm]}}{6} \quad (1)$$

4.15. PS Exposure

PS exposure was measured using the RealTime-Glo™ Annexin V Apoptosis Assay (Promega). Briefly, cells were seeded at 10^5 cells/100 µL medium in a white 96-multiwell plate with clear bottom and incubated O/N at 5% CO_2 and 37 °C. The reagent stock solution (1000×) was diluted in respective medium according to the manufacturer's protocol. Before the measurement 100 µL of the reagent solution was added to the cell suspension. Luminescence was measured using the Glomax Multi+ detection system (Promega, Madison, WI, USA).

4.16. Peptides

C-terminally amidated peptides R-DIM-P-LF11-322 (PFWRIRIRRPRRIRIRWFP-NH2, M = 2677.4 g/mol) and R-DIM-P-LF11-334 (PWRIRIRRPRRIRIRWP-NH2, M = 2382.4 g/mol) derived from Lactoferricin were purchased from PolyPeptide Group (San Diego, CA, USA). A purity of higher than 96% for all peptides had been determined by RP-HPLC. Peptide stock solutions were prepared in acetic acid (0.1%, v/v) to an approximate concentration of 3 mg/mL and treated by ultrasonication at 15 min for better solubility. Peptide concentration was determined by measurement of UV absorbance of tryptophan at a wavelength of 280 nm using a NanoDrop photometer (ND 1000, Peqlab, VWR International, Inc., Erlangen, Germany). All peptide stocks were stored at 4 °C until use.

4.17. PI Uptake Toxicity Assay

To determine peptide-induced cell death, cells were collected, resuspended in respective media and diluted to a concentration of 1×10^6 cells/mL. 100 µL aliquots (10^5 cells) were incubated with different peptide concentrations (5 µM to 100 µM) for up to 8 h in the presence of propidium iodide (PI; 2 µL/10^5 cells of 50 µg/mL, Molecular Probes Inc., Eugene, OR, USA) at RT in black 96-well plates. PI uptake was measured using the GloMax® Multi+ Detection System (Promega, Madison, WI, USA). Cytotoxicity was calculated from the percentage of PI-positive cells in media alone (P_0) and in the presence of peptide (P_X; Equation (2)). Triton-X-100 was used to determine 100% of PI positive cells (P_{100}).

$$\% \text{ PI} - \text{uptake} = \frac{100 \times (P_x - P_0)}{(P_{100} - P_0)} \quad (2)$$

Excitation and emission wavelengths were 536 nm and 617 nm, respectively. Each experiment was repeated at least 3 times.

4.18. Statistical Analysis

Statistical analyses were conducted in GraphPad Prism 9.2.0 (GraphPad Software, San Diego, CA, USA) using two-tailed student's t-tests to compare culture conditions (FBS or hPL) within cell line origins (pigmented or non-pigmented) or to compare the origins of cells within the same cultivation method, and two-tailed unpaired student's t-test to compare tumor weight and tumor growth of the four culture conditions on the day of sacrifice. A p-value ≤ 0.05 was considered statistically significant.

5. Conclusions

Melanoma in pregnancy is a vast, debated and complex field. Further research is urgently needed to provide evidence-based treatment and improve the management of melanoma in pregnancy. We have successfully established four cell lines from a melanoma metastasis from a pregnant woman. By using different tumor fractions and cultivation conditions, we were able to better mimic intratumoral heterogeneity as the four cell lines differ morphologically, genetically and phenotypically. In addition, we were able to demonstrate different cytotoxic effects by antitumor peptide treatments for all four cell lines, and thus may also consider them for new innovative treatment strategies.

Supplementary Materials: The following are available online at https://www.mdpi.com/article/10.3390/ijms222111318/s1.

Author Contributions: B.R. and S.S. designed the draft for this manuscript. S.S. performed main cell culture parts, Luminex Technology, collected and analyzed the data and wrote the manuscript with B.R. T.H. performed qPCR, S.A.W. and I.A. conducted the animal experiments. S.R. and D.Z. were responsible for the peptides. E.R. was responsible for the patient and patient history and acquired together with G.R. the tumor material. C.B. and P.S. prepared and supplied the hPL. A.A. provided the human specimen and was responsible for IHC stainings. W.B., B.H. and C.W. performed ICC stainings. M.G. and N.K. contributed to in vitro experiments. K.K. and E.H. were responsible for the genetic section of this manuscript. G.R. helped organizing the study. B.R. managed, supervised, and provided funding for this study. All authors contributed to the interpretation of the data, discussed the results, contributed to the manuscript, and approved the final version. All authors have read and agreed to the published version of the manuscript.

Funding: This research received no external funding.

Institutional Review Board Statement: Not applicable.

Informed Consent Statement: Not applicable.

Data Availability Statement: Not applicable.

Acknowledgments: We thank the DocSchool 'Molecular Medicine and Inflammation' at the Medical University of Graz for funding the publication costs.

Conflicts of Interest: The authors declare no conflict of interest.

References

1. Grzywa, T.M.; Paskal, W.; Włodarski, P.K. Intratumor and intertumor heterogeneity in melanoma. *Transl. Oncol.* **2017**, *10*, 956–975. [CrossRef]
2. Andor, N.; Graham, T.A.; Jansen, M.; Xia, L.C.; Aktipis, C.A.; Petritsch, C.; Ji, H.P.; Maley, C.C. Pan-cancer analysis of the extent and consequences of intratumor heterogeneity. *Nat. Med.* **2016**, *22*, 105–113. [CrossRef]
3. Pavlidis, N.A. Coexistence of pregnancy and malignancy. *Oncologist* **2002**, *7*, 279–287. [CrossRef]
4. Cottreau, C.M.; Dashevsky, I.; Andrade, S.E.; Li, D.-K.; Nekhlyudov, L.; Raebel, M.A.; Ritzwoller, D.P.; Partridge, A.H.; Pawloski, P.; Toh, S. Pregnancy-associated cancer: A U.S. population-based study. *J. Women's Health* **2019**, *28*, 250–257. [CrossRef] [PubMed]
5. Lawrence, M.S.; Stojanov, P.; Polak, P.; Kryukov, G.V.; Cibulskis, K.; Sivachenko, A.; Carter, S.L.; Stewart, C.; Mermel, C.H.; Roberts, S.A.; et al. Mutational heterogeneity in cancer and the search for new cancer-associated genes. *Nat. Cell Biol.* **2013**, *499*, 214–218. [CrossRef] [PubMed]
6. Govindarajan, B.; Bai, X.; Cohen, C.; Zhong, H.; Kilroy, S.; Louis, G.; Moses, M.; Arbiser, J.L. Malignant transformation of melanocytes to melanoma by constitutive activation of mitogen-activated protein kinase kinase (MAPKK) signaling. *J. Biol. Chem.* **2003**, *278*, 9790–9795. [CrossRef]
7. Richtig, G.; Hoeller, C.; Kashofer, K.; Aigelsreiter, A.; Heinemann, A.; Kwong, L.N.; Pichler, M.; Richtig, E. Beyond the BRAF^{V600E} hotspot: Biology and clinical implications of rare BRAF gene mutations in melanoma patients. *Br. J. Dermatol.* **2017**, *177*, 936–944. [CrossRef]
8. Vanni, I.; Tanda, E.T.; Spagnolo, F.; Andreotti, V.; Bruno, W.; Ghiorzo, P. The current state of molecular testing in the BRAF-mutated melanoma landscape. *Front. Mol. Biosci.* **2020**, *7*, 113. [CrossRef]

9. Seidel, J.; Otsuka, A.; Kabashima, K. Anti-PD-1 and Anti-CTLA-4 therapies in cancer: Mechanisms of action, efficacy, and limitations. *Front. Oncol.* **2018**, *8*, 86. [CrossRef] [PubMed]
10. Callahan, M.K.; Wolchok, J.D. At the bedside: CTLA-4- and PD-1-blocking antibodies in cancer immunotherapy. *J. Leukoc. Biol.* **2013**, *94*, 41–53. [CrossRef]
11. Winder, M.; Virós, A. Mechanisms of drug resistance in melanoma. In *New Approaches to Drug Discovery*; Springer: Berlin/Heidelberg, Germany, 2017; Volume 249, pp. 91–108. [CrossRef]
12. Cheng, L.; López-Beltrán, A.; Massari, F.; MacLennan, G.T.; Montironi, R. Molecular testing for BRAF mutations to inform melanoma treatment decisions: A move toward precision medicine. *Mod. Pathol.* **2018**, *31*, 24–38. [CrossRef]
13. Reuben, A.; Spencer, C.N.; Prieto, P.A.; Gopalakrishnan, V.; Reddy, S.; Miller, J.P.; Mao, X.; De Macedo, M.P.; Chen, J.; Song, X.; et al. Genomic and immune heterogeneity are associated with differential responses to therapy in melanoma. *NPJ Genom. Med.* **2017**, *2*, 10. [CrossRef]
14. Sarna, M.; Krzykawska-Serda, M.; Jakubowska, M.; Zadlo, A.; Urbanska, K.; Sarna, M.; Krzykawska-Serda, M.; Jakubowska, M.; Zadlo, A.; Urbanska, K. Melanin presence inhibits melanoma cell spread in mice in a unique mechanical fashion. *Sci. Rep.* **2019**, *9*, 9280. [CrossRef]
15. Laner-Plamberger, S.; Lener, T.; Schmid, D.; Streif, D.A.; Salzer, T.; Öller, M.; Hauser-Kronberger, C.; Fischer, T.; Jacobs, V.R.; Schallmoser, K.; et al. Mechanical fibrinogen-depletion supports heparin-free mesenchymal stem cell propagation in human platelet lysate. *J. Transl. Med.* **2015**, *13*, 354. [CrossRef] [PubMed]
16. Schallmoser, K.; Strunk, D. Preparation of pooled human platelet lysate (pHPL) as an efficient supplement for animal serum-free human stem cell cultures. *J. Vis. Exp.* **2009**, *32*, e1523. [CrossRef]
17. Riedl, S.; Leber, R.; Rinner, B.; Schaider, H.; Lohner, K.; Zweytick, D. Human lactoferricin derived di-peptides deploying loop structures induce apoptosis specifically in cancer cells through targeting membranous phosphatidylserine. *Biochim. Biophys. Acta Biomembr.* **2015**, *1848*, 2918–2931. [CrossRef] [PubMed]
18. Riedl, S.; Rinner, B.; Schaider, H.; Liegl-Atzwanger, B.; Meditz, K.; Preishuber-Pflügl, J.; Grissenberger, S.; Lohner, K.; Zweytick, D. In Vitro and In Vivo cytotoxic activity of human lactoferricin derived antitumor peptide R-DIM-P-LF11-334 on human malignant melanoma. *Oncotarget* **2017**, *8*, 71817–71832. [CrossRef] [PubMed]
19. Heitzer, E.; Groenewoud, A.; Meditz, K.; Lohberger, B.; Liegl-Atzwanger, B.; Prokesch, A.; Kashofer, K.; Behrens, D.; Haybaeck, J.; Kolb-Lenz, D.; et al. Human melanoma brain metastases cell line MUG-Mel1, isolated clones and their detailed characterization. *Sci. Rep.* **2019**, *9*, 4096. [CrossRef] [PubMed]
20. Wodlej, C.; Riedl, S.; Rinner, B.; Leber, R.; Drechsler, C.; Voelker, D.R.; Choi, J.-Y.; Lohner, K.; Zweytick, D. Interaction of two antitumor peptides with membrane lipids—influence of phosphatidylserine and cholesterol on specificity for melanoma cells. *PLoS ONE* **2019**, *14*, e0211187. [CrossRef] [PubMed]
21. Grissenberger, S.; Riedl, S.; Rinner, B.; Leber, R.; Zweytick, D. Design of human lactoferricin derived antitumor peptides-activity and specificity against malignant melanoma in 2D and 3D model studies. *Biochim. Biophys. Acta Biomembr.* **2020**, *1862*, 183264. [CrossRef]
22. Papo, N.; Shai, Y. Host defense peptides as new weapons in cancer treatment. *Cell. Mol. Life Sci.* **2005**, *62*, 784–790. [CrossRef]
23. Riedl, S.; Rinner, B.; Asslaber, M.; Schaider, H.; Walzer, S.M.; Novak, A.; Lohner, K.; Zweytick, D. In search of a novel target—phosphatidylserine exposed by non-apoptotic tumor cells and metastases of malignancies with poor treatment efficacy. *Biochim. Biophys. Acta Biomembr.* **2011**, *1808*, 2638–2645. [CrossRef]
24. Riedl, S.; Zweytick, D.; Lohner, K. Membrane-active host defense peptides—challenges and perspectives for the development of novel anticancer drugs. *Chem. Phys. Lipids* **2011**, *164*, 766–781. [CrossRef] [PubMed]
25. Willis, B.C.; Johnson, G.; Wang, J.; Cohen, C. SOX10: A useful marker for identifying metastatic melanoma in sentinel lymph nodes. *Appl. Immunohistochem. Mol. Morphol.* **2015**, *23*, 109–112. [CrossRef] [PubMed]
26. Shannan, B.; Perego, M.; Somasundaram, R.; Herlyn, M. Heterogeneity in melanoma. In *Melanoma*; Cancer Treatment and Research; Springer: Cham, Switzerland, 2016; Volume 167, pp. 1–15.
27. Kyrgidis, A.; Lallas, A.; Moscarella, E.; Longo, C.; Alfano, R.; Argenziano, G. Does pregnancy influence melanoma prognosis? A meta-analysis. *Melanoma Res.* **2017**, *27*, 289–299. [CrossRef] [PubMed]
28. Jhaveri, M.B.; Driscoll, M.S.; Grant-Kels, J.M. Melanoma in pregnancy. *Clin. Obstet. Gynecol.* **2011**, *54*, 537–545. [CrossRef] [PubMed]
29. Hepner, A.; Negrini, D.; Hase, E.A.; Exman, P.; Testa, L.; Trinconi, A.F.; Filassi, J.R.; Francisco, R.P.V.; Zugaib, M.; O'Connor, T.L.; et al. Cancer during pregnancy: The oncologist overview. *World J. Oncol.* **2019**, *10*, 28–34. [CrossRef] [PubMed]
30. Riker, A.I. *Melanoma: A Modern Multidisciplinary Approach*; Springer: Berlin/Heidelberg, Germany, 2018.
31. Tyler, K.H. Physiological skin changes during pregnancy. *Clin. Obstet. Gynecol.* **2015**, *58*, 119–124. [CrossRef]
32. Ding, L.; Kim, M.; Kanchi, K.L.; Dees, N.D.; Lu, C.; Griffith, M.; Fenstermacher, D.; Sung, H.; Miller, C.A.; Goetz, B.; et al. Clonal architectures and driver mutations in metastatic melanomas. *PLoS ONE* **2014**, *9*, e111153. [CrossRef]
33. Sanborn, J.Z.; Chung, J.; Purdom, E.; Wang, N.J.; Kakavand, H.; Wilmott, J.S.; Butler, T.; Thompson, J.F.; Mann, G.J.; Haydu, L.E.; et al. Phylogenetic analyses of melanoma reveal complex patterns of metastatic dissemination. *Proc. Natl. Acad. Sci. USA* **2015**, *112*, 10995–11000. [CrossRef]

34. Iida, J.; Pei, D.; Kang, T.; Simpson, M.A.; Herlyn, M.; Furcht, L.T.; McCarthy, J.B. Melanoma chondroitin sulfate proteoglycan regulates matrix metalloproteinase-dependent human melanoma invasion into type I collagen. *J. Biol. Chem.* **2001**, *276*, 18786–18794. [CrossRef] [PubMed]
35. Yang, J.; Price, M.A.; Neudauer, C.L.; Wilson, C.; Ferrone, S.; Xia, H.; Iida, J.; Simpson, M.A.; McCarthy, J.B. Melanoma chondroitin sulfate proteoglycan enhances FAK and ERK activation by distinct mechanisms. *J. Cell Biol.* **2004**, *165*, 881–891. [CrossRef] [PubMed]
36. Mrowietz, U.; Schwenk, U.; Maune, S.; Bartels, J.; Küpper, M.; Fichtner, I.; Schröder, J.-M.; Schadendorf, D. The chemokine RANTES is secreted by human melanoma cells and is associated with enhanced tumour formation in nude mice. *Br. J. Cancer* **1999**, *79*, 1025–1031. [CrossRef] [PubMed]
37. Payne, A.S.; Cornelius, L.A. The role of chemokines in melanoma tumor growth and metastasis. *J. Investig. Dermatol.* **2002**, *118*, 915–922. [CrossRef]
38. Tudrej, K.B.; Czepielewska, E.; Kozłowska-Wojciechowska, M. SOX10-MITF pathway activity in melanoma cells. *Arch. Med. Sci.* **2017**, *13*, 1493–1503. [CrossRef] [PubMed]
39. Redmer, T.; Welte, Y.; Behrens, D.; Fichtner, I.; Przybilla, D.; Wruck, W.; Yaspo, M.-L.; Lehrach, H.; Schäfer, R.; Regenbrecht, C.R.A. The nerve growth factor receptor CD271 is crucial to maintain tumorigenicity and stem-like properties of melanoma cells. *PLoS ONE* **2014**, *9*, e92596. [CrossRef]
40. Filipp, F.V.; Li, C.; Boiko, A.D. CD271 is a molecular switch with divergent roles in melanoma and melanocyte development. *Sci. Rep.* **2019**, *9*, 7696. [CrossRef]
41. Quintana, E.; Shackleton, M.; Foster, H.R.; Fullen, D.R.; Sabel, M.S.; Johnson, T.M.; Morrison, S.J. Phenotypic heterogeneity among tumorigenic melanoma cells from patients that is reversible and not hierarchically organized. *Cancer Cell* **2010**, *18*, 510–523. [CrossRef]
42. Manzano, J.L.; Layos, L.; Bugés, C.; de Los Llanos Gil, M.; Vila, L.; Martínez-Balibrea, E.; Martinez-Cardus, A. Resistant mechanisms to BRAF inhibitors in melanoma. *Ann. Transl. Med.* **2016**, *4*, 237. [CrossRef]
43. Somasundaram, R.; Villanueva, J.; Herlyn, M. Intratumoral heterogeneity as a therapy resistance mechanism: Role of melanoma subpopulations. *Adv. Pharmacol.* **2012**, *65*, 335–359. [PubMed]
44. Omholt, K.; Platz, A.; Kanter, L.; Ringborg, U.; Hansson, J. NRAS and BRAF mutations arise early during melanoma pathogenesis and are preserved throughout tumor progression. *Clin. Cancer Res.* **2003**, *9*, 6483–6488. [PubMed]
45. Harbst, K.; Lauss, M.; Cirenajwis, H.; Isaksson, K.; Rosengren, F.; Törngren, T.; Kvist, A.; Johansson, M.C.; Vallon-Christersson, J.; Baldetorp, B.; et al. Multiregion whole-exome sequencing uncovers the genetic evolution and mutational heterogeneity of early-stage metastatic melanoma. *Cancer Res.* **2016**, *76*, 4765–4774. [CrossRef]
46. Chiappetta, C.; Proietti, I.; Soccodato, V.; Puggioni, C.; Zaralli, R.; Pacini, L.; Porta, N.; Skroza, N.; Petrozza, V.; Potenza, C.; et al. BRAF and NRAS mutations are heterogeneous and not mutually exclusive in nodular melanoma. *Appl. Immunohistochem. Mol. Morphol.* **2015**, *23*, 172–177. [CrossRef]
47. Ziogas, D.C.; Diamantopoulos, P.; Benopoulou, O.; Anastasopoulou, A.; Bafaloukos, D.; Stratigos, A.J.; Kirkwood, J.M.; Gogas, H. Prognosis and management of BRAF V600E-mutated pregnancy-associated melanoma. *Oncologist* **2020**, *25*, e1209–e1220. [CrossRef]
48. Marcé, D.; Cornillier, H.; Denis, C.; Jonville-Bera, A.-P.; Machet, L. Partial response of metastatic melanoma to BRAF-inhibitor-monotherapy in a pregnant patient with no fetal toxicity. *Melanoma Res.* **2019**, *29*, 446–447. [CrossRef]
49. Schallmoser, K.; Bartmann, C.; Rohde, E.; Reinisch, A.; Kashofer, K.; Stadelmeyer, E.; Drexler, C.; Lanzer, G.; Linkesch, W.; Strunk, D. Human platelet lysate can replace fetal bovine serum for clinical-scale expansion of functional mesenchymal stromal cells. *Transfusion* **2007**, *47*, 1436–1446. [CrossRef] [PubMed]
50. Heitzer, E.; Ulz, P.; Belic, J.; Gutschi, S.; Quehenberger, F.; Fischereder, K.; Benezeder, T.; Auer, M.; Pischler, C.; Mannweiler, S.; et al. Tumor-associated copy number changes in the circulation of patients with prostate cancer identified through whole-genome sequencing. *Genome Med.* **2012**, *5*, 30. [CrossRef] [PubMed]

Communication

Polygodial and Ophiobolin A Analogues for Covalent Crosslinking of Anticancer Targets

Vladimir Maslivetc [1,*], Breana Laguera [1], Sunena Chandra [1], Ramesh Dasari [1], Wesley J. Olivier [2], Jason A. Smith [2], Alex C. Bissember [2], Marco Masi [3], Antonio Evidente [3], Veronique Mathieu [4,5] and Alexander Kornienko [1,*]

1. Department of Chemistry and Biochemistry, Texas State University, San Marcos, TX 78666, USA; breana.laguera@emory.edu (B.L.); sunena2293@yahoo.com (S.C.); rameshdiict2@gmail.com (R.D.)
2. School of Natural Sciences-Chemistry, University of Tasmania, Hobart, TAS 7001, Australia; wesley.olivier@utas.edu.au (W.J.O.); jason.smith@utas.edu.au (J.A.S.); alex.bissember@utas.edu.au (A.C.B.)
3. Dipartimento di Scienze Chimiche, Università di Napoli Federico II, Complesso Universitario Monte Sant'Angelo, Via Cintia 4, 80126 Napoli, Italy; marco.masi@unina.it (M.M.); evidente@unina.it (A.E.)
4. Department of Pharmacotherapy and Pharmaceutics, Faculté de Pharmacie, Université Libre de Bruxelles (ULB), 1050 Brussels, Belgium; Veronique.Mathieu@ulb.be
5. UCRC, ULB Cancer Research Center, Université Libre de Bruxelles (ULB), 1050 Brussels, Belgium
* Correspondence: vladmas@txstate.edu (V.M.); a_k76@txstate.edu (A.K.)

Abstract: In a search of small molecules active against apoptosis-resistant cancer cells, including glioma, melanoma, and non-small cell lung cancer, we previously prepared α,β- and γ,δ-unsaturated ester analogues of polygodial and ophiobolin A, compounds capable of pyrrolylation of primary amines and demonstrating double-digit micromolar antiproliferative potencies in cancer cells. In the current work, we synthesized dimeric and trimeric variants of such compounds in an effort to discover compounds that could crosslink biological primary amine containing targets. We showed that such compounds retain the pyrrolylation ability and possess enhanced single-digit micromolar potencies toward apoptosis-resistant cancer cells. Target identification studies of these interesting compounds are underway.

Keywords: anticancer activity; apoptosis resistance; ophiobolin A; polygodial; Wittig reaction

1. Introduction

Cancers with intrinsic resistance to apoptosis are characterized by the lack of responsiveness to the current chemotherapeutic agents that generally work by the induction of apoptosis in cancer cells. These cancers include tumors of the lung, melanoma and glioblastoma and they represent a major challenge in the clinic [1]. For example, patients afflicted by glioblastoma multiforme [2,3], have a median survival expectancy of less than 14 months when treated with the best available protocol that involves surgery, radiation and chemotherapy with temozolomide [4]. Further, the main cause of death of cancer patients are tumor metastases. Metastatic cells are resistant to anoikis, a type of apoptotic cell death triggered by the loss of contact with the extracellular matrix or neighboring cells [5]. Resistance to anoikis, and thus apoptosis, renders metastatic cells unresponsive to the large majority of proapoptotic agents as well [3,6]. Therefore, a search for novel anticancer agents that can overcome cancer cell resistance, including resistance to apoptosis, is an important pursuit.

In this connection, natural products represent a valuable source of not only cytotoxic compounds, but also those capable of overcoming the intrinsic resistance of cancer cells to apoptosis [7–10]. Half of the anticancer drugs developed since the 1940s either originated from natural sources, such as vinca alkaloids, taxanes, podophyllotoxin, and camptothecins; or synthetic derivatization of natural products, such as vinflunine (fluorinated vinca alkaloid derivative), cabazitaxel (derivative of a natural taxoid), and mifamurtide (derivative of muramyl dipeptide) [11,12].

In addition to their ability to infiltrate a variety of cell processes and locales, natural products are known to be able to modulate multiple molecular targets that are frequently deregulated in cancers. This can be especially useful for overcoming the resistance of cancer cells to single-target pharmaceutical drugs [13,14]. While apoptosis is the most widespread and well-studied mode of cell death induced by synthetic and naturally occurring anticancer agents, alternative mechanisms of cell death (including, autophagy, regulated necrosis, mitotic catastrophe, paraptosis, parthanatos, methuosis, and lysosomal membrane permeabilization) are prevalent among natural products [15–17].

In the pursuit of agents active against apoptosis- and multidrug-resistant cancer cells, our recent studies have focused on the synthetic derivatives of polygodial (Figure 1a), a sesquiterpenoid dialdehyde isolated from *Tasmannia lanceolata*, and ophiobolin A (Figure 1b), a plant toxin isolated from pathogenic fungi of the *Bipolaris* genus (*Drechslera gigantea* and *Bipolaris maydis*). For instance, we have shown that C12-Wittig derivatives of polygodial (1) exhibit antiproliferative activity mainly through cytostatic effects and in many cases display higher efficiency compared to the parent compound [18]. Additionally, both polygodial and compound 1 can readily undergo the pyrrole formation with primary amines to give 3 and 2, which can lead to pyrrolation of lysine residues in biological environment [18–20].

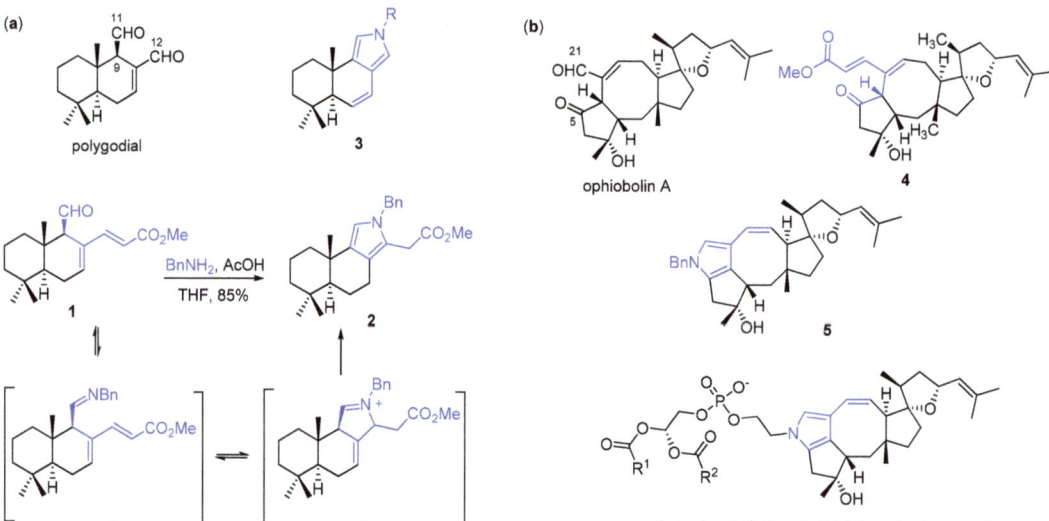

Figure 1. (a) Structures of polygodial, Wittig product 1, a proposed mechanism for pyrrole 2 formation involving intermediates I and II, unstable pyrrole 3; (b) Structures of ophiobolin A, Wittig product 4, pyrrole 5 and conjugated phosphatidylethanolamine 6.

Another natural anticancer product, ophiobolin A (Figure 1b), represents an interesting candidate for the treatment of apoptosis-resistant cancer cells due to a unique mechanism of action [21]. It was demonstrated to induce paraptosis in glioblastoma cells, offering an innovative strategy to combat this aggressive cancer. This cellular effect is understood to also originate from pyrrolylation of primary amine groups abundant in human cells (5) [22], such as intracellular lysine residues or phosphatidylethanolamine (6), which would lead to lipid bilayer destabilization [23]. Similarly to polygodial derivative 1, ophiobolin A can be converted into α,β-unsaturated ester 4 when treated with stabilized Wittig reagents [22].

The work reported in the current manuscript describes the early attempts of improving anticancer activity of polygodial and ophiobolin A derivatives through preparation of synthetic analogues that are capable of the covalent crosslinking of anticancer targets via the pyrrolylation of primary amines. The method that we envisioned involves the construction of dimeric conjugates of above-mentioned natural products. This approach has a proven record of success in oncology: such dimeric conjugates were often found to possess significantly improved anticancer activity compared to parent compounds (Figure 2) [24–31].

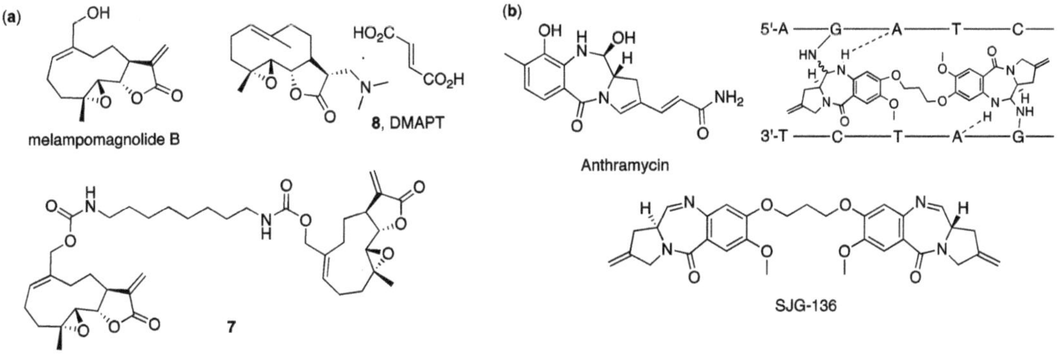

Figure 2. Dimeric anticancer conjugates of natural products: (**a**) Melampomagnolide B, its monomeric (**8**) and dimeric (**7**) derivatives; (**b**) Anthramycin, dimer SJG-136 and schematic representation of interstrand cross-link formation through DNA binding.

For instance, Crooks and coworkers [24] synthesized dimers of anticancer sesquiterpene lactone melampomagnolide B (Figure 2a) and assessed their activity against the NCI panel of 60 human hematological and solid tumor cell lines. Most active products exhibited potent growth inhibition (GI$_{50}$ = 0.16–0.99 µM) against the majority of cells. Notably, some of the compounds (**7**) displayed up to 1×10^6-fold higher cytotoxic effect against rat 9L-SF gliosarcoma cells when compared to previously developed monomeric melampomagnolide B analogs, such as DMAPT (**8**). This remarkable increase in activity for dimeric structures was attributed to the ability to provide multiple covalent interaction opportunities with multiple exposed surface cysteine residues of the GCLC and GCLM proteins via the formation of Michael addition adducts.

Another well-known example of rationally designed anticancer dimers involves the synthetic modification of the naturally occurring pyrrolobenzodiazepine (PBD) antitumor antibiotics, such as Anthramycin (Figure 2b) [25,26]. These compounds are sequence-selective DNA minor-groove binding agents. When monomeric PBD units were linked to form dimers containing two alkylating imine functionalities, it allowed them to form interstrand or intrastrand DNA cross-links in addition to mono-alkylated adducts, thus resulting in significantly greater cytotoxicity, antitumor activity and antibacterial activity compared to PBD monomers due to the different mode of DNA damage (Figure 2b) [25]. One of these dimers, SJG-136, has displayed potent cytotoxicity in the low nM range against human tumor cell lines and reached Phase II clinical trials in ovarian cancer and leukemia. Subsequent structural modification of the PBD dimers resulted in compounds with enhanced potency through increased interstrand DNA cross-linking ability, including agents with pM and in some cases sub-pM, activity in vitro [26]. These compounds were also employed as chemical payloads of antibody–drug conjugates, some of which are now in Phase III clinical trials [31].

2. Results and Discussion

In order to reduce the concept of multivalent anticancer conjugates to practice, in this work we focused on the preparation and assessment of dimeric polygodial and ophiobolin A α,β-unsaturated esters [32,33]. The synthesis of polygodial esters was initiated by the preparation of bromoacetates **9**, **10** and **11** with varied ethylene glycol chain lengths by heating diethylene, triethylene and tetraethylene glycols with bromoacetic acid under neat conditions (Figure 3). Bromoacetates **9**, **10** and **11** were then reacted with 2 eq of triphenylphosphine in toluene at room temperature to give phosphonium bromides **12**, **13** and **14**, respectively, which were conveniently purified with column chromatography on silica gel. Finally, the latter were subjected to Wittig olefination with polygodial, to yield dimeric α,β- and γ,δ-unsaturated esters **15**, **16** and **17** with varied linker chain lengths based on the number of incorporated ethylene glycol moieties. In a manner similar to the preparation of monomeric α,β- and γ,δ-unsaturated esters, such as **1** [18], polygodial reacted at the C-12 aldehyde only, possibly due to the sterically congested nature of the C-11 aldehyde group.

To confirm that the dimeric unsaturated esters retain the pyrrole forming ability upon the reaction with primary amines such as lysine residues in proteins, compound **15** was reacted with benzyl amine in the presence of AcOH in THF at room temperature and bis-pyrrole **18** was isolated in 70% yield (Figure 4). We also explored the preparation of a possible trimeric unsaturated ester. To this end, glycerol was reacted with bromoacetic acid under neat conditions to give tribromoacetate **19**. The latter was then subjected to a reaction with 3 eq of triphenylphosphine and phosphonium bromide **20** was isolated and purified with column chromatography on silica gel. The latter was treated with 3 eq of polygodial in the presence of Et₃N in THF at room temperature to give the desired trimeric α,β- and γ,δ-unsaturated ester **21** (Figure 5).

Ophiobolin A α,β- and γ,δ-unsaturated ester dimers **22** and **23**, containing one and two ethylene glycol unit linkers, respectively, were synthesized using bis(triphenylphosphonium) bromides **12** and **13** (Figure 6). The latter were stirred with excess of ophiobolin A in the presence of Et₃N in THF at room temperature for 48 h and yielded the desired dimers in good yields.

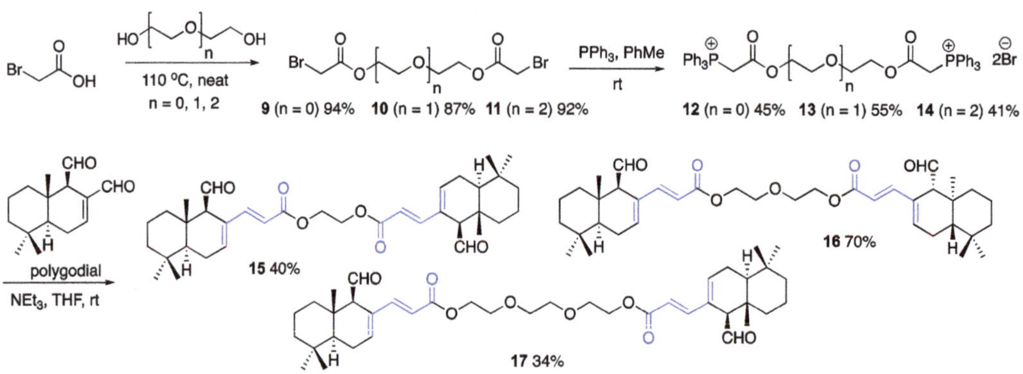

Figure 3. Synthesis of dimeric polygodial-based crosslinking agents **15**, **16** and **17**.

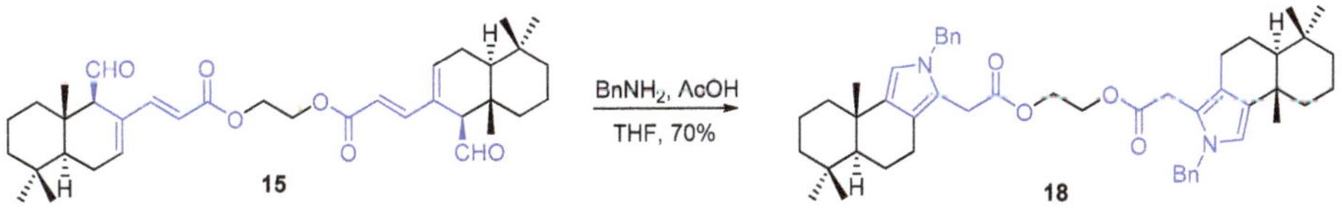

Figure 4. Demonstration of the feasibility of the formation of bis-pyrrole **18** from dimer **15**.

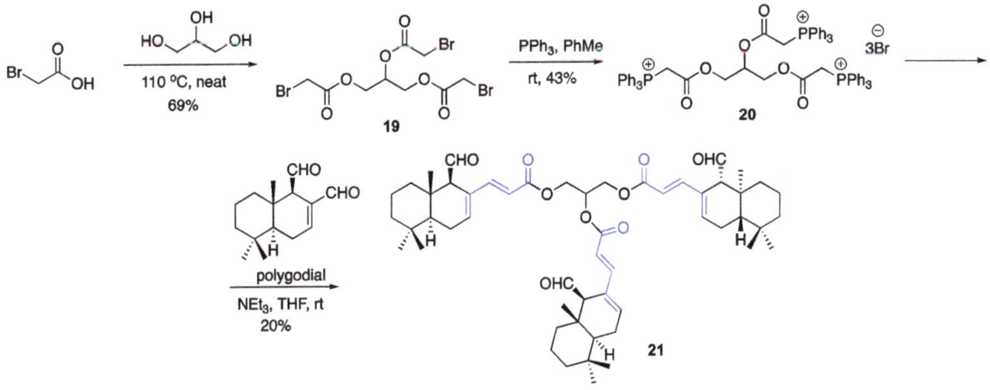

Figure 5. Synthesis of trimeric polygodial-based crosslinking agent **21**.

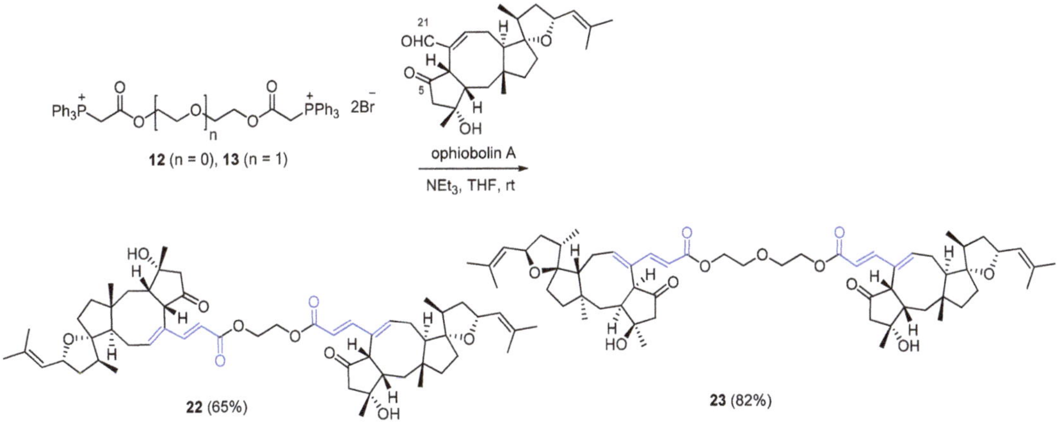

Figure 6. Synthesis of ophiobolin A-based crosslinking agents **22** and **23**.

The synthesized polygodial and ophiobolin A dimeric and trimeric unsaturated esters were evaluated for in vitro growth inhibition using the MTT colorimetric assay against a panel of six cancer cell lines. This included cells resistant to a number pro-apoptotic stimuli, such as human A549 non-small cell lung cancer (NSCLC) [34], human glioblastoma U373 [35,36], and human SKMEL-28 melanoma [37], as well as tumor models, which are largely susceptible to apoptosis-inducing stimuli, such as human Hs683 anaplastic oligodendroglioma [36], human MCF-7 breast adenocarcinoma [38] and mouse B16F10 melanoma (Table 1) [37]. Analysis of these data reveals that where monomeric α,β- and γ,δ-unsaturated esters of polygodial (**1**) and ophiobolin A (**4**) possess double digit micromolar antiproliferative potencies, dimeric polygodial unsaturated ester with one ethylene glycol linker **15** and both ophiobolin unsaturated esters **22** and **23** have single digit micromolar potencies. Thus, the conversion of monomeric covalently reacting polygodial and ophiobolin A unsaturated esters into their dimeric analogues is capable of enhancing the potency by an order of magnitude. Such potencies compare favorably with the clinically used cancer drug cisplatin (Table 1). It is noteworthy that the lengthening of the ethylene glycol linker between the polygodial molecules to two and to three units, i.e., **15→16→17**, progressively leads to a decrease in the potency of the dimers, pointing to a possible specificity of the interaction between these dimeric agents and their intracellular targets. This specificity is likely to be lost with the trimeric agent **21**, where the potency is in double digit micromolar region as well. Our data are similar to what has been seen in investigations of linker lengths in dimeric agents targeting G protein-coupled receptors [39], estrogen receptors [40], proteolysis targeting chimeras [39], among others [41].

Table 1. In vitro growth inhibitory effects of cross-linking agents in comparison with their monomeric counterparts.

Compound	GI$_{50}$ In Vitro Values (µM) [1]					
	Resistant to Apoptosis			Sensitive to Apoptosis		
	A549	SKMEL-28	U373	MCF7	Hs683	B16F10
1	30	42	41	36	29	7
15	7	14	9	4	4	5
16	17	22	28	9	8	9
17	55	29	79	36	50	24
21	25	24	28	23	17	6
4 [2]	59	86	86	43	59	ND
22	5	8	32	8	10	4
23	4	3	5	3	3	3
cisplatin	7	26	ND	35	7	8

[1] Mean concentration required to reduce the viability of cells by 50% after a 72 h treatment relative to a control, each experiment performed in sextuplicates, as determined by MTT assay. [2] Data are from ref. [22]. ND = not determined.

It is noteworthy that polygodial and ophiobolin A unsaturated esters displayed comparable potencies in cells both sensitive and resistant to apoptosis induction, indicating that this family of compounds is capable of overcoming apoptosis resistance in the clinic (Table 1) [18,22]. The comparison of the cellular effects of monomeric (**1** and **4**) and dimeric (**15** and **23**) polygodial and ophiobolin A esters on human glioblastoma U373 cells by phase contrast videomicroscopy did not reveal marked changes between the monomers and their respective dimers at their respective GI$_{50}$ concentrations (Figure 7). Phenotypically, it thus appears that the mode of action has not changed with the dimerization, despite the increase in potency.

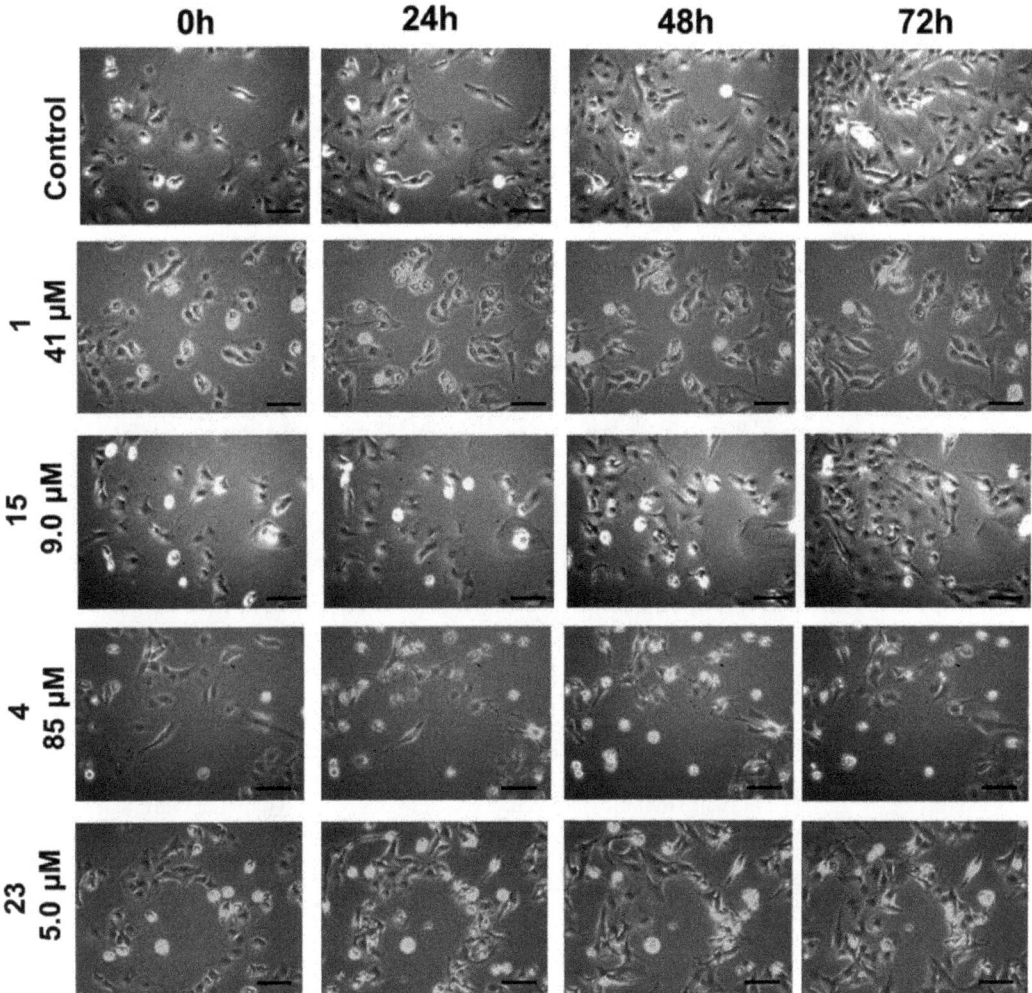

Figure 7. In vitro videomicroscopic analysis of the anticancer effects of monomeric Wittig products **1** and **4** as well as cross-linking agents **15** and **23**. The U373 human glioma cell line was treated with polygodial and ophiobolin A derivatives at their mean GI$_{50}$ concentrations (Table 1) or left untreated. The experiment was conducted once in triplicate. Scale bars correspond to 100 µm.

3. Materials and Methods

3.1. Human Cell Lines and Antiproliferative Effects

Breast carcinoma MCF-7 (DSMZ ACC107), oligodendroglioma Hs683 (ATCC HTB138), non-small cell lung cancer A549 (DSMZ ACC107), glioblastoma U373 (ECACC 08061901), melanoma SKMEL-28 (ATCC HTB72) and the murine melanoma B16F10 (ATCC CRL-6475) were cultured in RPMI supplemented with 10% heat-inactivated FBS (GIBCO code 10270106), 4 mM glutamine (Lonza code BE17-605E), 100 µg/mL gentamicin (Lonza code 17-5182), and penicillin-streptomycin (200 units/mL and 200 µg/mL) (Lonza code 17-602E). Cell lines were cultured in flasks, maintained and grown at 37 °C, 95% humidity, 5% CO_2. Antiproliferative effects of the compounds on these cell lines were evaluated through the colorimetric assay MTT. Briefly, cells were trypsinized and seeded in 96-well plates. Prior to treatment, compounds were dissolved in DMSO at a concentration of 10 mM. After 24 h, cells were treated with the compounds at different concentrations ranging from 10 nM to 100 µM or left untreated for 72 h. Cell viability was estimated by means of the MTT (3-(4,5-dimethylthiazol-2-yl)-2,5-diphenyl tetrazolium bromide, Sigma, Bornem, Belgium) mitochondrial reduction into formazan in living cells. The optical density of the untreated control was normalized as 100% of viable cells, allowing determination of the concentration that reduced their global growth by 50%.

3.2. Natural Product Isolation

Polygodial was isolated from *Tasmannia lanceolata* following a reported procedure [32].
Ophiobolin A was isolated from *Drechslera gigantea* following a reported procedure [33].

3.3. Selected Procedure for the Preparation of Polygodial-Based Cross-Linking Agent 16

To a solution of Wittig salt **13** (17.5 mg, 0.02 mmol) in THF (2 mL) was added triethylamine (18 µL, 0.13 mmol). The reaction was stirred for 20 min until the Wittig salt dissolved. A solution of polygodial (32.9 mg, 0.14 mmol) was dissolved in THF (0.5 mL) separately and added dropwise to the reaction mixture. The reaction mixture was stirred for 48 h and monitored by TLC (20% ethyl acetate/hexane). The reaction mixture was concentrated and purified by preparative TLC to yield 8.8 mg of **16** (70%). ^{1}H NMR (400 MHz, CDCl$_3$) δ 9.48 (d, J = 4.7 Hz, 2H), 7.33 (d, J = 10.6 Hz, 2H), 6.52 (dd, J = 8.6, 5.3 Hz, 2H), 5.53 (d, J = 16.3 Hz, 2H), 4.36–4.17 (m, 4H), 3.73–3.67 (m, 4H), 2.91–0.81 (m, 38H). ^{13}C NMR (500 MHz, CDCl$_3$) δ 205.6 (2C), 167.4 (2C), 147.4 (2C), 142.2 (2C), 130.9 (2C), 116.6 (2C), 69.6 (2C), 64.0 (2C), 63.2 (2C), 49.1 (2C), 42.2 (2C), 40.7 (2C), 37.9 (2C), 33.7 (2C), 33.6 (2C), 25.4 (2C), 22.7 (2C), 18.5 (2C), 16.0 (2C). HRMS m/z (ESI) calcd for C$_{38}$H$_{54}$O$_7$Na (M+Na) 645.3767, found 645.3769.

3.4. Selected Procedure for the Preparation of Ophiobolin A-Based Cross-Linking Agent 23

To a solution of Wittig salt **13** (3.1 mg, 3.6 µmol) in THF (2 mL) was added triethylamine (3.2 µL, 0.023 mmol). The reaction was stirred for 20 min until the Wittig salt dissolved. A solution of OpA (10 mg, 0.025 mmol) was dissolved in THF (1 mL) separately and added dropwise to the reaction mixture. The reaction mixture was stirred for 48 h and monitored by TLC (20% ethyl acetate/hexane). The reaction mixture was concentrated and purified by preparative TLC to yield 2.8 mg of **23** (82%). ^{1}H NMR (400 MHz, MeOD) δ 7.32 (dd, J = 15.6, 0.7 Hz, 2H), 6.43 (t, J = 8.5 Hz, 2H), 5.97 (d, J = 15.6 Hz, 2H), 5.23–5.16 (m, 2H), 4.55–4.45 (m, 2H), 4.27–4.18 (m, 4H), 3.75–3.69 (m, 4H), 3.67–3.62 (m, 2H), 3.60–3.55 (m, 2H), 2.58–2.45 (m, 4H), 2.45–2.36 (m, 2H), 2.25–2.21 (m, 2H), 2.19–2.06 (m, 6H), 1.87–1.77 (m, 2H), 1.74 (d, J = 1.2 Hz, 6H), 1.70 (d, J = 1.2 Hz, 6H), 1.67–1.63 (m, 2H), 1.62–1.39 (m, 8H), 1.37 (s, 6H), 1.33–1.26 (m, 2H), 1.09 (d, J = 7.2 Hz, 6H), 0.97–0.89 (m, 2H), 0.88 (s, 6H). ^{13}C NMR (100 MHz, MeOD) δ 167.3 (2C), 148.1 (2C), 137.1 (2C), 135.4 (2C), 133.6 (2C), 125.5 (2C), 115.9 (2C), 95.3 (2C), 77.1 (2C), 71.1 (2C), 68.9 (2C), 63.3 (2C), 61.3 (2C), 59.9 (2C), 54.5 (2C), 51.6 (2C), 50.3 (2C), 42.7 (2C), 42.6 (2C), 40.7 (2C), 36.5 (2C), 34.8 (2C), 30.1 (2C), 24.6 (2C), 24.1 (2C), 22.1 (2C), 17.0 (2C), 16.9 (2C), 16.8 (2C). HRMS m/z (ESI) calcd for C$_{58}$H$_{82}$O$_{11}$Na (M+Na) 977.5749, found 977.5164.

4. Conclusions

In conclusion, previously we synthesized α,β- and γ,δ-unsaturated esters of polygodial and ophiobolin A with potential covalent reactivity in cancer cells, i.e., pyrrolylation of lysine residues in proteins. In the present work we converted such covalently reacting compounds into dimeric and even trimeric species, capable of crosslinking proteins through such pyrrolylation reactions. Indeed, we showed that such dimeric compounds retain the ability to form pyrroles with primary amines. One selected polygodial dimer with the shortest linker length and both synthesized ophiobolin A dimers possess significantly enhanced antiproliferative potencies compared with their monomeric analogues, supporting this research premise. Mechanistic studies with a possible identification of intracellular targets for these molecules are underway.

Author Contributions: Conceptualization, A.K.; Methodology, B.L., S.C., R.D., W.J.O., and M.M.; Writing, V.M. (Vladimir Maslivetc) and A.K.; Supervision, J.A.S., A.C.B., A.E. and V.M. (Veronique Mathieu); Funding acquisition, A.K. All authors have read and agreed to the published version of the manuscript.

Funding: This research was funded by NIH, grants 5R21GM131717-02 and 1R15CA227680-01A1. The Belgian Brain Tumor Support is also acknowledged.

Institutional Review Board Statement: Not applicable.

Informed Consent Statement: Not applicable.

Data Availability Statement: Not applicable.

Acknowledgments: A.K. and Veronique Mathieu acknowledge Annelise De Carvalho, Aude Ingels and Vittoria Simioni for their contribution to the biological assays. The authors thank Maurizio Vurro and Angela Boari, Istituto di Scienze delle Produzioni Alimentari, CNR, Bari, Italy, for the supply of the culture filtrates of *D. gigantea*.

Conflicts of Interest: The authors declare no conflict of interest.

References

1. Brenner, H. Long-Term Survival Rates of Cancer Patients Achieved by the End of the 20th Century: A Period Analysis. *Lancet* **2002**, *360*, 1131–1135. [CrossRef]
2. Agnihotri, S.; Burrell, K.E.; Wolf, A.; Jalali, S.; Hawkins, C.; Rutka, J.T.; Zadeh, G. Glioblastoma, a Brief Review of History, Molecular Genetics, Animal Models and Novel Therapeutic Strategies. *Arch. Immunol. Ther. Exp.* **2013**, *61*, 25–41. [CrossRef] [PubMed]
3. Lefranc, F.; Sadeghi, N.; Camby, I.; Metens, T.; Dewitte, O.; Kiss, R. Present and Potential Future Issues in Glioblastoma Treatment. *Expert Rev. Anticancer Ther.* **2006**, *6*, 719–732. [CrossRef] [PubMed]
4. Stupp, R.; Hegi, M.E.; Mason, W.P.; Van Den Bent, M.J.; Taphoorn, M.J.B.; Janzer, R.C.; Ludwin, S.K.; Allgeier, A.; Fisher, B.; Belanger, K. Effects of Radiotherapy with Concomitant and Adjuvant Temozolomide versus Radiotherapy Alone on Survival in Glioblastoma in a Randomised Phase III Study: 5-Year Analysis of the EORTC-NCIC Trial. *Lancet Oncol.* **2009**, *10*, 459–466. [CrossRef]
5. Simpson, C.D.; Anyiwe, K.; Schimmer, A.D. Anoikis Resistance and Tumor Metastasis. *Cancer Lett.* **2008**, *272*, 177–185. [CrossRef] [PubMed]
6. Savage, P.; Stebbing, J.; Bower, M.; Crook, T. Why Does Cytotoxic Chemotherapy Cure Only Some Cancers? *Nat. Clin. Pract. Oncol.* **2009**, *6*, 43–52. [CrossRef] [PubMed]
7. Dasari, R.; Banuls, L.M.Y.; Masi, M.; Pelly, S.C.; Mathieu, V.; Green, I.R.; van Otterlo, W.A.L.; Evidente, A.; Kiss, R.; Kornienko, A. C1, C2-Ether Derivatives of the Amaryllidaceae Alkaloid Lycorine: Retention of Activity of Highly Lipophilic Analogues against Cancer Cells. *Bioorg. Med. Chem. Lett.* **2014**, *24*, 923–927. [CrossRef]
8. Kornienko, A.; Mathieu, V.; Rastogi, S.K.; Lefranc, F.; Kiss, R. Therapeutic Agents Triggering Nonapoptotic Cancer Cell Death. *J. Med. Chem.* **2013**, *56*, 4823–4839. [CrossRef]
9. Bury, M.; Girault, A.; Megalizzi, V.; Spiegl-Kreinecker, S.; Mathieu, V.; Berger, W.; Evidente, A.; Kornienko, A.; Gailly, P.; Vandier, C. Ophiobolin A Induces Paraptosis-like Cell Death in Human Glioblastoma Cells by Decreasing BKCa Channel Activity. *Cell Death Dis.* **2013**, *4*, e561. [CrossRef]
10. Mathieu, V.; Chantôme, A.; Lefranc, F.; Cimmino, A.; Miklos, W.; Paulitschke, V.; Mohr, T.; Maddau, L.; Kornienko, A.; Berger, W. Sphaeropsidin A Shows Promising Activity against Drug-Resistant Cancer Cells by Targeting Regulatory Volume Increase. *Cell. Mol. Life Sci.* **2015**, *72*, 3731–3746. [CrossRef]
11. Newman, D.J.; Cragg, G.M. Natural Products as Sources of New Drugs over the 30 Years from 1981 to 2010. *J. Nat. Prod.* **2012**, *75*, 311–335. [CrossRef] [PubMed]
12. Safarzadeh, E.; Shotorbani, S.S.; Baradaran, B. Herbal Medicine as Inducers of Apoptosis in Cancer Treatment. *Adv. Pharm. Bull.* **2014**, *4* (Suppl. 1), 421. [PubMed]
13. Wondrak, G.T. Redox-Directed Cancer Therapeutics: Molecular Mechanisms and Opportunities. *Antioxid. Redox Signal.* **2009**, *11*, 3013–3069. [CrossRef] [PubMed]
14. Ahmad, A.; Sakr, W.A.; Rahman, K.M. Novel Targets for Detection of Cancer and Their Modulation by Chemopreventive Natural Compounds. *Front. Biosci.* **2012**, *4*, 410–425. [CrossRef]
15. Lee, D.; Kim, I.Y.; Saha, S.; Choi, K.S. Paraptosis in the Anti-Cancer Arsenal of Natural Products. *Pharmacol. Ther.* **2016**, *162*, 120–133. [CrossRef]

16. Wang, X.; Feng, Y.; Wang, N.; Cheung, F.; Tan, H.Y.; Zhong, S.; Li, C.; Kobayashi, S. Chinese Medicines Induce Cell Death: The Molecular and Cellular Mechanisms for Cancer Therapy. *Biomed Res. Int.* **2014**, *2014*. [CrossRef]
17. Gali-Muhtasib, H.; Hmadi, R.; Kareh, M.; Tohme, R.; Darwiche, N. Cell Death Mechanisms of Plant-Derived Anticancer Drugs: Beyond Apoptosis. *Apoptosis* **2015**, *20*, 1531–1562. [CrossRef] [PubMed]
18. Dasari, R.; De Carvalho, A.; Medellin, D.C.; Middleton, K.N.; Hague, F.; Volmar, M.N.M.; Frolova, L.V.; Rossato, M.F.; Jorge, J.; Dybdal-Hargreaves, N.F. Wittig Derivatization of Sesquiterpenoid Polygodial Leads to Cytostatic Agents with Activity against Drug Resistant Cancer Cells and Capable of Pyrrolylation of Primary Amines. *Eur. J. Med. Chem.* **2015**, *103*, 226–237. [CrossRef]
19. Zhang, L.; Gavin, T.; DeCaprio, A.P.; LoPachin, R.M. γ-Diketone Axonopathy: Analyses of Cytoskeletal Motors and Highways in CNS Myelinated Axons. *Toxicol. Sci.* **2010**, *117*, 180–189. [CrossRef] [PubMed]
20. Graham, D.G.; Anthony, D.C.; Boekelheide, K.; Maschmann, N.A.; Richards, R.G.; Wolfram, J.W.; Shaw, B.R. Studies of the Molecular Pathogenesis of Hexane Neuropathy: II. Evidence That Pyrrole Derivatization of Lysyl Residues Leads to Protein Crosslinking. *Toxicol. Appl. Pharmacol.* **1982**, *64*, 415–422. [CrossRef]
21. Masi, M.; Dasari, R.; Evidente, A.; Mathieu, V.; Kornienko, A. Chemistry and Biology of Ophiobolin A and Its Congeners. *Bioorg. Med. Chem. Lett.* **2019**, *29*, 859–869. [CrossRef]
22. Dasari, R.; Masi, M.; Lisy, R.; Ferdérin, M.; English, L.R.; Cimmino, A.; Mathieu, V.; Brenner, A.J.; Kuhn, J.G.; Whitten, S.T. Fungal Metabolite Ophiobolin A as a Promising Anti-Glioma Agent: In vivo Evaluation, Structure–Activity Relationship and Unique Pyrrolylation of Primary Amines. *Bioorg. Med. Chem. Lett.* **2015**, *25*, 4544–4548. [CrossRef] [PubMed]
23. Chidley, C.; Trauger, S.A.; Birsoy, K.; O'Shea, E.K. The Anticancer Natural Product Ophiobolin A Induces Cytotoxicity by Covalent Modification of Phosphatidylethanolamine. *Elife* **2016**, *5*, e14601. [CrossRef] [PubMed]
24. Janganati, V.; Ponder, J.; Jordan, C.T.; Borrelli, M.J.; Penthala, N.R.; Crooks, P.A. Dimers of Melampomagnolide B Exhibit Potent Anticancer Activity against Hematological and Solid Tumor Cells. *J. Med. Chem.* **2015**, *58*, 8896–8906. [CrossRef] [PubMed]
25. Mantaj, J.; Jackson, P.J.M.; Rahman, K.M.; Thurston, D.E. From Anthramycin to Pyrrolobenzodiazepine (PBD)-containing Antibody–Drug Conjugates (ADCs). *Angew. Chemie Int. Ed.* **2017**, *56*, 462–488. [CrossRef]
26. Hartley, J.A. Antibody-Drug Conjugates (ADCs) Delivering Pyrrolobenzodiazepine (PBD) Dimers for Cancer Therapy. *Expert Opin. Biol. Ther.* **2021**, *21*, 931–943. [CrossRef]
27. Gong, Y.; Gallis, B.M.; Goodlett, D.R.; Yang, Y.; Lu, H.; Lacoste, E.; Lai, H.; Sasaki, T. Effects of Transferrin Conjugates of Artemisinin and Artemisinin Dimer on Breast Cancer Cell Lines. *Anticancer Res.* **2013**, *33*, 123–132.
28. Dury, L.; Nasr, R.; Lorendeau, D.; Comsa, E.; Wong, I.; Zhu, X.; Chan, K.-F.; Chan, T.-H.; Chow, L.; Falson, P. Flavonoid Dimers Are Highly Potent Killers of Multidrug Resistant Cancer Cells Overexpressing MRP1. *Biochem. Pharmacol.* **2017**, *124*, 10–18. [CrossRef]
29. Wong, I.L.K.; Zhu, X.; Chan, K.-F.; Law, M.C.; Lo, A.M.Y.; Hu, X.; Chow, L.M.C.; Chan, T.H. Discovery of Novel Flavonoid Dimers To Reverse Multidrug Resistance Protein 1 (MRP1, ABCC1) Mediated Drug Resistance in Cancers Using a High Throughput Platform with "Click Chemistry". *J. Med. Chem.* **2018**, *61*, 9931–9951. [CrossRef]
30. Zhu, X.; Wong, I.L.K.; Chan, K.-F.; Cui, J.; Law, M.C.; Chong, T.C.; Hu, X.; Chow, L.M.C.; Chan, T.H. Triazole Bridged Flavonoid Dimers as Potent, Nontoxic, and Highly Selective Breast Cancer Resistance Protein (BCRP/ABCG2) Inhibitors. *J. Med. Chem.* **2019**, *62*, 8578–8608. [CrossRef]
31. Kung Sutherland, M.S.; Walter, R.B.; Jeffrey, S.C.; Burke, P.J.; Yu, C.; Kostner, H.; Stone, I.; Ryan, M.C.; Sussman, D.; Lyon, R.P. SGN-CD33A: A Novel CD33-Targeting Antibody–Drug Conjugate Using a Pyrrolobenzodiazepine Dimer Is Active in Models of Drug-Resistant AML. *Blood J. Am. Soc. Hematol.* **2013**, *122*, 1455–1463. [CrossRef]
32. Just, J.; Jordan, T.B.; Paull, B.; Bissember, A.C.; Smith, J.A. Practical Isolation of Polygodial from Tasmannia Lanceolata: A Viable Scaffold for Synthesis. *Org. Biomol. Chem.* **2015**, *13*, 11200–11207. [CrossRef]
33. Bury, M.; Novo-Uzal, E.; Andolfi, A.; Cimini, S.; Wauthoz, N.; Heffeter, P.; Lallemand, B.; Avolio, F.; Delporte, C.; Cimmino, A. Ophiobolin A, a Sesterterpenoid Fungal Phytotoxin, Displays Higher in vitro Growth-Inhibitory Effects in Mammalian than in Plant Cells and Displays in vivo Antitumor Activity. *Int. J. Oncol.* **2013**, *43*, 575–585. [CrossRef]
34. Ingrassia, L.; Lefranc, F.; Dewelle, J.; Pottier, L.; Mathieu, V.; Spiegl-Kreinecker, S.; Sauvage, S.; El Yazidi, M.; Dehoux, M.; Berger, W. Structure− Activity Relationship Analysis of Novel Derivatives of Narciclasine (an Amaryllidaceae Isocarbostyril Derivative) as Potential Anticancer Agents. *J. Med. Chem.* **2009**, *52*, 1100–1114. [CrossRef]
35. Le Calvé, B.; Rynkowski, M.; Le Mercier, M.; Bruyère, C.; Lonez, C.; Gras, T.; Haibe-Kains, B.; Bontempi, G.; Decaestecker, C.; Ruysschaert, J.-M. Long-Term in vitro Treatment of Human Glioblastoma Cells with Temozolomide Increases Resistance in vivo through up-Regulation of GLUT Transporter and Aldo-Keto Reductase Enzyme AKR1C Expression. *Neoplasia* **2010**, *12*, 727–739. [CrossRef]
36. Branle, F.; Lefranc, F.; Camby, I.; Jeuken, J.; Geurts-Moespot, A.; Sprenger, S.; Sweep, F.; Kiss, R.; Salmon, I. Evaluation of the Efficiency of Chemotherapy in vivo Orthotopic Models of Human Glioma Cells with and without 1p19q Deletions and in C6 Rat Orthotopic Allografts Serving for the Evaluation of Surgery Combined with Chemotherapy. *Cancer Interdiscip. Int. J. Am. Cancer Soc.* **2002**, *95*, 641–655. [CrossRef]
37. Mathieu, V.; Pirker, C.; de Lassalle, E.M.; Vernier, M.; Mijatovic, T.; DeNeve, N.; Gaussin, J.; Dehoux, M.; Lefranc, F.; Berger, W. The Sodium Pump A1 Sub-unit: A Disease Progression–Related Target for Metastatic Melanoma Treatment. *J. Cell. Mol. Med.* **2009**, *13*, 3960–3972. [CrossRef] [PubMed]

38. Medellin, D.C.; Zhou, Q.; Scott, R.; Hill, R.M.; Frail, S.K.; Dasari, R.; Ontiveros, S.J.; Pelly, S.C.; van Otterlo, W.A.L.; Betancourt, T. Novel Microtubule-Targeting 7-Deazahypoxanthines Derived from Marine Alkaloid Rigidins with Potent in vitro and in vivo Anticancer Activities. *J. Med. Chem.* **2016**, *59*, 480–485. [CrossRef] [PubMed]
39. Perez-Benito, L.; Henry, A.; Matsoukas, M.-T.; Lopez, L.; Pulido, D.; Royo, M.; Cordomi, A.; Tresdern, G.; Pardo, L. The size matters? A computational tool to design bivalent ligands. *Bioinformatics* **2018**, *34*, 3857–3863. [CrossRef] [PubMed]
40. Knox, A.; Kalchschmid, C.; Schuster, D.; Gaggia, F.; Manzl, C.; Baecker, D.; Gust, R. Development of bivalent trialkene- and cyclofenil-derived dual estrogen receptor antagonist and downregulators. *Eur. J. Med. Chem.* **2020**, *192*, 112191. [CrossRef]
41. Paquin, A.; Reyes-Moreno, C.; Berube, G. Recent advances in the use of the dimerization strategy as a means to increase the biological potential of natural or synthetic molecules. *Molecules* **2021**, *26*, 2340. [CrossRef] [PubMed]

Review

Promising Antiviral Activities of Natural Flavonoids against SARS-CoV-2 Targets: Systematic Review

Ridhima Kaul [1,†], Pradipta Paul [1,†], Sanjay Kumar [2,3], Dietrich Büsselberg [4], Vivek Dhar Dwivedi [2] and Ali Chaari [1,*]

1. Weill Cornell Medicine-Qatar, Education City, Qatar Foundation, Doha 24144, Qatar; rik4001@qatar-med.cornell.edu (R.K.); prp4005@qatar-med.cornell.edu (P.P.)
2. Center for Bioinformatics, Computational and Systems Biology, Pathfinder Research and Training Foundation, Greater Noida 201308, India; sanjay93.sci@gmail.com (S.K.); vivek_bioinformatics@yahoo.com (V.D.D.)
3. School of Biotechnology, Jawaharlal Nehru University, New Delhi 110067, India
4. Department of Physiology and Biophysics, Weill Cornell Medicine-Qatar, Education City, Qatar Foundation, Doha 24144, Qatar; dib2015@qatar-med.cornell.edu
* Correspondence: alc2033@qatar-med.cornell.edu
† These authors contributed equally to this work.

Abstract: The ongoing COVID-19 pandemic, caused by the severe acute respiratory syndrome coronavirus 2 (SARS-CoV-2) became a globally leading public health concern over the past two years. Despite the development and administration of multiple vaccines, the mutation of newer strains and challenges to universal immunity has shifted the focus to the lack of efficacious drugs for therapeutic intervention for the disease. As with SARS-CoV, MERS-CoV, and other non-respiratory viruses, flavonoids present themselves as a promising therapeutic intervention given their success in silico, in vitro, in vivo, and more recently, in clinical studies. This review focuses on data from in vitro studies analyzing the effects of flavonoids on various key SARS-CoV-2 targets and presents an analysis of the structure-activity relationships for the same. From 27 primary papers, over 69 flavonoids were investigated for their activities against various SARS-CoV-2 targets, ranging from the promising 3C-like protease (3CLpro) to the less explored nucleocapsid (N) protein; the most promising were quercetin and myricetin derivatives, baicalein, baicalin, EGCG, and tannic acid. We further review promising in silico studies featuring activities of flavonoids against SARS-CoV-2 and list ongoing clinical studies involving the therapeutic potential of flavonoid-rich extracts in combination with synthetic drugs or other polyphenols and suggest prospects for the future of flavonoids against SARS-CoV-2.

Keywords: flavonoids; coronavirus; SARS-CoV-2; SARS-CoV; MERS-CoV

1. Introduction

In the last two decades, human coronaviruses caused three epidemics: Severe acute respiratory syndrome coronavirus (SARS-CoV-1) in 2003, the Middle East respiratory syndrome coronavirus (MERS-CoV) in 2012, and recently the SARS-CoV-2, responsible for the outbreak of the coronavirus disease 2019 (COVID-19) pandemic, from which the world is suffering since late 2019 [1–3]. COVID-19 continuously spreads at a high pace with new strains [4,5]. As of 16 August 2021, there were 207,173,086 confirmed cases of COVID-19, including 4,361,996 deaths, across the globe [4].

COVID-19 is associated with severe respiratory symptoms such as pneumonia and is commonly accompanied by fever; it is caused by CoV-2, which belongs to the Betacoronavirus genus in the family *Coronaviridae* of the order *Nidovirales* [6] (Figure 1). Members of this order share several distinctive characteristics. All of them are surrounded by an envelope that contains very large genomes (≥30 kilobases) characterized by a highly conserved genomic organization. They all express numerous nonstructural genes and replicate using a set of mRNAs [7]. The Beta-coronavirus genus has its ancestor in bat CoVs and includes other viruses such as the human coronaviruses HCoV-OC43 and HCoV-HKU1, which cause the common cold. The genus also includes MERS-CoV and SARS-CoV-1. The latter is closely related to the current CoV-2, though CoV-2 is reported to be more transmissible between individuals [8]. While MERS-CoV and SARS-CoV-1, like SARS-CoV-2, originate from bat CoVs, their intermediate hosts are probably dromedary camels (MERS), civet cats, or raccoon dogs (SARS-CoV-1) [9]. Phylogenetic studies of SARS-CoV-2 genomes and other coronaviruses reveal the phylogenetic relationship among SARS-CoV-2 and other β-CoVs, contributing to the fight with the actual pandemic [10–12]. Studies revealed that SARS-CoV-2 (GenBank: MN908947.3) present about 96% nucleotide sequence identity with bat coronavirus RaTG13 (GenBank: MN996532.1), 79.5% identity with SARS-CoV BJ01 (GenBank: AY278488.2) and 55% with MERS-CoV HCoV-EMC (GenBank: MH454272.1) [13]. Based on PubMed, since its emergence, over 100,000 papers have addressed the COVID-19 disease. However, there is still much to learn about SARS-CoV-2 to define efficient therapeutic and/or preventive strategies.

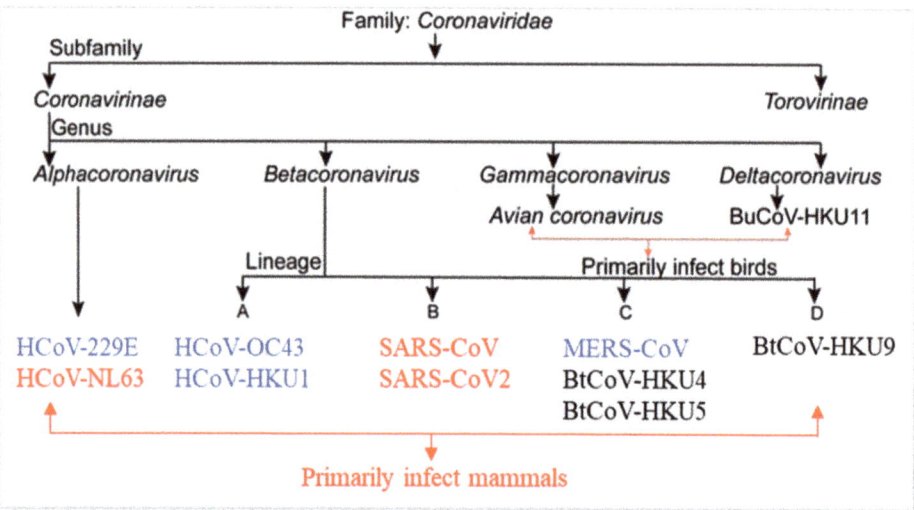

Figure 1. Classification of coronaviruses: the seven known HCoVs are in blue and red. Human coronaviruses (in red) bind the same host receptor, angiotensin-converting enzyme 2 (ACE2).

Despite the progress made in immunization and drug development, global infections keep rising. The highly unequal distribution of vaccines and the development of new COVID-19 strains raised a worldwide effort to find potential inhibitors of key viral processes. Moreover, most antiviral drugs today are single-target drugs designed against a unique viral enzyme. While developing new efficient therapeutic strategies and new drugs is a long process, natural substances are attractive therapeutic solutions in this context. They are proven to be the primary source of antimicrobial and antiviral drugs [14,15]. Many studies have demonstrated that targeting the virus-specific proteins is an effective strategy for drug discovery towards developing direct-acting antivirals [16–18]. Based on complementary approaches using both in vitro experiments and in silico virtual screenings, many drug discoveries are initiated by exploiting the effect of natural substances, including flavonoids, on essential coronavirus enzymes as a drug target (Figure 2) [19–23]. While in silico approaches such as molecular dynamics and structure-based virtual screening represent some of the early steps of drug discovery, in vitro studies utilize the findings to narrow down further and experiment with the most promising compounds in cell-based or cell-free methodologies.

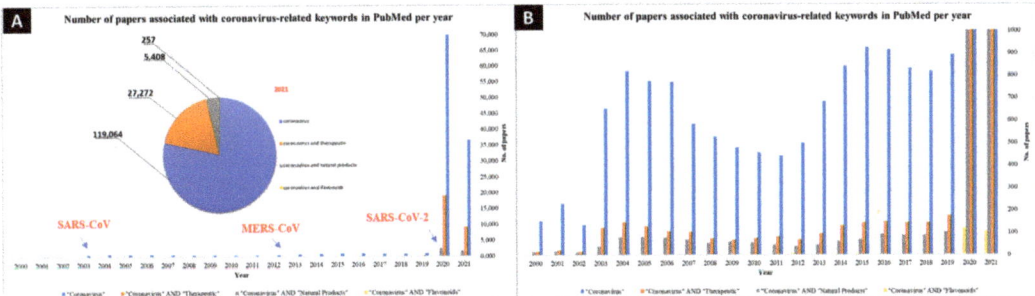

Figure 2. Comparison of the number of papers per year dealing with the words "coronavirus", "coronavirus and therapeutic", "coronavirus and natural products" and in the inset, total number of papers used keywords published from 2000 to 2021. (**A**) Bar graph to scale to account for all relevant papers published from 2020–2021, with indications at each major coronavirus outbreak and a pie chart showing the relative number of papers per key phrase. (**B**) Zoomed-in figure to account for low number of hits from 2000–2019 as well the low number of papers dealing with "coronavirus and flavonoids". (Source: PubMed Database (https://pubmed.ncbi.nlm.nih.gov/), accessed on 3 June 2021).

Flavonoids represent potential candidates to interfere with the coronavirus life cycle because of their safe administration with lack of systemic toxicity, their ability to work in synergy even with other drugs, and the capacity of their functional groups to interact with different cellular targets and intercept multiple pathways [24,25]. Flavonoids as natural substances represent potential candidates against the actual pandemic because of their biological availability and the participation of most countries (194 countries) on national-level policy for herbal medicines [26]. Moreover, the growing understanding of the efficiency of antiviral drug development based on flavonoids tested previously with other viruses and coronaviruses underlines the importance of exploring these natural products against SARS-CoV-2 [18,24,27].

Flavonoids include many secondary metabolites found in fruits, vegetables, and several plants [14,27]. From a chemical view, flavonoids are hydroxylated phenolic molecules characterized by their structural class and degree of hydroxylation. The hydroxyl functional groups of flavonoids are responsible for their antioxidant activity and are formed by two benzene rings (A and B rings), connected via a heterocyclic pyrene ring (C-ring) [28]. Based on their chemical structure, flavonoids are segregated into various classes, among which there are further subclasses [29,30]. They were extensively studied and recently gained increased interest among researchers and clinicians for their antimicrobial, antioxidant, anti-inflammatory, anti-cancer, and antiviral properties with numerous mechanisms to prevent infection and strengthen host immunity [31–37]. Among their many beneficial effects, antiviral properties can serve as a future therapeutic utility for drugging COVID-19.

Flavonoids as biologically active substances can affect coronaviruses at the stages of penetration and entry of the viral particle into the cell, replication of the viral nucleic acid, and release of the virion from the cell; they also can act on the host's cellular targets. These natural compounds could be a vital resource in the combat against coronaviruses, including the actual emerging pandemic. This review highlights the importance of flavonoids and their effects on various key SARS-CoV-2 targets and analyzes the structure–activity relationships for the same.

2. Methods

The review uses references from major databases such as Web of Science, PubMed, Scopus, Elsevier, Springer, and Google Scholar using keywords such as 'flavonoids', 'coronaviruses', or 'SARS-CoV', or 'MERS-CoV', or 'SARS-CoV-2'. An initial search in May 2021 was followed up with another on 17 August 2021 to include any new records that were published. After obtaining all reports from the databases, the papers were analyzed for relevant data; the search protocol is summarized in Figure 3. From 321 search hits,

266 review papers, in silico studies, duplicates, and non-English were removed. From the 55 papers qualified for further screening, 26 articles dealing with other viruses or different compounds were removed from the process.

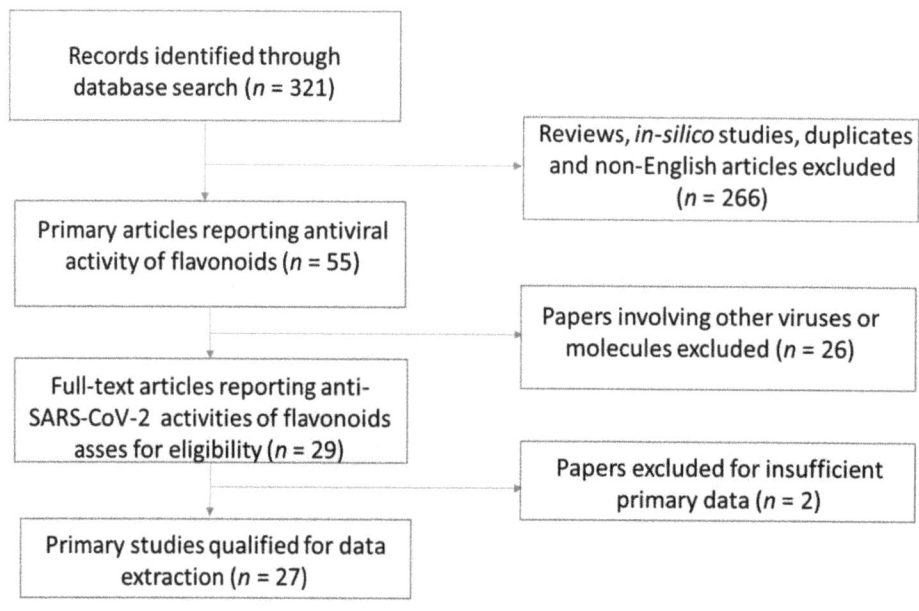

Figure 3. Flow chart of the search strategy.

Moreover, two publications without sufficient primary data on SARS-CoV-2 targets were excluded. Ultimately, 27 primary literature papers exploring the in vitro activities of over 69 flavonoids qualified for the final data extraction phase, with two independent reviewers extracting the data per study. Most studies detailed SARS-CoV-2 3C-like protease (3CLpro) as the most promising target, explaining the large number of flavonoids explored for their respective inhibitory activities against this protease. Other targets included the papain-like protease (PLpro), the spike (S) protein–ACE2 interaction, helicase, and the nucleocapsid (N) protein. Data were segregated according to SARS-CoV-2 target, class of flavonoid, and flavonoid (where multiple studies reported activities for the same flavonoid) and is summarized in tables under relevant sections. Wherever possible, mode of action, efficacy, cytotoxicity, and natural source efficacy of flavonoids are presented.

Nevertheless, other papers reporting general antiviral activities of flavonoids against CoV-2 without target specificity are also discussed. We compiled, evaluated, and analyzed the literature where flavonoids show inhibition against various SARS-CoV-2 targets through the use of various in vitro methodologies. We included both cell-free and cell-based methods, ultimately to suggest follow-up in vivo and clinical trials, focusing on a handful of effective molecules for the treatment of COVID-19. A search using the keywords "COVID-19", "flavonoids" and "polyphenols" was also conducted on ClinicalTrials.gov to compile and analyze ongoing real-world clinical trials reported until 17 August 2021 that explore the effect of flavonoids on patients suffering from COVID-19.

3. Results and Discussion

3.1. Coronaviruses Biology and Therapeutic Strategies for the Treatment of COVID-19 Infection

3.1.1. Genomic Characterization and Structure of SARS-CoV-2

SARS-CoV-2, like other coronaviruses, is enveloped with crown-like particles enclosing a positive-sense single-stranded RNA, which is characterized by a 5′ cap and a 3′ poly

(A) tail, which enables it to act as a messenger RNA for translation of replicase polyproteins once inside the host cell [38–40]. The single positive strand of SARS-CoV-2 genomic RNA is about 30 kb in size, and like those of SARS-CoV and MERS-CoV, its genome comprises 12 open reading frames (ORFs) in number (Figure 4B) [41].

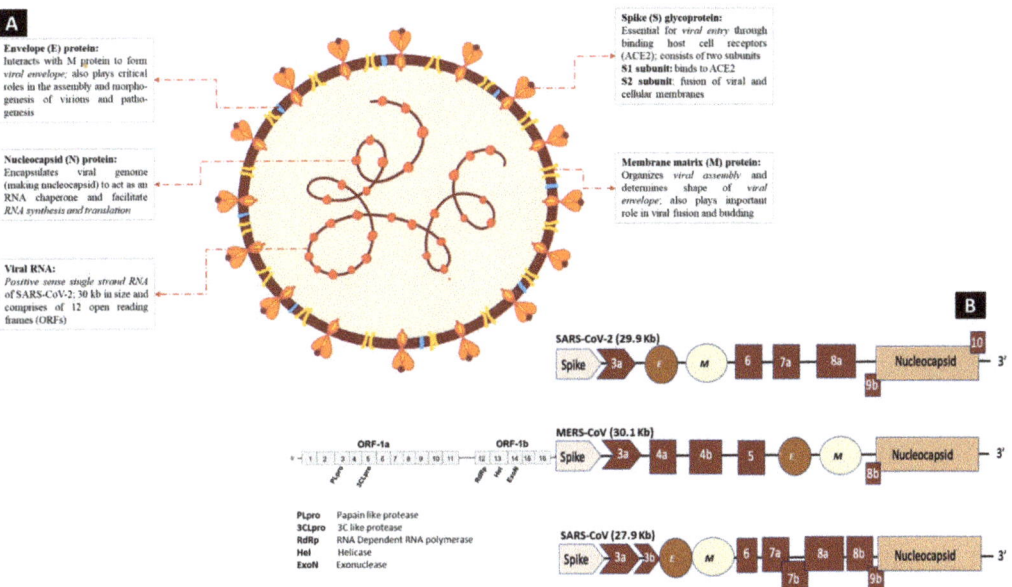

Figure 4. Schematic diagram of the SARS-CoV-2 structural and functional components. (A) The single-stranded viral RNA (ssRNA) is encapsulated at places with the nucleocapsid (N) protein, whereas the membrane consists of both the membrane matrix (M) and the envelope (E) proteins, from where the spike (S) protein protrudes outside and interacts with human ACE2 receptors during viral entry. (B) Consisting of two open reading frames (ORFs), ORF 1a and 1b, that code for Non-structural proteins (NSPs) as well as PLpro, 3CLpro, RdRp, exonuclease, the 29.9 kb ssRNA genome of SARS-CoV-2 observes similarities with the genomes of previous epidemic-causing coronaviruses, namely MERS-CoV (30.1 kb) and SARS-CoV (27.9 kb). In addition to coding for the polyproteins pp1a and pp1ab, the genome is responsible for translating numerous structural and functional molecules such as the S protein, E protein, M protein, and N protein.

The first ORF, ORF 1a and 1b, consists of 67% of the genome and encodes RNA polymerase and other nonstructural proteins (nsPs), including the main coronavirus protease chymotrypsin-like protease (3CLpro), RNA-dependent RNA polymerase (RdRp), and papain-like protease (PLpro). At the same time, the remaining ORFs generate several structural and accessory proteins [42,43]. The nonstructural proteins are essential for directing RNA synthesis and processing, cellular mRNA degradation, host immune response suppression, and double-membrane vesicle formation [44,45]. The four main structural proteins encoded by the genomic RNA consist of three main proteins surrounding the viral particle; the surface glycoprotein called the spike protein (S), membrane protein (M), an envelope protein (E), while the fourth nucleocapsid protein (N) is located internally and is intimately associated with the viral genomic RNA (Figure 4A) [41].

The S protein is a trimeric transmembrane glycoprotein responsible for binding with cellular receptors through its S1 subunit and for fusing the virus and the host cell before its entrance with its S2 subunit [46]. Both SARS-CoV and SARS-CoV-2 bind to the same functional host cell receptor, the angiotensin-converting enzyme 2 (ACE2), a membrane protein expressed in the lungs, heart, kidneys, and intestine of the host [13]. The M protein, a 25–30 kDa protein with three transmembrane domains, plays a central role in assembling new viral particles, where host factors and viruses come together to form new virus particles [47]. The E protein, a 9–12 kDa protein, presents an ion channel activity that plays a vital role in the virus life cycle, from assembly to release [47].

N protein, a 45 kDa RNA-binding protein, is the only protein that binds to the viral genome in a beads-on-a-string conformation. This protein is crucial in viral pathogenesis and can cooperate with some of the mentioned structural proteins such as the M protein, which helps to improve the efficiency of virus transcription and assembly [48].

3.1.2. Mechanism of Cell Entry and Life Cycle of the Virus

The virus cell cycle occurs in distinguished steps, including attachment, entry, induction of replicase proteins, replication, transcription, assembly, and discharge of mature viral particles [49] (Figure 5). Receptor binding and membrane fusion are the initial steps facilitated by the attachment of the virus to the host cells via binding of the spike protein (S1 region) to the receptor ACE2 for SARS-CoV and SARS-CoV-2 and the dipeptidyl peptidase 4 receptor (DPP4) for MERS-CoV [49]. This attachment is followed by a series of steps leading to the delivery of the viral genome into the cytoplasm.

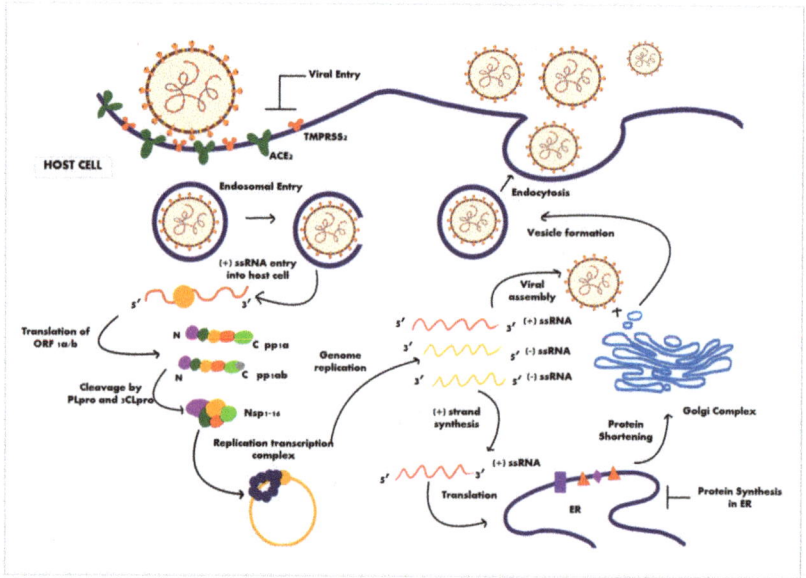

Figure 5. The SARS-CoV-2 viral life cycle with various stages as potential targets of therapeutic intervention.

Following the binding of the spike to ACE2, host proteases such as transmembrane protease serine 2 (TMPRSS2) are involved in the cleavage of the S protein, which helps the virion enter the host cell. Other proteases, such as Furin, then release the spike fusion peptide, allowing the cellular virus to pass through the endosomal pathway [49]. The virus RNA genome is released into the cytosol of the infected cell. The last step is favored by the low pH of the endosomal microenvironment and the S2 functional subunit of the S spike protein [50]. Once in the cytosol, the genomic RNA is translated into two polyproteins—pp1a and pp1ab—using the host ribosome machinery. The viral genome replication and the synthesis of functional and structural viral proteins occur. These polyproteins undergo proteolytic cleavage by two viral cysteine proteases, PLpro and 3CLpro, generating many nsPs, some of which subsequently assemble to form a replicase–transcriptase complex (RTC) [38,51–53]. The RTC, amongst various other domains, contains an RdRp domain which aids the replication of the positive-sense RNA to form a negative-strand RNA intermediate [52,53], which has two different fates. The negative RNA strand can undergo discontinuous transcription to create shorter, usually overlapping, sub-genomic RNA segments. Translation of this sub-genomic RNA produces essential structural proteins such as the M, E, and S proteins, which are then inserted into the secretory pathway of the host cell to be processed and packaged to form virion progeny. Another fate of the negative-strand viral RNA is to undergo further replication to give positive-sense RNA, which is inserted into the virion progeny being processed in the secretory pathway [52,53]. Once processing is completed, the new virion progeny is exocytosed by the cell, and the progeny can infect other host cells [52,53].

3.1.3. Promising Therapeutic Strategies for the Treatment of COVID-19 Infection

Although there is progress against the COVID-19 pandemic following the development of vaccines and some specific drugs that show minor effects against the disease, diverse events have limited this progress. They have raised a worldwide effort to find potential inhibitors relevant to mechanistic targets involved in SARS-CoV-2 infection. Early studies and pharmaceutical experience have shown that antivirals' de novo development is a time-, cost-, and effort-intensive endeavor [54,55]. Secondly, the safe use of natural compounds derived from natural sources against different viruses, including SARS-CoV and MERS-CoV, was acknowledged for several years, making them potential and powerful anti-COVID-19 drugs [24,51]. In this context, and besides the use and the assays, small molecule drugs, monoclonal antibodies, peptides, and interferon therapies to combat COVID-19, natural products including flavonoids emerged as a safe alternative therapeutic strategy against different targets for blocking the coronavirus life cycle at different stages of viral infection [55]. Flavonoids can directly target specific viral steps and enzymes or components at each phase of the virus life cycle. Figure 6 summarizes different viral and infected host targets that were revealed to be essential to develop potential therapeutics to inhibit the viral pathogenesis of SARS-CoV-2.

Figure 6. Viral and infected host cell targets important for potential therapeutics for inhibiting the viral pathogenesis of SARS-CoV-2.

The first therapeutic strategy targets the first step on the virus life cycle, which is the early entry of COVID-19 by interrupting spike–ACE2 protein–protein interaction, TMPRSS2 activity, and endocytic pathway-associated proteins such as clathrin and cathepsin L, preventing the internalization of SARS-CoV-2 in the host cell. The second therapeutic strategy involves the inhibition of the viral proteases 3CLpro and PLpro. The inhibition of these two key proteases blocks the production of non-structural proteins, including the RdRp and helicase, which directly block the transcription and replication of the virus [55]. Moreover, blocking directly viral replication enzymes may consist of the third potential target to treat COVID-19. Finally, the fourth therapeutic strategy targeting the release outside the infected cells of the new virions consists of reducing/silencing the expression and/or the activity of the ion channel viroporin 3a [55].

3.2. Flavonoids and Their Antiviral Properties

Flavonoids are the largest group of phenolic phytochemicals in higher plants. They also belong to secondary plant metabolites found in fruits, vegetables, seeds, roots, *propolis*, and other plant products such as tea and wine. With over 9000 structurally identified flavonoids, research has associated many of these compounds with multiple health-promoting effects, ranging from nutraceutical, pharmaceutical, medicinal, and cosmetic applications to their antioxidative, anti-inflammatory anti-mutagenic, and anti-carcinogenic properties [56,57]. In plants, they play an important role as components of cells to defend against pathogens, insects, and other stressful environments [58–62].

Produced by the phenylpropanoid pathway, flavonoids are hydroxylated phenolic molecules divided into classes by their structure, degree of hydroxylation, and polymerization. As mentioned before, their basic structure (flavan or 2-phenylchroman) consists of two benzene rings (A- and B-rings), connected through a heterocyclic pyrene ring (C-ring). The different subclasses of flavonoids include anthocyanins, chalcones, dihydrochalcones, dihydroflavonols, flavan-3-ols, flavanones, flavones, flavonols, flavanonols, and isoflavonoids (Figure 7).

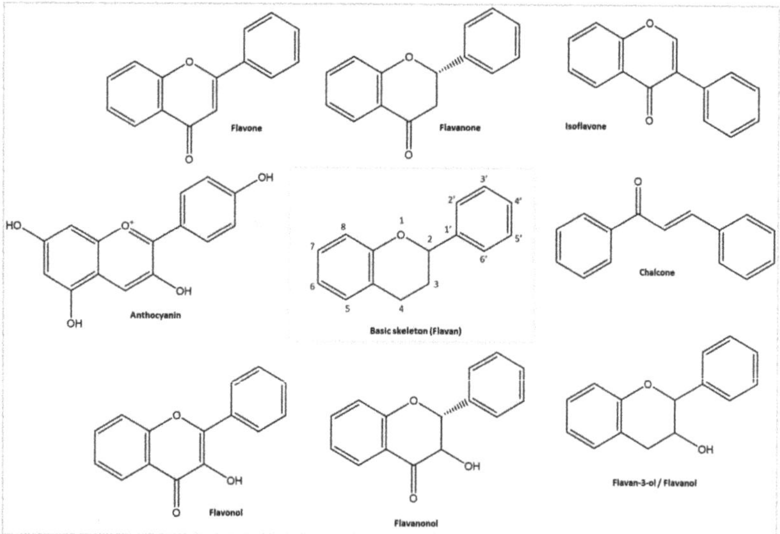

Figure 7. Basic structure of flavonoids (Flavan) and its different classes; image made using ChemDraw (https://perkinelmerinformatics.com/products/research/chemdraw/, accessed on 1 September 2021).

Unlike synthetic antiviral drugs with narrow antiviral activities and varying levels of patient-specific clinical efficacies, the potent bioactivity and the wide range of pharmacological and toxicological applications of flavonoids, make them be considered as potential therapeutics to both existing and novel public health concerns. In fact, their structures are often inspirations for synthetic therapeutic drugs [63–66]. Natural extracts and their isolated polyphenols are investigated in depth using in vitro, in vivo, and more recently, in silico studies for their antiviral effects against viral entry to release, as well as the intermediate cascades, such as R/DNA replication, protein translation, and post-translational modifications, of multiple viruses, both respiratory (such as H1N1) and non-respiratory (such as herpes simplex virus [HSV]) [24,49,67–72].

3.2.1. Antiviral Activities of Flavonoid against Non-Respiratory Viruses

In 2012, Johari et al. reported the flavone baicalein to have anti-adsorption, anti-replication, and direct virucidal effects against the Japanese encephalitis virus (JEV) [73]. Similar experiments by Ting et al. found that flavonol kaempferol and the isoflavone daidzein showed greater anti-JEV properties in pretreated cells [74]. Lani et al. reported that baicalein also repressed the Chikungunya virus (CHIKV) activity, while the glycosylated flavonol quercetagenin inhibited its replication, respectively. Along with the flavonol fisetin, all three inhibit different aspects of viral mechanism: extracellular stage, viral entry, gene replication, and other intracellular stages [75]. Independently, nobiletin and silymarin also were shown to act against CHIKV [76,77]. On the other hand, Zandi et al. showed quercetin to have direct antiviral activity against Dengue virus 2 (DENV-2) [78]. Rutin, baicalein, baicalin, daidzein, and naringenin also exhibited significant inhibition against DENV-2 [79–81]. Similarly, against enterovirus A71, the causative agent of encephalitis and hand, foot, and mouth disease (HFMD), apigenin, kaempferol, baicalein, and baicalin showed in vitro and in vivo promise [82–85].

Multiple flavonoids were investigated against the hepatitis C virus (HCV) in the last few decades, given its critical role in a wide array of liver diseases. The flavanone naringenin represses the release of HCV core dose-dependently and reduces its infectivity when pretreated with the flavonoid in vitro [86]. Apigenin, silybin, quercetin, ladanein, sorbifol, and pedalitin are also effective candidates against the same virus [87–90]. Similar promising concentration-dependent inhibitory effects were reported for epigallocatechin gallate (EGCG) against the Hepatitis B Virus (HBV) e antigen (HBeAg) [89]. Baicalin, genistein, and sodium rutin sulfate (SRS) all have inhibitory properties against the envelope fusion mediated viral entry of human immunodeficiency virus 1 (HIV-1) [89,91,92]. In addition to the above, multiple other naturally occurring and synthetic flavonoids inhibit non-respiratory viruses in the literature [49,72,93–95].

3.2.2. Antiviral Activities of Flavonoid against Respiratory Viruses

Imanisi et al. tested epigallocatechin (EGC) as a major constituent of green tea extract (GTE) against Madin-Darby canine kidney (MDCK) cells infected with various strains of the influenzas A and B viruses; the flavanol inhibited the early stages of viral infection in vitro [96]. Similarly, quercetin and its derivatives, silymarin, and multiple other flavonoids also have similar activities [92,97–99].

Another common viral pathogen causing respiratory diseases is rhinovirus (RV). Using BEAS-2B cell-based methodologies, quercetin decreases the levels of multiple strains of. Moreover, by pretreating the cells with quercetin before RV infection, viral endocytosis halted significantly, presumably by interaction with and inhibition of the cell enzyme PI-3-kinase; the flavonol potently inhibits the viral replication stage [100]. Additionally, Desideri et al. reported that the novel flavonoid derivatives 6-chloro-3-methoxy-flavone-4′-carboxylic acid, 6-chloro-4′-oxazolinyl, and 6-chloro-3-methoxy-4′-oxazolinyl flavone inhibited the activity of human rhino virus (HRV-1B) without having significant cytotoxic effects on the cells [101]. These examples show great promise in presenting applications of flavonoids against the novel coronavirus (n-CoV).

3.3. Antiviral Properties against Coronaviruses, Including SARS-CoV and MERS-CoV

Research exploring the antiviral activities of various natural flavonoids against animal and human coronaviruses span over three decades; some of this primary literature is compiled and reported in Table S1 (see Supplementary Material). In 1990, it was found that kaempferol reduces the replication of both human and bovine coronaviruses in vitro by ~65% at a concentration of just 10 µg/mL; against the same viruses, both chrysin and quercetin inhibited key stages in the replication and infectivity stages, although less effectively [72]. Later, theaflavin constituents (including those with galloyl moieties) of black tea synergistically inhibited bovine coronavirus (BCV) activity [102]. Perhaps most importantly, as priorly reported, quercetin 7-rhamnoside inhibits the non-respiratory coronavirus PEDV with a very potent IC50 = 0.014 µg/mL with high cellular cytotoxic tolerance; its structural analogs also showed promising activities (Table S1) [18].

3.3.1. SARS-CoV

SARS-CoV proteases, PLpro and 3CLpro, are the most investigated targets for flavonoid inhibition, some of these are listed in Table S1 [72,103–112]. In 2010, Ryu et al. demonstrated, using FRET assays, that the biflavonoid amentoflavone, a constituent of *Torreya nucifera*, non-competitively inhibits the SARS-CoV 3CLpro very effectively with an IC50 of 8.3 µM, while three other biflavonoids, namely, bilobetin, ginkgetin, and sciadopitysin, with methylation of 7-, 4′-, and 4′′′-hydroxyl groups, were less potent [103]. The flavonol herbacetin and the flavones rhoifolin and pectolinarin inhibit the protease; their hydrophobic aromatic rings and hydrophilic hydroxyl groups contributing to the binding affinity [104]. Among the seven flavonoids derived from *Pichia pastoris* explored by Nguyen et al., quercetin, ECGC, and GCG were most effective at inhibiting SARS-CoV 3CLpro in vitro; as the best inhibitor, GCG, a flavanol with two galloyl moieties, was further analyzed and was a competitive inhibitor showing a strong affinity for the protease (Table S1) [105].

Various polyphenols' inhibitory effects range from chalcones to flavonols, derived from *Broussonetia papyrifera* on both SARS-CoV cysteine proteases (3CLpro and PLpro). While PLpro inhibition was significantly more potent than 3CLpro overall, the chalcone Broussochalcone A was the most effective in the study [106]. Furthermore, twelve geranylated flavonoids from *Paulownia tomentosa* displayed dose-dependent SARS-CoV PLpro mixed inhibition using fluorogenic assay. While tomentin B, a reversible inhibitor that binds to the active site, was the most potent (lowest Ki; 3.5 µM, mixed), tomentin E had the lowest active concentration marker (IC50 = 5.0 ± 0.06 µM), whereas other constituents of the fruit also had effective concentrations (Table S1) [107].

The flavonol kaempferol's glycoside derivate, juglanin, inhibited viral production and released through interference with the Ba^{2+}-sensitive or cation-selective 3a channel protein of SARS-CoV potently with a very low IC50 of 2.3 µM [108]. Another SARS-CoV target explored is its helicase, nsP13. Scutellarein inhibits the ATPase activity of the helicase very effectively (IC50 = 0.86 ± 0.48 µM), while myricetin, myricitrin, amentoflavone, and Diosmetin-7-O-Glc-Xyl are less potent (Table S1) [109].

Moreover, the anthraquinone emodin, a constituent of the genus *Rheum* and *Polygonum*, inhibited the SARS-CoV S–ACE2 interaction concentration-dependent and also affected the infectivity of SARS-CoV spike-pseudotyped retrovirus to Vero cells [110]. Notably, the glycosylated flavone baicalin has general promising antiviral activity when tested against serum from patients infected with SARS-CoV at different periods post-incubation (Table S1) [111]. Similar experiments revealed that luteolin, procyanidin A2, procyanidin B1, and cinnamtannin B1 have antiviral activities of varying promising efficacies [27,113]. Antiviral activities of flavonoids against the SARS-CoV N protein and NTPase/helicase were also explored and published [114,115].

3.3.2. MERS-CoV

As for SARS-CoV, most flavonoids show anti-MERS-CoV activities targeted its proteases, namely 3CLpro and PLpro. Park et al., using fluorometric cleavage assay, also showed the inhibitory effects of multiple polyphenols derived from Broussonetia papyrifera on MERS-CoV cysteine proteases (3CLpro and PLpro), out of which the chalcone Broussochalcone B showed one of the most effective activities against 3CLpro (IC50 = 27.9 ± 1.2 µM). In contrast, Broussochalcone A was effective against PLpro (IC50 = 42.1 ± 5.0 µM) [106]. Jo et al., using FRET protease assays, reported anti-MERS-CoV 3CLpro activities of herbacetin, isobavachalcone, quercetin 3-β-d-glucoside, and helichrysetin [112]. The literature searches yielded that other flavonoids with lower effectiveness against MERS-CoV viral targets could be potential candidates for better therapeutic interventions against SARS-CoV-2 (Table S1) [116].

3.4. Antiviral Activity of Flavonoids against SARS-CoV-2

More than 69 flavonoids with inhibitory activities against specific SARS-CoV-2 targets were identified, most of whom belonged to the classes of flavonols and flavones, signifying their potent antiviral activities in general (Figure 8). Moreover, the most promising SARS-CoV-2 target was 3CLpro, followed closely by disrupting the S-ACE2 interaction and PLpro. A collection of structures of flavonoids showing anti-SARS-CoV-2 activities are reported in the Supplementary Material (Figures S1–S7), based on subclasses.

3.4.1. Antiviral Activity of Flavonoids against SARS-CoV-2 Proteases (3CLpro and PLpro)

SARS-CoV-2 proteases remain the most popular targets for the antiviral activity of flavonoids. Among these, 3CLpro was investigated against flavonols and flavones more than other subclasses of flavonoids. Tables 1 and 2 summarize the in vitro activities of various flavonoids against SARS-CoV-2 3CLpro, and PLpro reported in the literature, respectively [117–128].

Flavonols and Flavanonols

Flavonols have inhibitory activity in vitro against SARS-CoV-2 3CLpro. Of these, myricetin, quercetin, and their derivatives were the largest group of flavonols to have such activity. Among studies reporting the use of cell-based in vitro methodologies, the myricetin C7 derivative, 7-O-methyl-myricetin, was the most effective directly against SARS-CoV-2 3CLpro activity with an IC50 of 0.30 ± 0.00 µM. In contrast, another myricetin derivative, myricetin-7-yl diphenyl phosphate, was the most effective at inhibiting SARS-CoV-2 replication within cells with an EC50 of 3.15 ± 0.84 µM [117]. Among cell-free methods, base myricetin derived from black garlic and Polygoni avicularis is effective (IC50 = 43 ± 1 µM) against 3CLpro. Rutin has the lowest IC50 (32 µM); base quercetin also demonstrated a high affinity for the protease active site (Ki = 7.4 µM) [118,124,125].

Multiple flavonols derived from black garlic extract (IC50 = 137 ± 10 µg/mL), among whom quercetin and its derivatives showed remarkable efficacies for 3CLpro, likely due to the flavonols' interaction with the 3CLpro substrate-binding site as hypothesized and confirmed by independent researchers [117,118,121]. Others reported that quercetin acts as a competitive inhibitor of the 3CLpro active site and has a dose-dependent destabilizing effect on the thermal stability of the protease [124]. In general, the efficacies of most quercetin-derived glycosylated flavonols such as quercetin-4′-O-α-D-glucopyranoside and rutin (quercetin-3-O-rutinoside) against 3CLpro are significantly lower than that of base quercetin at 200 µM when assessed in the same study and under similar conditions [118]. However, other cell-free studies uphold rutin's status as a potent competitive inhibitor of the 3CLpro catalytic site due to its interaction with the catalytic dyad His41/Cys145 [125]. Similarly, varying cross-study data of quercetagenin, with cell-free and cell-based assays, report drastically different active concentration markers [118,119].

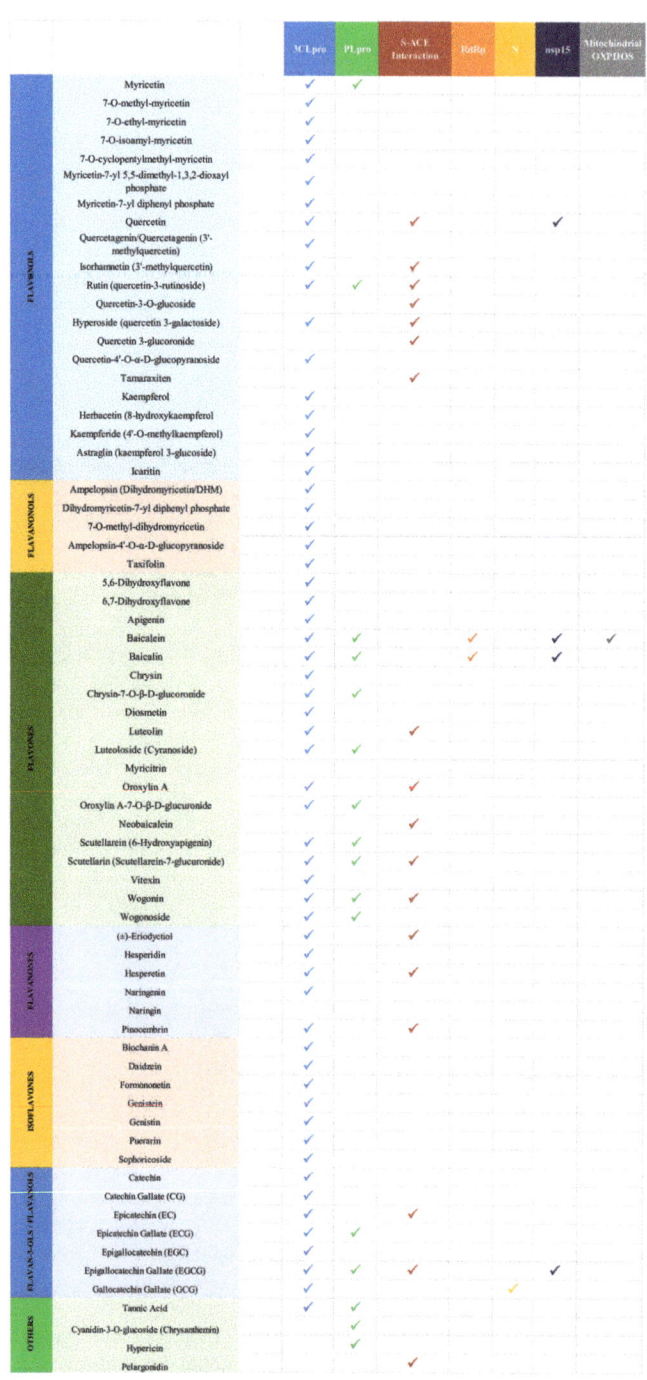

Figure 8. Summary of various flavonoids with anti-SARS-CoV-2 activities, segregated based on subclass and target of inhibition.

Table 1. Flavonoids with antiviral activities reported against SARS-CoV-2 3CLpro using in vitro methodologies segregated according to class.

Class	Class Flavonoid	Natural Source (N.S./Extract)	Efficacy of N.S. Extract	Mode of Action	Methods Used	IC50 (µM)	EC50 (µM)	% Inhibition	Reference
Flavonol	7-O-methyl-myricetin			Interacts with 3CLpro catalytic site	FRET assay, Vero E6 Cells, qRT-PCR	0.30 ± 0.00	12.59 ± 4.41		[117]
	7-O-ethyl-myricetin			Interacts with 3CLpro catalytic site	FRET assay, Vero E6 Cells, qRT-PCR	0.74 ± 0.06	51.01 ± 12.79		[117]
	7-O-isoamyl-myricetin			Interacts with 3CLpro catalytic site	FRET assay, Vero E6 Cells, qRT-PCR	1.92 ± 0.16	31.54 ± 0.74		[117]
	7-O-cyclopentylmethyl-myricetin			Interacts with 3CLpro catalytic site	FRET assay, Vero E6 Cells, qRT-PCR	2.45 ± 0.26	7.56 ± 2.34		[117]
	Astragalin (kaempferol 3-glucoside)	Black garlic extract	IC50: 137 ± 10 µg/mL, 100% inhibition at 0.5 mg/mL	Inhibits 3CLpro activity	FRET assay	143 ± 9		61 at 200 µM	[118]
	Herbacetin (8-hydroxykaempferol)	Flaxseed hulls, Rhodiola		Binds to the 3CLpro substrate binding site	Vero cells/RT PCR			59.1 ± 1.9 at 50 µM	[119]
	Hyperoside (quercetin 3-galactoside)	Nelumbo nucifera		Interacts with 3CLpro catalytic site	FRET assay, Vero E6 Cells, qRT-PCR			5.2 at 10 µM	[117]
	Icaritin	Black garlic extract	IC50: 137 ± 10 µg/mL 100% inhibition at 0.5 mg/mL	Inhibits 3CLpro activity	FRET assay			31 at 200 µM	[118]

Table 1. Cont.

Class	Class Flavonoid	Natural Source (N.S.)/Extract	Efficacy of N.S. Extract	Mode of Action	Methods Used	IC50 (µM)	EC50 (µM)	% Inhibition	Reference
	Isorhamnetin (3-methylquercetin)	Pears, olive oil, wine		Interacts with 3CLpro catalytic site	FRET assay, Vero E6 Cells, qRT-PCR			−2.6 at 10 µM	[117]
	Kaempferol	Black garlic extract	IC50: 137 ± 10 µg/mL 100% inhibition at 0.5 mg/mL	Inhibits 3CLpro activity	FRET assay			16 at 200 µM	[118]
		TCM		Binds to 3CLpro active site	Vero E6 Cells	34.46			[120]
	Kaempferide			Interacts with 3CLpro catalytic site	FRET assay, Vero E6 Cells, qRT-PCR			8.1 at 10 µM	[117]
		TCM		Binds to the 3CLpro substrate binding site.	FRET assay	>100			[121]
	Myricetin	Black garlic extract	IC50: 137 ± 10 µg/mL 100% inhibition at 0.5 mg/mL	Inhibits 3CLpro activity	FRET assay	43 ± 1		80 at 200 µM	[118]
		Polygoni avicularis		Binds to the 3CLpro substrate binding site	Vero cells/RT PCR	2.86 ± 0.23			[119]
				Positions itself in the 3CLpro binding pocket	FRET assay, BEAS-2B cells	3.684 ± 0.076		97.79 at 50 µM	[122]

Table 1. Cont.

Class	Class Flavonoid	Natural Source (N.S.)/Extract	Efficacy of N.S. Extract	Mode of Action	Methods Used	IC50 (µM)	EC50 (µM)	% Inhibition	Reference
				Binds at the catalytic site within the extended substrate-binding pocket	FRET assay, Vero E6 Cells, qRT-PCR	0.63 ± 0.01	8.00 ± 2.05	97.6 at 10 µM	[117]
		Ampelopsis grossedentata extract	99.74% inhibition at 100 µg/mL IC50 = 3.44 µg/mL	Modify key residue in domain III of 3CLpro	FRET assay	1.21 (60 min pre-incubation) 21.44 (0.5 min pre-incubation)			[123]
	Myricetin-7-yl 5,5-dimethyl-1,3,2-dioxayl phosphate			Interacts with 3CLpro catalytic site	FRET assay, Vero E6 Cells, qRT-PCR	6.62 ± 0.42	33.45 ± 11.96		[117]
	Myricetin-7-yl diphenyl phosphate			Interacts with 3CLpro catalytic site	FRET assay, Vero E6 Cells, qRT-PCR	3.13 ± 0.37	3.15 ± 0.84		[117]
	Quercetagenin/ quercetagenin	Black garlic extract	IC50: 137 ± 10 µg/mL 100% inhibition at 0.5 mg/mL	Inhibits 3CLpro activity	FRET assay	145 ± 6		58 at 200 µM	[118]
		Eriocaulon buergerianum		Binds to the 3CLpro substrate binding site	Vero cells/RT PCR	1.24 ± 0.14			[119]

Table 1. Cont.

Class	Class Flavonoid	Natural Source (N.S.)/Extract	Efficacy of N.S. Extract	Mode of Action	Methods Used	IC50 (µM)	EC50 (µM)	% Inhibition	Reference
	Quercetin	Black garlic extract	IC50: 137 ± 10 µg/mL 100% inhibition at 0.5 mg/mL	Inhibits 3CLpro activity	FRET assay	93 ± 5		74 at 200 µM	[118]
		TCM		Binds to the 3CLpro substrate binding site.	FRET assay	97.460 ± 2.263			[121]
				Interacts with 3CLpro catalytic site	FRET assay, Vero E6 Cells, qRT-PCR			41.3 at 10 µM	[117]
				Binds to SARS-CoV-2 3CLpro active site	FRET assay	Kiapp = 21 µM Ki = 7.4 µM Kd = 2.7 µM (no NaCl) Kd = 150 mM (150 mM NaCl)			[124]
	Quercetin-4′-O-α-D-glucopyranoside	Black garlic extract	IC50: 137 ± 10 µg/mL 100% inhibition at 0.5 mg/mL	Inhibits 3CLpro activity	FRET assay			26 at 200 µM	[118]
	Rutin (quercetin-3-O-rutinose)	Black garlic extract	IC50: 137 ± 10 µg/mL 100% inhibition at 0.5 mg/mL	Inhibits 3CLpro activity	FRET assay			45 at 200 µM	[118]
				Binds to the 3CLpro catalytic site.	FRET assay	32		43 at 30 µM 65 at 60 µM 80 at 120 µM	[125]

Table 1. Cont.

Class	Class Flavonoid	Natural Source (N.S.)/Extract	Efficacy of N.S. Extract	Mode of Action	Methods Used	IC50 (μM)	EC50 (μM)	% Inhibition	Reference
Flavanonol	Ampelopsin (dihydromyricetin/DHM)	Black garlic extract	IC50: 137 ± 10 μg/mL 100% inhibition at 0.5 mg/mL	Inhibits 3CLpro activity	FRET assay	128 ± 5		64 at 200 μM	[118]
		Ampelopsis japonica		Binds to the 3CLpro substrate binding site	Vero cells/RT PCR	1.20 ± 0.09			[119]
				Interacts with 3CLpro catalytic site	FRET assay, Vero E6 Cells, qRT-PCR	1.14 ± 0.03	13.56 ± 2.50	93.8 at 10 μM	[117]
		Ampelopsis grossedentata extract	99.74% inhibition at 100 μg/mL IC50 = 3.44 μg/mL	Modify key residue in domain III of 3CLpro	FRET assay	4.91 (60 min pre-incubation) 34.61 (0.5 min pre-incubation)			[123]
	Isodihydromyricetin	Ampelopsis grossedentata extract	99.74% inhibition at 100 μg/mL IC50 = 3.44 μg/mL	Modify key residue in domain III of 3CLpro	FRET assay	3.73 (60 min pre-incubation) 29.04 (0.5 min pre-incubation)			[123]
	Dihydromyricetin-7-yl diphenyl phosphate			Interacts with 3CLpro catalytic site	FRET assay, Vero E6 Cells, qRT-PCR	1.84 ± 0.22	9.03 ± 1.36		[117]
	7-O-methyl-dihydromyricetin			Interacts with 3CLpro catalytic site	FRET assay, Vero E6 Cells, qRT-PCR	0.26 ± 0.02	11.50 ± 4.57		[117]
	Ampelopsin-4′-O-α-D-glucopyranoside	Black garlic extract	IC50: 137 ± 10 μg/mL 100% inhibition at 0.5 mg/mL	Inhibits 3CLpro activity	FRET assay	195 ± 5		50 at 200 μM	[118]

Table 1. Cont.

Class	Class Flavonoid	Natural Source (N.S.)/Extract	Efficacy of N.S. Extract	Mode of Action	Methods Used	IC50 (µM)	EC50 (µM)	% Inhibition	Reference
	Taxifolin			Interacts with 3CLpro catalytic site	FRET assay, Vero E6 Cells, qRT-PCR			28.0 at 10 µM	[117]
		Ampelopsis grossedentata extract	99.74% inhibition at 100 µg/mL IC50 = 3.44 µg/mL	Inhibits 3CLpro activity	FRET assay	72.72 (60 min pre-incubation)			[123]
Flavanone	(±)-Eriodyctiol			Interacts with 3CLpro catalytic site	FRET assay, Vero E6 Cells, qRT-PCR			34.5 at 10 µM	[117]
	Hesperidin	Black garlic extract	IC50: 137 ± 10 µg/mL 100% inhibition at 0.5 mg/mL	Inhibits 3CLpro activity	FRET assay			22 at 200 µM	[118]
	Hesperetin			Interacts with 3CLpro catalytic site	FRET assay			13.8 at 10 µM	[117]
	Naringenin	Black garlic extract	IC50: 137 ± 10 µg/mL 100% inhibition at 0.5 mg/mL	Inhibits 3CLpro activity	FRET assay	150 ± 10		57 at 200 µM	[118]
		TCM		Binds to the 3CLpro substrate binding site.	FRET assay	>1000			[121]

Table 1. Cont.

Class	Class Flavonoid	Natural Source (N.S.)/Extract	Efficacy of N.S. Extract	Mode of Action	Methods Used	IC50 (µM)	EC50 (µM)	% Inhibition	Reference
Flavones	Naringin	Black garlic extract	IC50: 137 ± 10 µg/mL 100% inhibition at 0.5 mg/mL	Inhibits 3CLpro activity	FRET assay			18 at 200 µM	[118]
	5,6-dihydroxyflavone			Binds to the 3CLpro substrate binding site	Vero cells/RT PCR			26.6 ± 0.4 at 50 µM	[119]
	6,7-dihydroxyflavone			Binds to the 3CLpro substrate binding site	Vero cells/RT PCR			56.7 ± 2.0 at 50 µM	[119]
	Apigenin	Black garlic extract	IC50: 137 ± 10 µg/mL 100% inhibition at 0.5 mg/mL	Inhibits 3CLpro activity	FRET assay			25 at 200 µM	[118]
				Interacts with 3CLpro catalytic site	FRET assay			−1.0 at 10 µM	[117]
	Baicalein	Scutellaria baicalensis	IC50: 8.52 ± 0.54 µg/mL EC50: 0.74 ± 0.36 µg/mL CC50: > 500 µg/mL	Binds to the 3CLpro substrate binding site	Vero cells/RT PCR	0.39 ± 0.12	2.92 ± 0.06		[119]
				Binds to the core of the substrate-binding pocket, preventing substrate access to the active site	Vero E6 cells/CCK8 assays/qRT-PCR	0.94 ± 0.20	2.49 ± 1.19	99.4 at 100 µM 87 at 10 µM	[126]

Table 1. Cont.

Class	Class Flavonoid	Natural Source (N.S.)/Extract	Efficacy of N.S. Extract	Mode of Action	Methods Used	IC50 (μM)	EC50 (μM)	% Inhibition	Reference
	Baicalin	*Scutellaria baicalensis*	IC50: 8.52 ± 0.54 μg/mL EC50: 0.74 ± 0.36 μg/mL CC50: > 500 μg/mL	Binds to the 3CLpro substrate binding site	Vero cells/RT PCR	83.4 ± 0.9		41.5 ± 0.6 at 50 μM	[119]
				Binds to 3CLpro active site	Vero E6 cells/CCK8 assays/qRT-PCR	6.41 ± 0.95	27.87 ± 12.5	97.6 at 100 μM 68.9 at 10 μM	[126]
	Chrysin	Black garlic extract	IC50: 137 ± 10 μg/mL 100% inhibition at 0.5 mg/mL	Inhibits 3CLpro activity	FRET assay			9 at 200 μM	[118]
				Binds to the 3CLpro substrate binding site.	Vero cells/RT PCR			2.6 ± 1.1 at 50 μM	[119]
	Chrysin-7-O-β-D-glucoronide	*Scutellaria baicalensis*		Binds to 3CLpro active site	Vero E6 cells/CCK8 assays			50.6 at 100 μM 24.2 at 10 μM	[126]
	Diosmetin			Interacts with 3CLpro catalytic site	FRET assay			11.3 at 10 μM	[117]
	Luteolin	Black garlic extract	IC50: 137 ± 10 μg/mL 100% inhibition at 0.5 mg/mL	Inhibits 3CLpro activity	FRET assay			45 at 200 μM	[118]
		TCM		Binds to the 3CLpro substrate binding site.	FRET assay	89.670 ± 4.712			[121]

Table 1. Cont.

Class	Class Flavonoid	Natural Source (N.S.)/Extract	Efficacy of N.S. Extract	Mode of Action	Methods Used	IC50 (µM)	EC50 (µM)	% Inhibition	Reference
				Interacts with 3CLpro catalytic site	FRET assay			−4.1 at 10 µM	[117]
	Luteoloside (cyranoside)	L. japonica		Binds to 3CLpro active site	Vero E6 cells/CCK8 assays			65.4 at 100 µM 14.8 at 10 µM	[126]
	Myricitrin	Polygoni avicularis		Binds to the 3CLpro substrate binding site	Vero cells/RT PCR			30.8 ± 4.6 at 50 µM	[119]
		Ampelopsis grossedentata extract	99.74% inhibition at 100 µg/mL IC50 = 3.44 µg/mL	Modify key residue in domain III of 3CLpro	FRET assay	14.22 (60 min pre-incubation)			[123]
	Oroxylin A-7-O-β-D-glucuronide	Scutellaria baicalensis		Binds to 3CLpro active site	Vero E6 cells/CCK8 assays			33.0 at 100 µM	[126]
	Scutellarein (6-hydroxyapigenin)	Scutellaria, Erigerontis herba		Binds to the 3CLpro substrate binding site.	Vero cells/RT PCR	5.80 ± 0.22			[119]
		Scutellaria baicalensis		Binds to 3CLpro active site	Vero E6 cells/CCK8 assays	3.02 ± 0.11		101.6 at 100 µM 90.7 at 10 µM	[126]
	Scutellarin (scutellarein-7-glucuronide)	Scutellaria, Erigerontis herba		Binds to the 3CLpro substrate binding site.	Vero cells/RT PCR			28.9 ± 1.6 at 50 µM	[119]

Table 1. Cont.

Class	Class Flavonoid	Natural Source (N.S.)/Extract	Efficacy of N.S. Extract	Mode of Action	Methods Used	IC50 (µM)	EC50 (µM)	% Inhibition	Reference
		Scutellaria baicalensis		Binds to 3CLpro active site	Vero E6 cells/CCK8 assays			76.8 at 100 µM 18.9 at 10 µM	[126]
	Vitexin	Black garlic extract	IC50: 137 ± 10 µg/mL 100% inhibition at 0.5 mg/mL	Inhibits 3CLpro activity	FRET assay	180 ± 6		52 at 200 µM	[118]
	Wogonin	Scutellaria baicalensis	IC50: 8.52 ± 0.54 µg/mL EC50: 0.74 ± 0.36 µg/mL	Binds to the 3CLpro substrate binding site.	Vero cells/RT PCR			6.1 ± 0.8 at 50 µM	[119]
				Binds to 3CLpro active site	Vero E6 cells/CCK8 assays			3.6 at 100 µM	[126]
	Wogonoside	Scutellaria baicalensis	IC50: 8.52 ± 0.54 µg/mL EC50: 0.74 ± 0.36 µg/mL	Binds to the 3CLpro substrate binding site.	Vero cells/RT PCR			8.5 ± 3.3 at 50 µM	[119]
				Binds to 3CLpro active site	Vero E6 cells/CCK8 assays			20.4 at 100 µM	[126]
Isoflavones	Biochanin A			Interacts with 3CLpro catalytic site	FRET assay			5 at 10 µM	[117]
	Daidzein	Black garlic extract	IC50: 137 ± 10 µg/mL 100% inhibition at 0.5 mg/mL	Inhibits 3CLpro activity	FRET assay	56		100 at 200 µM	[118]
				Interacts with 3CLpro catalytic site	FRET assay			13.9 at 10 µM	[117]

Table 1. Cont.

Class	Class Flavonoid	Natural Source (N.S.)/Extract	Efficacy of N.S. Extract	Mode of Action	Methods Used	IC50 (μM)	EC50 (μM)	% Inhibition	Reference
	Formononetin			Interacts with 3CLpro catalytic site	FRET assay			16.0 at 10 μM	[117]
	Genistein			Interacts with 3CLpro catalytic site	FRET assay			15.0 at 10 μM	[117]
	Genistin	Black garlic extract	IC50: 137 ± 10 μg/mL 100% inhibition at 0.5 mg/mL	Inhibits 3CLpro activity	FRET assay			48 at 200 μM	[118]
				Interacts with 3CLpro catalytic site	FRET assay			25.5 at 10 μM	[117]
	Puerarin	Black garlic extract	IC50: 137 ± 10 μg/mL 100% inhibition at 0.5 mg/mL	Inhibits 3CLpro activity	FRET assay	42 ± 2		100 at 200 μM	[118]
	Sophoricoside			Interacts with 3CLpro catalytic site	FRET assay			10.3 at 10 μM	[117]
Flavan-3-ols/Flavanols	Catechin	Black garlic extract	IC50: 137 ± 10 μg/mL 100% inhibition at 0.5 mg/mL	Inhibits 3CLpro activity	FRET assay			9 at 200 μM	[118]
				Interacts with 3CLpro catalytic site	FRET assay			14.0 at 10 μM	[117]
	Catechin gallate (CG)	Black garlic extract	IC50: 137 ± 10 μg/mL 100% inhibition at 0.5 mg/mL	Inhibits 3CLpro activity	FRET assay			21 at 200 μM	[118]

Table 1. Cont.

Class	Class Flavonoid	Natural Source (N.S.)/Extract	Efficacy of N.S. Extract	Mode of Action	Methods Used	IC50 (µM)	EC50 (µM)	% Inhibition	Reference
	Epicatechin (EC)	Black garlic extract	IC50: 137 ± 10 µg/mL 100% inhibition at 0.5 mg/mL	Inhibits 3CLpro activity	FRET assay			8 at 200 µM	[118]
	Epicatechin gallate (ECG)	Black garlic extract	IC50: 137 ± 10 µg/mL 100% inhibition at 0.5 mg/mL	Inhibits 3CLpro activity	FRET assay			21 at 200 µM	[118]
	Epigallocatechin (EGC)	Black garlic extract	IC50: 137 ± 10 µg/mL 100% inhibition at 0.5 mg/mL	Inhibits 3CLpro activity	FRET assay			23 at 200 µM	[118]
	Epigallocatechin gallate (EGCG)	Black garlic extract	IC50: 137 ± 10 µg/mL 100% inhibition at 0.5 mg/mL	Inhibits 3CLpro activity	FRET assay	171 ± 5		53 at 200 µM	[118]
		TCM		Binds to the 3CLpro substrate binding site.	FRET assay	0.847 ± 0.005			[121]
	Gallocatechin gallate (GCG)	Black garlic extract	IC50: 137 ± 10 µg/mL 100% inhibition at 0.5 mg/mL	Inhibits 3CLpro activity	FRET assay			50 at 200 µM	[118]
Tannoid	Tannic acid	Black garlic extract	IC50: 137 ± 10 µg/mL, 100% inhibition at 0.5 mg/mL	Inhibits 3CLpro activity	FRET assay	9		100 at 200 µM	[118]
Others	Mixture of 11 flavonols	*Salvadora persica* L.		Inhibits 3CLpro activity	3CL protease assay, A549 cells	8.59 ± 0.3 µg mL^{-1}		85.56 ± 1.12%	[127]

Table 2. Flavonoids with antiviral activities reported against SARS-CoV-2 PLpro using in vitro methodologies segregated according to class.

Class	Flavonoid	Natural Source (N.S.)/Extract	Mode of Action	Methods Used	% Inhibition	Reference
Flavonol	Myricetin		Interacts with 3CLpro catalytic site	FRET assay	50 at 159.10 ± 38.33 μM	[117]
	Rutin		Binds to naphthalene inhibitor binding pocket	PLpro enzymatic inhibition assay	38 at 100 μM	[128]
Flavones	Baicalein	*Scutellaria baicalensis*	Binds to 3CLpro active site	Vero E6 cells/CCK8 assays	45.1 at 50 μM, 12.4 at 12.5 μM	[126]
	Baicalin	*Scutellaria baicalensis*	Binds to 3CLpro active site	Vero E6 cells/CCK8 assays	15.9 at 50 μM	[126]
	Chrysin-7-O-β-D-glucuronide	*Scutellaria baicalensis*	Binds to 3CLpro active site	Vero E6 cells/CCK8 assays	16.3 at 50 μM	[126]
	Luteoloside (cyranoside)	*L. japonica*	Binds to 3CLpro active site	Vero E6 cells/CCK8 assays	21.5 at 50 μM	[126]
	Oroxylin A-7-O-β-D-glucuronide	*Scutellaria baicalensis*	Binds to 3CLpro active site	Vero E6 cells/CCK8 assays	7.4 at 50 μM	[126]
	Scutellarein	*Scutellaria baicalensis*	Binds to 3CLpro active site	Vero E6 cells/CCK8 assays	65.7 at 50 μM, 14.4 at 12.5 μM	[126]
	Scutellarin	*Scutellaria baicalensis*	Binds to 3CLpro active site	Vero E6 cells/CCK8 assays	41.1 at 50 μM, 12.7 at 12.5 μM	[126]
	Wogonin	*Scutellaria baicalensis*	Binds to 3CLpro active site	Vero E6 cells/CCK8 assays	52.0 at 50 μM, 35.9 at 12.5 μM	[126]
	Wogonoside	*Scutellaria baicalensis*	Binds to 3CLpro active site	Vero E6 cells/CCK8 assays	14.4 at 50 μM	[126]
Flavan-3-ols/Flavanols	Epicatechin gallate (ECG)		Binds to naphthalene inhibitor binding pocket	PLpro enzymatic inhibition assay	20 at 100 μM	[128]
	Epigallocatechin gallate (EGCG)		Binds to naphthalene inhibitor binding pocket	PLpro enzymatic inhibition assay	13 at 100 μM	[128]
Anthocyanin	Cyanidin-3-O-glucoside (chrysanthemin)		Binds to naphthalene inhibitor binding pocket	PLpro enzymatic inhibition assay	20 at 100 μM	[128]
Others	Hypericin		Binds to naphthalene inhibitor binding pocket	PLpro enzymatic inhibition assay	97 at 100 μ M87 at 50 μM	[128]

In general, the efficacy of myricetin is higher than quercetin, with cell-free methods reporting generally weaker inhibitory capacities against 3CLpro than cell-based assays (Table 1) [117–119,122]. Moreover, myricetin is a promising compound given its inhibitory effect against viruses and weak cellular cytotoxicity [117]. Interestingly, many myricetin derivatives, such as ampelopsin (a.k.a. dihydromyricetin or DHM) and its other phosphorylated and glycosylated derivatives, demonstrate inhibition within similar thresholds [117,118]. Other flavonols and flavanonols target SARS-CoV-2 3CLpro less effectively, allowing for vital structure-efficacy relationship analysis (Table 1).

Lastly, myricetin and rutin possess anti-SARS-CoV-2 PLpro activities, with the latter interacting with the naphthalene inhibitor binding pocket of the protease; however, both of their actions are relatively weak (Table 2) [117,128]. No flavanonols had anti-PLpro activities.

Flavones, Flavanones, and Isoflavones

Flavones are another widely explored class of flavonoids for their anti-SARS-CoV-2 activities. *Scutellaria baicalensis* is a traditional Chinese plant that contains several flavonoid elements with antiviral activity. Liu et al. reported that its crude extract, with an IC50 of 8.52 ± 0.54 µg/mL and EC50 of 0.74 ± 0.36 µg/mL against SARS-CoV-2 3CLpro and replication, respectively, has low cellular cytotoxicity and consists of various active flavones [119]. Among these, baicalien was the most effective flavone inhibiting the protease mechanistically by binding to its substrate-binding site and consequently inhibiting viral replication. Two independent groups of researchers using cell-based assays confirm their activities; Liu et al. reported an IC50 value of 0.39 ± 0.12 µM and an EC50 value of 2.92 ± 0.06 µM, while Hai-Xia Su et al. demonstrated an IC50 of 0.94 ± 0.20 µM and an EC50 of 2.49 ± 1.19 µM. Both studies found very high cellular cytotoxic tolerance (>200 µM) to baicalein, showing promise as a therapeutic intervention [119,126]. Its glycosylated derivative, baicalin, and scetullarein are other flavones with promising activities against the SARS-CoV-2 3CLpro, albeit with slightly lower active concentration markers (Table 1).

A proposed mechanism suggests that there are several hydrogen bond interactions between the baicalein and SARS-CoV-2 3CLpro substrate-binding site; these include: the 6-OH of baicalein binds to the carbonyl group of L141, the 7-OH group of baicalein binds to the backbone amide group of G143 and the carbonyl group of baicalein binding to the backbone amide of E166. Additionally, the baicalein molecule covers up the H41 and C145 catalytic residues, adding its inhibitory effect. The higher activity of baicalein compared to baicalin may be attributed to the observation that the larger 7-glycosyl group is more prominent in baicalin, thus reducing effective interaction(s) with the binding site of the protease and consequently lower inhibitory activity. Liu et al. also studied other flavones derived from *Scutellaria baicalensis* extract, including Wogonin (lacks 6-OH and contains 8-methoxy) and Wogonoside (additionally contains C7 glycosylation); however, they showed drastically lower activities, allowing for significant structure-activity relationship analysis [119]. Other flavones extracted from *S. baicalensis* include oroxylin A-7-O-β-D-glucuronide and myricitrin. Overall, the inhibition of PLpro was generally weaker than that of 3CLpro between compounds; no IC50 was reported for this target (Table 2) [126].

Flavanones, as described before, can be characterized as flavones without a C2-C3 double bond on the C-ring. However, their activities against 3CLpro were limited. Nguyen et al. reported naringenin from black garlic extract as the most effective flavanone against SARS-CoV-2 3CLpro with an IC50 of 150 ± 10 µM using cell-free assays. In contrast, Du et al. reported a much higher active concentration for the same compound, likely due to differences in specific targets or methodologies [118,121]. (±)-Eriodictyol, hesperetin, hesperidin (glycosylated hesperetin), and naringin (glycosylated naringenin) are other flavanones that have only moderate to low efficacy against SARS-CoV-2 3CLpro (Table 1) [117,118]. No flavanones exhibited SARS-CoV-2 antiviral activity against PLpro.

Among all the flavonoids extracted from black garlic acid, the most efficacious isoflavone (similar to flavones, but with the B-ring attached to C3 rather than C2) was puerarin with an IC50 value of 42 ± 2 µM, determined using cell-free FRET assays. Daidzein and genistin from black garlic had moderate active concentration markers [116]. Other isoflavones were also studied for their anti-SARS-CoV-2 protease activities (Table 1) [117]. Isoflavones did not have anti-SARS-CoV-2 PLpro activities.

Flavan-3-ols/Flavanols

All reports on the activities of flavanols against SARS-CoV-2 3CLpro were performed on cell-free FRET assays. While both Nguyen et al. and Du et al. agree on the status of epigallocatechin gallate (EGCG) as the most promising flavanol against 3CLpro activity, their respective reports on the active concentration marker IC50 vary substantially. The former group reports an IC50 of 171 ± 5 µM for EGCG, whereas the latter group reported an IC50 of 0.847 ± 0.005 µM against the protease while reporting that the flavonoid binds to its substrate-binding site [118,121]. While both cell-free methods, this significant difference might be explained by the difference in material source, specific target, or methodologies. Other catechin derivatives have limited efficacy against 3CLpro; however, an increase in the presence of the galloyl moieties on the flavanol signaled increasing inhibitory activities (Table 1).

Pitsillou et al. reported that flavanols are also moderately effective against SARS-CoV-2 by binding to the naphthalene inhibitor binding pocket and blocking PLpro. The in vitro study used a PLpro enzymatic inhibition assay to investigate the effect of epicatechin gallate (ECG) and epigallocatechin gallate (EGCG) on PLpro. While neither of the compounds showed high activity levels, the study suggests that ECG is slightly more effective at inhibiting PLpro in SARS-CoV-2, revealing essential differences in the structural-activity relationship when comparing anti-3CLpro to anti-PLpro mechanisms (Table 2) [128].

Others

Tannic acid, a polyphenol similar to a flavonoid, has the most substantial inhibitory effect on SARS-CoV-2 3CLpro among black garlic constituents with an IC50 of 9 µM using cell-free FRET assay [118]. Like tannic acid, the special polyphenol hypericin has extraordinary potential for its action against the PLpro [128]. Furthermore, other classes of flavonoids, such as anthocyanins contain compounds such as cyanidin-3-O-β-glucoside (Chrysanthemin), which display antiviral activity inhibiting PLpro by binding to the naphthalene inhibitor binding site, although showing limited inhibition (Table 2) [128]. Lastly, Owis et al. upheld the status of a mixture of eleven flavonoids, constituting primarily of the flavonols kaempferol and isorhamnetin derivatives, as having potent cell-based inhibitory activities against 3CLpro using a protease assay and A549 cells. Furthermore, with a lipid encapsulation of the compound, the authors reported significantly higher inhibition (85% vs. 38%), owing to a higher uptake of these otherwise hydrophilic flavonoids through cell membranes (Table 1) [127].

3.4.2. Flavonoids against SARS-CoV-2 Spike RBD and hACE2 Interaction

Another promising target for anti-SARS-CoV-2 therapeutic intervention in vitro is the interaction between the viral S protein and the human ACE2 receptor during the viral entry phase. Table 3 summarizes extracted literature data concerning the in vitro activities of various flavonoids against this crucial interaction of the viral life cycle [129–132].

Table 3. Flavonoids with antiviral activities reported against SARS-CoV-2 spike protein–ACE2 interaction using in vitro methodologies segregated according to class.

Class	Flavonoid	Natural Source (N.S.)/Extract	Efficacy of N.S. Extract	Mode of Action	Methods Used	IC50 (µM)	% Inhibition	Reference
Flavonol	Isorhamnetin	Sea buckthorn berry		Binds to three residues involved in spike RBD–ACE2 interaction	HEK293 cells/SPR assay	$K_d = 2.51 \pm 0.68$ µM		[125]
				Inhibits rhACE2 activity	MCA Fluorescence, rhACE2 cells		14.7 ± 1.4 at 10 µM	[129]
	Quercetin	*Hippophae rhamnoides* L.		Binds to three residues involved in spike RBD–ACE2 interaction	HEK293 cells/SPR assay	$K_d = 5.92 \pm 0.92$ µM		[125]
				Inhibits rhACE2 activity	MCA Fluorescence, rhACE2 cells	At 2.5 min: 4.48 At 10.5 min: 29.5	66.2 ± 2.2 at 10 µM	[129]
	Quercetin-3-O-galactoside (hyperoside)			Inhibits rhACE2 activity	MCA Fluorescence, rhACE2 cells		34.2 ± 3.7 at 10 µM	[129]
	Quercetin-3-O-glucuronide			Inhibits rhACE2 activity	MCA Fluorescence, rhACE2 cells		33.1 ± 4.9 at 10 µM	[129]
	Quercetin-3-O-glucoside (isoquercetin)			Inhibits rhACE2 activity	MCA Fluorescence, rhACE2 cells		47.7 ± 3.7 at 10 µM	[129]
	Rutin (quercetin-3-)-rutinose)			Inhibits rhACE2 activity	MCA Fluorescence, rhACE2 cells		48.3 ± 4.7 at 10 µM	[129]
	Tamarixetin			Inhibits rhACE2 activity	MCA Fluorescence, rhACE2 cells		41.5 ± 5.0 at 10 µM	[129]

Table 3. Cont.

Class	Flavonoid	Natural Source (N.S.)/Extract	Efficacy of N.S. Extract	Mode of Action	Methods Used	IC50 (μM)	% Inhibition	Reference
Flavanone	(±)-Eriodictyol			Inhibits rhACE2 activity	MCA Fluorescence, rhACE2 cells		24.4 ± 1.4 at 10 μM	[129]
	Hesperetin	Anatolian Propolis	IC50: 1.14 μL		S1 colorimetric assay	16,880		[130]
	Pinocembrin	Anatolian Propolis	IC50: 1.14 μL		S1 colorimetric assay	29,530		[130]
Flavones	Luteolin			Inhibits rhACE2 activity	MCA Fluorescence, rhACE2 cells		37.1 ± 0.6 at 10 μM	[129]
	Neobaicalein	Radix Scutellariae		Binds to ACE2 receptor	CMC, HEK293T cells, CK8 assay, SPR assay	83.8		[120]
	Oroxylin A	Radix Scutellariae		Binds to ACE2 receptor	CMC, HEK293T cells, CK8 assay, SPR assay	164.6		[120]
	Scutellarin	Radix Scutellariae		Binds to ACE2 receptor	CMC, HEK293T cells, CK8 assay, SPR assay	170.9		[120]
	Wogonin	Radix Scutellariae		Binds to ACE2 receptor	CMC, HEK293T cells, CK8 assay, SPR assay	137.6		[120]
Flavan-3-ols/Flavanols	Epicatechin (EC)	Green tea		Interferes with SARS-CoV-2 spike RBD–ACE2 interaction	HEK293T-ACE2 cells/Huh7 cells/Vero cells	>20 μg/mL		[131]
				Inhibits rhACE2 activity	MCA Fluorescence, rhACE2 cells		27.4 ± 5.7 at 10 μM	[129]

Table 3. Cont.

Class	Flavonoid	Natural Source (N.S.)/Extract	Efficacy of N.S. Extract	Mode of Action	Methods Used	IC50 (µM)	% Inhibition	Reference
	Epigallocatechin gallate (EGCG)	Green tea		Interferes with SARS-CoV-2 spike RBD–ACE2 interaction	HEK293T-ACE2 cells/Huh7 cells/Vero cells/Plaque reduction assay	2.47 µg/mL		[131]
Anthocyanin	Pelargonidin			Binds to fatty acid binding pocket on spike RBD and attenuates spike–ACE2 interaction Reduces SARS-CoV-2 replication	ACE2-SARS-CoV-2 spike inhibitor screening assay Vero cells/Plaque assay		Screening Assay: >5 (10 µM), >15 (20 µM), >40 (50 µM)	[132]

Flavonols

Quercetin and its derivatives are the most predominant flavonols to display in vitro inhibition against the hACE2-spike RBD interaction. Using MCA Fluorescence and rhACE2 cells, Liu et al. reported an effective $66.2 \pm 2.2\%$ inhibition at 10 µM and an IC50 of 4.48 µM for base quercetin 2.5 min after incubation with the flavonol against rhACE2 activity, which increased to 29.5 µM after 10.5 min. Independent studies by Zhan et al. confirmed the high affinity of the flavonol through derivation of its Kd (5.92 ± 0.92 µM) and mode of inhibition (mixed) [129,133]. The activities of quercetin derivatives, although relatively limited, have also been explored, allowing for significant structure–activity relationship discussions (Table 3). Among these flavonols, the B-ring $3',4'$-dihydroxylation was responsible for the inhibitory activity; this group is also found in isorhamnetin, explaining similar, although limited, effects [129].

Flavones and Flavanones

While few flavones were investigated for disrupting the S-ACE2 interaction, these have shown fewer promising results than previous targets. Luteolin has a $37.1 \pm 0.6\%$ inhibition against rhACE2 activity at a low concentration of 10 µM using MCA Fluorescence [129]. Other *Radix Scutellariae*-derived flavones also prevent viral entry into cells through interaction ACE2 and other mechanisms (Table 3) [120]. The flavanones (±)-eriodictyol, hesperetin, and pinocembrin inhibit ACE2–spike interactions, although their effects are weak compared to previously mentioned compounds (Table 3).

Flavan-3-ols/Flavanols and Others

EGCG interferes with SARS-CoV-2 spike RBD–ACE2 interaction with the most potent IC50 concentration of IC50 of 2.47 µg/mL, while the flavanol EC, lacking two galloyl moieties, has a much lower inhibitory effect against ACE2 activity [131]. Similar reports are available for EGCG as an effective inhibitor against pseudotyped-SARS-CoV-2 cells with IC50 of 2.47 µg/mL and viral replication and receptor binding [131].

The anthocyanin pelargonidin binds to a fatty acid-binding pocket on spike RBD and attenuates spike–ACE2 interaction, thereby reducing SARS-CoV-2 spike–ACE2 interaction as well as viral replication in Vero cells (Table 3). Pelargonidin also reduces systemic inflammation in in vivo mouse models by interacting with the aryl hydrocarbon receptor (AHR) [132].

3.4.3. Antiviral Activities of Flavonoids against Other SARS-CoV-2 Targets

Table 4 enlists the activities of flavonoids against other less-investigated SARS-CoV-2 targets in various cell-based methodologies.

Flavones against SARS-CoV-2 RNA-Dependent RNA Polymerase (RdRp)

The replication stage is an essential step in the viral life cycle, yet few in vitro studies targeted the RdRp to explore the activities of flavonoids. A section of Table 4 summarizes data extracted from the literature concerning the auspicious inhibitory activities of two flavones against this critical enzyme responsible for viral replication. Baicalein and baicalin, flavone constituents of *Scutellaria baicalensis*, were shown to have anti-SARS-CoV-2 RdRp activities. Using Vero and Calu-3 cell-based in vitro methodologies, baicalein more effectively binds to the NiRAN domain and the palm subdomain on SARS-CoV-2 RdRp, thereby inhibiting its activity with an EC50 of 4.5 ± 0.2 µM (Vero cells) and 1.2 ± 0.03 µM (Calu-3 cells) [134]. Cytotoxicity assays showed CC50 values of 86-91 µM, signaling cellular cytotoxic tolerance to the flavone. Similar, although slightly less effective, was its glycosylated analog baicalin (Table 4) [134].

Table 4. Flavonoids with antiviral activities reported against other SARS-CoV-2 targets using in vitro methodologies segregated according to class.

Target	Class	Flavonoid	Natural Source (N.S.)/Extract	Mode of Action	Methods Used	IC50 (μM)	EC50 (μM)	% Inhibition	Reference
RdRp	Flavone	Baicalein	*Scutellaria baicalensis*	Binding to NiRAN domain and the palm subdomain	Vero CCL-81/Calu-3 cells/MTS assay/qRT-PCR assay/293T cells/Huh7.5 cells		Vero: 4.5 ± 0.2 Calu-3: 1.2 + 0.03	99.8 at 20 μM	[134]
	Flavone	Baicalin	*Scutellaria baicalensis*	Inhibits RdRp	Vero CCL-81/Calu-3 cells/MTS assay/qRT-PCR assay/293T cells/Huh7.5 cells		Vero: 9.0 ± 0.08 Calu-3: 8.0 ± 0.11	98 at 20 μM	[134]
N	Flavanol/flavan-3-ol	Gallocatechin Gallate (GCG)	Green Tea	Disrupts the LLPS of N by interfering with N-RNA binding	H1299 cells RT-qPCR	44.4			[1]
Mitochondrial OXPHOS	Flavone	Baicalein	*Scutellaria baicalensis*	Oxygen consumption inhibitor	Vero E6 cells	10			[135]
nsP15	Flavanol/flavan-3-ol	epigallocatechin gallate (EGCG)	Green Tea extract	Binds to nsp15 active site	Endoribonuclease assay, plaque assay, Vero cells	1.62 ± 0.36	PRNT50: 0.2 μM		[136]
	Flavone	Baicalin	*Scutellaria baicalensis*, *Scutellaria lateriflora*		Endoribonuclease assay, plaque assay, Vero cells	7.98 ± 1.46	PRNT50: 83.3 μM		[136]
	Flavone	Baicalein	*Scutellaria baicalensis*, *Scutellaria lateriflora*		Endoribonuclease assay, plaque assay, Vero cells	8.61			[136]
	Flavonol	Quercetin	Onion peels, red grapes, green leafy vegetables		Endoribonuclease assay, plaque assay, Vero cells	13.79			[136]

Flavan-3-ols/Flavanols against SARS-CoV-2 Nucleocapsid Protein

The nucleocapsid (N) protein is a rare target investigated by researchers to explore the inhibitory effects of flavonoids against SARS-CoV-2 given its location internally within the virus. Table 4 also summarizes data concerning the inhibitory activity of GCG against N protein. This flavanol extracted from green tea was studied in vitro with H1299 cells and qRT-PCR and was effective in blocking N by interfering with N-RNA binding and disrupting its liquid–liquid phase separation (LLPS) with a promising IC50 of 44.4 µM and a CC50 of 156.6 µM [1].

Flavonoids against Other SARS-CoV-2 Targets

A green tea extract containing EGCG, EGC, and ECG inhibited SARS-CoV-2 endoribonuclease nsP15 with an IC50 value of 2.54 µg/mL (Table 4). The study isolated the majority constituent of the extract, EGCG, which inhibited SARS-CoV-2 with an IC50 value of 1.62 µM or 0.74 µg/mL (three times lower than that of the extract) by binding to the nsP15 active site [1]. Furthermore, the authors reported anti-nsP15 activities of baicalin, baicalein, and quercetin in vitro with relatively limited, yet promising active concentration markers (Table 4).

Jang et al. studied the effects of EGCG on 3CLpro of human beta coronavirus (HCoV-OC43) and human alpha coronavirus (HCoV-229E) to substitute for SARS-CoV-2. Using a protease assay and qRT-PCR, the authors reported good IC50 values of 14.6 µM against viral production in the beta coronavirus model and 11.7 µM against the alphacoronavirus model. EGCG treatment decreased HCoV-OC43 induced cytotoxicity through plaque formation assays and measurement of changes in cell viability. Moreover, through qRT-PCR, EGCG had inhibitory effects against the beta coronavirus model by reducing the RNA levels of RdRp, membrane protein gene, and nucleocapsid protein gene when cells were conditioned in flavonoid media before and after infection. This points towards EGCG being responsible for inhibiting coronavirus production and transmission, potentially having similar effects on SARS-CoV-2 [137].

On the other hand, a cell-based in vitro study showed that the flavone baicalein inhibits viral replication of SARS-CoV-2 by blocking mitochondrial OXPHOS, which leads to a quick and robust decrease in the mitochondrial membrane potential (MMP) and acts as an oxygen consumption inhibitor [135] (Table 4).

The flavanone Naringenin, at a concentration of 62.5 µM, inhibits HCoV-OC43 by 100% in Vero cells by targeting the endo-lysosomal two-pore channel 2 (TPC2). SARS-CoV-2 infection was inhibited in a time- and dose-dependent manner by naringenin when analyzed for cytopathic effect in Vero cells, while not inducing toxicity on non-infected cells-m signaling towards the selectivity of the flavanone. The authors concluded that naringenin acted as a lysosomotropic active natural compound that exhibited human pan-Coronavirus antiviral activity [138].

Pitsillou et al. used enzymatic assays to explore the inhibition of PLpro deubiquitinase activity in vitro by various small molecules. At 100 µM, the anthraquinone derivative hypericin displayed the highest inhibition of >90%. The anthocyanin cyanidin-3-O-glucoside and the flavanols rutin, EGCG, and ECG followed in terms of their respective PLpro deubiquitinase activities [139].

Using Vero E6 and Calu-3 cells, Leal et al. recently demonstrated that *Siparuna cristata*-derived flavonols, 3,7-Di-O-methyl-kaempferol (kumatakenin) and 3,7,3′,4′-Tetra-O-methyl-quercetin (retusin), inhibited the general viral replication of SARS-CoV-2 with very promising effective concentrations. Kumatakein showed EC50 values of 10 ± 0.7 and 0.3 ± 0.02 µM in Vero and Calu cells, respectively; similarly, retusin also exhibited promise with EC50s of 0.4 ± 0.05 and 0.6 ± 0.06 µM, respectively [140]. This activity was credited to inhibiting the viral 3CLpro and PLpro due to the authors' in silico investigations but was not confirmed experimentally.

Song et al. investigated the effects of baicalein from *Scutellaria baicalensis* Georgi on SARS-CoV-2 induced infection parameters. The authors reported that the flavone inhibited cell damage induced by SARS-CoV-2 and improved the morphology of Vero E6 cells at concentrations of 0.1 µM and above. The compound was also studied using hACE2 transgenic mice infected with SARS-CoV-2. Baicalein improves respiratory function and inhibits pulmonary inflammatory cell infiltration while reducing inflammatory cytokines [141].

3.5. Structure-Activity Relationships (SARs) of Flavonoids

3.5.1. Effect of Flavonols and Flavanonols on SARS-CoV-2 3CLpro

Given the large number of studies exploring the effect of flavonoids on 3CLpro, it is feasible to compare different flavonoids within this class effectively and against this target. Using cell free assays, Nguyen et al. report that the most to least effective flavonol inhibitors of SARS-CoV-2 3CLpro were myricetin, quercetin, astragalin, quercetagenin, rutin, quercetin-4'-O-α-D-glucopyranoside, kaempferol (Figure 9) [118]. The presence of 3'-OH, 4'-OH, and 5'OH on the B-ring of myricetin accounts for its higher effectiveness compared to quercetin, with an absence of 5'-OH; and kaempferol, with a lack of both 3'-OH and 5'-OH. On the other hand, the presence of A-ring C6-OH accounted for a significant decrease in the inhibitory effect of quercetagenin (6-hydroxyquercetin) compared to quercetin. Further, the addition of carbohydrates to flavonols in place of the hydroxyl groups, as seen in the glycosylation of quercetin (at 4' on B-ring) to quercetin-4'-O-α-glucopyranoside and (at C3 on C-ring) to rutin (quercetin-3-O-rutinose), decreased the inhibitory effect of these compounds against 3CLpro in vitro. The higher position of astragalin on the spectrum given features such as an absence of B-ring 5'-OH and a presence of C-ring C3 glycosylation leads to the hypothesis that the presence of a hydroxyl on A-ring C6 is detrimental to a flavonol's activity against 3CLpro despite the presence of both B-ring 4' and 5'-OH as well as C-ring C3-OH (as seen in quercetagenin). It also suggests that a comparatively larger glycosylated group (such as in rutin) also reduces inhibitory activity significantly.

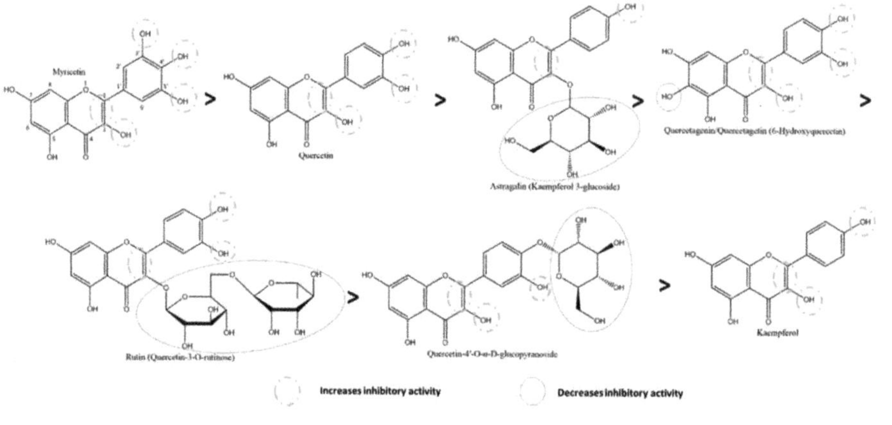

Figure 9. Structure–activity relationships of various flavonols against SARS-CoV-2 3CLpro, adapted from the experiments of [118] using cell-free based in vitro FRET assays. Substitutions and groups that result in an increase/decrease in inhibitory activities are shown in red and green, respectively.

Comparing the activities of myricetin and its a-ring C7-OH derivates highlights that all of them displayed potent in vitro activities against 3CLpro. The most effective of them was 7-O-methyl myricetin, followed by base myricetin and its ethyl-substituted derivative (Figure 10) [117]. Similarly, by following the trend, isoamyl- and cyclopentylmethyl-derivates show lower inhibition comparatively. In contrast, the largest substitutions, 7-yl diphenyl phosphate and 7-yl 5,5-dimethyl-1,3,2-dioxayl phosphate, were the least effective. This pattern of inhibition in myricetin derivatives shows that, in general, the larger substitutions have lower inhibitory activities against 3CLpro, suggesting a possible hint into the specificity of the binding mechanism.

Further, Owis et al. investigated the inhibitory effects of a mixture of the derivatives of kaempferol and isorhamnetin. The O-glycosylation at C3 of the C-ring in kaempferol glycosylated analogs was the source of the inhibitory activity despite the lower performance of base kaempferol [127].

By comparing the inhibition by myricetin to the flavanonol ampelopsin (also known as dihydromyricetin or DHM), it was deduced that the absence of the C-ring C2-C3 double bond further decreased the inhibitory effect, even more as expected, in the B-ring 4'-glycosylated flavanonol, ampelopsin-4'-O-α-D-glucopyranoside (Figure 11) [118]. On the other hand, 7-O-methyl-dihydromyricetin had relatively higher activity than base DHM. In contrast, a larger substitution, such as dihydromyricetin-7-yl diphenyl phosphate, reduced the activity [117]. However, all of these are derivatives of myricetin. Therefore, it is expected to have higher inhibitory activity against Taxifolin (or dihydroquercetin) due to the absence of the B-ring 3'-OH in the latter.

Figure 10. Structure–activity relationships of the flavonols myricetin and its analogs against SARS-CoV-2 3CLpro, adapted from the experiments of [117] using cell-based in vitro FRET assays. Substitutions and groups that result in increase/decrease in inhibitory activities are shown in red and green, respectively.

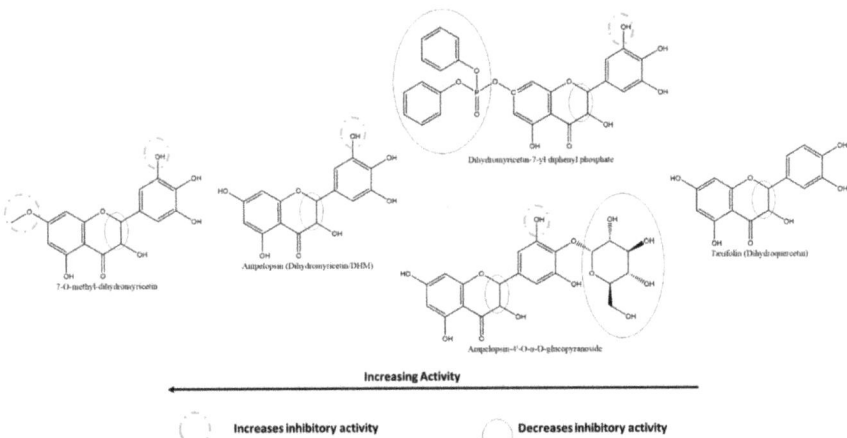

Figure 11. Structure–activity relationships of various flavanonols against SARS-CoV-2 3CLpro, adapted from the in vitro experiments of [117,118]. Substitutions and groups that result in increase/decrease in inhibitory activities are shown in red and green, respectively.

3.5.2. Effect of Flavones and Flavanones on SARS-CoV-2 3CLpro

Nguyen et al. also reported that at 200 µM, the most to least effective flavone inhibitors of SARS-CoV-2 3CLpro were vitexin > luteolin > apigenin > chrysin (Figure 12). Interestingly, although vitexin does not have a B-ring 5′-OH similar to luteolin, the presence of an A-ring C8-glycosylation resulted in a slightly higher inhibition of 3CLpro. The absence of the glycosylation (in apigenin) along with the absence of a B-ring 5′-OH resulted in a more than 50% decrease in inhibition compared to vitexin. The absence of all hydroxyl groups on the B-ring, including a 4′-OH, (as seen in chrysin) resulted in the lowest inhibition among the flavones in this cell-free based assay [118]. On the other hand, among the flavones investigated by Su et al., diosmetin (the 4′-methoxy derivatives of luteolin) was the only one that showed a slight inhibition at 10 µM. In contrast, apigenin and luteolin failed to show any inhibition, adding to the pattern in flavonols [117].

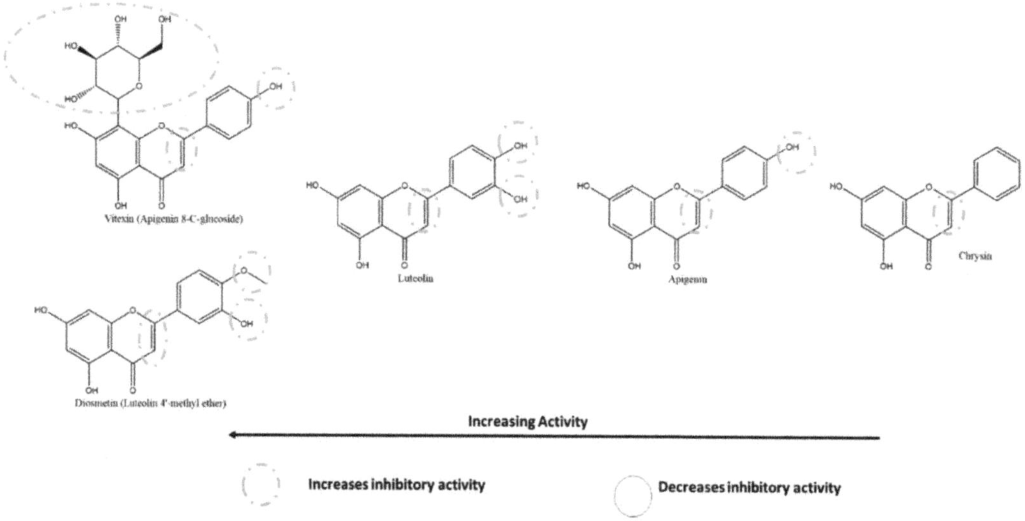

Figure 12. Structure–activity relationships of various flavones against SARS-CoV-2 3CLpro, adapted from the experiments of [118] using cell-free based in vitro FRET assays. Substitutions and groups that result in an increase/decrease in inhibitory activities are shown in red and green, respectively.

Liu et al. shed light on the activities of baicalein and its derivates against 3CLpro, and specifically, its substrate-binding site. As described earlier, the A-ring C6 and C7 hydroxyl groups increase their inhibitory effect given that these are responsible for the interaction with 3CLpro. The lower activity of its glycoside derivative, baicalin, is explained by the C7-OH glycosylation, which increases the size of the compound (Figure 13). The activity of scutellarein, slightly lower than baicalein, may be explained by their similarity in structures, although it is interesting to note that scutellarein, with its 4′-OH, has a higher IC50 value. However, its C7 glycoside derivative, scutellarin, has a lower activity than expected, given the reduction in inhibitory activities with glycosylation. The detrimental effects on the inhibitory activity of C-ring C3-glycosylation despite multiple continuous hydroxyl groups on the B-ring are seen in myricitrin. Comparing the actions of 5,6- and 6,7-dihydroxyflavone, one can notice that the presence of hydroxyl on the A-ring C7 position is important for the inhibitory function of flavones. Finally, the derivates wogonin and wogonoside lack the C6-OH, explaining their significantly lower activities [119].

Figure 13. Structure–activity relationships of various flavones against SARS-CoV-2 3CLpro, adapted from the experiments of [119] using cell-based in vitro FRET assays. Substitutions and groups that result in an increase/decrease in inhibitory activities are shown in red and green, respectively.

At 200 µM, the most to least effective flavanone inhibitors of SARS-CoV-2 3CLpro are naringenin > hesperidin > naringin (Figure 14) [117]. The A-ring C7-glycosylation of naringenin to naringin translates into more than a three-fold reduction in inhibitory activity. Moreover, being glycosylated at the same atom, hesperidin is much weaker than naringenin; however, it is slightly more inhibitory than naringin. This may be explained by substituting a methoxy group instead of a hydroxyl at the B-ring 4' position or the mere presence of a hydroxyl group at the B-ring 3' position in hesperidin rather than in naringin c. On the other hand, the results of Su et al. reveal that methoxy-group substitution at B-ring 4' position reduces the activity of hesperetin when compared to that of eriodyctiol (Figure 14) [117]. Overall, the differences in inhibition caused by methoxy-group substation at the hydroxyl place are slight and may be attributed to other factors; on the other hand, this variation in patterns provides a better insight into the binding mechanisms of flavanones and other compounds with 3CLpro.

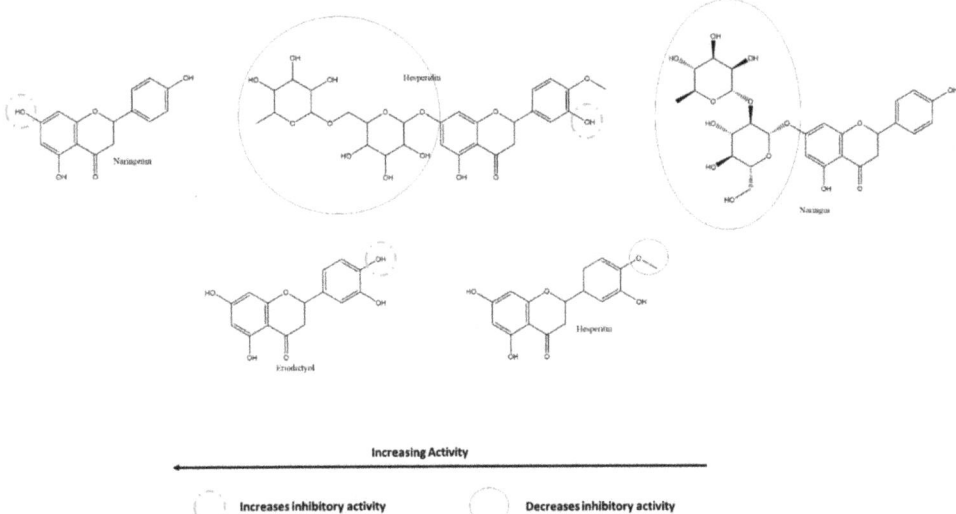

Figure 14. Structure–activity relationships of various flavanones against SARS-CoV-2 3CLpro, adapted from the experiments of [117,118] using in vitro FRET assays. Substitutions and groups that result in an increase/decrease in inhibitory activities are shown in red and green, respectively.

3.5.3. Effect of Flavan-3-ols/Flavanols on SARS-CoV-2 3CLpro

Among the flavan-3-ols/flavanols, Nguyen et al. reported that at 200 µM, the most to least effective inhibitors of SARS-CoV-2 3CLpro were in the order of EGCG ≈ GCG > EGC > CG = ECG > catechin ≈ EC. Akin to the relationship between myricetin, quercetin, and kaempferol, the presence of 3′, 4′, and 5′ hydroxyl groups on the B-ring of EGCG, GCG and EGC are responsible for their higher SARS-CoV-2 3CLpro inhibitory activity. The galloyl moiety on C-ring C3 of EGCG, GCG, CG, and ECG can also be inferred to increase the inhibitory activity compared to other flavanols such as EGC, catechin, and EC. The effect of 3′-OH is higher than that of the presence of C3-galloyl moiety (Figure 15). However, compared to flavanols, the absence of C-ring C2-C3 double bond and C4 carbonyl on the C-ring of flavanols explains their relative reduced activity [118].

Figure 15. Structure–activity relationships of various flavan-3-ols against SARS-CoV-2 3CLpro, adapted from the experiments of [118]. Substitutions and groups that result in an increase/decrease in inhibitory activities are shown in red and green, respectively.

3.5.4. Effect of Isoflavones on SARS-CoV-2 3CLpro

Nguyen et al. tested the inhibitory activity of three isoflavones, the order of whose activities against SARS-CoV-2 3CLpro were puerarin > daidzein > genistin using cell-free methods in vitro (Figure 16). The presence of A-ring C8-glycosylation in puerarin contributes to its high inhibitory activity. The C8-non-glycosylated daidzein and the C7-glycosylated genistin followed although all three activities were at least two times greater than their flavone analogs [118]. On the other hand, Su et al. also compared a few isoflavones against 3CLpro, revealing that B-ring 4′ glycosylation reduces the inhibitory effect. Substitution of the methoxy-group accounts for the higher inhibition by Formononetin compared to Daidzein and Genistein, whereas the presence of A-ring C5-hydroxyl appears to have the opposite effect [117].

Figure 16. Structure–activity relationships of various isoflavones against SARS-CoV-2 3CLpro, adapted from the experiments of [117,118] using in vitro FRET assays. Substitutions and groups that result in an increase/decrease in inhibitory activities are shown in red and green, respectively.

3.5.5. Effect of Flavones on SARS-CoV-2 PLpro

From the cell-based in vitro experiments of Su et al., one can also compare the activities of flavones on SARS-CoV-2 PLpro and consequently analyze the structure–activity relationships of the same (Figure 17) [126]. Scutellarein, with abundant A-ring hydroxyl groups and one on the B-ring, in addition to the absence of any glycosyl-substitutions, showed the highest efficacy against PLpro. As seen from the figure, in general, the presence of glycosylation in place of hydroxyl reduces the inhibitory activities of the flavones; however, this effect is more significant in the presence of hydroxyl groups on the B-ring, which appears to be the source of the anti-oxidative properties of flavonoids, in general.

Figure 17. Structure–activity relationships of various flavones against SARS-CoV-2 PLpro, adapted from the experiments of [126] using cell-based in vitro FRET assays. Substitutions and groups that result in an increase/decrease in inhibitory activities are shown in red and green, respectively.

3.5.6. Effect of Flavonols on SARS-CoV-2 Spike Protein and hACE2 Receptor Interaction

MCA fluorescence and rhACE2 cells determined the inhibitory activities of the flavonols quercetin and its derivates on S protein–ACE2 interaction in vitro. With its multiple hydroxyl groups at the A-, B-, and C-rings, base quercetin was the most potent inhibitor, followed closely by two C-ring C3 glycosides-rutin and isoquercetin, as well as Hyperoside and Quercetin-3-O-glucuronide (Figure 18). Contrary to what was seen for flavonoids against 3CLpro, the substitution of -OH by -OMe at the B-ring of flavonols against S–ACE2 interaction significantly reduces their respective activities [129].

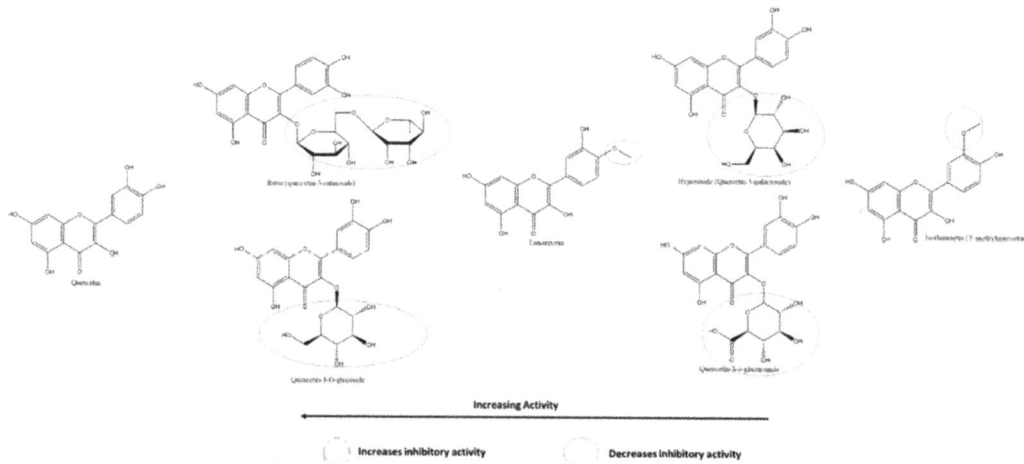

Figure 18. Structure–activity relationships of various flavonols against SARS-CoV-2 S-ACE2 interaction, adapted from the experiments of [129] using cell-based in vitro FRET assays. Substitutions and groups that result in increase/decrease in inhibitory activities are shown in red and green, respectively.

3.5.7. Effect of Flavones on SARS-CoV-2 Spike Protein and hACE2 Receptor Interaction

Using HEK293T cells and in vitro assays, Gao et al. showed that the flavones neobaicalein, wogonin, oroxylin A, and scutellarin inhibited the SARS-CoV-2 spike–ACE2 interaction in decreasing order (Figure 19). It can be deduced from their structures that methoxy- and hydroxyl groups on the 1′ and 5′ positions at the B-ring, respectively, increased the inhibitory action. In contrast, glycosylation at the A-ring C7 reduced it.

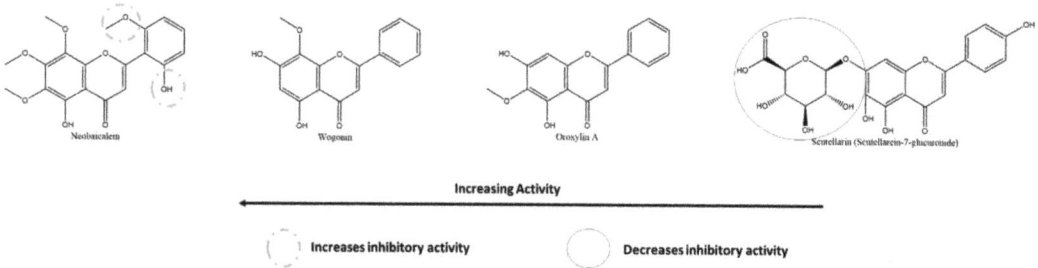

Figure 19. Structure–activity relationships of various flavones against SARS-CoV-2 S-ACE2 interaction, adapted from the experiments of [120] using cell-based in vitro FRET assays. Substitutions and groups that result in an increase/decrease in inhibitory activities are shown in red and green, respectively.

A few general patterns come to light from the analysis of the SARs of the various flavonoids reported in this study. Overall, the presence of hydroxyl groups on all rings, especially on B-ring, of flavonoids increase their respective activities. The most effective flavonoids against 3CLpro, the most promising target, were myricetin, with six hydroxyl groups spread over its three phenolic rings. Secondly, substituting these hydroxyl groups with sugars and other larger groups causes a decrease in inhibitory activity against all respective targets. Moreover, replacing hydroxyl with methoxy-groups resulted in increased activities for many flavonoids, leading to the hypothesis that the polar and electronegative nature of the hydroxyl groups is not the only responsible factor for their high effectiveness. Analysis of the SARs of flavonoids against various targets allows identifying the most effective molecule to test in further stages. Furthermore, it also guides pharmaceutical companies to develop synthetic drugs based on the most inhibiting structural groups in nature.

3.6. Flavonoids as Potential Inhibitors of SARS-CoV-2 Proteins: In Silico Studies

In silico approaches in the field of drug discovery such as structure-based virtual screening, molecular dynamics, and absorption, distribution, metabolism, excretion, and toxicity (ADMET) analysis have played an essential role in the screening and identification of various flavonoid inhibitors against the main targets of COVID-19: M^{pro}, spike glycoprotein, PL^{Pro}, RdRp, helicase and ACE2 within a brief period.

3.6.1. Flavonoids against SARS-CoV-2 M^{Pro}

Using molecular docking, Cherrak et al. identified quercetin-3-O-rhamnoside, myricetin 3-rutinoside, and rutin as the potential inhibitors of SARS-CoV-2 M^{pro} in decreasing order with binding energies of −9.7, −9.3, and −9.2 kcal.mol^{-1} respectively [142]. Another report also confirmed that rutin is a potential flavonoid against the SARS-CoV-2 M^{Pro} with a binding energy of −11.187 kcal/mol [143]. A recent study supported the inhibitory effect of rutin on SARS-CoV-2 Mpro via molecular docking with a binding energy −15.63 kcal/mol. Furthermore, ADMET analysis, combinatorial molecular simulations, and hybrid QM/MM approaches concluded that rutin binds very strongly at the active sites of SARS-CoV-2 M^{Pro} by forming three hydrogen bonds at His 163, Glu 166, Gln189 residues [144]. Rakshit et al. screened various flavonoids against M^{pro} and identified the top five potential flavonoids in order of rhoifolin, 5,7-dimethoxyflavanone-40-O-b-d-glucopyranoside, baicalin, luteolin, and kaempferol based on their binding energies of −9.28, −8.81, −8.29, −8.14, −8.11 kcal/mol [145]. The inhibitory effect of luteolin against the M^{pro} is recently confirmed [146]. Fayyaz et al. screened several flavonoids and identified three potentially active flavonoids, whose activities against SARS-CoV-2 M^{Pro} were in the order of rhodiolin > baicalin > silymarin based on their binding energy (−9.05, −8.85, −8.71 kcal/mol respectively) and dissociation constant (0.23, 0.33, 0.41 µm, respectively) using molecular docking and simulation studies [147]. Thioflavonol also inhibits Mpro [148] along with other flavonoids such as apigenin, daidzein, quercetin, kaempferol, luteolin, epigallocatechin, and kaempferol using molecular docking and simulation analysis [104].

The M^{Pro} is considered the most promising drug target for SARS-CoV-2 due to its proteolytic activity, cleaving viral polyprotein into independent functional proteins required for SARS-CoV-2 replication [149–151]. The other reason for its therapeutic importance is its dissimilarity to any human cell protease [152–154] and its similarity with the M^{Pro} of SARS-CoV [155]. The most common active site residues of M^{Pro} were Glu166, His163, and Met165, which were involved in the interaction with most of the flavonoids.

3.6.2. Flavonoids against SARS-CoV-2 Spike Glycoprotein

Rutin inhibits the SARS-CoV-2 spike glycoprotein but with less binding energy (−7.9 kcal/mol) [143]. On the other hand, naringin inhibits the spike glycoprotein more effectively with binding energy −9.8 kcal/mol compared to standard drug dexamethasone with the binding energy of −7.9 kcal/mol [156]. Fayyaz et al. described potentially active flavonoids, whose activities against SARS-CoV-2 spike protein were in the order of rhodiolin > hesperidin (with active site 1) > hesperidin (with active site 2) > silyhermin based on their binding energy (−8.68, −8.53, −8.18, −8.05 kcal/mol, respectively) and dissociation constant (0.43, 0.56, 1.01, 1.25 µm respectively) [147]. The authors showed that hesperidin could bind to two different active sites on the spike glycoprotein with different binding energies. Teli et al. highlighted that rutin could serve as a dual receptor inhibitor against the Mpro and spike glycoprotein of SARS-CoV-2 with improved ADMET parameters.

3.6.3. Flavonoids against SARS-CoV-2 PL^{Pro}

Potentially active flavonoids, whose activities against SARS-CoV-2 PL^{Pro} were in the order of baicalin > hesperidin > naringen > flemiflavanone D > Euchresta flavanone A on the basis of their binding energy (−10.82, −10.61, −10.17, −10.07, −9.95 kcal/mol, respectively) and dissociation constant (0.01, 0.02, 0.04, 0.04, 0.05 μm, respectively) using molecular docking and simulation studies [147]. The active site residues of PL^{Pro}, which are found common with almost all the flavonoids were Lys157, Leu162, Gly163, Asp164, and Glu167, and therefore, all these residues are critical for the ligand interaction.

3.6.4. Flavonoids against SARS-CoV-2 RdRp

An interesting protein–ligand blind docking approach proposed that the SARS-CoV-2 RNA replication can be inhibited by targeting its RdRp protein. Theaflavin inhibits the viral RNA replication by interfering with the RdRp catalytic pocket with the binding energy of −8.8 kcal/mol [157]. Shawan et al. identified luteolin as a potential inhibitor of ACE2 with the binding energy of −10.1 kcal/mol, which is very close to the binding energy of FDA-approved remdesivir (−10.0 kcal/mol) [158]. Fayyaz et al. identified three potentially active flavonoids against SARS-CoV-2 RdRp were in the order of hesperidin > baicalin > naringen based on their binding energy (−9.53, −9.01, −8.54 kcal/mol, respectively) and dissociation constant (0.1, 0.25, 0.55 μm, respectively) using molecular docking and simulation studies [147].

3.6.5. Flavonoids against SARS-CoV-2 Helicase

Along with Mpro, PL^{Pro}, RdRp, and spike protein, Fayyaz et al. identified two potential flavonoids: hesperidin and baicalin based on binding energies (−8.93 and −8.9 kcal/mol, respectively) and dissociation constant (0.283 and 0.29 μM, respectively). Hesperidin and baicalin are the only flavonoids that interact and inhibit all the main targets of SARS-CoV-2, such as M^{Pro}, PL^{Pro}, RdRp, and helicase with excellent binding energies; therefore, both of these flavonoids are considered as multi receptor/protein targets for COVID-19.

3.6.6. Flavonoids against SARS-CoV-2 ACE2

Luteolin is a potential inhibitor of ACE2 with the binding energy of −10.1 kcal/mol, which is very close to the binding energy of FDA-approved remdesivir (−10.0 kcal/mol) [158]. Using virtual screening via Autodock vina, studies identified various flavonoids such as Myritilin, myricitrin, δ-Viniferin, TaiwanhomoflavoneA, Afzelin Biorobin, and Nympholide A that can inhibit the ACE2 [159]. Similarly, Hesperetin, Baicalin, Scutellarin against ACE2 using virtual screening and molecular docking studies [160]. Tangeretin, Nobiletin, Naringenin, Brazilein, Brazilin, Galangin also inhibit ACE2 receptors [161].

3.7. Clinical Trials and Future Prospects

Many in vitro studies exploring the anti-SARS-CoV-2 action of flavonoids were published over the last two years since the advent of the COVID-19 pandemic. These studies were guided by in silico studies and the further in vitro or in vivo research of anti-SARS-CoV and anti-MERS-CoV activities of various flavonoids over the last decade owing to the two epidemics caused by these respective.

Until 17 August 2021, 13 clinical trials were reported and explored the effect and efficacy of various flavonoids and a few other polyphenols and their extracts on COVID-19 patients (Table 5 and Table S2). Of these, the most promising and popular flavonoid is the flavonol quercetin, furthering our findings. In particular, the study (NCT04401202) exploring the effect of *Nigella sativa* seed oil, which is rich in quercetin and kaempferol [162], found in its phase 2 trial that 62.1% of the intervention group which received *Nigella sativa* oil 500mg soft gel capsules orally twice daily recovered within 14 days, compared to only 36% in the control group. Following their progress can prove highly beneficial to clinicians around the globe in identifying potential COVID-19 therapeutic agents at the earliest.

Table 5. Some flavonoids and their natural source extracts are currently in clinical trials on COVID-19 patients. Extracted on 17 August 2021 from https://clinicaltrials.gov, accessed on 3 June 2021.

Study Title	Study Type	Number of Subjects Enrolled	Status
Nigella Sativa in COVID-19 (NCT04401202)	Prospective, Randomized, Open-label	183 COVID-19 positive participants *	Completed
Efficacy of Psidii Guava's Extract For COVID-19 (NCT04810728)	Experimental, randomized, double-blind clinical trial	90 COVID-19 positive participants b/w 13-59 yrs.	Phase 3
The Effectiveness of Phytotherapy in SARS-CoV 2 (COVID-19) (NCT04851821)	Randomized, double–masked, interventional clinical Trial with Parallel Assignment	80 COVID-19 positive participants *	Phase 1
Masitinib Combined with Isoquercetin and Best Supportive Care in Hospitalized Patients With Moderate and Severe COVID-19 (NCT04622865)	Randomized, double-blinded, triple-masked interventional clinical trial with Parallel Assignment	200 COVID-19 positive participants *	Phase 2
Quercetin In The Treatment of SARS-CoV 2 (NCT04853199)	Randomized, double-blinded, triple-masked interventional clinical trial with Parallel Assignment	200 COVID-19 positive participants *	Early Phase 1
Randomized Proof-of-Concept Trial to Evaluate the Safety and Explore the Effectiveness of Resveratrol, a Plant Polyphenol, for COVID-19 (NCT04400890)	Randomized placebo-controlled, double-blinded, quadruple-masked, interventional clinical trial	100 COVID-19 positive participants $\geq$ 45 yrs.	Phase 2
Tannin Specific Natural Extract for COVID-19 Infection (NCT04403646)	Double-blind, randomized, triple-masked	124 COVID-19 positive participants *	n/a
P2Et Extract in the Symptomatic Treatment of Subjects With COVID-19 (NCT04410510)	Double-blind, randomized, triple-masked, interventional clinical trial with parallel assignment	100 COVID-19 positive participants *	Phase 2/3
COVID-19, Hospitalized, Patients, Nasafytol (NCT04844658)	Standard-of-care comparative, open-labelled, parallel two-arms and randomized trial	50 COVID-19 positive participants *	Recruiting
Study to Investigate the Clinical Benefits of Dietary Supplement Quercetin for Managing Early COVID-19 Symptoms at Home (NCT04861298)	Open-labelled, randomized, parallel-assignment, interventional trial	142 COVID-19 positive participants *	Recruiting
Complementary Intervention for COVID-19 (NCT04487964)	Open-labelled, non-randomized	70 COVID-19 positive participants *	Recruiting
The Study of Quadruple Therapy Zinc, Quercetin, Bromelain and Vitamin C on the Clinical Outcomes of Patients Infected With COVID-19 (NCT04468139)	Open-labelled, single-assignment, interventional trial	60 COVID-19 positive participants *	Phase 4
Evaluation of the Effect of Anatolian Propolis on COVID-19 in Healthcare Professionals (NCT04680819)	Observational, prospective cohort study	50 HCWs at risk for developing COVID-19	Not yet recruiting

NLR: Neutrophil to Lymphocyte Ratio; hs-CRP: high-sensitivity C-reactive protein; ARDS: acute respiratory distress syndrome; PDI: Protein disulfide-isomerase; * participants $\geq$ 18 yrs.

4. Conclusions

The results of our review of in vitro and in silico studies are encouraging. Considering that no internationally accepted effective therapeutic intervention exists for COVID-19, there is an urgent need for more extended studies and further in vivo and clinical trials to confirm these results and promote the synthesis of more efficient drugs using the SARs outlined in previous sections.

In this review, we report that the flavonols quercetin, myricetin, and their derivatives, the flavones baicalin and baicalein, the flavan-3-ol EGCG, and finally, tannic acid, have the most promising scope for further evaluation using both in vivo and consequently clinical studies. Unfortunately, the tendency of flavonoids to aggregate and their limited bioavailability limit their therapeutic interventions. The use of flavonoids in combination with synthetic and commercially produced drugs showed promising results, but more research is needed to prove their synergistic effects. Looking at the growing concern of antiviral resistance, naturally occurring flavonoids are a promising alternative.

Supplementary Materials: The following are available online at https://www.mdpi.com/article/10.3390/ijms222011069/s1.

Author Contributions: A.C. designed the study, critically supervised the project, revised the text, and wrote some parts of the review; D.B. edited and reviewed the manuscript; V.D.D. and S.K. wrote a part of the manuscript; R.K. and P.P. equally carried out most of the study, wrote most of the manuscript and generated the tables and the graphs. All authors have read and agreed to the published version of the manuscript.

Funding: D.B. was supported by a National Priorities Research Program grant (NPRP 11S-1214-170101) from the Qatar National Research Fund (QNRF, a member of Qatar Foundation) Otherwise, this research received no external funding.

Institutional Review Board Statement: Not applicable.

Informed Consent Statement: Not applicable.

Data Availability Statement: No new data were created or analyzed in this study.

Acknowledgments: The publication of this article was funded by the Weill Cornell Medicine—Qatar Distributed eLibrary.

Conflicts of Interest: The authors declare no conflict of interest.

References

1. Zhao, M.; Yu, Y.; Sun, L.-M.; Xing, J.-Q.; Li, T.; Zhu, Y.; Wang, M.; Yu, Y.; Xue, W.; Xia, T.; et al. GCG inhibits SARS-CoV-2 replication by disrupting the liquid phase condensation of its nucleocapsid protein. *Nat. Commun.* **2021**, *12*, 1–14. [CrossRef]
2. Zhu, N.; Zhang, D.; Wang, W.; Li, X.; Yang, B.; Song, J.; Zhao, X.; Huang, B.; Shi, W.; Lu, R.; et al. A Novel Coronavirus from Patients with Pneumonia in China, 2019. *N. Engl. J. Med.* **2020**, *382*, 727–733. [CrossRef]
3. Wu, F.; Zhao, S.; Yu, B.; Chen, Y.-M.; Wang, W.; Song, Z.-G.; Hu, Y.; Tao, Z.-W.; Tian, J.-H.; Pei, Y.-Y.; et al. A new coronavirus associated with human respiratory disease in China. *Nature* **2020**, *579*, 265–269. [CrossRef] [PubMed]
4. World Health Organization. WHO Coronavirus (COVID-19) Dashboard. Available online: https://covid19.who.int/ (accessed on 17 August 2021).
5. "About Variants of the Virus that Causes COVID-19 | CDC". Available online: https://stacks.cdc.gov/view/cdc/104698 (accessed on 15 August 2021).
6. Loganathan, S.K.; Schleicher, K.; Malik, A.; Quevedo, R.; Langille, E.; Teng, K.; Oh, R.H.; Rathod, B.; Tsai, R.; Samavarchi-Tehrani, P.; et al. Rare driver mutations in head and neck squamous cell carcinomas converge on NOTCH signaling. *Science* **2020**, *367*, 1264–1269. [CrossRef]
7. Fehr, A.R.; Perlman, S. Coronaviruses: An Overview of Their Replication and Pathogenesis. In *Coronaviruses—Methods and Protocols*; Humana Press: New York, NY, USA, 2015; Volume 1282, pp. 1–23. [CrossRef]
8. Chan, J.F.-W.; Lau, S.K.P.; To, K.; Cheng, V.C.C.; Woo, P.C.Y.; Yuen, K.-Y. Middle East Respiratory Syndrome Coronavirus: Another Zoonotic Betacoronavirus Causing SARS-Like Disease. *Clin. Microbiol. Rev.* **2015**, *28*, 465–522. [CrossRef]
9. Walls, A.C.; Park, Y.-J.; Tortorici, M.A.; Wall, A.; McGuire, A.T.; Veesler, D. Structure, Function, and Antigenicity of the SARS-CoV-2 Spike Glycoprotein. *Cell* **2020**, *181*, 281–292. [CrossRef] [PubMed]

10. Kang, S.; Peng, W.; Zhu, Y.; Lu, S.; Zhou, M.; Lin, W.; Wu, W.; Huang, S.; Jiang, L.; Luo, X.; et al. Recent progress in understanding 2019 novel coronavirus (SARS-CoV-2) associated with human respiratory disease: Detection, mechanisms and treatment. *Int. J. Antimicrob. Agents* **2020**, *55*, 105950. [CrossRef]
11. Zhou, P.; Yang, X.-L.; Wang, X.-G.; Hu, B.; Zhang, L.; Zhang, W.; Si, H.-R.; Zhu, Y.; Li, B.; Huang, C.-L.; et al. A pneumonia outbreak associated with a new coronavirus of probable bat origin. *Nature* **2020**, *579*, 270–273. [CrossRef] [PubMed]
12. Lu, R.; Zhao, X.; Li, J.; Niu, P.; Yang, B.; Wu, H.; Wang, W.; Song, H.; Huang, B.; Zhu, N.; et al. Genomic characterisation and epidemiology of 2019 novel coronavirus: Implications for virus origins and receptor binding. *Lancet* **2020**, *395*, 565–574. [CrossRef]
13. Mittal, A.; Manjunath, K.; Ranjan, R.K.; Kaushik, S.; Kumar, S.; Verma, V. COVID-19 pandemic: Insights into structure, function, and hACE2 receptor recognition by SARS-CoV-2. *PLoS Pathog.* **2020**, *16*, e1008762. [CrossRef]
14. Mishra, C.B.; Pandey, P.; Sharma, R.D.; Malik, Z.; Mongre, R.K.; Lynn, A.M.; Prasad, R.; Jeon, R.; Prakash, A. Identifying the natural polyphenol catechin as a multi-targeted agent against SARS-CoV-2 for the plausible therapy of COVID-19: An integrated computational approach. *Brief. Bioinform.* **2020**, *22*, 1346–1360. [CrossRef]
15. O'Keefe, B.R.; Giomarelli, B.; Barnard, D.L.; Shenoy, S.R.; Chan, P.K.S.; McMahon, J.B.; Palmer, K.E.; Barnett, B.W.; Meyerholz, D.K.; Wohlford-Lenane, C.L.; et al. Broad-Spectrum In Vitro Activity and In Vivo Efficacy of the Antiviral Protein Griffithsin against Emerging Viruses of the Family Coronaviridae. *J. Virol.* **2010**, *84*, 2511–2521. [CrossRef]
16. Ertekin, S.S.; Morgado-Carrasco, D.; Forns, X.; Mascaró, J.M. Complete Remission of Hypertrophic Discoid Cutaneous Lupus Erythematosus after Treatment of Chronic Hepatitis C with Direct-Acting Antivirals. *JAMA Dermatol.* **2020**, *156*, 471. [CrossRef] [PubMed]
17. Holmes, J.A.; Chung, R.T. Shortening treatment with direct-acting antivirals in HCV-positive organ transplantation. *Lancet Gastroenterol. Hepatol.* **2020**, *5*, 626–627. [CrossRef]
18. Choi, H.-J.; Kim, J.-H.; Lee, C.-H.; Ahn, Y.-J.; Song, J.-H.; Baek, S.-H.; Kwon, D.-H. Antiviral activity of quercetin 7-rhamnoside against porcine epidemic diarrhea virus. *Antivir. Res.* **2009**, *81*, 77–81. [CrossRef] [PubMed]
19. Nunes, V.S.; Paschoal, D.F.S.; Costa, L.A.S.; Dos Santos, H.F. Antivirals virtual screening to SARS-CoV-2 non-structural proteins. *J. Biomol. Struct. Dyn.* **2021**, 1–15. [CrossRef] [PubMed]
20. Kumar, Y.; Singh, H.; Patel, C.N. In silico prediction of potential inhibitors for the main protease of SARS-CoV-2 using molecular docking and dynamics simulation based drug-repurposing. *J. Infect. Public Health* **2020**, *13*, 1210–1223. [CrossRef]
21. Gurung, A.B.; Ali, M.A.; Lee, J.; Farah, M.A.; Al-Anazi, K.M. Unravelling lead antiviral phytochemicals for the inhibition of SARS-CoV-2 Mpro enzyme through in silico approach. *Life Sci.* **2020**, *255*, 117831. [CrossRef] [PubMed]
22. Koulgi, S.; Jani, V.; Uppuladinne, V.N.M.; Sonavane, U.; Joshi, R.N. Natural plant products as potential inhibitors of RNA dependent RNA polymerase of Severe Acute Respiratory Syndrome Coronavirus-2. *PLoS ONE* **2021**, *16*, e0251801. [CrossRef]
23. Shah, B.; Modi, P.; Sagar, S.R. In silico studies on therapeutic agents for COVID-19: Drug repurposing approach. *Life Sci.* **2020**, *252*, 117652. [CrossRef]
24. Russo, M.; Moccia, S.; Spagnuolo, C.; Tedesco, I.; Russo, G.L. Roles of flavonoids against coronavirus infection. *Chem. Interact.* **2020**, *328*, 109211. [CrossRef]
25. Reynolds, D.; Huesemann, M.; Edmundson, S.; Sims, A.; Hurst, B.; Cady, S.; Beirne, N.; Freeman, J.; Berger, A.; Gao, S. Viral inhibitors derived from macroalgae, microalgae, and cyanobacteria: A review of antiviral potential throughout pathogenesis. *Algal Res.* **2021**, *57*, 102331. [CrossRef] [PubMed]
26. WHO Global Report on Traditional and Complementary Medicine. 2019. Available online: https://apps.who.int/iris/handle/10665/312342 (accessed on 15 August 2021).
27. Yi, L.; Li, Z.; Yuan, K.; Qu, X.; Chen, J.; Wang, G.; Zhang, H.; Luo, H.; Zhu, L.; Jiang, P.; et al. Small Molecules Blocking the Entry of Severe Acute Respiratory Syndrome Coronavirus into Host Cells. *J. Virol.* **2004**, *78*, 11334–11339. [CrossRef]
28. Chaari, A. Inhibition of human islet amyloid polypeptide aggregation and cellular toxicity by oleuropein and derivatives from olive oil. *Int. J. Biol. Macromol.* **2020**, *162*, 284–300. [CrossRef]
29. Thevarajan, I.; Nguyen, T.H.O.; Koutsakos, M.; Druce, J.; Caly, L.; van de Sandt, C.E.; Jia, X.; Nicholson, S.; Catton, M.; Cowie, B.; et al. Breadth of concomitant immune responses prior to patient recovery: A case report of non-severe COVID-19. *Nat. Med.* **2020**, *26*, 453–455. [CrossRef] [PubMed]
30. Zakaryan, H.; Arabyan, E.; Oo, A.; Zandi, K. Flavonoids: Promising natural compounds against viral infections. *Arch. Virol.* **2017**, *162*, 2539–2551. [CrossRef] [PubMed]
31. Fuzimoto, A.D.; Isidoro, C. The antiviral and coronavirus-host protein pathways inhibiting properties of herbs and natural compounds-Additional weapons in the fight against the COVID-19 pandemic? *J. Tradit. Complement. Med.* **2020**, *10*, 405–419. [CrossRef]
32. Abotaleb, M.; Liskova, A.; Kubatka, P.; Büsselberg, D. Therapeutic Potential of Plant Phenolic Acids in the Treatment of Cancer. *Biomolecules* **2020**, *10*, 221. [CrossRef] [PubMed]
33. Koklesova, L.; Liskova, A.; Samec, M.; Zhai, K.; Al-Ishaq, R.K.; Bugos, O.; Šudomová, M.; Biringer, K.; Pec, M.; Adamkov, M.; et al. Protective Effects of Flavonoids Against Mitochondriopathies and Associated Pathologies: Focus on the Predictive Approach and Personalized Prevention. *Int. J. Mol. Sci.* **2021**, *22*, 8649. [CrossRef] [PubMed]
34. Al-Ishaq, R.; Liskova, A.; Kubatka, P.; Büsselberg, D. Enzymatic Metabolism of Flavonoids by Gut Microbiota and Its Impact on Gastrointestinal Cancer. *Cancers* **2021**, *13*, 3934. [CrossRef]

35. Liskova, A.; Samec, M.; Koklesova, L.; Brockmueller, A.; Zhai, K.; Abdellatif, B.; Siddiqui, M.; Biringer, K.; Kudela, E.; Pec, M.; et al. Flavonoids as an effective sensitizer for anti-cancer therapy: Insights into multi-faceted mechanisms and applicability towards individualized patient profiles. *EPMA J.* **2021**, *12*, 155–176. [CrossRef]
36. Chaari, A.; Abdellatif, B.; Nabi, F.; Khan, R.H. Date palm (Phoenix dactylifera L.) fruit's polyphenols as potential inhibitors for human amylin fibril formation and toxicity in type 2 diabetes. *Int. J. Biol. Macromol.* **2020**, *164*, 1794–1808. [CrossRef] [PubMed]
37. Chaari, A.; Bendriss, G.; Zakaria, D.; McVeigh, C. Importance of Dietary Changes during the Coronavirus Pandemic: How to Upgrade Your Immune Response. *Front. Public Health* **2020**, *8*, 476. [CrossRef]
38. Shereen, M.A.; Khan, S.; Kazmi, A.; Bashir, N.; Siddique, R. COVID-19 infection: Emergence, transmission, and characteristics of human coronaviruses. *J. Adv. Res.* **2020**, *24*, 91–98. [CrossRef]
39. Piccolella, S.; Crescente, G.; Faramarzi, S.; Formato, M.; Pecoraro, M.T.; Pacifico, S. Polyphenols vs. Coronaviruses: How Far Has Research Moved Forward? *Molecules* **2020**, *25*, 4103. [CrossRef]
40. Satarker, S.; Nampoothiri, M. Structural Proteins in Severe Acute Respiratory Syndrome Coronavirus-2. *Arch. Med. Res.* **2020**, *51*, 482–491. [CrossRef] [PubMed]
41. Saxena, S.K. *Coronavirus Disease 2019 (COVID-19): Epidemiology, Pathogenesis, Diagnosis, and Therapeutics*; Springer Nature: Berlin/Heidelberg, Germany, 2020.
42. Güler, G.; Özdemir, H.; Omar, D.; Akdoğan, G. Coronavirus disease 2019 (COVID-19): Biophysical and biochemical aspects of SARS-CoV-2 and general characteristics. *Prog. Biophys. Mol. Biol.* **2021**, *164*, 3–18. [CrossRef] [PubMed]
43. Cui, J.; Li, F.; Shi, Z.-L. Origin and evolution of pathogenic coronaviruses. *Nat. Rev. Genet.* **2018**, *17*, 181–192. [CrossRef]
44. Al Adem, K.; Shanti, A.; Stefanini, C.; Lee, S. Inhibition of SARS-CoV-2 Entry into Host Cells Using Small Molecules. *Pharmaceuticals* **2020**, *13*, 447. [CrossRef]
45. Snijder, E.; Decroly, E.; Ziebuhr, J. The Nonstructural Proteins Directing Coronavirus RNA Synthesis and Processing. *Adv. Virus Res.* **2016**; *96*, 59–126. [CrossRef]
46. Chen, Y.; Guo, Y.; Pan, Y.; Zhao, Z.J. Structure analysis of the receptor binding of 2019-nCoV. *Biochem. Biophys. Res. Commun.* **2020**, *525*, 135–140. [CrossRef]
47. Arndt, A.L.; Larson, B.J.; Hogue, B.G. A Conserved Domain in the Coronavirus Membrane Protein Tail Is Important for Virus Assembly. *J. Virol.* **2010**, *84*, 11418–11428. [CrossRef] [PubMed]
48. Gordon, D.E.; Jang, G.M.; Bouhaddou, M.; Xu, J.; Obernier, K.; White, K.M.; O'Meara, M.J.; Rezelj, V.V.; Guo, J.Z.; Swaney, D.L.; et al. A SARS-CoV-2 protein interaction map reveals targets for drug repurposing. *Nature* **2020**, *583*, 459–468. [CrossRef] [PubMed]
49. Levy, E.; Delvin, E.; Marcil, V.; Spahis, S. Can phytotherapy with polyphenols serve as a powerful approach for the prevention and therapy tool of novel coronavirus disease 2019 (COVID-19)? *Am. J. Physiol. Metab.* **2020**, *319*, E689–E708. [CrossRef] [PubMed]
50. Belouzard, S.; Chu, V.C.; Whittaker, G.R. Activation of the SARS coronavirus spike protein via sequential proteolytic cleavage at two distinct sites. *Proc. Natl. Acad. Sci. USA* **2009**, *106*, 5871–5876. [CrossRef] [PubMed]
51. Hamid, S.; Mir, M.Y.; Rohela, G.K. Novel coronavirus disease (COVID-19): A pandemic (epidemiology, pathogenesis and potential therapeutics). *New Microbes New Infect.* **2020**, *35*, 100679. [CrossRef]
52. Guo, Y.-R.; Cao, Q.-D.; Hong, Z.-S.; Tan, Y.-Y.; Chen, S.-D.; Jin, H.-J.; Tan, K.-S.; Wang, D.-Y.; Yan, Y. The origin, transmission and clinical therapies on coronavirus disease 2019 (COVID-19) outbreak—An update on the status. *Mil. Med. Res.* **2020**, *7*, 1–10. [CrossRef]
53. Wang, Y.; Grunewald, M.; Perlman, S. Coronaviruses: An Updated Overview of Their Replication and Pathogenesis. *Coronaviruses* **2020**, *2203*, 1–29. [CrossRef]
54. Jeong, G.U.; Song, H.; Yoon, G.Y.; Kim, D.; Kwon, Y.-C. Therapeutic Strategies against COVID-19 and Structural Characterization of SARS-CoV-2: A Review. *Front. Microbiol.* **2020**, *11*, 1723. [CrossRef]
55. Das, G.; Ghosh, S.; Garg, S.; Ghosh, S.; Jana, A.; Samat, R.; Mukherjee, N.; Roy, R. An overview of key potential therapeutic strategies for combat in the COVID-19 battle. *RSC Adv.* **2020**, *10*, 28243–28266. [CrossRef]
56. Alzaabi, M.M.; Hamdy, R.; Ashmawy, N.S.; Hamoda, A.M.; Alkhayat, F.; Khademi, N.N.; Al Joud, S.M.A.; El-Keblawy, A.A.; Soliman, S.S.M. Flavonoids are promising safe therapy against COVID-19. *Phytochem. Rev.* **2021**, 1–22. [CrossRef]
57. Wang, T.-Y.; Li, Q.; Bi, K.-S. Bioactive flavonoids in medicinal plants: Structure, activity and biological fate. *Asian J. Pharm. Sci.* **2017**, *13*, 12–23. [CrossRef]
58. Panche, A.N.; Diwan, A.D.; Chandra, S.R. Flavonoids: An overview. *J. Nutr. Sci.* **2016**, *5*, e47. [CrossRef] [PubMed]
59. Cutting, W.C.; Dreisbach, R.H.; Azima, M.; Neff, B.J.; Brown, B.J.; Wray, J. Antiviral chemotherapy. V. Further report on flavonoids. *Stanf. Med. Bull.* **1951**, *9*, 236–242.
60. Gábor, M.; Eperjessy, E. Antibacterial Effect of Fisetin and Fisetinidin. *Nature* **1966**, *212*, 1273. [CrossRef]
61. Pusztai, R.; Béládi, I.; Bakai, M.; Mucsi, I.; Kukán, E. Study on the effect of flavonoids and related substances. I. The effect of quercetin on different viruses. *Acta Microbiol. Acad. Sci. Hung.* **1966**, *13*, 113–118.
62. Kincl, F.A.; Romo, J.; Rosenkranz, G.; Sondheimer, F. 804. The constituents of Casimiroa edulis llave et lex. Part I. The seed. *J. Chem. Soc.* **1956**, 4163–4169. [CrossRef]
63. Piccolella, S.; Crescente, G.; Candela, L.; Pacifico, S. Nutraceutical polyphenols: New analytical challenges and opportunities. *J. Pharm. Biomed. Anal.* **2019**, *175*, 112774. [CrossRef] [PubMed]

64. Piccolella, S.; Pacifico, S. Plant-Derived Polyphenols: A Chemopreventive and Chemoprotectant Worth-Exploring Resource in Toxicology. *Adv. Mol. Toxicol.* **2015**, *9*, 161–214. [CrossRef]
65. Martin, K.W.; Ernst, E. Antiviral agents from plants and herbs: A systematic review. *Antivir. Ther.* **2003**, *8*, 77–90. [CrossRef]
66. Helenius, A. Virus Entry: Looking Back and Moving Forward. *J. Mol. Biol.* **2018**, *430*, 1853–1862. [CrossRef] [PubMed]
67. Denaro, M.; Smeriglio, A.; Barreca, D.; De Francesco, C.; Occhiuto, C.; Milano, G.; Trombetta, D. Antiviral activity of plants and their isolated bioactive compounds: An update. *Phytotherapy Res.* **2019**, *34*, 742–768. [CrossRef] [PubMed]
68. Ahmad, A.; Kaleem, M.; Ahmed, Z.; Shafiq, H. Therapeutic potential of flavonoids and their mechanism of action against microbial and viral infections—A review. *Food Res. Int.* **2015**, *77*, 221–235. [CrossRef]
69. Béládi, I.; Pusztai, R.; Mucsi, I.; Bakay, M.; Gabor, M. Activity of some flavonoids against viruses. *Ann. N. Y. Acad. Sci.* **1977**, *284*, 358–364. [CrossRef]
70. Kaul, T.N.; Middleton, E.; Ogra, P.L. Antiviral effect of flavonoids on human viruses. *J. Med. Virol.* **1985**, *15*, 71–79. [CrossRef]
71. Vlietinck, A.J.; Berghe, D.A.V. Can ethnopharmacology contribute to the development of antiviral drugs? *J. Ethnopharmacol.* **1991**, *32*, 141–153. [CrossRef]
72. Debiaggi, M.; Tateo, F.; Pagani, L.; Luini, M.; Romero, E. Effects of propolis flavonoids on virus infectivity and replication. *Microbiologica* **1990**, *13*, 207–213.
73. Johari, J.; Kianmehr, A.; Mustafa, M.R.; Abubakar, S.; Zandi, K. Antiviral Activity of Baicalein and Quercetin against the Japanese Encephalitis Virus. *Int. J. Mol. Sci.* **2012**, *13*, 16785–16795. [CrossRef]
74. Zhang, T.; Wu, Z.; Du, J.; Hu, Y.; Liu, L.; Yang, F.; Jin, Q. Anti- Japanese-Encephalitis-Viral Effects of Kaempferol and Daidzin and Their RNA-Binding Characteristics. *PLoS ONE* **2012**, *7*, e30259. [CrossRef] [PubMed]
75. Lani, R.; Hassandarvish, P.; Shu, M.-H.; Phoon, W.H.; Chu, J.J.H.; Higgs, S.; Vanlandingham, D.; Abu Bakar, S.; Zandi, K. Antiviral activity of selected flavonoids against Chikungunya virus. *Antivir. Res.* **2016**, *133*, 50–61. [CrossRef] [PubMed]
76. Lin, S.-C.; Chen, M.-C.; Li, S.; Lin, C.-C.; Wang, T.T. Antiviral activity of nobiletin against chikungunya virus in vitro. *Antivir. Ther.* **2017**, *22*, 689–697. [CrossRef]
77. Lani, R.; Hassandarvish, P.; Chiam, C.W.; Moghaddam, E.; Chu, J.J.H.; Rausalu, K.; Merits, A.; Higgs, S.; VanLandingham, D.L.; Abu Bakar, S.; et al. Antiviral activity of silymarin against chikungunya virus. *Sci. Rep.* **2015**, *5*, 11421. [CrossRef]
78. Zandi, K.; Teoh, B.-T.; Sam, S.-S.; Wong, P.-F.; Mustafa, M.R.; AbuBakar, S. Antiviral activity of four types of bioflavonoid against dengue virus type-2. *Virol. J.* **2011**, *8*, 560. [CrossRef]
79. Chiow, K.; Phoon, M.; Putti, T.; Tan, B.K.; Chow, V.T. Evaluation of antiviral activities of Houttuynia cordata Thunb. extract, quercetin, quercetrin and cinanserin on murine coronavirus and dengue virus infection. *Asian Pac. J. Trop. Med.* **2015**, *9*, 1–7. [CrossRef]
80. Zandi, K.; Teoh, B.-T.; Sam, S.-S.; Wong, P.-F.; Mustafa, M.R.; AbuBakar, S. Novel antiviral activity of baicalein against dengue virus. *BMC Complement. Altern. Med.* **2012**, *12*, 214. [CrossRef]
81. Moghaddam, E.J.; Teoh, B.-T.; Sam, S.-S.; Lani, R.; Hassandarvish, P.; Chik, Z.; Yueh, A.; Abubakar, S.; Zandi, K. Baicalin, a metabolite of baicalein with antiviral activity against dengue virus. *Sci. Rep.* **2014**, *4*, 5452. [CrossRef]
82. Zhang, W.; Qiao, H.; Lv, Y.; Wang, J.; Chen, X.; Hou, Y.; Tan, R.; Li, E. Apigenin Inhibits Enterovirus-71 Infection by Disrupting Viral RNA Association with trans-Acting Factors. *PLoS ONE* **2014**, *9*, e110429. [CrossRef]
83. Dai, W.; Bi, J.; Li, F.; Wang, S.; Huang, X.; Meng, X.; Sun, B.; Wang, D.; Kong, W.; Jiang, C.; et al. Antiviral Efficacy of Flavonoids against Enterovirus 71 Infection in Vitro and in Newborn Mice. *Viruses* **2019**, *11*, 625. [CrossRef]
84. Li, X.; Liu, Y.; Wu, T.; Jin, Y.; Cheng, J.; Wan, C.; Qian, W.; Xing, F.; Shi, W. The Antiviral Effect of Baicalin on Enterovirus 71 In Vitro. *Viruses* **2015**, *7*, 4756–4771. [CrossRef]
85. Tsai, F.-J.; Lin, C.-W.; Lai, C.-C.; Lan, Y.-C.; Lai, C.-H.; Hung, C.-H.; Hsueh, K.-C.; Lin, T.-H.; Chang, H.C.; Wan, L.; et al. Kaempferol inhibits enterovirus 71 replication and internal ribosome entry site (IRES) activity through FUBP and HNRP proteins. *Food Chem.* **2011**, *128*, 312–322. [CrossRef]
86. Nahmias, Y.; Goldwasser, J.; Casali, M.; van Poll, D.; Wakita, T.; Chung, R.T.; Yarmush, M.L. Apolipoprotein B-dependent hepatitis C virus secretion is inhibited by the grapefruit flavonoid naringenin. *Hepatology* **2008**, *47*, 1437–1445. [CrossRef]
87. Shibata, C.; Ohno, M.; Otsuka, M.; Kishikawa, T.; Goto, K.; Muroyama, R.; Kato, N.; Yoshikawa, T.; Takata, A.; Koike, K. The flavonoid apigenin inhibits hepatitis C virus replication by decreasing mature microRNA122 levels. *Virology* **2014**, *462–463*, 42–48. [CrossRef]
88. Ferenci, P.; Scherzer, T.; Kerschner, H.; Rutter, K.; Beinhardt, S.; Hofer, H.; Schöniger-Hekele, M.; Holzmann, H.; Steindl-Munda, P. Silibinin Is a Potent Antiviral Agent in Patients With Chronic Hepatitis C Not Responding to Pegylated Interferon/Ribavirin Therapy. *Gastroenterology* **2008**, *135*, 1561–1567. [CrossRef]
89. Bachmetov, L.; Gal-Tanamy, M.; Shapira, A.; Vorobeychik, M.; Giterman-Galam, T.; Sathiyamoorthy, P.; Golan-Goldhirsh, A.; Benhar, I.; Tur-Kaspa, R.; Zemel, R. Suppression of hepatitis C virus by the flavonoid quercetin is mediated by inhibition of NS3 protease activity. *J. Viral Hepat.* **2011**, *19*, e81–e88. [CrossRef] [PubMed]
90. Shimizu, J.F.; Lima, C.S.; Pereira, C.M.; Bittar, C.; Batista, M.; Nazaré, A.C.; Polaquini, C.R.; Zothner, C.; Harris, M.; Rahal, P.; et al. Flavonoids from Pterogyne nitens Inhibit Hepatitis C Virus Entry. *Sci. Rep.* **2017**, *7*, 16127. [CrossRef] [PubMed]
91. Sauter, D.; Schwarz, S.; Wang, K.; Zhang, R.; Sun, B.; Schwarz, W. Genistein as Antiviral Drug against HIV Ion Channel. *Planta Med.* **2014**, *80*, 682–687. [CrossRef]

92. Tao, J.; Hu, Q.; Yang, J.; Li, R.; Li, X.; Lu, C.; Chen, C.; Wang, L.; Shattock, R.; Ben, K. In vitro anti-HIV and -HSV activity and safety of sodium rutin sulfate as a microbicide candidate. *Antivir. Res.* **2007**, *75*, 227–233. [CrossRef]
93. Badshah, S.L.; Faisal, S.; Muhammad, A.; Poulson, B.G.; Emwas, A.H.; Jaremko, M. Antiviral activities of flavonoids. *Biomed. Pharmacother.* **2021**, *140*, 111596. [CrossRef]
94. Ge, M.; Xiao, Y.; Chen, H.; Luo, F.; Du, G.; Zeng, F. Multiple antiviral approaches of (−)-epigallocatechin-3-gallate (EGCG) against porcine reproductive and respiratory syndrome virus infection in vitro. *Antivir. Res.* **2018**, *158*, 52–62. [CrossRef]
95. Mehany, T.; Khalifa, I.; Barakat, H.; Althwab, S.A.; Alharbi, Y.M.; El-Sohaimy, S. Polyphenols as promising biologically active substances for preventing SARS-CoV-2: A review with research evidence and underlying mechanisms. *Food Biosci.* **2021**, *40*, 100891. [CrossRef]
96. Imanishi, N.; Tuji, Y.; Katada, Y.; Maruhashi, M.; Konosu, S.; Mantani, N.; Terasawa, K.; Ochiai, H. Additional Inhibitory Effect of Tea Extract on the Growth of Influenza A and B Viruses in MDCK Cells. *Microbiol. Immunol.* **2002**, *46*, 491–494. [CrossRef]
97. Roschek, B.; Fink, R.; McMichael, M.D.; Li, D.; Alberte, R.S. Elderberry flavonoids bind to and prevent H1N1 infection in vitro. *Phytochemistry* **2009**, *70*, 1255–1261. [CrossRef] [PubMed]
98. Song, J.; Choi, H. Silymarin efficacy against influenza A virus replication. *Phytomedicine* **2011**, *18*, 832–835. [CrossRef] [PubMed]
99. Omrani, M.; Keshavarz, M.; Ebrahimi, S.N.; Mehrabi, M.; McGaw, L.J.; Abdalla, M.A.; Mehrbod, P. Potential Natural Products Against Respiratory Viruses: A Perspective to Develop Anti-COVID-19 Medicines. *Front. Pharmacol.* **2021**, *11*, 586993. [CrossRef] [PubMed]
100. Ganesan, S.; Faris, A.N.; Comstock, A.T.; Wang, Q.; Nanua, S.; Hershenson, M.B.; Sajjan, U.S. Quercetin inhibits rhinovirus replication in vitro and in vivo. *Antivir. Res.* **2012**, *94*, 258–271. [CrossRef]
101. Desideri, N.; Conti, C.; Sestili, I.; Tomao, P.; Stein, M.L.; Orsi, N. In vitro Evaluation of the Anti-Picornavirus Activities of New Synthetic Flavonoids. *Antivir. Chem. Chemother.* **1995**, *6*, 298–306. [CrossRef]
102. Clark, K.; Grant, P.; Sarr, A.; Belakere, J.; Swaggerty, C.; Phillips, T.; Woode, G. An in vitro study of theaflavins extracted from black tea to neutralize bovine rotavirus and bovine coronavirus infections. *Vet.-Microbiol.* **1998**, *63*, 147–157. [CrossRef]
103. Ryu, Y.B.; Jeong, H.J.; Kim, J.H.; Kim, Y.M.; Park, J.-Y.; Kim, D.; Naguyen, T.T.H.; Park, S.-J.; Chang, J.S.; Park, K.H. Biflavonoids from Torreya nucifera displaying SARS-CoV 3CLpro inhibition. *Bioorganic Med. Chem.* **2010**, *18*, 7940–7947. [CrossRef]
104. Jo, S.; Kim, S.; Shin, D.H.; Kim, M.-S. Inhibition of SARS-CoV 3CL protease by flavonoids. *J. Enzym. Inhib. Med. Chem.* **2019**, *35*, 145–151. [CrossRef]
105. Nguyen, T.T.H.; Woo, H.-J.; Kang, H.-K.; Nguyen, V.D.; Kim, Y.-M.; Kim, D.-W.; Ahn, S.-A.; Xia, Y.; Kim, D. Flavonoid-mediated inhibition of SARS coronavirus 3C-like protease expressed in Pichia pastoris. *Biotechnol. Lett.* **2012**, *34*, 831–838. [CrossRef]
106. Park, J.-Y.; Yuk, H.J.; Ryu, H.W.; Lim, S.H.; Kim, K.S.; Park, K.H.; Ryu, Y.B.; Lee, W.S. Evaluation of polyphenols from Broussonetia papyrifera as coronavirus protease inhibitors. *J. Enzym. Inhib. Med. Chem.* **2017**, *32*, 504–512. [CrossRef]
107. Cho, J.K.; Curtis-Long, M.J.; Lee, K.H.; Kim, D.W.; Ryu, H.W.; Yuk, H.J.; Park, K.H. Geranylated flavonoids displaying SARS-CoV papain-like protease inhibition from the fruits of Paulownia tomentosa. *Bioorganic Med. Chem.* **2013**, *21*, 3051–3057. [CrossRef]
108. Schwarz, S.; Sauter, D.; Wang, K.; Zhang, R.; Sun, B.; Karioti, A.; Bilia, A.R.; Efferth, T.; Schwarz, W. Kaempferol Derivatives as Antiviral Drugs against the 3a Channel Protein of Coronavirus. *Planta Med.* **2014**, *80*, 177–182. [CrossRef]
109. Yu, M.-S.; Lee, J.; Lee, J.M.; Kim, Y.; Chin, Y.-W.; Jee, J.-G.; Keum, Y.-S.; Jeong, Y.-J. Identification of myricetin and scutellarein as novel chemical inhibitors of the SARS coronavirus helicase, nsP13. *Bioorganic Med. Chem. Lett.* **2012**, *22*, 4049–4054. [CrossRef]
110. Ho, T.-Y.; Wu, S.-L.; Chen, J.-C.; Li, C.-C.; Hsiang, C.-Y. Emodin blocks the SARS coronavirus spike protein and angiotensin-converting enzyme 2 interaction. *Antivir. Res.* **2007**, *74*, 92–101. [CrossRef]
111. Chen, S.; Chan, K.; Jiang, Y.; Kao, R.Y.; Lu, H.; Fan, K.; Cheng, V.; Tsui, W.; Hung, I.F.N.; Lee, T. In vitro susceptibility of 10 clinical isolates of SARS coronavirus to selected antiviral compounds. *J. Clin. Virol.* **2004**, *31*, 69–75. [CrossRef] [PubMed]
112. Jo, S.; Kim, H.; Kim, S.; Shin, D.H.; Kim, M. Characteristics of flavonoids as potent MERS-CoV 3C-like protease inhibitors. *Chem. Biol. Drug Des.* **2019**, *94*, 2023–2030. [CrossRef]
113. Zhuang, M.; Jiang, H.; Suzuki, Y.; Li, X.; Xiao, P.; Tanaka, T.; Ling, H.; Yang, B.; Saitoh, H.; Zhang, L.; et al. Procyanidins and butanol extract of Cinnamomi Cortex inhibit SARS-CoV infection. *Antivir. Res.* **2009**, *82*, 73–81. [CrossRef]
114. Roh, C. A facile inhibitor screening of SARS coronavirus N protein using nanoparticle-based RNA oligonucleotide. *Int. J. Nanomed.* **2012**, *7*, 2173–2179. [CrossRef]
115. Lee, C.; Lee, J.M.; Lee, N.-R.; Kim, D.-E.; Jeong, Y.-J.; Chong, Y. Investigation of the pharmacophore space of Severe Acute Respiratory Syndrome coronavirus (SARS-CoV) NTPase/helicase by dihydroxychromone derivatives. *Bioorg. Med. Chem. Lett.* **2009**, *19*, 4538–4541. [CrossRef]
116. Solnier, J.; Fladerer, J.-P. Flavonoids: A complementary approach to conventional therapy of COVID-19? *Phytochem. Rev.* **2020**, *20*, 773–795. [CrossRef] [PubMed]
117. Su, H.; Yao, S.; Zhao, W.; Zhang, Y.; Liu, J.; Shao, Q.; Wang, Q.; Li, M.; Xie, H.; Shang, W.; et al. Identification of pyrogallol as a warhead in design of covalent inhibitors for the SARS-CoV-2 3CL protease. *Nat. Commun.* **2021**, *12*, 1–12. [CrossRef] [PubMed]
118. Nguyen, T.; Jung, J.-H.; Kim, M.-K.; Lim, S.; Choi, J.-M.; Chung, B.; Kim, D.-W.; Kim, D. The Inhibitory Effects of Plant Derivate Polyphenols on the Main Protease of SARS Coronavirus 2 and Their Structure–Activity Relationship. *Molecules* **2021**, *26*, 1924. [CrossRef] [PubMed]
119. Liu, H.; Ye, F.; Sun, Q.; Liang, H.; Li, C.; Li, S.; Lu, R.; Huang, B.; Tan, W.; Lai, L. Scutellaria baicalensis extract and baicalein inhibit replication of SARS-CoV-2 and its 3C-like protease in vitro. *J. Enzym. Inhib. Med. Chem.* **2021**, *36*, 497–503. [CrossRef]

120. Gao, J.; Ding, Y.; Wang, Y.; Liang, P.; Zhang, L.; Liu, R. Oroxylin A is a severe acute respiratory syndrome coronavirus 2-spiked pseudotyped virus blocker obtained from Radix Scutellariae using angiotensin-converting enzyme II/cell membrane chromatography. *Phytotherapy Res.* **2021**, *35*, 3194–3204. [CrossRef] [PubMed]
121. Du, A.; Zheng, R.; Disoma, C.; Li, S.; Chen, Z.; Li, S.; Liu, P.; Zhou, Y.; Shen, Y.; Liu, S.; et al. Epigallocatechin-3-gallate, an active ingredient of Traditional Chinese Medicines, inhibits the 3CLpro activity of SARS-CoV-2. *Int. J. Biol. Macromol.* **2021**, *176*, 1–12. [CrossRef]
122. Xiao, T.; Cui, M.; Zheng, C.; Wang, M.; Sun, R.; Gao, D.; Bao, J.; Ren, S.; Yang, B.; Lin, J.; et al. Myricetin Inhibits SARS-CoV-2 Viral Replication by Targeting Mpro and Ameliorates Pulmonary Inflammation. *Front. Pharmacol.* **2021**, *12*, 669642. [CrossRef]
123. Xiong, Y.; Zhu, G.-H.; Zhang, Y.-N.; Hu, Q.; Wang, H.-N.; Yu, H.-N.; Qin, X.-Y.; Guan, X.-Q.; Xiang, Y.-W.; Tang, H.; et al. Flavonoids in Ampelopsis grossedentata as covalent inhibitors of SARS-CoV-2 3CLpro: Inhibition potentials, covalent binding sites and inhibitory mechanisms. *Int. J. Biol. Macromol.* **2021**, *187*, 976–987. [CrossRef]
124. Abian, O.; Ortega-Alarcon, D.; Jimenez-Alesanco, A.; Ceballos-Laita, L.; Vega, S.; Reyburn, H.T.; Rizzuti, B.; Velazquez-Campoy, A. Structural stability of SARS-CoV-2 3CLpro and identification of quercetin as an inhibitor by experimental screening. *Int. J. Biol. Macromol.* **2020**, *164*, 1693–1703. [CrossRef]
125. Rizzuti, B.; Grande, F.; Conforti, F.; Jimenez-Alesanco, A.; Ceballos-Laita, L.; Ortega-Alarcon, D.; Vega, S.; Reyburn, H.; Abian, O.; Velazquez-Campoy, A. Rutin Is a Low Micromolar Inhibitor of SARS-CoV-2 Main Protease 3CLpro: Implications for Drug Design of Quercetin Analogs. *Biomedicines* **2021**, *9*, 375. [CrossRef]
126. Su, H.-X.; Yao, S.; Zhao, W.-F.; Li, M.-J.; Liu, J.; Shang, W.-J.; Xie, H.; Ke, C.-Q.; Hu, H.-C.; Gao, M.-N.; et al. Anti-SARS-CoV-2 activities in vitro of Shuanghuanglian preparations and bioactive ingredients. *Acta Pharmacol. Sin.* **2020**, *41*, 1167–1177. [CrossRef]
127. Owis, A.I.; El-Hawary, M.S.; El Amir, D.; Refaat, H.; Alaaeldin, E.; Aly, O.M.; Elrehany, M.A.; Kamel, M.S. Flavonoids of Salvadora persica L. (meswak) and its liposomal formulation as a potential inhibitor of SARS-CoV-2. *RSC Adv.* **2021**, *11*, 13537–13544. [CrossRef]
128. Pitsillou, E.; Liang, J.; Ververis, K.; Hung, A.; Karagiannis, T.C. Interaction of small molecules with the SARS-CoV-2 papain-like protease: In silico studies and in vitro validation of protease activity inhibition using an enzymatic inhibition assay. *J. Mol. Graph. Model.* **2021**, *104*, 107851. [CrossRef] [PubMed]
129. Liu, X.; Raghuvanshi, R.; Ceylan, F.D.; Bolling, B.W. Quercetin and Its Metabolites Inhibit Recombinant Human Angiotensin-Converting Enzyme 2 (ACE2) Activity. *J. Agric. Food Chem.* **2020**, *68*, 13982–13989. [CrossRef]
130. Güler, H.I.; Şal, F.A.; Can, Z.; Kara, Y.; Yildiz, O.; Beldüz, A.O.; Çanakçi, S.; Kolayli, S. Targeting CoV-2 spike RBD and ACE-2 interaction with flavonoids of Anatolian propolis by in silico and in vitro studies in terms of possible COVID-19 therapeutics. *Turk. J. Boil.* **2021**, *45*, 530–548. [CrossRef]
131. Henss, L.; Auste, A.; Schürmann, C.; Schmidt, C.; von Rhein, C.; Mühlebach, M.D.; Schnierle, B.S. The green tea catechin epigallocatechin gallate inhibits SARS-CoV-2 infection. *J. Gen. Virol.* **2021**, *102*, 001574. [CrossRef] [PubMed]
132. Biagioli, M.; Marchianò, S.; Roselli, R.; Di Giorgio, C.; Bellini, R.; Bordoni, M.; Gidari, A.; Sabbatini, S.; Francisci, D.; Fiorillo, B.; et al. Discovery of a AHR pelargonidin agonist that counter-regulates Ace2 expression and attenuates ACE2-SARS-CoV-2 interaction. *Biochem. Pharmacol.* **2021**, *188*, 114564. [CrossRef] [PubMed]
133. Zhan, Y.; Ta, W.; Tang, W.; Hua, R.; Wang, J.; Wang, C.; Lu, W. Potential antiviral activity of isorhamnetin against SARS-CoV-2 spike pseudotyped virus in vitro. *Drug Dev. Res.* **2021**. [CrossRef] [PubMed]
134. Zandi, K.; Musall, K.; Oo, A.; Cao, D.; Liang, B.; Hassandarvish, P.; Lan, S.; Slack, R.; Kirby, K.; Bassit, L.; et al. Baicalein and Baicalin Inhibit SARS-CoV-2 RNA-Dependent-RNA Polymerase. *Microorganisms* **2021**, *9*, 893. [CrossRef]
135. Huang, S.; Liu, Y.; Zhang, Y.; Zhang, R.; Zhu, C.; Fan, L.; Pei, G.; Zhang, B.; Shi, Y. Baicalein inhibits SARS-CoV-2/VSV replication with interfering mitochondrial oxidative phosphorylation in a mPTP dependent manner. *Signal Transduct. Target. Ther.* **2020**, *5*, 1–3. [CrossRef]
136. Hong, S.; Seo, S.H.; Woo, S.-J.; Kwon, Y.; Song, M.; Ha, N.-C. Epigallocatechin Gallate Inhibits the Uridylate-Specific Endoribonuclease Nsp15 and Efficiently Neutralizes the SARS-CoV-2 Strain. *J. Agric. Food Chem.* **2021**, *69*, 5948–5954. [CrossRef] [PubMed]
137. Jang, M.; Park, R.; Park, Y.-I.; Cha, Y.-E.; Yamamoto, A.; Lee, J.I.; Park, J. EGCG, a green tea polyphenol, inhibits human coronavirus replication in vitro. *Biochem. Biophys. Res. Commun.* **2021**, *547*, 23–28. [CrossRef]
138. Clementi, N.; Scagnolari, C.; D'Amore, A.; Palombi, F.; Criscuolo, E.; Frasca, F.; Pierangeli, A.; Mancini, N.; Antonelli, G.; Clementi, M.; et al. Naringenin is a powerful inhibitor of SARS-CoV-2 infection in vitro. *Pharmacol. Res.* **2020**, *163*, 105255. [CrossRef] [PubMed]
139. Pitsillou, E.; Liang, J.; Ververis, K.; Lim, K.W.; Hung, A.; Karagiannis, T.C. Identification of Small Molecule Inhibitors of the Deubiquitinating Activity of the SARS-CoV-2 Papain-Like Protease: In silico Molecular Docking Studies and in vitro Enzymatic Activity Assay. *Front. Chem.* **2020**, *8*, 623971. [CrossRef]
140. Leal, C.M.; Leitão, S.G.; Sausset, R.; Mendonça, S.C.; Nascimento, P.H.A.; Cheohen, C.F.d.A.R.; Esteves, M.E.A.; da Silva, M.L.; Gondim, T.S.; Monteiro, M.E.S.; et al. Flavonoids from Siparuna cristata as Potential Inhibitors of SARS-CoV-2 Replication. *Rev. Bras. Farm.* **2021**, 1–9. [CrossRef]
141. Song, J.; Zhang, L.; Xu, Y.; Yang, D.; Yang, S.; Zhang, W.; Wang, J.; Tian, S.; Yang, S.; Yuan, T.; et al. The comprehensive study on the therapeutic effects of baicalein for the treatment of COVID-19 in vivo and in vitro. *Biochem. Pharmacol.* **2020**, *183*, 114302. [CrossRef] [PubMed]

142. Cherrak, S.A.; Merzouk, H.; Mokhtari-Soulimane, N. Potential bioactive glycosylated flavonoids as SARS-CoV-2 main protease inhibitors: A molecular docking and simulation studies. *PLoS ONE* **2020**, *15*, e0240653. [CrossRef]
143. Teli, D.M.; Shah, M.B.; Chhabria, M.T. In silico Screening of Natural Compounds as Potential Inhibitors of SARS-CoV-2 Main Protease and Spike RBD: Targets for COVID-19. *Front. Mol. Biosci.* **2021**, *7*, 599079. [CrossRef]
144. Bharadwaj, S.; Dubey, A.; Yadava, U.; Mishra, S.K.; Kang, S.G.; Dwivedi, V.D. Exploration of natural compounds with anti-SARS-CoV-2 activity via inhibition of SARS-CoV-2 Mpro. *Brief. Bioinform.* **2021**, *22*, 1361–1377. [CrossRef] [PubMed]
145. Rakshit, G.; Dagur, P.; Satpathy, S.; Patra, A.; Jain, A.; Ghosh, M. Flavonoids as potential therapeutics against novel coronavirus disease-2019 (nCOVID-19). *J. Biomol. Struct. Dyn.* **2021**, 1–13. [CrossRef]
146. Yu, R.; Chen, L.; Lan, R.; Shen, R.; Li, P. Computational screening of antagonists against the SARS-CoV-2 (COVID-19) coronavirus by molecular docking. *Int. J. Antimicrob. Agents* **2020**, *56*, 106012. [CrossRef] [PubMed]
147. Akhter, S.; Batool, A.I.; Sclamoglu, Z.; Sevindik, M.; Eman, R.; Mustaqeem, M.; Aslam, M. Effectiveness of Natural Antioxidants against SARS-CoV-2? Insights from the In-Silico World. *Antibiotics* **2021**, *10*, 1011. [CrossRef]
148. Batool, F.; Mughal, E.U.; Zia, K.; Sadiq, A.; Naeem, N.; Javid, A.; Ul-Haq, Z.; Saeed, M. Synthetic flavonoids as potential antiviral agents against SARS-CoV-2 main protease. *J. Biomol. Struct. Dyn.* **2020**, 1–12. [CrossRef]
149. Ziebuhr, J.; Gorbalenya, A.; Snijder, E. Virus-encoded proteinases and proteolytic processing in the Nidovirales. *J. Gen. Virol.* **2000**, *81*, 853–879. [CrossRef] [PubMed]
150. Du, Q.-S.; Wang, S.-Q.; Zhu, Y.; Wei, D.-Q.; Guo, H.; Sirois, S.; Chou, K.-C. Polyprotein cleavage mechanism of SARS-CoV Mpro and chemical modification of the octapeptide. *Peptides* **2004**, *25*, 1857–1864. [CrossRef]
151. Zhang, L.; Lin, D.; Sun, X.; Curth, U.; Drosten, C.; Sauerhering, L.; Hilgenfeld, R. Conservation of substrate specificities among coronavirus main proteases. *J. Gen. Virol.* **2002**, *83*, 595–599. [CrossRef]
152. Zhang, L.; Lin, D.; Sun, X.; Curth, U.; Drosten, C.; Sauerhering, L.; Hilgenfeld, R. Crystal structure of SARS-CoV-2 main protease provides a basis for design of improved α-ketoamide inhibitors. *Science* **2020**, *368*, 409–412. [CrossRef] [PubMed]
153. Hilgenfeld, R. From SARS to MERS: Crystallographic studies on coronaviral proteases enable antiviral drug design. *FEBS J.* **2014**, *281*, 4085–4096. [CrossRef] [PubMed]
154. Zhang, L.; Lin, D.; Kusov, Y.; Nian, Y.; Ma, Q.; Wang, J.; Von Brunn, A.; Leyssen, P.; Lanko, K.; Neyts, J.; et al. α-Ketoamides as Broad-Spectrum Inhibitors of Coronavirus and Enterovirus Replication: Structure-Based Design, Synthesis, and Activity Assessment. *J. Med. Chem.* **2020**, *63*, 4562–4578. [CrossRef] [PubMed]
155. Qamar, M.T.U.; Alqahtani, S.M.; Alamri, M.A.; Chen, L.-L. Structural basis of SARS-CoV-2 3CLpro and anti-COVID-19 drug discovery from medicinal plants. *J. Pharm. Anal.* **2020**, *10*, 313–319. [CrossRef] [PubMed]
156. Jain, A.S.; Sushma, P.; Dharmashekar, C.; Beelagi, M.S.; Prasad, S.K.; Shivamallu, C.; Prasad, A.; Syed, A.; Marraiki, N.; Prasad, K.S. In silico evaluation of flavonoids as effective antiviral agents on the spike glycoprotein of SARS-CoV-2. *Saudi J. Biol. Sci.* **2020**, *28*, 1040–1051. [CrossRef]
157. Lung, J.; Lin, Y.; Yang, Y.; Chou, Y.; Shu, L.; Cheng, Y.; Liu, H.T.; Wu, C. The potential chemical structure of anti- SARS-CoV -2 RNA-dependent RNA polymerase. *J. Med. Virol.* **2020**, *92*, 693–697. [CrossRef] [PubMed]
158. Shawan, M.M.A.K.; Halder, S.K.; Hasan, A. Luteolin and abyssinone II as potential inhibitors of SARS-CoV-2: An in silico molecular modeling approach in battling the COVID-19 outbreak. *Bull. Natl. Res. Cent.* **2021**, *45*, 1–21. [CrossRef] [PubMed]
159. Joshi, R.S.; Jagdale, S.S.; Bansode, S.B.; Shankar, S.S.; Tellis, M.B.; Pandya, V.K.; Chugh, A.; Giri, A.P.; Kulkarni, M.J. Discovery of potential multi-target-directed ligands by targeting host-specific SARS-CoV-2 structurally conserved main protease. *J. Biomol. Struct. Dyn.* **2020**, 1–16. [CrossRef] [PubMed]
160. Chen, H.; Du, Q. Potential Natural Compounds for Preventing SARS-CoV-2 (2019-nCoV) Infection. Available online: https://www.preprints.org/manuscript/202001.0358/v3 (accessed on 3 June 2021).
161. Utomo, R.Y.; Ikawati, M.; Meiyanto, E. Revealing the Potency of Citrus and Galangal Con-stituents to Halt SARS-CoV-2 In-fection. *Preprints* **2020**, 2020030214. [CrossRef]
162. Topcagic, A.; Zeljkovic, S.C.; Karalija, E.; Galijasevic, S.; Sofic, E. Evaluation of phenolic profile, enzyme inhibitory and antimicrobial activities of Nigella sativa L. seed extracts. *Bosn. J. Basic Med. Sci.* **2017**, *17*, 286–294. [CrossRef]

International Journal of Molecular Sciences

Article

Synthesis of Demissidine Analogues from Tigogenin via Imine Intermediates †

Agnieszka Wojtkielewicz *, Urszula Kiełczewska, Aneta Baj and Jacek W. Morzycki *

Faculty of Chemistry, University of Białystok, K. Ciołkowskiego 1K, 15-245 Białystok, Poland; ulakielczewska@interia.eu (U.K.); aneta.baj@uwb.edu.pl (A.B.)
* Correspondence: a.wojtkielewicz@uwb.edu.pl (A.W.); morzycki@uwb.edu.pl (J.W.M.); Tel.: +48-857388043 (A.W.); +48-857388260 (J.W.M.)
† Dedicated to Prof. Dr. Ludger Wessjohann on the occasion of his 60th birthday.

Abstract: A five-step transformation of a spiroketal side chain of tigogenin into an indolizidine system present in solanidane alkaloids such as demissidine and solanidine was elaborated. The key intermediate in the synthesis was spiroimine **3** readily obtained from tigogenin by its RuO$_4$ oxidation to 5,6-dihydrokryptogenin followed by amination with aluminum amide generated in situ from DIBAlH and ammonium chloride. The mild reduction of spiroimine to a 26-hydroxy-dihydropyrrole derivative and subsequent mesylation resulted in the formation of 25-epidemissidinium salt or 23-sulfone depending on reaction conditions.

Keywords: steroidal alkaloids; solanidane alkaloids; demissidine; solanidine

1. Introduction

Demissidine and solanidine are the main representatives of the solanidane alkaloids that occur mainly as glycosides in potato species including *Solanum tuberosum*, *Solanum demissum*, and *Solanum acaule* (Figure 1) [1,2]. The various biological properties of these cholestane alkaloids have been reported in the literature [3]. Among these, α-solanine and α-chaconine, two main solanidine glycosides, are potent enough to inhibit proliferation and induce apoptosis in various types of cancer cells including cervical, liver, lymphoma, and stomach cancer cells [4]. The effectiveness of α-chaconine against hepatocellular cancer HepG2 cells is higher than the common anticancer agents doxorubicin and camptothecin [5]. Additionally, demissidine and its natural glycoside, commersonine, inhibit the growth of human colon and liver cancer cells in culture [5]. Apart from showing antitumor activity, solanidane-type alkaloids are known to act as natural insect deterrents, have antimicrobial and anti-inflammatory properties, inhibit acetylcholinesterase, and disrupt cell membranes [3,6–9]. Additionally, studies of solanidine and demissidine analogues confirm their potency for the design of new pharmacologically active agents [10–12].

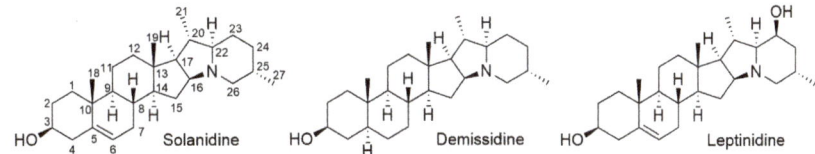

Figure 1. Steroidal alkaloids of solanidane type.

So far, eight syntheses of solanidine and demissidine have been described, four of them in the last decade, and the latest one was reported last year [13–20]. Although recently invented methods brought a significant improvement, the described methods suffer from several drawbacks, such as multi-step procedures or unsatisfactory yields.

Moreover, they cannot be easily adapted to the synthesis of demissidine or solanidine analogues. Therefore, the elaboration of an efficient route to demissidine congeners is still needed. An improved approach to the synthesis of different demissidine stereoisomers has been recently reported [21]. Here, we propose an alternative strategy toward demissidine analogues from an easily available steroid sapogenin—tigogenin.

2. Results and Discussion

We found that a convenient intermediate for the transformation of the spiroketal system present in steroidal sapogenins, e.g., tigogenin, into the solanidane framework of demissidine was spiroimine 3, shown in Scheme 1. This novel spirostane aza-analogue was obtained from tigogenin by a two-step protocol involving tigogenin oxidation to a 5,6-dihydrokryptogenin derivative and its reaction with aluminum amide as an aminating agent.

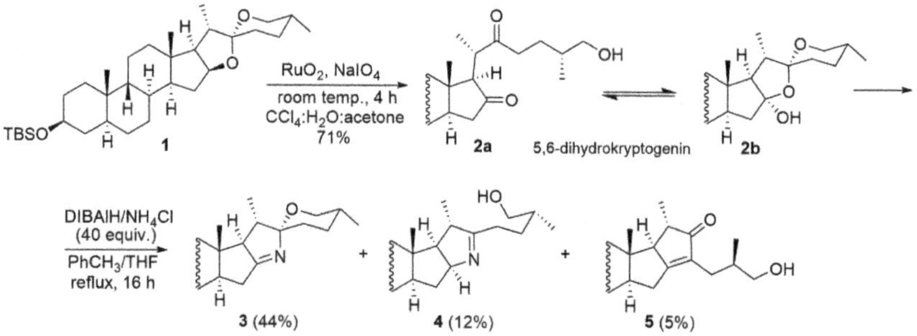

Scheme 1. Synthesis of spiroimine 3 from tigogenin 3-TBS ether (1).

The most convenient method for the oxidative cleavage of sapogenin spiroketal to hydroxy-diketone was chosen first. After perusing the known literature protocols [22–25], we employed the $RuO_4/NaIO_4$ catalytic system. The desired 5,6-dihydrokryptogenin derivative was obtained by the oxidation of tigogenin 3-TBS ether (1) as a mixture of two tautomers 2a and 2b in 71% yield. In the next step, the obtained product was subjected to a reaction with aluminum amide generated in situ from diisobutylaluminum hydride (DIBAlH) and ammonium salt. The use of various aminoalanes as aminating agents for such compounds as epoxides, ketones, carboxylic acids, and their derivatives (chlorides, esters) has previously been widely reported in the literature [26–31]. Our previous investigations have shown that the desired aminoalane might be readily synthesized by the treatment of DIBAlH with ammonium chloride under mild reaction conditions (0 °C – room temperature, THF, up to 2 h) [32]. However, the reagent proved to be unstable and its structure was not definitely determined. The reaction of the 2a/2b mixture with aminoalane prepared as described above was carried out in refluxing THF/toluene (Scheme 1). Spiroimine 3 was obtained as the main reaction product (44%) when using aluminum amide prepared in situ from 40 equivalents of DIBAlH and 42 equivalents of ammonium chloride. It is worth noting that this compound was not formed in the absence of DIBAlH. Compound 3 was accompanied by two minor products, imine 4 (12%) and enone 5 (5%). Ketone 5 was produced as a result of an aldol condensation probably due to enolization caused by aluminum amide playing a role as a Lewis acid. The formation of imine 4 in the experiment was unexpected and difficult to explain in terms of the substrate 2a/2b reaction with prepared aminoalane. It seems that some unreacted diisobutylaluminum hydride was still present in the reaction mixture, resulting in the reduction of initially produced spiroimine 3 to 4 (vide infra). The reaction of compounds 2a and 2b with aminoalane prepared from a lower amount of DIBAlH led to the incomplete conversion. For example, employing the aminating reagent produced from 20 equivalents of DIBAlH, imine 3 was obtained in 26% yield only, while the α,β-unsaturated ketone 5 was produced in 25% yield. In this case, compound 4 was not isolated. Both imines, 3 and 4, appeared to be convenient substrates for the synthesis of solasodine or solanidine derivatives. The mild reduction of spiroimine should provide hitherto unknown 'reverse' spirosolanes with the nitrogen atom in the pyrrolidine E-ring and the oxygen atom in the 'pyranose' F-ring. Moreover, the reductive cleavage of the spiroimine F ring may open a direct way to solanidane alkaloids possessing an indolizidine moiety.

First, the reduction of compound 3 under mild conditions was attempted. Interestingly, the expected hemiaminal 6 (Scheme 2) was not obtained, though various reducing agents were examined. Using an equimolar amount of various borohydrides, such as $NaBH_4$, $NaBH_4/I_2$, and $NaBH_3CN$, under different conditions (temperature, reaction times), the main isolated product was always imine 4 accompanied by small amounts of pyrrolidine 7. The other examined reducing agents (DIBAlH, H_2/PtO_2, H_2/Pd, Hantzsch ester/TFA [33], TESH/acid) proved less effective.

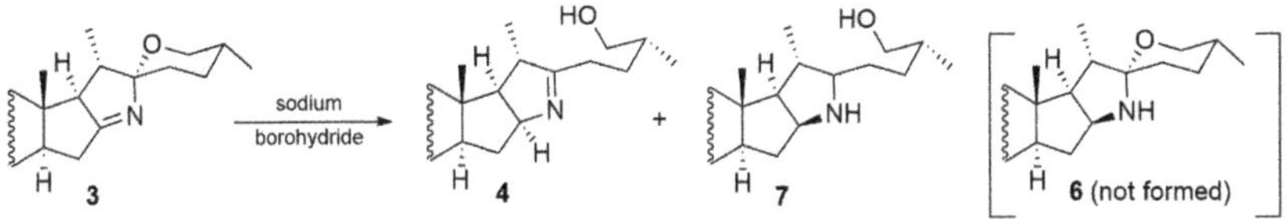

Scheme 2. Reduction of imine 3 with borohydrides.

The above-described results of the reduction experiments pointed out that compound 6 is less stable than its open-chain isomer 4 (confirmed by calculations). This explains unsuccessful attempts of imine 4 cyclization to 6 in the presence of acids. The observed behavior of imine 4 is clearly different from that of 'pseudosapogenins', which readily cyclize to spiroketals. The latter are relatively stable compounds, though their F-ring opening occurs when they are treated with strong Lewis acids. The natural aza-analogues of spirostanes (spirosolanes) containing the nitrogen atom in ring F, e.g., solasodine or tomatidine, are even more susceptible to an electrophilic attack than spiroketals. However, in the case of spirosolanes, the 'furanose' E-ring is readily opened [34]. This is because the cation resulting from the C–O bond cleavage is stabilized by electrons of the neighboring nitrogen atom. It seems that the isomeric compounds containing the nitrogen atom in the E-ring undergo the opening of the 'pyranose' F-ring for the same reason. The cleavage of the oxygen-containing ring in spirosolanes was also observed under the reducing conditions [35,36]. Despite the failure to obtain a 'reverse' spirosolane analogue from imine 3, it still seemed to be a convenient intermediate for the synthesis of solanidane alkaloids. A derivative of imine 4 was previously employed by Uhle and Tian to build an indolizidine system. In the solanidine analogue synthesis reported by Uhle [37], the imine was obtained in 20% yield from kryptogenin 16-(2,4-dinitrophenyl)hydrazone and transformed into 25-episolanidine by refluxing with KOH in ethylene glycol in 65% yield. In 2016, Tian and coworkers [18] developed a new way to synthesize solanidine and demissidine using diosgenin or tigogenin as a starting material, respectively. In the method proposed by the Chinese group, 26-methyl ester 22-imine was prepared in five steps and further transformed into the desired alkaloid by the selective reduction of the imine moiety to pyrrolidine, spontaneous intramolecular aminolysis of the obtained amino-ester to lactam, and reduction. The use of spiroimine 3 as an intermediate for the construction of an indolizidine unit allowed us to shorten the solanidane synthesis from tigogenin. The approach explored in our study involved the reduction of spiroimine 3 to dihydropyrrole 4 followed by its cyclization and reduction. As our initial studies on the imine reduction showed that only complex borohydrides were effective for this transformation, we went on to optimize the reduction reaction conditions using $NaBH_4$ and $NaBH_3CN$ as reducing agents. Our results are summarized in Table 1. Apart from compound 4, in most cases a small amount of amine 7 was formed. Imine 4 was obtained in the best yield in reaction with a $NaBH_4/I_2$ system (entry 5). With 1 equivalent of $NaBH_4$ (without any additives or with AcONa) at a low temperature and controlling the reaction time, we restrained imine over-reduction and obtained compound 4 in good yield (entry 2, 3, 4). $NaBH_3CN$ was less efficient (entry 6, 7). Additionally, when NaBH$_3$CN was used in the presence of AcOH, an imine–cyanoborane complex **8** was formed (Figure 2).

Table 1. The optimization of imine **3** reduction conditions.

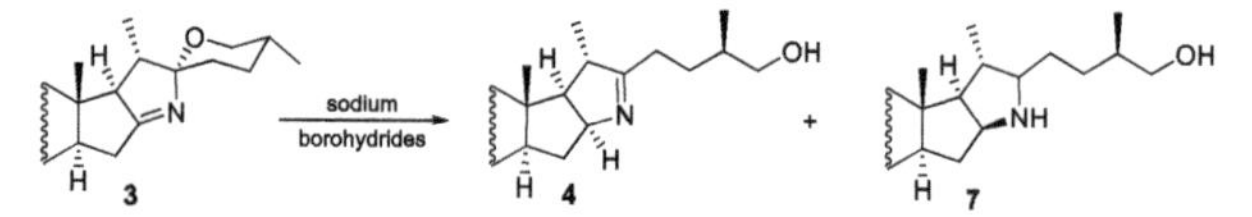

Entry	Reagent (equiv.)	Conditions	Product 4 Yield (%)	Product 7 Yield (%)	Substrate Recovery (%)
1	NaBH$_4$ (2)	MeOH/CH$_2$Cl$_2$, −18 °C, 1 h	64	30	<1
2	NaBH$_4$ (1)	MeOH/CH$_2$Cl$_2$, −18 °C, 1 h	58	<5	21
3	NaBH$_4$ (1)	MeOH/CH$_2$Cl$_2$, 0 °C, 1 h	67	13	<5
4	NaBH$_4$ (1)	NaOAc (1 equiv.), MeOH/CH$_2$Cl$_2$, 0 °C–room temp., 2 h	65	<5	30
5	NaBH$_4$ (4)	I$_2$ (4 equiv.), MeOH/CH$_2$Cl$_2$, 0 °C–reflux, 16 h	78	9	<1
6	NaBH$_3$CN (2)	AcOH (2 equiv.), THF, room temp., 1 h	complex 8 (48)	nd *	<5
7	NaBH$_3$CN (2)	THF/MeOH, room temp., 2 h	28	nd *	30

* nd—not detected.

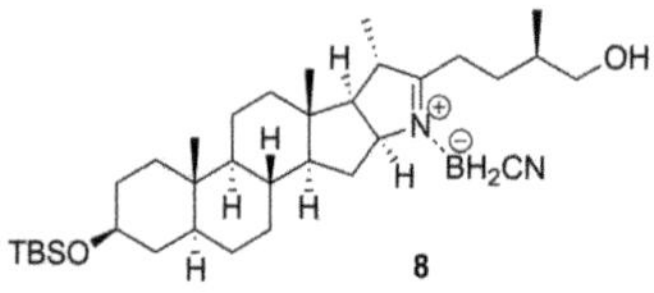

Figure 2. Complex **8** formed during the imine **3** reduction with NaBH$_3$CN/AcOH.

We envisaged that the activation of the 26-hydroxyl group in compound **4** would result in spontaneous ring closing to iminium salt. Therefore, we subjected compound **4** to a reaction with mesyl chloride. As examples of the successful chemoselective mesylation of the primary hydroxyl group in the presence of amine function could be found in the literature [38–40], we supposed that the chemoselective mesylation of hydroxy-imine should be reached under similar conditions. The initial mesylation of hydroxy-imine **4** carried out with 1.2 equivalents of mesyl chloride in the presence of triethylamine at −15 °C resulted mainly in a less polar product (26-mesyloxy-imine), which spontaneously cyclized after work-up to the desired iminium salt **9** (Scheme 3). Under mesylation conditions, TBS protection of the 3-OH group was also removed and the indolizinium salt **9** was isolated in 45% yield. Compound **9** was readily transformed into 25-epimissidine (**10**) by reduction with NaBH$_4$.

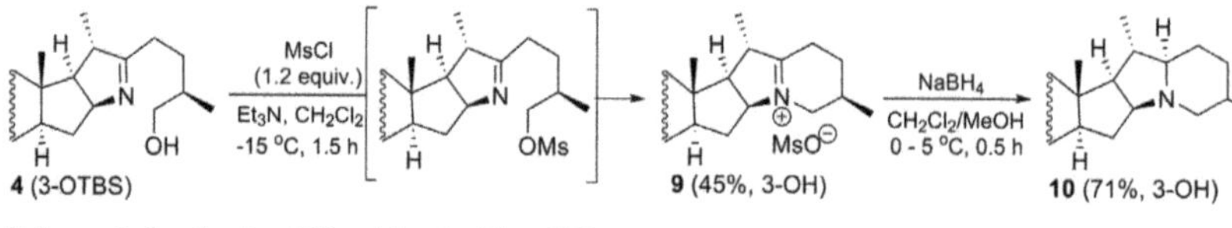

Scheme 3. Synthesis of 25-epimissidine (**10**).

Conducting the mesylation under slightly harsher conditions (1.2 equiv. of MsCl, Et$_3$N, 0 °C or 2 equiv. of MsCl, Py, DMAP(cat.), 0 °C–room temp.) led to a complex mixture of products. The iminium salt was formed only in 5% yield, while the main reaction product was identified as an enamine N,O-dimesyl derivative **11a** or **11b** (Figure 3).

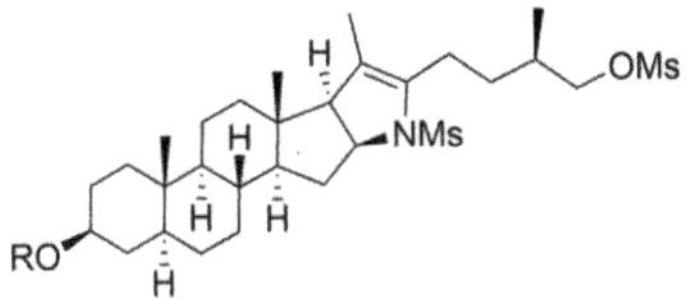

11a: R = Ms (for MsCl (2 equiv.), CH$_2$Cl$_2$/Py, DMAP cat.)
11b: R = TBS (for MsCl (1.2 equiv.), Et$_3$N, CH$_2$Cl$_2$, 0 °C)

Figure 3. Major products of imine **4** mesylation under harsh conditions.

As the changes made did not result in the yield improvement of the desired indolizinium salt, we also attempted to improve the chemoselectivity of O-mesylation by deactivating the imine nitrogen. For this purpose, hydroxy-imine **4** was reacted with hydrogen chloride (generated in situ from AcCl and MeOH) to obtain imine hydrochloride before mesylation. The crude salt without isolation was subjected to mesylation with 2 equivalents of MsCl in the presence of Et$_3$N at 0 °C–room temp. To our surprise, after basic work-up sulfone **12** (Scheme 4) was isolated, instead of the expected indolizinium salt **9**. The obtained solanidane seems to be a valuable intermediate for the synthesis of leptinidine analogues.

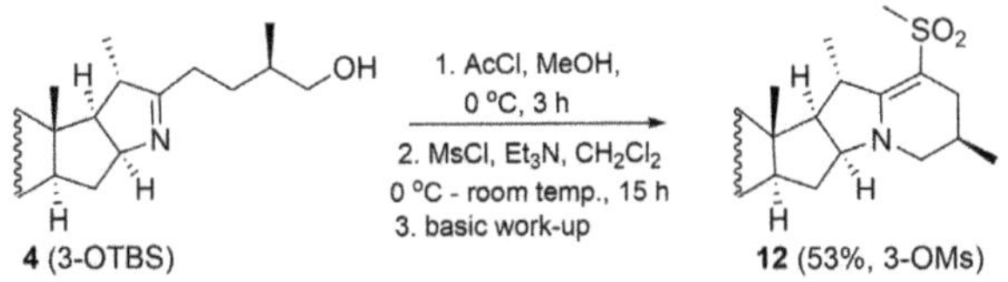

4 (3-OTBS)

1. AcCl, MeOH, 0 °C, 3 h
2. MsCl, Et$_3$N, CH$_2$Cl$_2$ 0 °C - room temp., 15 h
3. basic work-up

12 (53%, 3-OMs)

Scheme 4. Mesylation of imine **4** preceded by protonation with HCl.

The hypothetical mechanism of sulfone formation is outlined in Scheme 5. An addition of HCl caused the tautomerization of imine to enamine (I) via the in situ formation of iminium salt and simultaneous deprotection of TBS ether. The enamine (I) possessing three nucleophilic sites, primary OH group, secondary OH group, and enamine carbon atom, reacted further with mesyl chloride. Apart from alcohol mesylation (II), the mesylation of an enamine electron-rich carbon occurred, leading to sulfone formation with the reconstruction of imine in ring E (III). In the final step, the cyclization to indolizine took place via an intramolecular nucleophilic substitution of 26-mesylate with the imine nitrogen. The sequence of the last-mentioned transformations (the sulfone formation followed by the ring closing) is not obvious. The reverse order of transformations (with the cyclization first) is less likely but could not be excluded. It should be mentioned that a small amount of sulfone was also formed in the mesylation of imine **4** without pre-addition of HCl.

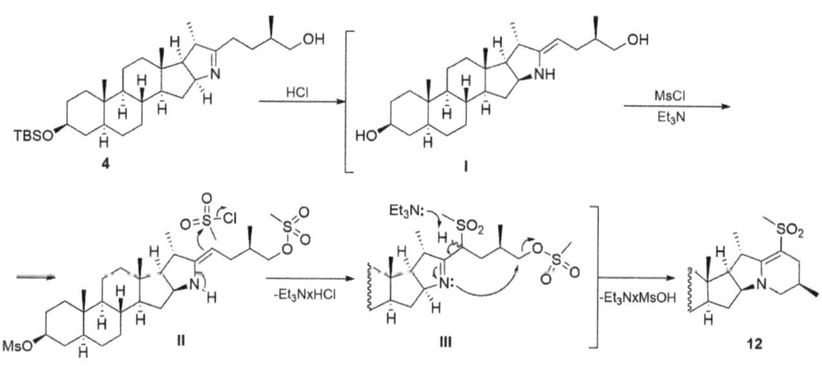

Scheme 5. Tentative mechanism of sulfone **12** formation from imine **4**.

The novel compounds prepared within the study, including the imine intermediates that frequently show antibiotic activity [41], will be subjected to biological activity evaluation in due course.

3. Materials and Methods

3.1. General

NMR spectra were recorded with Bruker Avance II 400 spectrometer operating at 400 MHz, using CDCl$_3$ solutions with TMS as the internal standard (only selected signals in the ^{1}H NMR spectra are reported). Coupling constants (J) are given in Hz. The spectra of compounds 3–10 and 12 are included in the Supplementary Materials. The FTIR spectra were obtained using Nicolet™ 6700 spectrometer (Thermo Scientific, Waltham, MA, USA). The spectra were recorded in the range between 4000 and 500 cm^{-1} with a resolution of 4 cm^{-1} and 32 scans using Attenuated Total Reflectance (ATR) techniques. ESI and ESI-HRMS spectra were obtained on the Agilent 6530 Accurate-Mass Q-TOF ESI and LC/MS system. Melting points were determined using MP70 Melting Point System (Mettler Toledo, Greifensee, Switzerland). Thin-layer chromatography (TLC) was performed on aluminum plates coated with silica gel 60 F254 (Merck, Darmstadt, Germany), by spraying with ceric ammonium molybdate (CAM) solution, followed by heating. The reaction products were isolated by column chromatography, performed using 70–230 mesh silica gel (J. T. Baker).

3.2. Chemical Synthesis

3.2.1. Oxidation of 3-TBS Tigogenin (1) with RuO$_4$/NaIO$_4$

Solution of NaIO$_4$ (1.8 g, 8.4 mmol) and RuO$_2$ (23 mg, 0.17 mmol) in the mixture of water (20 mL), acetone (10 mL), and tetrachloride (20 mL) was vigorously stirred until the yellow color of RuO$_4$ appeared. Then, a solution of 3-TBS tigogenin (**1**, 0.3 g, 0.57 mmol) in 8 mL of CCl$_4$ was added in three portions and the reaction mixture was stirred for 10 h at room temperature. After that time, the TLC control showed that no starting material remained. A few drops of isopropanol were added to quench RuO$_4$ and the resulting slurry was stirred for an additional 10 min at room temperature (yellow RuO$_4$ turned into black RuO$_2$). The reaction mixture was poured into water and product was extracted with CHCl$_3$. The extract was dried over anhydrous sodium sulfate, and the solvent was evaporated. Silica gel column chromatography afforded the product as an equilibrium mixture of two tautomers **2a** and **2b**, identical to that described in reference [42] in 73% total yield.

Compound **2a/2b**, eluted with 7.5% to 25% AcOEt/hexane: for main tautomer: ^{1}H NMR (400 MHz, CDCl$_3$) δ 3.58 (m, 2H), 3.47 (m, 1H), 2.57 (m, 1H), 2.62 (m, 1H), 1.02 (d, J = 7.0, 3H), 0.95 (d, J = 6.6, 3H), 0.89 (s, 9H), 0.81 (s, 3H), 0.74 (s, 3H), 0.06 (s, 6H); ESI-MS 547 [M+H]$^+$. HRMS calculated for C$_{33}$H$_{59}$O$_4$Si (M+H)$^+$, 547.4177; found 547.4230.

3.2.2. Synthesis of (25R)-3β-t-butyldimethylsililoxy-16-aza-spirost-16(N)-ene (3)

Preparation of the aminoalane reagent from DIBAlH and NH$_4$Cl

A solution of DIBAlH in toluene (1 M, 22 mL, 22 mmol, 40 equiv. relative to compounds **2a** and **2b**) was added to a cooled (0–5 °C) suspension of NH$_4$Cl (1.23 g, 23 mmol, 42 equiv.) in anhydrous THF (15 mL) under argon. The reaction was stirred for 15 min in an ice bath and then 1.5 h at room temperature. After this time, the obtained reagent solution was used directly for the reaction with compound **2a/2b**.

Synthesis of imine 3

The solution of aminoalane reagent (prepared from 40 equiv. of DIBAlH) was added dropwise to a solution of compound **2a** and **2b** (0.3 g, 0.549 mmol) in anhydrous THF (ca 6 mL) at room temperature. Then, stirring was continued for 16 h at reflux. After this time, the reaction mixture was cooled, quenched with aqueous solution of KHSO$_4$, and the product was extracted with ether. The extract was washed with water, dried over anhydrous sodium sulfate, and the solvent was evaporated. Silica gel column chromatography afforded three products: spiroimine **3** (44%) eluted with 10% AcOEt/hexane, α,β-unsaturated ketone **5** (5%) eluted with 15% AcOEt/hexane, and dihydropyrrole **4** (12%) eluted with 70% AcOEt/hexane.

Compound **3**: ^{1}H NMR (400 MHz, CDCl$_3$) δ 3.56 (m, 2H), 3.45 (dd, J = 11.0, 10.9, 1H), 2.56 (m, 1H), 2.45 (m, 1H), 1.01 (d, J = 6.9, 3H), 0.89 (s, 9H), 0.84 (d, J = 6.6, 3H), 0.83 (s, 3H), 0.61 (s, 3H), 0.06 (s, 6H); ^{13}C NMR (100 MHz, CDCl$_3$) δ 193.2 (C), 107.6 (C), 72.1 (CH), 70.6 (CH), 69.1 (CH$_2$), 56.8 (CH), 54.5 (CH), 45.0 (CH), 42.7 (CH), 39.8 (C), 38.6 (CH$_2$), 38.3 (CH$_2$), 37.0 (CH$_2$), 35.7 (C), 35.1 (CH), 33.7 (CH$_2$), 32.3 (CH$_2$), 31.9 (CH$_2$), 30.8 (CH), 29.3 (CH$_2$), 28.7 (CH$_2$), 28.5 (CH$_2$), 25.9 (3xCH$_3$), 20.9 (CH$_2$), 18.3 (C), 17.2 (CH$_3$), 13.5 (CH$_3$), 12.43 (CH$_3$), 12.39 (CH$_3$), −4.6 (2xCH$_3$); ESI-MS 528 [M+H]$^+$. HRMS calculated for C$_{33}$H$_{59}$NO$_2$Si (M+H)$^+$, 528.4231; found 528.4297; IR ATR, v_{max} (cm^{-1}): 1728, 1667, 1457, 1373, 1248, 1173, 1063.

Compound **4**: ^{1}H NMR (400 MHz, CDCl$_3$) δ 4.43 (m, 1H), 3.55 (m, 1H), 3.43 (dd, J = 11.1, 4.0, 1H), 3.30 (dd, J = 11.1, 6.0, 1H), 2.61 (q, J = 7.3, 1H), 2.29 (m, 2H), 2.22 (m, 1H), 1.08 (d, J = 7.3, 3H), 0.92 (d, J = 6.3, 3H), 0.89 (s, 9H), 0.79 (s, 3H), 0.51 (s, 3H), 0.05 (s, 6H); ^{13}C NMR (100 MHz, CDCl$_3$) δ 182.0 (C), 74.8 (CH), 72.1 (CH), 65.9 (CH$_2$), 61.6 (CH), 54.9 (CH), 54.6 (CH), 45.0 (CH), 44.5 (CH), 41.4 (C), 39.2 (CH$_2$), 38.6 (CH$_2$), 37.2 (CH$_2$), 35.9 (CH), 35.6 (C), 35.1 (CH), 32.4 (CH$_2$), 32.0 (CH$_2$), 31.9 (CH$_2$), 28.7 (CH$_2$), 27.6 (CH$_2$), 27.5 (CH$_2$), 25.9 (3xCH$_3$), 20.8 (CH$_2$), 18.8 (CH$_3$), 18.3 (C), 17.0 (CH$_3$), 14.0 (CH$_3$), 12.4 (CH$_3$), −4.6 (2xCH$_3$); ESI-MS 530 [M+H]$^+$. HRMS calculated for C$_{33}$H$_{60}$NO$_2$Si (M+H)$^+$, 530.4388; found 530.4398; IR ATR, v_{max} (cm^{-1}): 3235, 1631, 1454, 1372, 1250, 1095, 1062.

Compound **5**: ^{1}H NMR (400 MHz, CDCl$_3$) δ 3.57 (m, 1H), 3.47 (m, 1H), 3.37 (m, 1H), 3.12 (m, 1H), 2.58 (m, 1H), 2.39 (dd, J = 13.5, 5.6, 1H), 2.24 (bs, 1H), 2.17 (q, J = 7.2, 1H), 1.19 (d, J = 7.2, 3H), 0.89 (s, 9H), 0.83 (s, 3H), 0.80 (d, J = 6.8, 3H), 0.54 (s, 3H), 0.06 (s, 6H); ^{13}C NMR (100 MHz, CDCl$_3$) δ 214.9 (C), 183.9 (C), 134.1 (C), 72.0 (CH), 66.0 (CH$_2$), 65.8 (CH), 55.9 (CH), 54.7 (CH), 45.0 (CH), 42.8 (CH), 41.2 (C), 38.6 (CH$_2$), 37.9 (CH$_2$), 37.1 (CH$_2$), 35.7 (C), 35.2 (CH), 35.0 (CH), 32.2 (CH$_2$), 31.9 (CH$_2$), 28.5 (CH$_2$), 28.0 (CH$_2$), 26.2 (CH$_2$), 25.9 (3xCH$_3$), 20.9 (CH$_2$), 18.3 (C), 16.4 (CH$_3$), 14.5 (CH$_3$), 12.4 (CH$_3$), 12.3 (CH$_3$), -4.6 (2xCH$_3$); ESI-MS 529 [M+H]$^+$. HRMS calculated for C$_{33}$H$_{57}$O$_3$Si (M+H)$^+$, 529.4071; found 529.4062; IR ATR, v_{max} (cm^{-1}): 3431, 1697, 1654, 1456, 1373, 1248, 1080, 834, 772.

3.2.3. General Procedure for Imine 3 Reduction with Complex Sodium Hydride

To the stirred solution of imine **3** (1 equiv.) in the proper solvent, reducing agents (NaBH$_4$, NaBH$_3$CN) and additives (NaOAc, I$_2$, AcOH) were added. The detailed reaction conditions are indicated in Table 1. The reaction mixture was monitored by TLC. The reaction mixture was poured into water and extracted with CHCl$_3$. The extract was washed with water, dried over anhydrous sodium sulfate, and the solvent was evaporated. The crude products (**4, 7, 8**) were isolated by silica gel column chromatography.

Compound **7**, eluted with 8% MeOH/CHCl$_3$: ^{1}H NMR (400 MHz, CDCl$_3$) δ 3.75 (m, 1H), 3.51 (m, 2H), 3.37 (m, 1H), 2.88 (m, 1H), 2.01 (m, 1H), 1.01 (d, J = 6.4, 3H), 0.88 (s, 9H), 0.864 (s, 3H), 0.859 (d, J = 6.3, 3H), 0.80 (s, 3H), 0.05 (s, 6H); ^{13}C NMR (100 MHz,

CDCl$_3$) δ 72.1 (CH), 70.4 (CH), 67.3 (CH$_2$), 63.2 (CH), 62.2 (CH), 57.6 (CH), 54.3 (CH), 45.0 (CH), 41.2 (C), 40.0 (CH$_2$), 38.6 (CH$_2$), 38.4 (CH), 37.1 (CH$_2$), 35.6 (C), 34.8 (CH), 34.7 (CH), 32.3 (CH$_2$), 31.9 (CH$_2$), 30.8 (CH$_2$), 29.6 (CH$_2$), 28.6 (CH$_2$), 27.9 (CH$_2$), 25.9 (3×CH$_3$), 20.9 (CH$_2$), 18.2 (CH$_3$), 18.1 (C), 17.0 (CH$_3$), 16.0 (CH$_3$), 12.3 (CH$_3$), −4.6 (2×CH$_3$); ESI-MS 532 [M+H]$^+$. HRMS calculated for C$_{33}$H$_{62}$NO$_2$Si (M+H)$^+$, 532.4544; found 532.4559; IR ATR, ν$_{max}$ (cm^{-1}): 3288, 1454, 1368, 1247, 1092.

Compound 8 (obtained by reduction with NaBH$_3$CN, Table 1, entry 6), eluted with 45% AcOEt/hexane: ^{1}H NMR (400 MHz, CDCl$_3$) δ 4.66 (m, 1H), 3,62–3.50 (m, 3H), 3.07 (q, J = 7.4, 1H), 2.91 (dd, J = 12.4, 4.3, 1H), 2.48–2.35 (m, 2H), 1.83 (d, J = 8.4, 1H), 1.22 (d, J = 7.4, 3H), 1.00 (d, J = 6.6, 3H), 0.89 (s, 9H), 0.79 (s, 3H), 0.57 (s, 3H), 0,05 (ε, 6H); ^{13}C NMR (100 MHz, CDCl$_3$) δ 191. 4 (C), 76.9 (CH), 72.0 (CH), 66.7 (CH$_2$), 57.6 (CH), 54.4 (CH), 54.0 (CH), 44.9 (CH), 43.6 (CH), 42.3 (C), 38.52 (CH$_2$), 38.49 (CH$_2$), 37.2 (CH$_2$), 35.8 (CH), 35.6 (C), 34.9 (CH), 32.2 (CH$_2$), 31.9 (CH$_2$), 31.2 (CH$_2$), 28.7 (CH$_2$), 28.4 (CH$_2$), 27.6 (CH$_2$), 25.9 (3×CH$_3$), 20.5 (CH$_2$), 18.5 (CH$_3$), 18.2 (C), 16.3 (CH$_3$), 14.6 (CH$_3$), 12.3 (CH$_3$), −4.57 (CH$_3$), −4.59 (CH$_3$); ^{11}B NMR (128 MHz, CDCl$_3$) δ −24.52; ESI-MS 1159 [2M+Na]$^+$. IR ATR, ν$_{max}$ (cm^{-1}): 3468, 2401, 1636, 1458, 1249, 1093, 1056.

3.2.4. Synthesis of Compound 9

To a solution of 4 (19 mg, 0.036 mmol) in dichloromethane (2 mL) at −15 °C, Et$_3$N (0.01 mL, 7.3 mg, 0.072 mmol) and 0.22 mL of solution of MsCl (0.03 mL) in dichloromethane (2 mL) were added, successively. The reaction mixture was continuously stirred at −15 °C for 1.5 h and quenched by adding aqueous NaHCO$_3$, and the layers were separated and the aqueous layer was extracted with chloroform. The organic layers were combined, dried over Na$_2$SO$_4$, and evaporated under reduced pressure. The crude product was purified by column chromatography (20% MeOH/CHCl$_3$) to obtain compound 9 (45%).

Compound 9: ^{1}H NMR (400 MHz, CDCl$_3$) δ 5.45 (m, 1H), 4.25 (m, 1H), 3.61 (m, 1H), 3.13-3.00 (m, 3H), 2.75 (s, 3H), 2.72 (m, 1H), 2.62 (m, 1H), 2.32 (m, 1H), 2.18 (d, J = 6.1, 1H), 1.57 (d, J = 7.1, 3H), 1.08 (d, J = 5.9, 3H), 0.81 (s, 3H), 0.62 (s, 3H); δ ^{13}C NMR (100 MHz, CDCl$_3$) δ 190.6 (C), 75.7 (CH), 71.0 (CH), 57.0 (CH), 54.3 (CH), 54.0 (CH), 52.7 (CH$_2$), 44.6 (CH), 44.5 (CH), 42.0 (C), 39.4 (CH$_3$), 38.2 (CH$_2$), 38.0 (CH$_2$), 36.9 (CH$_2$), 35.5 (C), 35.0 (CH), 32.0 (CH$_2$), 31.3 (CH$_2$), 29.2 (CH$_2$), 28.3 (CH$_2$), 25.84 (CH), 25.82 (CH$_2$), 24.6 (CH$_2$), 20.5 (CH$_2$), 18.2 (CH$_3$), 18.1 (CH$_3$), 14.6 (CH$_3$), 12.3 (CH$_3$); ESI-MS 398 [M$^+$]; IR ATR, ν$_{max}$ (cm^{-1}): 3377, 1664, 1628, 1456, 1195, 1043.

3.2.5. Synthesis of 25-epidemissidine (10)

To the stirred ice-cooled solution of compound 9 (20 mg, 0.04 mmol) in MeOH (2 mL)/DCM (2 mL), NaBH$_4$ (4.6 mg, 0.12 mmol) was added. The stirring of the reaction mixture was continued at −10 – 0 °C for 0.5 h. The reaction mixture was poured into water and extracted with CHCl$_3$. The extract was washed with water, dried over anhydrous sodium sulfate, and the solvent was evaporated. The crude product was purified by column chromatography (20% AcOEt/hexane) to obtain compound 10 (71%), identical to that described in ref. [21].

Compound 10: ^{1}H NMR (400 MHz, CDCl$_3$): δ 3.60 (m, 1H), 2.59 (m, 2H), 1.03 (d, J = 7.0, 3H), 0.92 (d, J = 6.7, 3H), 0.85 (s, 3H), 0.82 (s, 3H).

3.2.6. Synthesis of Compound 12

Acetyl chloride (0.027 mL, 29 mg, 0.38 mmol) was added to the stirred, ice-cold solution of compound 4 (20 mg, 0.038 mmol) in dry MeOH (3 mL). The reaction mixture was stirred for 3 h and allowed to warm up to room temperature. Then, the solvent was evaporated under reduced pressure and the residue was dissolved in dichloromethane (2 mL) and THF (2 mL). To the obtained suspension, triethylamine (0.02 mL) and 0.58 mL of a solution of MsCl (0.02 mL) in CH$_2$Cl$_2$ (2 mL) were added. The reaction mixture was stirred overnight, allowing it to warm up to room temperature. After this time, the mixture was poured into aqueous NaHCO$_3$ and extracted with CHCl$_3$. The extract was washed with water, dried over anhydrous sodium sulfate, and the solvent was evaporated. The crude product was purified by column chromatography (35% AcOEt/hexane) to afford compound 12 (53%).

Compound 12: ^{1}H NMR (400 MHz, CDCl$_3$) δ 4.62 (m, 1H), 4.04 (m, 1H), 3.61 (q, J = 7.0, 1H), 3.02 (m, 1H), 3.00 (s, 3H), 2.83 (s, 3H), 2.74 (m, 1H), 2.50 (d, J = 12.5, 1H), 1.24 (d, J = 7.0, 3H), 1.04 (d, J = 6.1, 3H), 0.83 (s, 3H), 0.61 (s, 3H); ^{13}C NMR (100 MHz, CDCl$_3$) δ 160.9 (C), 88.9 (C), 81.6 (CH), 65.4 (CH), 60.5 (CH), 55.0 (CH), 54.1 (CH), 49.0 (CH$_2$), 44.8 (CH), 43.0 (CH$_3$), 41.8 (C), 38.9 (CH$_3$), 38.2 (CH$_2$), 36.7 (CH$_2$), 36.6 (CH), 35.3 (C), 35.1 (CH$_2$), 34.9 (CH), 32.0 (CH$_2$), 30.8 (CH$_2$), 30.7 (CH$_2$), 28.6 (CH$_2$), 28.3 (CH$_2$), 26.8 (CH), 23.0 (CH$_3$), 20.5 (CH$_2$), 18.7 (CH$_3$), 13.5 (CH$_3$), 12.1 (CH$_3$); ESI-MS 554 [M+H]$^+$, 1129 [2M+Na]$^+$. HRMS calculated for C$_{29}$H$_{48}$NO$_5$S$_2$ (M+H)$^+$ 554.2968; found 554.2965; IR ATR, ν$_{max}$ (cm^{-1}): 1658, 1453, 1333, 1212, 1163, 1036, 925.

4. Conclusions

In summary, we developed a novel, concise synthesis of solanidanes from the spirostane sapogenin tigogenin. The indolizidine moiety present in solanidane-type alkaloids was constructed from spirostane in five steps involving tigogenin oxidation, amination, reduction, mesylation, and reduction again. The key intermediate for the proposed approach was spiroimine obtained in the reaction of a 5,6-dihydrokrytogenin derivative with aminoalane generated in situ from DIBAlH and NH$_4$Cl. Depending on mesylation conditions, two different solanidanes were obtained: the indolizinium salt 9a, which was readily converted into 25-epimissidine (10), and the 23-sulfone derivative 12, a convenient intermediate for further derivatization.

Supplementary Materials: The following are available online at https://www.mdpi.com/article/10.3390/ijms221910879/s1, ^{1}H NMR, ^{13}C NMR spectra of compounds 3–10 and 12.

Author Contributions: Conceptualization, A.W. and J.W.M.; investigation, A.W. and U.K.; methodology, A.W. and A.B.; formal analysis, A.W.; writing—original draft preparation, A.W.; writing—review, editing, and supervising, J.W.M. All authors have read and agreed to the published version of the manuscript.

Funding: The authors acknowledge the financial support from the National Science Centre, Poland (Grant 2015/17/B/ST5/02892).

Conflicts of Interest: The authors declare no conflict of interest.

References

1. Attaur, R.; Choudhary, M.I. Diterpenoid and steroidal alkaloids. *Nat. Prod. Rep.* **1997**, *14*, 191–203. [CrossRef]
2. Li, H.J.; Jiang, Y.; Li, P. Chemistry, bioactivity and geographical diversity of steroidal alkaloids from the *Liliaceae* family. *Nat. Prod. Rep.* **2006**, *23*, 735–752. [CrossRef]
3. Jiang, Q.-W.; Chen, M.-W.; Cheng, K.-J.; Yu, P.-Z.; Wei, X.; Shi, Z. Therapeutic potential of steroidal alkaloids in cancer and other diseases. *Med. Res. Rev.* **2016**, *36*, 119–143. [CrossRef] [PubMed]
4. Friedman, M.; Lee, K.R.; Kim, H.J.; Lee, I.S.; Kozukue, N. Anticarcinogenic effects of glycoalkaloids from potatoes against human cervical, liver, lymphoma, and stomach cancer cells. *J. Agric. Food Chem.* **2005**, *53*, 6162–6169. [CrossRef] [PubMed]
5. Lee, K.R.; Kozukue, N.; Han, J.S.; Park, J.H.; Chang, E.Y.; Baek, E.J.; Chang, J.S.; Friedman, M. Glycoalkaloids and metabolites inhibit the growth of human colon (HT29) and liver (HepG2) cancer cells. *J. Agric. Food. Chem.* **2004**, *52*, 2832–2839. [CrossRef]
6. Tingey, W.M. Glycoalkaloids as pest resistance factors. *Am. J. Potato Res.* **1984**, *61*, 157–167. [CrossRef]
7. Schreiber, K. Steroid alkaloids: The *Solanum* group. In *The Alkaloids: Chemistry and Physiology*; Manske, R.H.F., Ed.; Academic Press: New York, NY, USA, 1968; Volume 10, pp. 1–192.
8. Roddick, J.G. The acetylcholinesterase-inhibitory activity of steroidal glycoalkaloids and their aglycones. *Phytochemistry* **1989**, *28*, 2631–2634. [CrossRef]
9. Roddick, J.G.; Weissenberg, M.; Leonard, A.L. Membrane disruption and enzyme inhibition by naturally-occurring and modified chacotriose-containing *Solanum* steroidal glycoalkaloids. *Phytochemistry* **2001**, *56*, 603–610. [CrossRef]

10. Mótyán, G.; Kovács, F.; Wölfling, J.; Gyovai, A.; Zupkó, I.; Frank, É. Microwave-assisted stereoselective approach to novel steroidal ring D-fused 2-pyrazolines and an evaluation of their cell-growth inhibitory effects in vitro. *Steroids* **2016**, *112*, 36–46. [CrossRef] [PubMed]
11. Abdalla, M.M.; Al-Omar, M.A.; Bhat, M.A.; Amr, A.-G.E.; Al-Mohizeae, A.M. Steroidal pyrazolines evaluated as aromatase and quinone reductase-2 inhibitors for chemoprevention of cancer. *Int. J. Biol. Macromol.* **2012**, *50*, 1127–1132. [CrossRef]
12. Schuster, D.; Wolber, G. Identification of bioactive natural products by pharmacophore-based virtual screening. *Curr. Pharm. Des.* **2010**, *16*, 1666–1681. [CrossRef] [PubMed]
13. Kuhn, R.; Low, I.; Trischmann, H. Uberfuhrung von tomatidin in demissidin. *Angew. Chem.* **1952**, *64*, 397–397. [CrossRef]
14. Sato, Y.; Latham, H.G. Chemistry of dihydrotomatidines. *J. Am. Chem. Soc.* **1956**, *78*, 3146–3150. [CrossRef]
15. Schreiber, K.; Ronsch, H. Solanum-alkaloide-LIV. Synthese von solanidin und 22-iso-solanidin aus tomatid-5-en-3β-ol. *Tetrahedron* **1965**, *21*, 645–650. [CrossRef]
16. Adam, G.; Schreiber, K. Synthese des steroid alkaloids demissidin aus 3β-acetoxy-pregn-5-en-20-on; aufbau des solanidine-gerustes durch Hofmann-Loffler-Freytag-cyclisierung. *Tetrahedron Lett.* **1963**, *4*, 943–948. [CrossRef]
17. Zhang, Z.; Giampa, G.M.; Draghici, C.; Huang, Q.; Brewer, M. Synthesis of demissidine by a ring fragmentation 1,3-dipolar cycloaddition approach. *Org. Lett.* **2013**, *15*, 2100–2103. [CrossRef]
18. Zhang, Z.-D.; Shi, Y.; Wu, J.-J.; Lin, J.-R.; Tian, W.-S. Synthesis of demissidine and solanidine. *Org. Lett.* **2016**, *18*, 3038–3040. [CrossRef]
19. Hou, L.-L.; Shi, Y.; Zhang, Z.-D.; Wu, J.-J.; Yang, Q.-X.; Tian, W.-S. Divergent synthesis of solanidine and 22-epi-solanidine. *J. Org. Chem.* **2017**, *82*, 7463–7469. [CrossRef]
20. Wang, Y.; Huang, G.; Shi, Y.; Tian, W.-S.; Zhuang, C.; Chen, E.-F. Asymmetric synthesis of (−)-solanidine and (−)-tomatidenol. *Org. Biomol. Chem.* **2020**, *18*, 3169–3176. [CrossRef]
21. Rivas-Loaiza, J.A.; Baj, A.; López, Y.; Witkowski, S.; Wojtkielewicz, A.; Morzycki, J.W. Synthesis of solanum alkaloid demissidine stereoisomers and analogues. *J. Org. Chem.* **2021**, *86*, 1575–1582. [CrossRef]
22. Cheng, M.S.; Wang, Q.L.; Tian, Q.; Song, Y.H.; Liu, Y.X.; Li, Q.; Xu, X.; Miao, H.D.; Yao, X.S.; Yang, Z. Total synthesis of methyl protodioscin: A potent agent with antitumor activity. *J. Org. Chem.* **2003**, *68*, 3658–3662. [CrossRef] [PubMed]
23. Lee, J.S.; Cao, H.; Fuchs, P.L. Ruthenium-catalyzed mild C–H oxyfunctionalization of cyclic steroidal ethers. *J. Org. Chem.* **2007**, *72*, 5820–5823. [CrossRef] [PubMed]
24. Yu, B.; Liao, J.; Zhang, J.; Hui, Y. The first synthetic route to furostan saponins. *Tetrahedron Lett.* **2001**, *42*, 77–79. [CrossRef]
25. Bovicelli, P.; Lupettelli, P.; Fracas, D. Sapogenins and dimethyldioxirane: A new entry to cholestanes functionalized at the side chain. *Tetrahedron Lett.* **1994**, *35*, 935–938. [CrossRef]
26. Overman, L.E.; Sugai, S. A Convenient Method for obtaining *trans*-2-aminocyclohexanol and *trans*-2-aminocyclopentanol in enantiomerically pure form. *J. Org. Chem.* **1985**, *50*, 4154–4155. [CrossRef]
27. Overman, L.E.; Flippin, L.A. Facile aminolysis of epoxides with diethylaluminum amides. *Tetrahedron Lett.* **1981**, *22*, 195–198. [CrossRef]
28. Ashworth, I.W.; Bowden, M.C.; Dembofsky, B.; Levin, D.; Moss, W.; Robinson, E.; Szczur, N.; Virica, J. A new route for manufacture of 3-cyano-1-naphthalenecarboxylic acid. *Org. Process. Res. Dev.* **2003**, *7*, 74–81. [CrossRef]
29. Veerasamy, N.; Carlson, E.C.; Collett, N.D.; Saha, M.; Carter, R.G. Enantioselective approach to quinolizidines: Total synthesis of cermizine D and formal syntheses of senepodine G and cermizine C. *J. Org. Chem.* **2013**, *78*, 4779–4800. [CrossRef]
30. Bakunova, S.M.; Bakunov, S.A.; Wenzler, T.; Barszcz, T.; Werbovetz, K.A.; Brun, R.; Hall, J.E.; Tidwell, R.R. Synthesis and in vitro antiprotozoal activity of bisbenzofuran cations. *J. Med. Chem.* **2007**, *50*, 5807–5823. [CrossRef]
31. Huang, S.-Q.; Zheng, X.; Deng, X.-M. DIBAL-H-H_2NR and DIBAL-H-HNR_1R_2·HCl complexes for efficient conversion of lactones and esters to amides. *Tetrahedron Lett.* **2001**, *42*, 9039–9041. [CrossRef]
32. Wojtkielewicz, A.; Łotowski, Z.; Morzycki, J.W. One-step synthesis of nitriles from acids, esters and amides using DIBAL-H and ammonium chloride. *Synlett* **2015**, *26*, 2288–2292. [CrossRef]
33. Bechara, W.S.; Charette, A.B.; Na, R.; Wang, W.; Zheng, C. Diethyl 1,4-dihydro-2,6-dimethyl-3,5-pyridinedicarboxylate. In *Encyclopedia of Reagents for Organic Synthesis 2020*; John Wiley & Sons, Ltd.: Hoboken, NJ, USA, 2020. [CrossRef]
34. Czajkowska-Szczykowska, D.; Jastrzebska, I.; Rode, J.E.; Morzycki, J.W. Revision of the structure of N,O-diacetylsolasodine. Unusual epimerization at the spiro carbon atom during acetylation of solasodine. *J. Nat. Prod.* **2019**, *82*, 59–65. [CrossRef]
35. Chagnon, F.; Guay, I.; Bonin, M.-A.; Mitchell, G.; Bouarab, K.; Malouin, F.; Marsault, É. Unraveling the structure-reactivity relationship of tomatidine, a steroid alkaloid with unique antibiotic properties against persistent forms of *Staphylococcus aureus*. *Eur. J. Med. Chem.* **2014**, *80*, 605–620. [CrossRef]
36. Zha, X.M.; Zhang, F.R.; Shan, J.Q.; Zhang, Y.H.; Liu, J.O.; Sun, H.B. Synthesis and evaluation of in vitro anticancer activity of novel solasodine derivatives. *Chin. Chem. Lett.* **2010**, *21*, 1087–1090. [CrossRef]
37. Uhle, F.C.; Sallmann, F. The synthesis and transformations of a steroid pyrroline Derivative. *J. Am. Chem. Soc.* **1960**, *82*, 1190–1199. [CrossRef]
38. Reddy, C.R.; Ramesh, P.; Latha, B. Formal syntheses of 5,8-disubstituted indolizidine alkaloids (−)-205A, (−)-207A, and (−)-235B. *Synlett* **2016**, *27*, 481–484. [CrossRef]

39. Donohoe, T.J.; Sintim, H.O.; Hollinshead, J. A Noncarbohydrate Based Approach to Polyhydroxylated. *J. Org. Chem.* **2005**, *70*, 7297–7304. [CrossRef]
40. Mátyus, P.; Szilágyi, G.; Kasztreiner, E.; Sohár, P. Studies on pyridazine derivatives. IV. Mesylation reaction of pyridazinylpyrazoles. *J. Heterocycl. Chem.* **1980**, *17*, 781–783. [CrossRef]
41. Back, T.G.; Hu, N.-X. Synthesis of Antibiotic A25822 B. *Tetrahedron* **1993**, *49*, 337–348. [CrossRef]
42. Cheng, S.; Du, Y.; Mac, B.; Tanc, D. Total synthesis of a furostan saponin, timosaponin BII. *Org. Biomol. Chem.* **2009**, *7*, 3112–3118. [CrossRef]

Review

Natural Products with Activity against Lung Cancer: A Review Focusing on the Tumor Microenvironment

Yue Yang, Ning Li *, Tian-Ming Wang and Lei Di *

Inflammation and Immune Mediated Diseases Laboratory of Anhui Province, School of Pharmacy, Anhui Medical University, Hefei 230032, China; yangyue9288@163.com (Y.Y.); wtm1818@163.com (T.-M.W.)
* Correspondence: 1993500019@ahmu.edu.cn (N.L.); dilei@ahmu.edu.cn (L.D.); Tel.: +86-551-6516-1115 (N.L.)

Abstract: Lung cancer is one of the most prevalent malignancies worldwide. Despite the undeniable progress in lung cancer research made over the past decade, it is still the leading cause of cancer-related deaths and continues to challenge scientists and researchers engaged in searching for therapeutics and drugs. The tumor microenvironment (TME) is recognized as one of the major hallmarks of epithelial cancers, including the majority of lung cancers, and is associated with tumorigenesis, progression, invasion, and metastasis. Targeting of the TME has received increasing attention in recent years. Natural products have historically made substantial contributions to pharmacotherapy, especially for cancer. In this review, we emphasize the role of the TME and summarize the experimental proof demonstrating the antitumor effects and underlying mechanisms of natural products that target the TME. We also review the effects of natural products used in combination with anticancer agents. Moreover, we highlight nanotechnology and other materials used to enhance the effects of natural products. Overall, our hope is that this review of these natural products will encourage more thoughts and ideas on therapeutic development to benefit lung cancer patients.

Keywords: natural products; tumor microenvironment (TME); lung cancer; phytochemicals; botanical agents

1. Introduction

Lung cancer, which is classified into non-small cell lung cancer (NSCLC) and small cell lung cancer (SCLC), is one of the most prevalent malignancies worldwide in terms of both incidence and mortality (18.0% of total cancer deaths) [1]. Therapeutic options for this cancer are surgery, radiation therapy, and systemic treatments including chemotherapy, targeted therapy, hormonal therapy, and immunotherapy. Because the diagnosis of SCLC is rarely localized, surgical resection plays a small role in its treatment. Most patients with SCLC receive chemotherapy. Approximately 56% of patients with stage I and II NSCLC undergo surgery. For patients with stage III NSCLC, only 18% are treated with surgery, while most (62%) undergo chemotherapy or radiotherapy. Immune and targeted therapeutic drugs are used for the treatment of advanced NSCLC, but some drugs are only used for the treatment of cancers with specific gene mutations [2]. Despite the tremendous efforts in research on the treatment of lung cancer, the incidence and mortality rates of lung cancer have not decreased significantly [3]. Therefore, it is necessary to find more treatment strategies and drugs for lung cancer.

The tumor microenvironment (TME) is a complex ecosystem consisting of the vasculature, extracellular matrix (ECM), cytokines and growth factors, and many different populations of stromal cells, such as myeloid-derived suppressor cells (MDSCs), tumor-associated macrophages (TAMs), and tumor-associated fibroblasts (TAFs) [4]. Over the past decade, the TME has been recognized as playing key roles in lung cancer initiation and progression [5,6]. As the composition of the TME is heterogeneous, interactions between cancer and stromal cells in the microenvironment regulate the main characteristics of cancer, including its immune suppression, angiogenesis, and inflammation properties [7]. These properties support the growth, invasion, and metastasis of cancer cells. Therefore, the role of the TME in lung cancer has received increasing attention. The TME is regarded as a target-rich environment in the development of new anticancer drugs. Strategies that target cancer cells are considered as treatment avenues. Moreover, different from cancer cells, the stromal cells in the TME are genetically stable, therefore, they are attractive therapeutic targets that are associated with reductions in drug resistance and tumor recurrence [8]. More and more studies are focusing on the TME as a target for drug research and development (Figure 1).

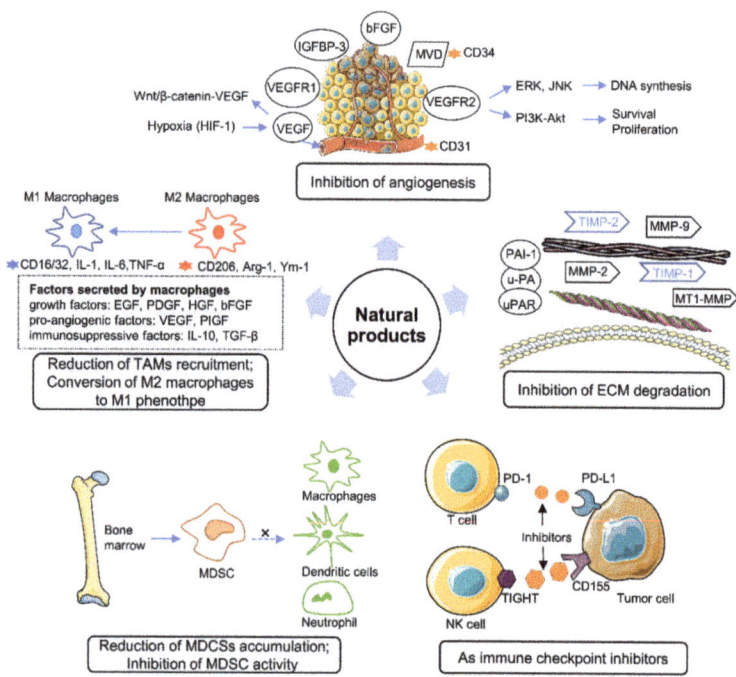

Figure 1. Modulation of the TME by natural products.

Natural products are precious gifts from nature to mankind. They include extracts of animals and plants, metabolites of insects, marine organisms, and microorganisms, as well as many chemical components found endogenously in humans and animals. In addition, traditional Chinese medicine (TCM) is based on the combination of natural products and TCM theory. Natural products have always been an important source of drug discovery. According to the latest statistics on drugs approved by the Food and Drug Administration (FDA) in the United States, many prescription medicines used for treatment originate from natural products. From 1946 to 2019, more than 50% of newly approved drugs were natural small molecules [9]. Plant preparations and Chinese medicines are multi-component, multi-channel, and multi-target products. Due to their diverse structures and activities, natural products continue to attract researchers' attention. Although the TME has been widely studied, the natural products that target and regulate the TME of lung cancer have not been systematically summarized. In this review, we describe the antitumor effect of natural products on the TME in lung cancer. We summarize relevant natural products, including descriptions of their anti-tumor actions in terms of modulating the TME in lung cancer when given alone (Table 1), in combination with anticancer drugs (Table 2), and in combination with materials such as nanomaterials.

Table 1. The effects of natural products on modulation of the TME.

No.	Natural Products	Common Source	Cell Lines or Animal Models or Patients	Function or Molecular Mechanism	Ref.
			Targeting angiogenesis		
1	Jolkinolide A (1)	*Euphorbia fischeriana*	A549, HUVEC; A549 cell xenograft mice	Inhibition of the Akt-STAT3-mTOR signaling pathway and reduction of VEGF protein expression; inhibition of HUVEC migration	[10]
2	Jolkinolide B (2)				
3	Parthenolide (3)	*Tanacetum parthenium*	A549, H526	Inhibition of A549 and H526 cell proliferation in the presence and absence of nicotine; induction of apoptosis; inhibition of angiogenesis; down-regulation of Bcl-2 expression and up-regulation of E2F1, p53, GADD45, Bax, Bim, and caspase 3,7,8,9 expression	[11]
4	Galbanic acid (4)	*Ferula assafoetida*	LLC, HUVEC; LLC-bearing mice	Inhibition of VEGF-inducible HUVECs and LLC proliferation; tube formation, migration/invasion inhibition in HUVECs; decreased phosphorylation of p38MAPK, JNK and Akt, and decreased expression of VEGFR targeting eNOS; inhibition of tumor-induced angiogenesis and tumor growth in mice; reduction of CD34 and Ki67	[12]
5	Salvicine (5)	*Salvia prionitis*	A549, HMEC	Inhibition of A549 cell viability; inhibition of the migration and tube formation of HMECs; reduced mRNA expression levels of bFGF; VEGF mRNA expression unchanged	[13]
6	Tanshinone IIA (6)	*Salvia miltiorrhiza*	A549	Proliferation inhibition; apoptosis induction; cell cycle arrest at the S stage; downregulation of protein expression of VEGF and VEGFR2	[14]
7	β-hydroxyisovalerylshikonin (7)	*Lithospermum erythrorhizon*	HUVEC, VPC; LLC xenograft mice	Inhibition of blood vessel formation by chicken chorioallantoic membrane assay; suppression of in vivo angiogenesis; suppression of VEGFR2 and Tie2; inhibition of HUVECs and VPC growth; suppression of MAPK and Sp1-dependent VEGFR2 and Tie2 mRNA expression	[15]
8	Isogambogenic acid (8)	*Gamboge hanburyi*	A549, HUVEC; transgenic FLK-1 promoter EGFP zebrafish; A549 xenograft mice	More effective for inhibiting HUVEC proliferation than A549; angiogenesis inhibition in zebrafish embryos; suppression of angiogenesis and tumor growth in mice; inhibition of vessel sprouting ex vivo; inhibition of VEGF-induced migration, invasion, and tube formation, morphological changes in HUVECs	[16]
9	Tubeimoside-1 (9)	*Bolbostemma paniculatum*	H460, A549, eEND2; H460 xenograft mice	Inhibition of tumor growth and vascularization; vascular sprouting; eEND2, H460, and A549 cell viability; eEND2 cell migration; VEGFR2 and Tie2 expression; and the Akt/mTOR pathway	[17]
10	Decursin (10)	*Angelica gigas*	HUVEC, LLC; LLC xenograft mice	Inhibition of VEGF-induced HUVEC proliferation, angiogenesis, and blood vessel formation; suppression of VEGF-induced phosphorylation of p42/44 ERK and JNK MAPK in endothelial cells and VEGF-induced MMP-2 activation; suppression of tumor growth and angiogenesis in nude mice	[18]
11	Decursinol angelate (11)				
12	Farnesiferol C (12)	*Ferula assafoetida*	HUVEC, LLC; LLC allograft tumor mice	Inhibition of VEGF-induced proliferation, migration, invasion, and MMP-2 secretion of HUVECs; inhibition of VEGF-induced vessel sprouting ex vivo; inhibition of tumor growth, angiogenesis, proliferation in vivo; inhibition of VEGF-induced phosphorylation of p125 FAK (pY861), Src (pY416), ERK1/2, p38MAPK, and JNK	[19]
13	Ergosterol (13)	*Agaricus blazei*	s180, LLC; s180 and LLC-bearing mice; mice inoculated with matrigel	Prevention of neovascularization induced by both LLC cells and matrigel; inhibition of tumor growth	[20]
14	Grape seed proanthocyanidins (GSP)		A549, H1299, H226, H460, H157, H1975, H1650, H3255, HCC827, BEAS-2B; A549 and H1299 xenograft mice	Inhibition of the proliferation of human lung cancer cells but not normal human bronchial epithelial cells; GSP-induced inhibition of proliferation is blocked by IGFBP-3 knockdown; inhibition of tumor growth and neovascularization; upregulation of IGFBP-3 protein levels in plasma and lung tumors; inhibition of VEGF expression	[21–25]
15	Pomegranate fruit extract		B(a)P-induced lung tumorigenesis in A/J mouse	Inhibition of the activation of NF-κB, IKKα, PI3k and mTOR, and phosphorylation of IκBα, MAPKs, Akt and c-met; down-regulation of Ki-67, PCNA, CD31, VEGF, and iNOS expression	[26]
16	Extracts of *Astragali Mongolici* and *Rhizoma Curcumae*		LLC-bearing mice	Decreases in tumor weight and tumor MVD; down-regulation of p38 MAPK, p-p38 MAPK, ERK1/2, p-ERK1/2, JNK, p-JNK, and VEGF expression	[27]

Table 1. Cont.

No.	Natural Products	Common Source	Cell Lines or Animal Models or Patients	Function or Molecular Mechanism	Ref.
17	*Scutellaria barbata* extract		CL1-5, HEL299, 293T, LL2, HMEC-1; LL2 lung metastatic mice	Decrease in the transcriptional activity of HIF-1α by inactivation of AKT; inhibition of VEGF expression in CL1-5 cells, migration and proliferation of HMEC-1 cells; inhibition of tumor growth	[28]
18	*Ginkgo biloba* exocarp extracts		LLC; LLC transplanted tumor mice	Inhibition of LLC cell proliferation; downregulation of CD34, Wnt3α, β-catenin, p-AKT/AKT, VEGF, and VEGFR2 expression	[29]
19	Green tea extract		NNK-induced lung tumorigenesis in mouse	Decreases in CD31, MVD and VEGF expression; apoptosis	[30]
20	An-te-xiao capsule	Chinese medicine	A549, H460, H520, LLC, HUVEC; LLC xenograft mouse; H460 and H520 xenograft mice	No acute oral toxicity; prolongation of survival time; inhibition of tumor growth; decreases in MVD, CD31, and the blood vessel number; inhibition of Td-EC migration, invasion, and tube formation in the presence or absence of VEGF; inhibition of VEGF secretion and VEGFR2 phosphorylation	[31]
21	Erbanxiao solution	Chinese medicine	Patients with lung cancer	Inhibition of tumor angiogenesis by changing the levels of VEGF, bFGF, and TNF-α	[32]
22	Ka-mi-kae-kyuk-tang	Korean herbal cocktail	HUVEC, LLC; LLC-bearing mice; PC-3 xenograft mice	Suppression of bFGF stimulated endothelial membrane receptor-tyrosine kinase signaling to ERK1/2, cell motility, capillary differentiation, and to a less extent mitogenesis; inhibition of HIF1α and VEGF; suppression of tumor growth;	[33]
23	Qingzaojiufei decoction	Chinese medicine	LLC	Inhibition of LLC proliferation and growth; up-regulation of p53, and down-regulation of c-myc and Bcl-2; reduction of MMP-9, VEGF, VEGFR, p-ERK1/2	[34]
24	Yiqichutan formula	Chinese medicine	A549, H460, H446; A549-bearing rats, H460-bearing rats, H446-bearing rats	Inhibition of tumor growth; reductions in CD31 expression and the number of blood vessels; decreases in VEGF, HIF-1, DLL4, and Notch-1 protein expression and VEGF mRNA expression	[35]
			Targeting the ECM		
25	Curcumin (**14**)	*Curcuma logna*	A549; A549 xenograft mice	Attenuation of the GLUT1/MT1-MMP/MMP2 pathway	[36]
26	Honokiol (**15**)	*Magnolia officinalis*	H1299	Disruption of HDAC6-mediated Hsp90/MMP-9 interaction; MMP-9 protein degradation; inhibition of migration, invasion, and MMP-9 proteolytic activity and expression; regulation of ubiquitin proteasome system	[37]
27	Theaflavin (**16**)	Black tea	LL2-Lu3	Inhibition of cell invasion, MMP-2 and MMP-9 secretion, and type IV collagenases	[38]
28	Theaflavin digallate (**17**)				
29	EGCG (**18**)	Green tea	CL1-5, CL1-0	Cell cycle G2/M arrest; inhibition of cell invasion and migration; repression of MMP-2 and -9 activities; reduction of nuclear translocation of NF-κB and Sp1; JNK signaling pathway	[39,40]
			A549, HUVEC; A549 xenograft mice	Inhibition of nicotine-induced migration and invasion; down-regulation of HIF-1α, VEGF, COX-2, p-Akt, p-ERK, and vimentin protein levels; up-regulation of p53 and β-catenin protein levels; suppression of HIF-1α and VEGF protein expression	[41]
30	Steroidal saponins extracted from *Paris polyphylla*		A549	Suppression of cell proliferation, adhesion, migration, and invasion; downregulation of MMP-2 and -9 protein levels; inhibition of MMP-2 and -9 activity	[42]
31	Methanolic extract of *Euchelus asper*		A549; chick chorio-allantoic membrane model	Cell cycle subG1 phase arrest; reduction of A549 proliferation, MMP-2 and -9; decrease of the branching points of the 1st order blood vessels or capillaries of the chorio-allantoic membrane	[43]
32	*Phyllanthus urinaria* extract		A549, LLC; LLC-bearing mice	Cytotoxicity; inhibition of invasion and migration; inhibition of u-PA, and MMP-2 and -9 activity; inhibition of TIMP-2 and PAI-1 protein expression; inhibition of transcriptional activity of MMP-2 promoter; inhibition of p-Akt, NF-κB, c-Jun, and c-Fos; decrease in lung metastases	[44]
33	*Rosa gallica* petal extract		A549	Downregulation of the PCNA, cyclin D1, and c-myc; suppression of cell migration and invasion; inhibition of the expression and activity of MMP-2 and -9; regulation of EGFR-MAPK and mTOR-Akt signaling pathways	[45]
34	*Viola Yedoensis* extract		A549, LLC	Cytotoxicity; inhibition of cell invasiveness and migration; suppression of MMP-2, -9 and u-PA; decreases in TIMP-2 and TIMP-1 protein levels; increase in PAI-1 protein expression; decrease in NF-κB DNA binding activity	[46]

Table 1. Cont.

No.	Natural Products	Common Source	Cell Lines or Animal Models or Patients	Function or Molecular Mechanism	Ref.
35	*Cinnamomum cassia* extract		A549, H1299	Cytotoxicity; inhibition of cell migration, invasiveness and motility; inhibition of MMP-2 u-PA, and RhoA protein levels; decrease of cell-matrix adhesion to gelatin and collagen; decrease of FAK and ERK1/2 phosphorylation;	[47]
36	*Duchesnea indica* extracts		A549, H1299	Inhibition of MMP-2 and u-PA activity; cell invasion and metastasis inhibition; downregulation of the expression of p-ERK, p-FAK Tyr397, p-paxillin Tyr118, c-Jun, c-Fos, and TGF-b1 induced-vimentin; inhibition of tumor growth	[48]
37	*Fructus phyllanthi* tannin fraction		H1703, H460, A549, HT1080	Cytotoxicity; inhibition of cell migration and invasion; down-regulation of p-ERK1/2, MMP-2 and -9 expression level, up-regulation of p-JNK expression; regulation of the MAPK pathway	[49]
38	Butanol fraction extract of *Psidium cattleianum* leaf		H1299	Suppression of activities, protein and mRNA expression levels of MMP-2 and -9; inhibition of adhesion, migration and invasion; downregulation of the mRNA level of uPAR; suppression of the ERK1/2 signaling pathway	[50]
39	*Terminalia catappa* leaf extract		A549, LLC	Absence of cytotoxicity; inhibition of invasion, metastasis and motility; inhibition of u-PA, MMP-2, and -9 activity; inhibition of protein levels of TIMP-2 and PAI-1	[51]
40	*Rhizoma Paridis* saponins		LA795 xenograft mice	Inhibition of tumor growth; downregulation of mRNA expression of MMP-2 and -9 and ascendance of TIMP-2	[52]
41	*Selaginella tamariscina* extract		A549, LLC; LLC-bearing mice	Inhibition of invasion and motility; reduced activity of u-PA MMP-2 and -9; increase of protein levels of TIMP-2 and PAI-1 in A549 cells; decrease in lung metastases in mice	[53]
42	Ethanol extract of *Ocimum sanctum*		LLC; LLC lung metastasis mouse model	Cytotoxicity; inhibition of cell adhesion and invasion; inhibition of nodules mediated by LLC cells; decrease in the activity of the enzymes SOD, CAT, and GSH-Px	[54]
43	Rhubarb serum metabolites		A549; A549 lung metastatic mouse model	Suppression of the activity and expression level of MMP-2; inhibition of NF-κB/c-Jun pathway; inhibition of u-PA expression; inhibition of cell motility in vitro and lung metastasis *in vivo*	[55]
44	Fuzheng Kang-Ai decoction	Chinese medicine	A549, PC9, H1650	Suppression of cell proliferation; inhibition of cell migration and invasion; downregulation of MMP-9 activity and protein expression; downregulation of EMT related protein N-cadherin and vimentin	[56]
45	Yifei Tongluo	Chinese medicine	LLC-bearing mice	Inhibition of tumor growth; prolonged survival; fewer nodules on the lung surface; inhibition of MVD, CD34, VEGF, MMP-2, MMP-9, N-cadherin, and vimentin; increases in E-cadherin expression and NK cytotoxic activity; increased percentages of CD4$^+$, CD8$^+$ T, and NK cells; upregulation of Th1-type cytokines and levels of IFN-γ and IL-2 in the serum and reduction in IL-10 and TGF-β1; downregulation of PI3K/AKT, MAPK, and TGFβ/Smad2 pathways; upregulation of the JNK and p38 pathways	[57]
			Targeting MDSCs		
46	Resveratrol (**19**)	Grape skin and seeds	LLC; LLC-bearing mice	Decrease in G-MDSC accumulation by triggering its apoptosis and decreasing recruitment; promotion of CD8$^+$IFN-γ$^+$ cell expansion; impairing the suppressive capability of G-MDSC on CD8$^+$ T cells; G-MDSC differentiation into CD11c$^+$ or F4/80$^+$ cells	[58]
47	Silymarin (**20**)	*Silybum marianum*	LLC-bearing mice	Inhibition of tumor growth; apoptosis; increases in the infiltration and function of CD8$^+$ T cells; increases in IFN-γ and IL-2 levels, decrease in the IL-10 level	[59]
48	Vitamin D (**21**)	Sea fish, animal liver, etc.	COVID-19 patients	Beneficial effects come from reducing the macrophage and MDSC hyperinflammatory response	[60]
49	Polysaccharide from *Ganoderma lucidum*		LLC-bearing mice	Inhibition of tumor growth; reduction of MDSC accumulation in spleen and tumor tissue; increase in the percentage of CD4$^+$, CD8$^+$ T cells and IFN-γ and IL-12 production in the spleen; decreases in arginase activity and NO production; increase in IL-12 production in tumor tissue; regulation of the CARD9-NF-κB-IDO pathway in MDSCs	[61]
50	Curdlan produced by *Alcaligenes faecalis*		LLC-bearing mice	Promotion of MDSC differentiation; impairment of the suppressive capability of MDSCs; inhibition of tumor progression by reducing MDSCs and enhancing the CTL and Th1 responses;	[62]

Table 1. *Cont.*

No.	Natural Products	Common Source	Cell Lines or Animal Models or Patients	Function or Molecular Mechanism	Ref.
51	Ze-Qi-Tang formula	Chinese medicine	LLC; orthotopic lung cancer mouse model	G-MDSC apoptosis through the STAT3/S100A9/Bcl-2/Caspase-3 signaling pathway; elimination of MDSCs and enrichment of antitumor T cells; inhibition of tumor growth; prolongation of survival	[63]
			Targeting TAMs		
52	Pterostilbene (**22**)	*Pterocarpus santalinus*	A549, H441	Decrease in the induction of stemness by M2-TAMs; prevention of M2-TAM polarization and decrease in side-population cells; suppression of the self-renewal ability in M2-TAMs-co-cultured lung cancer cells accompanied by down-regulation of MUC1, NF-κB, CD133, β-catenin, and Sox2 expression	[64]
53	Dihydroisotanshinone I (**23**)	*Salvia miltiorrhiza*	A549, H460	Inhibition of cell motility and migration; blockage of the macrophage recruitment ability of lung cancer cells; inhibition of CCL2 secretion; blockage of p-STAT3	[65]
54	Ginsenoside Rh2 (**24**)	Ginseng	A549, H1299	Inhibition of cell proliferation and migration; decrease in the secretion and mRNA and protein levels of VEGF-C, MMP-2, and -9; decreases in VEGF-C and CD206 expression by tumor tissues	[66]
55	Sea fare hydrolysate		A549, HCC-366, RAW264.7,	Polarization of M1 macrophages in RAW264.7 cells; reduction of IL-4-induced M2 polarization in mouse peritoneal macrophages with reductions in the M2 markers Arg-1 and Ym-1; suppression of M2 macrophage polarization in human TAMs with reductions in the M2 markers CCL18, CD206, CD209, fibronectin-1, and IL-10 and increases in the M1 markers IL-1, IL-6, and TNF-α; reductions in the activity of STAT3 and p38 in TAMs; cytotoxicity and G2/M arrest	[67]
56	Yu-Ping-Feng	Chinese medicine	orthotopic lung tumor-bearing mice	Survival prolongation; increases in the CD4⁺ T Cell and M1 macrophage populations; cytotoxicity of CD4⁺ T cells; enhancement of the Th1 immunity response; STAT1 activation in M1 macrophages	[68]
57	Bu-Fei- Decoction	Chinese medicine	A549, H1975; A549 and H1975 xenograft mice	Suppression of cell proliferation, migration, and invasion in TAM conditioned medium; reduced expression of IL-10 and PD-L1; suppression of A549 and H1975 tumor growth, and PD-L1, IL-10 and CD206 protein expression in xenograft mice	[69]
			Targeting immune checkpoint		
58	Green tea extract and EGCG		A549, Lu99; NNK-induced lung tumor mice	Downregulation of IFN-γ-induced PD-L1 protein; inhibition of STAT1 and Akt phosphorylation; inhibition of the IFNR/JAK2/STAT1 and EGFR/Akt signaling pathway; reduction of PD-L1-positive cells and inhibition of tumor growth in mice	[70]
59	Berberine (**25**)	*Coptis chinensis*	A549, H157, H358, H460, H1299, H1975, LLC, Jurkat; Lewis tumor xenograft mice	Negative regulator of PD-L1; decrease in PD-L1 expression; recovery of the sensitivity of cancer cells to T-cell killing; suppression of xenograft tumor growth; tumor-infiltrating T-cell enhancement; PD-L1 destabilization by binding to and inhibition of CSN5 activity; inhibition of CSN5 activity by directly binding to CSN5 at Glu76	[71]
60	Ginsenoside Rk1 (**26**)	Ginseng	A549, PC9; A549 xenograft mice	Inhibition of cell proliferation and tumor growth; cell cycle arrest in the G1 phase; apoptosis via the NF-κB signaling pathway; downregulation of PD-L1 and NF-κB expression	[72]
61	Platycodin D (**27**)	*Platycodon grandifloras*	H1975, H358	Decrease in the PD-L1 protein level; increase in IL-2 secretion; extracellular release of PD-L1 independent of the hemolytic mechanism	[73]
62	Rediocide A (**28**)	*Trigonostemon rediocides*	A549, H1299	Blockage of cell immuno-resistance; increase in granzyme B release and IFN-γ secretion; down-regulation of CD155 expression	[74]

Table 2. The effects of natural products combined with chemotherapy drugs on modulation of the TME.

Natural Products	Combined Clinical Drugs	Cell Lines or Animal Models	Function or Molecular Mechanism	Ref.
Brucea Javanica oil	Anlotinib	H446; H460 liver-metastasis mouse model	Enhancement of anlotinib efficacy against liver metastasis from SCLC; reduction of anlotinib-induced weight loss in mice; enhancement of the anti-angiogenic effect (inhibition of tumor microvessels growth) of amlotinib	[75]
Mahonia aquifolium extract	Doxorubicin	A549	Increased cytotoxicity; cell cycle arrest in the subG1 phase; pronounced DOX retention; lower migratory ability and colony formation potential; decrease in MMP-9 expression	[76]
Fei-Liu-Ping ointment	Cyclophosphamide	A549, THP-1; LLC xenograft mice	Enhancement of tumor growth inhibition; down-regulation of the inflammatory cytokines TNF-α, IL-6 and IL-1β levels; increase in E-cadherin expression and decrease in N-cadherin and MMP-9, expression; inhibition of cell proliferation and invasion; inhibition of NF-κB activity, expression, and nuclear translocation	[77]
	Celecoxib	LL/2-luc-M38; LLC xenograft mice	Enhancement of tumor growth inhibition; inhibition of Cox-2, mPGES-1, VEGF, PDGFRβ, MMP-2, and -9 expression; down-regulation of E-cadherin expression, upregulation of N-cadherin and Vimentin expression	[78]
Resveratrol	Dasatinib, 5-fluorouridine	A549, Bm7	Inhibition of cell migration; ADAM9 degradation via the ubiquitin-proteasome pathway; synergistic anticancer effects to inhibit cell proliferation	[79]
Carnosic acid	Cisplatin	LLC-bearing mice	Enhancement of tumor growth suppression and apoptosis; reduction of side effects (body weight loss) of cisplatin; promotion of CD8$^+$ T cells-mediated antitumor immune response; function and accumulation decrease of MDSCs; downregulation of CD11b$^+$ Gr1$^+$ MDSCs, Arg-1, iNOS-2, and MMP-9 levels	[80]
Ginsenoside Rh2	Cisplatin	A549, H1299	Enhancement of cisplatin-induced cell apoptosis by repressing autophagy; scavenging of cisplatin-induced superoxide autophagy generation; inhibition of cisplatin-induced EGFR-PI3K-AKT pathway activation; inhibition of the cisplatin-induced PD-L1 expression	[81]
Water extract of ginseng	Cisplatin	A549, THP-1; LLC-bearing mice	Increase in the expression of the M1 macrophage marker iNOS, decrease in the expression of the M2 marker Arg-1; regulation of TAMs polarization; reductions in tumor growth and cisplatin-induced immunosuppression	[82]

2. Effects of Natural Products on the Tumor Microenvironment

2.1. Natural Products Involved in Angiogenesis Inhibition

Angiogenesis refers to the formation of new blood vessels from the pre-existing vasculature, and is an important hallmark in the development of malignant tumors [83]. Tumor growth and subsequent metastasis require nutrients and oxygen supplied by an elaborate network of blood vessels [84]. Endothelial cells are the main cells that are directly involved in angiogenesis. Cytokines in the TME, such as vascular endothelial growth factor (VEGF), can induce angiogenesis through different mechanisms. Angiogenesis inhibitors targeting various components of the VEGF pathway have been developed. These inhibitors were found and designed to inhibit VEGF ligand, VEGFR (vascular endothelial growth factor receptor), and downstream signal elements. VEGFR1 and VEGFR2 are VEGF receptors, of which VEGFR2 (KDR/Flk) has stronger tyrosine kinase activity [85]. VEGFR2 is the major signal transducer in angiogenesis, and its actions include the activation of ERK and JNK for DNA synthesis and the activation of PI3K-Akt for survival and proliferation. The VEGF antibody bevacizumab, the VEGFR2 antibody ramucirumab, and the tyrosine kinase inhibitor nintedanib have been approved for clinical use [5]. However, in some cases, it is difficult for these compounds to penetrate into the smallest blood vessels of tumor tissue. Due to the great difference in penetration of angiogenesis inhibitors in different tumor tissues, the drug concentration can only reach an effective concentration in some tumors. Therefore, the clinical benefit of such compounds is limited [86]. Researchers are continuously searching for natural antiangiogenic agents. It is hoped that natural products could block the formation of new blood vessels to prevent or slow down the growth or metastasis of cancer.

Jolkinolide A (**1**) and jolkinolide B (**2**), diterpenoids isolated from the roots of *Euphorbia fischeriana*, were shown to decrease the expression of VEGF and inhibit the Akt/STAT3/mTOR signaling pathway in A549 cells. The inhibitory effect of jolkinolide B is more obvious than that of jolkinolide A. The medium of A549 cells stimulated with either jolkinolide A or jolkinolide B was found to inhibit the proliferation and migration of human umbilical vein endothelial cells (HUVECs) in a concentration dependent manner [10]. Parthenolide (**3**), a sesquiterpene lactone extracted from *Tanacetum parthenium*, was shown to inhibit the proliferation of A549 cells in the absence and presence of nicotine. The activity of capsase-3, a key enzyme and indicator in apoptosis induction pathways, was found to be enhanced with parthenolide treatment. Essential protein VEGF levels of angiogenesis were also significantly reduced [11]. The LLC mouse model was established by Kim et al. to show that galbanic acid (**4**) extracted from *Ferula assafoetida* inhibits angiogenesis and tumor growth and reduces microvessel density (MVD) index CD34 expression in vivo. Galbanic acid was shown to disrupt the VEGF-induced tube formation of HUVECs. The phosphorylation of downstream signaling compounds such as p38MAPK, JNK and Akt was found to be decreased by galbanic acid treatment in VEGF-treated HUVECs in vitro [12]. Salvicine (**5**) was shown to decrease mRNA expression of basic fibroblast growth factor (bFGF), an enhancer of angiogenesis, but it was not associated with a change in VEGF expression [13]. Natural compounds targeting VEGFR2 include tanshinone IIA (**6**) [14], β-hydroxyisovalerylshikonin (**7**) [15], isogambogenic acid (**8**) [16], tubeimoside-1 (**9**) [17], decursin (**10**) and decursinol angelate (**11**) [18]. Farnesiferol C (**12**) was shown to exert antitumor activity and antiangiogenic actions by targeting VEGFR1 [19]. Takaku et al. reported that ergosterol (**13**) or its metabolites might be involved in neovascularization inhibition using an LLC model [20]. The chemical structures of antiangiogenic compounds identified in recent research are displayed in Figure 2.

Plant preparations are a promising choice for the development of more effective chemoprevention and chemotherapy strategies. Grape seed proanthocyanidins (GSPs), a mixture of flavanols/polyphenols, mainly containing proanthocyanidins (89%), can be used as dietary supplements with antioxidant and anticancer properties [21,22]. Akhtar et al. showed that GSPs inhibit the proliferation of a variety of human NSCLC cells and mouse Lewis lung carcinoma (LLC) cells in a dose- and time-dependent manner in vitro. GSPs were not found to inhibit the proliferation of BEAS-2B normal human bronchial epithelial cells. GSPs were also shown to inhibit the tumor growth of A549 and H1299 xenografts in vivo. The results of a tumor tissue immunohistochemical assay showed a reduction in VEGF protein expression with GSPs treatment. CD31 contributes to the formation of neovascularization and is therefore a biomarker of angiogenesis [23]. Immunofluorescence staining showed that the expression trend of CD31 is consistent with that of VEGF after treatment with GSPs. This further verified that GSPs could inhibit angiogenesis. Moreover, the protein level of IGFBP-3 with antiangiogenic antitumor activity in lung tumor tissues and plasma increased with GSPs treatment [24,25]. Khan et al. studied the antitumor effect of pomegranate fruit extract on growth, progression, angiogenesis, and signaling pathways in a primary lung tumor mouse model. Pomegranate fruit extract was found to effectively inhibit the incidence of lung cancer and reduce the activation of PI3K/Akt, MAPK, NF-κB, mTOR signaling, and c-met. Additionally, the expression of markers of cell proliferation or angiogenesis, including Ki67, PCNA, VEGF, iNOS, and CD31, was also reduced. Thus, pomegranate fruit extract exerts tumor growth inhibition and angiogenesis effects through multiple signaling pathways [26]. Extracts of *Astragali Mongolici* and *Rhizoma Curcumae* were found to inhibit LLC growth and angiogenesis in a xenograft mouse model through the reduction of tumor MVD; decreased expression of VEGF; and activation inhibition of p38MAPK, ERK1/2, and JNK [27]. Hypoxia can promote angiogenesis by increasing the expression and secretion of VEGF [87]. Thus, hypoxia-inducible factor-1 (HIF-1) plays a key role in tumor angiogenesis. *Scutellaria barbata* extract was reported to inhibit angiogenesis by decreasing the expression of VEGF, HIF-1α, and the phosphorylated upstream signal mediator Akt in lung tumors [28]. *Ginkgo biloba* exocarp extracts were found to inhibit angiogenesis by blocking the Wnt/β-catenin-VEGF signaling pathway in LLC. mRNA expression levels of VEGF and VEGFR2 and protein expression levels of CD34, Wnt3a, and β-catenin were all decreased [29]. Green tea was shown to inhibit angiogenesis through reductions of MVD and VEGF in A/J mice [30].

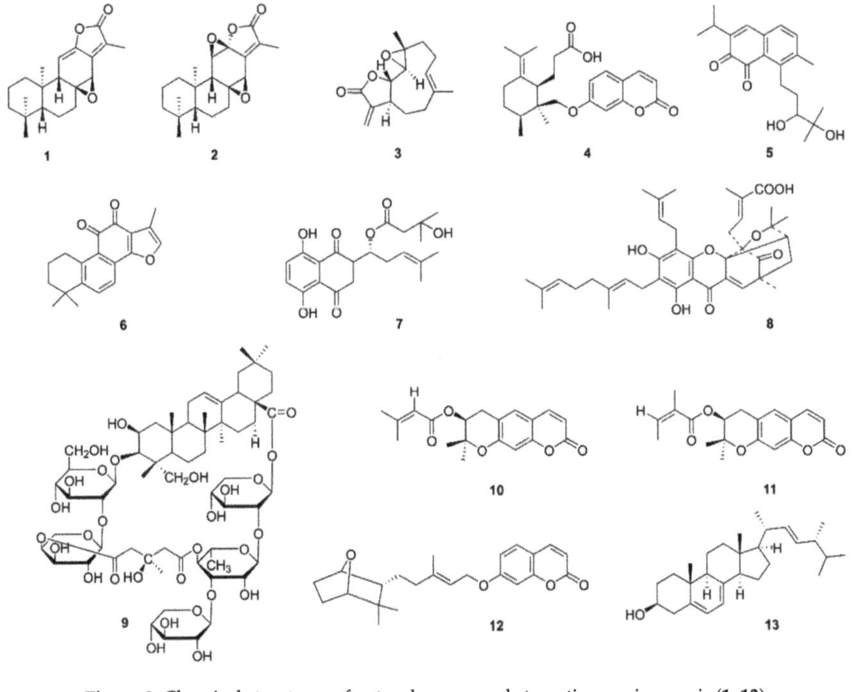

Figure 2. Chemical structures of natural compounds targeting angiogenesis (1–13).

The an-te-xiao capsule, which accounts for all alkaloids in *Solanum lyratum*, is used as an antineoplastic medicine in China. The an-te-xiao capsule was shown to prolong the survival time of Lewis tumor mice with no acute oral toxicity. In the presence or absence of VEGF, the migration, invasion, and tube formation of tumor-derived vascular endothelial cells (Td-ECs) were shown to be suppressed in A549, H460, and H520 cells when an an-te-xiao capsule was taken. Secretion of VEGF and phosphorylation of VEGFR2 were also inhibited [31]. Another Chinese medicine, the erbanxiao solution, was shown to significantly inhibit tumor angiogenesis in lung cancer patients, possibly by changing levels of VEGF, bFGF, and TNF-α [32]. Some other natural products that have shown growth inhibition or anti angiogenesis effects by regulating the expression of VEGF and related signaling molecules include the Korean herbal cocktail ka-mi-kae-kyuk-tang [33] and Chinese medicines Qingzaojiufei decoction [34] and Yiqichutan formula [35].

2.2. Natural Products Control ECM Degradation

The ECM, consisting of collagen, glycosaminoglycans, proteoglycans, and laminin, is the primary component of the TME and is found in the interstitial and epithelial vessels. On the one hand, the ECM mediates interactions between cancer cells and stromal cells, promoting carcinogenesis. On the other hand, the ECM is an important barrier against tumor metastasis in tissues [88]. Degradation of the ECM promotes cancer cells to traverse the ECM and migrate into blood vessels. Then, under the activity of some cytokines, cancer cells pass through the vessel wall and extravasate to secondary sites where they continue to proliferate and form metastatic lesions [89]. Different types of proteases can cause the degradation of the ECM, the most important of which are the matrix metalloproteinases (MMPs). MMPs, a family of zinc-dependent endopeptidases produced by fibroblasts, epithelial cells, and immune cells, degrade various subtypes of collagen as well as other elements of the ECM. The urokinase-type plasminogen activator (u-PA), a key proteolytic enzyme, is known to involve the degration of ECM and convert proMMPs to active MMPs including MMP-2, which is a member of the MMP family. Membrane type 1–matrix metalloproteinase (MT1-MMP), the expression of which is abnormal in tumors, is involved in the regulation of MMP-2 activity [90]. MMP-9, another member of the MMP family, has certain value as a biomarker of various cancers [91]. Thus, inhibiting the activity or expression of MMPs may help to suppress tumor invasion and metastasis.

In many cancers, including lung cancer, GLUT1 (glucose transporter 1) is overexpressed and regarded as a prognostic indicator. Curcumin (**14**) is a widely studied natural product with diverse activities [92]. Liao et al. reported that protein and mRNA expression of GLUT1, MT1-MMP, and MMP2 reduced in A549 cells following curcumin treatment at concentrations of 15 and 30 μM. Moreover, the anti-migration and anti-invasion effects of curcumin were shown to be damaged and MT1-MMP and MMP2 expression was up-regulated in GLUT1-overexpressed A549 cells. Consistent with the in vitro results, following curcumin treatment, the metastatic rate in nude mice that were untreated and transfected with empty vector A549 cells was shown to be about 50%, while the metastatic rate was 84% in nude mice bearing A549 cells and transfected with pcDNA3.1-GLUT1. The results showed that the overexpression of GLUT1 hinders the anti-metastasis effect of curcumin. Curcumin was shown to suppress migration and invasion by modulating the GLUT1/MT1-MMP/MMP2 pathway in A549 cells [36]. Another active compound, honokiol (**15**), derived from *Magnolia officinalis*, was found to inhibit migration and invasion by disrupting the expression of MMP-9 and Hsp90/MMP-9 interactions mediated by HDAC6 in H1299 cells. Honokiol was shown to promote ubiquitin–proteasome degradation of MMP-9 rather than inhibiting its transcription process. HDAC6 is a special deacetylase that regulates protein stability of Hsp90. Its actions are associated with the activation of MMP-2/9 through protein–protein interactions. Honokiol was shown to inhibit the expression of acetyl-α-tubulin, which is a specific substrate of HDAC6. Using a cell model, it was further proven that MMP-9 is regulated by HDAC6 [37]. In some earlier studies about natural active components, theaflavin (**16**) and theaflavin digallate (**17**), which are biologically active derivatives from black tea, were found to exert anti-metastasis effects through inhibiting type IV collagenase in LL2-Lu3 mouse LLC cells [38]. The green tea polyphenol (-)-Epigallocatechin-3-gallate (EGCG, **18**), which has a variety of activities, has been widely studied [39]. Deng et al. reported that EGCG exerts an anti-invasion effect by inhibiting mRNA and protein levels of MMP-2 via the JNK pathway in CL1-5 cells, which are highly invasive. EGCG was shown to suppress the activity of the MMP-2 promoter in a dose-dependent manner. Moreover, EGCG was found to enhance the anticancer effects of docetaxel and reduce MMP-2 expression [40]. Shi et al. reported that EGCG suppresses migration and invasion through inhibition of the epithelial-mesenchymal transition (EMT) and angiogenesis induced by nicotine [41]. The chemical structures of compounds targeting the ECM are displayed in Figure 3.

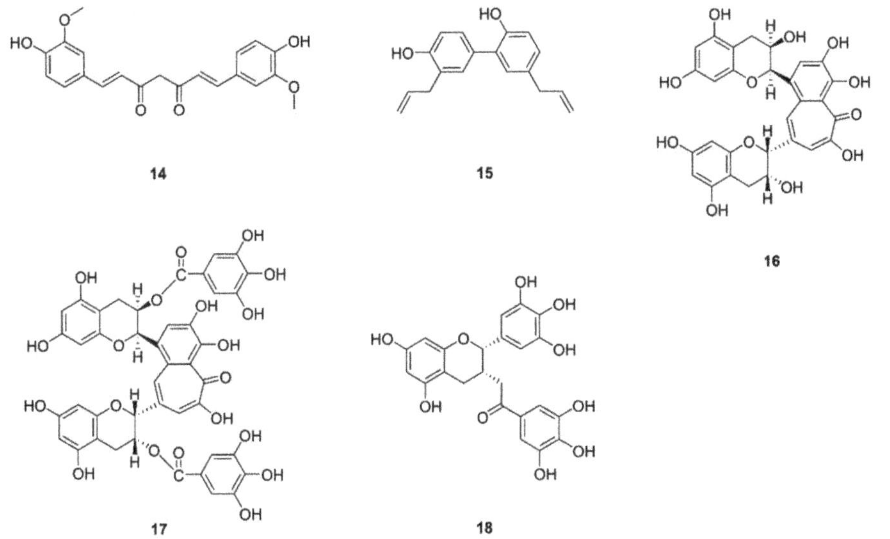

Figure 3. Chemical structures of natural compounds targeting the ECM (**14–18**).

Steroidal saponins extracted from *Paris polyphylla* (PPSS) were shown to inhibit A549 cell growth, adhesion, migration and invasion in a concentration-dependent manner. The anti-invasive mechanism underlying these processes is that PPSS reduces the protein expression and activity of MMP-2 and MMP-9 [42]. Methanolic extract of *Euchelus asper* was also shown to exert anti-proliferative activities by decreasing the expression of MMP-2 and MMP-9 in vitro [43]. Another study reported that *Phyllanthus urinaria* extracts (PUE) suppress the migration and invasion of A549 and LLC cells through reduced expression of MMP-2, MMP-9, and u-PA [44]. Rose is one of the most important ornamental plants. A previous study reported that *Rosa gallica* petal extract (RPE) inhibits the proliferation, metastasis, and invasion of A375 cells by reducing the expression and activity of MMP-2 and -9. Different from the PUE, RPE was also shown to modulate the EGFR-MAPK and mTOR-Akt signaling pathways [45]. *Viola Yedoensis* extract (VYE) was not only found to inhibit the activity of MMP-2, -9, and u-PA, it was also shown to suppress the protein expression levels of TIMP-2, TIMP-1, and PAI-1 in A549 and LLC cells. Further research showed that VYE inhibits the binding of NF-κB to DNA. Thus, the inhibition of NF-κB suppresses MMP-2 and u-PA expression and lung cancer cell invasion [46]. Focal adhesion kinase (FAK) was found to be overexpressed and activated in some late-stage cancers. Activated p-FAK has been shown to promote migration and invasion and modulate u-PA and MMPs [93]. Active ERK1/2 was also shown to promote MMP production [94].

Wu et al. reported that *Cinnamomum cassia*, a traditional food and medicinal plant, exhibits anti-metastasis ability through reduced phosphorylation of FAK and ERK1/2 as well as downregulating MMP-2 and u-PA in A549 and H1299 cells [47]. Chen et al. reported that *Duchesnea indica* extracts (DIE) inhibit the expression of p-ERK1/2 and p-FAK in A549 and H1299 cells, subsequently reducing the expression of u-PA and MMP-2 mediated by p-ERK1/2 and the expression of p-paxillin, vimentin, fibronectin, and N-cadherin. Additionally, the expression of the epithelial marker E-cadherin was found to increase. These changes in signal molecules by DIE were found to inhibit cell adhesion, migration, invasion and the epithelial–mesenchymal transition (EMT). In an A549-bearing nude mouse xenograft, tumor growth was shown to be efficiently retarded by DIE treatment compared with a control group. A higher level of E-cadherin and lower levels of MMP-2 and N-cadherin were examined in tumor tissues in DIE-treated mice [48]. A number of studies have shown that various botanical agents, such as fructus phyllanthi tannin fraction and butanol fraction extract of *Psidium cattleianum* leaf, influence the ECM by downregulating the expression and activity of MMP-2 and -9 as well as the activation of ERK1/2. Fructus phyllanthi, the dried ripened fruit of *Phyllanthus emblica*, which has been used for thousands of years in the Tibetan area, was shown to modulate the MAPK pathway by dose-dependently upregulating the expression of p-JNK in H1703 cells [49]. The butanol fraction extract of *Psidium cattleianum* leaves was shown to suppress the urokinase plasminogen activator receptor (uPAR) and MAPK signaling pathway [50]. *Terminalia catappa* leaf extract was found to inhibit the activity of MMP-2, -9 and u-PA and up-regulate the expression of the proteins TIMP-2 and PAI-1 [51]. The main ingredients in *Paris polyphylla* are steroid saponins known as *Rhizoma Paridis* saponins (RPS). An experiment in which T739 mice were injected subcutaneously with LA795 mouse lung adenocarcinoma cells showed that after RPS treatment, mRNA levels of MMP-2 and -9 were reduced and TIMP-2 was upregulated in tumor tissues [52]. *Selaginella tamariscina* extracts were shown to not only downregulate the activity of MMP2/9 and u-PA but also decrease the protein levels of TIMP and PAI in A549 and LLC cells [53]. There is evidence showing that metastasis can also be inhibited by regulating antioxidant enzymes [95]. *Ocimum sanctum* is generally known as "Holy basil" and is used in Ayurvedic medicine in India [96]. Ethanol extract of *Ocimum sanctum* (EEOS) has been shown to play roles in adhesion and invasion in LLC cells by inhibiting MMP-9 rather than MMP-2 activation. Meanwhile, antioxidant enzyme activity, including the activity of including SOD, CAT, and GSH-Px, was found to decrease in lung cancer tissues in LLC bearing mice treated with EEOS. Additionally, the ratio of GSH/GSSG was shown to decrease [54]. A unique experiment using an ex vivo approach demonstrated the suppression of metastasis and investigated the mechanisms underlying the actions of serum metabolites from rhubarb, the dried root and rhizome of *Rheum palmatum*. First, serum metabolites were extracted from rats that had been administered rhubarb by gavage. Then, A549 cells were cocultured with rhubarb serum metabolites. The results of a wound healing assay, zymography, RT-PCR, and Western blot analysis showed that rhubarb serum metabolites suppress the activity and expression of MMP-2 and u-PA. Protein expression levels of phosphorylated NF-κB and c-Jun were reduced. Many studies have shown that transcription factors such as NF-κB, c-Jun, and AP-1 are involved in the transcriptional regulation of MMP-2 and -9 [97–100]. It has been indicated that some active components of rhubarb serum metabolism inhibit the activity of MMP-2 by inhibiting the u-PA and NF-κB/c-Jun pathway. These components were shown to block motility and inhibit the metastasis of A549 cells in vitro. Using a lung metastatic mouse model, the number of metastatic nodules in the lungs of rhubarb-treated mice was shown to be reduced compared with a control group [55].

Fuzheng Kang-Ai decoction (FZKA), a classic Chinese herbal medicine, is used to treat cancers. Li et al. reported that FZKA inhibits the metastasis of H1650, A549, and PC-9 cells through inhibition of the STAT3/MMP-9 pathway and EMT. After FZKA treatment, the activity and expression of MMP-9 were shown to be reduced. Additionally, activation of signal transducer and activator of transcription 3 (STAT3) was inhibited. However, the overexpression of STAT3 was demonstrated to rescue the activity of MMP-9. On the other hand, the expression of the mesenchymal markers N-cadherin and vimentin was found to reduce following FZKA treatment [56]. Another Chinese herbal formula, Yifei Tongluo (YFTL), was found to inhibit tumor growth, metastasis, and angiogenesis; prolong survival; and improve immunity through multiple signaling pathways in Lewis lung cancer bearing mice. The expression of the major angiogenesis-associated protein VEGF was found to be inhibited in both tumor tissues and serum. YFTL was shown to induce significant reductions in MMP-2, MMP-9, N-cadherin, and vimentin expression levels as well as increasing E-cadherin expression. CD4$^+$ and CD8$^+$ T lymphocytes are the major components of T cell-mediated anti-tumor immunity [101]. Natural killer (NK) cells also play a role in antitumor immunity [102]. CD4$^+$, CD8$^+$, and NK cells were found to increase following treatment with YFTL. The cytokine levels of IL-2, IFN-γ, IL-10, and TGF-β1, components that promote antitumor immunity through proinflammatory actions, were found to increase in the serum of tumor-bearing mice following treatment with YFTL. In addition, experimental results showed that YFTL inhibited the ERK1/2, TGF1/Smad2, and PI3K/Akt, pathways and upregulated the p38 and JNK pathways. Taken together, these YFTL-regulated factors have been shown to suppress tumor growth and metastasis in Lewis-tumor-bearing mice [57].

2.3. Natural Products Reduce the Accumulation of MDSCs

MDSCs, as regulators of the immune system, are a heterogeneous population of immature myeloid progenitors and precursors of dendritic cells, granulocytes, and macrophages [103,104]. Uncontrolled expansion of MDSCs disrupts the normal homeostasis of the immune system and eventually lead to tumor progression and the development of other diseases. In cancer patients, an increase in MDSCs has been shown to induce intense immunosuppressive activity through inhibiting the functions of NK cells and cytotoxic T lymphocytes (CTLs) [105,106]. MDSCs are classified into two subsets: monocytic-MDSC (M-MDSC) and granulocytic-MDSC (G-MDSC). Anti-MDSC treatments, such as abrogating suppressive activity or reducing the number of MDSCs in the tumor microenvironment, have been successfully used as anticancer therapy [107].

Resveratrol (**19**) is a polyphenol derived from grape skin and seeds that provides an abundance of health benefits [108]. Zhao et al. reported that resveratrol reduces G-MCSD accumulation by triggering its apoptosis and promotes CD8$^+$IFN-γ^+ cell expansion in Lewis lung carcinoma bearing mice. Resveratrol was also shown to impair the activity of CD8$^+$ T cells, which are suppressed by MDSCs. Moreover, resveratrol was shown to enhance the differentiation of MDSCs separated from mice into CD11c$^+$ (pro-inflammatory macrophage- or dendritic cell-like) and F4/80$^+$ (macrophage marker) cells. Taken together, these results show that resveratrol inhibits the development of tumors in mice with Lewis lung carcinoma [58]. In another study of the same mouse model, Wu et al. provided evidence that silymarin (**20**) reduces the proportion of MDSCs in tumor tissues. Additionally, the mRNA expression levels of iNOS2, Arg-1 and MMP-9 were found to be reduced in tumor tissues, indicating that the function of MDSCs was suppressed. Mature T lymphocytes are mainly divided into CD4$^+$ and CD8$^+$ cells. CD8$^+$ T cells are activated by CD4$^+$ cells and migrate to the tumor site, exerting a cytotoxic effect. The study showed that silymarin increased the expression of CD8+ T cells in the tumor tissues compared to the control group [59]. Vitamin D (**21**) is a fat-soluble vitamin that regulates calcium, phosphate, and magnesium homeostasis, and influences elements of human health, including the immune response and tumorigenesis [109]. A study on COVID-19 patients showed that the absence of vitamin D increased the severity of the acute respiratory distress syndrome (ARDS) induced by cytokine storm. Vitamin D supplementation could affect the inflammatory responses of macrophages and MDSCs, inhibiting a strong inflammatory response and reducing the ARDS of COVID-19 patients [60].

Polysaccharides are the main components of herbs. Polysaccharides extracted from herbal medicine have important medicinal value with their actions including immunity improvement and anti-tumor effects [110]. A homogeneous polysaccharide from *Ganoderma lucidum* (GLP) was reported to induce differentiation and reduce the accumulation of MDSCs through the CARD9-NF-κB-IDO signaling pathway, preventing lung cancer development in an LLC mouse model [61,111]. Polysaccharides come not only from plants but also from bacteria. Curdlan, composed of linear b-(1,3)-glycosidic linkages, is a bacterial polysaccharide produced by *Alcaligenes faecalis*. Rui et al. reported that curdlan can promote the differentiation of MDSCs into a more mature state. A reduction of MDSCs was found to downregulate immunosuppression in LLC-bearing mice, thus having an antitumor effect [62]. Besides polysaccharides, Ishiguro et al. reported that water extract from *Euglena gracilis* induced G MDSC apoptosis and the differentiation of M-MDSCs into macrophages. Thus, the water extract stimulated host antitumor immunity and inhibited lung tumor growth [112]. Chemical structures of compounds targeting MDSCs are displayed in Figure 4.

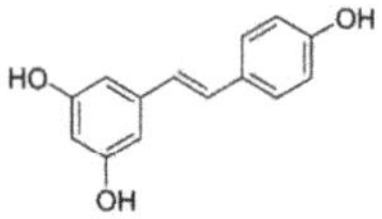

Figure 4. Chemical structures of natural compounds targeting MDSCs (**19–21**).

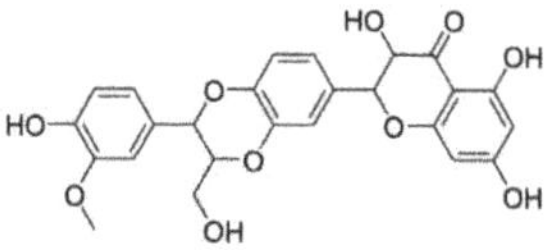

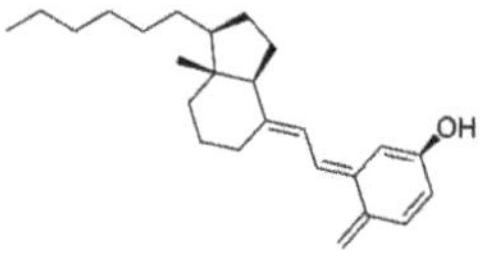

The Ze-Qi-Tang formula (ZQT) has been used traditionally to treat respiratory diseases. Xu et al. firstly illustrated the immunomodulatory effect of ZQT in NSCLC. ZQT was found to induce G-MDSC apoptosis through the STAT3/S100A9/Bcl-2/caspase-3 signaling pathway, and it significantly reduced the number of MDSCs. ZQT was shown to remodel the immune tumor microenvironment by eliminating MDSCs and enriching T cells, prolonging survival and inhibiting tumor growth in a orthotopic mouse model of lung cancer [63].

2.4. Natural Products Regulate TAMs

Macrophages acquire diverse phenotypes in response to different stimuli generated by activated stromal cells or cancer cells in the microenvironment [113]. M1 macrophages promote antitumor responses, while M2 macrophages drive tumor progression. In the tumor microenvironment, TAMs are generally M2-polarized [114]. TAMs are one of the major cell populations within the stroma of various cancers. It was reported that having a high TAMs number is closely related to the presence of advanced cancer [115]. TAMs stimulate cell proliferation and promote angiogenesis and tumor metastasis by secreting various cytokines. For example, the proliferation of tumor cells is promoted by growth factors such as EGF, PDGF, HGF, and bFGF, which are secreted by macrophages. TAMs promote angiogenesis through the release of pro-angiogenic factors such as VEGF and PIGF. Tumor invasion is promoted by substrates and tumor aggregation factors, such as EGF. Immunosuppression is achieved by immunosuppressive factors, such as IL-10 and TGF-β [116]. Moreover, TAMs produce a series of proteases including u-PA and MMPs to degrade ECM and promote cancer cell invasion and migration [117]. It appears that a reduction in TAM recruitment and the conversion of M2 macrophages to M1 phenotype are effective cancer treatment strategies.

MUC1, a kind of pro-oncogenic mucin, is overexpressed in different cancer types and is an indicator of a poorer prognosis [118]. Cancer stem cells (CSCs), key drivers of tumor progression, have the same properties as normal stem cells, such as their self-renewal ability and the potential to transition epithelial into mesenchymal cells [119]. Huang et al. examined the role of MUC1 in TAMs and its connection with the generation of lung CSCs. Significantly increased MUC1 transcription was identified in the lung tissues of lung adenocarcinoma patients compared to those with normal lung tissues. In an experiment involving the coculture of CSCs and M2-TAMs, MUC1 and cancer stemness genes were shown to significantly increase. Pterostilbene (**22**), a polyphenol isolated from the heartwood of red sandalwood (*Pterocarpus santalinus*), is an analog of resveratrol. Huang et al. provided evidence that pterostilbene suppresses M2-polarization through MUC1 inhibition and reduces M2-TAM-induced CSC generation [64]. A study found that dihydroisotanshinone I (DT, **23**), an active ingredient of *Salvia miltiorrhiza*, improves the survival rate of advanced lung cancer patients. This research also indicated that DT has the capability to suppress the migration of A549 and H460 cells, cells cultured with the macrophage medium, and lung cancer/macrophage coculture. DT was shown to inhibit the macrophage recruitment of lung cancer cells by decreasing the expression of chemokine (C-C motif) ligand 2 (CCL2) secreted from both lung cancer cells and macrophages. CCL2 has been recognized as the strongest chemoattractant involved in macrophage recruitment and is a powerful initiator of inflammation [120]. Notably, the CCL2 signaling pathway is a prominent mechanism through which TAMs promote the growth and metastasis of lung cancer cells through bidirectional interactions between lung cancer cells and macrophages [121]. DT was shown to interrupt the crosstalk between macrophages and lung cancer cells [65]. It may be an attractive strategy for transforming TAMs from surface M2 to M1 in the tumor microenvironment [122]. Ginsenoside Rh2 (**24**), an active component of ginseng, has been proven to have the potential to convert M2 macrophages into the M1 subset. A549 and H1299 cells were induced to secrete and express high levels of VEGF-C, MMP-2, and MMP-9 following coculture with M2 macrophages derived from RAW264.7 cells in vitro. In contrast, treatment with ginsenoside Rh2 decreased the secretion and expression of VEGF-C, MMP-2 and MMP-9 and inhibited the proliferation and migration of lung cancer cells. A flow cytometry assay showed a decrease in the M2 phenotype marker CD206 but an increase in the M1 macrophage marker CD16/32 following ginsenoside Rh2 treatment. Furthermore, in C57BL/6 mice subcutaneously injected LLC, ginsenoside Rh2 treatment reduced the expression of the VEGF-C and M2 macrophage marker CD206 [66]. Chemical structures of compounds targeting TAMs are displayed in Figure 5.

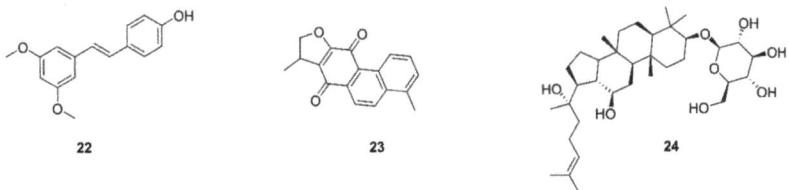

Figure 5. Chemical structures of natural compounds targeting TAMs (**22**–**24**).

The sea hare is a marine organism with various active secondary metabolites [123]. Sea fare hydrolysate was shown to induce the polarization of M1 macrophages and decrease M2 polarization in both RAW264.7 cells and mouse peritoneal macrophages. Sea fare hydrolysate was found to upregulate M1 markers (IL-1β, IL-6, and TNF-α) and downregulate M2 markers (CD206, CD209 and FN-1) in human macrophages and TAMs. In addition, sea fare hydrolysate was shown to inhibit A549 cell growth when cocultured with either M1 cells or M2 cells. Different from most natural products, sea fare hydrolysate was shown to induce M1 and M2 polarization in macrophages, not only in one direction. It might be an effective cancer therapy [67].

Two traditional Chinese medicines were used to treat lung disease in ancient China. Yu-Ping-Feng (YPF), consisting of *Astragalus membranaceus*, *Atractylodes macrocephala*, and *Saposhnikovia divaricate*, has been shown to prolong the survival of orthotopic LLC mice. The percentages of M1 macrophages and CD4$^+$ T cells in spleen and tumor tissues were found to increase following YPF administration. YPF was also shown to enhance the cytotoxicity of CD4$^+$ T cells and macrophage-mediated LLC lysis [68]. Another Chinese medicine formula, Bu-Fei-Decoction (BFD) was shown to dose-dependently inhibit the ability of

A549 and H1975 cells, which were increased by exposure to a conditioned medium from TAMs, to undergo proliferation, migration, and invasion. PD-L1 expression was found to be promoted by IL-10 secreted from TAMs. BFD was shown to decrease the expression of CD206, PD-L1, and IL-10 in lung cancer cells. Thus, BFD has been shown to block crosstalk between TAMs and cancer cells by inhibiting the expression of PD-L1 and IL-10 in vivo and in vitro [69].

2.5. Natural Products as Important Immune Checkpoint Inhibitors

The immune checkpoint pathway is a series of cell–cell interactions that function to inhibit the hyperactive effector T cells under normal conditions and prevent attacks on normal cells. However, cancer cells can escape immune destruction by dysregulating the immune response [124]. The most frequently studied immune checkpoint molecules related to lung cancer are PD-1 (programmed cell death protein 1) and CTLA4 [125]. Immune checkpoint inhibitors are monoclonal antibodies developed for corresponding immune checkpoints. Their main function is to block interactions between tumor cells expressing immune checkpoints and immune cells to block the inhibitory effect of tumor cells on immune cells [126,127]. The development of checkpoint-blockade-based immunotherapies has provided more options and attractive weapons in the battle against cancer [128]. However, the use of immunotherapeutic drugs for the treatment of lung cancer can lead to immune-mediated toxicity conditions such as pneumonia, endocrine disease, nephritis, colitis, and pulmonary toxicity [2,129]. Most natural products have low toxicity and high effectiveness. Natural small active molecules have better permeability than monoclonal antibody-based drugs. Due to their multi-component and multi-target characteristics, plant preparations can be combined with monoclonal antibody targeted drugs to achieve an increased efficiency and reduced toxicity.

Many scholars have reported that various natural products have inhibitory effects on programmed cell death ligand 1 (PD-L1) within the TME. Rawangkan et al. reported that green tea extract reduces the percentage of PD-L1 positive cells in lung tumor tissues and the average number of tumors per mouse treated with 4-(methylnitrosamino)-1-(3-pyridyl)-1-butanone, a tobacco-specific carcinogen. In an in vitro experiment, EGCG and green tea extract were shown to downregulate IFN-γ-induced PD-L1 protein expression through JAK2/STAT1 and EGFR/Akt pathways in A549 cells. They were also shown to inhibit EGF-induced PD-L1 expression in Lu99 cells. In a coculture experiment, EGCG was shown to reduce the mRNA expression of PD-L1 in F10-OVA cells and partially restore the mRNA expression of IL-2 in tumor specific T cells [70]. IL-2 is considered as a key molecule in the promotion of T cell proliferation and differentiation, so it has also been called T cell growth factor for decades [130]. Unexpected and interestingly, a recent study showed that IL-2 regulates tumor-reactive $CD8^+$ T cell exhaustion in the middle and late tumor stages [131]. Another widely studied active natural compound is berberine (**25**), which is selectively bound to glutamate and inhibits the PD-1/PD-L1 axis through its deubiquitination effect, leading to the ubiquitination and degradation of PD-L1 [71]. Existing evidence shows that the activation of NF-κB promotes the proliferation of regulatory T cells (Treg), leading to the transcription of PD-L1 by binding to its promoter [132]. Ginsenoside Rk1 (**26**), a bioactive ingredient of ginseng, has been shown to downregulate the protein expression of PD-L1 by targeting the NF-κB signaling pathway in A549 and PC9 cells as well as in an A549-xenograft nude mouse model. In addition, ginsenoside Rk1 was shown to induce apoptosis and cell cycle arrest in lung cancer cells [72]. Platycodin D (**27**), a triterpenoid saponin isolated from the southeast Asian functional food *Platycodon grandifloras*, was reported to trigger the extracellular release of PD-L1, leading to a reduction in the inhibition of immunity [73]. However, exosomes carrying PD-L1 in the tumor microenvironment secreted by tumor cells were shown to transfer to distant places where they exerted immunosuppression effects [133]. The targeted inhibition of PD-L1 is particularly important.

A newly discovered human cancer immune checkpoint, CD155, is a cell surface adhesion molecule. The T cell immunoreceptor with Ig and ITIM domains (TIGIT) is an inhibitory receptor that is mainly expressed on NK, Treg, CD4$^+$, and CD8+ T cells. The presence of CD155 on the tumor surface combined with TIGIT was shown to inhibit the function of NK and other immune cells [134]. A small molecule, rediocide A (**28**), isolated from *Trigonostemon rediocides*, was shown to reduce the expression of CD155 in A549 and H1299 cells by 11% and 14%, respectively, thus blocking the tumor immune resistance to NK cells [74]. Chemical structures of compounds that exert immune checkpoint inhibitor effects are displayed in Figure 6.

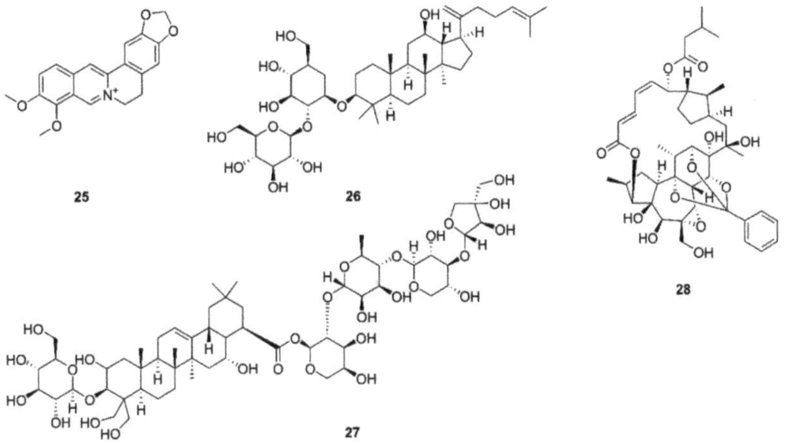

Figure 6. Chemical structures of natural compounds targeting immune checkpoints (**25–28**).

3. Combination of Natural Products and Anticancer Drugs

Tumor drug resistance is an obstacle to tumor treatment that may lead to tumor recurrence or treatment failure. More and more evidence is supporting the idea that the combination of natural products and anticancer drugs can have better therapeutic benefits. Natural products have increased efficiency, reduced toxicity, and can induce improved immunity by regulating various signal pathways.

Compared with anlotinib monotherapy, the combination of anlotinib and the traditional Chinese medicine *Brucea Javanica* oil was shown to inhibit the growth and angiogenesis of SCLC liver metastases more significantly. *Brucea Javanica* oil also reduced weight loss in model and normal mice following anotinib treatment [75]. *Mahonia aquifolium* extract was shown to promote the antitumor effects of doxorubicin. The extract was shown to prolong the action time of doxorubicin in A549 cells. The combined application of doxorubicin/extract was shown to decrease MMP-9 expression. A549 cells treated with the extract and doxorubicin combination displayed lower colony and migratory formation potential than untreated cells or cells only treated with doxorubicin. The application of this combination was found to reduce the dosage of doxorubicin required, thereby reducing the toxicity to normal tissues [76]. The combination of Fei-Liu-Ping ointment and cyclophosphamide was shown to suppress lung cancer growth and invasion by inhibiting the tumor inflammatory environment. This suggests that Fei-Liu-Ping ointment can be used alone or in combination with the routine treatment of inflammation-related pneumonia [77]. Liu et al. also proved that the combination of Fei-Liu-Ping ointment with celecoxib inhibits the tumor inflammatory microenvironment in an LLC xenograft model [78]. Disintegrin and a metalloproteinase (ADAM9), a type I transmembrane protein, are overexpressed in various cancers, including lung cancer [135–137]. Lin et al. proved that a secreted form of ADAM9 promotes cancer invasion through tumor-stromal interactions [137]. Subsequently, they indicated that resveratrol inhibits the protein expression of ADAM9 in A549 and Bm7 cells through the ubiquitin–proteasome pathway. A synergistic anticancer effect was shown when resveratrol was used in combination with dasatinib or 5-fluorouridine [79]. Studies have shown that the application of carnosic acid, ginsenoside Rh2, and water extract of ginseng enhances the antitumor effects of cisplatin by decreasing PD-L1 expression, inhibiting MDSCs, or regulating macrophage polarization [80–82].

4. Combination of Natural Products with Nanotechnology or Other Materials for Targeting the TME

Technological advances have led to the development of innovative drug delivery systems [138,139]. Various natural and synthetic materials have been used as potential biomaterial carriers of therapeutic agents in cancer therapy. Tumor treatment requires the localization of active substances in tumor cells. The combination of drugs and materials aids in achieving better localization of active substances and minimizes the impact of active substances on normal cells or maximizes the impact on tumor cells. Some materials themselves have anti-tumor effects, and their combination with drugs enhances the effects of the drugs. Nano-, micro-, and macroscale drug delivery systems are used to improve the bioavailability of drugs [140]. In addition, the combination of polysaccharides with more polymers to improve their required functional properties, such as encapsulation, stability, and release of drugs, is a common practice. Pectin is a natural excellent macromolecule polymer with biocompatible and biodegradable properties that is used for targeted drug delivery [141,142]. Moreover, some experiments have shown that pectin has the ability to inhibit tumor growth in cancers [143,144]. Poly (vinyl pyrrolidone) (PVP) is used in the development of biomedical applications as a complexing agent or cross-linker with excellent biocompatibility and solubility characteristics. Gaikwad et al. prepared pectin-PVP based curcumin particulates of different ratios and evaluated their localized delivery to lung cancer tumors. The results showed the optimal ratio of particles and indicated that it could be used for inhalation in lung cancer treatment. Spray-dried pectin-PVP curcumin was shown to enhance curcumin solubility. It was also shown to inhibit cancer cell proliferation and angiogenesis more than curcumin treatment alone [145]. Singh et al. reported that nanoparticles encapsulating polyphenols, EGCG and theaflavin and combined with cisplatin exhibited more biological effectiveness and stronger inhibition of cell proliferation, metastasis, and angiogenesis biomarkers than EGCG/theaflavin alone.

5. Conclusions

Natural products are important sources of new drugs. In this review, we focused on natural products that have been reported to have anticancer activity targeting the TME in lung cancer. Our findings are of great significance for the development of new plant-derived chemotherapy agents for the treatment of lung cancer. Most studies in this area have been related to angiogenesis, MDSCs, and TAMs, and research on some other aspects is lacking. The use of appropriate phytochemicals, medicinal plants, or other natural substances in combination with immune checkpoint inhibitors for lung cancer treatment may be a better choice than using monotherapies. However, there are few reports on combined use in the literature. Moreover, the activity of some natural products is not very high due to problems with their stability and bioavailability. This can be solved by structural modifications or by combining these compounds with material technologies such as nanotechnologies. Moreover, nanoparticle-mediated delivery of natural compounds may limit the unwanted toxicity of chemotherapeutic agents. Unfortunately, few studies have been done on the effects of natural products in combination with other materials on the TME in lung cancer. In addition, the determination of the components of botanical agents and traditional Chinese medicine extracts has been a problem requiring resolution for a long time. Intensified technology is needed to identify natural products and active derivatives and to research potential antitumor effects.

Author Contributions: Investigation, writing—original draft preparation, Y.Y.; methodology, funding acquisition, N.L.; writing—review and editing T.-M.W.; validation, formal analysis, supervision, L.D. All authors have read and agreed to the published version of the manuscript.

Funding: This research was funded by the Key Research and Development Projects of Anhui Province, grant number 202004a07020035.

Institutional Review Board Statement: Not applicable.

Informed Consent Statement: Not applicable.

Conflicts of Interest: The authors declare no conflict of interest.

References

1. Sung, H.; Ferlay, J.; Siegel, R.L.; Laversanne, M.; Soerjomataram, I.; Jemal, A.; Bray, F. Global Cancer Statistics 2020: GLOBOCAN Estimates of Incidence and Mortality Worldwide for 36 Cancers in 185 Countries. *CA Cancer J. Clin.* **2021**, *71*, 209–249. [CrossRef]
2. Miller, K.D.; Nogueira, L.; Mariotto, A.B.; Rowland, J.H.; Yabroff, K.R.; Alfano, C.M.; Jemal, A.; Kramer, J.L.; Siegel, R.L. Cancer treatment and survivorship statistics, 2019. *CA Cancer J. Clin.* **2019**, *69*, 363–385. [CrossRef]
3. Bray, F.; Ferlay, J.; Soerjomataram, I.; Siegel, R.L.; Torre, L.A.; Jemal, A. Global cancer statistics 2018: GLOBOCAN estimates of incidence and mortality worldwide for 36 cancers in 185 countries. *CA Cancer J. Clin.* **2018**, *68*, 394–424. [CrossRef]
4. Mittal, V.; El Rayes, T.; Narula, N.; McGraw, T.E.; Altorki, N.K.; Barcellos-Hoff, M.H. The Microenvironment of Lung Cancer and Therapeutic Implications. *Adv. Exp. Med. Biol.* **2016**, *890*, 75–110. [PubMed]
5. Altorki, N.K.; Markowitz, G.J.; Gao, D.; Port, J.L.; Saxena, A.; Stiles, B.; McGraw, T.; Mittal, V. The lung microenvironment: An important regulator of tumour growth and metastasis. *Nat. Rev. Cancer* **2019**, *19*, 9–31. [CrossRef] [PubMed]
6. Hanahan, D.; Coussens, L.M. Accessories to the crime: Functions of cells recruited to the tumor microenvironment. *Cancer Cell* **2012**, *21*, 309–322. [CrossRef] [PubMed]
7. Chen, Z.; Fillmore, C.M.; Hammerman, P.S.; Kim, C.F.; Wong, K.K. Non-small-cell lung cancers: A heterogeneous set of diseases. *Nat. Rev. Cancer* **2014**, *14*, 535–546. [CrossRef]
8. Quail, D.F.; Joyce, J.A. Microenvironmental regulation of tumor progression and metastasis. *Nat. Med.* **2013**, *19*, 1423–1437. [CrossRef]
9. Newman, D.J.; Cragg, G.M. Natural Products as Sources of New Drugs over the Nearly Four Decades from 01/1981 to 09/2019. *J. Nat. Prod.* **2020**, *83*, 770–803. [CrossRef]
10. Shen, L.; Zhang, S.Q.; Liu, L.; Sun, Y.; Wu, Y.X.; Xie, L.P.; Liu, J.C. Jolkinolide A and Jolkinolide B Inhibit Proliferation of A549 Cells and Activity of Human Umbilical Vein Endothelial Cells. *Med. Sci. Monit.* **2017**, *23*, 223–237. [CrossRef]
11. Talib, W.H.; Al Kury, L.T. Parthenolide inhibits tumor-promoting effects of nicotine in lung cancer by inducing P53-dependent apoptosis and inhibiting VEGF expression. *Biomed. Pharmacother.* **2018**, *107*, 1488–1495. [CrossRef]
12. Kim, K.H.; Lee, H.J.; Jeong, S.J.; Lee, H.J.; Lee, E.O.; Kim, H.S.; Zhang, Y.; Ryu, S.Y.; Lee, M.H.; Lu, J.; et al. Galbanic acid isolated from Ferula assafoetida exerts in vivo anti-tumor activity in association with anti-angiogenesis and anti-proliferation. *Pharm. Res.* **2011**, *28*, 597–609. [CrossRef]
13. Zhang, Y.; Wang, L.; Chen, Y.; Qing, C. Anti-angiogenic activity of salvicine. *Pharm. Biol.* **2013**, *51*, 1061–1065. [CrossRef]
14. Xie, J.; Liu, J.; Liu, H.; Liang, S.; Lin, M.; Gu, Y.; Liu, T.; Wang, D.; Ge, H.; Mo, S.L. The antitumor effect of tanshinone IIA on anti-proliferation and decreasing VEGF/VEGFR2 expression on the human non-small cell lung cancer A549 cell line. *Acta Pharm. Sin. B* **2015**, *5*, 554–563. [CrossRef] [PubMed]
15. Komi, Y.; Suzuki, Y.; Shimamura, M.; Kajimoto, S.; Nakajo, S.; Masuda, M.; Shibuya, M.; Itabe, H.; Shimokado, K.; Oettgen, P.; et al. Mechanism of inhibition of tumor angiogenesis by beta-hydroxyisovalerylshikonin. *Cancer Sci.* **2009**, *100*, 269–277. [CrossRef] [PubMed]
16. Fan, Y.; Peng, A.; He, S.; Shao, X.; Nie, C.; Chen, L. Isogambogenic acid inhibits tumour angiogenesis by suppressing Rho GTPases and vascular endothelial growth factor receptor 2 signalling pathway. *J. Chemother* **2013**, *25*, 298–308. [CrossRef] [PubMed]
17. Gu, Y.; Korbel, C.; Scheuer, C.; Nenicu, A.; Menger, M.D.; Laschke, M.W. Tubeimoside-1 suppresses tumor angiogenesis by stimulation of proteasomal VEGFR2 and Tie2 degradation in a non-small cell lung cancer xenograft model. *Oncotarget* **2016**, *7*, 5258–5272. [CrossRef]
18. Jung, M.H.; Lee, S.H.; Ahn, E.M.; Lee, Y.M. Decursin and decursinol angelate inhibit VEGF-induced angiogenesis via suppression of the VEGFR-2-signaling pathway. *Carcinogenesis* **2009**, *30*, 655–661. [CrossRef]
19. Lee, J.H.; Choi, S.; Lee, Y.; Lee, H.J.; Kim, K.H.; Ahn, K.S.; Bae, H.; Lee, H.J.; Lee, E.O.; Ahn, K.S.; et al. Herbal compound farnesiferol C exerts antiangiogenic and antitumor activity and targets multiple aspects of VEGFR1 (Flt1) or VEGFR2 (Flk1) signaling cascades. *Mol. Cancer Ther.* **2010**, *9*, 389–399. [CrossRef]
20. Takaku, T.; Kimura, Y.; Okuda, H. Isolation of an antitumor compound from Agaricus blazei Murill and its mechanism of action. *J. Nutr.* **2001**, *131*, 1409–1413. [CrossRef]
21. Nandakumar, V.; Singh, T.; Katiyar, S.K. Multi-targeted prevention and therapy of cancer by proanthocyanidins. *Cancer Lett.* **2008**, *269*, 378–387. [CrossRef]

22. Sharma, S.D.; Meeran, S.M.; Katiyar, S.K. Dietary grape seed proanthocyanidins inhibit UVB-induced oxidative stress and activation of mitogen-activated protein kinases and nuclear factor-kappaB signaling in in vivo SKH-1 hairless mice. *Mol. Cancer Ther.* **2007**, *6*, 995–1005. [CrossRef]
23. Risau, W. Differentiation of endothelium. *FASEB J.* **1995**, *9*, 926–933. [CrossRef] [PubMed]
24. Kim, J.H.; Choi, D.S.; Lee, O.H.; Oh, S.H.; Lippman, S.M.; Lee, H.Y. Antiangiogenic antitumor activities of IGFBP-3 are mediated by IGF-independent suppression of Erk1/2 activation and Egr-1-mediated transcriptional events. *Blood* **2011**, *118*, 2622–2631. [CrossRef] [PubMed]
25. Akhtar, S.; Meeran, S.M.; Katiyar, N.; Katiyar, S.K. Grape seed proanthocyanidins inhibit the growth of human non-small cell lung cancer xenografts by targeting insulin-like growth factor binding protein-3, tumor cell proliferation, and angiogenic factors. *Clin. Cancer Res.* **2009**, *15*, 821–831. [CrossRef] [PubMed]
26. Khan, N.; Afaq, F.; Kweon, M.H.; Kim, K.; Mukhtar, H. Oral consumption of pomegranate fruit extract inhibits growth and progression of primary lung tumors in mice. *Cancer Res.* **2007**, *67*, 3475–3482. [CrossRef]
27. Xu, C.; Wang, Y.; Feng, J.; Xu, R.; Dou, Y. Extracts from Huangqi (Radix Astragali Mongoliciplus) and Ezhu (Rhizoma Curcumae Phaeocaulis) inhibit Lewis lung carcinoma cell growth in a xenograft mouse model by impairing mitogen-activated protein kinase signaling, vascular endothelial growth factor production, and angiogenesis. *J. Tradit. Chin. Med.* **2019**, *39*, 559–565. [PubMed]
28. Shiau, A.L.; Shen, Y.T.; Hsieh, J.L.; Wu, C.L.; Lee, C.H. Scutellaria barbata inhibits angiogenesis through downregulation of HIF-1 alpha in lung tumor. *Environ. Toxicol.* **2014**, *29*, 363–370. [CrossRef]
29. Han, D.; Cao, C.; Su, Y.; Wang, J.; Sun, J.; Chen, H.; Xu, A. Ginkgo biloba exocarp extracts inhibits angiogenesis and its effects on Wnt/beta-catenin-VEGF signaling pathway in Lewis lung cancer. *J. Ethnopharmacol.* **2016**, *192*, 406–412. [CrossRef]
30. Liao, J.; Yang, G.Y.; Park, E.S.; Meng, X.; Sun, Y.; Jia, D.; Seril, D.N.; Yang, C.S. Inhibition of lung carcinogenesis and effects on angiogenesis and apoptosis in A/J mice by oral administration of green tea. *Nutr. Cancer* **2004**, *48*, 44–53. [CrossRef] [PubMed]
31. Han, L.; Wang, J.N.; Cao, X.Q.; Sun, C.X.; Du, X. An-te-xiao capsule inhibits tumor growth in non-small cell lung cancer by targeting angiogenesis. *Biomed. Pharmacother.* **2018**, *108*, 941–951. [CrossRef]
32. Liu, J.; Liu, Y. Influence of erbanxiao solution on inhibiting angiogenesis in stasis toxin stagnation of non-small cell lung cancer. *J. Tradit. Chin. Med.* **2013**, *33*, 303–306. [CrossRef]
33. Lee, H.J.; Lee, E.O.; Rhee, Y.H.; Ahn, K.S.; Li, G.X.; Jiang, C.; Lu, J.; Kim, S.H. An oriental herbal cocktail, ka-mi-kae-kyuk-tang, exerts anti-cancer activities by targeting angiogenesis, apoptosis and metastasis. *Carcinogenesis* **2006**, *27*, 2455–2463. [CrossRef]
34. Xie, B.; Xie, X.; Rao, B.; Liu, S.; Liu, H. Molecular Mechanisms Underlying the Inhibitory Effects of Qingzaojiufei Decoction on Tumor Growth in Lewis Lung Carcinoma. *Integr. Cancer Ther.* **2018**, *17*, 467–476. [CrossRef]
35. Li, J.; Han, R.; Li, J.; Zhai, L.; Xie, X.; Zhang, J.; Chen, Y.; Luo, J.; Wang, S.; Sun, Z.; et al. Analysis of Molecular Mechanism of YiqiChutan Formula Regulating DLL4-Notch Signaling to Inhibit Angiogenesis in Lung Cancer. *Biomed. Res. Int.* **2021**, *2021*, 8875503. [PubMed]
36. Liao, H.; Wang, Z.; Deng, Z.; Ren, H.; Li, X. Curcumin inhibits lung cancer invasion and metastasis by attenuating GLUT1/MT1-MMP/MMP2 pathway. *Int. J. Clin. Exp. Med.* **2015**, *8*, 8948–8957. [PubMed]
37. Pai, J.T.; Hsu, C.Y.; Hsieh, Y.S.; Tsai, T.Y.; Hua, K.T.; Weng, M.S. Suppressing migration and invasion of H1299 lung cancer cells by honokiol through disrupting expression of an HDAC6-mediated matrix metalloproteinase 9. *Food Sci. Nutr.* **2020**, *8*, 1534–1545. [CrossRef] [PubMed]
38. Sazuka, M.; Imazawa, H.; Shoji, Y.; Mita, T.; Hara, Y.; Isemura, M. Inhibition of collagenases from mouse lung carcinoma cells by green tea catechins and black tea theaflavins. *Biosci. Biotechnol. Biochem.* **1997**, *61*, 1504–1506. [CrossRef] [PubMed]
39. Chakrawarti, L.; Agrawal, R.; Dang, S.; Gupta, S.; Gabrani, R. Therapeutic effects of EGCG: A patent review. *Expert Opin. Ther. Pat.* **2016**, *26*, 907–916. [CrossRef] [PubMed]
40. Deng, Y.T.; Lin, J.K. EGCG inhibits the invasion of highly invasive CL1-5 lung cancer cells through suppressing MMP-2 expression via JNK signaling and induces G2/M arrest. *J. Agric. Food Chem.* **2011**, *59*, 13318–13327. [CrossRef] [PubMed]
41. Shi, J.; Liu, F.; Zhang, W.; Liu, X.; Lin, B.; Tang, X. Epigallocatechin-3-gallate inhibits nicotine-induced migration and invasion by the suppression of angiogenesis and epithelial-mesenchymal transition in non-small cell lung cancer cells. *Oncol. Rep.* **2015**, *33*, 2972–2980. [CrossRef] [PubMed]
42. He, H.; Zheng, L.; Sun, Y.P.; Zhang, G.W.; Yue, Z.G. Steroidal saponins from Paris polyphylla suppress adhesion, migration and invasion of human lung cancer A549 cells via down-regulating MMP-2 and MMP-9. *Asian Pac. J. Cancer Prev.* **2014**, *15*, 10911–10916. [CrossRef] [PubMed]
43. Agrawal, S.; Chaugule, S.; More, S.; Rane, G.; Indap, M. Methanolic extract of Euchelus asper exhibits in-ovo anti-angiogenic and in vitro anti-proliferative activities. *Biol. Res.* **2017**, *50*, 41. [CrossRef]
44. Tseng, H.H.; Chen, P.N.; Kuo, W.H.; Wang, J.W.; Chu, S.C.; Hsieh, Y.S. Antimetastatic potentials of Phyllanthus urinaria L on A549 and Lewis lung carcinoma cells via repression of matrix-degrading proteases. *Integr. Cancer Ther.* **2012**, *11*, 267–278. [CrossRef]
45. Lim, W.C.; Choi, H.K.; Kim, K.T.; Lim, T.G. Rose (Rosa gallica) Petal Extract Suppress Proliferation, Migration, and Invasion of Human Lung Adenocarcinoma A549 Cells through via the EGFR Signaling Pathway. *Molecules* **2020**, *25*, 5119. [CrossRef] [PubMed]
46. Huang, S.F.; Chu, S.C.; Hsieh, Y.H.; Chen, P.N.; Hsieh, Y.S. Viola Yedoensis Suppresses Cell Invasion by Targeting the Protease and NF-kappaB Activities in A549 and Lewis Lung Carcinoma Cells. *Int. J. Med. Sci.* **2018**, *15*, 280–290. [CrossRef]

47. Wu, H.C.; Horng, C.T.; Lee, Y.L.; Chen, P.N.; Lin, C.Y.; Liao, C.Y.; Hsieh, Y.S.; Chu, S.C. Cinnamomum Cassia Extracts Suppress Human Lung Cancer Cells Invasion by Reducing u-PA/MMP Expression through the FAK to ERK Pathways. *Int. J. Med. Sci.* **2018**, *15*, 115–123. [CrossRef] [PubMed]
48. Chen, P.N.; Yang, S.F.; Yu, C.C.; Lin, C.Y.; Huang, S.H.; Chu, S.C.; Hsieh, Y.S. Duchesnea indica extract suppresses the migration of human lung adenocarcinoma cells by inhibiting epithelial-mesenchymal transition. *Environ. Toxicol.* **2017**, *32*, 2053–2063. [CrossRef]
49. Zhao, H.J.; Liu, T.; Mao, X.; Han, S.X.; Liang, R.X.; Hui, L.Q.; Cao, C.Y.; You, Y.; Zhang, L.Z. Fructus phyllanthi tannin fraction induces apoptosis and inhibits migration and invasion of human lung squamous carcinoma cells in vitro via MAPK/MMP pathways. *Acta Pharmacol. Sin.* **2015**, *36*, 758–768. [CrossRef]
50. Im, I.; Park, K.R.; Kim, S.M.; Kim, M.C.; Park, J.H.; Nam, D.; Jang, H.J.; Shim, B.S.; Ahn, K.S.; Mosaddik, A.; et al. The butanol fraction of guava (Psidium cattleianum Sabine) leaf extract suppresses MMP-2 and MMP-9 expression and activity through the suppression of the ERK1/2 MAPK signaling pathway. *Nutr. Cancer* **2012**, *64*, 255–266. [CrossRef]
51. Chu, S.C.; Yang, S.F.; Liu, S.J.; Kuo, W.H.; Chang, Y.Z.; Hsieh, Y.S. In vitro and in vivo antimetastatic effects of Terminalia catappa L. leaves on lung cancer cells. *Food Chem. Toxicol.* **2007**, *45*, 1194–1201. [CrossRef]
52. Man, S.; Gao, W.; Zhang, Y.; Yan, L.; Ma, C.; Liu, C.; Huang, L. Antitumor and antimetastatic activities of Rhizoma Paridis saponins. *Steroids* **2009**, *74*, 1051–1056. [CrossRef]
53. Yang, S.F.; Chu, S.C.; Liu, S.J.; Chen, Y.C.; Chang, Y.Z.; Hsieh, Y.S. Antimetastatic activities of Selaginella tamariscina (Beauv.) on lung cancer cells in vitro and in vivo. *J. Ethnopharmacol.* **2007**, *110*, 483–489. [CrossRef]
54. Kim, S.C.; Magesh, V.; Jeong, S.J.; Lee, H.J.; Ahn, K.S.; Lee, H.J.; Lee, E.O.; Kim, S.H.; Lee, M.H.; Kim, J.H.; et al. Ethanol extract of Ocimum sanctum exerts anti-metastatic activity through inactivation of matrix metalloproteinase-9 and enhancement of anti-oxidant enzymes. *Food Chem. Toxicol.* **2010**, *48*, 1478–1482. [CrossRef]
55. Shia, C.S.; Suresh, G.; Hou, Y.C.; Lin, Y.C.; Chao, P.D.; Juang, S.H. Suppression on metastasis by rhubarb through modulation on MMP-2 and uPA in human A549 lung adenocarcinoma: An ex vivo approach. *J. Ethnopharmacol.* **2011**, *133*, 426–433. [CrossRef]
56. Li, L.; Wang, S.; Yang, X.; Long, S.; Xiao, S.; Wu, W.; Hann, S.S. Traditional Chinese medicine, Fuzheng KangAi decoction, inhibits metastasis of lung cancer cells through the STAT3/MMP9 pathway. *Mol. Med. Rep.* **2017**, *16*, 2461–2468. [CrossRef] [PubMed]
57. Qi, Q.; Hou, Y.; Li, A.; Sun, Y.; Li, S.; Zhao, Z. Yifei Tongluo, a Chinese Herbal Formula, Suppresses Tumor Growth and Metastasis and Exerts Immunomodulatory Effect in Lewis Lung Carcinoma Mice. *Molecules* **2019**, *24*, 731. [CrossRef] [PubMed]
58. Zhao, Y.; Shao, Q.; Zhu, H.; Xu, H.; Long, W.; Yu, B.; Zhou, L.; Xu, H.; Wu, Y.; Su, Z. Resveratrol ameliorates Lewis lung carcinoma-bearing mice development, decreases granulocytic myeloid-derived suppressor cell accumulation and impairs its suppressive ability. *Cancer Sci.* **2018**, *109*, 2677–2686. [CrossRef] [PubMed]
59. Wu, T.; Liu, W.; Guo, W.; Zhu, X. Silymarin suppressed lung cancer growth in mice via inhibiting myeloid-derived suppressor cells. *Biomed. Pharmacother.* **2016**, *81*, 460–467. [CrossRef] [PubMed]
60. Kloc, M.; Ghobrial, R.M.; Lipinska-Opalka, A.; Wawrzyniak, A.; Zdanowski, R.; Kalicki, B.; Kubiak, J.Z. Effects of vitamin D on macrophages and myeloid-derived suppressor cells (MDSCs) hyperinflammatory response in the lungs of COVID-19 patients. *Cell Immunol.* **2021**, *360*, 104259. [CrossRef]
61. Wang, Y.; Fan, X.; Wu, X. Ganoderma lucidum polysaccharide (GLP) enhances antitumor immune response by regulating differentiation and inhibition of MDSCs via a CARD9-NF-kappaB-IDO pathway. *Biosci. Rep.* **2020**, *40*, BSR20201170. [CrossRef]
62. Rui, K.; Tian, J.; Tang, X.; Ma, J.; Xu, P.; Tian, X.; Wang, Y.; Xu, H.; Lu, L.; Wang, S. Curdlan blocks the immune suppression by myeloid-derived suppressor cells and reduces tumor burden. *Immunol. Res.* **2016**, *64*, 931–939. [CrossRef]
63. Xu, Z.H.; Zhu, Y.Z.; Su, L.; Tang, X.Y.; Yao, C.; Jiao, X.N.; Hou, Y.F.; Chen, X.; Wei, L.Y.; Wang, W.T.; et al. Ze-Qi-Tang Formula Induces Granulocytic Myeloid-Derived Suppressor Cell Apoptosis via STAT3/S100A9/Bcl-2/Caspase-3 Signaling to Prolong the Survival of Mice with Orthotopic Lung Cancer. *Mediators Inflamm.* **2021**, *2021*, 8856326. [CrossRef] [PubMed]
64. Huang, W.C.; Chan, M.L.; Chen, M.J.; Tsai, T.H.; Chen, Y.J. Modulation of macrophage polarization and lung cancer cell stemness by MUC1 and development of a related small-molecule inhibitor pterostilbene. *Oncotarget* **2016**, *7*, 39363–39375. [CrossRef] [PubMed]
65. Wu, C.Y.; Cherng, J.Y.; Yang, Y.H.; Lin, C.L.; Kuan, F.C.; Lin, Y.Y.; Lin, Y.S.; Shu, L.H.; Cheng, Y.C.; Liu, H.T.; et al. Danshen improves survival of patients with advanced lung cancer and targeting the relationship between macrophages and lung cancer cells. *Oncotarget* **2017**, *8*, 90925–90947. [CrossRef] [PubMed]
66. Li, H.; Huang, N.; Zhu, W.; Wu, J.; Yang, X.; Teng, W.; Tian, J.; Fang, Z.; Luo, Y.; Chen, M.; et al. Modulation the crosstalk between tumor-associated macrophages and non-small cell lung cancer to inhibit tumor migration and invasion by ginsenoside Rh2. *BMC Cancer* **2018**, *18*, 579. [CrossRef] [PubMed]
67. Nyiramana, M.M.; Cho, S.B.; Kim, E.J.; Kim, M.J.; Ryu, J.H.; Nam, H.J.; Kim, N.G.; Park, S.H.; Choi, Y.J.; Kang, S.S.; et al. Sea Hare Hydrolysate-Induced Reduction of Human Non-Small Cell Lung Cancer Cell Growth through Regulation of Macrophage Polarization and Non-Apoptotic Regulated Cell Death Pathways. *Cancers* **2020**, *12*, 726. [CrossRef] [PubMed]
68. Wang, L.; Wu, W.; Zhu, X.; Ng, W.; Gong, C.; Yao, C.; Ni, Z.; Yan, X.; Fang, C.; Zhu, S. The Ancient Chinese Decoction Yu-Ping-Feng Suppresses Orthotopic Lewis Lung Cancer Tumor Growth Through Increasing M1 Macrophage Polarization and CD4(+) T Cell Cytotoxicity. *Front. Pharmacol.* **2019**, *10*, 1333. [CrossRef] [PubMed]

69. Pang, L.; Han, S.; Jiao, Y.; Jiang, S.; He, X.; Li, P. Bu Fei Decoction attenuates the tumor associated macrophage stimulated proliferation, migration, invasion and immunosuppression of non-small cell lung cancer, partially via IL-10 and PD-L1 regulation. *Int. J. Oncol.* **2017**, *51*, 25–38. [CrossRef]
70. Rawangkan, A.; Wongsirisin, P.; Namiki, K.; Iida, K.; Kobayashi, Y.; Shimizu, Y.; Fujiki, H.; Suganuma, M. Green Tea Catechin Is an Alternative Immune Checkpoint Inhibitor that Inhibits PD-L1 Expression and Lung Tumor Growth. *Molecules* **2018**, *23*, 2071. [CrossRef] [PubMed]
71. Liu, Y.; Liu, X.; Zhang, N.; Yin, M.; Dong, J.; Zeng, Q.; Mao, G.; Song, D.; Liu, L.; Deng, H. Berberine diminishes cancer cell PD-L1 expression and facilitates antitumor immunity via inhibiting the deubiquitination activity of CSN5. *Acta Pharm. Sin. B* **2020**, *10*, 2299–2312. [CrossRef]
72. Hu, M.; Yang, J.; Qu, L.; Deng, X.; Duan, Z.; Fu, R.; Liang, L.; Fan, D. Ginsenoside Rk1 induces apoptosis and downregulates the expression of PD-L1 by targeting the NF-kappaB pathway in lung adenocarcinoma. *Food Funct.* **2020**, *11*, 456–471. [CrossRef]
73. Huang, M.Y.; Jiang, X.M.; Xu, Y.L.; Yuan, L.W.; Chen, Y.C.; Cui, G.; Huang, R.Y.; Liu, B.; Wang, Y.; Chen, X.; et al. Platycodin D triggers the extracellular release of programed death Ligand-1 in lung cancer cells. *Food Chem. Toxicol.* **2019**, *131*, 110537. [CrossRef] [PubMed]
74. Ng, W.; Gong, C.; Yan, X.; Si, G.; Fang, C.; Wang, L.; Zhu, X.; Xu, Z.; Yao, C.; Zhu, S. Targeting CD155 by rediocide-A overcomes tumour immuno-resistance to natural killer cells. *Pharm. Biol.* **2021**, *59*, 47–53. [CrossRef] [PubMed]
75. Peng, S.; Dong, W.; Chu, Q.; Meng, J.; Yang, H.; Du, Y.; Sun, Y.U.; Hoffman, R.M. Traditional Chinese Medicine Brucea Javanica Oil Enhances the Efficacy of Anlotinib in a Mouse Model of Liver-Metastasis of Small-cell Lung Cancer. *In Vivo* **2021**, *35*, 1437–1441. [CrossRef] [PubMed]
76. Damjanovic, A.; Kolundzija, B.; Matic, I.Z.; Krivokuca, A.; Zdunic, G.; Savikin, K.; Jankovic, R.; Stankovic, J.A.; Stanojkovic, T.P. Mahonia aquifolium Extracts Promote Doxorubicin Effects against Lung Adenocarcinoma Cells In Vitro. *Molecules* **2020**, *25*, 5233. [CrossRef] [PubMed]
77. Li, W.; Chen, C.; Saud, S.M.; Geng, L.; Zhang, G.; Liu, R.; Hua, B. Fei-Liu-Ping ointment inhibits lung cancer growth and invasion by suppressing tumor inflammatory microenvironment. *BMC Complement Altern. Med.* **2014**, *14*, 153. [CrossRef]
78. Liu, R.; Zheng, H.; Li, W.; Guo, Q.; He, S.; Hirasaki, Y.; Hou, W.; Hua, B.; Li, C.; Bao, Y.; et al. Anti-tumor enhancement of Fei-Liu-Ping ointment in combination with celecoxib via cyclooxygenase-2-mediated lung metastatic inflammatory microenvironment in Lewis lung carcinoma xenograft mouse model. *J. Transl. Med.* **2015**, *13*, 366. [CrossRef] [PubMed]
79. Lin, Y.S.; Hsieh, C.Y.; Kuo, T.T.; Lin, C.C.; Lin, C.Y.; Sher, Y.P. Resveratrol-mediated ADAM9 degradation decreases cancer progression and provides synergistic effects in combination with chemotherapy. *Am. J. Cancer Res.* **2020**, *10*, 3828–3837.
80. Liu, W.; Wu, T.C.; Hong, D.M.; Hu, Y.; Fan, T.; Guo, W.J.; Xu, Q. Carnosic acid enhances the anti-lung cancer effect of cisplatin by inhibiting myeloid-derived suppressor cells. *Chin. J. Nat. Med.* **2018**, *16*, 907–915. [CrossRef]
81. Chen, Y.; Zhang, Y.; Song, W.; Zhang, Y.; Dong, X.; Tan, M. Ginsenoside Rh2 Improves the Cisplatin Anti-tumor Effect in Lung Adenocarcinoma A549 Cells via Superoxide and PD-L1. *Anticancer Agents Med. Chem.* **2020**, *20*, 495–503. [CrossRef]
82. Chen, Y.; Bi, L.; Luo, H.; Jiang, Y.; Chen, F.; Wang, Y.; Wei, G.; Chen, W. Water extract of ginseng and astragalus regulates macrophage polarization and synergistically enhances DDP's anticancer effect. *J. Ethnopharmacol.* **2019**, *232*, 11–20. [CrossRef]
83. Hanahan, D.; Weinberg, R.A. The hallmarks of cancer. *Cell* **2000**, *100*, 57–70. [CrossRef]
84. Pandya, N.M.; Dhalla, N.S.; Santani, D.D. Angiogenesis-A new target for future therapy. *Vascul. Pharmacol.* **2006**, *44*, 265–274. [CrossRef]
85. Simons, M.; Gordon, E.; Claesson-Welsh, L. Mechanisms and regulation of endothelial VEGF receptor signalling. *Nat. Rev. Mol. Cell Biol.* **2016**, *17*, 611–625. [CrossRef] [PubMed]
86. Torok, S.; Rezeli, M.; Kelemen, O.; Vegvari, A.; Watanabe, K.; Sugihara, Y.; Tisza, A.; Marton, T.; Kovacs, I.; Tovari, J.; et al. Limited Tumor Tissue Drug Penetration Contributes to Primary Resistance against Angiogenesis Inhibitors. *Theranostics* **2017**, *7*, 400–412. [CrossRef]
87. Zimna, A.; Kurpisz, M. Hypoxia-Inducible Factor-1 in Physiological and Pathophysiological Angiogenesis: Applications and Therapies. *Biomed. Res. Int.* **2015**, *2015*, 549412. [CrossRef]
88. Lu, P.; Weaver, V.M.; Werb, Z. The extracellular matrix: A dynamic niche in cancer progression. *J. Cell Biol.* **2012**, *196*, 395–406. [CrossRef] [PubMed]
89. Kai, F.; Drain, A.P.; Weaver, V.M. The Extracellular Matrix Modulates the Metastatic Journey. *Dev. Cell* **2019**, *49*, 332–346. [CrossRef] [PubMed]
90. Lee, H.; Chang, K.W.; Yang, H.Y.; Lin, P.W.; Chen, S.U.; Huang, Y.L. MT1-MMP regulates MMP-2 expression and angiogenesis-related functions in human umbilical vein endothelial cells. *Biochem. Biophys. Res. Commun.* **2013**, *437*, 232–238. [CrossRef]
91. Huang, H. Matrix Metalloproteinase-9 (MMP-9) as a Cancer Biomarker and MMP-9 Biosensors: Recent Advances. *Sensors* **2018**, *18*, 3249. [CrossRef]
92. Hewlings, S.J.; Kalman, D.S. Curcumin: A Review of Its Effects on Human Health. *Foods* **2017**, *6*, 92. [CrossRef]
93. Sulzmaier, F.J.; Jean, C.; Schlaepfer, D.D. FAK in cancer: Mechanistic findings and clinical applications. *Nat. Rev. Cancer* **2014**, *14*, 598–610. [CrossRef] [PubMed]
94. Chen, L.C.; Shibu, M.A.; Liu, C.J.; Han, C.K.; Ju, D.T.; Chen, P.Y.; Viswanadha, V.P.; Lai, C.H.; Kuo, W.W.; Huang, C.Y. ERK1/2 mediates the lipopolysaccharide-induced upregulation of FGF-2, uPA, MMP-2, MMP-9 and cellular migration in cardiac fibroblasts. *Chem. Biol. Interact.* **2019**, *306*, 62–69. [CrossRef] [PubMed]

95. Nishikawa, M.; Hashida, M. Inhibition of tumour metastasis by targeted delivery of antioxidant enzymes. *Expert Opin. Drug Deliv.* **2006**, *3*, 355–369. [CrossRef]
96. Singh, S.; Majumdar, D.K.; Rehan, H.M. Evaluation of anti-inflammatory potential of fixed oil of Ocimum sanctum (Holybasil) and its possible mechanism of action. *J. Ethnopharmacol.* **1996**, *54*, 19–26. [CrossRef]
97. Zhang, G.; Luo, X.; Sumithran, E.; Pua, V.S.; Barnetson, R.S.; Halliday, G.M.; Khachigian, L.M. Squamous cell carcinoma growth in mice and in culture is regulated by c-Jun and its control of matrix metalloproteinase-2 and -9 expression. *Oncogene* **2006**, *25*, 7260–7266. [CrossRef] [PubMed]
98. Vayalil, P.K.; Katiyar, S.K. Treatment of epigallocatechin-3-gallate inhibits matrix metalloproteinases-2 and -9 via inhibition of activation of mitogen-activated protein kinases, c-jun and NF-kappaB in human prostate carcinoma DU-145 cells. *Prostate* **2004**, *59*, 33–42. [CrossRef]
99. Cheung, L.W.; Leung, P.C.; Wong, A.S. Gonadotropin-releasing hormone promotes ovarian cancer cell invasiveness through c-Jun NH2-terminal kinase-mediated activation of matrix metalloproteinase (MMP)-2 and MMP-9. *Cancer Res.* **2006**, *66*, 10902–10910. [CrossRef] [PubMed]
100. Shieh, J.M.; Chiang, T.A.; Chang, W.T.; Chao, C.H.; Lee, Y.C.; Huang, G.Y.; Shih, Y.X.; Shih, Y.W. Plumbagin inhibits TPA-induced MMP-2 and u-PA expressions by reducing binding activities of NF-kappaB and AP-1 via ERK signaling pathway in A549 human lung cancer cells. *Mol. Cell Biochem.* **2010**, *335*, 181–193. [CrossRef]
101. Vesely, M.D.; Kershaw, M.H.; Schreiber, R.D.; Smyth, M.J. Natural innate and adaptive immunity to cancer. *Annu. Rev. Immunol.* **2011**, *29*, 235–271. [CrossRef]
102. Childs, R.W.; Carlsten, M. Therapeutic approaches to enhance natural killer cell cytotoxicity against cancer: The force awakens. *Nat. Rev. Drug Discov.* **2015**, *14*, 487–498. [CrossRef]
103. Chioda, M.; Peranzoni, E.; Desantis, G.; Papalini, F.; Falisi, E.; Solito, S.; Mandruzzato, S.; Bronte, V. Myeloid cell diversification and complexity: An old concept with new turns in oncology. *Cancer Metastasis Rev.* **2011**, *30*, 27–43. [CrossRef] [PubMed]
104. Gabrilovich, D.I.; Nagaraj, S. Myeloid-derived suppressor cells as regulators of the immune system. *Nat. Rev. Immunol.* **2009**, *9*, 162–174. [CrossRef] [PubMed]
105. Kumar, V.; Patel, S.; Tcyganov, E.; Gabrilovich, D.I. The Nature of Myeloid-Derived Suppressor Cells in the Tumor Microenvironment. *Trends Immunol.* **2016**, *37*, 208–220. [CrossRef] [PubMed]
106. Groth, C.; Hu, X.; Weber, R.; Fleming, V.; Altevogt, P.; Utikal, J.; Umansky, V. Immunosuppression mediated by myeloid-derived suppressor cells (MDSCs) during tumour progression. *Br. J. Cancer* **2019**, *120*, 16–25. [CrossRef] [PubMed]
107. Law, A.M.K.; Valdes-Mora, F.; Gallego-Ortega, D. Myeloid-Derived Suppressor Cells as a Therapeutic Target for Cancer. *Cells* **2020**, *9*, 561. [CrossRef]
108. Galiniak, S.; Aebisher, D.; Bartusik-Aebisher, D. Health benefits of resveratrol administration. *Acta Biochim. Pol.* **2019**, *66*, 13–21. [CrossRef]
109. Bouillon, R.; Carmeliet, G.; Verlinden, L.; van Etten, E.; Verstuyf, A.; Luderer, H.F.; Lieben, L.; Mathieu, C.; Demay, M. Vitamin D and human health: Lessons from vitamin D receptor null mice. *Endocr. Rev.* **2008**, *29*, 726–776. [CrossRef]
110. Zeng, P.; Li, J.; Chen, Y.; Zhang, L. The structures and biological functions of polysaccharides from traditional Chinese herbs. *Prog. Mol. Biol. Transl. Sci.* **2019**, *163*, 423–444.
111. Qu, J.; Liu, L.; Xu, Q.; Ren, J.; Xu, Z.; Dou, H.; Shen, S.; Hou, Y.; Mou, Y.; Wang, T. CARD9 prevents lung cancer development by suppressing the expansion of myeloid-derived suppressor cells and IDO production. *Int. J. Cancer* **2019**, *145*, 2225–2237. [CrossRef]
112. Ishiguro, S.; Upreti, D.; Robben, N.; Burghart, R.; Loyd, M.; Ogun, D.; Le, T.; Delzeit, J.; Nakashima, A.; Thakkar, R.; et al. Water extract from Euglena gracilis prevents lung carcinoma growth in mice by attenuation of the myeloid-derived cell population. *Biomed. Pharmacother.* **2020**, *127*, 110166. [CrossRef] [PubMed]
113. Shapouri-Moghaddam, A.; Mohammadian, S.; Vazini, H.; Taghadosi, M.; Esmaeili, S.A.; Mardani, F.; Seifi, B.; Mohammadi, A.; Afshari, J.T.; Sahebkar, A. Macrophage plasticity, polarization, and function in health and disease. *J. Cell Physiol.* **2018**, *233*, 6425–6440. [CrossRef] [PubMed]
114. Chanmee, T.; Ontong, P.; Konno, K.; Itano, N. Tumor-associated macrophages as major players in the tumor microenvironment. *Cancers* **2014**, *6*, 1670–1690. [CrossRef]
115. Condeelis, J.; Pollard, J.W. Macrophages: Obligate partners for tumor cell migration, invasion, and metastasis. *Cell* **2006**, *124*, 263–266. [CrossRef]
116. Wan, L.Q.; Tan, Y.; Jiang, M.; Hua, Q. The prognostic impact of traditional Chinese medicine monomers on tumor-associated macrophages in non-small cell lung cancer. *Chin. J. Nat. Med.* **2019**, *17*, 729–737. [CrossRef]
117. Pollard, J.W. Tumour-educated macrophages promote tumour progression and metastasis. *Nat. Rev. Cancer* **2004**, *4*, 71–78. [CrossRef]
118. Nath, S.; Mukherjee, P. MUC1: A multifaceted oncoprotein with a key role in cancer progression. *Trends Mol. Med.* **2014**, *20*, 332–342. [CrossRef]
119. Ayob, A.Z.; Ramasamy, T.S. Cancer stem cells as key drivers of tumour progression. *J. Biomed. Sci.* **2018**, *25*, 20. [CrossRef]
120. Jin, J.; Lin, J.; Xu, A.; Lou, J.; Qian, C.; Li, X.; Wang, Y.; Yu, W.; Tao, H. CCL2: An Important Mediator Between Tumor Cells and Host Cells in Tumor Microenvironment. *Front. Oncol.* **2021**, *11*, 722916. [CrossRef]

121. Schmall, A.; Al-Tamari, H.M.; Herold, S.; Kampschulte, M.; Weigert, A.; Wietelmann, A.; Vipotnik, N.; Grimminger, F.; Seeger, W.; Pullamsetti, S.S.; et al. Macrophage and cancer cell cross-talk via CCR2 and CX3CR1 is a fundamental mechanism driving lung cancer. *Am. J. Respir. Crit. Care Med.* **2015**, *191*, 437–447. [CrossRef]
122. Han, S.; Wang, W.; Wang, S.; Yang, T.; Zhang, G.; Wang, D.; Ju, R.; Lu, Y.; Wang, H.; Wang, L. Tumor microenvironment remodeling and tumor therapy based on M2-like tumor associated macrophage-targeting nano-complexes. *Theranostics* **2021**, *11*, 2892–2916. [CrossRef]
123. Pereira, R.B.; Andrade, P.B.; Valentao, P. Chemical Diversity and Biological Properties of Secondary Metabolites from Sea Hares of Aplysia Genus. *Mar. Drugs* **2016**, *14*, 39. [CrossRef]
124. Hanahan, D.; Weinberg, R.A. Hallmarks of cancer: The next generation. *Cell* **2011**, *144*, 646–674. [CrossRef]
125. Jain, P.; Jain, C.; Velcheti, V. Role of immune-checkpoint inhibitors in lung cancer. *Ther. Adv. Respir. Dis.* **2018**, *12*, 1753465817750075. [CrossRef]
126. Iams, W.T.; Porter, J.; Horn, L. Immunotherapeutic approaches for small-cell lung cancer. *Nat. Rev. Clin. Oncol.* **2020**, *17*, 300–312. [CrossRef]
127. Leonetti, A.; Wever, B.; Mazzaschi, G.; Assaraf, Y.G.; Rolfo, C.; Quaini, F.; Tiseo, M.; Giovannetti, E. Molecular basis and rationale for combining immune checkpoint inhibitors with chemotherapy in non-small cell lung cancer. *Drug Resist. Updat.* **2019**, *46*, 100644. [CrossRef] [PubMed]
128. Yang, Y. Cancer immunotherapy: Harnessing the immune system to battle cancer. *J. Clin. Investig.* **2015**, *125*, 3335–3337. [CrossRef] [PubMed]
129. Suresh, K.; Naidoo, J.; Lin, C.T.; Danoff, S. Immune Checkpoint Immunotherapy for Non-Small Cell Lung Cancer: Benefits and Pulmonary Toxicities. *Chest* **2018**, *154*, 1416–1423. [CrossRef] [PubMed]
130. Oppenheim, J.J. IL-2: More than a T cell growth factor. *J. Immunol.* **2007**, *179*, 1413–1414. [CrossRef]
131. Liu, Y.; Zhou, N.; Zhou, L.; Wang, J.; Zhou, Y.; Zhang, T.; Fang, Y.; Deng, J.; Gao, Y.; Liang, X.; et al. IL-2 regulates tumor-reactive CD8(+) T cell exhaustion by activating the aryl hydrocarbon receptor. *Nat. Immunol.* **2021**, *22*, 358–369. [CrossRef]
132. Antonangeli, F.; Natalini, A.; Garassino, M.C.; Sica, A.; Santoni, A.; Di Rosa, F. Regulation of PD-L1 Expression by NF-kappaB in Cancer. *Front Immunol.* **2020**, *11*, 584626. [CrossRef]
133. Chen, G.; Huang, A.C.; Zhang, W.; Zhang, G.; Wu, M.; Xu, W.; Yu, Z.; Yang, J.; Wang, B.; Sun, H.; et al. Exosomal PD-L1 contributes to immunosuppression and is associated with anti-PD-1 response. *Nature* **2018**, *560*, 382–386. [CrossRef] [PubMed]
134. Liu, L.; You, X.; Han, S.; Sun, Y.; Zhang, J.; Zhang, Y. CD155/TIGIT, a novel immune checkpoint in human cancers (Review). *Oncol. Rep.* **2021**, *45*, 835–845. [CrossRef]
135. Zhang, J.; Qi, J.; Chen, N.; Fu, W.; Zhou, B.; He, A. High expression of a disintegrin and metalloproteinase-9 predicts a shortened survival time in completely resected stage I non-small cell lung cancer. *Oncol. Lett.* **2013**, *5*, 1461–1466. [CrossRef] [PubMed]
136. Zhou, R.; Cho, W.C.S.; Ma, V.; Cheuk, W.; So, Y.K.; Wong, S.C.C.; Zhang, M.; Li, C.; Sun, Y.; Zhang, H.; et al. ADAM9 Mediates Triple-Negative Breast Cancer Progression via AKT/NF-kappaB Pathway. *Front. Med.* **2020**, *7*, 214. [CrossRef] [PubMed]
137. Mazzocca, A.; Coppari, R.; De Franco, R.; Cho, J.Y.; Libermann, T.A.; Pinzani, M.; Toker, A. A secreted form of ADAM9 promotes carcinoma invasion through tumor-stromal interactions. *Cancer Res.* **2005**, *65*, 4728–4738. [CrossRef]
138. Liu, J.; Huang, Y.; Kumar, A.; Tan, A.; Jin, S.; Mozhi, A.; Liang, X.J. pH-sensitive nano-systems for drug delivery in cancer therapy. *Biotechnol. Adv.* **2014**, *32*, 693–710. [CrossRef] [PubMed]
139. Cai, M.; Chen, G.; Qin, L.; Qu, C.; Dong, X.; Ni, J.; Yin, X. Metal Organic Frameworks as Drug Targeting Delivery Vehicles in the Treatment of Cancer. *Pharmaceutics* **2020**, *12*, 232. [CrossRef]
140. Huang, P.; Wang, X.; Liang, X.; Yang, J.; Zhang, C.; Kong, D.; Wang, W. Nano-, micro-, and macroscale drug delivery systems for cancer immunotherapy. *Acta Biomater.* **2019**, *85*, 1–26. [CrossRef]
141. Sriamornsak, P. Application of pectin in oral drug delivery. *Expert Opin. Drug Deliv.* **2011**, *8*, 1009–1023. [CrossRef]
142. Wang, S.Y.; Meng, Y.J.; Li, J.; Liu, J.P.; Liu, Z.Q.; Li, D.Q. A novel and simple oral colon-specific drug delivery system based on the pectin/modified nano-carbon sphere nanocomposite gel films. *Int. J. Biol. Macromol.* **2020**, *157*, 170–176. [CrossRef] [PubMed]
143. Fang, T.; Liu, D.D.; Ning, H.M.; Dan, L.; Sun, J.Y.; Huang, X.J.; Dong, Y.; Geng, M.Y.; Yun, S.F.; Yan, J.; et al. Modified citrus pectin inhibited bladder tumor growth through downregulation of galectin-3. *Acta Pharmacol. Sin.* **2018**, *39*, 1885–1893. [CrossRef] [PubMed]
144. Zhang, S.L.; Mao, Y.Q.; Zhang, Z.Y.; Li, Z.M.; Kong, C.Y.; Chen, H.L.; Cai, P.R.; Han, B.; Ye, T.; Wang, L.S. Pectin supplement significantly enhanced the anti-PD-1 efficacy in tumor-bearing mice humanized with gut microbiota from patients with colorectal cancer. *Theranostics* **2021**, *11*, 4155–4170. [CrossRef]
145. Gaikwad, D.; Shewale, R.; Patil, V.; Mali, D.; Gaikwad, U.; Jadhav, N. Enhancement in in vitro anti-angiogenesis activity and cytotoxicity in lung cancer cell by pectin-PVP based curcumin particulates. *Int. J. Biol. Macromol.* **2017**, *104*, 656–664. [CrossRef] [PubMed]

Article

The Natural Pigment Violacein Potentially Suppresses the Proliferation and Stemness of Hepatocellular Carcinoma Cells In Vitro

Yu Jin Kim [1,†], Nayeong Yuk [2,†], Hee Jeong Shin [1] and Hye Jin Jung [1,2,3,*]

1. Department of Life Science and Biochemical Engineering, Sun Moon University, Asan 31460, Korea; petaldew17@naver.com (Y.J.K.); gmlwjd903@naver.com (H.J.S.)
2. Department of Pharmaceutical Engineering and Biotechnology, Sun Moon University, Asan 31460, Korea; nayeong7249@naver.com
3. Genome-Based BioIT Convergence Institute, Sun Moon University, Asan 31460, Korea
* Correspondence: poka96@sunmoon.ac.kr; Tel.: +82-41-530-2354; Fax: +82-41-530-2939
† These authors contributed equally to this work.

Abstract: Hepatocellular carcinoma (HCC) is a malignant type of primary liver cancer with high incidence and mortality, worldwide. A major challenge in the treatment of HCC is chemotherapeutic resistance. It is therefore necessary to develop novel anticancer drugs for suppressing the growth of HCC cells and overcoming drug resistance for improving the treatment of HCC. Violacein is a deep violet-colored indole derivative that is produced by several bacterial strains, including *Chromobacterium violaceum*, and it possesses numerous pharmacological properties, including antitumor activity. However, the therapeutic effects of violacein and the mechanism underlying its antitumor effect against HCC remain to be elucidated. This study is the first to demonstrate that violacein inhibits the proliferation and stemness of Huh7 and Hep3B HCC cells. The antiproliferative effect of violacein was attributed to cell cycle arrest at the sub-G1 phase and the induction of apoptotic cell death. Violacein induced nuclear condensation, dissipated mitochondrial membrane potential (MMP), increased generation of reactive oxygen species (ROS), activated the caspase cascade, and upregulated p53 and p21. The anticancer effect of violacein on HCC cells was also associated with the downregulation of protein kinase B (AKT) and extracellular signal-regulated kinase (ERK)1/2 signaling. Violacein not only suppressed the proliferation and formation of tumorspheres of Huh7 and Hep3B cancer stem-like cells but also reduced the expression of key markers of cancer stemness, including CD133, Sox2, Oct4, and Nanog, by inhibiting the signal transducer and activator of transcription 3 (STAT3)/AKT/ERK pathways. These results suggest the therapeutic potential of violacein in effectively suppressing HCC by targeting the proliferation and stemness of HCC cells.

Keywords: violacein; hepatocellular carcinoma; proliferation; apoptosis; stemness

1. Introduction

Hepatocellular carcinoma (HCC) is a primary malignancy of the liver, and it accounts for approximately 75% of all cases of liver cancer [1]. In 2018, HCC was the seventh most prevalent cancer worldwide and the second most common cause of cancer-related mortality [2]. In particular, the incidence of HCC is high in Asia and Africa [3]. The primary causes of HCC include chronic hepatitis B virus (HBV) or chronic hepatitis C virus (HCV) infections, alcoholic cirrhosis, and non-alcoholic steatohepatitis (NASH) [4]. To date, the early diagnosis and effective treatment of HCC continues to be a challenge. Most patients are asymptomatic and develop symptoms at an advanced stage of the disease. The long-term prognosis of treatment is good when detected very early; however, for the majority of patients with HCC, the disease is detected at a stage where surgical treatment is no longer feasible [5]. Therefore, most patients with HCC require chemotherapy, which involves the use of chemical agents for destroying cancer cells and inhibiting the growth of new cancer cells [6]. Chemotherapy, along with radiation therapy, surgery, and immunotherapy, have been used for the treatment of HCC for many years, but the progress is poor due to the high recurrence rate and frequent cirrhosis [7]. Sorafenib, a tyrosine kinase inhibitor, inhibits the activities of Raf kinase and vascular endothelial growth factor receptor (VEGFR), and it is the most commonly used targeted chemotherapy drug for the treatment of HCC. However, it has been demonstrated that sorafenib improves patient survival by only about 7–10 months [8,9]. Other kinase inhibitors that have been recently approved for the treatment of HCC include regorafenib and lenvatinib. However, the treatment benefits of these drugs have not been shown to be significantly superior to those of sorafenib [10,11]. Therefore, for the effective treatment of HCC, continued studies are necessary for identifying novel candidate drugs to overcome drug resistance.

Apoptosis, a genetically encoded cell death program, is associated with characteristic morphological and biochemical changes, including chromatin condensation, DNA fragmentation, and loss of mitochondrial membrane potential (MMP) [12,13]. Apoptosis is activated by intrinsic and extrinsic pathways and is modulated by a multitude of factors, including intracellular mediators of signal transduction and nuclear proteins that regulate gene expression, DNA replication, and the cell cycle [13]. Representatively, the activation of caspases, the increased expression of tumor suppressor p53, and the generation of reactive oxygen species (ROS) are major factors for the initiation of apoptosis and are molecular targets of numerous anticancer drugs [14,15]. As the main goal of cancer therapy is to eliminate cancer cells while minimizing damage to normal cells, the specific apoptosis of cancer cells can serve as a promising therapeutic strategy against cancer [16]. Therefore, drugs that induce the specific apoptosis of HCC cells can be considered as a potential treatment option for the disease.

Cancer stem cells (CSCs), also known as tumor-initiating cells (TICs), have stem cell-like properties and are defined as a small subpopulation of cancer cells that contributes to tumor diversity and heterogeneity [17,18]. CSCs are more tumorigenic and resistant to anticancer therapies than non-stem cancer cells [18,19]. Therefore, CSCs are closely related to tumor carcinogenesis, recurrence, and metastasis [18,19]. Mature hepatocytes, hepatoblasts, and bile cells can transform into liver CSCs (LCSCs) during liver injury and trigger hepatic regeneration or lead to a state of oncogenic dedifferentiation [20]. LCSCs are being increasingly recognized as responsible for the initiation, relapse, metastasis, and chemoresistance of HCC [21]. Therefore, the suppression of LCSCs is a promising strategy for the effective treatment of HCC and the prevention of relapse.

Natural products are a highly useful source of bioactive molecules and serve as an important and valuable resource for drug development [22]. Several preclinical and research findings have demonstrated that bioactive compounds derived from natural products show considerable potential in the prevention and treatment of several types of cancer [22,23]. Violacein, an indole derivative produced as a secondary metabolite of *Chromobacterium violaceum* (Figure 1A), is a purple-colored natural pigment that exhibits various biological properties, including antimicrobial, antiparasitic, anti-inflammatory, and antitumor activities [24]. Previous studies have demonstrated that violacein exhibits anticancer activity by inducing apoptosis in various types of cancer, including breast cancer, colon cancer, lung cancer, and leukemia [25–27]. However, the mechanism underlying the anticancer effect of violacein in HCC remains to be elucidated.

In this study, the anticancer effects and molecular mechanism underlying the anticancer property of violacein in Huh7 and Hep3B HCC cells were investigated. The results of our study demonstrate that violacein effectively inhibited the proliferation of HCC cells by upregulating the apoptotic pathways and also eradicated the stem-like features by downregulating major cancer stemness regulators in HCC cells. We therefore suggest that the natural compound violacein can serve as a potential therapeutic alternative for HCC.

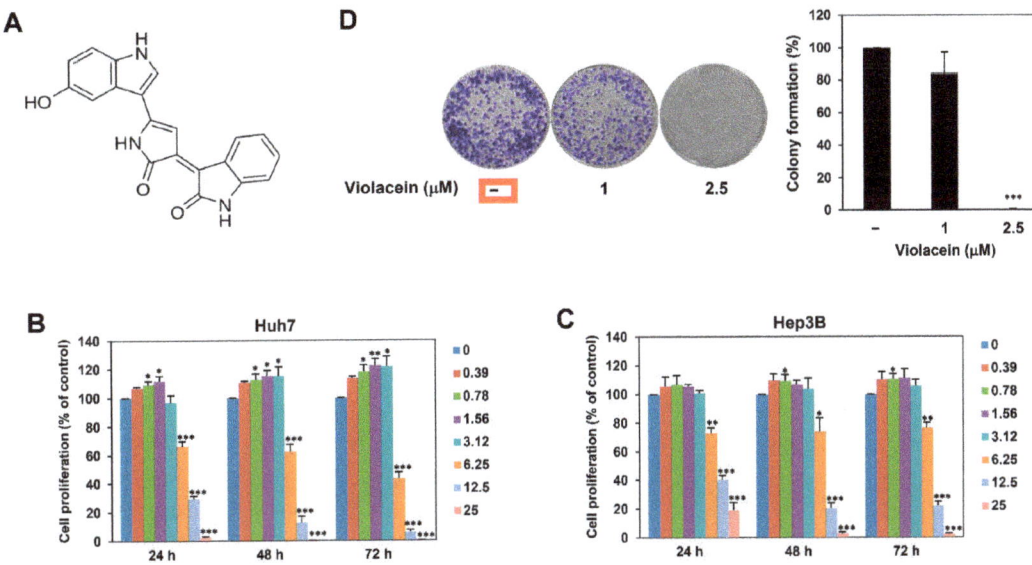

Figure 1. Violacein inhibits the proliferation of Huh7 HCC cells. (**A**) Chemical structure of violacein. (**B,C**) Effect of violacein on the proliferation of Huh7 and Hep3B cells. The cells were treated with violacein at various concentrations (0–25 µM) for 24, 48, or 72 h. Cell proliferation was measured using the CellTiter-Glo® luminescent assay system. (**D**) Effect of violacein on the colony-forming ability of Huh7 cells. The cells were incubated in the absence or presence of violacein (1 and 2.5 µM) for 12 days. The cell colonies were detected by crystal violet staining. * $p < 0.05$, ** $p < 0.01$, *** $p < 0.001$ vs. the control.

2. Results

2.1. Violacein Inhibits the Proliferation of Huh7 HCC Cells

In order to examine whether violacein affects the proliferation of HCC cells, Huh7 cells were treated with violacein at various concentrations (0–25 µM) for 24, 48, or 72 h. Cell proliferation was subsequently measured using the ATP-monitoring luminescence assay. As depicted in Figure 1B, treatment with violacein showed a biphasic dose–response on the proliferation of Huh7 cells. At low concentrations (0.39–3.12 µM), violacein induced the proliferation of Huh7 cells. However, at higher concentrations (6.25–25 µM), it inhibited the Huh7 cell proliferation, with IC_{50} values of 7.97, 6.71, and 6.10 µM at 24, 48, and 72 h, respectively. A similar pattern of dose–response was observed in different HCC cells, Hep3B (Figure 1C). The IC_{50} values of violacein for Hep3B cells were determined to be 8.01, 8.41, and 8.23 µM at 24, 48, and 72 h, respectively. These data indicate that violacein may have a biphasic effect on the proliferation of HCC cells by modulating the action of molecular targets involved in HCC cell proliferation in a concentration-dependent biphasic manner.

We next evaluated the effect of violacein on the formation of Huh7 cell colonies. Colony formation was observed on day 12 after treatment with violacein. As depicted in Figures 1D and S1, 2.5 µM violacein completely suppressed the colony-forming ability of Huh7 cells. Taken together, these results demonstrate that violacein has inhibitory potential on the proliferation of HCC cells.

2.2. Violacein Promotes Apoptotic Characteristics in Huh7 HCC Cells

In order to further investigate the mechanisms underlying the antiproliferative effect of violacein in Huh7 HCC cells, we determined by 4′,6-diamidino-2-phenylindole (DAPI) staining whether violacein causes nuclear morphological changes in Huh7 cells. As depicted in Figure 2A, treatment with violacein caused nuclear condensation, a prominent hallmark of apoptosis.

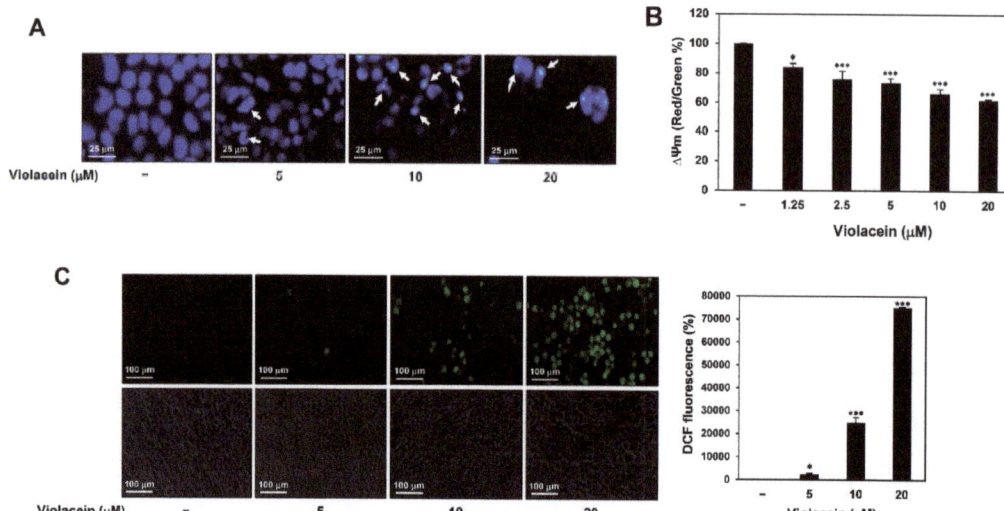

Figure 2. Violacein promotes apoptotic characteristics in Huh7 HCC cells. (**A**) Effect of violacein on the nuclear morphology. Huh7 cells were treated with violacein (5, 10, and 20 µM) for 24 h. Changes in nuclear morphology were monitored by DAPI staining under a fluorescence microscope. The condensed nuclei are indicated by white arrows. (**B**) Effect of violacein on the MMP. Huh7 cells were treated with violacein (0–20 µM) for 24 h and stained with JC-1. The fluorescence intensity of J-aggregates and JC-1 monomers was detected by a multimode microplate reader. (**C**) Effect of violacein on the generation of intracellular ROS. Huh7 cells were treated with violacein (5, 10, and 20 µM) for 6 h. The levels of ROS were detected with DCFH-DA using a fluorescence microscope and were further quantified by densitometry. * $p < 0.05$, *** $p < 0.001$ vs. the control.

The loss of MMP ($\Delta\psi$m) can induce early apoptosis [12,13]. We therefore evaluated the effect of treatment with violacein on the MMP in Huh7 cells using the cationic 5,5′,6,6′-tetra-chloro-1,1′,3,3′-tetraethylbenzimidazol-carbocyanine iodide (JC-1) fluorescent dye. As depicted in Figure 2B, violacein significantly decreased the ratio of intensity of red/green fluorescence in a dose-dependent manner, indicating that violacein caused the loss of MMP in Huh7 cells.

Intracellular ROS play a central role in biological processes, especially in the induction of apoptosis [12,15]. In order to assess whether violacein affects the generation of ROS in Huh7 cells, the level of intracellular ROS was measured using 2′,7′-dichlorofluorescein diacetate (DCFH-DA). Violacein significantly increased the accumulation of ROS in a dose-dependent manner (Figure 2C). Taken together, these results suggest that violacein may inhibit the proliferation of Huh7 HCC cells by inducing apoptotic cell death.

2.3. Violacein Induces Cell Cycle Arrest at the Sub-G1 Phase and Cellular Apoptosis in Huh7 HCC Cells

In order to further explore the role of apoptosis in the inhibitory effect of violacein on the proliferation of HCC cells, we investigated the effect of violacein on the progression of the cell cycle in Huh7 cells by flow cytometric analysis. DNA fragmentation can be measured by flow cytometry using the sub-G1 assay. The small DNA fragments generated during apoptosis leak out of cells, decreasing the total DNA content of apoptotic cells. By staining DNA with PI, hypodiploid apoptotic cells can be detected as a "sub-G1" population of the PI histogram. As depicted in Figure 3A, treatment with violacein significantly increased the population of cells in the sub-G1 phase and decreased the population of cells in the G0/G1, S, and G2/M phases, compared with those of the untreated control group. These data imply that violacein caused nuclear fragmentation in the HCC cells and consequently arrested cell cycle progression at the sub-G1 phase, which represents apoptotic cells.

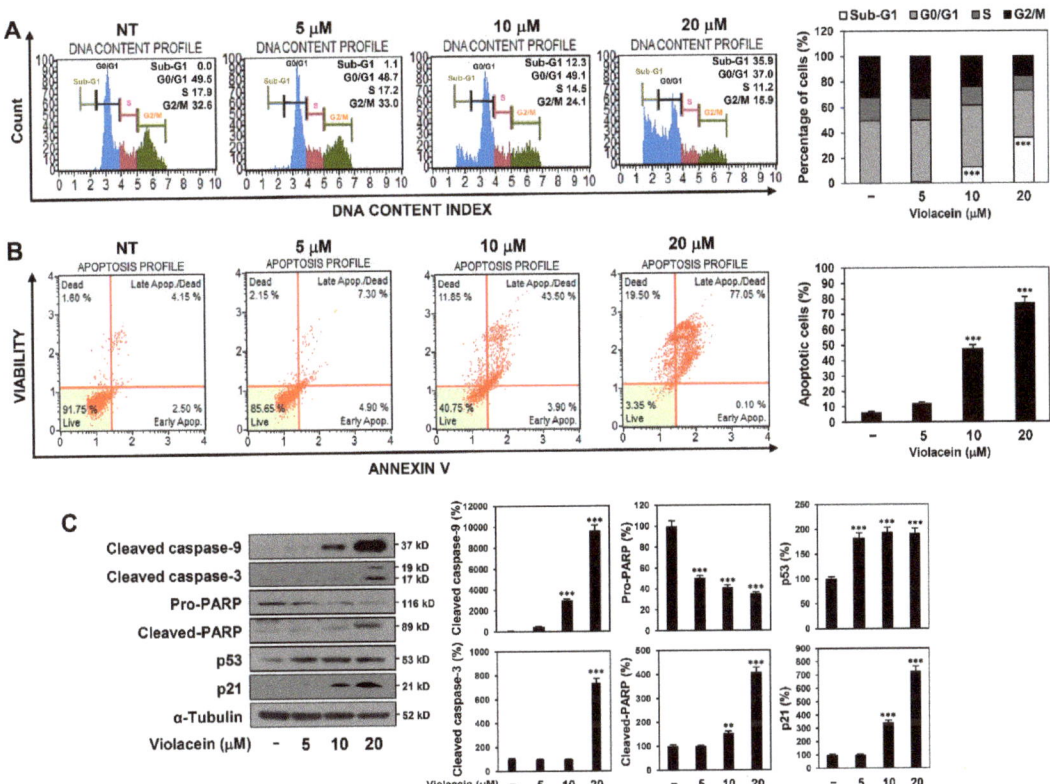

Figure 3. Violacein induces cell cycle arrest at the sub-G1 phase and cellular apoptosis in Huh7 HCC cells. (A–C) Huh7 cells were treated with violacein (5, 10, and 20 µM) for 24 h. (A) Effect of violacein on the cell cycle. The cell cycle distribution was evaluated using a Muse Cell Analyzer with Muse® Cell Cycle kit. (B) Effect of violacein on the apoptotic cell death. Apoptotic cells were detected using a Muse Cell Analyzer with Muse® Annexin V & Dead Cell kit. (C) Effect of violacein on the expression of apoptosis regulators. Protein levels were detected by Western blot analysis using specific antibodies and were further quantified by densitometry. α–tubulin levels were used as an internal control. ** $p < 0.01$, *** $p < 0.001$ vs. the control.

Next, cellular apoptosis was analyzed by flow cytometry by Annexin V-FITC and PI dual labeling. Treatment with violacein significantly increased the proportion of apoptotic cells in a dose-dependent manner compared with that in the untreated control group (Figure 3B). These data demonstrate that violacein induced apoptosis in Huh7 HCC cells.

In order to further elucidate the molecular mechanism underlying violacein-induced apoptosis, we assessed the effect of violacein on the expression of crucial mediators of apoptosis in Huh7 cells. Caspases play an important role in apoptosis and are activated by proteolytic cleavage [12,13]. Upon induction of apoptosis, cytochrome c released from mitochondria associates with caspase-9 and apoptotic protease activating factor-1 (Apaf-1). The complex processes caspase-9 (47 kD) into several subunits, including a p37 fragment (37 kD). The cleaved caspase-9 further processes caspase-3 (35 kD) that is a critical executioner of apoptosis. The cleaved caspase-3 (17/19 kD) is responsible for the proteolytic cleavage (89 kD) of downstream substrates involved in apoptotic changes, including a 116 kD nuclear poly (ADP-ribose) polymerase (PARP). As depicted in Figure 3C, treatment with violacein increased the levels of cleaved caspase-9, cleaved caspase-3, and cleaved PARP. Additionally, violacein upregulated the expression of the tumor suppressor p53 and its transcriptional target, p21, which are implicated in both cell cycle arrest and apoptosis. These results suggest that violacein may induce cellular apoptosis via the activation of p53- and caspase-mediated apoptotic pathways in Huh7 HCC cells.

2.4. Violacein Downregulates Protein Kinase B (AKT) and Extracellular Signal-Regulated Kinase (ERK) Signaling in Huh7 HCC Cells

Aberrantly activated AKT and ERK signaling contributes to the survival and proliferation of several types of cancers, including HCC [28]. AKT is a serine/threonine kinase that promotes cell survival by inhibiting apoptosis. ERK1 and ERK2 isoforms are 44 and 42 kD serine/threonine kinases, respectively, belonging to the mitogen-activated protein kinase family. We therefore examined whether violacein affects the key signaling pathways in Huh7 cells. In particular, to clarify whether the inhibitory effect of violacein on HCC cell proliferation is due to inactivation of the signaling pathways, the effect of the compound was evaluated at the early time point of 1 h. As depicted in Figure 4, violacein more effectively inhibited the expression levels of phosphorylated forms compared with unphosphorylated AKT and ERK1/2 proteins. These data indicate that the antiproliferative effect of violacein on Huh7 cells could be mediated via the downregulation of AKT and ERK signaling pathways.

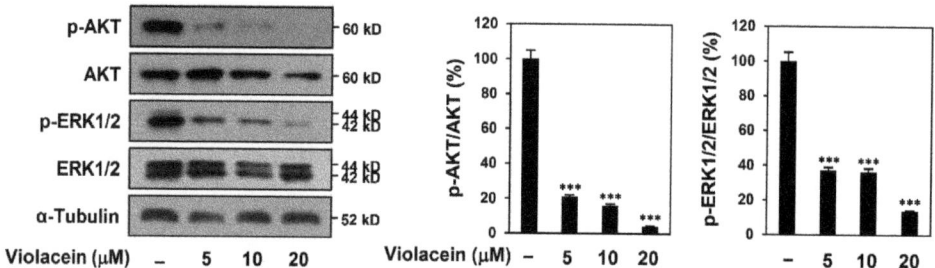

Figure 4. Violacein downregulates AKT and ERK signaling in Huh7 HCC cells. The cells were treated with violacein (5, 10, and 20 μM) for 1 h. Protein levels were detected by Western blot analysis using specific antibodies and were further quantified by densitometry. α–tubulin levels were used as an internal control. *** $p < 0.001$ vs. the control.

2.5. Violacein Inhibits the Proliferation and Formation of Tumorspheres of Huh7 Cancer Stem-Like Cells

Accumulating evidence has revealed that CSCs, a subpopulation of tumor cells, facilitate metastasis, recurrence, and resistance to chemotherapy or radiotherapy in HCC [17]. We therefore further investigated the effect of violacein on the stemness of Huh7 cells. In order to propagate the cancer stem-like cells, Huh7 cells were grown in serum-free spheroid suspension culture [29]. The 3D spheroid cell culture is known to stimulate in vivo cellular conditions better in comparison with 2D cell culture, and it is widely used to increase a subpopulation of cancer cells with stem cell-like properties, thereby providing new insights into cancer treatment and CSC research [29]. As depicted in Figure S2, the expression levels of several key stemness-related markers were remarkably increased in the Huh7 tumorsphere cells cultured in serum-free media containing EGF and bFGF compared to the Huh7 adherent cells cultured in 10% FBS-supplemented media. These data indicate that the serum-free spheroid culture can expand the CSC population from HCC cell lines.

The Huh7 cancer stem-like cells were treated with violacein at various concentrations (0–100 μM) for 7 days, and cell proliferation was measured using the ATP-monitoring luminescence assay. As depicted in Figure 5A, treatment with violacein inhibited the proliferation of Huh7 cancer stem-like cells, and the IC_{50} value was determined to be 16.47 µM. Furthermore, both the size and number of tumorspheres formed by the Huh7 cancer stem-like cells were effectively reduced following treatment with violacein (Figure 5B,C). In particular, the tumorsphere-forming ability was remarkably suppressed by violacein, even at concentrations 10 times lower than the IC_{50} value for inhibiting the proliferation of Huh7 cancer stem-like cells. We further identified that violacein potently inhibited the tumorsphere-forming ability of Hep3B cancer stem-like cells (Figure S3). These data suggest the possible therapeutic potential of violacein in eradicating CSCs in HCC.

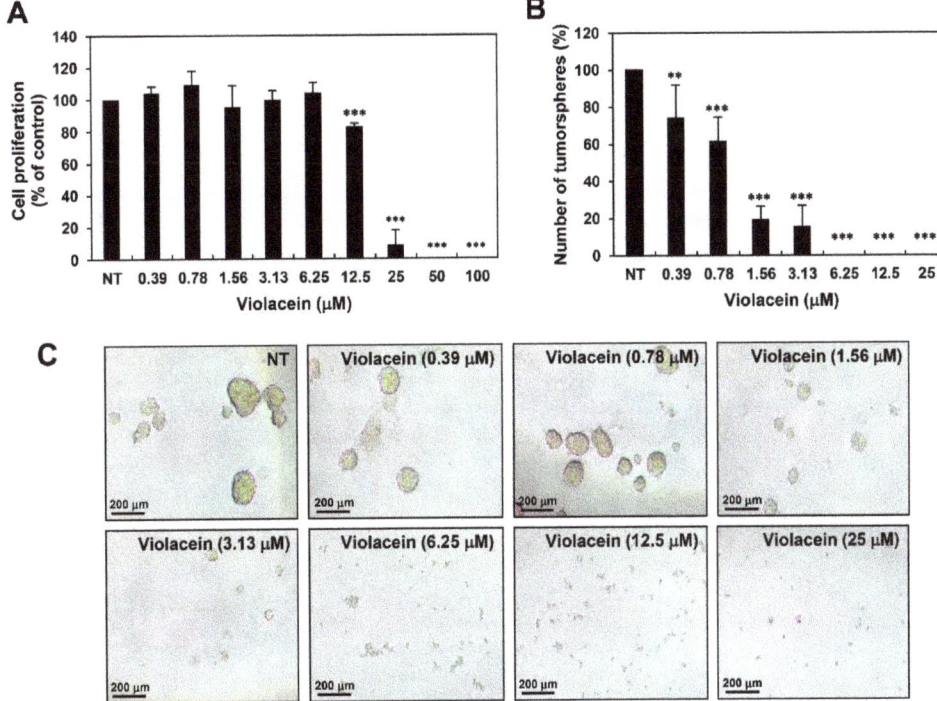

Figure 5. Violacein inhibits the proliferation and formation of tumorspheres of Huh7 cancer stem-like cells. (**A**) Effect of violacein on the proliferation of Huh7 cancer stem-like cells. The cells were treated with violacein at various concentrations (0–100 µM) and were incubated with the CSC culture media for 7 days. Cell proliferation was measured using the CellTiter-Glo® luminescent assay system. (**B,C**) Effect of violacein on the tumorsphere-forming ability of Huh7 cancer stem-like cells. The cells were treated with violacein at various concentrations (0–25 µM) and were incubated with the CSC culture media for 7 days. The number of tumorspheres in each well was counted under an optical microscope. ** $p < 0.01$, *** $p < 0.001$ vs. the control.

2.6. Violacein Downregulates Cancer Stemness-Related Markers and Signal Transducer and Activator of Transcription 3 (STAT3)/AKT/ERK Signaling in Huh7 Cancer Stem-Like Cells

We next evaluated whether violacein regulates the expression of key stemness-related markers in the CSCs in HCC. The results demonstrate that violacein markedly suppressed the expression of CD133, Sox2, Oct4, and Nanog in Huh7 cancer stem-like cells (Figure 6A). However, it slightly inhibited the expression of aldehyde dehydrogenase 1A1 (ALDH1A1) at concentrations of 1 and 5 µM, while it did not reduce the protein levels at 2.5 µM. Meanwhile, violacein suppressed the expression of integrin α6 (125 kD reduced and 150 kD nonreduced forms) at 1 and 2.5 µM treatment, but not at 5 µM.

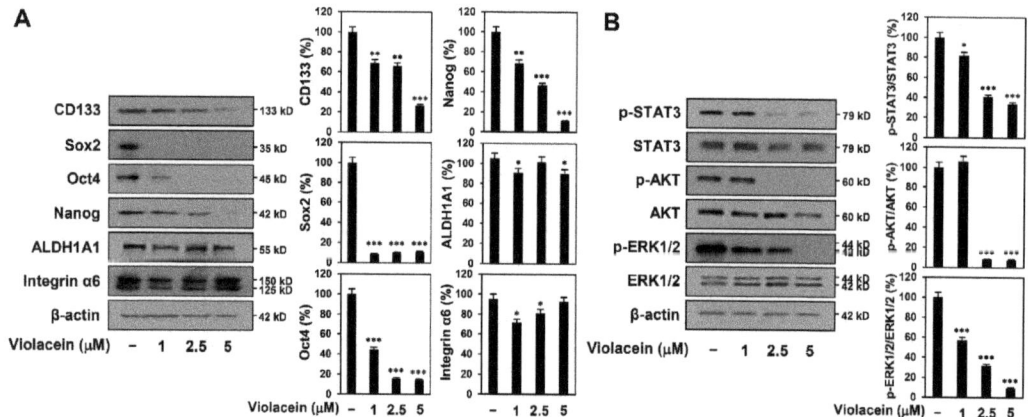

Figure 6. Violacein downregulates cancer stemness-related markers and STAT3/AKT/ERK signaling in Huh7 cancer stem-like cells. (**A,B**) The cells were treated with violacein (1, 2.5, and 5 µM) and were incubated with the CSC culture media for 72 h. Protein levels were detected by Western blot analysis using specific antibodies and were further quantified by densitometry. β-actin levels were used as an internal control. * $p < 0.05$, ** $p < 0.01$, *** $p < 0.001$ vs. the control.

It has been reported that STAT3 signaling regulates the expression of several CSC markers, including CD133 and Nanog, in HCC [30]. The AKT and ERK signaling pathways are also involved in the maintenance of self-renewal in CSCs in HCC [31,32]. Treatment with violacein inhibited the phosphorylation of STAT3, AKT, and ERK1/2 without affecting the total protein expression (Figure 6B). These results demonstrate that violacein may inhibit the stem-like features of Huh7 HCC cells by inactivating the STAT3/AKT/ERK pathway.

3. Discussion

HCC is the most common primary malignancy of the liver, with a high mortality rate that is attributed to difficulties in early diagnosis, high recurrence rate, and frequent cirrhosis [1,7]. Therefore, the continued exploration and development of novel anticancer agents is necessary for reducing the rate of recurrence of HCC and achieving effective chemotherapy.

Violacein is a natural pigment with numerous biological activities, and it is obtained from several bacterial strains, including *C. violaceum* [24]. Several studies have demonstrated that violacein can induce apoptosis in a variety of cancer cells, including breast cancer, colon cancer, lung cancer, and leukemia [25–27]. It has been demonstrated that violacein induces the accumulation of ROS in tumors by increasing the expression of genes related to oxidative stress, including GSK3β and TNF-α, and induces apoptosis by upregulating the expression of apoptosis-inducing genes, including p53, caspase-3, caspase-8, caspase-9, and PARP [25,26]. However, the anticancer activity and mechanism underlying the antitumor effect of violacein in HCC are yet to be investigated. This study was the first to evaluate the anticancer effect of violacein in Huh7 and Hep3B HCC cells.

The evasion of apoptosis is the hallmark of cancer and results in cancer progression and drug resistance [33]. Therefore, the induction of apoptosis through the reactivation of major apoptotic pathways and the inhibition of cell survival signaling cascades can offer a potential strategy for efficiently suppressing the growth of HCC cells. The results of our study demonstrate that violacein effectively inhibited the proliferation of HCC cells by inducing apoptosis. Violacein causes cell cycle arrest at the sub-G1 phase and activates the key regulatory mechanisms of apoptosis, including nuclear condensation, dissipation of MMP, increased generation of ROS, activation of the caspase cascade, and upregulation of p53 and p21. Additionally, the antiproliferative effect of violacein on HCC cells was mediated via the downregulation of AKT and ERK1/2 signaling pathways, which play an important role in sustaining proliferation and preventing the apoptosis of tumor cells [28]. These findings demonstrate that violacein potently inhibits the proliferation of HCC cells via activation of apoptotic pathways and deactivation of survival pathways.

CSCs are a small subpopulation of cancer cells with stem cell properties, including self-renewal and multilineage differentiation, and they have been identified in many tumors, including liver cancer [17–19]. It has been demonstrated that LCSCs, also known as hepatic cancer stem cells, contribute to the initiation, relapse, metastasis, and chemoresistance of HCC [20,21]. Therefore, LCSCs are a vital target for the successful treatment of HCC. Notably, the results of our study demonstrate that violacein significantly inhibited the proliferation and formation of tumorspheres of Huh7 and Hep3B cancer stem-like cells. Additionally, violacein effectively suppressed the key stemness-related markers and signaling pathways in cells.

Accumulating evidence has revealed that several cell surface markers, transcription factors, and molecular signaling pathways are activated in LCSCs, which mediate the development and maintenance of LCSCs [34,35]. The CD133 cell surface marker has been detected in numerous types of CSCs, and CD133-positive HCC has stronger carcinogenicity and proliferative capacity [34]. The key regulatory transcription factors, including Sox2, Oct4, and Nanog, not only play a critical role in the maintenance and self-renewal of LCSCs but also contribute to the malignancy of HCC via mechanisms related to drug resistance, including epithelial–mesenchymal transition (EMT) [21,34–36]. ALDH1A1, an isoform of the aldehyde dehydrogenase enzyme, partakes in the clearance of toxic aldehydes and reduces oxidative stress, which consequently leads to the survival of CSCs and poor prognosis [37]. Integrin $\alpha 6$, the receptor for the extracellular matrix (ECM) protein laminin, plays an important role in maintaining the CSC niche and regulating homeostasis in CSCs [36]. In this study, violacein noticeably reduced the expression levels of CD133, Sox2, Oct4, and Nanog in Huh7 cancer stem-like cells. However, it differently regulated the expression levels of ALDH1A1 and integrin $\alpha 6$ at each treatment concentration.

Additionally, several molecular signaling pathways, including the STAT3, AKT, and ERK pathways, are aberrantly activated in HCC, leading to the dysregulation of downstream target genes that control proliferation, survival, invasion, and stemness [30–32]. Activated STAT3 translocates to the nucleus and induces the transcription of several markers of CSCs in HCC, including CD133 and Nanog, thereby promoting the maintenance and self-renewal of LCSCs [38,39]. The AKT and ERK signaling pathways are also implicated in the proliferation, metabolism, and differentiation of LCSCs via upregulation of the expression of Sox2, Oct4, and Nanog [31,32,40]. The results of our study demonstrate that violacein significantly inhibited the phosphorylation of STAT3, AKT, and ERK1/2 without reducing the total protein levels in Huh7 cancer stem-like cells. These findings suggest that violacein may suppress the cancer stem-like features of HCC cells by inhibiting the expression of key cancer stemness markers, including CD133, Sox2, Oct4, and Nanog, via downregulation of the STAT3/AKT/ERK signaling pathways.

Taken together, the study demonstrates the therapeutic potential of violacein in effectively suppressing HCC by targeting both the proliferation and stemness of HCC cells. However, further in vivo experiments with animal models and identification of the primary molecular target of violacein are necessary before violacein can be used for the clinical treatment of HCC.

4. Materials and Methods

4.1. Materials

Violacein was kindly provided by Prof. Jae Kyung Sohng (Sun Moon University, Asan, Korea) and dissolved in dimethyl sulfoxide (DMSO) at a concentration of 100 mM. Dulbecco's modified Eagle's medium (DMEM) and Roswell Park Memorial institute-1640 (RPMI-1640) were purchased from Corning Cellgro (Manassas, VA, USA). DMEM/F12 and trypsin were purchased from HyClone (Marlborough, MA, USA). Epidermal growth factor (EGF) and basic fibroblast growth factor (bFGF) were obtained from Prospecbio (East

Brunswick, NJ, USA). Fetal bovine serum (FBS), B-27 serum-free supplement, L-glutamine, and penicillin/streptomycin were obtained from Gibco (Grand Island, NY, USA). Penicillin-streptomycin-amphotericin B and Accutase were obtained from Lonza (Walkersville, MD, USA) and EMD Millipore (Temecula, CA, USA), respectively. Heparin, DAPI, DCFH-DA, and JC-1 were purchased from Sigma-Aldrich (St. Louis, MO, USA). The CellTiter-Glo® 2.0 Cell Viability Assay kit was purchased from Promega (Madison, WI, USA). The Muse® Annexin V & Dead Cell and Muse® Cell Cycle kits were purchased from Luminex (Austin, TX, USA). The antibodies against cleaved caspase-9 (cat. no. 9501), cleaved caspase-3 (cat. no. 9661), PARP (cat. no. 9542), p53 (cat. no. 2524), p21 (cat. no. 2947), ALDH1A1 (cat. no. 12035), Integrin α6 (cat. no. 3750), Sox2 (cat. no. 3579), Oct4 (cat. no. 2750), Nanog (cat. no. 3580), CD133 (cat. no. 64326), phospho-STAT3 (cat. no. 9145), STAT3 (cat. no. 9139), phospho-AKT (cat. no. 4060), AKT (cat. no. 9272), phospho-ERK1/2 (cat. no. 9101), ERK1/2 (cat. no. 9102), α-tubulin (cat. no. 2144), rabbit IgG (cat. no. 7074), and mouse IgG (cat. no. 7076) were purchased from Cell Signaling Technology (Danvers, MA, USA). Anti-β-actin antibody (cat. no. ab6276) was purchased from Abcam (Cambridge, UK).

4.2. Cell Culture

Huh7 and Hep3B human HCC cells were purchased from the Korean Cell Line Bank (Seoul, Korea). The Huh7 and Hep3B cells were grown in DMEM supplemented with 10% FBS and 1% penicillin–streptomycin–amphotericin B, respectively. The Huh7 and Hep3B cancer stem-like cells were cultured in DMEM/F12 containing $1 \times$ B-27, 5 μg/mL heparin, 2 mM L-glutamine, 20 ng/mL EGF, 20 ng/mL bFGF, and 1% penicillin/streptomycin. All the cells were maintained at 37 °C in a humidified CO_2 incubator with 5% CO_2 (Thermo Scientific, Vantaa, Finland).

4.3. Cell Proliferation Assay

Cell proliferation was quantitatively evaluated using the CellTiter-Glo® 2.0 Cell Viability Assay kit. Briefly, the cells (5×10^3 cells/well) were seeded in a 96-white-well culture plate and treated with violacein at various concentrations. Following incubation for the indicated durations, 20 μL of substrate solution was added to each well, and the culture plate was shaken for 2 min and subsequently incubated for 8 min in the dark. Luminescence was detected using a multimode microplate reader (BioTek, Inc., Winooski, VT, USA). The IC_{50} values determined from the obtained data were analyzed by the curve-fitting program in GraphPad Prism, version 5 (GraphPad Software, La Jolla, CA, USA).

4.4. Colony Formation Assay

Huh7 cells (1×10^3 cells/well) were seeded in 6-well culture plates and treated with violacein. After 12 days of incubation, the colonies that had formed were fixed with 3.7% formaldehyde by incubating for 20 min, and stained with 0.5% crystal violet reagent by incubating for 15 min. The stained colonies were washed with phosphate-buffered saline (PBS), and the number of visible colonies in each well was counted.

4.5. Nuclear Fluorescent Staining with DAPI

Huh7 cells (7×10^4 cells/well) were seeded in 24-well culture plates and treated with violacein for 24 h. The cells were fixed with 3.7% formaldehyde by incubating for 15 min and stained with 5 μg/mL DAPI by incubating for 15 min. The fluorescent images were obtained using a fluorescence microscope (Optinity KI-2000F, Korea Lab Tech, Seong Nam, Korea).

4.6. Measurement of MMP

Huh7 cells (5×10^4 cells/well) were seeded in 96-black-well culture plates and treated with violacein for 24 h. The cells were subsequently treated with 10 μg/mL of the JC-1 fluorescent dye and were incubated for 20 min. The fluorescence intensity of the J-aggregates that formed and the JC-1 monomers was detected at wavelengths of 530/590 nm (excitation/emission) and 485/528 nm, respectively, using a multimode microplate reader (BioTek, Winooski, VT, USA). Mitochondrial depolarization was indicated by a decrease in the ratio of intensity of red/green fluorescence.

4.7. Measurement of ROS

Huh7 cells (5×10^4 cells/well) were seeded in a 24-well culture plate and treated with violacein for 6 h. The cells were stained with 15 μM DCFH-DA by incubating for 10 min. The fluorescent images were obtained using a fluorescence microscope (Optinity KI-2000F, Korea Lab Tech, Seong Nam, Korea), and the fluorescence density was analyzed using ImageJ software, version 1.5 (NIH).

4.8. Analysis of Apoptosis

Huh7 cells (2×10^5 cells/well) were seeded in a 60 mm cell culture dish and treated with violacein for 24 h. The cells were harvested, washed with PBS, and stained with 100 μL of Muse® Annexin V & Dead Cell reagent, according to the manufacturer's instructions. The stained cells were analyzed using a Guava® Muse® Cell Analyzer (MuseSoft_V1.8.0.3; Luminex Corporation, Austin, TX, USA).

4.9. Cell Cycle Analysis

Huh7 cells (1×10^5 cells/well) were seeded in a 60 mm cell culture dish and treated with violacein for 24 h. The cells were harvested, washed with PBS, and fixed with 70% ethanol at $-20\ °C$ for 3 h. After washing with PBS, the cells were stained with 200 μL of Muse® Cell Cycle reagent, according to the manufacturer's instructions. The stained cells were analyzed using a Guava® Muse® Cell Analyzer.

4.10. Western Blot Analysis

The cells were lysed using RIPA buffer, supplemented with protease and phosphatase inhibitors (ATTO, Tokyo, Japan). The concentrations of the proteins were determined using the Pierce® BCA Protein Assay Kit (Thermo Fisher Scientific, Inc., Rockford, IL, USA). Equal amounts of cell lysates were separated by 7.5–15% sodium dodecyl sulfate-polyacrylamide gel electrophoresis (SDS-PAGE) and subsequently transferred to polyvinylidene difluoride (PVDF) membranes (EMD Millipore, Hayward, CA, USA) using standard electroblotting procedures. The blots were blocked with 5% skim milk in Tris-buffered saline with Tween-20 (TBST) at room temperature for 1 h, and immunolabeled with the primary antibodies against cleaved capase-3, cleaved caspase-3, PARP, p21, p53, phospho-ERK1/2, ERK1/2, phospho-AKT, AKT, phospho-STAT3, STAT3, ALDH1A1, integrin α6, Sox2, Oct4, Nanog, CD133, α-tubulin (dilution 1:2000), and β-actin (dilution 1:10000) by incubating overnight at 4 °C. After washing thrice with TBST, the membranes were incubated with horseradish peroxidase-conjugated anti-rabbit or anti-mouse (dilution 1:3000) secondary antibody for 1 h at room temperature. Immunolabeling was detected using an enhanced chemiluminescence (ECL) kit (Bio-Rad Laboratories, Hercules, CA, USA), according to the manufacturer's instructions. The density of the bands was analyzed using ImageJ software, version 1.5 (NIH).

4.11. Tumorsphere Forming Assay

Huh7 and Hep3B cancer stem-like cells (3×10^3 cells/well) were seeded in 96-well culture plates using serum-free media with EGF and bFGF, and they were treated with violacein at various concentrations. After 7 days of incubation, the size and number of tumorspheres were determined using an optical microscope (Olympus, Tokyo, Japan).

4.12. Statistical Analyses

The results are presented as the mean ± standard deviation (SD) of data obtained from at least three independent experiments. The differences among the groups were analyzed by analysis of variance (ANOVA), calculated using the SPSS software, version 9.0 (SPSS

Inc., Chicago, Ill., USA). Post hoc analysis was performed using Tukey's test. Statistical significance was considered at $p < 0.05$.

5. Conclusions

This study is the first to demonstrate the anticancer effect of violacein against HCC cells and elucidate the molecular mechanism underlying its antitumor activity. The results demonstrate that violacein effectively inhibited the proliferation of HCC cells by inducing cell cycle arrest at the sub-G1 phase and triggering apoptosis. The apoptosis induced by violacein was associated with nuclear condensation; loss of MMP; increased generation of ROS; activation of caspase-9, caspase-3, and PARP; upregulation of p53 and p21; and down regulation of AKT and ERK1/2 signaling. Furthermore, violacein significantly suppressed the proliferation and formation of tumorspheres of HCC stem-like cells by reducing the expression of HCC stemness markers, including CD133, Sox2, Oct4, and Nanog, and by inhibiting the STAT3/AKT/ERK signaling pathways. In conclusion, these findings suggest that violacein has the chemotherapeutic potential to effectively suppress HCC by targeting both the proliferation and stemness of HCC cells.

Supplementary Materials: The following are available online at https://www.mdpi.com/article/10.3390/ijms221910731/s1.

Author Contributions: Conceptualization, H.J.J.; methodology, Y.J.K. and N.Y.; software, H.J.J.; validation, Y.J.K.; formal analysis, Y.J.K. and N.Y.; investigation, Y.J.K., N.Y. and H.J.S.; resources, H.J.J.; data curation, Y.J.K. and N.Y.; writing—original draft preparation, Y.J.K.; writing—review and editing, H.J.J.; visualization, Y.J.K., N.Y. and H.J.S.; supervision, H.J.J.; project administration, H.J.J.; funding acquisition, H.J.J. All authors have read and agreed to the published version of the manuscript.

Funding: This research was supported by the Basic Science Research Program through the National Research Foundation of Korea funded by the Ministry of Science and ICT (NRF-2019R1A2C1009033) and the Ministry of Education (NRF-2021R1I1A3050093). This work was also supported by the Brain Korea 21 Project, Republic of Korea.

Institutional Review Board Statement: Not applicable.

Informed Consent Statement: Not applicable.

Data Availability Statement: The data that support the findings of this study are available from the corresponding author upon reasonable request.

Acknowledgments: We are very grateful to Jae Kyung Sohng (Sun Moon University, Asan, Korea) for providing violacein.

Conflicts of Interest: The authors declare no conflict of interest.

References

1. Altekruse, S.F.; Devesa, S.S.; Dickie, L.A.; McGlynn, K.A.; Kleiner, D.E. Histological classification of liver and intrahepatic bile duct cancers in SEER registries. *J. Regist. Manag.* **2011**, *38*, 201–205.
2. Bray, F.; Ferlay, J.; Soerjomataram, I.; Siegel, R.L.; Torre, L.A.; Jemal, A. Global cancer statistics 2018: GLOBOCAN estimates of incidence and mortality worldwide for 36 cancers in 185 countries. *CA Cancer J. Clin.* **2018**, *68*, 394–424. [CrossRef]
3. Petrick, J.L.; Florio, A.A.; Znaor, A.; Ruggieri, D.; Laversanne, M.; Alvarez, C.S.; Ferlay, J.; Valery, P.C.; Bray, F.; McGlynn, K.A. International trends in hepatocellular carcinoma incidence, 1978–2012. *Int. J. Cancer* **2020**, *147*, 317–330. [CrossRef]
4. Medavaram, S.; Zhang, Y. Emerging therapies in advanced hepatocellular carcinoma. *Exp. Hematol. Oncol.* **2018**, *7*, 17. [CrossRef] [PubMed]
5. Dimitroulis, D.; Damaskos, C.; Valsami, S.; Davakis, S.; Garmpis, N.; Spartalis, E.; Athanasiou, A.; Moris, D.; Sakellariou, S.; Kykalos, S.; et al. From diagnosis to treatment of hepatocellular carcinoma: An epidemic problem for both developed and developing world. *World J. Gastroenterol.* **2017**, *23*, 5282–5294. [CrossRef] [PubMed]
6. Ogunwobi, O.O.; Harricharran, T.; Huaman, J.; Galuza, A.; Odumuwagun, O.; Tan, Y.; Ma, G.X.; Nguyen, M.T. Mechanisms of hepatocellular carcinoma progression. *World J. Gastroenterol.* **2019**, *25*, 2279–2293. [CrossRef] [PubMed]
7. Liu, C.Y.; Chen, K.F.; Chen, P.J. Treatment of liver cancer. *Cold Spring Harb. Perspect. Med.* **2015**, *5*, a021535. [CrossRef]
8. Ikeda, M.; Morizane, C.; Ueno, M.; Okusaka, T.; Ishii, H.; Furuse, J. Chemotherapy for hepatocellular carcinoma: Current status and future perspectives. *Jpn. J. Clin. Oncol.* **2018**, *48*, 103–114. [CrossRef]

9. Keating, G.M. Sorafenib: A review in hepatocellular carcinoma. *Target. Oncol.* **2017**, *12*, 243–253. [CrossRef]
10. Personeni, N.; Pressiani, T.; Santoro, A.; Rimassa, L. Regorafenib in hepatocellular carcinoma: Latest evidence and clinical implications. *Drugs Context* **2018**, *7*, 212533. [CrossRef]
11. Spallanzani, A.; Orsi, G.; Andrikou, K.; Gelsomino, F.; Rimini, M.; Riggi, L.; Cascinu, S. Lenvatinib as a therapy for unresectable hepatocellular carcinoma. *Expert Rev. Anticancer Ther.* **2018**, *18*, 1069–1076. [CrossRef] [PubMed]
12. Cooper, E.H. The biology of cell death in tumours. *Cell Tissue Kinet.* **1973**, *6*, 87–95. [CrossRef] [PubMed]
13. Elmore, S. Apoptosis: A review of programmed cell death. *Toxicol. Pathol.* **2007**, *35*, 495–516. [CrossRef] [PubMed]
14. Kaufmann, S.H.; Earnshaw, W.C. Induction of apoptosis by cancer chemotherapy. *Exp. Cell Res.* **2000**, *256*, 42–49. [CrossRef] [PubMed]
15. Guzik, T.J.; Harrison, D.G. Vascular NADPH oxidases as drug targets for novel antioxidant strategies. *Drug Discov. Today* **2006**, *11*, 524–533. [CrossRef]
16. Fulda, S. Targeting apoptosis for anticancer therapy. *Semin. Cancer Biol.* **2015**, *31*, 84–88. [CrossRef] [PubMed]
17. Chiba, T.; Kamiya, A.; Yokosuka, O.; Iwama, A. Cancer stem cells in hepatocellular carcinoma: Recent progress and perspective. *Cancer Lett.* **2009**, *286*, 145–153. [CrossRef]
18. Lee, G.; Hall, R.R.; Ahmed, A.U. Cancer stem cells: Cellular plasticity, niche, and its clinical relevance. *J. Stem Cell Res. Ther.* **2016**, *6*, 363. [CrossRef]
19. Jordan, C.T.; Guzman, M.L.; Noble, M. Cancer stem cells. *N. Engl. J. Med.* **2006**, *355*, 1253–1261. [CrossRef]
20. Vu, N.B.; Nguyen, T.T.; Tran, L.C.D.; Do, C.D.; Nguyen, B.H.; Phan, N.K.; Pham, P.V. Doxorubicin and 5-fluorouracil resistant hepatic cancer cells demonstrate stem-like properties. *Cytotechnology* **2013**, *65*, 491–503. [CrossRef]
21. Nio, K.; Yamashita, T.; Kaneko, S. The evolving concept of liver cancer stem cells. *Mol. Cancer* **2017**, *16*, 4. [CrossRef] [PubMed]
22. Millimouno, F.M.; Dong, J.; Yang, L.; Li, J.; Li, X. Targeting apoptosis pathways in cancer and perspectives with natural compounds from mother nature. *Cancer Prev. Res.* **2014**, *7*, 1081–1107. [CrossRef]
23. Shin, H.J.; Han, J.M.; Choi, Y.S.; Jung, H.J. Pterostilbene suppresses both cancer cells and cancer stem-like cells in cervical cancer with superior bioavailability to resveratrol. *Molecules* **2020**, *25*, 228. [CrossRef]
24. Andrighetti-Fröhner, C.R.; Antonio, R.V.; Creczynski-Pasa, T.B.; Barardi, C.R.M.; Simões, C.M.O. Cytotoxicity and potential antiviral evaluation of violacein produced by *Chromobacterium violaceum*. *Mem. Inst. Oswaldo Cruz* **2003**, *98*, 843–848. [CrossRef]
25. Alshatwi, A.A.; Subash-Babu, P.; Antonisamy, P. Violacein induces apoptosis in human breast cancer cells through up regulation of BAX, p53 and down regulation of MDM2. *Exp. Toxicol. Pathol.* **2016**, *68*, 89–97. [CrossRef]
26. Carvalho, D.D.; Costa, F.T.M.; Duran, N.; Haun, M. Cytotoxic activity of violacein in human colon cancer cells. *Toxicol. In Vitro* **2006**, *20*, 1514–1521. [CrossRef]
27. Ferreira, C.V.; Bos, C.L.; Versteeg, H.H.; Justo, G.Z.; Durán, N.; Peppelenbosch, M.P. Molecular mechanism of violacein-mediated human leukemia cell death. *Blood* **2004**, *104*, 1459–1464. [CrossRef]
28. Scchmitz, K.J.; Wohlschlaeger, J.; Lang, H.; Sotiropoulos, G.C.; Malago, M.; Steveling, K.; Reis, H.; Cicinnati, V.R.; Schmid, K.W.; Baba, H.A. Activation of the ERK and AKT signalling pathway predicts poor prognosis in hepatocellular carcinoma and ERK activation in cancer tissue is associated with hepatitis C virus infection. *J. Hepatol.* **2008**, *48*, 83–90. [CrossRef] [PubMed]
29. Kim, S.M.; Han, J.M.; Le, T.T.; Sohng, J.K.; Jung, H.J. Anticancer and antiangiogenic activities of novel α-mangostin glycosides in human hepatocellular carcinoma cells via downregulation of c-Met and HIF-1α. *Int. J. Mol. Sci.* **2020**, *21*, 4043. [CrossRef] [PubMed]
30. Shih, P.C.; Mei, K.C. Role of STAT3 signaling transduction pathways in cancer stem cell-associated chemoresistance. *Drug Discov. Today* **2021**, *56*, 1450–1458. [CrossRef]
31. Ma, S.; Lee, T.K.; Zheng, B.J.; Chan, K.W.; Guan, X.Y. CD133+ HCC cancer stem cells confer chemoresistance by preferential expression of the Akt/PKB survival pathway. *Oncogene* **2008**, *27*, 1749–1758. [CrossRef]
32. Chen, H.A.; Kuo, T.C.; Tseng, C.F.; Ma, J.T.; Yang, S.T.; Yen, C.J.; Yang, C.Y.; Sung, S.Y.; Su, J.L. Angiopoietin-like protein 1 antagonizes MET receptor activity to repress sorafenib resistance and cancer stemness in hepatocellular carcinoma. *Hepatology* **2016**, *64*, 1637–1651. [CrossRef]
33. Tsuruo, T.; Naito, M.; Tomida, A.; Fujita, N.; Mashima, T.; Sakamoto, H.; Haga, N. Molecular targeting therapy of cancer: Drug resistance, apoptosis and survival signal. *Cancer Sci.* **2003**, *94*, 15–21. [CrossRef]
34. Lan, X.; Wu, Y.Z.; Wang, Y.; Wu, F.R.; Zang, C.B.; Tang, C.; Cao, S.; Li, S.L. CD133 silencing inhibits stemness properties and enhances chemoradiosensitivity in CD133-positive liver cancer stem cells. *Int. J. Mol. Med.* **2013**, *31*, 315–324. [CrossRef]
35. Ma, S.; Chan, K.W.; Hu, L.; Lee, T.K.W.; Wo, J.Y.H.; Ng, I.O.L.; Zheng, B.J.; Guan, X.Y. Identification and characterization of tumorigenic liver cancer stem/progenitor cells. *Gastroenterology* **2007**, *132*, 2542–2556. [CrossRef] [PubMed]
36. Barclay, W.W.; Axanova, L.S.; Chen, W.; Romero, L.; Maund, S.L.; Soker, S.; Lees, C.J.; Cramer, S.D. Characterization of adult prostatic progenitor/stem cells exhibiting self-renewal and multilineage differentiation. *Stem Cells* **2008**, *26*, 600–610. [CrossRef] [PubMed]
37. Tomita, H.; Tanaka, K.; Tanaka, T.; Hara, A. Aldehyde dehydrogenase 1A1 in stem cells and cancer. *Oncotarget* **2016**, *7*, 11018–11032. [CrossRef] [PubMed]
38. Ghoshal, S.; Fuchs, B.C.; Tanabe, K.K. STAT3 is a key transcriptional regulator of cancer stem cell marker CD133 in HCC. *Hepatobiliary Surg. Nutr.* **2016**, *5*, 201–203. [CrossRef] [PubMed]

39. Sun, C.; Sun, L.; Jiang, K.; Gao, D.M.; Kang, X.N.; Wang, C.; Zhang, S.; Huang, S.; Qin, X.; Li, Y.; et al. NANOG promotes liver cancer cell invasion by inducing epithelial–mesenchymal transition through NODAL/SMAD3 signaling pathway. *Int. J. Biochem. Cell Biol.* **2013**, *45*, 1099–1108. [CrossRef] [PubMed]
40. Huang, P.; Qiu, J.; Li, B.; Hong, J.; Lu, C.; Wang, L.; Wang, J.; Hu, Y.; Jia, W.; Yuan, Y. Role of Sox2 and Oct4 in predicting survival of hepatocellular carcinoma patients after hepatectomy. *Clin. Biochem.* **2011**, *44*, 582–589. [CrossRef]

Article

Incomptine A Induces Apoptosis, ROS Production and a Differential Protein Expression on Non-Hodgkin's Lymphoma Cells †

Emmanuel Pina-Jiménez [1,2], Fernando Calzada [3,*], Elihú Bautista [4], Rosa María Ordoñez-Razo [2], Claudia Velázquez [5], Elizabeth Barbosa [6] and Normand García-Hernández [2,*]

1. Posgrado en Ciencias Biológicas, Universidad Nacional Autónoma de México, Ciudad Universitaria 3000, Coyoacán, Mexico City CP 04510, Mexico; epinaj@hotmail.com
2. Unidad de Investigación Médica en Genética Humana, UMAE Hospital Pediatría 2° Piso, Centro Médico Nacional Siglo XXI, Instituto Mexicano del Seguro Social, Av. Cuauhtémoc 330, Col. Doctores, Mexico City CP 06725, Mexico; romaorr@yahoo.com.mx
3. Unidad de Investigación Médica en Farmacología, UMAE Hospital de Especialidades, 2° Piso CORSE, Centro Médico Nacional Siglo XXI, Instituto Mexicano del Seguro Social, Av. Cuauhtémoc 330, Col. Doctores, Mexico City CP 06725, Mexico
4. CONACYT—Consorcio de Investigación, Innovación y Desarrollo para las Zonas Áridas, Instituto Potosino de Investigación Científica y Tecnológica A.C., San Luis Potosí CP 78216, Mexico; francisco.bautista@ipicyt.edu.mx
5. Área Académica de Farmacia, Instituto de Ciencias de la Salud, Universidad Autónoma del Estado de Hidalgo, Km 4.5, Carretera Pachuca-Tulancingo, Unidad Universitaria, Pachuca CP 42076, Mexico; cvg09@yahoo.com
6. Sección de Estudios de Posgrado e Investigación, Escuela Superior de Medicina, Instituto Politécnico Nacional, Salvador Díaz Mirón esq. Plan de San Luis S/N, Miguel Hidalgo, Casco de Santo Tomas, Mexico City CP 11340, Mexico; rebc78@yahoo.com.mx
* Correspondence: fercalber10@gmail.com (F.C.); normandgarcia@gmail.com (N.G.-H.)
† This work is taken the Ph.D. thesis from Emmanuel Pina-Jiménez.

Abstract: Sesquiterpene lactones are of pharmaceutical interest due their cytotoxic and antitumor properties, which are commonly found within plants of several genera from the Asteraceae family such as the *Decachaeta* genus. From *Decachaeta incompta* four heliangolide, namely incomptines A–D have been isolated. In this study, cytotoxic properties of incomptine A (**IA**) were evaluated on four lymphoma cancer cell lines: U-937, Farage, SU-DHL-2, and REC-1. The type of cell death induced by **IA** and its effects on U-937 cells were analyzed based on its capability to induce apoptosis and produce reactive oxygen species (ROS) through flow cytometry with 4′,6-diamidino-2-phenylindole staining, dual annexin V/DAPI staining, and dichlorofluorescein 2′,7′-diacetate, respectively. A differential protein expression analysis study was carried out by isobaric tags for relative and absolute quantitation (iTRAQ) through UPLC-MS/MS. Results reveal that **IA** exhibited cytotoxic activity against the cell line U-937 (CC$_{50}$ of 0.12 ± 0.02 μM) and the incubation of these cells in presence of **IA** significantly increased apoptotic population and intracellular ROS levels. In the proteomic approach 1548 proteins were differentially expressed, out of which 587 exhibited a fold-change ≥ 1.5 and 961 a fold-change ≤ 0.67. Most of these differentially regulated proteins are involved in apoptosis, oxidative stress, glycolytic metabolism, or cytoskeleton structuration.

Keywords: incomptine A; sesquiterpene lactone; *Decachaeta incompta*; cytotoxic activity; iTRAQ; apoptosis; ROS production

1. Introduction

The non-Hodgkin's lymphomas (NHL) are a heterogeneous group of illnesses that arise primarily in the lymph nodes or other lymph tissues due to malignant transformation of B, T, and NK lymphocytes. In 2015, nearly 4.3 million people around the world were reported with NHL, which accounted for 231,400 deaths. In Mexico until 2017, NHL were the fourth cause of mortality between patients with cancer [1]. The anticancer drugs commonly used alone or combined for the treatment of NHL include cyclophosphamide, prednisone, cisplatin, methotrexate, and doxorubicin as well as vincristine and etoposide. All these drugs show strong side effects. In Mexico, methotrexate is regarded as the preferred drug, however, in most cases, high dose methotrexate administration required for lymphoma treatment leads to genotoxic damage and the appearance of side effects including acute kidney injury, nephrotoxicity, myelosuppression, mucositis, and hepatotoxicity as well as dermatologic toxicity [1,2]. Therefore, there is a need to search and develop new, safer, and more effective anticancer agents.

Incomptine A (**IA**) is a heliangolide-type sesquiterpene lactone isolated from leaves of *Decachaeta incompta* [3]. The study of sesquiterpene lactones (**SL**) has led to discover the molecular mechanisms involved in the cytotoxicity of these compounds which include reduced glutathione depletion, prevented NF-κB activation, increased intra-cellular reactive oxygen species levels, and downregulation of Bcl-2 anti-apoptotic proteins, as well as intrinsic apoptosis activation [4]. Furthermore, the SL can inhibit glycolytic enzymes [5], leading to energy metabolism disruption and compromising cell viability [6]. **IA** has a 3-methylenedihydrofuran-2(3H)-one moiety, to which the cytotoxic properties of sesquiterpene lactones have been associated, due to its capability to act as a Michael acceptor and react with sulfhydryl residues of proteins. Previously, it was reported that **IA** (Figure 1) induces energy metabolism disruption as result of downregulation of the glycolytic enzymes enolase and fructose biphosphate aldolase in *Entamoeba histolytica* trophozoites [7], and exhibit a broad range of biological properties including antiprotozoal, antibacterial, trypanocidal, phytotoxic, and spermatic activities as well as anti-propulsive properties [3,8,9]. Herein, we examined the cytotoxic effects of **IA** against U-937 cells including apoptosis induction, and reactive oxygen species (ROS) production, as well as the differential protein expression.

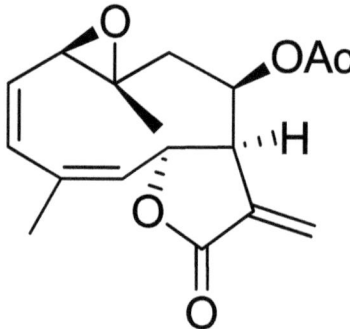

Figure 1. Structure of incomptine A (**IA**).

2. Results and Discussion

Non-Hodgkin's lymphoma represents a wide spectrum of illnesses that are a significant cause of morbidity and mortality around the world. Chemotherapy is an effective treatment against NHL that is used alone or combined including methotrexate, cyclophosphamide and doxorubicin. However, until today all the drugs used induced various side effects such is the case of methotrexate that is the drug of choice in Mexico [1,2,7]. In an effort to improve the therapy of cancer, specifically NHL, the study of specialized metabolites isolated from medicinal plants such as sesquiterpene lactones [4] constitute a source of potential new anti-lymphoma compounds with low toxicity. Based on these facts, the above prompted to assay the cytotoxic effects of **IA** and explore its mechanism of action.

In the present work, the cytotoxic activities of incomptine A (**IA**) against four subtypes of NHL cell lines, U-937 (diffuse histiocytic lymphoma), Farage (diffuse large B-cell non-Hodgkin's lymphoma), SU-DHL-2 (diffuse large B-cell lymphoma), and REC-1 (mantle cell lymphoma), were evaluated by DAPI staining for 24 h using methotrexate as a drug control and flow cytometry analysis. The results (Table 1) revealed that all subtypes of NHL human cells used were susceptible to incomptine A (**IA**) with CC_{50} values ranging from 0.12 ± 0.02 to 3.5 ± 0.01 µM, being most cytotoxic against the U-937 cell line. In this case, incomptine A (**1**) was four-fold more active than methotrexate, used as drug control. Additionally, the use of high doses of methotrexate has been reported to be toxic to humans [2] and mice [10]. These results suggest that **IA** has antitumoral potential [4,11–13].

Table 1. Cytotoxic activity of incomptine A in CC_{50} (mM) ± SEM against four sub-types of NHL human cell lines.

NHL Cell Line	Incomptine A (IA) CC_{50} (mM) [a]	Methotrexate CC_{50} (mM) [a]
U-937	0.12 ± 0.02	0.5 ± 0.004
Farage	2.3 ± 0.55	1.5 ± 0.01
SU-DHL-2	3.2 ± 0.04	1.3 ± 0.01
REC-1	3.5 ± 0.01	8.1 ± 0.05

[a] Cytotoxic concentration required to kill 50% of the cells (CC_{50}). Data represent the means ± SEM, were analyzed using Graph Pad Prism, $n = 3$. NHL, non-Hodgkin's lymphoma.

In addition, because DAPI staining was used, the strong cytotoxic properties of **IA** can be associated with an apoptotic process [11,12]. To explore the possible mechanism of action of cell death induced on U-937 cell lines. Cells were incubated with a concentration equivalent to CC_{50} value of **IA** (0.12 µM) or MTX (0.5 µM) for 24 h, and were then harvested, stained with annexin V/DAPI and further analyzed by flow cytometry. The distribution of cells resulting from this experiment is shown in Figure 2A. U-937 cells treated with **IA** or MTX showed a decrease in the percentage of viable cells with 39.9% and 49.4%, respectively, compared to control (84.3%). Additionally, the results showed an important early apoptotic effect on U-937 cells caused by incomptine A (**IA**), it was closer than MTX (Figure 2B).

The production of reactive oxygen species (ROS) has been associated with apoptosis induction under physiologic and pathological conditions [13–15]. Previous studies carried out with other SL indicate that cytotoxic, antitumor, and anticancer activities are associated to the presence of a 3-methylenedihydrofuran-2 (3H)-one moiety [4]. Therefore, interaction of this moiety with reduced glutathione (GSH), leads to its depletion and in consequence, the accumulation of reactive oxygen species that damage cell biomolecules such as lipids, structural proteins and DNA, as well as cause the initiation of mitochondria-dependent apoptosis pathway [4,16]. Based on the above, the production of ROS induced by **IA** was also measured. The results showed (Figure 3A) that DCF oxidation shifts the fluorescence peak towards the right, which is proportional to ROS production. Compared to positive control (H_2O_2), compound **IA** slightly shifts the DCF fluorescence peak, and MTX was inactive. Mean fluorescence intensity of DCF was recorded for each treatment and subsequently the DCF index was calculated with respect to control (Figure 3B). We observed that compound **IA** and H_2O_2 had significant differences in DCF oxidation, whereas MTX showed no changes. The cytotoxic activity, apoptosis induction, and increase of reactive oxygen species suggest that **IA** induces oxidative stress that finally leads to apoptosis on U-937 cells [4,16], unlike methotrexate where endogenous ROS levels did not increase, suggesting that incomptine A (**IA**) and MTX have a different action mechanism.

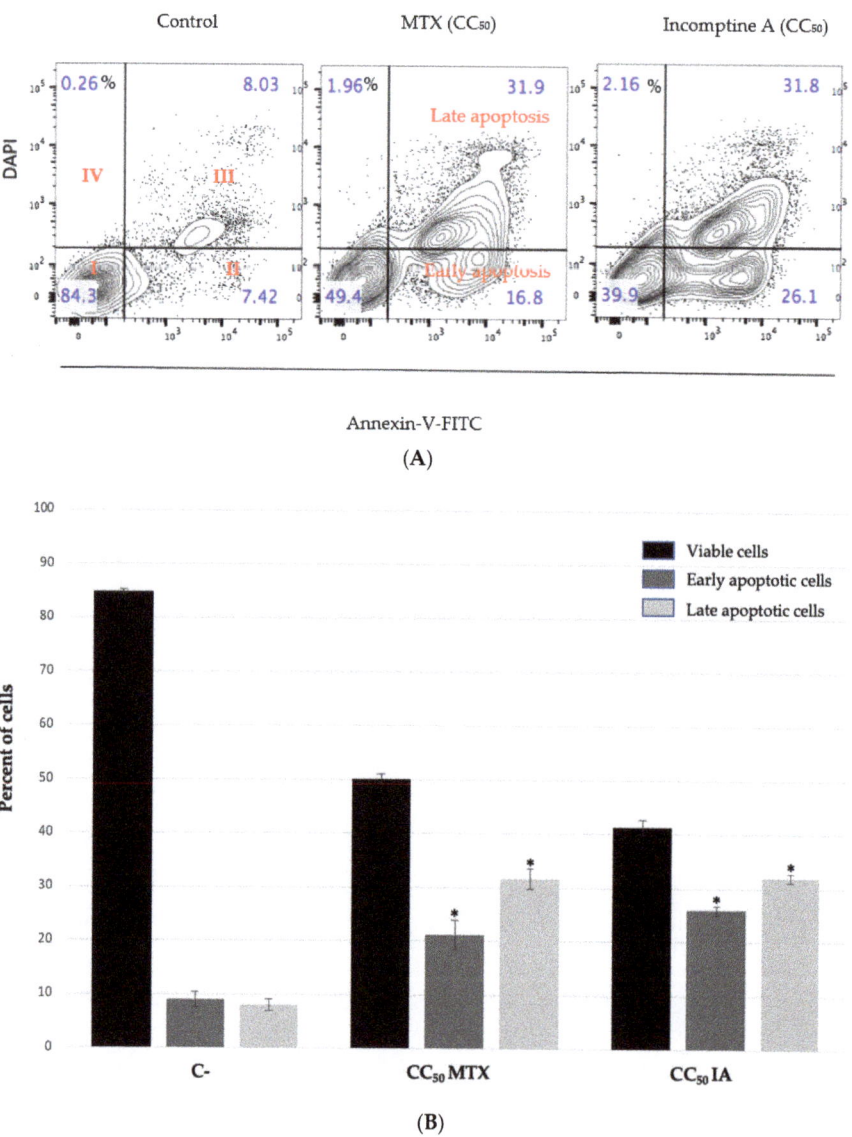

Figure 2. Apoptosis induced on U-937 cells by incomptine A (**IA**) and methotrexate (MTX). U-937 cells were treated with CC_{50} (mM) of compounds for 24 h. (**A**) Representative images of contour maps from evaluated treatments with relative death percentage for each quadrant (red numbers). Quadrant I annexin V−/DAPI−; quadrant II annexin V+/DAPI− early apoptosis; quadrant III annexin V+/DAPI+ late apoptosis, and quadrant IV annexin V−/DAPI−. Apoptotic population increases in IA and MTX CC_{50} treatments, revealing apoptosis induction. (**B**) Histograms of apoptosis were calculated respectively to negative control, finding significant differences in incomptine A and MTX treatments. Data is expressed as means ± SEM, $n = 3$; * $p < 0.05$.

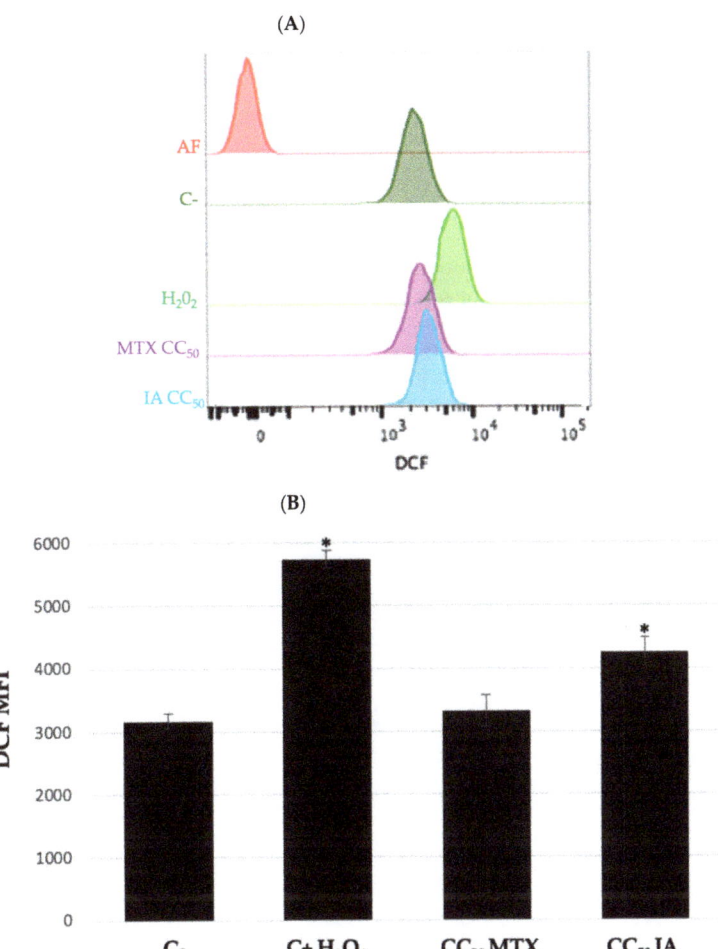

Figure 3. Production of ROS. Quantification was performed by luminescence using DCF reagent. Incomptine A (**IA**) increases oxygen reactive species in U-937 cells. (**A**) Representative histogram from evaluated treatments. Fluorescence peak of DCF moves towards the right depending on DCF oxidation by increasing intracellular reactive oxygen species, positive control H_2O_2 showed the maximum shift of fluorescence peak. (**B**) DCF index was calculated, respectively, to negative control, finding significant differences in positive control H_2O_2 and incomptine A treatments. Data is expressed as means ± SEM, $n = 3$; * $p < 0.05$. MTX, methotrexate; IA, incomptine A; DCF, dichlorofluorescein 2′,7′-diacetate.

To understand the cytotoxic mechanism of **IA** on U-937 cells, a proteomic approach was carried out, in which we analyzed the differential protein expression by isobaric tags for relative and absolute quantitation (iTRAQ) coupled with liquid chromatography–tandem mass spectrometry (LC-MS/MS) [17]. U-937 cells were treated with incomptine A (0.12 µM) or methotrexate (0.5 µM) for 24 h and non-treated cells were used as control. We identified and quantified 3222 proteins for all evaluated conditions. Normalized reporter values were used to calculate the ratios of each treatment with respect to control for all proteins [17,18]. Fold change criteria for upregulated proteins were established for those with ratios $\geq$ 1.5 and downregulated for proteins with ratios $\leq$ 0.67. Differentially expressed proteins were analyzed using STRING Functional Enrichment Analysis to identify biological processes and pathways that were altered by administration of incomptine A (**IA**) [17–19]. Enrichment analysis revealed 82 significantly enriched biological processes and signaling pathways ($p < 0.05$), but only those with percentages of proteins involved in each process greater than 5% were plotted in Figure 4. We observed that IA modifies the expression of proteins involved in apoptosis, oxidative stress response, DNA repair, cell cycle checkpoints, glycolysis, cytoskeleton organization, NF-κB signaling and autophagy (Table 2); whereas, the MTX alter the expression of proteins implicated in purine nucleotide biosynthesis related pathways, G2/M checkpoint and intrinsic apoptosis. Proteomic analysis was validated by identifying the reported antitumor mechanism of methotrexate, which depends on dihydrofolate reductase inhibition to prevent catalysis of dihydrofolate to tetrahydrofolate, an essential cofactor in purine and thymidylate biosynthesis, which are crucial events for cell division and proliferation [20]. Exhausted nucleotide biosynthesis is associated with a poor prognosis of neoplasms of breast cancer by action of guanosine-5-triphosphate (GTP), that generates guanosine monophosphate (GMP), which is required for cell proliferation by MAPK pathway [21]. Our data confirmed that biological processes such as GMP, nucleobase, nucleoside and purine ribonucleoside monophosphate biosynthesis were differentially expressed by MTX treatment on U-937 cells, intrinsic apoptosis signaling, DNA synthesis and G2/M checkpoint [22]. The activity of MTX seems to vary among different cell groups, since in lymphoid cell lines it increases reactive oxygen species, leading to DNA damage, activation of cell cycle arrest and DNA repair mechanisms whereas oxidative stress induction occurs to a lesser extent in monocytic cells [23,24], which was confirmed in our studied U-937 histiocytic lymphoma monocytes.

The treatment with incomptine A (**IA**) differentially expressed changes in the amounts of proteins involved in process cells such as apoptotic mitochondrial fragmentation, apoptosis, detoxification of ROS, cell cycle checkpoints, DNA repair, autophagy, canonical glycolysis, pyruvate metabolism and cytoskeleton organization. Upregulation of pro-apoptotic proteins BAX, PDCD4 and NDRG1 in our U-937 cells indicates modification of mitochondrial membrane associated with activation of intrinsic apoptosis [25]. Glutathione transferases are involved in chemoresistance and have been studied as molecular targets. These enzymes can reduce glutathione levels or inhibit their function to prevent detoxification generated by chemotherapy [26], leading to generation of oxidative stress which induces DNA damage and apoptosis [27].

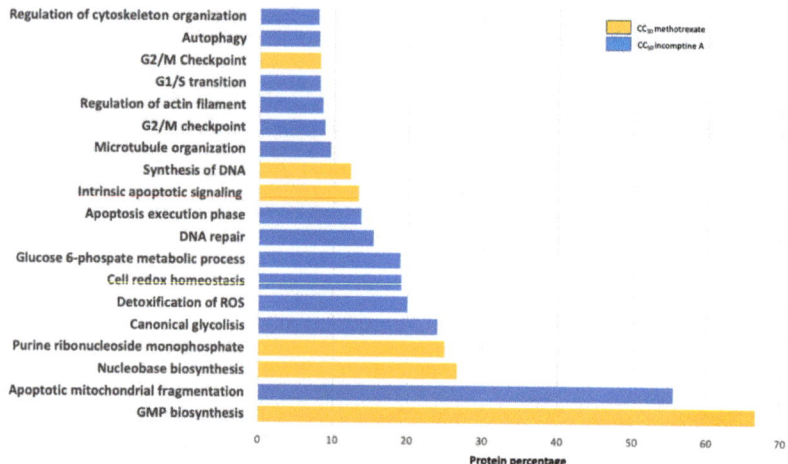

Figure 4. Functional enrichment analysis. Enrichment analysis revealed 82 significantly enriched biological processes and signaling pathways ($p < 0.05$). Intrinsic apoptosis can be found in incomptine A and methotrexate treatments, but different biologic processes are differentially expressed.

Table 2. Protein expression dysregulation involved in U-937 cell death induced by incomptine A (**IA**) and identified by iTRAQ. Fold change (FC) values are shown for all identified key proteins for each treatment.

Protein ID	Protein Name	Gene Name	MTX (FC)	IA (FC)
Q07812	Apoptosis regulator BAX	BAX	1.12	1.75
Q53EL6	Programmed cell death protein 4	PDCD4	1.98	2.80
Q92597	NDRG1 protein	NDRG1	1.53	1.64
Q16548	Bcl-2-related protein A1	BCL2A1	0.77	0.53
Q9NYF8	Bcl-2-associated transcription factor 1	BCLAF1	0.56	0.26
P00374	Dihydrofolate reductase	DHFR	1.42	1.77
P11586	Tetrahydrofolate synthase	MTHFD1	1.29	2.1
P60891	Ribose phosphate pyrophosphokinase 1	PRPS1	1.65	2.24
P10620	Microsomal glutathione S-transferase 1	MGST1	0.76	0.59
Q9Y2Q3	Glutathione S-transferase kappa 1	GSTK1	0.65	0.49
Q06830	Peroxiredoxin-1	PRDX1	0.93	2.74
P32119	Peroxiredoxin-2	PRDX2	1.21	1.95
P30041	Peroxiredoxin-6	PRDX6	1.27	2.29
O14920	Inhibitor of nuclear factor kappa-B kinase subunit beta	IKBKB	1.23	1.73
P19838	Nuclear factor NF-kappa-B p105 subunit	NFKB1	1.11	1.68
Q04206	Nuclear factor NF-kappa-B p65 subunit	RELA	1.79	1.62
Q04864	c-Rel	REL	1.45	1.9
P42224	Signal transducer and activator of transcription 1-alpha/beta	STAT1	1.50	1.97
P30281	G1/S-specific cyclin-D3	CCND3	1.21	0.57
P43246	DNA mismatch repair protein Msh2	MSH2	1.39	1.95
O95352	Autophagy protein 7	ATG7	1.40	2.73
Q9NT62	Autophagy protein 3	ATG3	1.31	2.21
P11279	Lysosomal associated membrane protein 1	LAMP1	0.91	1.60
P13473	Lysosomal associated membrane protein 2	LAMP2	1.47	1.86
P00338	L-lactate dehydrogenase A chain	LDHA	0.97	0.54
P07195	L-lactate dehydrogenase B chain	LDHB	0.78	0.51
P04075	Fructose-bisphosphate aldolase A	ALDOA	1.06	0.34
P18669	Phosphoglycerate mutase 1	PGAM1	1.25	0.62
O43707	Alpha actinin 4	ACTN4	1.27	0.51
Q6IRU2	Tropomyosin alpha-4 chain	TPM4	0.96	0.43

We identified downregulation of glutathione transferases caused by incomptine A (**IA**), and some of them were microsomal glutathione S-transferase and glutathione S-transferase kappa 1, upregulation of peroxiredoxin oxidative stress response transferases PRDX1, PRDX2 and PRDX6. Oxidative stress induced by **IA**, also led to upregulation of DNA repair protein MSH2, involved in 8-oxoguanine repair [27,28], and autophagic proteins ATG3, ATG7, LAMP1 and LAMP2 on U-937 cells, coinciding with previously reported autophagic activity of sesquiterpene lactones [29].

According to the results obtained, the oxidative stress observed in U-937 cells by ROS production, and the subsequent proteomic analysis suggested that a mechanism involving NF-κB is not possible due to NF-κB signaling p105 and p65 were found upregulated. NF-κB is known for being upregulated in cancer, conferring tumor cells the ability to evade apoptosis through the expression of Bcl-2 family anti-apoptotic proteins [30,31]. Anti-apoptotic Bcl-2 family members were not detected dysregulated, only the Bcl-2 associated transcription factor 1 was found downregulated, which confirms that **IA** does not interact with NF-κB on U-937 cells.

On the other hand, aerobic glycolysis is known to be abnormally activated in non-Hodgkin's lymphoma, conferring drug resistance to tumor cells, consequently glycolytic enzymes are also considered as drug development targets [32]. Interestingly, the proteomic analysis revealed differential expression of glycolysis pathway, being lactate dehydrogenase A (LDHA), lactate dehydrogenase B (LDHB) and fructose-biphosphate aldolase A (ALDOA), the identified downregulated glycolytic enzymes by (**IA**) treatment. ALDOA is a key glycolytic enzyme, the high expression of which has been associated to tumor progression and poor prognosis in hepatocellular carcinoma [33]. Glycolytic enzymes play a physiological role that is not limited to catalytic function and can participate as regulators of cellular processes, such as actin filaments organization, p53 signaling, and cell cycle progression, considered as necessary cellular events for survival and proliferation of cancer cells. When ALDOA-actin cytoskeleton interaction is perturbed, significant elevation of ROS production leads to decreased ATP synthesis, increased calcium levels and activates caspases [34]. Since **IA** also downregulated cytoskeleton elements such as alpha actinin-4 and tropomyosin alpha-4 chain, additional studies must be realized to know how compound **IA** treatment triggers the activation of this mechanism related to regulatory function of ALDOA, and how the downregulation of ALDOA and its glycolytic activity is associated to the induction of oxidative stress. To our knowledge, this is the first report that the inhibition of glycolytic enzymes (ALDOA, LDHA, and LDHB) and cytoskeleton proteins (alpha actinin-4 and tropomyosin alpha-4 chain) may be part of the mechanism by which SL produce their antitumor effects.

In order to visualize the different clusters involved in pro-apoptotic activity of **IA**, we performed a Protein–Protein Interaction Analysis (PPI) of differentially expressed proteins (Figure 5) and the results obtained suggest that the possible action mechanism may be the apoptosis activation, where glycolytic enzymes ALDOA, LDHA and LDHB were biologically connected to pro-apoptotic proteins as BAX and BID, as well to the glutathione transferases MGST1 and GSTK1 [33,34].

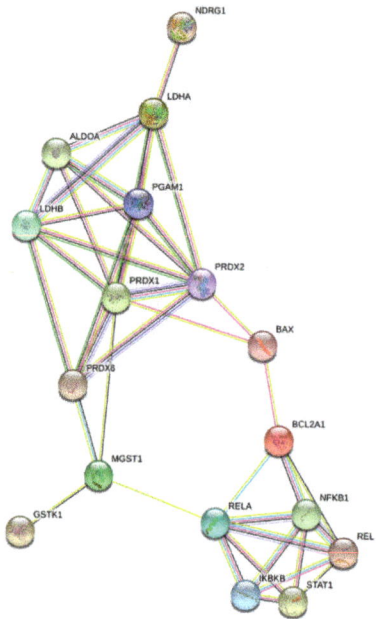

Figure 5. Protein–protein interaction (PPI) networks of differentially expressed proteins. Line shape indicates the predicted molecular mechanism of action. PPI presented 17 nodes and 33 edges with a PPI enrichment p-value < 1.0 meaning that proteins are biologically connected as a group.

3. Conclusions

This study used cytotoxic effects of incomptine A (**IA**) against U-937 cells including apoptosis induction, and reactive oxygen species (ROS) production, followed by iTRAQ labeling approach to determinate its potential as an antitumor agent. Our findings suggest that **IA** is a sesquiterpene lactone with antitumor potential that induces apoptosis by redox imbalance [4] on U-937 cells. In addition, antitumor potential could be associated with inhibition of glycolytic enzymes such as LDHA, LDHB and ALDOA. Future systematic studies will investigate how **IA** regulates the expression of glycolytic enzymes and its correlation with apoptosis induction and ROS production. Research should be directed to confirm the antitumor potential of **IA** in non-Hodgkin's lymphoma using animal models [1].

4. Materials and Methods

4.1. Incomptine A (IA) Isolation

Compound **IA** was isolated from the aerial parts of *Decachaeta incompta* (syn.: *Eupatorium incomptum*; Asteraceae) collected in the State of Oaxaca, Mexico. The plant was identified by the MS Abigail Aguilar Contreras taxonomist of the Instituto Mexicano del Seguro Social (IMSS). A voucher specimen (15311) was deposited in the Herbarium IMSSM. The extraction and isolation procedure were performed according the protocol previously described [3]. Identification of **IA** was made by comparison (NMR, TLC, and HPLC-DAD) with an authentic sample, having a purity near to 99%. Dimethyl sulfoxide (DMSO) was used to dissolve **IA** and methotrexate (PISA pharmaceutical).

4.2. Chemicals

TMT10plex Isobaric Label Reagent Set, Pierce Quantitative Colorimetric Peptide Assay (Thermo Fisher Science, Waltham, MA, USA). Triethylammonium bicarbonate buffer (1.0 M, pH 8.5 ± 0.1), Tris (2-carboxyethyl) phosphine hydrochloride solution (0. M, pH 7.0), iodoacetamide (IAA), formic acid (FA), acetonitrile (MeCN), methanol (MeOH), and trypsin from bovine pancreas (Promega, Madison, WI, USA) were used. Ultrapure water was obtained from a Millipore purification system.

4.3. Cell Cultures and Reagents

NHL cell lines: U-937 (histiocytic lymphoma, cat. CRL-1593.2); SU-DHL-2 (large B-cell lymphoma, cat. CRL-2956); Farage (B-cell lymphoma, cat. CRL-2630), and REC-1 (mantle B-cell lymphoma, cat. CRL-3004) were obtained from the American Type Culture Collection (ATCC). Cells were cultured in RPMI 1640 culture medium (GIBCO Cat: 11875–093), added with 5% fetal bovine serum (GIBCO Cat: 16000044) in a 37 °C incubator and 5% CO_2. Cell cultures were maintained at a density of 2×10^6 cells in T75 flasks.

4.3.1. Cytotoxic Activity

Concentration that kills 50% cell population (CC_{50}) was determined in NHL cell lines after 24 h upon exposure to **IA** by flow cytometry viability analysis. In total, 50,000 cells were seeded in 500 µL of RPMI 1640 medium supplemented with fetal bovine serum in 48-well culture plates. Cells were allowed to grow for 4 h prior to exposure to different concentrations (0.01, 0.1, 1, and 10 µM) of **IA** or methotrexate (0.01, 0.1, 1, and 10 µM). DMSO treatments were used as solvent control to rule out that cell death is due to its effect. A working solution of **IA** was prepared with RPMI 1640 and DMSO. A total of 10,000 events per sample were registered after 24 h of treatment on a FACS Verse cytometer (Beckton Dickinson) equipped with filters 404, 488, and 561 nm. DAPI was excited by the 404 nm laser and emission light was collected with filter 450/50 BP [35,36]. Percentage of mortality (DAPI positive cells) was recorded for each treatment and linearized on Excel software to obtain CC_{50}.

4.3.2. Annexin V/DAPI Apoptosis Assay

Apoptotic cell death index induced on U-937 cell line by **IA** was determined by flow cytometry using the Annexin V/DAPI apoptosis detection kit (BD Bio-sciences), according to the manufacturer instructions. In total, 50,000 cells were seeded in 500 µL of RPMI 1640 medium supplemented with fetal bovine serum in 48-well culture plates. Cells were allowed to grow for 4 h prior to exposure to the concentrations of **IA** (0.12 µM) or methotrexate (0.5 µM). Methotrexate was used as a drug control. DMSO was used as solvent control to rule out that cell death. A total of 10,000 events of each treatment were registered after 24 h of treatment administration on a FACS Verse cytometer (Beckton Dickinson). Relative percentages of apoptotic cells were recorded for each treatment [35,36].

4.3.3. Reactive Oxygen Species Assay

Oxidative stress induced on U-937 cell line by **IA** was evaluated by flow cytometry. The increase of reactive oxygen species levels was detected based on the 2′7′dichlorofluorescin di-acetate (DCF) oxidation to 2′7′dichlorofluorescin assay. In total, 50,000 cells were seeded in 500 µL of RPMI 1640 medium supplemented with fetal bovine serum in 48-well culture plates and treated for 30 min with **IA** (0.12 µM) or methotrexate (0.5 µM). DMSO treatment was used as solvent control to rule out that ROS increased levels is due to its effect at equivalent concentration of incomptine A (**IA**) CC_{50}. A total of 50 µM hydrogen peroxide (H_2O_2) was used as a positive control of increasing ROS levels agent and a negative control without treatment were used. DCF mean fluorescence intensity (MFI) of DCF were recorded to perform comparisons between treatments [36].

4.3.4. Cell Lysis and iTRAQ-Based Proteomic Analysis

First, 1×10^8 U-937 cells were harvested in 50 mL of RPMI 1640 medium added with 5% fetal bovine serum (GIBCO Cat: 16000044) at 37 °C and 5% CO_2. Amounts necessary to obtain final concentration of **IA** (0.12 µM) or methotrexate (0.5 µM) were added and the resulting mixtures were incubated for 24 h. Non-treated cells were used as negative control. After the treatment, cells were centrifuged at 2500 rpm × 3 min and resulting pellets were washed with 1 mL of PBS and lysed with 300 µL of lysis buffer (8 M urea, 0.8 M NH_4HCO_3, pH 8.0) and TissueLyser. After cell lysis the tubes were centrifuged at 12,000 rpm for 15 min at 4 °C. The supernatants were collected, and protein concentration was measured using the Pierce BCA protein assay kit. Then, 100 µg of proteins per sample were reduced with 5 µL of 10 mM TCEP for 1 h at 56 °C, and subsequently alkylated with 20 µL of 20 mM IAA for 1 h at room temperature in the dark. Then, free trypsin was added into the protein solution at a ratio of 1:50. The resulting solution was incubated at 37 °C overnight and then lyophilized. The residues were re-dissolved with TEAB (100 mM). After, 20 µL of anhydrous MeCN was added to each tube and shaken for 5 min at room temperature. Each sample was centrifugated to obtain a supernatant, then the samples were transferred to the iTRAQ (TMT10 plex reagent, Thermo Fisher Science) reagent vial as follows: C, TMT10-127C; MTX, TMT10-128C, and IA, TMT10-130N. The reaction mixtures were incubated 1 h at room temperature. The reaction was quenched by addition of 8 mL of hydroxylamine 5% and incubated again for 15 min and transferred to a new microcentrifuge tube

4.3.5. Nano LC-MS/MS Analyses

The iTRAQ labeled peptides were analyzed by LC-MS/MS on an Ultimate 3000 nano UHPLC system (Thermo Fisher Scientific) coupled to a Q Exactive HF mass spectrometer (Thermo Scientific) equipped with a nano spray flex ion source (Thermo Scientific). Briefly, 2 µg dissolved of each sample were injected onto a trap (PepMap C18, 100 Å, 100 µm × 2 cm, 5 µm) column, followed by fractionation on an analytical (PepMap C18, 100 Å, 75 µm × 50 cm, 2 µm) column. Linear gradients of 5–7% solvent B in 2 min, from 7% to 20% solvent B in 80 min, from 20% to 40% solvent B in 35 min, then from 40% to 90% solvent B in 4 min at 200 nL/min flow rate. Solvent A consisted of 0.1% formic acid in water and solvent B contained 0.1% formic acid in 80% Me CN. For TMT-labeled samples, the full scan was performed between the window m/z 350–1650 m/z at the resolution 120,000 at 200 Th. The automatic gain control target for the full scan was set to 3 to 6. The MS/MS scan was operated in Top 15 mode using the following settings: resolution 30,000 at 200 Th; automatic gain control target 1e5; normalized collision energy at 32%; isolation window of 1.2 Th; charge sate exclusion: unassigned, 1, >6; dynamic exclusion 40 s.

4.3.6. Protein Identification

The raw MS files of six replicates were analyzed and compared against the human protein database using Maxquant (1.6.2.6). The parameters were set as follows: the protein modifications were carbamidomethylation (C) (fixed), oxidation (M) (variable), TMT-10Plex; the enzyme specificity was set to trypsin; the maximum missed cleavages were set to 2; the precursor ion mass tolerance was set to 10 ppm, and MS/MS tolerance was 0.6 Da.

4.3.7. Differential Protein Analysis

Generated data were exported to an Excel file containing the 3222 identified and quantified proteins. Normalized TMT reporter values C (TMT10-127C), MTX (TMT10-128C), and IA (TMT10-130N) were used to determine protein differential expression between treatments, establishing the following ratios: MTX/C, and IA/C. The fold change criteria for upregulated proteins were established for those with ratios greater than ≥ 1.5 and downregulated for those proteins with ratios lower than ≤ 0.67. Differentially expressed proteins for each comparison were analyzed using STRING Functional Enrichment Analysis, Protein–Protein Interaction Analysis (PPI) and Kyoto Encyclopedia of Genes and Genomes (KEGG).

4.3.8. Statistical Analysis

Results are expressed as the mean values ± standard error of the mean. Statistical analyses were performed by using Excel version 16.46 for Macintosh. The statistical evaluation was carried out through an analysis of variance followed by Dunnet's test for multiple comparisons. $p < 0.05$ (*) was considered a statistically significant difference between the group means.

Author Contributions: Conceptualization, F.C. and N.G.-H.; methodology, E.P.-J., R.M.O.-R.; software, E.P.-J.; formal analysis, F.C. and N.G.-H.; investigation, E.P.-J., C.V. and E.B. (Elizabeth Barbosa) and E.B. (Elihú Bautista); resources, F.C. and N.G.-H.; writing—original draft preparation, F.C. and E.P.-J.; writing—review and editing, F.C., N.G.-H. and R.M.O.-R.; visualization; project administration, N.G.-H. and F.C. All authors have read and agreed to the published version of the manuscript.

Funding: This research received no external funding.

Informed Consent Statement: Not applicable.

Data Availability Statement: The additional data on this study is available on request from corresponding author.

Acknowledgments: This research was funded by Instituto Mexicano del Seguro Social, contract grant sponsor Fondo de Investigación en Salud: FIS/IMSS/PROT/EMER18/1846 (Approval no: R-2018-785-111). This work is from the PhD. Thesis of Emmanuel Pina-Jiménez. We would like to thank Consejo Nacional de Ciencia y Tecnologia for the scholarship (CVU 631725) and IMSS for the institutional scholarship (2015-070 99096818). We want to thank María Antonieta Chávez González for facilitating the use of the flow cytometer and reagents used in viability tests, apoptosis and reactive oxygen species. We want to thank David Douterlungne Rotsaert for his suggestions.

Conflicts of Interest: The authors declare no conflict of interest.

References

1. Calzada, F.; Ramirez-Santos, J.; Valdes, M.; Garcia-Hernandez, N.; Pina-Jiménez, E.; Ordoñez-Razo, R.M. Evaluation of acute oral toxicity, brine shrimp lethality, and antilymphoma activity of geranylgeraniol and *Annona macroprophyllata* Leaf extracts. *Rev. Bras. Farmacogn.* **2020**, *30*, 301–304. [CrossRef]
2. Howard, S.C.; McCormick, J.; Pui, C.H.; Buddington, R.K.; Harvey, R.D. Preventing and managing toxicities of high-dose methotrexate. *Oncologist* **2016**, *21*, 1471–1482. [CrossRef] [PubMed]
3. Calzada, F.; Yepez-Mulia, L.; Tapia-Contreras, A.; Alfredo, O. Antiprotozoal and antibacterial properties of *Decachaeta incompta*. *Rev. Latinoamer. Quim.* **2009**, *37*, 97–103.
4. Babaei, G.; Aliarab, A.; Abroon, S.; Rasmi, Y.; Aziz, S.G. Application of sesquiterpene lactone: A new promising way for cancer therapy based on anticancer activity. *Biomed. Pharmacother.* **2018**, *106*, 239–246. [CrossRef] [PubMed]

5. Baer-Dubowska, W.; Gnojkowski, J.; Chmiel, J. Inhibicja enzymów przemiany glikolitycznej przez laktony seskwiterpe nowe w limfocytach stymulowanych fitohemaglutynina [Inhibition of glycolytic enzymes by sesquiterpene lactones in phytohemagglutinin-stimulated lymphocytes]. *Folia Med. Cracov.* **1980**, *22*, 393–402. [PubMed]
6. Ritterson, L.C.; Tolan, D.R. Targeting of several glycolytic enzymes using RNA interference reveals aldolase affects cancer cell proliferation through a non-glycolytic mechanism. *J. Biol. Chem.* **2012**, *287*, 42554–42563.
7. Velazquez-Dominguez, J.; Marchat, L.A.; Lopez-Camarillo, C.; Mendoza-Hernandez, G.; Sanchez Espindola, E.; Calzada, F.; Ortega-Hernandez, A.; Sanchez-Monroy, V.; Ramirez-Moreno, E. Effect of the sesquiterpene lactone incomptine A in the energy metabolism of Entamoeba histolytica. *Exp. Parasitol.* **2013**, *135*, 503–510.
8. Guerrero, C.; Taboada, J.; Diaz, J.B.; Oliva, A.; Ortega, A. Incomptinas A y B dos heliangolidas aisladas de *Decachaeta incompta*. Estudio preliminar de las actividades biologicas de incomptina B. *Rev. Latinoamer. Quim.* **1994**, *23*, 142–147.
9. Calzada, F.; Valdes, M.; Barbosa, E.; Velazquez, C.; Bautista, E. Evaluation of antipropulsive activity of *Decachaeta in compta* (DC) King and Robinson and its sesquiterpene lactones on induced hyperperistalsis in rats. *Phcog. Mag.* **2020**, *16*, S272–S275. [CrossRef]
10. Calzada, F.; Solares-Pascasio, J.I.; Valdes, M.; Garcia-Hernandez, N.; Velazquez, C.; Ordoñez-Razo, R.M.; Barbosa, E. Antilymphoma potential of the ethanol extract and rutin obtained of the leaves from *Schinus molle* Linn. *Phcog. Res.* **2018**, *10*, 119–123. [CrossRef]
11. Abdolmohamma, M.H.; Fouladdel, S.; Shafiee, A.; Amin, G.; Ghaffari, S.; Azizi, E. Anticancer effects and cell cycle analysis on human breast cancer T47D cells treated with extracts of Astrodaucus persicus (Boiss.) Drude in comparison to doxorubicin. *J. Pharm. Sci.* **2008**, *16*, 112–118.
12. Perumal, A.; AlSalhi, M.S.; Kanakarajan, S.; DEvanesan, S.; Selvaraj, R.; Tamizhazhagan, V. Phytochemical evaluation and anticancer activity of rambutan (*Nephelium lappaceum*) fruit endocarp extracts against human hepatocellular carcinoma (HepG-2) cells. *Saudi J. Biol. Sci.* **2021**, *28*, 1816–1825. [CrossRef] [PubMed]
13. Suffness, M.; Douros, J. Current Current status of the NCI plant and animal product program. *J. Nat. Prod.* **1982**, *45*, 1–14. [CrossRef] [PubMed]
14. Simon, H.U.; Haj-Yehia, A.; Levi-Schaffer, F. Role of reactive oxygen species (ROS) in apoptosis induction. *Apoptosis* **2000**, *5*, 415–418. [CrossRef] [PubMed]
15. Halasi, M.; Wang, M.; Chavan, T.S.; Gaponeko, V.; Hay, N.; Gartel, A. ROS inhibitor N-acetyl-L-cysteine antagonizes the activity of proteasome inhibitors. *Biochem. J.* **2013**, *454*, 201–208. [CrossRef] [PubMed]
16. Zunino, S.J.; JMDucore Storms, D.H. Parthenolide induces significant apoptosis and production of reactive oxygen species in high-risk pre-B leukemia cells. *Cancer Lett.* **2007**, *254*, 119–127. [CrossRef]
17. Gautam, P.; Nair, S.C.; Gupta, M.K.; Rakesh, S.; Polisetty, R.V.; Uppin, M.S.; Sundaram, C.; Puligopu, A.K.; Ankathi, P.; Purohit, A.K.; et al. Proteins with Altered Levels in Plasma from Glioblastoma Patients as Revealed by iTRAQ-Based Quantitative Proteomic Analysis. *PLoS ONE* **2012**, *7*, e46153.
18. Li, S.; Su, X.; Jin, Q.; Li, G.; Sun, Y.; Abdullah, M.; Cai, Y.; Lin, Y. iTRAQ-Based identification of proteins related to lignin synthesis in the pear pollinated with pollen from different varieties. *Molecules* **2018**, *23*, 548. [CrossRef]
19. Zhang, Y.; Wang, Y.; Li, S.; Zhang, X.; Li, W.; Luo, S.; Sun, Z.; Nie, R. iTRAQ-based quantitative proteomic analysis of processed *Euphorbia lathyrism* L. for reducing the intestinal toxicity. *Proteome Sci.* **2018**, *16*, 8. [CrossRef]
20. Tran, P.N.; Tate, C.J.; Ridgway, M.; Saliba, K.J.; Kirk, K. Human dihydrofolate reductase influences the sensitivity of the malaria parasite Plasmodium falciparum to ketotifen-A cautionary tale in screening transgenic parasites. *Int. J. Parasitol. Drugs Drug Resist.* **2016**, *6*, 179–183. [CrossRef]
21. Lv, Y.; Wang, X.; Li, X.; Xu, G.; Bai, Y.; Wu, J.; Piao, Y.; Shi, Y.; Xiang, R.; Wang, L. Nucleotide de novo synthesis increases breast cancer stemness and metastasis via cGMP-PKG-MAPK signaling pathway. *PLoS Biol.* **2020**, *18*, e3000872. [CrossRef]
22. Spurlock, C.F., III; Tossberg, J.T.; Fuchs, H.A.; Olsen, N.J.; Aune, T.M. Methotrexate increases expression of cell cycle checkpoint genes via JNK activation. *Arthritis Rheum.* **2012**, *64*, 1780–1789. [CrossRef]
23. Herman, S.; Zurgil, N.; Deutsch, M. Low dose methotrexate induces apoptosis with reactive oxygen species involvement in T lymphocytic cell lines to a greater extent than in monocytic lines. *Inflamm. Res.* **2005**, *54*, 273–280. [CrossRef]
24. Canellos, G.P.; Skarin, A.T.; Rosenthal, D.S.; Moloney, W.C.; Frei, E. III. Methotrexate as a single agent and in combination chemotherapy for the treatment of non-Hodgkin's lymphoma of unfavorable histology. *Cancer Treat. Rep.* **1981**, *65*, 125–129.
25. Kuwana, T.; King, L.E.; Cosentino, K.; Suess, J.; Garcia-Saez, A.J.; Gilmore, A.P.; Newmeyer, D.D. Mitochondrial residence of the apoptosis inducer BAX is more important than BAX oligomerization in promoting membrane permeabilization. *J. Biol. Chem.* **2020**, *295*, 1623–1636. [CrossRef]
26. Pljesa-Ercegovac, M.; Savic-Radojevic, A.; Matic, M.; Coric, V.; Djukic, T.; Radic, T.; Simic, T. Glutathione transferases: Potential targets to overcome chemoresistance in solid tumors. *Int. J. Mol. Sci.* **2018**, *19*, 3785. [CrossRef]
27. Lang, J.Y.; Ma, K.; Guo, J.X.; Sun, H. Oxidative stress induces B lymphocyte DNA damage and apoptosis by upregulating p66shc. *Eur. Rev. Med. Pharmacol. Sci.* **2018**, *22*, 1051–1060. [PubMed]
28. Bridge, G.; Rashid, S.; Martin, S.A. DNA mismatch repair and oxidative DNA damage: Implications for cancer biology and treatment. *Cancers* **2014**, *6*, 1597–1614. [CrossRef] [PubMed]
29. D'Anneo, A.; Carlisi, D.; Lauricella, M.; Puleio, R.; Martinez, R.; Bella, S.D.; Marco, P.D.; Emanuele, S.; Fiore, R.D.; Guercio, A.; et al. Parthenolide generates reactive oxygen species and autophagy in MDA-MB231 cells. A soluble parthenolide analogue inhibits tumour growth and metastasis in a xenograft model of breast cancer. *Cell Death Dis.* **2013**, *4*, e891. [CrossRef] [PubMed]

30. Kordes, U.; Krappmann, D.; Heissmeyer, V.; Ludwing, W.D.; Scheidereit, C. Transcription factor NF-κB is constitutively activated in acute lymphoblastic leukemia cells. *Leukemia* **2000**, *14*, 399–402. [CrossRef] [PubMed]
31. Luo, J.L.; Kamata, H.; Karin, M. IKK/NF-κB signaling: Balancing life and death—A new approach to cancer therapy. *J. Clin. Investig.* **2005**, *115*, 2625–2632. [CrossRef]
32. Xu, L.; Xu, M.; Tong, X. Effects of aerobic glycolysis on pathogenesis and drug resistance of non-Hodgkin lymphoma. *Zhejiang Da Xue Xue Bao Yi Xue Ban* **2019**, *48*, 219–223. [PubMed]
33. Tang, Y.; Yang, X.; Feng, K.; Hu, C.; Li, S. High expression of aldolase A is associated with tumor progression and poor prognosis in hepatocellular carcinoma. *J. Gastrointest. Oncol.* **2021**, *12*, 174–183. [CrossRef] [PubMed]
34. Gizak, A.; Wisniewski, J.; Heron, P.; Mamczur, P.; Sygusch, J.; Rakus, D. Targeting a moonlighting function of aldolase induces apoptosis in cancer cells. *Cell Death Dis.* **2019**, *10*, 712. [CrossRef] [PubMed]
35. Wallberg, F.; Tenev, T.; Meier, P. Analysis of Apoptosis and Necroptosis by Fluorescence-Activated Cell Sorting. *Cold Spring Harb. Protoc.* **2016**, *2016*, 087387. [CrossRef] [PubMed]
36. Flores-Lopez, G.; Moreno-Lorenzana, D.; Ayala-Sanchez, M.; Aviles-Vazquez, S.; Torres-Martinez, H.; Crooks, P.A.; Guzman, M.L.; Mayani, H.; Chavez-Gonzalez, A. Parthenolide and DMAPT induce cell death in primitive CML cells through reactive oxygen species. *J. Cell. Mol. Med.* **2018**, *22*, 4899–4912. [CrossRef]

International Journal of Molecular Sciences

Article

Access to New Cytotoxic Triterpene and Steroidal Acid-TEMPO Conjugates by Ugi Multicomponent-Reactions †

Haider N. Sultani [1], Ibrahim Morgan [1], Hidayat Hussain [1], Andreas H. Roos [2], Haleh H. Haeri [2], Goran N. Kaluđerović [1,3], Dariush Hinderberger [2] and Bernhard Westermann [1,4,*]

1. Department of Bioorganic Chemistry, Leibniz-Institute of Plant Biochemistry, Weinberg 3, 06120 Halle, Germany; haidersoltani@yahoo.com (H.N.S.); Ibrahim.morgan@ipb-halle.de (I.M.); Hidayat.Hussain@ipb-halle.de (H.H.); goran.kaluderovic@hs-merseburg.de (G.N.K.)
2. Physical Chemistry—Complex Self-Organizing Systems, Institute of Chemistry, Martin Luther University Halle-Wittenberg, von-Danckelmann-Platz 4, 06120 Halle, Germany; Andreas.roos@chemie.uni-halle.de (A.H.R.); haleh.hashemi-haeri@chemie.uni-halle.de (H.H.H.); dariush.hinderberger@chemie.uni-halle.de (D.H.)
3. Department of Engineering and Natural Sciences, University of Applied Sciences Merseburg, Eberhard-Leibnitz-Strasse 2, 06217 Merseburg, Germany
4. Organic Chemistry, Institute of Chemistry, Martin-Luther University Halle-Wittenberg, Kurt-Mothes-Strasse 2, 06120 Halle, Germany
* Correspondence: Bernhard.Westermann@ipb-halle.de; Tel.: +49-345-5582-1340; Fax: +49-345-5582-1309
† This paper is dedicated to L. A. Wessjohann on the occasion of his 60th birthday.

Abstract: Multicomponent reactions, especially the Ugi-four component reaction (U-4CR), provide powerful protocols to efficiently access compounds having potent biological and pharmacological effects. Thus, a diverse library of betulinic acid (BA), fusidic acid (FA), cholic acid (CA) conjugates with TEMPO (nitroxide) have been prepared using this approach, which also makes them applicable in electron paramagnetic resonance (EPR) spectroscopy. Moreover, convertible amide modified spin-labelled fusidic acid derivatives were selected for post-Ugi modification utilizing a wide range of reaction conditions which kept the paramagnetic center intact. The nitroxide labelled betulinic acid analogue **6** possesses cytotoxic effects towards two investigated cell lines: prostate cancer PC3 (IC$_{50}$ 7.4 ± 0.7 µM) and colon cancer HT29 (IC$_{50}$ 9.0 ± 0.4 µM). Notably, spin-labelled fusidic acid derivative **8** acts strongly against these two cancer cell lines (PC3: IC$_{50}$ 6.0 ± 1.1 µM; HT29: IC$_{50}$ 7.4 ± 0.6 µM). Additionally, another fusidic acid analogue **9** was also found to be active towards HT29 with IC$_{50}$ 7.0 ± 0.3 µM (CV). Studies on the mode of action revealed that compound **8** increased the level of caspase-3 significantly which clearly indicates induction of apoptosis by activation of the caspase pathway. Furthermore, the exclusive mitochondria targeting of compound **18** was successfully achieved, since mitochondria are the major source of ROS generation.

Keywords: multi-component reaction; fusidic acid; TEMPO-conjugate; electron paramagnetic resonance (EPR) spectroscopy; caspase-3

1. Introduction

Reactive oxygen species (ROS) are involved in numerous processes, which mediate physiological and pathophysiological signal transductions. Upon unregulated increased ROS production, redox imbalances occur, which cause atherosclerosis, cardiovascular diseases, hypertension, diabetes mellitus, neurodegenerative and immune-inflammatory diseases. In addition, the impact of oxidants and antioxidants in tumor cell proliferation is observed frequently. On the molecular level, ROS causes oxidative stress, which is responsible for damaging cell structures by acting on lipids, membranes, proteins, and DNA. This behavior of ROS in cancer cells, in particular, offers a basis for the prevention of tumor progression and metastasis by ROS scavengers [1,2]. Therefore, antioxidant therapies are sought to selectively inhibit the growth of tumor cells to induce cellular differentiation and to alter the intracellular redox state [3]. Besides this, antioxidative food supplements enhance the positive outcome of conventional cancer treatments. In the context of cancer therapy, nitroxide-modified natural products have been shown to be useful due to their ability to remove superoxide anions, trap carbon-centered radicals, or to terminate chain reactions [4]. Furthermore, these derivatizations have been applied in bioreductive drugs to initiate additional cytotoxic events, which make them useful as antitumor drugs [5].

Betulinic acid (BA, **1**, Figure 1) and its congeners are well known for their abilities to act as natural cytotoxic products [6–10]. These triterpenes have been modified quite extensively providing products with enhanced biological activities [11–15]. Modifications of these lupane triterpenoids with nitroxyl radicals have been shown to produce a positive outcome on the cytotoxic activity on several cell lines (e.g., CEM13, U937, MT4) [16,17]. Fusidic acid (FA, **2**) is a triterpene acid that belongs to the family of tetracyclic fusidane nor-triterpenes and has been clinically employed as an antibiotic for staphylococcal infections [18]. Additionally, it has been reported that fusidic acid sodium salt, an approved bacteriostatic antibiotic, showed significant cytotoxic effects (in vitro and in vivo) towards various colon cancer cells alone or coupled with 5-fluorouracil [19,20]. Moreover, various studies have been reported that cholic acid (CA, **3**), which is steroidal acid, can be used for the prevention and treatment of colon cancer [21,22].

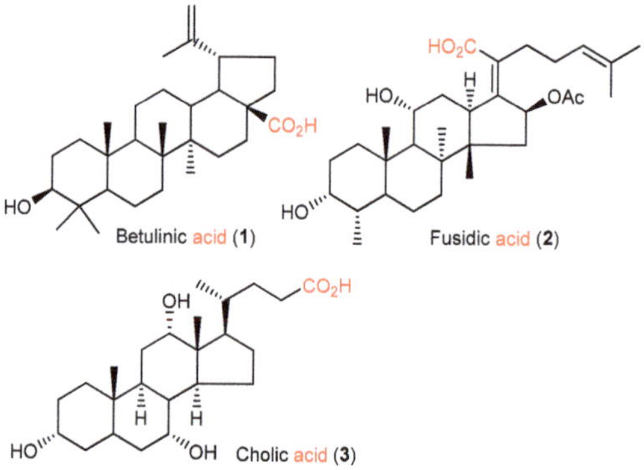

Figure 1. Structures of Betulinic acid (**1**), Fusidic acid (**2**), and Cholic acid (**3**).

Based on these promising cytotoxic effects of the natural products **1–3**, we prepared a library of BA (**1**), FA (**2**), CA (**3**) conjugates with TEMPO (nitroxide) by utilizing an Ugi multicomponent reaction approach (U-4CR) with the aim of enhancing the cytotoxic potentials of our conjugates. Although only a few reports have been published about the U-4CR modifications on BA (**1**) as anti-inflammatory agents [23,24], none of them investigated the fusion to nitroxide. In a previous communication, we demonstrated that the U-4CR strategy is very well suited to achieve spin-labelled products [25]. In the present study, we use an amino spin-label viz.: 4-amino-2,2,6,6-tetramethylpiperidine-1-oxyl (**4**, 4-NH_2-TEMPO) as a U-4CR counterpart (Scheme 1) allowing for the preparation of the natural acid-TEMPO adducts. The spin-labelled FA derivative **8** acts strongly against two investigated cancer cell lines of prostate cancer (PC3) and colon cancer (HT29) and induces apoptosis by a caspase-dependent mechanism.

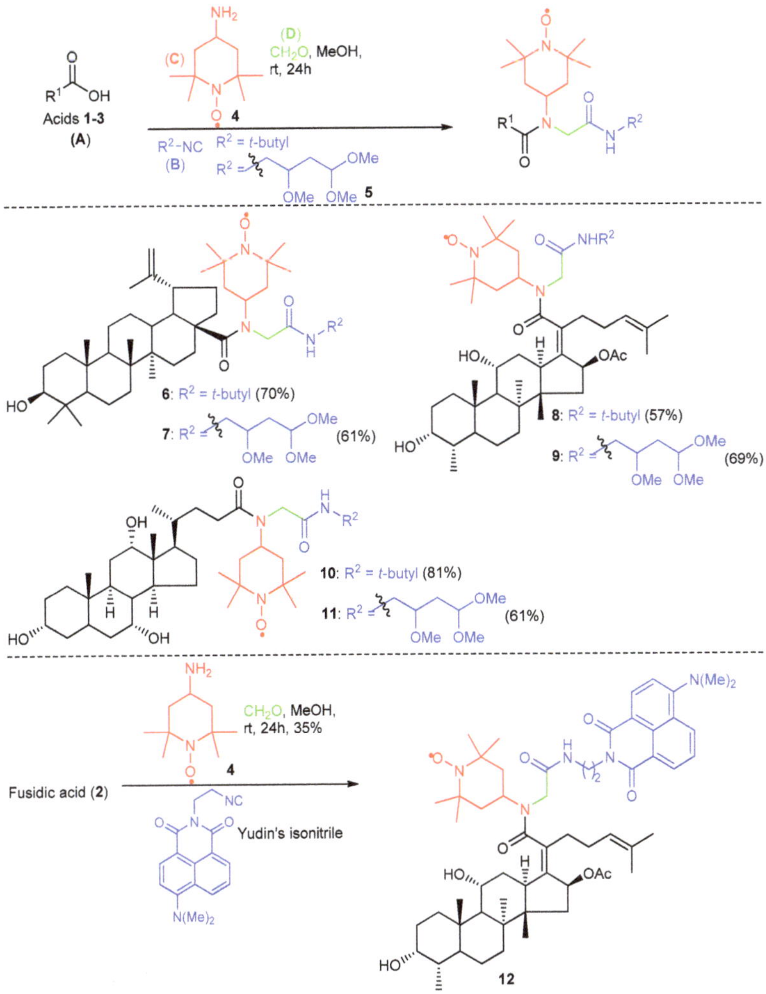

Scheme 1. Synthesis of terpenoic acid-TEMPO adducts 6–12.

2. Results and Discussion

2.1. Chemistry

The general synthetic pathway for the preparation of natural acids 1–3-based Ugi products 6–12 is outlined in Scheme 1. Natural acid-TEMPO adducts 6–11 were synthesized in a single step operation by utilizing the Ugi four-component reaction. These compounds were prepared in moderate to good yields (57–81%) by the reaction of BA (1), FA (2), or CA (3) as the acid component (A), convertible IPB isonitrile 5 [26], or *t*-butyl isocyanide as isonitrile component (B), 4-NH$_2$-TEMPO as amine (C), and paraformaldehyde (D) in the presence of MeOH. Encouraged by our previous results [25] that spin-label TEMPO is not affected under the reaction conditions of the U-4CR, we plan to couple nitroxide comprising amine viz.: 4-NH$_2$-TEMPO (4; as amine component) to enhance the cytotoxicity of natural acids 1–3. Moreover, BA (1), FA (2), and CA (3) have a tertiary carboxylic acid, vinyl carboxylic acid, and secondary carboxylic acid groups respectively. We found that the alteration of these acids did not play any significant role in the product yields. To introduce further chemical diversity via the Ugi synthetic procedure, we additionally prepared fusidic acid-based Ugi product **12** by utilizing Yudin's fluorescent isocyanide [27].

In order to expand the diversity of the Ugi products synthesized, the advantage was taken of the isonitrile functionality as illustrated for compounds **7** and **9** (Scheme 2). As demonstrated earlier [26], the secondary amide can be transformed upon acidic treatment to acyl pyrroles, which can easily be cleaved by nucleophiles. Thus, in the presence of camphor sulfonic acid (CSA) both Ugi-products **7** and **9** were transformed to the corresponding acyl pyrroles **13** and **15**, which upon treatment with KOH were converted into the corresponding carboxylates **14** and **16**. However, no selectivity could be obtained for the acetyl moiety in the fusidic acid derivative **15** since the ester moiety (C-16 acetyl group) was cleaved as well, as expected under these conditions. To achieve selectivity in the displacement of the acyl pyrrole, an alternative procedure (DMAP, H_2O/t-BuOH) was successful and furnished the C-16 acetyl fusidic acid analog **17**. For biological evaluation, we envisioned the preparation of a conjugate with a triphenylphosphine moiety, since this moiety is known to selectively bind to mitochondria membranes. Again, the U-4CR proved to be the synthetic protocol of choice, since in a single step not only the triphenylphosphine moiety but a dye (Yudin's dye/Yudin's isonitrile) can be assembled to form the product in the same synthetic process yielding the fusidic acid analog **18**.

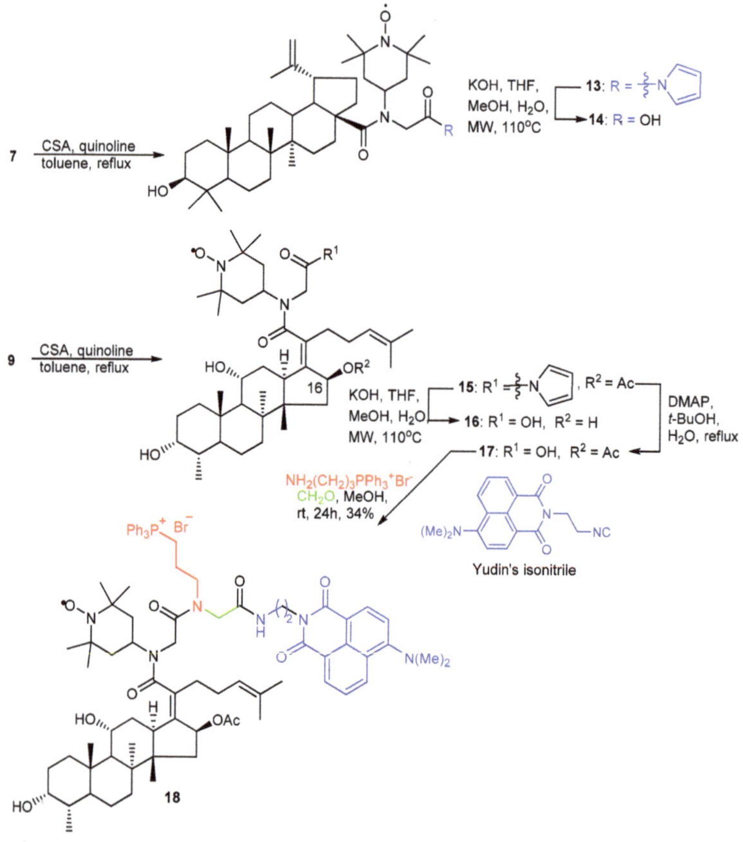

Scheme 2. Synthesis of compounds **13–18**.

2.2. Characterization of the Nitroxide Conjugated Compounds by Electron Paramagnetic Resonance (EPR) Spectroscopy

The radical nature of the synthesized nitroxide conjugates were verified by continuous wave (CW) EPR spectroscopy. Figure 2 shows the CW EPR spectra of the nitroxide conjugated compounds **6–12, 14, 16, 17,** and **18**. Conventional triplet pattern of TEMPO nitroxide with relative spectral intensities of 1:1:1 can be seen in Figure 2 due to the coupling of the unpaired electron to the N-atom which indicates that the nitroxide was intact during the EPR measurements.

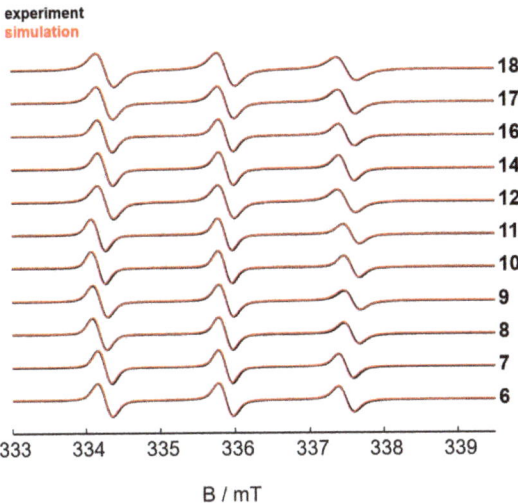

Figure 2. Experimental room temperature CW EPR (X-band, microwave frequency ≈ 9.4 GHz) spectra (black) and the corresponding simulations (red) of the synthesized nitroxide conjugates are given.

The EPR characteristics, mainly the isotropic hyperfine coupling (A_{iso}) and isotropic rotational correlation times (τ_c), dominate solution-state EPR spectra as a measure of the line spacing and line shape, were obtained by simulations using the Easyspin software package [28]. All synthesized nitroxide adducts show hyperfine couplings (A_{iso}~45–47 MHz) that are indicative of a water-exposed nitroxide moiety. The isotropic rotational correlation times (τ_c), as a simple measure of nitroxide rotational dynamics were monitored and were found to be between 1–2 ns for the synthesized nitroxide adducts, in good agreement with what can be expected when attached to medium-sized molecules as in this case. During the simulations, the g-values were kept constant at g_{iso}~2.005, the commonly found value for piperidine-based nitroxide radicals [29–31]. The numerical values are summarized in Table S1. Altogether, one can state that the EPR parameters clearly show that none of the spin-labelled natural products seems to be aggregated/micellized or non-homogeneously dissolved in aqueous solution.

2.3. Cytotoxic Activity

The first set of synthesized spin-labelled adducts **6–11** were subjected to fast screening by MTT and CV assays to have an overall view on their potential activity against human cancer cell lines viz.: PC3 (prostate cancer) and HT29 (colon cancer). Two concentrations were employed viz.: 0.1 and 10 µM and compared to the activity of unmodified BA (**1**), FA (**2**), and CA (**3**). As shown in Figure 3 both the betulinic acid derivatives **6** and **7** and the fusidic acid derivatives **8** and **9** showed a significant reduction in cell viability when compared to cholic acid derivatives **10** and **11**. The low anticancer activity of cholic acid and its derivatives can be attributed to the high lipophilic character (log p ~ 2.02) and low water solubility, which makes it difficult to cross membranes effectively to be present in high concentrations in the cytosol of the cancer cells [32].

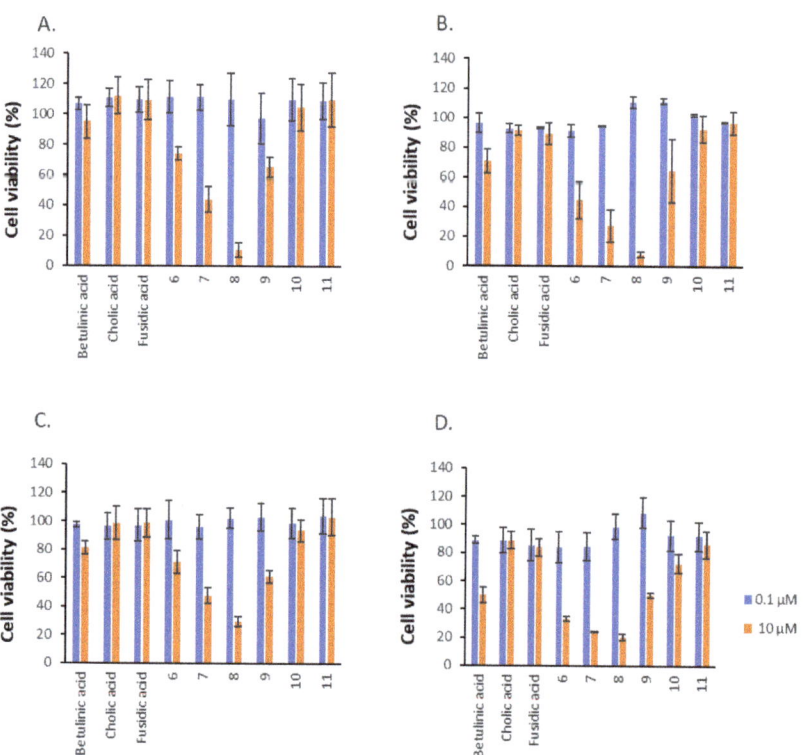

Figure 3. Cell viability of HT29 (**A**,**B**) and PC3 (**C**,**D**); cell lines were treated by the investigated compounds for 48 h. Cell viability was determined using CV assay (**A**,**C**) and MTT assay (**B**,**D**).

As expected, FA (**2**) failed to show any activity against the tested cancer cell lines. However, the unanticipated very high activity displayed by its derivatives **8** and **9**, especially when compared to the well-known anticancer activity of BA (**1**) was encouraging [33]. The IC$_{50}$ values were also determined for the most active compounds against both cell lines used in our appraisal. The IC$_{50}$ values are illustrated in Table 1, where it is evident that fusidic acid analogue **8** is the most active compound against both cell lines tested. Additionally, another fusidic acid analogue **9** was also active towards HT29 cells with an IC$_{50}$ of 6.98 ± 0.25 µM (CV). Moreover, the betulinic acid analogue **6** possesses cytotoxic effects towards PC3 (IC$_{50}$: 7.43 ± 0.72 µM) and HT29 cells (IC$_{50}$: 8.98 ± 0.43 µM). In addition, it is worthy to note that FA alone possesses no activity when compared to its spin-labelled adducts. Thus, it may clearly be noted that the structure modification provided by the Ugi multicomponent reaction dramatically enhances the anticancer activity of these classes of terpenes.

Table 1. IC$_{50}$ values (µM) for the most active compounds against HT29 and PC3 cell lines determined by MTT and CV assays.

Compound	PC3		HT29	
	CV	MTT	CV	MTT
1	24.64 ± 1.78	25.43 ± 4.35	24.97 ± 0.57	19.02 ± 2.26
6	13.69 ± 0.80	7.43 ± 0.72	13.16 ± 0.97	8.98 ± 0.43
7	10.59 ± 0.85	10.54 ± 0.91	13.82 ± 0.29	11.87 ± 0.94
8	7.44 ± 0.80	6.00 ± 1.09	8.10 ± 0.43	7.41 ± 0.56
9	15.26 ± 1.01	13.85 ± 2.04	6.98 ± 0.25	12.94 ± 1.03
18	9.27 ± 0.73	6.19 ± 0.20	16.30 ± 0.87	12.23 ± 0.67

Compound **8**, the most promising conjugate, was selected to determine its mode of action against the PC3 cell line based on its cytotoxic effects. To determine the mode of cell death induced by compound **8**, the AnnV/PI assay was performed since the degree of induction of apoptosis by compound **8** can be effectively measured. The assay determines the expression of phosphatidylserine on the cell surface by annexin V (AnnV) stain and the DNA fragmentation by propidium iodide (PI) (Figure 4). Compound **8** was tested at two different concentrations (IC$_{50}$, 2 × IC$_{50}$) for 48 h and it was analysed using flow cytometry. It is clearly shown, that this compound increases both early and late apoptosis, only when the PC3 cancer cells were treated with 2 × IC$_{50}$ value with a total apoptotic event of 68% compared to the control of 16%. To study the impact of compound **8** on the cell cycle distribution, the DAPI assay was performed as outlined in Figure 5. Based on the results obtained, compound **8** caused a dose-dependent increase in the entrapment of cells in sub G1-phase. This accumulation of the cells in the sub G1-phase of the cell cycle indicates the fragmentation of the DNA that has occurred due to the induction of apoptosis by compound **8**.

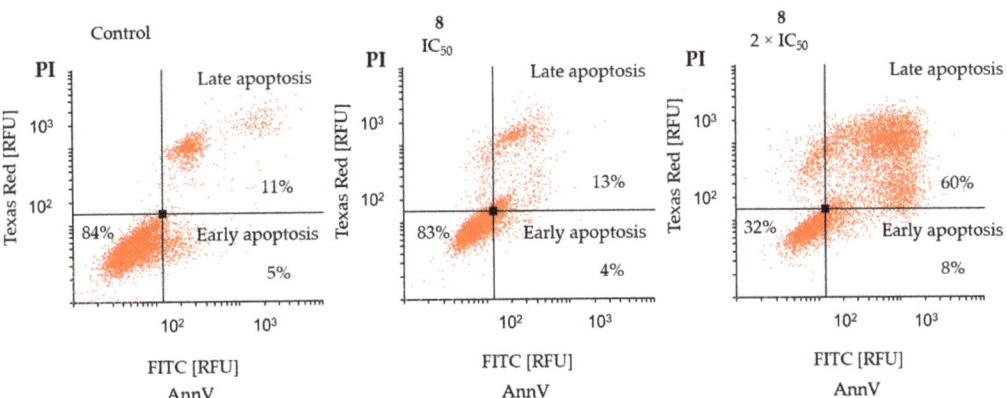

Figure 4. The impact of compound **8** on the apoptosis induction in PC3 cells. **8** was tested using IC$_{50}$, 2 × IC$_{50}$ concentrations for 48 h (AnnV/PI assay).

Reyes et al. reported that natural triterpenoic acids induce caspase-dependent apoptosis and in particular, caspase-3 [34]. Furthermore, recent reports showed that treating tumor cells with nitroxides can also induce apoptosis by a caspase activation mechanism [35]. These studies inspired us to investigate the possibility of caspase-3 being involved in the mechanism of action of conjugate **8** to explain the apoptotic mode of cell death, passivate the level of protein expression of the anti-apoptotic protein Bcl-XL and the housekeeping proteins (β-actin and α/β-tubulin). Results illustrated in Figure 6 show that **8** increased the level of caspase-3 significantly after 48 h of incubation, which clearly indicates the induction of apoptosis by activation of the caspase pathway. Expression of the anti-apoptotic protein

Bcl-XL which is a transmembrane molecule in the mitochondria was also measured. After 48 h of incubation, it is clearly evident that the level of Bcl-XL decreases which supports the apoptosis by triggering the caspase-3 activation pathway.

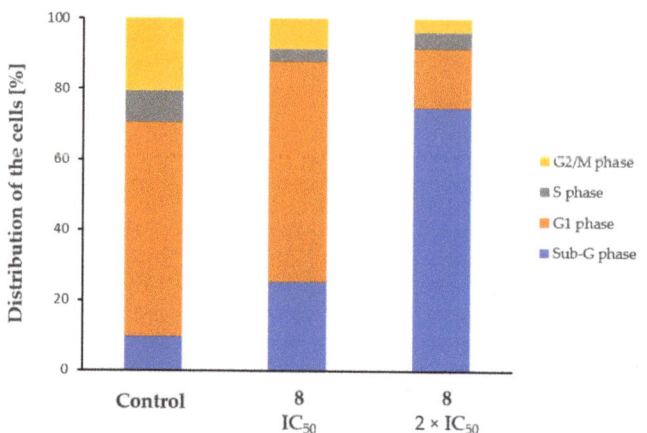

Figure 5. The impact of compound **8** on the distribution of PC3 cells in the cell cycle phases. Compound **8** was tested using IC_{50}, $2 \times IC_{50}$ concentrations for 48 h (DAPI assay).

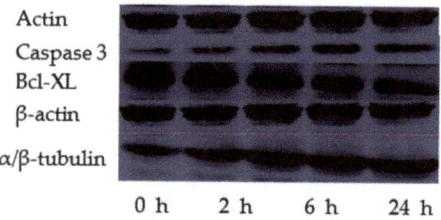

Figure 6. Effect of compound **8** on protein expression in PC3 cells (relative to β-actin).

β-Actin is a housekeeping protein that is involved in the restriction of the cell motility, structure integrity, and in addition to its resistance to different cellular treatment, which makes it a good choice as a housekeeping protein for western blot analysis [36]. Indeed, after 48 h of incubation, no change in its expression was detected. α/β-Tubulins is another housekeeping protein control used since their expression should remain unchanged. Surprisingly, the behavior of this protein was manifested by a strong elevation of the expression level being obvious after 48 h of incubation as shown in Figure 6. Recent studies report that microtubulin increases during apoptosis and functions as a physical barrier preventing caspase from spreading into the cellular cortex. In addition, it increases phosphatidylserine (PS) externalization which helps the macrophage for efficient clearance [37].

The influence of compound **8** (IC_{50} and $2 \times IC_{50}$) on the ROS production in PC3 cells was monitored using dihydrorhodamine (DHR) assay for 48 h and the data were analysed with flow cytometry.

As shown in Figure 7 compound **8** indeed reduced the level of ROS as anticipated in a dose-dependent manner.

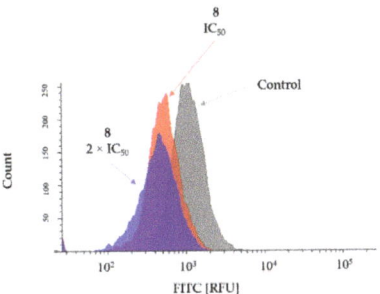

Figure 7. The impact of compound **8** on ROS produced by PC3. **8** was tested using IC_{50}, $2 \times IC_{50}$ concentrations for 48 h using DHR assay, fluorescence was detected in FITC channel (ex/em: 488/520 nm).

2.4. Fluorescent Imaging Study

Fluorescent conjugate **12** was initially used to determine if this dye-tagged analogue of fusidic acid conjugate **8** can target the mitochondria, since mitochondria are the major source of ROS generation and therefore is more sensible for ROS manipulation. Unfortunately, after PC3 cancer cells were incubated with **12** (depicted as green color in Figure 8), no mitochondrial targeting was observed. Therefore, we turned our attention to test the mitochondrial targeting of TPP-conjugate **18**. After 24 h of incubation of compound **18** with PC3 cells, a clear mitochondrial targeting was successfully achieved as shown in Figure 9. Additionally, cytotoxic activity of **18** against PC3 and HT29 was found to be with significant effects towards PC3 cancer cell line (IC_{50}: 6.18 ± 0.20 μM, MTT).

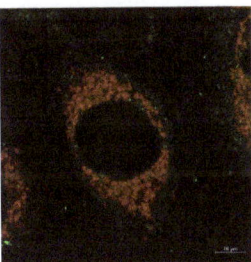

Figure 8. Fluorescent imaging with fluorescent conjugate **12** (ex/em: 488/520 nm) double-stained with MitoTracker™ Deep Red (ex/em: 596/615 nm) in PC3 cancer cell line.

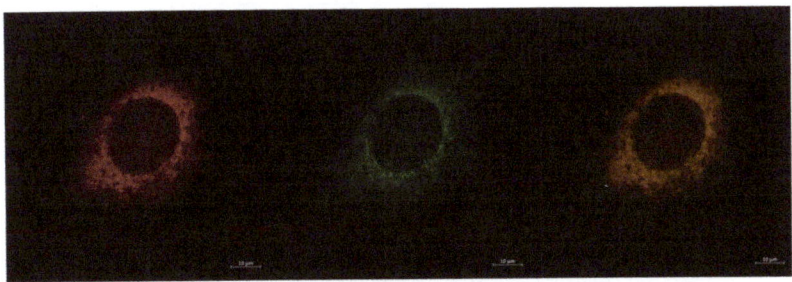

MitoTracker™ Deep Red   TPP-conjugate **18**   Overlay

Figure 9. Fluorescent imaging of PC3 cancer cell line stained with TPP-conjugate **18** (ex/em: 488/520 nm) and with MitoTracker™ Deep Red (ex/em: 596/615 nm).

3. Conclusions

TEMPO radical conjugation to natural products can serve as a potential strategy to obtain new hybrid compounds with novel mechanisms of action. Conjugate **8** revealed a high activity against both PC3 and HT29 cancer cell lines (PC3: IC$_{50}$ 6.0 ± 1.1 μM; HT29: IC$_{50}$ 7.4 ± 0.6 μM), furthermore, apoptosis was induced through the caspase activation mechanism. In addition, targeting mitochondria (the major source of ROS generation) was successfully achieved with **18**. Moreover, it was clearly demonstrated that utilizing Ugi multicomponent reactions is a powerful synthetic tool that gives access to a wide variety of different analogues via a fairly easy synthetic effort. We envisioned that utilizing the power of MCR, large interesting libraries of natural product TEMPO conjugates for the treatment of cancer can be generated.

4. Materials and Methods

4.1. Chemistry

4.1.1. Materials

All commercially available reagents were purchased and used without further purification. Convertible isocyanide 2-isocyano-2-methylpropyl phenyl carbonate "IPB" was synthesized following reported procedures [27]. (3-Aminopropyl)triphenylphosphonium bromide (TPP-NH$_2$) was also synthesized following reported protocols [38]. HPLC grade methanol was used in all Ugi reactions. Analytical thin layer chromatography (TLC) was performed using silica gel 60 F$_{254}$ aluminum sheets (Merck, Darmstadt, Germany) and the visualization of the spots has been done under UV light (254 nm) or by developing with a solution of cerium sulfate. Flash column chromatography was performed using silica gel (0.040–0.063 mm). ^{1}H- and ^{13}C-NMR spectra were recorded in solutions on a 400 NMR Varian MERCURY-VX 400 at 22 °C at 400 MHz and 100 MHz, or an Agilent (Varian, Santa Clara, CA, USA) VNMRS 600 NMR spectrometer at 599.83 MHz and 150.83 MHz respectively. Chemical shifts (δ) are reported in ppm relative to TMS (^{1}H-NMR) and to the solvent signal (^{13}C NMR spectra). The positive-ion high-resolution ESI mass spectra were obtained with an Orbitrap Elite mass spectrometer (Thermo Fisher Scientific, Schwerte, Germany) equipped with HESI electrospray ion source (positive spray voltage 4 kV, capillary temperature 275 °C, source heater temperature 80 °C, FTMS resolution 60,000). Nitrogen was used as sheath gas. The instrument was externally calibrated using the Pierce LTQ Velos ESI positive ion calibration solution (product number 88323, ThermoFisher Scientific, Rockford, IL, USA). The data were evaluated using the software Xcalibur 2.7 SP1. Analytical RP-HPLC analysis was performed with an 1100 system (Agilent, Santa Clara, CA, USA) on a reverse-phase C18 column (4.6 × 150 mm, 5 μm) with a PDA detector. A linear gradient from 5% to 100% of solvent B in solvent A over 15–30 min at a flow rate of 0.8 mL min^{-1}. Detection was accomplished at 210 nm. Solvent A: 0.1% (v/v) formic acid (FA) in water. Solvent B: 0.1% (v/v) FA in acetonitrile.

PBS, RPMI 1640, and Trypsin EDTA were from Capricorn Scientific (Ebsdorfergrund, Germany). β-mercaptoethanol was from Bio-Rad (Hercules, CA, USA). Anti-rabbit IgG HRP-linked antibody, α/β-tubulin rabbit Ab, caspase-3 rabbit Ab were purchased from Cell Signalling Technology (Frankfurt/Main, Germany), while Bcl-XL rabbit antibody was obtained from Abcam (Cambridge, UK). DMSO was bought from Duchefa Biochemie (Harleem, The Netherlands). ECL Prime Western Blotting System was supplied by GE Healthcare (Braunschweig, Germany). AnnV/PI, PAGE Ruler, EDTA Solution, Trypan blue, MitoTracker™ Deep Red and Halt Protease Inhibitor Cocktail were obtained from ThermoFisher Scientific, Schwerte, Germany). Ethanol, Na$_2$HPO$_4$, NaH$_2$PO$_4$, and BSA were bought from Merck, Darmstadt, Germany). Digitonin was from Riedel De Haen (Seelze, Germany). Acetic acid, APS, FCS, glycerol, glycine, methanol, NaOH, penicillin/streptomycin, Roti-quant "5x", TEMED and TRIS were from Roth (Karlsruhe, Germany). Acrylamide/Bisacrylamide was bought from Serva (Heidelberg, Germany). Finally, bromophenol blue, CV, DAPI, MTT, Triton X-100, and Tween-20 were from Sigma Aldrich (St. Louis, MO, USA).

4.1.2. General Procedure A for the Ugi-4CR

To a stirred solution of TEMPO amine **4** (0.1 mmol) in methanol (250 µL, 0.4 M) was added paraformaldehyde (0.1 mmol) and the mixture was stirred for 2 h. After this time the acid (0.1 mmol) and isonitrile (0.1 mmol) were added before stirring was continued for 18 h. The solvent was removed under reduced pressure and the crude material purified by column chromatography to afford the desired products.

Note: due to the paramagnetism of nitroxide moieties, NMR cannot provide information useful for structural elucidation of nitroxide-containing products, therefore, reduction of the paramagnetic center was performed with phenylhydrazine or hydrazobenzene [39,40].

(4-{[2-(tert-Butylamino)-2-oxoethyl][(1R,3aS,5aR,5bR,9S,11aR)-9-hydroxy-5a,5b,8,8,11a-pentamethyl-1-(prop-1-en-2-yl)icosahydro-3aH-cyclopenta[a]chrysene-3a-carbonyl]amino}-2,2,6,6-tetramethylpiperidin-1-yl)oxidanyl (**6**)

Obtained using the general method **A**, N-oxyl amine **4** and paraformaldehyde were used, then betulinic acid and *t*-butyl isonitrile were added. The crude reaction product was purified by silica gel column chromatography (ethyl acetate/hexane 8:2) to yield compound **6** (51 mg, 0.070 mmol, 70%) as red solid. R_F 0.77 (ethyl acetate/hexane 8:2). NMR of the corresponding hydroxylamine after phenylhydrazine reduction. ^{1}H-NMR (600 MHz, CDCl$_3$) δ 4.72 (s, 1H), 4.59 (s, 1H), 4.35 (d, *J* = 12.6 Hz, 1H), 3.65 (s, 2H), 3.16 (dd, *J* = 11.3, 4.8 Hz, 2H), 2.98 (m, 1H), 2.87–2.81 (m, 1H), 2.24–2.18 (m, 1H), 2.06–1.20 (m, 49H), 0.96 (s, 3H), 0.95 (s, 3H), 0.92 (s, 3H), 0.80 (s, 3H), 0.74 (s, 3H). ^{13}C-NMR (151 MHz, CDCl$_3$): δ 176.0 (CO), 109.5, 78.9, 55.5, 54.9, 53.0, 50.9, 50.8, 49.3, 47.7, 45.8, 42.8, 42.3, 42.1, 40.8, 38.9, 38.8, 37.3, 36.9, 36.1, 34.3, 32.2, 32.0, 31.6, 29.9, 29.8, 28.7 (4 × CH$_3$ (TEMPO)), 28.1 (CH$_3$ (*t*Bu)), 27.5, 25.7, 21.1, 20.1, 19.9, 19.6, 18.3, 16.2, 15.5, 14.8. HRMS (ESI) *m/z*: 723.5890 [M + H]$^+$, calcd. for [C$_{45}$H$_{77}$N$_3$O$_4$]$^+$ 723.5914.

[4-({(1R,3aS,5aR,5bR,9S,11aR)-9-Hydroxy-5a,5b,8,8,11a-pentamethyl-1-(prop-1-en-2-yl)icosahydro-3aH-cyclopenta[a]chrysene-3a-carbonyl}{2-oxo-2-[(2,4,4-trimethoxybutyl)amino]ethyl}-amino)-2,2,6,6-tetramethylpiperidin-1-yl]oxidanyl (**7**)

Obtained using the general method **A**, N-oxyl amine **4** and paraformaldehyde were used, then betulinic acid and IPB isonitrile **5** were added. The crude reaction product was purified by silica gel column chromatography (ethyl acetate/hexane 8:2) to yield compound **7** (50 mg, 0.061 mmol, 61%) as red solid. R_F 0.35 (ethyl acetate/hexane 8:2). NMR of the corresponding hydroxylamine after phenylhydrazine reduction. ^{1}H-NMR (600 MHz, CDCl$_3$) δ 4.72 (s, 1H), 4.59 (s, 1H), 4.54–4.49 (m, 1H). 4.38 (t, *J* = 12.5 Hz, 1H (ipb)), 3.66 (s, 2H), 3.36–3.30 (m, 12H). (IPB Ugi moiety), 3.16 (dd, *J* = 11.3, 4.8 Hz, 2H), 2.98 (m, 1H), 2.87–2.81 (m, 1H), 2.27–2.15 (m, 1H), 2.11–1.09 (m, 42H), 0.96 (s, 3H), 0.95 (s, 3H), 0.91 (s, Hz, 3H), 0.81 (s, 3H), 0.75 (s, 3H). ^{13}C-NMR (151 MHz, CDCl$_3$) δ 176.08 (CO), 109.50, 79.08, 57.04, 56.98, 56.49, 55.57, 55.47, 53.24, 53.07, 50.93, 45.82, 45.77, 42.59, 42.09, 41.51, 41.30, 40.86, 38.99, 38.85, 38.37, 37.34, 36.10, 35.20, 34.39, 32.21, 31.61, 31.52, 31.44, 31.01, 29.97, 29.82, 28.11(4 x CH$_3$ (TEMPO)), 27.53, 25.73, 21.21, 21.17, 20.32, 20.17, 19.71, 16.31, 16.08, 16.05, 15.48, 14.83. HRMS (ESI) *m/z*: 812.6138 [M]$^+$, calcd. for [C$_{48}$H$_{82}$N$_3$O$_7$]$^+$ 812.6153.

[4-({(2E)-2-[(2S,3aS,3bS,6S,7R,9aS,10R,11aR)-2-(Acetyloxy)-7,10-dihydroxy-3a,3b,6,9a-tetramethylhexadecahydro-1H-cyclopenta[a]phenanthren-1-ylidene]-6-methylhept-5-enoyl}[2-(tert-butylamino)-2-oxoethyl]amino)-2,2,6,6-tetramethylpiperidin-1-yl]oxidanyl (**8**)

Obtained using the general method **A**, N-oxyl amine **4** and paraformaldehyde were used, then fusidic acid and *t*-butyl isonitrile were added. The crude reaction product was purified by silica gel column chromatography (DCM/MeOH 9:1) to yield compound **8** (51 mg, 0.057 mmol, 57%) as red solid. R_F 0.65 (DCM/MeOH 9:1). NMR of the corresponding hydroxylamine after phenylhydrazine reduction. ^{1}H-NMR (600 MHz, CDCl$_3$) δ 5.69 (d, *J* = 8.6 Hz, 1H), 5.07 (m, 1H), 4.33–4.22 (m, 3H), 3.71 (s, 2H), 3.26 (d, *J* = 14.6 Hz, 1H),

3.05–3.02 (m, 1H), 2.80–2.75 (m, 2H), 2.32–2.28 (m, 1H), 2.22–2.02 (m, 5H), 1.89 (s, 3H), 1.86–1.82 (m, 2H), 1.76–1.71 (m, 4H), 1.68 (s, 3H), 1.61 (s, 3H), 1.60–1.53 (m, 4H), 1.34 (d, J = 7.0 Hz, 3H), 1.28 (s, 9H), 1.26–1.14 (m, 15H), 1.13–1.08 (m, 2H), 0.96 (s, 3H), 0.91–0.88 (m, 6H). ^{13}C-NMR (151 MHz, CDCl$_3$) δ 173.21, 169.87, 169.06, 133.15, 132.45, 122.56, 74.81, 73.10, 71.29, 68.10, 50.95, 50.84, 49.76, 49.63, 49.42, 44.51, 41.63, 39.45, 39.26, 37.09, 36.22, 36.09, 35.22, 32.28, 32.28, 30.28, 30.24, 28.69, 28.57 (4 x CH$_3$ (TEMPO)), 28.52, 25.86, 22.97, 20.86, 20.51, 20.10, 18.05, 17.88, 16.03. HRMS (ESI) *m/z*: 783.5744 [M + H]$^+$, calcd. for [C$_{46}$H$_{77}$N$_3$O$_7$]$^+$ 783.5762.

[4-({(2E)-2-[(2S,3aS,3bS,6S,7R,9aS,10R,11aR)-2-(Acetyloxy)-7,10-dihydroxy-3a,3b,6,9a-tetramethylhexadecahydro-1H-cyclopenta[a]phenanthren-1-ylidene]-6-methylhept-5-enoyl}{2-oxo-2-[(2,4,4-trimethoxybutyl)amino]ethyl}amino)-2,2,6,6-tetramethylpiperidin-1-yl]oxidanyl (**9**)

Obtained using the general method **A**, *N*-oxyl amine **4** and paraformaldehyde were used, then fusidic acid and IPB isonitrile **5** were added. The crude reaction product was purified by silica gel column chromatography (DCM/MeOH 9:1) to yield compound **9** (60 mg, 0.068 mmol, 69%) as red solid. R_F 0.72 (DCM/MeOH 9:1). NMR of the corresponding hydroxylamine after phenylhydrazine reduction. ^{1}H-NMR (600 MHz, CDCl$_3$) δ 5.69 (d, J = 8.4 Hz, 1H), 5.08 (m, 1H), 4.51 (m, 1H), 4.34–4.31 (m, 1H), 3.76–3.73 (m, 1H), 3.65 (s, 2H), 3.39–3.29 (m, 12H), 3.14 (m, 1H), 3.07–3.02 (m, 1H), 2.75 (m, 1H), 2.37–2.27 (m, 1H), 2.23–2.08 (m, 5H), 2.04 (s, 3H), 1.92 (s, 3H), 1.88–1.80 (m, 4H), 1.76–1.71 (m, 4H), 1.68 (s, 3H), 1.62 (s, 3H), 1.60–1.48 (m, 4H), 1.36 (s, 3H), 1.31–1.17 (m, 15H), 1.15–1.10 (m, 2H), 0.97 (s, 3H), 0.93–0.88 (m, 6H). ^{13}C-NMR (151 MHz, CDCl$_3$) δ 173.21, 169.87, 169.06, 133.15, 132.45, 122.56, 74.81, 71.29, 68.10, 50.95, 49.63, 49.42, 44.51, 42.68, 39.45, 39.26, 37.09, 36.37, 36.22, 36.09, 35.22, 32.28, 30.28, 30.24, 30.07, 28.57 (4 x CH$_3$ (TEMPO)), 28.52, 27.81, 27.73, 25.86, 23.85, 22.97, 21.43, 21.21, 20.86, 20.51, 20.10, 18.19, 18.05, 17.88, 16.03.15.68. HRMS (ESI) *m/z*: 873.6062 [M + H]$^+$, calcd. for [C$_{49}$H$_{83}$N$_3$O$_{10}$]$^+$ 873.6078.

[4-({2-(tert-Butylamino)-2-oxoethyl}{(4R)-4-[(1R,3aS,4R,7R,9aS,9bS,11S,11aR)-4,7,11-trihydroxy-9a,11a-dimethylhexadecahydro-1H-cyclopenta[a]phenanthren-1-yl]pentanoyl}amino)-2,2,6,6-tetramethylpiperidin-1-yl]oxidanyl (**10**)

Obtained using the general method **A**, *N*-oxyl amine **4** and paraformaldehyde were used, then cholic acid and *t*-butyl isonitrile were added. The crude reaction product was purified by silica gel column chromatography (DCM/MeOH 9:1) to yield compound **10** (55 mg, 0.081 mmol, 81%) as red solid. R_F 0.62 (DCM/MeOH 9:1). NMR of the corresponding hydroxylamine after phenylhydrazine reduction. ^{1}H-NMR (600 MHz, CDCl$_3$) δ 4.14–4.08 (m, 1H), 4.06–3.98 (m, 2H), 3.97–3.91 (m, 1H), 3.85–3.75 (m, 1H), 2.51–2.42 (m, 1H), 2.36–2.28 (m, 1H), 2.27–2.13 (m, 4H), 1.94–1.80 (m, 5H), 1.79–1.60 (m, 7H), 1.61–1.46 (m, 7H), 1.37–1.15 (m, 25H), 1.13–1.00 (m, 3H), 0.97 (s, 3H), 0.87 (s, 3H), 0.67 (d, J = 7.5 Hz, 3H). ^{13}C-NMR (151 MHz, CDCl$_3$) δ 173.21, 169.87, 169.06, 133.15, 132.45, 122.56, 74.81, 73.10, 71.29, 68.10, 50.95, 50.84, 49.76, 49.63, 49.42, 44.51, 41.63, 39.45, 39.26, 37.09, 36.22, 36.09, 35.22, 32.28, 32.28, 30.28, 30.24, 28.57, 28.52, 28.52, 25.86, 22.97, 20.86, 20.51, 20.10, 18.05, 17.88, 16.03. HRMS (ESI) *m/z*: 675.5164 [M + H]$^+$, calcd. for [C$_{39}$H$_{69}$N$_3$O$_6$]$^+$ 675.5186.

[2,2,6,6-Tetramethyl-4-({2-oxo-2-[(2,4,4-trimethoxybutyl)amino]ethyl}{(4R)-4-[(1R,3aS,4R,7R,9aS,9bS,11S,11aR)-4,7,11-trihydroxy-9a,11a-dimethylhexadecahydro-1H-cyclopenta[a]phenanthren-1-yl]pentanoyl}amino)piperidin-1-yl]oxidanyl (**11**)

Obtained using the general method **A**, *N*-oxyl amine **4** and paraformaldehyde were used, then cholic acid and IPB isonitrile **5** were added. The crude reaction product was purified by silica gel column chromatography (DCM/MeOH 9:1) to yield compound **11** (47 mg, 0.061 mmol, 61%) as red solid. R_F 0.55 (DCM/MeOH 9:1). NMR of the corresponding hydroxylamine after phenylhydrazine reduction. ^{1}H-NMR (600 MHz, CDCl$_3$) δ 4.55–4.48 (m, 1H), 4.12–4.00 (m, 1H), 3.95–3.81 (m, 2H), 3.45 (s, 3H), 3.37–3.28 (m, 12H), 3.28–3.22 (m, 1H), 2.48 (m, 2H), 2.36–2.11 (m, 5H), 1.98–1.80 (m, 6H), 1.80–1.69 (m, 5H), 1.69–1.39 (m, 7H), 1.40–1.16 (m, 15H), 1.15–1.00 (m, 4H), 0.96 (s, 3H), 0.88 (s, 3H), 0.68 (s, 3H). ^{13}C-NMR (151 MHz, CDCl$_3$) δ 174.58, 170.07, 151.16, 101.75, 76.18, 72.86, 71.70, 68.28,

59.33, 59.28, 57.03, 53.33, 53.06, 52.96, 50.55, 46.99, 46.39, 45.61, 41.80, 41.38, 41.28, 35.57, 35.20, 35.09, 34.64, 32.16, 31.54, 30.85, 30.39, 28.20, 27.46, 26.51, 23.12, 22.42, 19.88, 17.44, 14.62, 12.48. HRMS (ESI) m/z: 764.5410 [M]$^+$, calcd. for $[C_{42}H_{74}N_3O_9]^+$ 764.5425.

[4-({(2E)-2-[(2S,3aS,3bS,6S,7R,9aS,10R,11aR)-2-(Acetyloxy)-7,10-dihydroxy-3a,3b,6,9a-tetramethylhexadecahydro-1H-cyclopenta[a]phenanthren-1-ylidene]-6-methylhept-5-enoyl}[2-({2-[6-(dimethylamino)-1,3-dioxo-1H-benzo[de]isoquinolin-2(3H)-yl]ethyl}amino)-2-oxoethyl]amino)-2,2,6,6-tetramethylpiperidin-1-yl]oxidanyl (**12**)

Obtained using the general method **A**, N-oxyl amine **4** and paraformaldehyde were used, then fusidic acid and Yudin's isonitrile were added. The crude reaction product was purified by silica gel column chromatography (DCM/MeOH 9:1) to yield compound **12** (35 mg, 0.035 mmol, 35%) as yellow powder. R_F 0.1 (DCM/MeOH 9:1). NMR of the corresponding hydroxylamine after the addition of hydrazobenzene. ^{1}H-NMR (400 MHz, DMSO-d_6) δ 8.49 (dd, J = 8.5, 2.1 Hz, 1H), 8.45 (d, J = 7.5 Hz, 1H), 8.34 (dd, J = 8.2, 1.9 Hz, 1H), 7.78–7.72 (m, 1H), 7.20 (d, J = 8.6 Hz, 1H), 5.65 (d, J = 8.5 Hz, 1H), 5.09–5.01 (m, 1H), 4.91–4.76 (m, 2H), 4.18–4.11 (m, 1H), 4.07 (d, J = 3.9 Hz, 1H), 3.99 (d, J = 3.9 Hz, 1H), 3.90–3.81 (m, 2H), 3.53 (s, 2H), 3.50–3.47 (m, 1H), 3.07 (s, 6H), 2.99–2.85 (m, 1H), 2.70–2.60 (m, 1H), 2.40–2.17 (m, 3H), 2.16–1.93 (m, 5H), 1.85 (d, J = 13.3 Hz, 2H), 1.62 (dd, J = 9.8, 5.2 Hz, 7H, (Fusidic acid; 5H+TEMPO; 2H)), 1.57–1.50 (m, 6H), 1.49–1.28 (m, 6H), 1.25 (s, 3H), 1.08–0.95 (m, 17H, (fusidic acid 3H, TEMPO 14H)), 0.86 (s, 3H), 0.81–0.75 (m, 6H). ^{13}C-NMR (101 MHz, DMSO-d_6) δ 171.61, 169.91, 169.10, 164.32, 163.64, 147.18, 133.72, 132.31, 131.95, 131.88, 130.94, 130.21, 125.46, 124.76, 123.96, 123.48, 114.04, 113.47, 74.34, 69.68, 66.23, 58.57, 58.38, 51.63, 49.42, 49.07, 48.98, 48.65, 44.84, 43.22, 39.05, 36.90, 36.81, 35.67, 35.55, 33.31, 33.05, 32.37, 32.25, 30.69, 29.93, 28.82, 28.49, 25.92, 23.94, 23.18, 20.67, 20.13, 18.01, 17.97, 16.74, 16.71. HRMS (ESI) m/z: 993.6145 [M + H]$^+$, calcd. for $[C_{58}H_{83}N_5O_9]^+$ 993.6191.

4.1.3. General Procedure **B** for the Conversion of Ugi Products **7** and **9** to Corresponding Spin-Labelled N-Acylpyrroles **13** and **15**

To a solution of Ugi products **7** and **9** (0.05 mmol) in toluene (10 mL) was added 10-camphorsulfonic acid (10 mol%) and quinoline (10 mol%). The mixture was stirred for 1 min at room temperature and then refluxed for at least 30 min until TLC showed complete conversion. The mixture was cooled to room temperature, transferred to a separatory funnel and washed with 1M aqueous HCl (2 × 30 mL). The acidic aqueous phase was further extracted with ethyl acetate (1 × 20 mL). The organic layers were combined, washed with NaHCO$_3$ and brine (2 × 20 mL), dried over anhydrous Na$_2$SO$_4$, filtered and evaporated under reduced pressure to obtain the N-acyl pyrrole derivatives **13** and **15**, which were used in the next step without further purification.

4.1.4. General Procedure **C** for the Conversion of the N-Acylpyrroles **13** and **15** into Their Corresponding Carboxylic Acids **14** and **16**
Method **C1**

To a solution of intermediates **13** and **15** (0.025 mmol) in a mixture of THF (2 mL), methanol (2 mL), water (2 mL), potassium hydroxide (0.5 mmol) was added. This mixture was heated to 110 °C for 30 min in a microwave (90 W heating, 6 W keeping temperature). The reaction mixture was diluted with methanol (10 mL) and the pH value was set to pH = 2 by the addition of saturated aqueous NaHSO$_4$ solution. The aqueous phase was extracted with ethyl acetate (3 × 100 mL) and it was dried over Na$_2$SO$_4$. After filtration and evaporation of the solvent, the crude residue was purified by column chromatography (DCM/MeOH 8:2). By this method compounds, **14** and **16** were obtained.

(4-{(Carboxymethyl)[(1R,3aS,5aR,5bR,9S,11aR)-9-hydroxy-5a,5b,8,8,11a-pentamethyl-1-(prop-1-en-2-yl)icosahydro-3aH-cyclopenta[a]chrysene-3a-carbonyl]amino}-2,2,6,6-tetramethylpiperidin-1-yl)oxidanyl (**14**)

Obtained using the general method **C1**, the reaction crude reaction product was purified by silica gel column chromatography (DCM/MeOH 8:2) to yield compound **14**

(11.8 mg, 0.014 mmol, 70%) as orange powder. R_F 0.12 (DCM/MeOH 8:2). NMR of the corresponding hydroxylamine after the addition of hydrazobenzene. ^{1}H-NMR (400 MHz, DMSO-d_6) δ 11.92 (s, 1H), 4.64 (d, J = 2.7 Hz, 1H), 4.57–4.51 (m, 1H), 4.26 (d, J = 5.1 Hz, 1H), 3.80–3.62 (m, 1H), 3.52 (s, 2H), 2.97 (m, 2H), 2.80 (m, 1H), 2.68 (m, 1H), 2.36–2.31 (m, 1H), 1.96 (m, 2H), 1.78 (dd, J = 13.0, 9.0 Hz, 2H), 1.64 (s, 3H), 1.55 (t, J = 12.7 Hz, 5H), 1.51–1.38 (m, 6H), 1.38–1.21 (m, 8H), 1.08 (d, J = 3.1 Hz, 14H), 0.92 (s, 3H), 0.88 (s, 3H), 0.84 (s, 3H), 0.77 (s, 3H), 0.66 (s, 3H). ^{13}C-NMR (101 MHz, DMSO-d_6) δ 173.71, 171.03, 151.04, 109.22, 76.75, 58.20, 54.97, 53.73, 52.04, 50.17, 48.12, 45.40, 43.82, 41.50, 38.48, 38.28, 38.23, 36.73, 35.18, 33.92, 32.46, 30.63, 29.07, 27.15, 27.12, 25.15, 20.65, 19.78, 19.63, 19.07, 17.93, 15.97, 15.85, 15.78, 14.31. HRMS (ESI) m/z: 666.4996 [M − H]$^-$, calcd. for [C$_{41}$H$_{66}$N$_2$O$_5$]$^-$ 666.4972.

{4-[(Carboxymethyl){(2E)-6-methyl-2-[(2S,3aS,3bS,6S,7R,9aS,10R,11aR)-2,7,10-trihydroxy-3a, 3b,6,9a-tetramethylhexadecahydro-1H-cyclopenta[a]phenanthren-1-ylidene]hept-5-enoyl}-amino]- 2,2,6,6-tetramethylpiperidin-1-yl}oxidanyl **(16)**

Obtained using the general method **C1**, the reaction crude reaction product was purified by silica gel column chromatography (DCM/MeOH 8:2) to yield compound **16** (10 mg, 0.014 mmol, 58%) as orange oil. R_F 0.10 (DCM/MeOH 8:2). NMR of the corresponding hydroxylamine after the additon of hydrazobenzene. ^{1}H-NMR (400 MHz, DMSO-d_6) δ 10.22 (s, 1H), 5.46 (d, J = 8.6 Hz, 1H), 5.14–5.05 (m, 1H), 4.48 (d, J = 8.3 Hz, 1H), 4.17–4.07 (m, 1H), 4.00 (d, J = 3.6 Hz, 1H), 3.94 (d, J = 3.9 Hz, 1H), 3.51 (t, J = 3.2 Hz, 1H), 3.17 (dd, J = 15.1, 9.9 Hz, 1H), 2.99–2.75 (m, 1H), 2.71–2.55 (m, 1H), 2.40–1.93 (m, 8H), 1.93–1.66 (m, 4H), 1.63 (s, 3H), 1.55 (t, J = 2.2 Hz, 3H), 1.52–1.29 (m, 6H), 1.26 (s, 3H), 1.24–1.07 (m, 3H), 1.07–0.91 (m, 14H), 0.89 (d, J = 6.2 Hz, 6H), 0.79 (d, J = 6.7 Hz, 3H). ^{13}C-NMR (101 MHz, DMSO-d_6) δ 174.42, 171.52, 147.95, 147.48, 135.99, 124.70, 69.98, 69.76, 66.50, 58.62, 58.55, 56.47, 49.67, 49.07, 48.89, 43.28, 41.71, 39.22, 36.97, 35.86, 35.67, 33.37, 33.13, 30.75, 29.57, 27.60, 25.89, 23.95, 23.06, 21.21, 19.01, 18.17, 16.77, 16.00. HRMS (ESI) m/z: 684.4732 [M − H]$^-$, calcd. for [C$_{40}$H$_{64}$N$_2$O$_7$]$^-$ 684.4714.

Method **C2**

Method **C2** was established to keep the C-16 acetyl group intact on position 16 of the fusidic acid skeleton. N-acylpyrrole **15** (0.025 mmol) was dissolved in a mixture of t-BuOH (10 mL) and H$_2$O (5 mL). Then, DMAP (0.015 mmol) was added and the reaction mixture was heated at reflux for 5 h, after which TLC (DCM/MeOH 8:2) indicated the saponification into the carboxylic acid **17**. The reaction mixture was concentrated to a volume of 10 mL in a rotary evaporator. Saturated NaHCO$_3$ solution (10 mL) and CH$_2$Cl$_2$ (20 mL) were added. After the separation of the organic layer, the water layer was extracted with CH$_2$Cl$_2$ (2 × 30 mL). Then the water layer was acidified with NaHSO$_4$ (2 M) and extracted with EtOAc (3 × 20 mL). The combined organic solutions of the acidic extraction were dried over Na$_2$SO$_4$, filtered, and evaporated to give carboxylic acid derivative, which was further purified by column chromatography (DCM/MeOH 8:2).

{4-[{(2E)-2-[(2S,3aS,3bS,6S,7R,9aS,10R,11aR)-2-(Acetyloxy)-7,10-dihydroxy-3a,3b,6,9a- tetramethylhexadecahydro-1H-cyclopenta[a]phenanthren-1-ylidene]-6-methylhept-5-enoyl} (carboxymethyl)amino]-2,2,6,6-tetramethylpiperidin-1-yl}oxidanyl **(17)**

Obtained using the general method **C2**, the reaction crude reaction product was purified by silica gel column chromatography (DCM/MeOH 8:2) to yield compound **17** (12 mg, 0.016 mmol, 66%) as orange oil. R_F 0.15 (DCM/MeOH 8:2). NMR of the corresponding hydroxylamine after the addition of hydrazobenzene. ^{1}H-NMR (400 MHz, DMSO-d_6) δ 8.89 (s, 1H), 5.44 (d, J = 8.5 Hz, 1H), 5.10 (t, J = 7.0 Hz, 1H), 4.18–4.13 (m, 1H), 3.98 (d, J = 3.7 Hz, 1H), 3.95 (d, 1H), 3.87 (d, J = 17.0 Hz, 2H), 3.53–3.49 (m, 1H), 2.97–2.87 (m, 1H), 2.73–2.66 (m, 1H, TEMPO), 2.26–2.17 (m, 3H), 2.10–1.97 (m, 5H), 1.94–1.82 (m, 5H), 1.81–1.70 (m, 2H, TEMPO), 1.64 (s, 3H), 1.57 (s, 3H), 1.53–1.29 (m, 6H), 1.27 (s, 3H), 1.26–1.12 (m, 3H), 1.10–0.95 (m, 14H, TEMPO), 0.88 (s, 3H), 0.84 (s, 3H), 0.79 (d, J = 6.7 Hz, 3H). ^{13}C-NMR (101 MHz, DMSO-d_6) δ 175.94, 171.54, 170.02, 154.96, 147.92, 133.53, 123.91,

74.27, 69.74, 66.27, 58.63, 58.47, 54.20, 49.43, 49.10, 48.97, 43.02, 39.25, 38.90, 36.94, 36.69, 36.09, 35.71, 31.77, 29.78, 27.52, 26.07, 23.86, 23.22, 21.37, 20.72, 20.18, 18.21, 18.15, 16.78, 14.43. HRMS (ESI) m/z: 726.4807 [M − H]$^−$, calcd. for [C$_{42}$H$_{66}$N$_2$O$_8$]$^−$ 726.4819.

{4-[{(2E)-2-[(2S,3aS,3bS,6S,7R,9aS,10R,11aR)-2-(Acetyloxy)-7,10-dihydroxy-3a,3b,6,9a-tetramethylhexadecahydro-1H-cyclopenta[a]phenanthren-1-ylidene]-6-methylhept-5-enoyl}(2-{[2-({2-[6-(dimethylamino)-1,3-dioxo-1H-benzo[de]isoquinolin-2(3H)-yl]ethyl}amino)-2-oxoethyl][3-(triphenylphosphaniumyl)propyl]amino}-2-oxoethyl)amino]-2,2,6,6-tetramethylpiperidin-1-yl}oxidanyl bromide (**18**)

Obtained using the general method **A** in 0.013 mmol scale (3-aminopropyl) triphenylphosphonium bromide and paraformaldehyde were used, then fusidic acid derivative **17** and Yudin's isonitrile were added. The crude reaction product was purified by silica gel column chromatography (DCM/MeOH 8:2) to yield compound **18** (6 mg, 0.004 mmol, 34%) as yellow powder. R_F 0.12 (DCM/MeOH 8:2). NMR of the corresponding hydroxylamine after the additon of hydrazobenzene. ^{1}H-NMR (400 MHz, DMSO-d_6) δ 8.57 (d, J = 8.8 Hz, 1H), 8.54 (d, J = 7.5 Hz, 1H), 8.51 (d, J = 8.3 Hz, 1H), 7.87–7.72 (m, 15H), 7.18–7.15 (m, 1H), 5.44 (q, J = 15.3, 12.0 Hz, 1H), 5.10 (d, J = 10.6 Hz, 1H), 4.62 (t, J = 7.3 Hz, 2H), 4.20–4.14 (m, 1H), 4.08–3.97 (m, 6H), 3.62–3.52 (m, 4H), 3.37 (d, J = 7.9 Hz, 2H), 3.16 (s, 4H), 3.10 (s, 6H), 2.95–2.88 (m, 1H), 2.69–2.62 (m, 1H), 2.46–2.11 (m, 8H), 2.11–1.92 (m, 7H), 1.50 (s, 3H), 1.47–1.32 (m, 6H), 1.30 (s, 3H), 1.24 (s, 3H), 1.15–0.95 (m, 17H), 0.89 (s, 3H), 0.85 (s, 3H), 0.80 (d, J = 6.7 Hz, 3H). ^{13}C-NMR (101 MHz, DMSO-d_6) δ 171.12, 169.83, 169.46, 168.61, 163.83, 163.15, 156.51, 148.71, 135.13, 135.10, 133.60, 133.49, 132.17, 131.82, 131.51, 130.42, 130.29, 129.73, 128.76, 128.46, 124.98, 123.47, 122.99, 118.40, 117.82, 113.56, 112.99, 73.58, 69.19, 65.74, 58.08, 57.89, 54.62, 51.14, 48.94, 48.58, 48.16, 44.36, 43.54, 42.08, 38.59, 36.66, 36.42, 36.33, 36.26, 36.16, 35.84, 31.78, 30.21, 27.01, 25.55, 23.46, 23.33, 21.07, 20.80, 20.67, 20.08, 19.64, 18.37, 17.84, 17.71, 17.49, 16.25. HRMS (ESI) m/z: 1352.7621 [M]$^+$, calcd. for [C$_{81}$H$_{105}$N$_6$O$_{10}$P]$^+$ 1352.7624.

4.2. EPR Spectroscopy and Sample Preparation

X-Band (~9.43 GHz) room temperature CW-EPR measurements were performed on a Magnettech MiniScope MS400 benchtop spectrometer (Magnettech, Berlin, Germany). Spectra were recorded with a microwave power of 3.16 mW, 100 kHz modulation frequency, modulation amplitude of 0.1mT and 4096 points. The final spectra were accumulations of 10 scans, each took 60 s. The samples were dissolved in methanol. Therefore, to reduce the line broadening effect due to the dissolved oxygen in the solvent, all samples were flushed with argon before EPR measurements.

4.3. Biology

4.3.1. Cell Lines and Cultivation

PC3 and HT29 cell lines were supplied by the Leibniz Institute of Plant Biochemistry. The cells were grown in RPMI 1640 completed medium (supplemented with 10% FCS, 1% glutamine, and 1% penicillin/streptomycin) at 37 °C and 5% CO$_2$. Cells were seeded at 5 × 10^3 cells/well in 96-well plates for viability determination and 1.5 × 10^5 cells/well in 6-well plates for flow cytometry and western blotting.

4.3.2. MTT and CV Assays

For the fast screening the two cell lines were treated with 0.1 and 10 µM of the synthesized compounds 6-12, and 18 for 48 h. The compounds which showed anticancer activity was further analyzed to determine their IC$_{50}$, in which, each compound was tested in 7 different concentrations (100, 50, 25, 12.5, 6.25, 3.125, 1.56 µM) for 48 h. Afterward, for the CV assay, the cells were fixed by 4% paraformaldehyde for 15 min at RT and then the cells were stained with 0.1% CV solution for 15 min. Subsequently, the cells were washed with dd H$_2$O, dried overnight and the dye was dissolved using 33% acetic acid. For MTT assay, the cells were incubated with MTT (0.5 mg/mL) for 20 min. Then, the

MTT solution was removed and the dye was dissolved using DMSO. The dissolved dyes were measured using an automated microplate reader (Spectramax, Molecular Devices, San José, CA, USA) at 570 nm with a background wavelength of 670 nm. The IC_{50} values were calculated using the four-parameter logistic function and presented as the mean and all assays were performed in three biological replicates. The cell viability was expressed as a percentage compared to a negative control which was cells treated with complete medium and a positive control which was cells treated with digitonin (125 µM) [41,42].

4.3.3. Apoptosis Analysis

The PC3 cells were prepared in a 6 well plate, treated with IC_{50} and 2 × IC_{50} of compound **8** (7.4 and 14.9 µM), and incubated for 48 h at 37 °C and 5% CO_2. After the incubation, cells were stained by AnnV and PI (5 µL of AnnV, 2 µL of PI in 100 µL PBS) to determine apoptosis using flow cytometry (FACSAria III, BD Biosciences, Franklin Lakes, NJ, USA). The procedure was carried out according to the manufacturer's supplied instructions [42].

4.3.4. Cell Cycle Analysis

The PC3 cells were prepared in a 6 well-plate and treated with IC_{50} and 2 × IC_{50} of compound **8** (7.4 and 14.9 µM) and incubated for 48 h at 37 °C and 5% CO_2. Afterward, the cells were fixed in 70% ethanol overnight at 2 °C and then, stained with 1 µg/mL of DAPI at room temperature for 10 min. At last, the cells were analyzed by flow cytometry (FACSAria III) [42].

4.3.5. Western Blot Analysis

PC3 cells were cultivated with an IC_{50} dose of **8** for 2 h, 6 h, 12 h, 24 h, and 48 h. The cell lysis was performed using protein lysis buffer (62.5 mMTris–HCl (pH 6.8), 2% (w/v) SDS, 10% glycerol, and 50 mM dithiothreitol). The proteins were electrically separated using 12% SDS-polyacrylamide gels where a PageRuler prestained ladder was used as a protein molecular weight marker. The proteins were electrically transferred to nitrocellulose membranes by western blot system (Owl HEP-1, ThermoFisher Scientific, Schwerte, Germany). The membranes were blocked by 5% (w/v) BSA in PBS with 0.1% Tween 20 for 1 h at RT. Afterwards, blots were incubated overnight at 4 °C with α/β-Tubulin rabbit Ab, Caspase-3 rabbit Ab, β-actin rabbit Ab, and Bcl-XL rabbit Ab. As a secondary antibody Anti-rabbit IgG, HRP-linked Antibody was used. Bands were visualized using an ECL Prime Western Blotting System.

4.3.6. Investigation of ROS Production

For the detection of reactive oxygen and nitrogen species, PC3 cells were stained with 1 µM of DHR solution in 0.1% PBS for 10 min. Afterwards, the cells were treated with IC_{50} and 2 × IC_{50} of compound **8** for 48 h. After 48 h, cells were trypsinized, washed with PBS, and then analyzed with flow cytometry [43,44].

4.4. Microscopy

Fluorescent Microscopy

PC3 cells were seeded in a 6-well plate for 24 h at 37 °C and 5% CO_2. Afterward, cells were stained with 0.1 µM of MitoTracker™ Deep Red in a complete medium for 15 min (based on the manufacturer's protocol). The cells were washed twice with PBS. After washing, cells were treated with the IC_{50} of the tested compound for 24 h. The cells were washed twice with PBS, upon which 1 mL of medium was added. Finally, the cells were observed using GFP and Texas Red channels using LSM700 (Carl Zeiss, Jena, Germany) and EVOS FL AUTO (ThermoFisher, Schwerte, Germany).

Supplementary Materials: The following are available online at https://www.mdpi.com/article/10.3390/ijms22137125/s1. Table S1. EPR charactersitics for nitroxides 6-12, 14, 16-18; ^{1}H-, ^{13}C-NMR spectra and HPLC-profiles of products 6-12, 14, 16-18.

Author Contributions: H.N.S. synthesized and analyzed the compounds, wrote the initial manuscript, I.M. carried out the biological experiments and assisted in the biological data compilation, H.H. co-analyzed the spectroscopic data and assisted in the initial manuscript, A.H.R. and H.H.H. measured and analyzed the EPR-spectra, G.N.K. oversaw the biological assay data and assisted in the final manuscript, D.H. oversaw the EPR-analysis and co-designed the project, B.W. designed and coordinated the project, finalized the manuscript. All authors have read and agreed to the published version of the manuscript.

Funding: This research received no external funding.

Acknowledgments: Hidayat Hussain gratefully acknowledges support by the AvH-foundation.

Conflicts of Interest: The authors declare no competing financial interest.

References

1. Kumari, S.; Badana, A.K.; Murali Mohan, G.; Shailender, G.; Malla, R.R. Reactive Oxygen Species: A Key Constituent in Cancer Survival. *Biomark. Insights* **2018**, *13*, 1–9. [CrossRef]
2. Liou, G.-Y.; Storz, P. Reactive oxygen species in cancer. *Free Radic. Res.* **2010**, *44*, 479–496. [CrossRef]
3. Firuzi, O.; Miri, R.; Tavakkoli, M.; Saso, L. Antioxidant therapy: Current status and future prospects. *Curr. Med. Chem.* **2011**, *18*, 3871–3888. [CrossRef]
4. Krasowska, A.; Piasecki, A.; Murzyn, A.; Sigler, K. Assaying the antioxidant and radical scavenging properties of aliphatic mono- and di-N-oxides in superoxide dismutase-deficient yeast and in a chemiluminescence test. *Folia Microbiol.* **2007**, *52*, 45–51. [CrossRef]
5. Anderson, R.F.; Shinde, S.S.; Hay, M.P.; Denny, W.A. Potentiation of the cytotoxicity of the anticancer agent tirapazamine by benzotriazine N-oxides: the role of redox equilibria. *J. Am. Chem. Soc.* **2006**, *128*, 245–249. [CrossRef]
6. Hordyjewska, A.; Ostapiuk, A.; Horecka, A.; Kurzepa, J. Betulin and betulinic acid: Triterpenoids derivatives with a powerful biological potential. *Phytochem. Rev.* **2019**, *18*, 929–951. [CrossRef]
7. Bednarczyk-Cwynar, B. An overwiew on the chemistry and biochemistry of triterpenoids. *Mini-Rev. Org. Chem.* **2014**, *11*, 251–252. [CrossRef]
8. Amiria, S.; Dastghaibb, S.; Ahmadic, M.; Mehrbodd, P.; Khademe, F.; Behroujb, H.; Aghanoorif, M.R.; Machajg, F.; Ghamsaric, M.; Rosikg, J.; et al. Betulin and its derivatives as novel compounds with different pharmacological effects. *Biotechnol. Adv.* **2020**, *38*, 107409. [CrossRef]
9. Fulda, S.; Kroemer, G. Targeting mitochondrial apoptosis by betulinic acid in human cancers. *Drug Discov. Today* **2009**, *14*, 885–890. [CrossRef]
10. Kumar, P.; Bhadauria, A.S.; Singh, A.K.; Saha, S. Betulinic acid as apoptosis activator: Molecular mechanisms, mathematical modeling and chemical modifications. *Life Sci.* **2018**, *209*, 24–33. [CrossRef]
11. Csuk, R.; Barthel, A.; Schwarz, S.; Kommera, H.; Paschke, R. Synthesis and biological evaluation of antitumor-active γ-butyrolactone substituted betulin derivatives. *Bioorganic Med. Chem.* **2010**, *18*, 2549–2558. [CrossRef] [PubMed]
12. Sommerwerk, S.; Heller, L.; Kerzig, C.; Kramell, A.E.; Csuk, R. Rhodamine B conjugates of triterpenoic acids are cytotoxic mitocans even at nanomolar concentrations. *Eur. J. Med. Chem.* **2017**, *127*, 1–9. [CrossRef] [PubMed]
13. Wiemann, J.; Heller, L.; Perl, V.; Kluge, R.; Ströhl, D.; Csuk, R. Betulinic acid derived hydroxamates and betulin derived carbamates are interesting scaffolds for the synthesis of novel cytotoxic compounds. *Eur. J. Med. Chem.* **2015**, *106*, 194–210. [CrossRef] [PubMed]
14. Wolfram, R.K.; Fischer, L.; Kluge, R.; Ströhl, D.; Al-Harrasi, A.; Csuk, R. Homopiperazine-rhodamine B adducts of triterpenoic acids are strong mitocans. *Eur. J. Med. Chem.* **2018**, *155*, 869–879. [CrossRef] [PubMed]
15. Wolfram, R.K.; Heller, L.; Csuk, R. Targeting mitochondria: Esters of rhodamine B with triterpenoids are mitocanic triggers of apoptosis. *Eur. J. Med. Chem.* **2018**, *152*, 21–30. [CrossRef]
16. Antimonova, A.N.; Petrenko, N.I.; Shults, E.E.; Polienko, I.F.; Shakirov, M.M.; Irtegova, I.G.; Pokrovskii, M.A.; Sherman, K.M.; Grigor'ev, I.A. Synthetic transformations of higher terpenoids. XXX. Synthesis and cytotoxic activity of betulonic acid amides with a piperidine or pyrrolidine nitroxide moiety. *Bioorganicheskaia khimiia* **2013**, *39*, 206–211.
17. Popov, S.A.; Shpatov, A.V.; Grigor'ev, I.A. Synthesis of substituted esters of ursolic, betulonic, and betulinic acids containing the nitroxyl radical 4-Amino-2,2,6,6-tetramethylpiperidine-1-oxyl. *Chem. Nat. Compd.* **2015**, *51*, 87–90. [CrossRef]
18. Zhao, M.; Gödecke, T.; Gunn, J.; Duan, J.A.; Che, C.T. Protostane and fusidane triterpenes. *Molecules* **2013**, *18*, 4054–4080. [CrossRef] [PubMed]
19. Zykova, T.; Zhu, F.; Wang, L.; Li, H.; Lim, D.Y.; Yao, K.; Roh, E.; Yoon, S.P.; Kim, H.G.; Bae, K.B.; et al. Targeting PRPK function blocks colon cancer metastasis. *Mol. Cancer Ther.* **2018**, *17*, 1101–1113. [CrossRef]

20. Ni, J.; Guo, M.; Cao, Y.; Lei, L.; Liu, K.; Wang, B.; Lu, F.; Zhai, R.; Gao, X.; Yan, C.; et al. Discovery, synthesis of novel fusidic acid derivatives possessed amino-terminal groups at the 3-hydroxyl position with anticancer activity. *Eur. J. Med. Chem.* **2019**, *162*, 122–131. [CrossRef]
21. Magnuson, B.A.; Carr, I.; Bird, R.P. Ability of aberrant crypt foci characteristics to predict colonic tumor incidence in rats fed cholic acid. *Cancer Res.* **1993**, *53*, 4499–4504.
22. Peterlik, M. Role of bile acid secretion in human colorectal cancer. *Wien Med. Wochenschr.* **2008**, *158*, 539–541. [CrossRef]
23. Govdi, A.I.; Sokolova, N.V.; Sorokina, I.V.; Baev, D.S.; Tolstikova, T.G.; Mamatyuk, V.I.; Fadeev, D.S.; Vasilevsky, S.F.; Nenajdenko, V.G. Synthesis of new betulinic acid–peptide conjugates and in vivo and in silico studies of the influence of peptide moieties on the triterpenoid core activity. *Med. Chem. Commun.* **2015**, *6*, 230–238. [CrossRef]
24. Govdi, A.I.; Vasilevsky, S.F.; Sokolova, N.V.; Sorokina, I.V.; Tolstikova, T.G.; Nenajdenko, V.G. Betulonic acid–peptide conjugates: Synthesis and evaluation of anti-inflammatory activity. *Mendeleev Commun.* **2013**, *23*, 260–261. [CrossRef]
25. Sultani, H.N.; Haeri, H.H.; Hinderberger, D.; Westermann, B. Spin-labelled diketopiperazines and peptide–peptoid chimera by Ugi-multi-component-reactions. *Org. Biomol. Chem.* **2016**, *14*, 11336–11341. [CrossRef] [PubMed]
26. Neves Filho, R.A.W.; Stark, S.; Morejon, M.C.; Westermann, B.; Wessjohann, L.A. 4-Isocyanopermethylbutane-1,1,3-triol (IPB): A convertible isonitrile for multicomponent reactions. *Tetrahedron Lett.* **2012**, *53*, 5360–5363. [CrossRef]
27. Rotstein, B.H.; Mourtada, R.; Kelley, S.O.; Yudin, A.K. Solvatochromic reagents for multicomponent reactions and their utility in the development of cell-permeable macrocyclic peptide vectors. *Chem. Eur. J.* **2011**, *17*, 12257–12261. [CrossRef] [PubMed]
28. Stoll, S.; Schweiger, A. EasySpin, a comprehensive software package for spectral simulation and analysis in EPR. *J. Magn. Reson.* **2006**, *178*, 42–55. [CrossRef] [PubMed]
29. Ondar, M.A.; Grinberg, O.Y.; Dubinskii, A.A.; Lebedev, Y.S. Study of the effect of the medium on the magnetic-resonance parameters of nitroxyl radicals by high-resolution EPR spectroscopy. *Sov. J. Chem. Phys.* **1985**, *3*, 781–792.
30. Snipes, W.; Cupp, J.; Cohn, G.; Keith, A. Electron spinal resonance analysis of the nitroxide spin label 2,2,6,6-tetramethylpipidone-N-oxyl (Tempone) in single crystals of the reduced tempone matrix. *Biophys. J.* **1974**, *14*, 20–32. [CrossRef]
31. Kawamura, T.; Matsunami, S.; Yonezawa, T. Solvent effects on the g-value of di-t-butyl nitric oxide. *Bull. Chem. Soc. Jpn.* **1967**, *40*, 1111–1115. [CrossRef]
32. Roda, A.; Minutello, A.; Angellotti, M.A.; Fini, A. Bile acid structure-activity relationship: Evaluation of bile acid lipophilicity using 1-octanol/water partition coefficient and reverse phase HPLC. *J. Lipid Res.* **1990**, *31*, 1433–1443. [CrossRef]
33. Fulda, S. Betulinic acid is a natural product with a range of biological effects. *Int. J. Mol. Sci.* **2008**, *9*, 1096–1107. [CrossRef] [PubMed]
34. Reyes, F.J.; Centelles, J.J.; Lupiáñez, J.A.; Cascante, M. (2α,3β)-2,3-Dihydroxyolean-12-en-28-oic acid, a new natural triterpene from *Olea europea*, induces caspase dependent apoptosis selectively in colon adenocarcinoma cells. *FEBS Lett.* **2006**, *580*, 6302–6310. [CrossRef] [PubMed]
35. Suy, S.; Mitchell, J.B.; Samuni, A.; Mueller, S.; Kasid, U. Nitroxide tempo, a small molecule, induces apoptosis in prostate carcinoma cells and suppresses tumor growth in athymic mice. *Cancer* **2005**, *103*, 1302–1313. [CrossRef]
36. Nie, X.; Li, C.; Hu, S.; Xue, F.; Kang, Y.J.; Zhang, W. An appropriate loading control for western blot analysis in animal models of myocardial ischemic infarction. *Biochem. Biophys. Rep.* **2017**, *12*, 108–113. [CrossRef]
37. Oropesa-Ávila, M.; Fernández-Vega, A.; de La Mata, M.; Maraver, J.G.; Cordero, M.D.; Cotán, D.; Calero, C.P.; Paz, M.V.; Pavón, A.D.; Sánchez, M.A.; et al. Apoptotic microtubules delimit an active caspase free area in the cellular cortex during the execution phase of apoptosis. *Cell Death Dis.* **2013**, *4*, e527. [CrossRef]
38. Xu, J.; Zeng, F.; Wu, H.; Wu, S. A mitochondrial-targeting and NO-based anticancer nanosystem with enhanced photo-controllability and low dark-toxicity. *J. Mater. Chem. B* **2015**, *3*, 4904–4912. [CrossRef]
39. Lee, T.D.; Keana, J.F.W. In situ reduction of nitroxide spin labels with phenylhydrazine in deuteriochloroform solution. Convenient method for obtaining structural information on nitroxides using nuclear magnetic resonance spectroscopy. *J. Org. Chem.* **1975**, *40*, 3145–3147. [CrossRef]
40. Li, Y.; Lei, X.; Li, X.; Lawler, R.G.; Murata, Y.; Komatsu, K.; Turro, N.J. Indirect 1H NMR characterization of H2@C60 nitroxide derivatives and their nuclear spin relaxation. *Chem. Commun.* **2011**, *47*, 12527–12529. [CrossRef]
41. Sladowski, D.; Steer, S.J.; Clothier, R.H.; Balls, M. An improved MTT assay. *J. Immunol. Methods* **1993**, *157*, 203–207. [CrossRef]
42. Kaluđerović, G.N.; Krajnović, T.; Momcilovic, M.; Stosic-Grujicic, S.; Mijatović, S.; Maksimović-Ivanić, D.; Hey-Hawkins, E. Ruthenium(II) p-cymene complex bearing 2,2′-dipyridylamine targets caspase 3 deficient MCF-7 breast cancer cells without disruption of antitumor immune response. *J. Inorg. Biochem.* **2015**, *153*, 315–321. [CrossRef] [PubMed]
43. Krajnović, T.; Kaluđerović, G.N.; Wessjohann, L.A.; Mijatović, S.; Maksimović-Ivanić, D. Versatile antitumor potential of isoxanthohumol: Enhancement of paclitaxel activity in vivo. *Pharmacol. Res.* **2016**, *105*, 62–73. [CrossRef] [PubMed]
44. Maksimovic-Ivanic, D.; Mijatovic, S.; Harhaji, L.; Miljkovic, D.; Dabideen, D.; Fan Cheng, K.; Mangano, K.; Malaponte, G.; Al-Abed, Y.; Libra, M.; et al. Anticancer properties of the novel nitric oxide-donating compound (S,R)-3-phenyl-4,5-dihydro-5-isoxazole acetic acid-nitric oxide in vitro and in vivo. *Mol. Cancer Ther.* **2008**, *7*, 510–520. [CrossRef] [PubMed]

MDPI
St. Alban-Anlage 66
4052 Basel
Switzerland
Tel. +41 61 683 77 34
Fax +41 61 302 89 18
www.mdpi.com

International Journal of Molecular Sciences Editorial Office
E-mail: ijms@mdpi.com
www.mdpi.com/journal/ijms

www.ingramcontent.com/pod-product-compliance
Lightning Source LLC
LaVergne TN
LVHW070233100526
838202LV00015B/2124